SECURITY IN
RFID AND SENSOR
NETWORKS

SECURITY IN RFID AND SENSOR NETWORKS

Edited by
Yan Zhang
Paris Kitsos

CRC Press
Taylor & Francis Group
Boca Raton   London   New York

CRC Press is an imprint of the
Taylor & Francis Group, an **informa** business
AN AUERBACH BOOK

Auerbach Publications
Taylor & Francis Group
6000 Broken Sound Parkway NW, Suite 300
Boca Raton, FL 33487-2742

International Standard Book Number-13: 978-1-4200-6839-9 (Hardcover)

Library of Congress Cataloging-in-Publication Data

Zhang, Yan, 1977-
　　Security in RFID and sensor networks / Yan Zhang and Paris Kitsos.
　　　　p. cm. -- (Wireless networks and mobile communications)
　　Includes bibliographical references and index.
　　ISBN 978-1-4200-6839-9 (alk. paper)
　　1. Sensor networks--Security measures. 2. Radio frequency identification systems--Security measures. I. Kitsos, Paris. II. Title.

TK7872.D48Z42 2009
681'.2--dc22
　　　　　　　　　　　　　　　　　　　　　　　　　　　　　　　　　2008037932

Visit the Taylor & Francis Web site at
http://www.taylorandfrancis.com

and the Auerbach Web site at
http://www.auerbach-publications.com

Contents

PART I   Security in RFID

PART II   *Security in Wireless Sensor Networks*

PART III   Security in Integerated RFID and WSN

Editors

Yan Zhang received his BS in communication engineering from Nanjing University of Post and Telecommunications, China; his MS in electrical engineering from Beijing University of Aeronautics and Astronautics, China; and his PhD in the School of Electrical & Electronics Engineering, Nanyang Technological University, Singapore. He is an associate editor on the editorial board of *Wiley Wireless Communications and Mobile Computing (WCMC)*; *Security and Communication Networks* (Wiley); *International Journal of Network Security; International Journal of Ubiquitous Computing*; *Transactions on Internet and Information Systems (TIIS)*; *International Journal of Autonomous and Adaptive Communications Systems (IJAACS)*; *International Journal of Ultra Wideband Communications and Systems (IJUWBCS)*; and *International Journal of Smart Home (IJSH)*. He is currently serving as the book series editor for Wireless Networks and Mobile Communications book series (Auerbach Publications, CRC Press, Taylor & Francis Group). He serves as guest coeditor for the following: *IEEE Intelligent Systems*, special issue on "Context-Aware Middleware and Intelligent Agents for Smart Environments"; *Wiley Security and Communication Networks* special issue on "Secure Multimedia Communication"; *Springer Wireless Personal Communications* special issue on selected papers from ISWCS 2007; *Elsevier Computer Communications* special issue on "Adaptive Multicarrier Communications and Networks"; *International Journal of Autonomous and Adaptive Communications Systems (IJAACS)* special issue on "Cognitive Radio Systems"; *The Journal of Universal Computer Science (JUCS)* special issue on "Multimedia Security in Communication"; *Springer Journal of Cluster Computing* special issue on "Algorithm and Distributed Computing in Wireless Sensor Networks"; *EURASIP Journal on Wireless Communications and Networking (JWCN)* special issue on "OFDMA Architectures, Protocols, and Applications"; and *Springer Journal of Wireless Personal Communications* special issue on "Security and Multimodality in Pervasive Environments."

He is serving as coeditor for several books: *Resource, Mobility and Security Management in Wireless Networks and Mobile Communications*; *Wireless Mesh Networking: Architectures, Protocols and Standards*; *Millimeter-Wave Technology in Wireless PAN, LAN and MAN*; *Distributed Antenna Systems: Open Architecture for Future Wireless Communications*; *Security in Wireless Mesh Networks*; *Mobile WiMAX: Toward Broadband Wireless Metropolitan Area Networks*; *Wireless Quality-of-Service: Techniques, Standards and Applications*; *Broadband Mobile Multimedia: Techniques and Applications*; *Internet of Things: From RFID to the Next-Generation Pervasive Networked Systems*; *Unlicensed Mobile Access Technology: Protocols, Architectures, Security, Standards and Applications*; *Cooperative Wireless Communications*; *WiMAX Network Planning and Optimization*; *RFID Security: Techniques, Protocols and System-On-Chip Design*; *Autonomic Computing and Networking*; *Security in RFID and Sensor Networks*; *Handbook of Research on Wireless Security*; *Handbook of Research on Secure Multimedia Distribution*; *RFID and Sensor Networks*; *Cognitive Radio Networks*; *Wireless Technologies for Intelligent Transportation Systems*; *Vehicular Networks: Techniques, Standards and Applications*; *Orthogonal Frequency Division Multiple Access (OFDMA)*; *Game Theory for Wireless Communications and Networking*; and *Delay Tolerant Networks: Protocols and Applications*.

Dr. Zhang serves as symposium cochair for the following: ChinaCom 2009; program cochair for BROADNETS 2009; program cochair for IWCMC 2009; workshop cochair for ADHOC-NETS 2009; general cochair for COGCOM 2009; program cochair for UC-Sec 2009; journal liasion chair for IEEE BWA 2009; track cochair for ITNG 2009; publicity cochair for SMPE 2009; publicity cochair for COMSWARE 2009; publicity cochair for ISA 2009; general cochair for WAMSNet 2008; publicity cochair for TrustCom 2008; general cochair for COGCOM 2008; workshop cochair for IEEE APSCC 2008; general cochair for WITS-08; program cochair for PCAC 2008; general cochair for CONET 2008; workshop chair for SecTech 2008; workshop chair for

SEA 2008; workshop co-organizer for MUSIC'08; workshop co-organizer for 4G-WiMAX 2008; publicity cochair for SMPE-08; international journals coordinating cochair for FGCN-08; publicity cochair for ICCCAS 2008; workshop chair for ISA 2008; symposium cochair for ChinaCom 2008; industrial cochair for MobiHoc 2008; program cochair for UIC-08; general cochair for CoNET 2007; general cochair for WAMSNet 2007; workshop cochair FGCN 2007; program vice-cochair for IEEE ISM 2007; publicity cochair for UIC-07; publication chair for IEEE ISWCS 2007; program cochair for IEEE PCAC'07; special track cochair for "Mobility and Resource Management in Wireless/Mobile Networks" in ITNG 2007; special session co-organizer for "Wireless Mesh Networks" in PDCS 2006; and a member of technical program committee for numerous international conferences, including ICC, GLOBECOM, WCNC, PIMRC, VTC, CCNC, AINA, and ISWCS. He received the Best Paper Award in the IEEE 21st International Conference on Advanced Information Networking and Applications (AINA-07).

Since August 2006, he has been working with Simula Research Laboratory, Norway (http://www.simula.no/). His research interests include resource, mobility, spectrum, data, energy, and security management in wireless networks and mobile computing. He is a member of IEEE and IEEE ComSoc.

Paris Kitsos received his BS in physics in 1999 and his PhD in 2004 from the Department of Electrical and Computer Engineering, both at the University of Patras. Since June 2005 he has been a research fellow with the Digital Systems & Media Computing Laboratory, School of Science & Technology, Hellenic Open University (HOU), Greece (http://dsmc.eap.gr/en/main.php). He is an associate editor of *Computer and Electrical Engineering* (an international journal [Elsevier]) and a member on the editorial board of *International Journal of Reconfigurable Computing* (Hindawi). He is also the coeditor of *RFID Security. Techniques, Protocols and System-On-Chip Design* published by Springer in 2008. He has participated as a program and technical committee member in more than 40 conferences and workshops in the area of his research. He has also participated as a guest coeditor in the following special issues: *Computer and Electrical Engineering* (an international journal [Elsevier Ltd]) "Security of Computers and Networks"; *Wireless Personal Communications* (an international journal [Springer]) "Information Security and Data Protection in Future Generation Communication and Networking"; and *Security and Communication Network* (SCN) (Wiley journal) "Secure Multimedia Communication."

Dr. Kitsos' research interests include VLSI design and efficient hardware implementations of cryptographic algorithms and security protocols for wireless communication systems, and hardware implementations of RFID cryptography algorithms. He is an adjunct lecturer in the Department of Computer Science and Technology, University of Peloponnese. He has published more than 60 publications in international journals, books, and technical reports, and also reviews manuscripts for books, international journals, and conferences/workshops in the areas of his research. He is also a member of the Institute of Electrical and Electronics Engineers (IEEE).

Contributors

Hossam Afifi
Mobility and Security Group
Department of Wireless Networks and
 Multimedia Services
Institut National des Télécommunications
Evry, France

Slobodan Bojanić
Department of Electronic Engineering
Universidad Politecnica de Madrid
Madrid, Spain

Leonid Bolotnyy
Department of Computer Science
School of Engineering and Applied Science
University of Virginia
Charlottesville, Virginia

Trevor Burbridge
British Telecommunications plc
Adastral Park, Martlesham Heath
Ipswich, United Kingdom

Aldar C.-F. Chan
Department of Computer Science
National University of Singapore
Singapore

Ioannis Chatzigiannakis
Department of Computer Engineering
 and Informatics
University of Patras
Patras, Greece

and

Computer Technology Institute
Patras, Greece

Hao Chen
Department of Electrical and Computer
 Engineering
Thomas J. Watson School of Engineering
State University of New York
Binghamton, New York

Yu Chen
Department of Electrical and Computer
 Engineering
Thomas J. Watson School of Engineering
State University of New York
Binghamton, New York

Xiaowen Chu
Department of Computer Science
Hong Kong Baptist University
Kowloon, Hong Kong

Tassos Dimitriou
Athens Information Technology
Athens, Greece

Alberto Peinado Domínguez
Escuela Técnica Superior de Ingenieros
 de Telecomunicación
University of Málaga
Málaga, Spain

Saar Drimer
Computer Laboratory
University of Cambridge
Cambridge, United Kingdom

Juan M. Estevez-Tapiador
Computer Science Department
Carlos III University of Madrid
Madrid, Spain

Jorge Munilla Fajardo
Escuela Técnica Superior de Ingenieros
 de Telecomunicación
University of Málaga
Málaga, Spain

Elgar Fleisch
Department of Management, Technology,
 and Economics
Eidgenössische Technische Hochscule
Zurich, Switzerland

and

Institute of Technology Management
University of St. Gallen
Saint Gallen, Switzerland

Sepideh Fouladgar
Mobility and Security Group
Department of Wireless Networks and
  Multimedia Services
Institut National des Télécommunications
Evry, France

Felix C. Freiling
Department of Informatics
University of Mannheim
Mannheim, Germany

Filippo Gandino
Dipartimento di Automatica e Informatica
Politecnico di Torino
Torino, Italy

Damianos Gavalas
Department of Cultural Technology
  and Communication
University of Aegean
Lesvos, Greece

Thanassis Giannetsos
Athens Information Technology
Athens, Greece

Gerhard P. Hancke
Smart Card Centre
Information Security Group
Royal Holloway University of London
London, United Kingdom

Julio Cesar Hernandez-Castro
Computer Science Department
Carlos III University of Madrid
Madrid, Spain

Jaap-Henk Hoepman
TNO Information and Communication
  Technology
Groningen, the Netherlands

and

Digital Security Group
Faculty of Science
Radboud University
Nijmegen, the Netherlands

Dijiang Huang
Computer Science and Engineering Department
Arizona State University
Tempe, Arizona

Alexander Ilic
Department of Management, Technology,
  and Economics
Eidgenössische Technische Hochschule
Zurich, Switzerland

José L. Imaña
Department of Computer Architecture
  and Systems Engineering
Faculty of Physics
Complutense University of Madrid
Madrid, Spain

Yixin Jiang
Department of Electrical and Computer
  Engineering
University of Waterloo
Waterloo, Ontario, Canada

Elisavet Konstantinou
Department of Information and Communication
  Systems Engineering
University of the Aegean
Karlovassi, Greece

Charalampos Konstantopoulos
Department of Informatics
University of Piraeus
Piraeus, Greece

Ioannis Krontiris
Athens Information Technology
Athens, Greece

Wei-Shinn Ku
Department of Computer Science
  and Software Engineering
Auburn University
Auburn, Alabama

Shiguo Lian
France Telecom R&D Beijing
Beijing, China

Augusto Lima
School of Information Technology
and Engineering
University of Ottawa
Ottawa, Ontario, Canada

Chuang Lin
Department of Computer Science
and Technology
Tsinghua University
Beijing, China

Zhongxuan Liu
France Telecom R&D Beijing
Beijing, China

Florian Michahelles
Department of Management, Technology,
and Economics
Eidgenössische Technische Hochschule
Zurich, Switzerland

Ali Miri
School of Information Technology
and Engineering

and

Department of Mathematics and Statistics
University of Ottawa
Ottawa, Ontario, Canada

Bartolomeo Montrucchio
Dipartimento di Automatica e Informatica
Politecnico di Torino
Torino, Italy

Aristides Mpitziopoulos
Department of Cultural Technology
and Communication
University of Aegean
Lesvos, Greece

Monica Nevins
Department of Mathematics and Statistics
University of Ottawa
Ottawa, Ontario, Canada

Ilker Onat
School of Information Technology
and Engineering
University of Ottawa
Ottawa, Ontario, Canada

Goran Pantelić
Network Security Technology
Belgrade, Serbia

Grammati Pantziou
Department of Informatics
Technological Educational Institution
of Athens
Athens, Greece

Pedro Peris-Lopez
Computer Science Department
Carlos III University of Madrid
Madrid, Spain

Maurizio Rebaudengo
Dipartimento di Automatica e Informatica
Politecnico di Torino
Torino, Italy

Arturo Ribagorda
Computer Science Department
Carlos III University of Madrid
Madrid, Spain

Gabriel Robins
Department of Computer Science
School of Engineering and Applied Science
University of Virginia
Charlottesville, Virginia

Erwing R. Sanchez
Dipartimento di Automatica e Informatica
Politecnico di Torino
Torino, Italy

Raymond Sbrusch
Division of Computing and Mathematics
University of Houston–Clear Lake
Houston, Texas

Xuemin (Sherman) Shen
Department of Electrical and Computer
Engineering
University of Waterloo
Waterloo, Ontario, Canada

Minghui Shi
Department of Electrical and Computer
Engineering
University of Waterloo
Waterloo, Ontario, Canada

Andrea Soppera
British Telecommunications plc
Martlesham Heath
Ipswich, United Kingdom

Violeta Tomašević
School of Applied Informatics
Singidunum University
Belgrade, Serbia

Thijs Veugen
TNO Information and Communication
 Technology
Delft, the Netherlands

and

Department of Mediamatics
Faculty of Electrical Engineering,
 Mathematics, and Computer Science
 (EEMCS)
Delft University of Technology
Delft, the Netherlands

Qinghua Wang
Department of Information Technology
 and Media
Mid Sweden University
Sundsvall, Sweden

T. Andrew Yang
Division of Computing and Mathematics
University of Houston–Clear Lake
Houston, Texas

Tingting Zhang
Department of Information Technology
 and Media
Mid Sweden University
Sundsvall, Sweden

Zhibin Zhou
Computer Science and Engineering Department
Arizona State University
Tempe, Arizona

Part I

Security in RFID

1 Multi-Tag RFID Systems

Leonid Bolotnyy and Gabriel Robins

CONTENTS

Radio-frequency identification (RFID) is a promising technology for automated object identification that does not require line of sight, and accurate object identification is the primary objective of RFID. However, many factors such as object occlusions, metal/liquid opaqueness, and environmental conditions (e.g., radio noise) impede object detection, thus degrading the overall availability, reliability, and dependability of RFID systems. For example, a recent major study by Wal-Mart has shown that object detection probability can be as low as 66 percent. To improve the accuracy of object identification, we propose the tagging of objects with multiple tags. We show that this strategy dramatically improves the efficacy of RFID systems, even in the face of (radiopaque) metals and liquids, radio noise, and other interfering factors. We define different types of multi-tags and examine their benefits using analytics, simulations, and experiments with commercial RFID equipment. We investigate the effects of multi-tags on anticollision algorithms, and develop several techniques that enable multi-tags to enhance RFID security. We suggest new promising applications of multi-tags, ranging from improving patient safety to preventing illegal deforestation. We analyze the economics of multi-tag RFID systems and argue that the benefits of multi-tags can substantially outweigh the costs in many current applications, and that this trend will become even more pronounced in the future.

1.1   INTRODUCTION

Bar code scanners require a line of sight to the bar codes, and they usually have to be close to the objects being identified. Moreover, bar codes are scanned one at a time, and bar code scanners (or the bar codes themselves) must physically move between successive reads. This mechanical process limits the bar-code read rate to at most a few bar codes per second. On the other hand, RFID readers can read hundreds of tags per second and they do not require line of sight, thus allowing for fast automation of the reading process, and therefore making RFID-based identification very appealing commercially. However, as the identification process is automated, we must ensure the successful reading of all the tags within the readers' field to detect all objects.

Object detection is impeded by ubiquitous background radio noise. Moreover, metals and liquids reflect or absorb radio signals, further degrading the readers' ability to achieve accurate and complete tag identification. Missed items, even at a relatively low rate of 1 percent, can result in large financial losses for businesses with low profit margins that rely on RFID-enabled automatic checkout stations. This situation is real and serious, because milk, water, juices, and canned/metal-foil-wrapped (i.e., Faraday caged) goods are commonly stocked in markets. Experiments by Wal-Mart in 2005 showed 90 percent tag detection at case level, 95 percent tag detection on conveyor belts, and only 66 percent detection rate of individual items inside fully loaded pallets [1].

A report by the Defense Logistics Agency [2] showed that only 3 percent of the tags attached to objects moving through the Global Transportation Network (GTN) did not reach the destination (165 single-tagged objects were tracked in this study). However, the same report shows that only 20 percent of the tags were recorded in the GTN at every checkpoint, and at one of the checkpoints fewer than 2 percent of tags of one particular type were detected. In addition, some of the tags were registered on arrival, but not on departure. As a result of these low object detection rates, accurate real-time tracking of objects moving through the GTN network was not possible. This report underscores the unreliability of object detection using a single RFID tag per object.

Cardinal Health, a multi-billion-dollar healthcare company, conducted RFID trials in 2006 which showed mixed results [3]. Several product lines were automatically tagged, programmed, and later tracked. The company reported that only 94.8 percent/97.7 percent of tags were encoded correctly. The accuracy of product tracking varied widely from ~8 to 100 percent, depending on the product, the tracking location, and the tracking stage. Most product detection rates were in the low-80 percent to mid-90 percent range. These results show the inadequacy of standard RFID solutions for healthcare applications.

In addition to ambient radio noise, environmental conditions such as temperature and humidity can also adversely affect the success of object detection [4]. Moreover, objects moving at high speeds can have significantly reduced detection rates. The number of objects stacked together, variation in tag receptivity (even among tags from the same manufactured batch), and tag aging (and degradation in general) can diminish the object detection probabilities as well. From the security standpoint, objects tagged with a single tag are easier to steal (a simple metal foil placed over the tag can block detection). In addition, RFID systems used in healthcare pose a special dependability challenge, because RFID system deployment will directly affect patients' welfare.

To address the problems discussed above, we propose attaching multiple RFID tags to each object, as opposed to using only a single tag per object [5]. Multi-tags will greatly improve object detection probabilities and increase reader–multi-tag communication distances, even in the presence of metallics, liquids, radio noise, and adverse environmental conditions. Multi-tags will greatly benefit theft deterrence and prevention applications, as well as dependable computing applications such as healthcare, where higher reliability and safety are required. All these benefits can be achieved at a reasonable cost, as we discuss below.

1.2 MULTI-TAG APPROACH

We base our analysis of multi-tags on the expected angle of incidence of the radio signal from the reader to the tag. We perform the analysis for inductive coupling as well as for far-field propagation. In the case of inductive coupling, Figure 1.1 depicts the angle α of the tag relative to the perpendicular direction of the signal transmitted from the reader, and gives the formula of the voltage induced in the tag by the received signal [6]. We analyze the expected voltage in one tag, as well as in ensembles of two, three, and four identical tags, assuming a fixed frequency, signal strength, and antenna geometry (i.e., loop area and number of antenna coil turns). In other words, we focus on the parameter that induces many of the benefits of multi-tags, namely the expected incidence angle of the arriving signal.

We define the angle β to be the angle between the tag and the direction of the arriving signal (rather than focusing on the angle between the tag and the perpendicular orientation of the tag to the B-field). We therefore replace $\cos(\alpha)$ with $\sin(\beta)$ in the voltage equation in Figure 1.1. Our goal is to maximize $\sin(\beta)$ in the voltage equation to maximize the induced voltage and thus the strength of the received signal. Also, because power $\sim$ voltage^2 on board a tag, we obtain power $\sim \sin^2(\beta)$.

Similarly, for far-field propagation, the power induced in the antenna by the signal is proportional to the gain of the antenna, which in turn is proportional to Poynting's vector $p = E \times H$ where E is the instantaneous electric field intensity and H is the instantaneous magnetic field intensity. We

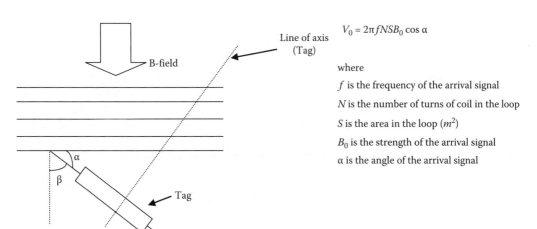

$V_0 = 2\pi f N S B_0 \cos \alpha$

where

f is the frequency of the arrival signal
N is the number of turns of coil in the loop
S is the area in the loop (m^2)
B_0 is the strength of the arrival signal
α is the angle of the arrival signal

FIGURE 1.1 Reader-induced voltage on board a tag.

also have $E \sim \sin(\beta)$ and $H \sim \sin(\beta)$. So, we obtain power $\sim \sin^2(\beta)$ [7–9]. Therefore, to improve object detection for both inductive coupling and far-field propagation, we seek to bring the expected incidence angle β closer to 90°.

Besides improvements in expected power generated on board a tag, multi-tags improve object detection because even if one tag is occluded/damaged, another tag may still be detectable. In our theoretical analysis of improvements in object detection using multi-tags, we ignore environmental conditions (e.g., radio noise), object occlusions, presence of metals and liquids in the vicinity of an object, number of objects stacked together, etc. We will explore the effect of detection impeding factors in our extensive experimental studies discussed in Section 1.3.

1.2.1 OPTIMAL PLACEMENT OF MULTI-TAGS

The first question is how to orient the tags relative to each other to maximize the expected angle of incidence of the radio wave with respect to one of the tag antennas. In our analysis, we assume a uniform distribution for the direction of the arriving signal. Indeed, in many RFID applications the orientation of a tag's antenna to the arriving signal can be arbitrary (e.g., products in a shopping cart, cell phone in a pocket). In the case of a single tag, the tag can be positioned arbitrarily, because its orientation would not affect the expected (uniformly distributed) signal arrival angle. For two tags, it is optimal to position them perpendicular to one another in the x–y and x–z planes. Similarly, for three tags, tags should be positioned pair-wise perpendicular in the x–y, x–z, and y–z planes. For four tags, it turns out that to maximize the expected signal incidence angle to at least one of the tags, it is best to position them parallel to the faces of a tetrahedron, a platonic solid.*

To validate our conjecture of optimal multi-tag placement for two and three tags, we computed the expected largest grazing angle of the radio signal to one of the tags analytically and using a simulation. The optimal placement of four tags was validated using simulations only, due to the considerable complexity of the corresponding analytical expressions. We first compute the expected largest grazing angle for two tags. Let α be the angle between two tags. Without loss of generality, let $0 \leq \alpha \leq \frac{\pi}{2}$. Then the average grazing angle ξ is

$$\xi = \frac{180}{\pi} \cdot \frac{1}{2\pi} \int\limits_{0}^{2\pi} \int\limits_{0}^{\frac{\pi}{2}} \mathrm{Max}[\Delta_1, |\Delta_2|] \sin(\phi) d\theta \, d\phi$$

where
$\Delta_1 = \frac{\pi}{2} - \phi$
$\Delta_2 = \mathrm{Arcsin}[\sin(\alpha)\sin(\theta)\sin(\phi) + \cos(\alpha)\cos(\phi)]$ [10]

To determine the optimal positioning of two tags, we want to maximize ξ subject to constraint that $0 \leq \alpha \leq \frac{\pi}{2}$. We performed computations of ξ using numerical integration in MATHEMATICA. The computations showed monotonic increase in ξ as the angle α increases from 0° to 90°. Therefore, when two tags are perpendicular to each other the expected largest grazing angle to one of the tags is maximal and equals $\sim$47.98°. We can use the above ξ equation for two planes to compute the expected grazing angle for one plane by setting $\alpha = 0$. We obtain the following:

$$\xi = \frac{180}{\pi} \cdot \frac{1}{2\pi} \int\limits_{0}^{2\pi} \int\limits_{0}^{\frac{\pi}{2}} \left(\frac{\pi}{2} - \phi\right) \sin(\phi) d\theta \, d\phi \approx 32.7°$$

---

* For five or more tags, it becomes more complicated to analytically determine the optimal relative positioning of the tags, except for specific special cases, such as for $N = 6$ where the tags should ideally be placed parallel to the faces of a dodecahedron, and $N = 10$ where the tags should be parallel to the faces of an icosahedron.

We performed similar computations for three tags. Let α be the angle between the first tag and the second tag. Let α_1 and α_2 be angles between tag 1/tag 3 and tag 2/tag 3, respectively. We obtain an almost identical expected angle formula as for two tags, except that additional coefficient are present in the body of the integral:

$$\xi = \frac{180}{\pi} \cdot \frac{1}{2\pi} \int_0^{2\pi} \int_0^{\frac{\pi}{2}} \text{Max}[\Delta_1, |\Delta_2|, |\Delta_3|] \sin(\phi) d\theta \, d\phi$$

where

$\Delta_1 = \frac{\pi}{2} - \phi$

$\Delta_2 = \text{Arcsin}[\sin(\alpha) \sin(\theta) \sin(\phi) + \cos(\alpha) \cos(\phi)]$

$\Delta_3 = \text{Arcsin}[x \cdot \cos(\theta) \sin(\phi) + y \cdot \sin(\theta) \sin(\phi) + z \cdot \cos(\phi)]$

$z = \cos(\alpha_1)$

$y = \dfrac{\cos(\alpha_2) - \cos(\alpha) \cos(\alpha_1)}{\sin(\alpha)}$

$x = \sqrt{1 - y^2 - z^2}$

Note, the above average angle ξ equation contains singularities, as it is not defined for some geometrically impossible (α_1, α_2) pairs [10]. The average angle ξ equation for three tags is much more complicated than the equation for two tags, and we computed the equation using numerical methods in MATHEMATICA for only select values of inter-tag angles α, α_1, α_2. The expected grazing angle values were computed extensively for all valid discrete (1°–90°) inter-tag angles using the simulation, which we describe below. The values computed using the simulation agreed closely with the values computed using MATHEMATICA, giving us confidence in the correctness of the formulas. The simulation showed that the optimal positioning of three tags is the mutually perpendicular positioning, resulting in the expected largest grazing angle $\sim$58.11°.

To corroborate our analytical computations of the largest average grazing angle, we developed a software simulator that computes the expected grazing angle for an arbitrary number of tags. The simulator enumerates all possible multi-tag orientations and for each orientation it calculates the average value of the maximum angle to any tag over many randomly generated simulated signals. The result of the simulation is the largest of the average maximum angles over all possible multi-tag positions. The simulator also records average maximum grazing angles for all orientations for future comparison with analytical computations.

To calculate the expected angle of incidence for a given multi-tag positioning, our simulator generates a random uniformly distributed point on the surface of a sphere [11]. This point determines the direction of a random uniformly distributed radio signal relative to the origin, and calculates the angle to every tag in the multi-tag ensemble, while recording the largest of these angles. For one- and two-tag ensembles our simulation generates 10 million such random trials and averages the induced maximum angles. For three-tag ensembles, it generates 1 million random trials, and for four tags, it generates 100,000 trials. The runtime of the simulator is $\Theta(k \cdot n^{2m-1})$ where k is the number of random reader signals, n is the number of possible angles between tags, and m is the number of tags (size of multi-tag ensemble), $m \geq 1$. Some of our simulations ran for several weeks on a single machine. Also, we ran week-long decomposed parallel computations on a cluster of 64 dual-processor Alpha PCs through portable batch system job scheduling software.

The results obtained from the analytical computations agree with the experimental results for one, two, and three tags at many points within a reasonably small error bound. We show the close match between the analytical and simulation computations for two tags in Figure 1.2. Similar results, although with smaller precision (due to fewer random trials) were obtained for three tags. The near-identical analytical and simulation results raise our confidence level in the correctness of the average angle computations. Accurate computations for one, two, and three tags allowed us to use only

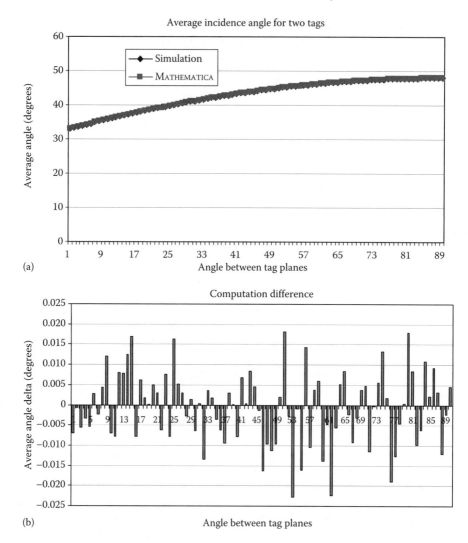

FIGURE 1.2 Accuracy of angle computation for two tags. Analytical and simulation computations are in tight agreement with at most several hundredth of a degree difference. (a) Comparison of analytical and simulation computations. (b) Computational error.

the simulator to compute the average angle for a larger tag ensemble (i.e., four tags), because the complex geometries involved make it intractable to analytically compute this quantity. For four tags, the average maximum grazing angle is ~61.86°. Figure 1.3 shows the simulation results of the expected largest incidence angle for one-, two-, three-, and four-tag configurations.

We note that there is a two-digit increase in the expected angle as we move from one tag to two tags, and also as we go from two tags to three tags, but only a 3° improvement as we move from three tags to four tags. This suggests that adding an extra tag or two may be beneficial for the purpose of increasing the induced voltage (and thus improving the communication range), but using four or more tags will not garner substantial additional benefit in that respect. Nevertheless, even though the benefit of having more than three tags per object to increase the reader–tag communication range may be relatively small, there are other benefits to using more than three tags. For example, if an alternate benefit of multi-tags (e.g., theft prevention or human safety) is the primary goal, we may still benefit

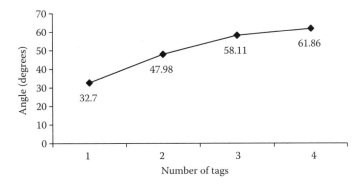

FIGURE 1.3 Expected largest reader signal's angle to any tag.

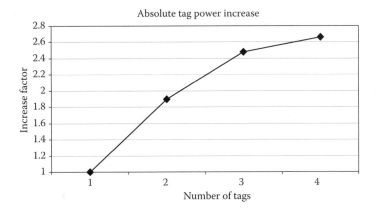

FIGURE 1.4 Absolute tag power increase for various number of tags per object.

from using more than three tags per object (and we can achieve further detection improvements by optimizing the tags' positioning).

Having computed the expected incidence angle improvements using multi-tags, we can determine the absolute and relative tag power gains for various multi-tag ensembles. Recalling that power $\sim \sin^2(\beta)$, Figure 1.4 shows the expected improvements in tag power. An increase in the expected tag power boosts the expected reader-to-tag communication distance. Figure 1.5 depicts the expected communication range increase for inductive coupling technologies as the number of tags is increased, and also for far-field propagation scenarios. These values were computed based on the relation of the distance between a reader and a tag, and the tag power generated by the reader. For backscattering technology, the effective communication distance varies as $\sim \sqrt{\text{power}}$; for inductive coupling, the maximum communication distance varies as $\sim \sqrt[6]{\text{power}}$ [12].

Our incidence angle-based analysis assumes that the signal can come from any direction with equal likelihood, which is realistic for many applications (e.g., goods randomly piled inside a shopping cart, cell phone arbitrarily placed inside a pocket). However, for some applications where the position/orientation of the object is known in advance or may only span a narrow range of possibilities, the optimal positioning of the tags may be different from the assumption-free ones suggested above. Similarly, the number of tags may vary among objects, to further optimize overall detection. Also, in our analysis of optimal multi-tag positioning we considered tag orientations only. However, to further improve objects detection, multi-tags should be spaced apart to reduce the likelihood of multiple tags being occluded simultaneously.

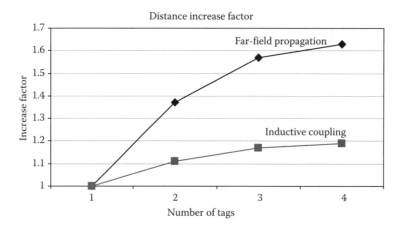

FIGURE 1.5 Expected factor of communication range increase for inductive coupling and far-field propagation as a function of the number of tags per object.

1.3 EXPERIMENTAL EQUIPMENT AND SETUP

To validate our analytical and simulation studies, we conducted an extensive experimental evaluation of multi-tags. Our experiments were performed using commercial FCC-compliant equipment (Figure 1.6), namely ultrahigh frequency (UHF) readers manufactured by Alien Technology (model ALR-9800, four antennas, multi-protocol, 915 MHz) and ThingMagic (model Mercury 4). We utilized sets of linear and circular reader antennas from Alien Technology, and circular reader antennas from ThingMagic. A single Alien Technology reader antenna can either broadcast or receive signals, whereas the more versatile ThingMagic antenna can both send and receive signals. We used several types of tags from UPM Raflatac, the world's leading RFID tag manufacturer. In particular, we picked unipolar (dipole) UPM Rafsec UHF "Impinj 34 × 54 ETSI/FCC" tags and bipolar (quadrupole) UPM Rafsec UHF "Impinj 70 × 70 ETSI/FCC" tags for our experiments.

We performed the experiments in an otherwise empty room to minimize radio interference and signal reflection anomalies. We placed multiple tags on a diverse set of 20 solid nonmetallic objects[*]

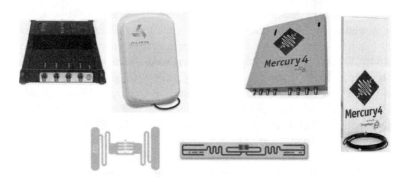

FIGURE 1.6 Multi-tag experimental equipment: Alien Technology Corporation reader and antenna (top left), ThingMagic reader and antenna (top right), UPM Rafsec and Alien Technology Corporation tag samples (bottom row).

---

[*] Solid nonmetallic multi-tagged objects included soap bars, cereal boxes, paper plates, plastic boxes, packaged foods, clothing items, etc.

using four tags per object, and a set of 20 metal and liquid-containing objects* using three tags per object. We positioned tags perpendicular to each other whenever possible, and spread the tags as far apart in space across an object as possible, to minimize tag occlusions by other tags or objects. The experiments with solid nonmetallic objects used sets of both unipolar and bipolar tags. The experiments containing metallic and liquid objects were performed only with unipolar UPM Rafsec UHF "Impinj 34 × 54 ETSI/FCC" tags.

We positioned Alien Technology reader antennas side-by-side in pairs, with each pair consisting of a sending and a receiving antenna. Each pair of antennas was equidistant to the center of a plastic bag containing objects, placed 20.5 in. above the floor, and aligned perpendicularly toward the center of the bag. We allowed sufficient time for the reader to read all the tags within its range by performing many tag reads and maintaining adequate time-outs between reads to make sure that the effects of the environmental noise were minimized. We performed our experiments for linear as well as for circular antennas using seven different power levels ranging from 25.6 to 31.6 dBm, in increments of 1 dBm.

In a separate set of experiments, circular ThingMagic antennas were equidistant and perpendicular to the bag containing the objects, located 33 in. above the floor, in the rectangular "gate" formation. Each ThingMagic antenna was both sending and receiving signals. As with the Alien Technology hardware, we allowed sufficient reader time for object identification. We randomly (re)shuffled the tagged objects multiple times to change the tags' spatial orientations with respect to the reader's antennas, to improve the statistical significance of the results (the values reported in the tables and graphs below are averages over all random object shufflings). We also varied the power emitted by the antennas, keeping in mind that the communication distance is proportional to $\sqrt{power}$.

We will mostly describe our experiments involving the Alien Technology hardware, because this equipment allowed us to collect data for both circular and linear antennas. Linear antennas have smaller angular coverage than circular (omnidirectional) antennas in exchange for greater signal strength in the specific direction. In the discussions and graphs below, we will implicitly assume that the Alien Technology equipment was used in each experiment, unless explicitly stated that the ThingMagic hardware was used instead. Similarly, all the experiments discussed in Section 1.4 have used the unipolar UPM Rafsec UHF tags "Impinj 34 × 54 ETSI/FCC," unless explicitly stated that bipolar tags were used.

1.4 EXPERIMENTAL RESULTS

1.4.1 LINEAR ANTENNAS

Our experiments show that multi-tags considerably improve object detection probabilities for linear antennas. The detection probabilities for different numbers of tags per object, different numbers of reader antennas, and various reader power levels are summarized in Table 1.1. This table shows that switching from one to two tags per object produces a high double-digit increase in tag detection probability, and a low double-digit increase when moving from two to three tags, but only single-digit increase from three to four tags. These results corroborate our theoretical expectations [13].

Figure 1.7a graphically shows the increase in object detection probability for each object (the objects are sorted along the x-axis according to their detection probabilities). Again, we observe significant separations between the first three curves. In Figure 1.7b, we compare object detection improvements between two tags per object versus two reader antennas. From this data we can see a

---

* The multi-tagged metallic and liquid objects included cans of tomato sauce, canned vegetables, canned and bottled soda, bottled water, etc.

TABLE 1.1

Detailed Statistics Showing the Average Detection Probability for Linear Antenna(s) as a Function of the Power Level for Different Antenna Configurations and for Different Numbers of Tags Per Object

	Antenna Pair #1				Antenna Pair #2				Antenna Pairs #1 and #2			
	1 Tag	2 Tags	3 Tags	4 Tags	1 Tag	2 Tags	3 Tags	4 Tags	1 Tag	2 Tags	3 Tags	4 Tags
Power: 31.6 dBm	0.5800	0.7930	0.8945	0.9385	0.5715	0.7970	0.9010	0.9570	0.6495	0.8450	0.9300	0.9695
Power: 30.6 dBm	0.5280	0.7500	0.8575	0.9070	0.4730	0.6980	0.8210	0.8950	0.5890	0.7970	0.8930	0.9380
Power: 29.6 dBm	0.4645	0.6895	0.8110	0.8760	0.4220	0.6545	0.7925	0.8885	0.5370	0.7555	0.8635	0.9195
Power: 28.6 dBm	0.4140	0.6360	0.7645	0.8390	0.4350	0.6615	0.7920	0.8695	0.4920	0.7155	0.8295	0.8880
Power: 27.6 dBm	0.3425	0.5435	0.6770	0.7645	0.3765	0.5940	0.7340	0.8200	0.4380	0.6620	0.7880	0.8565
Power: 26.6 dBm	0.3275	0.5345	0.6740	0.7695	0.3235	0.5255	0.6635	0.7580	0.3985	0.6195	0.7540	0.8380
Power: 25.6 dBm	0.2575	0.4410	0.5790	0.6895	0.2785	0.4615	0.5825	0.6580	0.3430	0.5565	0.6975	0.7880

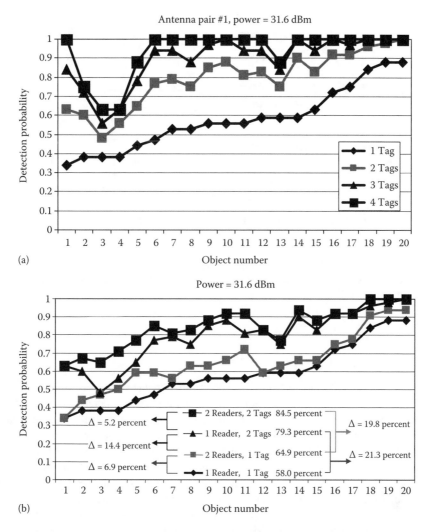

FIGURE 1.7 (a) Average object detection probability improvements for linear antennas as the number of tags per object increases. (b) Comparisons of multi-tags with multiple readers for linear antennas. Note that attaching multiple tags to an object yields higher average object detection probabilities than adding more readers. Objects are sorted based on single-tag detection probability.

dramatic double-digit improvement from adding a second tag to each object, and only a low single-digit improvement from adding a second reader, yielding almost a factor of 4 improvement in object detection probability using multi-tags as compared to multiple readers.

1.4.2 CIRCULAR ANTENNAS

As with linear antennas, experiments with circular antennas show a dramatic double-digit average improvement in object detection as the number of tags per object increases. However, the detection probabilities for circular antennas are higher than for linear ones, because the orientation of objects with respect to the reader antennas varies widely. From the comparisons of different numbers of multi-tags and multiple readers (Figure 1.8), we can see that for circular antennas the advantage of adding a tag is on par with that of adding a reader. We also observed that the average object detection probabilities decrease more rapidly for circular than for linear antennas, as a function of decreasing antenna power [10].

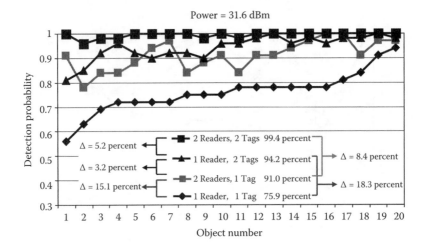

FIGURE 1.8 Comparing multi-tags with multiple readers for circular antennas. Attaching multiple tags to an object produces higher object detection probability than adding more readers. Objects are sorted based on single-tag detection probability.

1.5   IMPORTANCE OF TAG ORIENTATION

In our analytical analysis of multi-tags [13], we determined that (two or three) multi-tags should be oriented perpendicular to each other to obtain the most benefits in object detection. We experimentally confirm this claim by varying multi-tags orientation, collecting tag identification data, and calculating object detection probabilities for different multi-tag orientations. Then, we analyze these data and draw appropriate conclusions. We performed experiments with unipolar tags (UPM Rafsec UHF tag Impinj 34×54 ETSI/FCC) whose tag-plane orientation matters, and with bipolar tags (UPM Rafsec UHF tag Impinj 70×70 ETSI/FCC) whose tag-plane orientation has no effect on tag detection.

With unipolar tags, we ran experiments comparing differently oriented pairs of tags. One orientation which we call *180-same* refers to two tags positioned on the same plane and having identical orientation. The second orientation *180-diff* refers to two tags positioned on the same plane, but one of the tags is rotated 90° relative to the orientation of the other tag. The third orientation *90-same* refers to two tags having identical orientation, but positioned on perpendicular planes. Finally, the forth tag orientation *90-diff* refers to two tags positioned on perpendicular planes with one tag rotated 90° relative to the other tag. In our experiments we compared these four different tag orientations, and the results are presented in Figure 1.9a. The results show that tags perpendicular to each other yield a higher probability of detecting at least one of them than tags that have identical orientation. In addition, to increase detection probability, it is better to position tags on perpendicular planes, rather than to place all the tags in the same plane.

With bipolar tags we compared two possible tag orientations—180, where tags are positioned on parallel planes, and 90, where tags are positioned on perpendicular planes. These are the only possibilities because tag orientations within the plane have no effect on (ideal) bipolar tag detection. The results of the experiments shown in Figure 1.9b demonstrate no difference between tag orientations for omnidirectional/circular antennas, but a drastic advantage for perpendicular 90 tags over parallel 180 tags for directional/linear antennas. These results show that multi-tags improve object detection not only because they increase the total antenna size per object and decrease the probability of antenna occlusions but also because the expected grazing angle between the signal from the reader and one of the tags increases, which in turn raises the expected power on board one of the tags. These findings confirm our theoretical expectations.

	Circular		Linear	
	1 Tag	2 Tags	1 Tag	2 Tags
180-same		0.5500		0.3700
180-diff	0.4784	0.7454	0.3311	0.5272
90-same		0.6727		0.5272
90-diff		0.8000		0.6363

(a)

	Circular			Linear		
	1 Tag	2 Tags	3 Tags	1 Tag	2 Tags	3 Tags
180	0.75	1	1	0.53	0.57	0.70
90		0.93	1		0.97	1

(b)

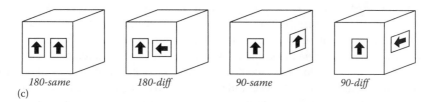

180-same   180-diff   90-same   90-diff

(c)

FIGURE 1.9 The two tables comparing object detection probabilities for unipolar and bipolar tags for different multi-tag orientations. The results show the significance of perpendicular multi-tag orientation, especially for directional/linear antennas. In Figure 1.9a, *180-same* refers to identically oriented tags positioned on parallel planes; *180-diff* refers to perpendicularly oriented tags positioned on parallel planes; *90-same* refers to identically oriented tags positioned on perpendicular planes; *90-diff* refers to perpendicularly oriented tags positioned on perpendicular planes. In Figure 1.9b, 180 refers to tags positioned on parallel planes; 90 refers to tags positioned on perpendicular planes.

1.6 CONTROLLING EXPERIMENTAL VARIABLES

It is important in RF experiments to carefully isolate and control the variables to ensure the accuracy of the results. In our multi-tag experiments, we controlled the effects of radio noise, reader variability, tag variability, the number and type of reader antennas, reader power level, and the distance from the reader antennas to the objects. To control the effect of ambient radio noise, we ran our experiments multiple times, sometimes even across multiple days to ensure that statistical properties of the data are stable. To accurately calculate improvements in object detection with multi-tags, we allowed sufficient time for the reader to read the tags. The reader parameters were carefully selected to ensure that all tags within a reader's detectability range were read. To ensure that our results are independent of the particular reader and antenna manufacturer/brand, we ran our experiments with readers and antennas from two different manufacturers. In all of our experiments, we used consistent tag types and ensured that tag variability does not affect our experiments. We will discuss tag variability further in Section 1.6.1. The reader and identical reader antennas were carefully selected and objects were placed on a rotating platform at a fixed distance from the reader. The reader power levels were carefully controlled via a parameter in the software driver.

1.6.1 TAG VARIABILITY

To determine tag properties and control tag variability, we performed multiple tag variability tests. It is widely believed that RFID tags with different chip manufacturers and antenna geometries have different detectability/receptivity properties [14]. The importance of tag receptivity and its use as a tag performance metric is addressed in Ref. [15]. Similarly, no two chips are truly identical due to inherent very-large-scale integration (VLSI) manufacturing variations [16]. Indeed, we found differences in tag detectability among tags of the same type, even among ones coming from the very same tag roll. In fact, these inherent tag receptivity differences were surprisingly high, with up to an order-of-magnitude difference in detectability between the "best" and "worst" tags. These findings provide yet another incentive for deploying multi-tags to ensure consistent object detection.

In our tag variability experiments, we used a ThingMagic reader, one circular ThingMagic antenna, and "UPM Rafsec UHF tag Impinj 34 × 54 ETSI/FCC" tags. Tags were elevated 26 in. from the floor, and positioned perpendicular to the antenna at a distance of 59.5 in. from the antenna center. The reader power level was set to 31.6 dBm. Each tag was read 200 times and the number of successful reads was recorded. We paused for 50 ms between reads to allow tags sufficient time to lose power and initialize their state. The reader was allotted 10 ms to read a tag. In this way, we computed the detectability/receptivity of 75 seemingly identical tags. To ensure data consistency, each experiment was performed twice and repeated the next day with the tags rotated 180°.

The smallest number of successful reads out of 200 was 8 and the largest was 91. The average was 43.44 and the standard deviation was 23.92. The Pearson product–moment correlation coefficient between two reads of each tag on the same day was 0.99 and the correlation between reads across two days was 0.98. Figure 1.10b shows the distribution of the number of successful tag reads, and Figure 1.10a compares the number of successful reads for each tag across the two sets of experiments conducted on consecutive days. To magnify the visual spread between tags, we show the number of successful tag reads out of 400 by summing the detectability across the two runs of each day. Similarly, high tag detectability variations were found in other UPM Rafsec tag types.

1.6.2 READER VARIABILITY

To ensure that our results are not dependent on the reader/antenna manufacturers, we repeated our experiments using ThingMagic readers and ThingMagic circular antennas.[*] Because the tag detection algorithms used by ThingMagic and their implementations are different from those of Alien Technology, and because ThingMagic antennas are much bigger than those by Alien Technology, the detection probabilities we obtained differed between these two systems. However, the percentage improvements of multi-tags versus single-tagged objects were similar for both systems, supporting our hypothesis that the percentage improvements in object detection using multi-tags is mostly independent from the specific equipment used. Table 1.2 shows the statistics of object detection improvements using circular ThingMagic antennas for a different number of tag ensembles per object. In addition to providing the second set of data, the ThingMagic equipment enabled the collection of data for three and four antennas, whereas the Alien Technology readers work with only one and two antennas.

1.7 OBJECT DETECTION IN THE PRESENCE OF METALS AND LIQUIDS

So far we have discussed multi-tags experiments with only solid nonmetallic objects [17]. In some practical scenarios, however, the items to be identified can contain mixtures of nonmetallic objects, as well as metallic and liquid materials, making reliable object identification more problematic. It is more difficult to detect metallics and liquids because they tend to interfere with and occlude radio signals, thus preventing readers from receiving accurately decodable tag responses [18]. Metallic and liquid objects can also occlude other nonmetallic objects and thus interfere with the detection of these as well.

When metals and liquids are present, the detection probabilities for solid and nonmetallic objects decrease due to radio interference from the metallics and liquids. In our experiments, we observed a 4–10 percent decrease in the detection probability of solid objects, depending on the antenna type and the number of tags per object, as compared to situations where no liquids or metallics are

---

[*] Experimental results similar to ours using equipment from Symbol Technologies (now Motorola, Inc.) were reported [19].

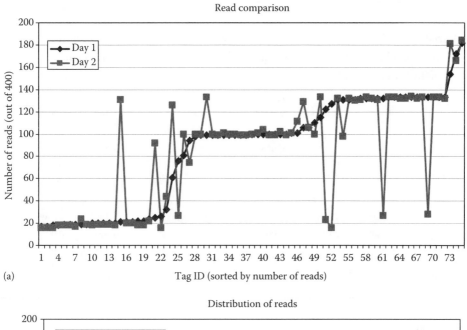

(a)

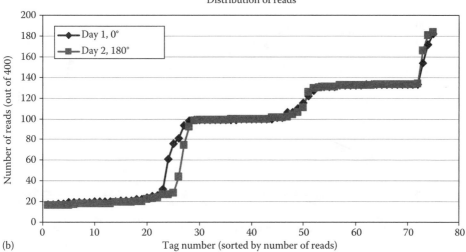

(b)

FIGURE 1.10 Characterization of tag detectability/receptivity. (a) Comparison of the number of successful reads per tag across two days. The tags are sorted based on the number of successful reads to better illustrate the data. (b) Distribution of successful tag reads across two days and two tag orientations. The number of successful reads shown is out of a total of 400 attempted. We observe a significant separation between several "clusters" of tag performance levels.

present. Figure 1.11 shows the average object detection probability for solid nonmetallic objects for circular reader antennas. In the graph, the top curve represents the detection probabilities of solid nonmetallic objects when metallics and liquids are absent, and the bottom curve represents the detection probabilities of solid nonmetallic objects when metallics and liquids are present.

To detect metallic and liquid objects in our experiments, we had to considerably reduce the distance from the objects to the readers to ensure that tags are actually detectable at that range. Specifically, we reduced the approximate reader-to-tag distance to 32 in., from the 55 in. range used for solid and nonmetallic objects. In addition, we had to operate readers at high power levels only. To avoid using special tags that are specifically designed for metals and liquids, and be able to compare

TABLE 1.2

Detection Probability Statistics for Circular ThingMagic Antennas as a Function of the Power Level for Different Antenna Configurations and for a Different Number of Tags Per Object

	1 Antenna				2 Antennas			
	1 Tag	2 Tags	3 Tags	4 Tags	1 Tag	2 Tags	3 Tags	4 Tags
Power: 31.6 dBm	0.6528	0.8511	0.9291	0.9662	0.8335	0.9580	0.9874	0.9979
Power: 30.6 dBm	0.5668	0.7775	0.8761	0.9257	0.7567	0.9129	0.9537	0.9667
Power: 29.6 dBm	0.4813	0.6932	0.8033	0.8653	0.6755	0.8630	0.9233	0.9485
Power: 28.6 dBm	0.3818	0.5778	0.6960	0.7736	0.5614	0.7702	0.8588	0.9105

	3 Antennas				4 Antennas			
	1 Tag	2 Tags	3 Tags	4 Tags	1 Tag	2 Tags	3 Tags	4 Tags
Power: 31.6 dBm	0.8847	0.9782	0.9958	1	0.8910	0.9800	0.9970	1
Power: 30.6 dBm	0.8176	0.9442	0.9686	0.9750	0.8255	0.9465	0.9690	0.9750
Power: 29.6 dBm	0.7476	0.9100	0.9492	0.9615	0.7600	0.9160	0.9515	0.9630
Power: 28.6 dBm	0.6355	0.8323	0.9025	0.9400	0.6535	0.8450	0.9100	0.9445

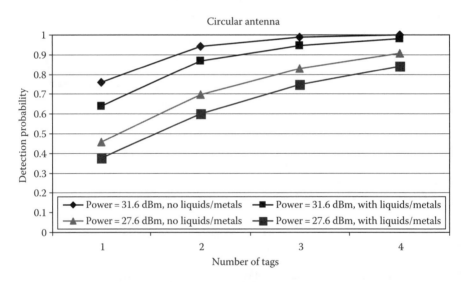

FIGURE 1.11 Comparison of average detection probabilities using circular antennas for solid nonmetallic objects when metallic/liquid objects are present and absent.

relative improvements of multi-tags for solid/nonmetallic objects with liquids and metallics, we used a few millimeter thin spacers between the objects and the tags. The space between the objects and the tags enabled bouncing radio signals to detect tags, yet kept the tags close enough to the metallic and liquid objects to retain the signal-interfering absorption and reflection characteristics of the liquids and metals.

Based on our experimental results, multi-tags are highly effective in improving object detection in the presence of metallics and liquids. We observed an almost linear improvement in metallic and liquid objects detection when the number of tags per object is increased, as compared to the rapidly increasing and then leveling detection probability curve for solid nonmetallic objects. Figure 1.12 shows detection probability for several power levels and antenna configurations. The results of separate experiments using the ThingMagic hardware show rapidly vanishing improvements in object

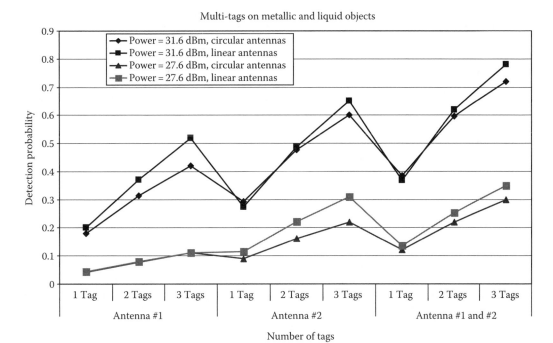

FIGURE 1.12   Comparison of average detection probabilities of metallic and liquid objects using one and two linear and circular antennas for various power levels.

detection probabilities as the number of antennas increases, yet an almost linear improvement in object detection probabilities as the number of tags per object is increased [10].

1.8   EFFECT OF OBJECT QUANTITY ON DETECTION

Aside from environmental conditions such as temperature, humidity, radio noise, and the presence of metallics and liquids in the objects' vicinity, the mere number of objects stacked together affects the average detection probability of an object. This occurs because the objects to be identified act as radio signal occluders, shielding other objects' tags from the readers. To better understand the effect of the number of objects on the average object detection probability, we conducted several experiments. The results of these experiments confirmed our expectations and revealed interesting patterns that we describe next.

We performed two back-to-back experiments to determine the effect of the number of objects on the average object detection probability. In these experiments, we used circular ThingMagic antennas and unipolar tags. In the first experiment, we grouped 15 solid nonmetallic and 15 metallic and liquid objects and determined the average object detection probabilities for liquids and metallics, and separately for solid, nonmetallic objects. In the second experiment, we grouped 20 solid nonmetallic and 20 liquid metallic objects, and again determined the average object detection probabilities. To ensure that the reader has sufficient time to detect all reader-visible tags in both experiments, we allocated 3 s for the reader to detect tags in the 15/15 experiment and (proportionally) 4 s for the 20/20 experiment. The detection probability statistics were calculated for various numbers of tags per object, as well as different numbers of reader antennas. For accurate comparison, in calculating the statistics in the second experiment, we used a subset of 15 solid nonmetallic and 15 liquid metallic objects that matched the objects in the first experiment.

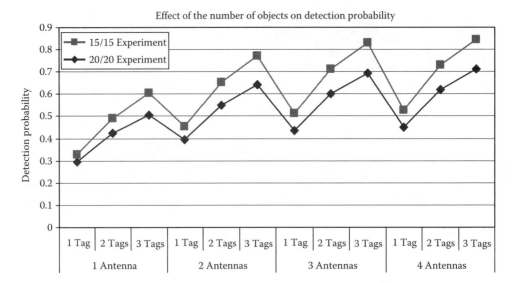

FIGURE 1.13 Effect of the number of objects on the average object detection probability. In the 15/15 experiment, we used 15 metallic and liquid objects, and 15 solid nonmetallic objects. Similarly, in the 20/20 experiment, we used 20 metallic and liquid objects, and 20 solid nonmetallic objects.

We compared the average object detection probabilities between two experiments, varying the number of tags per object and the number of reader antennas. Figure 1.13 shows the results of this comparison for metallic and liquid objects. Observe that the average detection probability of an object in a 15/15 experiment is greater than in a 20/20 experiment, as expected (because higher numbers of objects increase the likelihood of occlusions). The difference is more dramatic and vivid for metallic and liquid objects than for solid nonmetallic ones because the reader is operating at a high power level to detect metallic and liquid objects.

Note that the difference in object detection probabilities between the two experiments is greater when more tags are attached to an object, and when multiple readers are used for object identification. This occurs due to an overall improvement in object detection when multi-tags and multiple readers are used. These experiments clearly illustrate that multi-tags have a more positive influence than multiple readers on detection probabilities, especially in the presence of metallics and liquids, and when identifying larger groups of objects.

1.9 EFFECT OF MULTI-TAGS ON ANTICOLLISION ALGORITHMS

Anticollision algorithms enable a reader to uniquely identify tags while minimizing the number of tag broadcasting collisions (i.e., simultaneous interfering transmissions by the tags). Multi-tags have no effect on two variants of Binary Tree-Walking [8,20], and may at most double/triple the total read time for double/triple-tags over single tag for Slotted Aloha [8] and for Randomized Tree-Walking [21–23]. Our theoretical and experimental study of multi-tags addressed how multi-tags improve object detection. It is worth noting, however, that because not all tags are detected, the time required to identify all reader-visible tags is considerably less than double (or triple) the time needed to identify single-tagged objects by some anticollision protocols.

In particular, from our experiments we observed that 25–75 percent of all tags on solid/ nonmetallic objects are detected with one reader antenna, depending on its type and power level. The percentages are much lower for metallic and liquid objects. Therefore, attaching two tags to each object may not add any significant overall time delay for object identification. Moreover, current RFID technology can read hundreds of tags per second, making the increase in the number of tags insignificant, even in real-time systems. Finally, in many scenarios the benefits of successfully identifying all the objects certainly justifies a modest increase in identification time. Based on the above observations, RFID system designers should select an appropriate anticollision algorithm based on the number of objects that may have to be identified near-simultaneously, the number of tags attached to each object, and the expected objects' velocities (if the objects to be identified are moving).

1.10   MULTI-TAGS AS SECURITY ENHANCERS

1.10.1   CHAFFING AND WINNOWING

Multi-tags can enhance RFID security using the idea of "chaffing and winnowing" [24]. Chaffing creates messages with phony message authentication codes (MACs), and winnowing filters fake messages by comparing the MAC received along with the message against the MAC computed by the recipient. The achieved confidentiality can be made arbitrarily strong with smaller packet sizes. Sending chaff probabilistically, or controlling the amount of chaff sent will hide the real number of tags in the reader's interrogation zone [23]. This relatively low-cost technique is especially useful in preventing adversaries from performing accurate inventorization. For example, a business may want to perform rapid covert inventorization of a competitor, relying on RFID tagged inventory, and thus gain valuable information about a competitor's business practices. Prevention of unauthorized inventorization is a very interesting problem that merits future research.

1.10.2   PREVENTING SIDE-CHANNEL ATTACKS

Multi-tags can prevent certain side-channel attacks (e.g., "power analysis" attacks). An adversary can use power analysis attack to learn the kill password* of an electronic product code (EPC) tag, as demonstrated by Oren and Shamir in Ref. [25]. They showed that when an EPC compliant tag receives a kill password from the reader one bit at a time, the tag's power operation changes, allowing an adversary to detect power spikes when the tag receives an invalid bit. In a multi-tag scenario, one tag can counterbalance the power budget of the other tag by operating in an "opposite" mode, thus preventing simple power analysis, and consequently preventing the discovery of a kill password by an adversary.

1.10.3   SPLITTING ID AMONG MULTI-TAGS

Another technique to prevent accurate adversarial inventorization is the splitting of the tag ID/data into several parts, and distribute these parts among multi-tags. The multi-tags can transmit the data to the readers at different frequencies using code division multiple access, making it difficult for an adversary to reconstruct the complete signal (tag ID/data). This technique was used by the British during World War II to prevent the Germans from jamming Allied transmissions [26]. Note that the data splitting technique is unlikely to prevent adversarial tracking on its own because the tag's data is sent in the clear, but in conjunction with privacy preserving techniques it can be a powerful security mechanism. Splitting the data between tags may lower the overall cost of the system.

1.11   APPLICATIONS OF MULTI-TAGS

Multi-tags can be deployed in a variety of useful applications and serve many purposes. They can be used for specific tasks such as determining the location and orientation of objects, as well as ensuring system reliability, availability, and even safety. In addition, multi-tags can be a considerable deterrent to illegal activities such as theft and forgery, and they can enhance RFID security and privacy. For example, multi-tags can speed up the execution of some algorithms through parallel computation. Below, we give examples of scenarios and systems where multi-tags can be effective. These examples

---

* When a tag receives the correct kill password from a reader, it stops responding to future reader queries.

do not cover all possible applications; rather, they serve mainly to illustrate the wide range of uses and applications of multi-tags.

1.11.1 RELIABILITY

There are many RFID applications where system reliability is critical. For example, in a store scenario, checkout RFID readers should reliably detect all items purchased by the consumer. Missed items, even at a relatively low rate of 1 percent, can incur huge losses to a typical low-profit-margin business, thus significantly affecting the store's bottom line. Also, objects moving through a supply chain should be detected reliably to enable accurate real-time inventory control and early theft detection. In general, in most applications where goods change hands or objects move through an RFID checkpoint, all objects should be detected and identified accurately. Multi-tags attached to objects will greatly increase objects' detection probabilities at a reasonable cost.

1.11.2 AVAILABILITY

One example where multi-tags can improve system availability is in "yoking-proof" scenarios, where a potentially adversarial reader communicates with a group of tags and generates a proof that the tags were identified near-simultaneously [27,28]. The constructed proof is later verified by an off-line verifier. The integrity of the system hinges on the tags of all objects being detectable by the reader when required, because otherwise no valid proof can be created, even by an honest reader. The problem is exacerbated because of the tight timing constraints of the protocol, and the inherent variations in tag receptivity [17]. In such "yoking-proof" scenarios, multi-tags can be attached to each object, thus greatly increasing the probability of at least one tag per object being detectable. Note that here multi-tags may need to be physically connected to each other, so that they can consistently share their states with each other to prevent the possible forgery of a yoking proof. Another example of an application where availability is important is the real-time tracking of critical household or business objects such as remote controls, car keys, firearms, and important documents, among others.

1.11.3 SAFETY

Another, perhaps unexpected, area where multi-tags can be of great benefit is safety. Specifically, multi-tags can be used in healthcare to track medical instruments (e.g., gauze sponges). For example, surgical sponges, among other foreign objects, are sometimes left inside humans during operations, causing highly undesirable consequences that adversely affect the patients. Recent medical studies [29] have shown surprisingly good results in detecting RFID-equipped surgical gauze sponges during operations. However, to accurately detect all the sponges requires very careful and precise positionings of the reader. If the distance between the reader and the tags is increased even slightly, the tags may go undetected and thus the object may be inadvertently left inside the patient. In addition, the sponges may be located amid bodily fluids, further decreasing the detection probabilities. Finally, the tags on the sponges may break or malfunction, causing readers to miss tags, which may result in serious human injury. Attaching multi-tags to surgical sponges will greatly increase the probability of all sponges being detected and accounted for, which would translate into improved patient safety and reduced hospital liability.

1.11.4 OBJECT LOCATION

The location of a multi-tagged object can be more accurately determined than that of a single-tagged one. Well-known location triangulation methods can be utilized to determine the position of each tag, thus reducing the error in computing a multi-tagged object's location coordinates. A carefully engineered multi-tag RFID system can be used to determine not only an object's position but also its spatial orientation [30]. Directional antennas and orientation-sensitive RFID tags can be deployed to make such a system highly effective. Creating a working prototype of such a system and applying it in real-world scenarios is an interesting area for future research.

1.11.5 PACKAGING

Many RFID tag types are delivered to the customer on a continuous paper roll, and the customer later programs the tags with unique IDs. We envision that tags will soon be cheap enough to embed into, e.g., adhesive packaging tape used to wrap packages and containers, thus simplifying the multi-tagging of boxed objects, and enabling automatic tag diversity and orientation selection to greatly improve object detection at negligible cost. With higher tag ubiquity and the multi-tagging of objects, the testing of RFID tags will be obviated, because even a low tag production yield will enable the overall system to function properly. The acceptability of lower tag manufacturing yields will further reduce the production costs, while ensuring high object detection probabilities as well as improved dependability and reliability of RFID systems.

1.11.6 THEFT PREVENTION

Another useful set of applications of multi-tags is in theft prevention. Increasing the number of tags attached to (or embedded in) an object will make it much more difficult for a thief to shield or remove all of the tags, thereby increasing the probability of him getting caught. For example, one intriguing application of this could be the prevention of illegal deforestation* by embedding tags in the trunks of living trees [13]. Because tags are very cheap compared to the cost of lumber (especially for rare or legally protected trees such as Redwoods), the economics of such applications are financially viable. When logs are shipped and sold, they can be scanned for tags whose presence will determine the origin of the wood (and possibly convey other useful information, such as weather and environmental statistics tracked over the tree's lifetime). It would be prohibitively expensive for illegal loggers to detect and remove all of the tags from a given tree trunk, thus substantially increasing the cost and risk of illegal deforestation, at a relatively low cost to the protection agencies.

The attachment of the radio antenna(s) to the silicon chip, and tag packaging itself incur the majority of the cost in RFID tag manufacturing [31]. However, if we use multi-tags for theft prevention as described above, we neither need to package the tags nor be particularly precise or careful when attaching antenna(s) to chips. The mere large number of tags per object will guarantee that enough tags are still detectable, and will thus deter theft. The simpler process of producing unpackaged tags will considerably streamline the tag manufacturing process and consequently reduce their cost. In addition, in such scenarios, manufacturing yields are no longer required to remain high, and tag testing steps may be skipped as well, further contributing to significant tag cost reductions. We discuss the economics of multi-tag RFID in more detail in the next section.

1.11.7 TAGGING BULK MATERIALS

Cheap redundant multi-tags can be embedded into bulk materials (e.g., fertilizers, explosives, chemicals, propellants, crude oil, etc.) to prevent their unintended acquisition, transportation, and possible misuse. If tags are embedded into certain bulk materials at a reasonably small proportion to the size/quantity/weight of a substance, they will not adversely affect the normal use of these materials (e.g., crude oil can be tagged at the rate of ten multi-tags per barrel, and these tags can be removed during the final stages of the refinement process). If required, the tags can have limited lifespans or even be (bio)degradable. The RFID tagging of fertilizers/explosives can help law enforcement agencies trace the producer or buyer. The tagging of bulk materials can also directly prevent criminals/terrorists from causing damage by enabling law enforcement agencies to detect the presence of dangerous substances in proximity (or ominously en route) to sensitive locations or particular sites of interest, hopefully before an illegal act transpires.

---

* Illegal deforestation is not a hypothetical problem (e.g., see a recent news article in Ref. [32]).

1.12 ECONOMICS OF MULTI-TAGS

Based on RFID trials by corporations and government agencies (see Section 1.1), and our experimental results [17], it is clear that object detection probabilities are far from perfect, even when multiple readers/antennas are used. Multi-tags, potentially in conjunction with multiple readers, can help address the object detection problem. The cost of RFID tags in 2007 is around 8¢ each, making the multi-tagging of high-cost items viable today. In addition, the cost of tags is decreasing at an exponential rate following Moore's law, and this trend will enable the cost-effective tagging of even low-cost objects in the near future. Also, the cost of RFID tags is decreasing substantially faster than the cost of RFID readers, due to improving manufacturing yields and an economy of scale driven by massive deployments. Moreover, this price gap is expected to continue to widen due to the increasing demand for cheap RFID tags. The anticipated future omnipresence and ubiquity of RFID tags is expected to eventually reduce the cost of RFID tags into the subpenny level.

1.12.1 COSTS AND BENEFITS OF MULTI-TAGS

The cost of passive RFID tags has been decreasing rapidly over the last decade. From 2001 to 2006, the cost of passive tags has speedily dropped from $1.15 to $0.08 a piece, when at least 1 million units were purchased [33–35]. Based on this historical data, we predict that tags will cost $0.06 by the end of 2007, and 5¢ in 2008. A 5¢ price point for tags was considered the threshold for supporting a strong business case for item-level tagging [36], and now this target price is just around the corner. Based on the efforts of some companies and researchers working on RFID tag technology [31,37], we believe that ~1 penny tags will become a reality around the year 2011. Eventually, tags will be printed directly onto objects and cost less than a penny to produce. This cost milestone will make RFID a truly ubiquitous and affordable technology. Figure 1.14 depicts the historical (and our projected) decreasing cost trends for tags.

When considering the cost of RFID tags or even the cost of an entire RFID system, it is critical to also analyze the benefits that RFID brings to an application. A complete business analysis of deploying RFID should be performed, because the benefit of deploying RFID in an application can considerably outweigh the cost, even at today's prices. Specifically, the business analyses of RFID systems should take into account the direct savings that RFID deployment will enable, such as higher employee productivity, automated business processes, workforce reductions, and the valuable information collected through RFID.

In supply chain management scenarios, the benefits of RFID deployment are tremendous. First, the merchandize can be tracked in real-time, allowing more efficient scheduling of operations. RFID may also allow reductions in the number of workers, because many currently manual processes can be automated. RFID can also prevent theft of goods, which are stolen predominantly by insiders. According to National Association for Shoplifting Prevention (NASP), insider thieves outnumber outsider thieves six to one [38,39]. It has been documented that over 1 percent of goods in retail

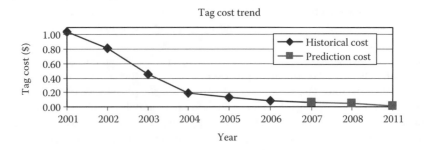

FIGURE 1.14 Cost trend of passive RFID tags over time, and our cost prediction for the future. The price per tag is based on the purchase of at least 1 million tags.

stores are stolen [39], and the real losses due to theft are likely to be much higher, as companies tend to underreport theft statistics. Multi-tag technology enables objects to be tracked more effectively, not only during transport or checkout but also during manufacturing and warehousing, which can significantly reduce theft rates and thereby increase profits.

1.12.2 TAG MANUFACTURING YIELD ISSUES

Manufacturing yield is one of the main criteria that influence the cost of VLSI chips. This is because customers have to pay not only for the good chips delivered to them but also for the defective chips that never made it out of the fabrication facility, as well as for the labor-intensive separation of the good ones from the defective ones. For example, according to recent research by RFID vendors, as many as 30 percent of RFID chips are damaged during production when chips are attached to their antennas, and an additional 10–15 percent are damaged during the printing process [4].

Due to the redundancy built into our proposed multi-tag RFID systems, we can often ignore the manufacturing yield. Some manufactured RFID tags may be defective, while others may fail in the field, but if multiple tags are attached to each object, the probability that all the tags fail is still quite small. This considerably increases the overall reliability of a multi-tag RFID system, and also decreases the tag manufacturing costs (e.g., expensive manufacturing steps such as testing may be dispensed with).

The failure rate of deployed RFID tags in the field is estimated to be as high as 20 percent [40]. This large failure rate induces an additional cost pressure on RFID tag manufacturing, because individual tags must be made more reliable, or extensively tested after manufacturing. Even after packaging, tags may become defective. For example, 5 percent of the tags that we purchased for our experiments were marked by the manufacturer as defective; moreover, we discovered several additional inoperable tags during the tag programming phase of our experiments. As with the yield issue, multi-tags allow us to ignore damaged tags and statistically rely on the promise that enough multi-tags will remain operational to satisfy an application's requirements. This property of multi-tag systems helps to improve the overall reliability and cost of deployed multi-tag RFID systems.

1.12.3 RFID DEMAND DRIVERS

A strong driver of cost in RFID systems is the scope of the demand for this technology. With increases in demand, the number of produced RFID units will increase, which drives the amortized development costs down. However, many companies are hesitant to deploy RFID technology because the business case is not entirely clear or proven. This classic "chicken-and-egg" dilemma has inhibited the massive deployments of RFID systems so far. With improvements in RFID technology, the cost of RFID systems should decrease, creating a more convincing business case for companies and accelerating the demand for the technology, which will in turn reduce the amortized cost of RFID tags even further. The demand for RFID will be driven by many companies with a wide range of specializations and fields, led by major players such as Wal-Mart and DoD, and the desire to remain competitive in rapidly evolving marketplaces. Consequently, companies will experience mounting pressures to adopt RFID technology, and multi-tag-based strategies will help bootstrap undecided companies into this technology and help propel them into the RFID age.

1.12.4 COST-EFFECTIVE TAG DESIGN TECHNIQUES

Overall tag cost can be reduced by developing better and cheaper tag components and assembling them in a more cost-effective manner. We give some practical examples of advanced memory design, antenna design, and assembly technologies to illustrate how technological developments drive down RFID costs. The cost of RFID tags can be reduced through innovative lower-cost memory design technologies. For example, the chip manufacturer Impinj, Inc., uses "self-adaptive silicon," which enables the low-cost reliable analog storage of bits in floating gates [41]. Another way to decrease the tag cost is to speed up the tag manufacturing and packaging processes. For example, Alien Technology has developed fluidic self-assembly (FSA), which allows for the placement of a large number of very small components across the surface in a single operation, significantly speeding up tag assembly. This technology involves flowing tiny microchips in a special fluid over a base containing holes shaped to catch the chips [42]. In addition to designing antennas with improved receptivity and orientation, measures can be taken to lower antenna costs. For example, Symbol Technologies reduced the cost of antennas by manufacturing them out of aluminum rather than silver. The company also compressed antennas into small, low-powered inlay, thus reducing tag area and cost [43].

1.12.5 SUMMARY OF MULTI-TAG ECONOMICS

RFID technology leverages Moore's law in the positive direction. RFID tags are getting both smaller and cheaper over time, resulting in a multiplicative corresponding reduction in tag cost. In addition, RFID tag yields are improving, further compounding the effect of these trends on cost reduction. Also, engineering and manufacturing tolerances for RFID chips are much larger than for high-end chips (e.g., RFID chips can operate at low clock speeds, extreme miniaturization is not a prominent problem in RFID production, etc.). Moreover, the VLSI manufacturing equipment for RFID tags does not have to be cutting edge, which reduces the cost pressure when constructing tag fabrication facilities. Rapidly increasing demand for RFID, along with cheaper manufacturing techniques and improving yields, is expected to rapidly bring the cost of RFID tags into the subpenny levels in the near future, making multi-tags ever more affordable. In short, multi-tags are clearly economically viable, and their benefits are bound to become even more dramatic over time.

1.13 CONCLUSION

There are many obstacles to reliable RFID-based object identification. Environmental conditions such as temperature, humidity, ambient radio noise, object quantities/geometries, etc. can significantly interfere with object detection and accurate identification. Dramatic variations in tag receptivity and detectability, even among tags of the same type and production batch, reduce the reliability of tag detection. The metals and liquids present in or around objects (or the environment) can reflect or absorb radio signals, thus preventing accurate signal decoding. In addition, objects density, concentration, and placement geometry can adversely influence object detection, thus affecting the availability, reliability, and even safety of an RFID system.

To significantly improve object detection, we proposed to attach multiple RFID tags (multi-tags) to each object. We defined different types of multi-tags. Through analytics, simulations, and experiments, we showed that multi-tags should be positioned perpendicular to each other whenever possible, and separated from each other to reduce chances of occlusion. Our experiments showed that multiple readers improve object detection only moderately, yet multi-tags provide much more dramatic gains in average object detection probability. We showed that multi-tags are very effective in dealing with radio noise, tag variability, and the presence of metallics and liquids among objects, as well as high object densities. We gave examples of numerous applications that could greatly benefit from multi-tags. We proposed several techniques to enhance RFID security using multi-tags. We analyzed the economics of multi-tags and argued that multi-tags are cost-effective even today for many cost-sensitive, safety-critical, and security-oriented applications. We predicted that multi-tags will become cost-justifiable for many more applications in the near future, as the cost of passive tags continues to rapidly drop. We also stressed the importance of careful RFID system design to ensure the desired operation and performance.

It is important to note that although multi-tags considerably improve object detection, especially in conjunction with multiple readers, they do not guarantee 100 percent object detection. Given the numerous obstacles to reliable object identification, it is very difficult to provide detection guarantees. In practical RFID deployments, the deployment site should be carefully analyzed for radio interfering phenomenon, allowing the system engineers to make appropriate design decisions. More research in the areas of RFID chip and antenna design, as well as RF technology, is needed to further improve the reliability of RFID-tagged object detection.

Neglecting to carefully consider the benefits and costs of multi-tags in a specific deployment may result in financial loses, degraded overall system performance, and other unintended consequences. For example, improving object detection might unexpectedly aid thieves in locating valuable items, thus hurting object owners. Similarly, the overuse of multi-tags may create additional interferences in the operation of anticollision algorithms, thus degrading object detection. It is also possible that for some applications multi-tags are not economically viable because they may require unjustifiable investment in extra tags and equipment to optimize tag placement. In general, RFID deployments require careful planning and testing on a case-by-case basis.

In summary, we believe that multi-tag RFID technology promises many benefits to numerous applications, and will expedite reductions in tag manufacturing cost. This will positively tip the cost-benefit scale in favor of massive RFID deployments, and encourage many companies, organizations, and communities to join the age of ubiquitous RFID.

ACKNOWLEDGMENT

This research was supported by grant CNS-0716635 from the National Science Foundation.

REFERENCES

[1] IDTechEx. RFID progress at Wal-Mart. www.idtechex.com/products/en/articles/00000161.asp, October 2005.

[2] SRA International, Inc. Resolute ordinance movement evaluation. http://www.dla.mil/j-6/ait/Documents/Reports/Resolute_Ordinance_May2001.as%px, May 2001. Appendix F—GTN Tag History.

[3] Cardinal Health releases RFID pilot results. http://www.cardinal.com/content/news/11142006_91731.asp, November 2006. Cardinal Health, Dublin, OH.

[4] Gao. Key considerations related to federal implementation of radio frequency identification technology. http://www.gao.gov/new.items/d05849t.pdf, June 2005. Testimony before the Subcommittee on Economic Security, Infrastructure Protection, and Cybersecurity, House Committee on Homeland Security.

[5] L. Bolotnyy and G. Robins. Multi-tag rfid systems. *International Journal of Internet Protocol Technology, Special Issue on RFID: Technologies, Applications, and Trends*, 2(3/4):218–231, 2007.

[6] Y. Lee. RFID coil design. Technical Report AN678, Microchip Technology, Inc., Chandler, AZ, 1998. ww1.microchip.com/downloads/en/AppNotes/00678b.pdf.

[7] C. A. Balanis. *Antenna Theory Analysis and Design*. John Wiley & Sons, New York, 1997.

[8] K. Finkenzeller. *RFID Handbook*. John Wiley & Sons, West Sussex, England, 2003.

[9] T. Scharfeld. An analysis of the fundamental constraints on low cost passive radio-frequency identification system design. Master's thesis, MIT, Cambridge, MA, 2001.

[10] L. Bolotnyy. New directions in reliability, security and privacy in radio frequency identification systems. PhD thesis, University of Virginia, Charlottesville, VA, 2007.

[11] G. Marsaglia. Choosing a point from the surface of a sphere. *Annals of Mathematical Statistics*, 43(2):645–646, 1972.

[12] M. Reynolds. The physics of RFID, 2003. http://www.media.mit.edu/events/movies/video.php?id=rfid privacy-2003-11-15-1.

[13] L. Bolotnyy and G. Robins. Multi-tag radio frequency identification systems. In *Proceedings of the IEEE Workshop on Automatic Identification Advanced Technologies (Auto-ID)*, pp. 83–88, Buffalo, NY, October 2005.

[14] E. Schuster, T. Scharfeld, P. Kar, D. Brock, and S. Allen. Analyzing the rfid tag read rate issue. http://mitdatacenter.org/CutterITAdvisor.pdf, 2004.

[15] Impinj. Receptivity—A tag performance metric. www.impinj.com/files/MR_MZ_WP_00005_Tag Receptivity.pdf, December 2005.

[16] Y. Chen, A. B. Kahng, G. Robins, and A. Zelikovsky. Area fill synthesis for uniform layout density. *IEEE Transactions on Computer-Aided Design*, 21(10):1132–1147, 2002.

[17] L. Bolotnyy, S. Krize, and G. Robins. The practicality of multi-tag rfid systems. In *Proceedings of the International Workshop on RFID Technology—Concepts, Applications, Challenges (IWRT 2007)*, pp. 100–113, Madeira, Portugal, June 2007.

[18] L. Bolotnyy and G. Robins. The case for multi-tag rfid systems. In *International Conference on Wireless Algorithms, Systems and Applications (WASA)*, Chicago, IL, August 2007.

[19] A. Rahmati, L. Zhong, M. Hiltunen, and R. Jana. Reliability techniques for rfid-based object tracking applications. In *Annual IEEE/IFIP International Conference on Dependable Systems and Networks (DSN)*, pp. 113–118, Edinburgh, U.K., June 2007.

[20] A. Juels, R. Rivest, and M. Szedlo. The blocker tag: Selective blocking of RFID tags for consumer privacy. In V. Atluri (ed.), *Proceedings of the ACM Conference on Computer and Communications Security*, pp. 103–111, Washington, DC, October 2003.

[21] Auto-ID Center. Draft Protocol Specification for a 900 MHz Class 0 Radio Frequency Identification Tag, 2003.

[22] L. Bolotnyy and G. Robins. Randomized pseudo-random function tree walking algorithm for secure radio-frequency identification. In *Proceedings of the IEEE Workshop on Automatic Identification Advanced Technologies (Auto-ID)*, pp. 43–48, Buffalo, NY, October 2005.

[23] S. Weis. Security and privacy in radio-frequency identification devices. Master's thesis, MIT, Cambridge, MA, May 2003.

[24] R. Rivest. Chaffing and winnowing: Confidentiality without encryption. *CryptoBytes*, 4(1):12–17, 1998.

[25] Y. Oren and A. Shamir. Power analysis of rfid tags, 2006. http://www.wisdom.weizmann.ac.il/~yossio/rfid/.

[26] D. Nolan. Internet technologies in a converged network environment. *NCS Technical Information Bulletin* 04-2, 2004.

[27] L. Bolotnyy and G. Robins. Generalized 'yoking proofs' for a group of radio frequency identification tags. In *International Conference on Mobile and Ubiquitous Systems (Mobiquitous)*, San Jose, CA, July 2006.

[28] A. Juels. "yoking-proofs" for RFID tags. In R. Sandhu and R. Thomas (Eds.), *International Workshop on Pervasive Computing and Communication Security*, pp. 138–143, Orlando, FL, March 2004.

[29] A. Macario, D. Morris, and S. Morris. Initial clinical evaluation of a handheld device for detecting retained surgical gauze sponges using radiofrequency identification technology. *Archives of Surgery*, 141:659–662, 2006.

[30] S. Hinske. Determining the position and orientation of multi-tagged objects using rfid technology. In *Fifth Annual IEEE International Conference on Pervasive Computing and Communications Workshops (PerComW'07)*, pp. 377–381, White Plains, NY, March 2007.

[31] P. Peumans. Monolithic, low-cost rfid tags. http://peumans-pc.stanford.edu/research/monolithic-low-cost-rfid-tags, 2006. Project at Stanford Organic Electronics Lab.

[32] KIROTV.com. *3 Accused of Felling Old-Growth Trees.* http://www.kirotv.com/news/14210458/detail.html, September 2007.

[33] R. Moscatiello. Forecasting the unit cost of RFID tags. http://www.mountainviewsystems.net/Forecasting%20the%20Unit%20Cost%20of%20RFID%20Tags.pdf, July 2003.

[34] M. O'Connor. Alien drops tag price to 12.9 cents. *RFID Journal.* http://www.rfidjournal.com/article/articleview/1870/1/1/, September 2005.

[35] M. Roberti. A 5-cent breakthrough. *RFID Journal.* http://www.rfidjournal.com/article/articleview/2295/1/128/, May 2006.

[36] S. Sarma. Towards the five-cent tag. Technical Report MIT-AUTOID-WH-006, 2001. Auto-ID Labs.

[37] OrganicID, Inc. Printable, plastic RFID tags. In *Printed Electronics*, OrganicID, Inc., New Orleans, LA, December 2004.

[38] National Association for Shoplifting Prevention (NASP). Shoplifting statistics. www.shopliftingprevention.org.

[39] M. Vargas. Shoplifting and employee theft recovery up last year. http://retailindustry.about.com/od/lp/a/bl_hayes_theft.htm, 2005. From 17th Annual Retail Theft Survey by Jack L. Hayes International, Inc.

[40] *RFID Journal.* RFID system components and costs, 2005. http://www.rfidjournal.com/article/articleview/1336/3/129/.

[41] Impinj, Inc. Our technology. http://www.impinj.com/advantage/our-technology.aspx.

[42] FSA manufacturing. http://www.alientechnology.com/technology/fsa_manufacturing.php. Alien Technology, Inc., Morgan Hill, CA.

[43] Symbol Technologies. Symbol technologies launches portfolio of RFID generation 2 and specialty tag inlays, May 2006. http://news.thomasnet.com/companystory/484018.

2 Attacking RFID Systems

*Pedro Peris-Lopez, Julio Cesar Hernandez-Castro,
Juan M. Estevez-Tapiador, and Arturo Ribagorda*

CONTENTS

A great number of hackers end up working in the security departments of IT and telecommunications companies. In other words, the best way of making a system secure is knowing how it can be attacked. Radio-frequency identification (RFID) is no different from any other technology, so the possible attacks on it should be studied in depth. The extent of an attack can vary considerably; some attacks focus on a particular part of the system (e.g., the tag) whereas others target the whole system. Although there are references to such attacks in a number of publications, a rigorous study has not been made of the subject until now. We examine, in this chapter, the main threats to RFID security. First, we look at data and location privacy. Although these are the risks most often referred to in the literature, there are other equally important problems to consider too. RFID systems are made up of three main components (tag, reader, and back-end database), so we have grouped the threats according to the unit involved in the attack. First, we examine those related to tags and readers such as eavesdropping, cloning, replay, and relay attacks. Then we look at the threats to the back-end database (e.g., object name service [ONS] attack, virus). By the end of this chapter (and with the opportunity to consult the extensive bibliography for further details), we hope the reader will have acquired a basic understanding of the principal security risks in RFID.

2.1 INTRODUCTION

2.1.1 BACKGROUND

Press stories about radio-frequency identification (RFID) often give inaccurate descriptions of the possibilities that exist for abuse of this technology. They predict a world where all our possessions will have a unique identification tag: clothes, books, electronic items, medicines, etc. For example, an attacker outside your house equipped with a commercial reader would be able to draw up an inventory of all your possessions, and particular information such as your health and lifestyle could also be revealed. Also, it is said that this technology allows "Big Brother" to know when you are in public places (office, cinemas, stores, pubs, etc.), tracking all your movements and compromising your privacy in terms of your whereabouts (location).

RFID technology is a pervasive technology, perhaps one of the most pervasive in history. While security concerns about the possibility of abuse of this pervasive technology are legitimate, misinformation, and hysteria should be avoided. One should be aware that ways of collecting, storing, and analyzing vast amounts of information about consumers and citizens existed before the appearance of RFID technology. For example, we usually pay with credit cards, give our names and address for merchandizing, use cookies while surfing the Internet, etc.

In this chapter we give an overview of the risks and threats related to RFID technology, helping the reader to become better acquainted with this technology. Although the privacy issues are the main focus in literature [1–12], there are other risks that should be considered when a RFID system is designed.

2.1.2 ATTACK OBJECTIVES

The objectives of each attack can be very different. It is important to identify the potential targets to understand all the possible attacks. The target can be the complete system (i.e., disrupt the whole of a business system) or only a section of the entire system (i.e., a particular item).

A great number of information systems focus solely on protecting the transmitted data. However, when designing RFID systems, additional objectives, such as tracking or data manipulation should be considered. Imagine the following example in a store: an attacker modifies the tag content of an item reducing its price from 100 € to 9.90 €. This leads to a loss of 90 percent for the store. In this scenario, the data may be transmitted in secure form and the database has not been manipulated. However, fraud is carried out because part of the system has been manipulated. Therefore, to make a system secure, all of its components should be considered. Neglecting one component, whatever the security level of the remaining components, could compromise the security of the whole system.

The objectives of the attacks are very different. As we see in the above example, the attack may be perpetrated to steal or reduce the price of a single item, while other attacks could aim to prevent all sales at a store. An attacker may introduce corrupt information in the database to render it inoperative. Some attacks, such as the faraday cage or active jamming, are inherent in the wireless technology employed. Other attacks are focused on eliminating physical access control, and ignore the data. Other attacks even involve fraudulent border crossings, identity stealing from legitimate e-passports, etc.

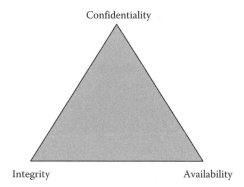

FIGURE 2.1   Three pillars of security: the CIA triad.

2.1.3 SECURITY NEEDS

As any other mission-critical system, it is important to minimize the threats to the confidentiality, integrity, and availability (CIA) of data and computing resources. These three factors are often referred to as "The Big Three." Figure 2.1 illustrates the balance between these three factors.

However, not all systems need the same security level. For example, not all systems need 99.999 percent availability or require that its users be authenticated via retinal scans. Because of this, it is necessary to analyze and evaluate each system (sensitivity of the data, potential loss from incidents, criticality of the mission, etc.) to determine the CIA requirements. To give another example, the security requirements of tags used in e-passports should not equal those employed in the supply chain (i.e., tag compliant to EPC Class-1 Generation-2).

Confidentiality: The information is accessible only to those authorized for access. Privacy information, such as the static identifiers transmitted by tags, fits into the confidentiality dimension. Both users and companies consider this issue of utmost importance. Furthermore, RFID technology allows the tracking of items. From a user perspective, tracking should be avoided. However, companies may control the movements of materials in the supply chains, increasing the productivity of their processes.

Integrity: The assurance that the messages transmitted between two parties are not modified in transit. Additionally, some systems provide the authenticity of messages. The receipt is able to prove that a message was originated by the purported sender and is not a forgery (nonrepudiation). An example of this kind of attack is the spoofing attack.

Availability: System availability is whether (or how often) a system is available for use by its intended users. This factor will determine the performance and the scalability level of the system. Denial-of-service (DoS) attacks are usual threats for availability (i.e., active jamming of the radio channel or preventing the normal operation of vicinity tags by using some kind of blocker tag).

Each time a new technology is implanted, contingency plans for various points of failure should be designed. We recommend periodical security audits to review the security polices, procedures, and IT infrastructures. As has been frequently mentioned, RFID technology may be a replacement for bar-code technology. Nevertheless, new risk scenarios should be considered with its implantation. For example, consider the repercussions of a bar-code reader failing or an RFID reading going down. When a bar-code reader fails, an operator can manually enter the codes into the terminal and the system works, albeit with relatively slowness. On the other hand, if the RFID reader is processing high volumes of items and these items are moving at high speed, the consequences will be much worse. Security needs should therefore be considered a priority.

2.2   MAIN SECURITY CONCERNS

2.2.1   PRIVACY

No one shall be subjected to arbitrary interference with his privacy, family, home, or correspondence, nor to attacks upon his honor and reputation. Everyone has the right to the protection of the law against such interference or attacks [13].

Whereas data-processing systems are designed to serve man; whereas they must, whatever the nationality or residence of individuals, respect their fundamental rights and freedoms, notably the right to privacy, and contribute to economic and social progress, trade expansion and the well-being of individuals [14].

Privacy has no definite boundaries and its meaning is not the same for different people. In general terms, it is the ability of an individual or group to keep their lives and personal affairs out of public view, or to control the flow of information about themselves.

The invasion of privacy by governments, corporations, or individuals is controlled by a country's laws, constitutions, or privacy laws. For example, taxation processes normally require detailed private information about earnings. The EU Directive 95/46/EC [14] on the protection of individuals with regard to the processing of personal data and the free movement of this, limits and regulates the collection of personal information. Additionally, Article 8 of the European Convention of Human Rights identifies the right to have private and family life respected. Within this framework, monitoring the use of e-mails, Internet, or phones in the workplace, without notifying employees or obtaining their consent can result in legal action.

RFID technology is a pervasive technology, and seems destined to become more and more so. As Weiser already predicted in 1991, one of the main problems that ubiquitous computing has to solve is privacy [15]. Leakage of information is a problem that occurs when data sent by tags reveals sensitive information about the labeled items. Products labeled with insecure tags reveal their memory contents when queried by readers. Usually, readers are not authenticated and tags answer in a transparent and indiscriminate way.

As an example of the threat this could pose, consider the pharmaceutical sector where tagged medication is planned for the immediate future. Imagine that when you leave the chemist's with a given drug—say an antidepressive or AIDS treatment, an attacker standing by the door equipped with a reader could find out what kind of medication you have just bought. In a similar scenario, thieves equipped with tag readers could search people, selecting those with multiple tagged bank bills to rob, and they would know how much they would earn with each robbery.

Advanced applications, where personal information is stored in the tags, have appeared recently. E-passports are a good example of this sort of application. As part of its U.S.-VISIT program, the U.S. government mandated the adoption of e-passports by the 27 nations in its Visa-Waiver Program. A combination of RFID technology and biometric technology is employed [7,16,17]. The RFID tags store the same information that is printed on its first page (name, date of birth, passport number, etc.) as well as biometric information (facial image). In phase-2 of the European e-passport project [18], the biometric data from two fingerprints, which is very sensitive information, will also be stored.

Several organizations like CASPIAN [19] and FOEBUD [20] are strongly against the massive deployment of RFID technology. They believe that RFID technology will lead to a significant loss of citizens' privacy. Some of CASPIAN's activities include successful boycott campaigns against important companies like Benetton [21,22], Tesco [23], and Gillette [24], to name but a few. Additionally, a book titled "*SPYCHIPS: How Major Corporations and Government Plan to Track your Every Move with RFID*" and published in 2005 [25], has contributed to promoting suspicion about RFID technology.

Another example of objection to RFID technology is the case of California State Senator Joe Simitian (Senate Bill 682), who planned to restrict the use of identification systems based on RFID technology: "The act would prohibit identity documents created, mandated, or issued by various public entities from containing a contactless integrated circuit or other device that can broadcast personal information or enable personal information to be scanned remotely" [26]. Due to significant industry opposition, Bill 682 was stalled in the Assembly Appropriations Committee and an important missed deadline resulted in the expiry of the Bill. Legislative maneuvring allowed the resurrection of the case by means of Bill 768 [27]. This bill was finally vetoed by California Governor Arnold Schwarzenegger. In particular, Bill 768 proposed to

1. Criminalize the "skimming" of personal data from RFID-enable identification documents.
2. Implement specific provisions to ensure the security of data contained in such identification documents.
3. Impose a three-year moratorium on the use of RFID technology in certain types of government-issued identification documents.

In 2002, Garfinkel proposed a set of rights that should be upheld by any system that uses RFID technology [28]. Consumers should have:

1. Right to know whether products contain RFID tags.
2. Right to have RFID tags removed o deactivated when they purchase products.
3. Right to use RFID-enabled services without RFID tags.
4. Right to access an RFID tag's stored data.
5. Right to know when, where and why the tags are being read.

These rights are not necessarily considered as the basis for a new law, but as a framework for voluntary guidelines that companies wishing to deploy this technology may adopt publicly.

2.2.2 TRACKING

Location information is a set of data describing an individual's location over a period of time [29]. The resolution of the system (time and localization) depends on the technology used to collect data.

Indeed, location privacy can be viewed as a particular type of privacy information [30]. A secondary effect of wireless communication is that information can be made public and collected. In a mobile phone context, the regions are divided up into cells. Each time a phone enters a new cell, the mobile is registered. Mobile phone operators record handset location information and supply it to third parties (i.e., police, the company that subscribed the localization service, etc.). Other techniques such as triangulation can be used to increase the precision of the system. The new localization services (i.e., third-generation mobile phones) allow an accuracy of a few meters by means of the incorporation of a global positioning system (GPS) receiver. In data network context, Wireless 802.11 Ethernet cards obtain connectivity by registering with access points which could be used to locate a network device.

RFID technology is not a high-tech bugging device. It does not possess GPS functionality or the ability to communicate with satellites. RFID tags do not have the storage and transmission capability for large quantities of information. An RFID system is normally composed of three components: tags, readers, and a back-end database. Readers are connected, using a secure channel, to the database. When a database is present in the system, tags might only transmit an identifier. This identifier is used as a index-search in the database to obtain all the information associated with the tag. Therefore, only people with access to the database can obtain the information about the labeled item.

Most of the time, tags provide the same identifier. Although an attacker cannot obtain the information about the tagged item, an association between the tag and its holder can easily be established. Even where individual tags only contain product codes rather than a unique serial number, tracking is still possible using an assembly of tags (constellations) [31]. To clarify the potential risks of tracking, some examples are given:

Wall-Mart: It is an American public corporation, currently one of the world's largest. It has concentrated on streamlining the supply chain, which is why it encourages all its suppliers to incorporate RFID technology in their supply chains. The substitution of bar codes by RFID tags allows an increase in the reading-rate of the pallets as they move along the conveyor belt. RFID readers can automatically scan these as they enter or leave the warehouse, saving time and improving product flow. Right now, RFID technology is used at pallet level. Individual packaging is the next logical step.

Individual product packaging: Imagine that your Tag Heuer bifocals possess a tag, and this tag stores a 96 bit static identifier, allowing an attacker to establish a link between the identifier and you. On association, an attacker could know when you passed through a given place, for example when you enter or leave your home, when you arrive at or leave your office, etc. Even worse, the attacker could locate several readers in your favorite mall. He could collect data over a long time (data, time, shop, etc.) acquiring a consumer profile of you. Finally, he could send you personalized advertising information depending on your shopping habits.

E-passports: Since October 2006, the United States required the adoption of e-passports by all the countries in its Visa-Waiver Program. The International Civil Aviation Organization (ICAO) standard specifies one mandatory cryptographic feature (passive authentication) and two optional cryptographic features (basic access control and active authentication). Passive authentication only demonstrates that tag content is authentic but it does not prove that the data container is secure. Basic authentication ensures that tag content can only be read by an authorized reader. Additionally, a session key is established, encrypting all the information exchanged between the tag and the reader. Active authentication is an anticloning feature, but it does not prevent unauthorized readings. Independent of the security mechanism used, tracking is possible. The electronic chip required by the ICAO must conform to ISO/IEC 14443 A/B already adopted in other applications [32,33]. The collision avoidance in ISO 14443 uses unique identifiers that allow readers to distinguish one tag from another [17]. However, this identifier will allow an attacker to unequivocally identify an e-passports's holder. One simple countermeasure is to generate a new random identifier each time the tag is read.

As has been shown, RFID technology is not the only one that permits the tracking of people (i.e., video surveillance, mobile phone, Wireless 802.11 Ethernet cards, GPS, etc.). Nevertheless, the equipment used to track people holding RFID tags is not very expensive. If we return to the example of tracking in a mall, we will understand one of the principal differences between RFID and other localization technologies. The great majority of malls have a video surveillance system. You can be filmed in all the supermarket sections in which you buy an item. Then, the information obtained by the system (images) has to be processed to obtain your consumer profile. However, if RFID technology was employed, data could be automatically collected without the need for subsequent data processing as in video systems.

2.3 TAGS AND READERS

2.3.1 OPERATING FREQUENCIES AND READING DISTANCES

RFID tags operate in four primary frequency bands [34]:

1. Low frequency (LF) (120–140 kHz)
2. High frequency (HF) (13.56 MHz)
3. Ultrahigh frequency (UHF) (860–960 MHz)
4. Super high frequency/microwave (μW) (2.45 GHz and above)

The characteristics of different frequencies are summarized in Table 2.1.

TABLE 2.1

Tag Frequencies and Reading Distances

Frequency Band	Frequency	Distance	Energy Transfer
Low (LF)	125 kHz	1–90 cm, typically around 45 cm	Inductive coupling
High (HF)	13.56 MHz	1–75 cm, typically under 40 cm	Inductive coupling
Ultrahigh (UHF)	865–868 MHz (Europe) 902–928 MHz (United States) 433 MHz (active tags)	Up to 9 m	Electromagnetic coupling
Microwave (μW)	2.45 GHz 5.8 GHz	Typically 0.3–0.9 m	Electromagnetic coupling

LF tags: These tags operate at 120–140 kHz. They are generally passive and use near-field inductive coupling. So they are suited for applications reading small amounts of data at relatively slow speeds and at short distances. Their read range varies from 1 to 90 cm, typically below 45 cm. LF tags do not support simultaneous tag reads. LF tags are relatively costly because they require a longer, more expensive copper antenna. They penetrate materials such as water, tissue, wood, and aluminum. Their common applications are in animal identification, automobile security, electronic article surveillance, commerce, and other areas.

HF tags: These tags operate at 13.56 MHz. They are typically passive and typically use inductive coupling. HF tags penetrate materials well, such as water, tissue, wood, aluminum, etc. Their data rates are higher than LF tags and their cost is lower due to the simple antenna design. Their read ranges varies from 1 to 75 cm, typically under 40 cm. HF tags are used in smart shelf, smart cards, libraries, baggage handling, and other applications.

UHF tags: UHF active and passive tags can operate at different frequencies. UHF active tags operate at 433 MHz, and UHF passive tags usually operate at 860–960 MHz. Generally, passive UHF tags are not very effective around metals and water. They perform well at distances greater than 90 cm. UHF passive tags usually reach about 9 m. UHF tags have good non-line-of-sight communication, a high data rate, and can store relatively large amounts of data.

Super high frequency/microwaves tags: These tags operate at frequencies of 2.45 GHz and above (also 5.8 GHz) and can be either active or passive. Their characteristics are similar to those of UHF tags. However, they have faster read rates and are less effective around metals and liquids than tags of lower frequencies. These tags can be smaller in size compared to LF, HF, and UHF tags and are used for electronic toll collection as well as for the tracking of shipping containers, trains, commercial vehicles, parking, etc. The read range varies from 0.3 to 0.9 m for passive tags and is very dependent on design. Active systems also use microwave frequency.

2.3.2 Eavesdropping

RFID technology operates through radio, so communication can be surreptitiously overheard. In Ref. [35], the possible distances at which an attacker can listen to the messages exchanged between a tag and a reader are categorized (see Figure 2.2).

Forward channel eavesdropping range: In the reader-to-tag channel (forward channel) the reader broadcasts a strong signal, allowing its monitoring from a long distance.

Backward channel eavesdropping range: The signal transmitted in the tag-to-reader (backward channel) is relatively weak, and may only be monitored in close proximity to the tag.

Operating range: The read ranges shown in Section 2.3.1 are the operating read range using sales-standard readers.

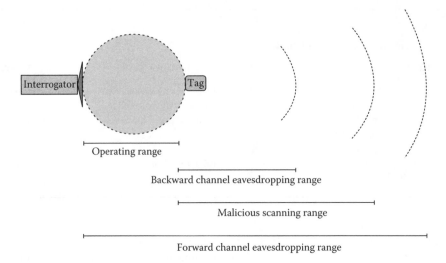

FIGURE 2.2 Eavesdropping range classification. (From Ranasinghe, D.C. and Cole, P.H., Confronting security and privacy threats in modern RFID systems. In *Proceedings of ACSSC 06*, 2006, pp. 2058–2064. With permission.)

Malicious scanning range: An adversary may build his own reader-archiving longer read ranges, especially if regulations about radio devices are not respected. A conversation between a reader and a tag can be eavesdropped over a greater distance than is possible with direct communication. For example, tags compliant to ISO 14443 have a reading distance of around 10 cm (using standard equipment). However, Kfir et al. showed that this distance can be increased to 55 cm employing a loop antenna and signal processing [36].

Eavesdropping is particular problematic for two reasons:

1. Feasibility: it can be accomplished from long distances.
2. Detection difficulty: it is purely passive and does not imply power signal emission.

Eavesdropping attacks are a serious threat mainly when sensitive information is transmitted on the channel. To give an example, we consider the use of RFID technology in payments cards (RFID credit cards) [37]. In an eavesdropping attack, information exchanged between the credit card reader and the RFID credit card is captured. Heydt-Banjamin et al. showed how this attack can be carried out [38]. An antenna was located next to an off-the-shelf RFID credit card reader. The radio signal picked up by the antenna was processed to translate it into human readable form. In particular, the following pieces of data were captured: cardholder name, complete credit card number, credit card expiry date, credit card type, and finally information about software version and supported communications protocols. As the above example shows, eavesdropping attacks should therefore be considered and treated seriously.

2.3.3 AUTHENTICATION

Entity authentication allows the verification of the identity of one entity by another. The authenticity of the claimed entity can only be ascertained for the instant of the authentication exchange. A secure means of communication should be used to provide authenticity of the subsequent data exchanged. To prevent replay attacks, a time-variant parameter, such as a time stamp, a sequence number, or a challenge may be used. The messages exchanged between entities are called tokens. At least one token has to be exchanged for unilateral authentication and at least two tokens for mutual authentication. An additional token may be needed if a challenge has to be sent to initiate the protocol.

In RFID context, the first proposals found in literature are based on unilateral authentication [39–41]. However, the necessity of mutual authentication has been confirmed in many publications [42–45]. In ISO/IEC 9784, the different mechanisms for entity authentication are described [46]:

- Part 1: General model
- Part 2: Entity authentication using symmetric techniques
- Part 3: Entity authentication using a public key algorithm
- Part 4: Entity authentication using a cryptographic check function

Use of a cryptographic check function seems to be the most precise solution for RFID. Due to the fact that standard cryptographic primitives exceed the capabilities of a great number of tags, the design of lightweight primitives is imperative, at least for low-cost RFID tags.

The two entities (claimant/verifier) share a secret authentication key. An entity corroborates its identity by demonstrating knowledge of the shared key. This is accomplished by using a secret key with a cryptographic check function applied to specific data to obtain a cryptographic check value. This value can be recalculated by the verifier and compared with the received value. The following mechanisms, as shown in Figure 2.3, are possible.

2.3.4 SKIMMING

Takashimaya, one of the largest retailers in Japan, now sells antiskimming cards called "Sherry" at their department stores. Consumers can just put the cards in their wallets to prevent their RFID-chipped train passes, etc. from skimming attacks.

The antiskimming card functions by creating a reverse electromagnetic field like Taiyo's technology [47].

One pass—unilateral authentication

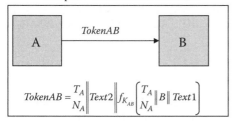

Two pass—unilateral authentication

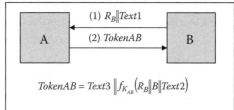

Two pass—mutual authentication

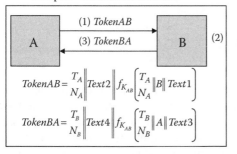

Three pass—mutual authentication

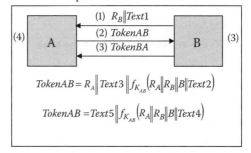

FIGURE 2.3   Entity authentication mechanisms.

Eavesdropping is the opportunistic interception of information exchanged between a legitimate tag and legitimate reader. However, skimming occurs when the data stored on the RFID tag is read without the owner's knowledge or consent. An unauthorized reader interacts with the tag to obtain the data. This attack can be carried out because most of the tags broadcast their memory content without requiring authentication.

One interesting project is the Adam Laurie's RFIDIOt project [48]. Specifically, RFIDIOt is an open source library for exploring RFID devices. Several experiments with readers operating at 13.56 MHz and 125/134.2 kHz are shown. The number of standards supported by the library is around 50. Some examples of the attacks carried out are the following:

Nonauthentication example: In 2004, Verichip received approval to develop a human-implant RFID microchip [49]. About twice the length of a grain of rice, the device is typically implanted above the triceps of an individual's right arm. Once scanned at the proper frequency, the Verichip answers with a unique 16 digit number which can correlate the user to the information stored on a database. The type of tag used by Verichip appears to be an EM4x05. This kind of tag can be read simply with the program "readlfx.py," obtaining the following information: card ID, tag type, application identifier, country code, and national ID.

Password authentication example: Since 2003, the Oyster card has been used on Transport for London and National Rail services. The Oyster card is a contactless smart card, with a claimed proximity range of about 8 cm, and based on Philips's MIFARE® standard [50]. A code for attacking this kind of card is included. The sample program "bruteforce.py" can be run against it, and it will try to log in the sector 0 by choosing random numbers as the key.

Nowadays, the security of e-passports have aroused a great interest [16,17,51,52]. Skimming is problematic because e-passports possess sensitive data. The mandatory passive authentication mechanism demands the use of digital signatures. A reader will be able to verify that the data came from the correct passport-issuing authority. However, digital signatures do not link data to a specific passport. Additionally, if only passive authentication is supported, an attacker equipped with a reader could obtain sensitive information such as your name, birthday, or even your facial photograph. This is possible because readers are not authenticated—in other words, the tag answers indiscriminately. Certain projects exist which give the code needed to read e-passports: RFIDIOt (Adam Laurie) [48], OpenMRTD (Harald Welte) [53], and JMRTD (SoS group, ICIS, Radbound University) [54].

2.3.5 CLONING AND PHYSICAL ATTACKS

Symmetric-key cryptography can be used to avoid tag cloning attacks. Specifically, a challenge–response like the following can be employed. First, the tag is singulated from many by means of a collision-avoidance protocol like the binary tree walking protocol. The tag (T_i) shares the key (K_i) with the reader. Afterward, the following messages are exchanged:

1. The reader generates a fresh random number (R) and transmits it to the tag.
2. The tag computes $H = g(K_i, R)$ and sends back to the reader.
3. The reader computes $H' = g(K_i', R)$ and checks its equality with H.

The g function can be implemented by a hash function or, alternatively, by an encryption function. Note that if the g function is well constructed and appropriately deployed, it is infeasible for an attacker to simulate the tag. Because standard cryptographic primitives (hash functions, message authentication codes, block/stream ciphers, etc.) are extravagant solutions for low-cost RFID tags on account of their demand for circuit size, power consumption, and memory size [55], the design of new lightweight primitives is pressing.

For some kinds of tags, resources are not so restricted. However, their cost is much higher than low-cost RFID tags (i.e., tags used in supply chain). An example of these sort of tags are e-passports. The active authentication method is an anticloning feature. The mechanism relies on public cryptography. It works by having e-passports prove possession of a private key:

1. The tag generates an 8 bytes nonce and sends it to the tag.
2. The tag digitally signs this value using its private key and transmits it to the reader.
3. The reader can verify the correctness of the response with the public key supposedly associated with the passport.

Tamper-resistant microprocessors are used to store and process private and sensitive information, such as private keys or electronic money. The attacker should not be able to retrieve or modify this information. To achieve this objective, chips are designed so that the information is not accessible using external means and can only be accessed by the embedded software, which should contain the appropriate security measures.

Making simple electronic devices secure against tampering is very difficult, as a great number of attacks are possible, including [56]:

- Mechanical machining
- Laser machining
- Energy attacks
- Temperature imprinting
- Probe attacks
- Active or injector probes
- Energy probes
- Manual material removal
- Clock glitching
- Electronic beam read/write
- Imaging technology
- Water machining
- Shaped charge technology
- Radiation imprinting
- High-voltage imprinting
- Passive probes
- Pico probes
- Matching methods
- High or low voltage
- Circuit disruption
- IR laser read/write
  ⋮

As sensitive information such as cryptographic keys are stored on the chips, tamper-resistant devices may be designed to erase this information when penetration of their security encapsulation or out-of-specification environmental parameters is detected. Some devices are even able to erase all their information after their power supply has been interrupted.

In the RFID context, we have to distinguish between low-cost RFID tags and tags used in applications without severe price restrictions. Low-cost RFID tags are very constrained resources (storing, computing, and energy consumption). These kinds of tags are usually nonresistant to physical attacks. An example of these kinds of tags are tags compliant with the EPC Class-1 Generation-2 specification [57]. High-cost tags, sometimes called contactless chips or smart cards, are not so restrictive regarding resources. However, price increases from 0.05 € to several euros. For example, the chips used in e-passports have an EAL 5+ security level, the highest security level for chips [58]. Therefore, an attacker will not be able to acquire the private key used in private authentication to avoid cloning attacks. The plusID tag, manufactured by Bradcom, is another example of tamper-resistant tags [59]. Initially, its security level was 2 (tamper evidence) according to Federal Information Processing Standards (FIPS), but it was finally increased to level 3 (tamper resistant).

2.3.6 REPLAY AND RELAY ATTACKS

A replay attack copies a stream of messages between two parties and replays it to one or more of two parties. A generalized definition of a replay attack could be the following: an attack on a security protocol using replay of messages from a different context into the intended (or original and expected) context, thereby fooling the honest participant(s) into thinking they have successfully completed the protocol run [60]. An exhaustive classification of replay attacks can be found in Ref. [61].

Common techniques to avoid replay attacks are incremental sequence number, clock synchronization, or a nonce. In Ref. [62], a set of design principles for avoiding replays attacks in cryptographic protocols is presented. In a RFID context, clock synchronization is not feasible because passive RFID tags cannot make use of clocks, as these kind of tags do not have an on-board power source. Incremental sequence such as session token may be a straightforward solution if tracking is not considered a threat. Therefore, the use of nonce is the most suitable option for RFID tags.

A number of factors combine to make relay attacks on RFID technology is possible. Tags are read over a distance and activated automatically when close to a reader. Therefore, an attacker could communicate with a tag without the knowledge of its owner.

Two devices, as shown in Figure 2.4, are involved in the relay attack: the ghost and the leech [36]. The ghost is a device which fakes a card to the reader, and the leech is a device which fakes a reader to the card. A fast communication channel between the legitimate reader and the victim card is created by the ghost and the leech:

1. Legitimate reader sends a message (A) to the ghost.
2. Ghost receives it and forwards this message (A) to the leech through the fast communication channel (minimum delay).
3. Leech fakes the real reader, and sends the message (A) to the legitimate tag.
4. Legitimate tag computes a new message (B) and transmits it to the leech.

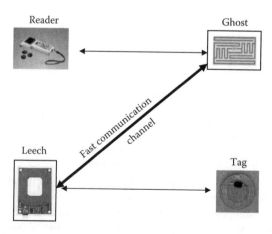

FIGURE 2.4   Relay attacks.

5. Leech receives it and forwards this message (B) to the ghost through the fast communication channel.
6. Ghost forwards this message (B) to the real reader.

This sort of attack dispels the assumption that readers and tags should be very close to communicate. Additionally, even if communications are encrypted, the attack is feasible because messages are only relayed through a fast communication channel, without requiring knowledge of its content. In Ref. [63], a practical relay attack against ISO 14443 compliant tags is described.

2.3.7 HIDING

RFID technology uses electromagnetic radio waves. Labeled items can be therefore protected by insulating them from any kind of electromagnetic radiation:

Faraday cage: A faraday cage or shield is a container made of conducting material, or a mesh of such material. This blocks out radio signals of certain frequencies. There are currently a number of companies that sell this type of solution [64,65].
Passive jamming: Each time a reader wants to interact with a single tag, the tag will have to be singulated from a population of tags. A collision-avoidance protocol such as Aloha or binary tree walking protocol may be employed. To conceal the presence of a particular tag, this could simulate the full spectrum of possible tags in the singulation phase, hiding its presence. This concept was first introduced by Juels et al. as the "Blocker tag" [66]. In 2004, a variant of the blocker concept, named "soft blocking," was introduced [67]. This involves software (or firmware) modules that offer a different balance of characteristics from ordinary blockers.
Active jamming: Another way of achieving isolation from electromagnetic waves is disturbing the radio channel known as active jamming of RF signals. This disturbance may be effected with a device that actively broadcasts radio signals, so as to completely disrupt the radio channel, thus preventing the normal operation of RFID readers. However, in most cases, government regulations on radio emissions (power and bandwidth) are violated [68].

2.3.8 DEACTIVATING

Some methods exist for deactivating tags and rendering them unreadable. The most common method consists of generating a high-power RF field that induces sufficient current to burn out a weak section of the antenna. The connection between the chip and the antenna is cut off, rendering it useless. This method is usually chosen to address privacy concerns and to deactivate tags that are used to label individual items or prevent thefts in stores.

The benefits of using RFID technology in a store are clear. However, the deactivation of tags may be malicious. The necessary technology can be available to anyone. The usual range of a "kill" signal is only a few centimeters. However, designing and building a high-gain antenna with a high-power transmitter is easy. Using batteries, it could probably fit into a backpack. Then an attacker entering a store could kill all the tags, causing widespread retail chaos. A practical implementation of this sort of attack is the RFID-Zapper project [69,70].

Karjoth et al. proposed the use of physical RFID structures that permit a consumer to disable a tag by mechanically altering the tag [71]. In "clipped tags," the consumer can physically separate the body (chip) from the head (antenna) in an intuitive way. Such separation provides visual confirmation that the tag has been deactivated. Then the tag can be reactivated by means of physical contact. The reactivation requires deliberate actions on the part of its owner. Indeed, reactivation cannot be carried out without the owner's knowledge unless the item was stolen.

To avoid wanton deactivation of tags, the use of kill passwords has been proposed. Tags compliant to the EPC Class-1 Generation-2 implement this feature [57]. When an electronic product code (EPC) tag receives the "kill" command, it renders itself permanently inoperative. However, to protect tags from malicious deactivation, the kill command is PIN protected. One of the main problems linked to solutions based on password is its management. Employing the same password for all tags could be a naive solution. Nevertheless, if a tag is compromised, all the tags would be at risk. Another straightforward solution is that each tag has a different password with the associated management and scalability problems.

The potential benefits of RFID technology usage are reduced if tags are permanently deactivated. Instead of killing tags, they could be put to sleep, rendering them only temporarily inoperative. As with the killing process, sleeping/waking up tags will not offer real protection if anyone is able to accomplish these operations. So some form of access control, such a PINs, will be needed. To sleep/wake up a tag, a PIN has to be transmitted.

2.3.9 CRYPTOGRAPHIC VULNERABILITIES

In nineteenth century, Kerckhoffs sets out the principles to the public known of cryptography systems [72]:

1. The system must be practically, if not mathematically, indecipherable.
2. It must not be required to be secret, and it must be able to fall into the hands of the enemy without inconvenience.
3. Its key must be communicable and retainable without the help of written notes, and changeable or able to be modified at the will of the correspondents.
4. It must be applicable to telegraphic correspondence.
5. It must be portable, and its usage and function must not require the concourse of several people.
6. Finally, it is necessary, given the circumstances that command its application, that the system be easy to use, requiring neither mental strain nor the knowledge of a long series of rules to observe.

RFID tags are very constrained devices, with restrictions in power consumption, storage, and circuitry. Due to these severe limitations, some commercial RFID tags support weak cryptographic primitives, and thus vulnerable authentication protocols. Additionally, some of these cryptographic primitives are proprietary. The use of proprietary solutions is not really inadequate if algorithms are published to be analyzed by the research community. However, time has shown, the security of an algorithm cannot reside in "obscurity." A system relying on security through obscurity may have theoretical or actual security vulnerabilities, but its owners or designers believe that the flaws are unknown, and that attackers are unlikely to find them [72].

Texas Instruments manufacture a low-frequency tag, named digital signature transponder (DST). The DST executes a challenge–response protocol. The reader and the DST share a secret key K_i. The reader sends a challenge R to the DST. The DST computes an encryption function of the challenge $C = e_{K_i}(R)$ and sends this value to the reader. The reader computes $C' = e_{K_i'}(R)$ and compares this value with the received value. The challenge is 40 bits in length, and the output of the encryption function is 24 bits in length. The length of the K_i is only 40 bits. It is a very short length. The National Institute of Standards and Technology [73] and the ECRYPT EU Network of Excellence on cryptography [74] recommended in 2005 a key length of 80 bits for a minimal level of general-purpose protection, and 112 bits for the following ten years.

The most common uses of DST are the following:

1. DST is employed as a theft-deterrent (immobilizer keys) in automobiles, such as Ford and Toyota vehicles.
2. DST serves as a wireless payment device (speedpass), which can be used by more than seven million individuals in around 10,000 Exxon and Mobile gas stations.

Texas Instruments has not published details of the encryption algorithm, basing itself on security through algorithm obscurity. A team of researchers at Johns Hopkins University and RSA Laboratories discovered security vulnerabilities in the DST [75]. In particular, a successful reverse engineering of the DST encryption algorithm was accomplished. First, a rough schematic of the cipher was obtained from a published Texas Instruments presentation. With the reverse engineering of the cipher, they showed that a 40 bit key length was inadequate, the cipher being not only vulnerable to brute-force attacks as known by cryptographers. The attack can be divided into three phases:

Reverse engineering: They were equipped with a DST reader and some blank DST tags. With the reader and the blank tags, the output of the encryption function, with any key and challenge, could be obtained. Using specific key/challenge pairs and centering on the schematic of the encryption, operational details of the algorithm were derived.

Key cracking: After determining the encryption algorithm, a programmed hardware "key cracker" was implemented to recover the unique cryptographic key of the DST. The cracker operated by brute force (full space of 2^{40}). Given two input–output, in about 30 minutes the secret key was recovered.

Simulation: They programmed a hardware device with the key recovered from the DST. This device could impersonate the original DST.

The research on the DST exemplifies Kerckhoffs principles. Another significant example is the proprietary CRYPTO1 encryption algorithm used in Philips Mifare cards, which has been recently reverse engineered [76]. We recommend the publication of any algorithms. Open algorithms can be analyzed and refined by the scientific community, bolstering confidence in their security.

2.4 BACK-END DATABASE

2.4.1 TAG COUNTERFEITING AND DUPLICATION

Because the incorporation of RFID technology in sensitive applications such as passports [77] or pharmaceutical pedigrees [78], the possibility of creating counterfeiting tags has unleashed some concerns.

Here are some arguments that may dissuade users from alarmist attitudes [79]:

1. Usually, each tag has a unique identifier (ID) that allows its unequivocal identification. To counterfeit a tag, one would have to modify the identity of an item, which generally implies tag manipulation. The tag (ID) implementation may vary in each manufacturer as well as in each product. The major manufacturers first program the tag and then lock it. So resistance to using attacks lies in the lock. In most cases, it is not possible to unlock the tag without using invasive techniques. These techniques are not commonly available to the general industry.
2. RFID tags are generally sold preprogrammed with their identifiers, this being one of the phases of the normal production process. The ID format usually accords with a standard. The nonavailability of blank tags will therefore reduce the possibility of counterfeiting.
3. Another alternative is the design of blank tags. However, even with the equipment necessary for IC fabrication, designing these kind of chips is not an easy task.

Despite the difficulty of counterfeiting tags, on some occasions tags are duplicated. It is a similar problem to that of credit card fraud where a card is duplicated and possibly used in multiple places at the same time. As duplicate tags cannot be operatively distinguished, the back-end database should detect rare conditions. An example of a rare condition is the following: a tag cannot be in the toll gate on the Madrid–Barcelona motorway and 15 minutes later in the toll gate of Valencia–Barcelona motorway. The design of back-end database should be considered case by case [80].

2.4.2   EPC NETWORK: ONS ATTACKS

The EPCglobal network is made up of three key elements, as displayed in Figure 2.5:

1. EPC information services (EPC-IS)
2. EPC discovery services
3. Object name service

When an RFID tag is manufactured with an EPC, the EPC is registered within the ONS. The RFID tag is attached to a product, and the EPC becomes a part of that product as it moves through the supply chain. The particular product information is added to the manufacturer's EPC-IS, and the knowledge that this data exists within the manufacturer's EPC-IS is passed to the EPC discovery service.

The ONS is a distributed but authoritative directory service that routes request for information about EPCs. Existing or new network resources can be employed to route the requests. The ONS is similar to domain name service (DNS) both technologically and functionally. When a query is sent to the ONS including the EPC code, one or more localizations (uniform resource locator or URL) where information about items reside, are returned. The ONS service is divided in two layers. First, the Root ONS, which is the authoritative directory of manufacturers whose products may have information on the EPC Network. Second, the Local ONS, which is the directory of products for that particular manufacturer.

As the ONS can be considered a subset of the DNS, the same security risks are applicable. In 2004, a threat analysis of the domain name system was published as RFC 3833 [81]. Some of the principal threats identified are the following:

1. Packet interception: manipulating Internet Protocol (IP) packets carrying DNS information.
2. Query prediction: manipulating the query/answer schemes of the DNS protocol.
3. Cache poisoning: injecting manipulated information into DNS caches.
4. Betrayal by trusted server: attacker controlling DNS servers in use.
5. DoS: DNS is vulnerable to DoS as happens in any other network service. Additionally, the DNS itself might be used to attack third parties.

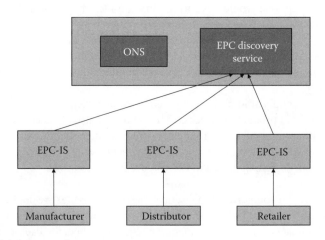

FIGURE 2.5   EPCglobal network.

However, there are some risks that are particular to the ONS service [82]:

1. Privacy: There are many situations where the EPC of an RFID tag can be considered highly sensitive information. Sensitive information can be obtained even knowing just part of the EPC. For example, knowing only the class of the identifier, you can find out the kind of object. To obtain the information associated with a tag, the EPC-IS has to be located. Even if the connections to the EPC-IS are secured (i.e., Secure Sockets Layer [SSL]/Transport Layer Security [TLS] protocol), the initial ONS look-up process is not authenticated nor encrypted in the first place. Therefore, sensitive information passes in clear on the channel (middleware-networks-DNS server).

2. Integrity: The correctness and the completeness of the information should be guaranteed. An attacker controlling intermediate DNS servers or launching a successful man-in-the-middle attack could forge the list of URLs (i.e., a fraudulent server). To prevent this attack, an authentication mechanism should be used for the EPC-IS.

3. Availability: If the adoption of the EPC network is widespread, there will be a great number of companies dependent on network services. ONS will become a service highly exposed to attacks. These could include distribute denial-of-service (DDoS) attacks that reduce the functioning of the server or its network connection by issuing countless and intense queries, or targeted exploits that shut down the server software or its operating system.

2.4.3 VIRUS ATTACKS

The RFID tag memory contains a unique identifier, but additional data may be stored. The data size varies from a few bytes to several kilobytes. The memory where this additional information is stored is rewritable. The information sent by the tags is implicity trusted, which implies some security threats [80,83]:

1. Buffer overflow: It is one of the most frequent sources of security vulnerabilities in software. Programming languages, such as C or C++, are not memory safe. In other words, the length of the inputs are not checked. An attacker could introduce an input that is deliberately longer, writing out of the buffer. As program control data is often located in memory areas adjacent to data buffers, the buffer overflow may lead the program to execute an arbitrary code. As a great number of tags have severe storage limitations, resource-rich tag simulating devices could be utilized [84].

2. Code insertion: An attacker might inject malicious code into an application, using any script language (i.e., common gateway interface, Java, Perl, etc.). RFID tags with data written in a script language could perform an attack of this kind. Imagine that the tags used for tracking baggage in the airport contain the airport destination in its data field. Each time a tag is read, the back-end system fires the query, "select * form location_table where airport =<tag data>." Imagine that an attacker stores in one piece of baggage "MAD;shutdown." When this data is read, the database will be shutdown and the baggage system will crashed.

3. Structure Query Language (SQL) injection: It is a type of code insertion attack, executing SQL codes in the database that were not intended. The main objectives of these attacks are the following: enumerate the database structure, retrieve authorized data, make unauthorized modifications or deletions, etc. RFID tags could contain data for a SQL injection attack. Storage limitation is not a problem, as it is possible to do a lot of harm with a very small amount of SQL. For example, the SQL "drop table <*tablename*>" will delete a specified database table.

Summarizing, an RFID tag is an unsecured and untrusted database. So the information obtained from such devices should be analyzed until there is sufficient evidence that the data is accurate. However, this is not a new concept, as in all information systems the imput data should be examined to ensure that it will not cause problems.

REFERENCES

[1] L. Bolotnyy and G. Robins. Physically unclonable function-based security and privacy in RFID systems. In *Proceedings of PerCom'07*, pp. 211–220. IEEE Computer Society Press, Washington, DC, 2007.

[2] J. Cichon, M. Klonowski, and M. Kutylowski. Privacy protection in dynamic systems based on RFID tags. In *Proceedings of PerSec'07*, pp. 235–240. IEEE Computer Society Press, Washington, DC, 2007.

[3] T. Heydt-Benjamin, H.-J. Chae, B. Defend, and K. Fu. Privacy for public transportation. In *Proceedings of PET'06, LNCS*, 4258, pp. 1–19. Springer-Verlag, Cambridge, U.K., 2006.

[4] T. Hjorth. Supporting privacy in RFID systems. Master thesis, Technical University of Denmark, Lyngby, Denmark, 2004.

[5] A. Juels and R. Pappu. Squealing euros: Privacy protection in RFID-enabled banknotes. In *Proceedings of FC'03, LNCS*, 2742, pp. 103–121. Springer-Verlag, Guadeloupe, French West Indies, 2003.

[6] S. Kinoshita, M. Ohkubo, F. Hoshino, G. Morohashi, O. Shionoiri, and A. Kanai. Privacy enhanced active RFID tag. In *Proceedings of ECHISE'05*, Munich, Germany, 2005.

[7] E. Kosta, M. Meints, M. Hensen, and M. Gasson. An analysis of security and privacy issues relating to RFID enabled epassports. In *Proceedings of Sec'07, IFIP*, 232, pp. 467–472. Springer, Sandton, South Africa, 2007.

[8] D. Ranasinghe, D. Engels, and P. Cole. Security and privacy: Modest proposals for low-cost RFID systems. In *Proceedings of Auto-ID Labs Research Workshop*, Zurich, Switzerland, 2004.

[9] M. Rieback, B. Crispo, and A. Tanenbaum. Uniting legislation with RFID privacy-enhancing technologies. In *Security and Protection of Information*, Brno, Czech Republic, 2005.

[10] M. Rieback, G. Gaydadjiev, B. Crispo, R. Hofman, and A. Tanenbaum. A platform for RFID security and privacy administration. In *Proceedings of LISA'06*, Washington, DC, 2006.

[11] S.E. Sarma, S.A. Weis, and D.W. Engels. RFID systems and security and privacy implications. In *Proceedings of CHES'02, LNCS*, 2523, pp. 454–470. Springer-Verlag, Redwood City, CA, 2002.

[12] S. Spiekermann and H. Ziekow. RFID: A 7-point plan to ensure privacy. In *Proceedings of ECIS'05*, Regensburg, Germany, 2005.

[13] Universal declaration of human rights, Article 12, 1948.

[14] EU Directive 95/46/EC—Data Protection Directive. Official Journal of the European Communities, November 23, 1995.

[15] M. Weiser. The computer for the 21st century. *Scientific American*, 265(3):94–104, September 1991.

[16] J.-H. Hoepman, E. Hubbers, B. Jacobs, M. Oostdijk, and R. Wichers Schreur. Crossing borders: Security and privacy issues of the European e-passport. In *Proceedings of IWSEC'06, LNCS*, 4266, pp. 152–167. Springer-Verlag, Kyoto, Japan, 2006.

[17] A. Juels, D. Molnar, and D. Wagner. Security and privacy issues in e-passports. In *Proceedings of SecureComm'05*. IEEE Computer Society, Athens, Greece, 2005.

[18] Advanced security mechanisms for machine readable travel documents—extended access control (EAC) version. 1.0.1. Technical guideline TR-03110, Federal Office of Information Security, Bonn, Germany, 2006.

[19] CASPIAN. http://www.nocards.org/, October 1, 2005.

[20] FoeBuD. http://www.foebud.org/rfid, October 5, 2005.

[21] Boycott Benetton. http://www.boycottbenetton.com/, April 9, 2003.

[22] *RFID Journal*. Behind the benetton brouhaha. http://www.rfidjournal.com, April 14, 2003.

[23] Boycott Tesco. http://www.boycotttesco.com/, Janurary 26, 2005.

[24] Boycott Guillette. http://www.boycottguillette.com/, September 2, 2003.

[25] K. Albrecht and L. McIntyre. *SPYCHIPS: How Major Corporations and Government Plan to Track your Every Move with RFID*. Nelson Communications, Inc., Nashville, TN, 2005.

[26] California Senate Bill 682. http://www.epic.org/privacy/rfid/, February 22, 2005.

[27] What's in California's proposed RFID Bill? http://www.rfidproductsnew.com, January 20, 2006.

[28] S. Garfinkel. Bill of Rights. http://www.technologyreview.com, October 2002.

[29] G. Danezis, S. Lewis, and R. Anderson. How much is location privacy worth. In *Proceedings of Workshop of Economics of IS'05*, Cambridge, MA, 2005.

[30] A. Beresfor and F. Stajano. Location privacy in pervasive computing. *IEEE Pervasive Computing*, 2(1):1536–1268, 2003.

[31] S. Weis, S. Sarma, R. Rivest, and D. Engels. Security and privacy aspects of low-cost radio frequency identification systems. In *Proceedings of SPC'03, LNCS*, 2802, pp. 454–469. Springer-Verlag, 2003.

[32] Identification cards–contactless integrated circuits cards–proximity cards. http://www.wg8.de/sdi.html, 2001.

[33] Machine readable travel documments, Doc. 9303. http://www.mrtd.icao.int, July 8, 2006.

[34] M. Brown, E. Zeisel, and R. Sabella. *RFID+Exam Cram*. Que Publishing, Indianapolis, IN, 2006.

[35] D.C. Ranasinghe and P.H. Cole. Confronting security and privacy threats in modern RFID systems. In *Proceedings of ACSSC 06*, pp. 2058–2064, Pacific Grove, CA, 2006.

[36] Z. Kfir and A. Wool. Picking virtual pockets using relay attacks on contactless smartcard systems. In *Proceedings of SecureComm'05*. IEEE Computer Society, Athens, Greece, 2005.

[37] J. Atkinson. Contactless credit card consumer report. http://www.findcreditcards.org, April 3, 2006.

[38] T.S Heydt-Benjamin, D.V. Bailey, K. Fu, A. Juels, and T. Ohare. Vulnerabilities in first-generation RFID-enabled credit cards. In *Proceedings of FC'07*, LNCS. Springer-Verlag, Lowlands, Scarborough, Trinidad/Tobago, 2007.

[39] M. Feldhofer. A proposal for an authentication protocol in a security layer for RFID smart tags. In *Proceedings of MELECON'04*, vol. 2. IEEE Computer Society, Dubrovnik, Croatia, 2004.

[40] I. Vajda and L. Buttyán. Lightweight authentication protocols for low-cost RFID tags. In *Proceedings of UBICOMP'03*, Seattle, WA, 2003.

[41] M. Ohkubo, K. Suzuki, and S. Kinoshita. Cryptographic approach to "privacy-friendly" tags. In *Proceedings of RFID Privacy Workshop*, MIT, Cambridge, MA, 2003.

[42] P. Peris-Lopez, J.C. Hernandez-Castro, J. Estevez-Tapiador, and A. Ribagorda. M2AP: A minimalist mutual-authentication protocol for low-cost RFID tags. In *Proceedings of UIC'06, LNCS*, 4519, pp. 912–923. Springer-Verlag, Wuhan and Three Gorges, China, 2006.

[43] H.Y. Chien and C.H. Chen. Mutual authentication protocol for RFID conforming to EPC class-1 generation-2 standards. *Computer Standards and Interfaces*, 29(2):254–259, 2007.

[44] A. Juels. Minimalist cryptography for low-cost RFID tags. In *Proceedings of SCN'04, LNCS*, 3352, pp. 149–164. Springer-Verlag, Amalfi, Italy, 2004.

[45] D. Molnar and D. Wagner. Privacy and security in library RFID: Issues, practices, and architectures. In *Proceedings of ACM CCS'04*, pp. 210–219. ACM Press, Washington, DC, 2004.

[46] ISO/IEC 9798 Information Technology—Security techniques—Entity authentication. http://www.iso.org, 1995.

[47] Anti-skimming in Japan. http://www.future.iftf.org/index.html, August 10, 2005.

[48] A. Laurie. RFIDIOt project. http://www.rfidiot.org, August 5, 2007.

[49] Verichip corporation. http://www.verichipcorp.com, August 15, 2007.

[50] Easing traveling in London's congested public transport network. http://www.mifare.net/showcases/london.asp, August 10, 2007.

[51] M. Halváč, Martin, and T. Rosa. A note on the relay attacks on e-passports: The case of Czech e-passports. In *Cryptology ePrint Archive, Report 2007/244*, IACR, 2007.

[52] D. Carluccio, K. Lemke, and C. Paar. Electromagnetic side channel analysis of a contactless smart card: First results. In *Handout of Workshop on RFID Security*, Graz, Austria, 2006.

[53] H. Welte. OpenMRTD project. http://www.openmrtd.org, August 7, 2007.

[54] SoSGroup, ICIS, and Radbound University. JMRTD project. http://www.jmrtd.sourceforge.net/, August 9, 2007.

[55] A Juels. RFID security and privacy: A research survey. Manuscript, 2005.

[56] S.H. Weingart. Physical security devices for computer subsystems: A survey of attacks and defenses. In *Proceedings of CHES'00, LNCS*, 1965, pp. 302–317. Springer-Verlag, Worcester, MA, 2000.

[57] Class-1 Generation-2 UHF air interface protocol standard version 1.0.9: "Gen-2". http://www.epcglobalinc.org/standards/, 2005.

[58] C. Lee, D. Houdeau, and R. Bergmann. Evolution of the e-passport. http://www.homelandsecurityasia.com, September 3, 2007.

[59] C. Swedberg. Broadcom introduces secure RFID chip. RFID Journal. http://www.rfidjournal.com, June 29, 2006.

[60] S. Malladi, S. Alves-Foss, and R. Heckendorn. On preventing replay attacks on security protocols. In *Proceedings of SM'02*, pp. 77–83, CSREA Press, Las Vegas, NV, 2003.

[61] P. Syverson. A taxonomy of replay attacks. In *Proceedings of CSF'94*, pp. 187–191. IEEE Computer Society, Franconia, NH, 1994.

[62] T. Aura. Strategies against replay attacks. In *Proceedings of CSF'97*. IEEE Computer Society, Rockport, MA, 1997.

[63] G. Hancke. Practical attacks on proximity identification systems (short paper). In *Proceedings of SP'06*. IEEE Computer Society, Oakland, CA, 2000.

[64] mCloak for RFID tags. http://www.mobilecloak.com/rfidtag/rfid.tag.html, September 10, 2005.

[65] Envelope to help you do it with your security, privacy, and discretion intact. http://www.emvelope.com, August 13, 2007.

[66] A. Juels, R. Rivest, and M. Szydlo. The blocker tag: Selective blocking of RFID tags for consumer privacy. In *ACM CCS'03*, pp. 103–111. ACM Press, Washington, DC, 2003.

[67] A. Juels and J. Brainard. Soft blocking: Flexible blocker tags on the cheap. In *WPES'04*, pp. 1–7. ACM Press, Washington, DC, 2004.

[68] RSA Laboratories. Faq on RFID and RFID privacy. http://www.rsa.com/rsalabs/node.asp?id=2120, October 4, 2007.

[69] J. Collins. RFID-Zapper shoots to kill. *RFID Journal*, January 23, 2006.

[70] MiniMe and Mahajivana. RFID-Zapper project. http://www.events.ccc.de/congress/2005/static/r/f/i/RFID-Zapper(EN)_77f3.html, 2006.

[71] G. Karjoth and P.A. Moskowitz. Disabling RFID tags with visible confirmation: Clipped tags are silenced. In *Proceedings of WPES'05*. ACM Press, Alexandria, VA, 2005.

[72] A. Kerckhoffs. La cryptographie militaire. *Journal des Sciencies*, 9:161–191, 1983.

[73] Recommendation for key management. Technical Report Special Publication 800-57 Draft, National Institute of Technology, 2005.

[74] Year report on algorithms and keysizes. Technical Report IST-2002-507932, ECRYPT, 2006.

[75] S. Bono, M. Greem, A. Stubblefield, A. Juels, A. Rubin, and M. Syzdlo. Security analysis of a cryptographically-enabled device. In *Proceedings of SSYM'05*. Usenix Association, Alexandria, VA, 2005.

[76] N. Karten and H. Plotz. Mifare little security, despite obscurity. http://events.ccc.de/congress/2007/Fahrplan/events/2378.en.html, 2007.

[77] P. Prince. United States sets date for e-passports. *RFID Journal*, October 25, 2005.

[78] E. Wasserman. Purdue Pharma to run pedigree pilot. *RFID Journal*, May 31, 2005.

[79] M. Guillory. Analysis: Counterfeit tags. http://www.aimglobal.org, June 30, 2005.

[80] F. Thornton, B. Haines, A. Das, H. Bhargava, A. Campbell, and J. Kleinschmidt. *RFID Security*. Syngress Publishing, 2006.

[81] D. Atkins and R. Austein. Threat analysis of the domain name system (DNS). In *Request for Comments—RFC 3833*, Berkeley, CA, 2004.

[82] B. Fabian, G. Oliver, and S. Spiekermann. Security analysis of the object name service for RFID. In *Proceedings of SecPerU'05*. IEEE Computer Society, Santorini Island, Greece, 2005.

[83] M. Rieback, C. Bruno, and A. Tanenbaum. Is your car infected with a computer virus? In *Proceedings of PerCom'06*. IEEE Computer Society, Pisa, Italy, 2006.

[84] B. Jamali, P.H. Cole, and D. Engels. In *Networked RFID Systems and Lightweight Cryptography, chapter RFID Tag Vulnerabilities in RFID Systems*, pp. 147–155. Springer, 2007.

3 RFID Relay Attacks: System Analysis, Modeling, and Implementation

Augusto Lima, Ali Miri, and Monica Nevins

CONTENTS

Radio-frequency identification (RFID) is fast becoming a ubiquitous tool for secure, anonymous identification of individuals. Every day, people use RFID tags to gain access to otherwise restricted areas or equipment, or to quickly process their transactions at the gas pump. RFID technologies are even used for highly sensitive identification documents, including passports. However, with this ever-increasing use come the inevitable questions of security and authenticity. RFID systems have many potential vulnerabilities; of these, let us focus on the simplest physical-layer attack: the relay attack.

In this chapter, we focus on practical aspects of the implementation of relay attacks on RFID systems, while providing the mathematical analysis of such attacks. An overview of ISO 14443 standards and requirements is given to provide a basis for understanding the details behind the communication between tags (also known as proximity integrated circuity cards [PICCs]) and tag readers (proximity coupling devices [PCDs]). The details provided are sufficient to allow a reader with a basic telecommunications and electronics background to be able to mount such attacks, and thus acquire the knowledge necessary to develop countermeasures. We analyze the security of some proposed solutions to counter relay attacks, and conclude with suggestions on how the security of these devices can be improved by using mathematical algorithms or by applying different techniques with today's hardware technology.

3.1   INTRODUCTION

Radio-frequency identifications (RFIDs) are widely used in devices such as access cards and speed passes. The growth in use of these devices worldwide has been exponential.

In this chapter, we introduce the principal concepts of several schemes used to attack the systems that use these types of devices, focusing primarily on so-called *relay attacks*. We analyze relay attacks in detail, and describe details of their implementation. The purpose of this is twofold: first, to provide a deeper understanding of RFID vulnerabilities; and second to serve as a background for the further analysis of security solutions needed to counter such attacks.

Relay attacks pose a serious threat to the security of RFID systems, in part because they are at the physical layer, and so can circumvent application-layer cryptographic countermeasures. In many instances, a victim will not be even aware of the attack. Moreover, relay attacks are simple, and can be achieved with low-cost equipment. The range of these types of attacks, and their success rate, can be improved using various techniques and equipment, some of which are discussed in Section 3.4.4.

Figure 3.1 depicts a typical RFID communication system, which consists of a tag, a reader, and a server. Note that in an RFID system, one needs to consider not only the communication system itself but also the power needed by each device in the system. In particular, there are two types of tags used—*passive* and *active*—each with different power requirements. Active tags are self-powered. Passive tags do not have their own power source, and need to obtain their energy via the reader

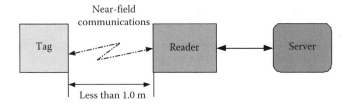

FIGURE 3.1 RFID communication system.

TABLE 3.1

Frequency Spectrum Usage of RFID

Frequency Band	Characteristics	Applications
100–500 KHz	Short to medium range, very low throughput	Access control, inventory information, animal control
12–13 MHz	Short range	Access control, smart cards
900 MHz	Medium reading speed	Access control, smart cards
2.45 and 5.8 GHz	Long range, higher speeds	Vehicle identification, toll collection

carrier. As a result, passive tags can usually communicate with readers no further than 1 m away. This restricted range of operation is often considered an addition to the security of the underlying RFID system. Table 3.1 provides an overview of the frequency spectrum usage for typical RFID applications.

The first goal of this chapter is to present the typical setup and hardware needed to implement these attacks. This discussion demonstrates the vulnerability of the RFID-enabled devices in the marketplace today and the need for better solutions with respect to security.

We begin in Section 3.2 by outlining some of the common physical-layer attacks which threaten the security of RFID systems, with special attention to describing relay attacks. With these attacks in mind, we proceed to give a detailed overview of a typical RFID system in Section 3.3. We go on to consider the issue of the channel link budget in Section 3.4, which helps to define the basic system features of the relay attack.

Also in Section 3.4, we consider how multiple protocols, beamforming, and increased channel capacity can be exploited to improve relay attacks (and, on the flip side, security). Enhancements of existing attacks using such state-of-the-art technologies provide better attack ranges and success rates at the cost of more complex algorithms and more powerful implementation hardware. The issue of system delay and its effects on the successful implementation of an attack is addressed in Section 3.5. We present implementation details of relay attacks in Section 3.6. Corresponding aspects of security issues of both near-field communications (NFC) (passive) devices and far-field communications (active) devices are presented in Section 3.7. Our conclusions are presented in Section 3.8 and some standards and specifications for RFID systems are included in the Appendix.

3.2 COMMON ATTACKS ON RFID DEVICES

This section covers the most common threats and attacks on RFID-type devices, with a focus on relay attacks.

3.2.1 REPLAY ATTACKS

In a replay attack, an adversary can simply store and replay a previous communication between a tag and reader to impersonate that tag. Examples of this type of attack already shown in the literature are cloning of SpeedPasses and credit card purchases [1]. The attacker can make use of RF signature and baseband analysis. The attacker learns about the victim's system by analyzing the envelope of the RF signal transmitted by the victim. This analysis is done without the need to demodulate the signal. It is worth mentioning that encrypting the communication between the reader and the tag will not circumvent this type of attack. Some countermeasures suggested include the use of time stamps, or challenge-and-response protocols. Such countermeasures are nevertheless still susceptible to replay attacks; they are discussed in Section 3.7.2.

3.2.2   POWER ANALYSIS ATTACKS

One side-channel attack that can be used against RFID tags is the *power analysis* attack. The aim of such an attack is to find the secret information used by a device by making a correlation between the power consumption of the device and changes in the state of the cryptosystem used in the device. The power consumption of most devices is proportional to internal bit transitions over a given time, which can then provide information about secret information used at different stages of the cryptosystem. To perform a side-channel attack, it is assumed that the attacker has some knowledge of the internal hardware and software of the device, or that such knowledge can be obtained using a reverse engineering approach. We will not address power analysis attacks in this chapter.

3.2.3   RELAY ATTACKS

The basic premise of a relay attack is similar to that of a man-in-the-middle attack. Given that a (passive) tag is powered and responds to a reader, an adversary can use a rogue reader to initiate and establish a communication with a legitimate tag. Information learned can then be relayed to a fake tag (possibly at a great distance away) that can communicate with the legitimate reader as a clone of the legitimate tag. This type of attack can be performed without the legitimate tag holder's knowledge. However, the attack needs to be done without introducing significant system delay, as this could foil the attack.

As relay attacks are the focus of this chapter, let us now give a broad overview of various issues with regard to the modeling and implementation of this type of attack, as well as its implications to the users of RFID-enabled devices.

The relay attack is slightly different from the replay attack, in the sense that the attacker does not store previous messages nor does he replay them. Instead, the attacker intercepts the communication between the tag and the reader and tries to relay the messages between the two. If the attacker can act quickly enough to pass the information to the legitimate tag and to respond to the legitimate reader, he can impersonate the legitimate tag, or the legitimate reader. Because the attacker does not have to be able to decipher the content of the communication between the legitimate tag and reader, this type of attack is difficult to protect against.

An example of this type of attack could be a situation where someone—the attacker—wants to gain access to a restricted location for which another—the victim—has access rights. For instance, research companies and government offices with limited or very restricted access often use electronic access tags. The victim's access information is stored in the victim's tag. In this situation, the false reader must be placed close enough to the victim—from a few centimeters to a few meters depending on the technique used—and it must fake the valid reader located far away.

The attacker's system in this case is depicted in Figure 3.2. Here, the system is composed of a false reader, a "long-distance" communication system, and possibly a fake tag. The selection of communication system depends on which distance the relay needs to take place. In today's technologies, for short distances, that attacker could either use ZigBee or WiFi, whereas for longer distances, WiMax or other cellular technologies would be more appropriate. That said, for longer distances delay issues may have to be addressed.

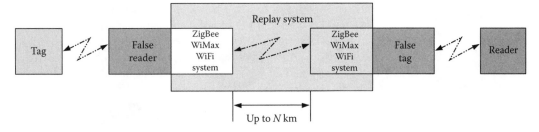

FIGURE 3.2 Relay attack system.

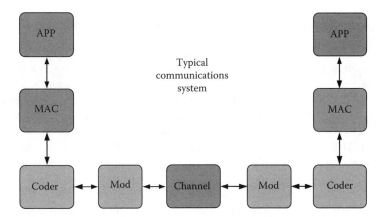

FIGURE 3.3 Typical communication system.

3.3 RFID SYSTEM MODEL

We begin discussing the requirements of a possible attack by first considering a general communications system, shown in Figure 3.3. Using this general system and all its building blocks, we can then explain some of the existing RFID protocols currently available in the marketplace. Here, we are going to focus on the NFC systems, which use passive tags, and provide analysis for this type of system. Other types of systems yield to similar analysis, with some modifications to the requirements and assumptions. This section discusses modeling of RFID systems, using a typical environment such as MATLAB or Simulink. This model is used as a basis for determining the hardware requirements, as well as to uncover implementation issues of relay attacks.

3.3.1 APPLICATION SOFTWARE (APP)

The application software is a simple piece of software designed to provide an interface to the end user to communicate with the tags, as well as possibly other software applications that may be used at a high level during implementation. It is typically stored in the reader only, as the cost of tags and their simplicity make it prohibitive for them to support any software. Tags typically support only very simple protocols, which are implemented in hardware. Passive devices are designed with a very low power level requirement, and operate without the need of an internal battery. This implies a higher power requirement for a valid reader transmitter, although this must not be higher than the limits stated in Ref. [2].

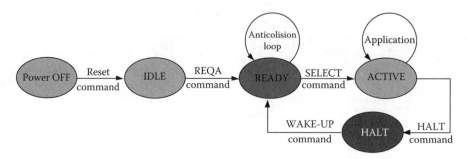

FIGURE 3.4 PICC type A state diagram.

3.3.2 INITIALIZATION AND ANTICOLLISION PROTOCOL (MAC)

MAC is also a simple layer that takes care of the initialization and anticollision protocols. Figure 3.4 shows the state diagram for an ISO 14443 type tag. When a tag is placed close to a reader, it receives power and it changes its state from OFF to IDLE. From time to time, the reader sends a Request for Acknowledgment (REQA) command, and upon the acknowledgment of this command, the tag will go into the READY state. If more than one card is present in the vicinity of the reader, the anticollision algorithm SELECTs one of the cards, using a tree search algorithm, to put in an ACTIVE state. The tag will go into HALT state either due to a HALT command [3] or due to an application-specific command. The tag will need a WAKE-UP command to get out of HALT state, at a later time.

Figure 3.5 shows the initialization process that includes an anticollision protocol. To illustrate how the process works, let us assume that the tag is close to the reader and it is already in the IDLE state. At this state, the tag is energized and searching for a REQA from the reader. As long as it does not receive the REQA but is receiving energy from the reader, it continues in this state. Once the REQA is received the tag goes into READY state and sends back the channel condition to the reader using an answer to request type A (ATQA) message. If the channel condition is such that the communication can be started, the reader will check to see whether it has received more than one ATQA message, in which case it has to select one of the present tags to continue the communication process. The selection message is then broadcasted and only one tag will be selected. If a particular tag receives the SELECT message and it is not the one that has been selected by the reader, it will go back to IDLE state. The chosen tag will acknowledge the reader by sending back a Select Acknowledge (SAK) message, and will go into ACTIVE state. The reader then will send a Request to Answer Selected (RATS) message to the chosen tag, asking for an Answer To Select (ATS) message with its setting. The reader will also send to the tag a protocol parameter selection message, if there is more than one protocol available to be used.

3.3.3 CODER

In this section, we discuss the advantages of using a coded signal instead of using pure Return to Zero (RZ) bit streams. From Figure 3.3, we see that the MAC block will generate a serial bitstream to be transmitted to the tag or reader. This RZ serial data contains energy starting at zero DC to the maximum frequency to be transmitted. This maximum frequency is generally bounded by the baseband low-pass filter. In some cases this low-pass filter is a root-raised cosine filter. There are several problems in modulating a signal with spectral energy very close to zero [4] and to avoid this implementation problem, designers and system engineers take advantage of applying coding to the incoming serial data stream.

For instance, during the communication from the tag to the reader, and before up-converting the serial stream, the serial bitstream is coded using the modified Miller for ISO 14443 type A and the Non Return to Zero (NRZ) for type B. Figure 3.6 shows a sequence [0011010] coded using RZ,

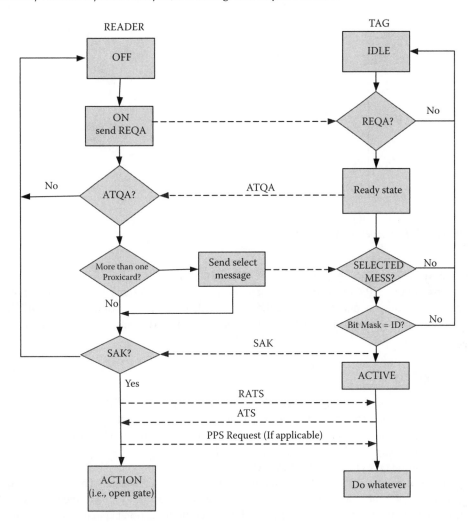

FIGURE 3.5 Initialization and anticollision flowchart.

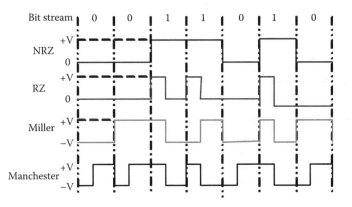

FIGURE 3.6 Coded serial data stream.

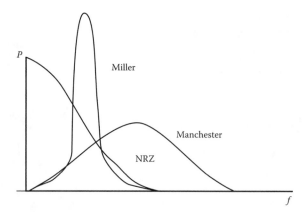

FIGURE 3.7   Spectrum density for coded signals.

Miller and Manchester codings, and Figure 3.7 shows their respective spectrum density functions. One can observe that NRZ has components starting at zero but Miller and Manchester do not have low-frequencies components. This is due the fact that there is no DC component for Miller and Manchester-coded waveforms but there are DC components for NRZ and RZ streams, as shown in Figure 3.6.

3.3.4 MODULATOR

The modulation scheme is usually of very robust modulation type. Even if there is no information to be sent, the reader usually transmits the pure carrier to provide power for the tag. For ISO 14443A/B, modulation and coding are described in Table 3.2. In this table, the reader is referred to as the proximity coupling device (PCD) and the tag as the PICC.

The modulation types are described in detail below.

Figure 3.8 depicts a typical modulator model and two steps up-converters. The first mixer moves the serial stream $p(t)$ into the subcarriers' frequencies and the second step is to move the signal to the carrier frequency (in this case, $f_c = 13.56$ MHz). For communication between the reader and tag for either type A or B, Ref. [2] states that the subcarrier frequencies f_{sc} shall satisfy $f_{sc} = f_c/16$ (about 847 kHz).

The use of the mixer is only for modeling purposes. In an actual design, the mixer function may be designed not as explicitly as in the model depicted in Figure 3.8. Also notice that there is no baseband filter or amplifier on this diagram; this will be added in final stages of the design.

TABLE 3.2
ISO 14443 A and B: Differences

Direction	Characteristic	Type A	Type B
PCD to PICC	Modulation	ASK	ASK
	Modulation index	100 percent	10 percent
	Coding	Modified Miller	NRZ-L
	Transmission rate	106 kbps	106 kbps
PICC to PCD	Modulation	OOK	BPSK
	Modulation index	100 percent	N/A
	Coding	Manchester	NRZ-L
	Transmission rate	106 kbps	106 kbps

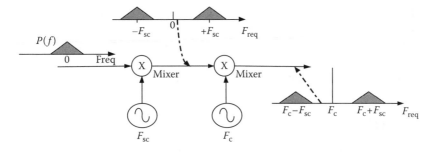

FIGURE 3.8 Modulator model.

Let us now give an overview of the different modulation types used in ISO 14443.

3.3.4.1 On–Off Keying (OOK)

This type of modulation is implemented by turning the carrier on and off in accordance with the data to be transmitted. It is very robust to signal-to-noise ratio (SNR), phase noise, and RF nonlinearity effects that usually generate degradation of communication systems not using robust modulation schemes [4], or are imposed due to design cost constraints. Figure 3.9 depicts this type of modulation constellation. Notice that in the signal space, there are only two positions in which the carrier will be landing (indicated by dark dots in Figure 3.9): either at the origin or at maximum amplitude.

Mathematically, the modulated OOK signal can be represented by

$$s(t) = p(t) \cos(2\pi f_c t + \Phi), \tag{3.1}$$

where
  f_c denotes the carrier frequency
  Φ is a given arbitrary constant representing the phase
  $p(t)$ is the multiplicative coefficient function of the coded bitstream which will define the value
      of the carrier frequency for a given bit to be transmitted

This multiplicative factor will assume only two different values during the transmission before the baseband filter and it is defined as

$$p(t) = \begin{cases} 1 & \text{if logic state} = 0; \\ 0 & \text{if logic state} = 1. \end{cases} \tag{3.2}$$

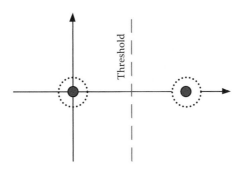

FIGURE 3.9 OOK modulation constellation.

For random transmitted signals, using Binary Symmetric Channel type communications systems and under Additive White Gaussian Noise (AWGN), the threshold value used by the decoder is usually half of the bit energy transmitted. The detector at the receive side usually uses a hard decision metric which has simple implementation, thus helping to reduce the cost of the device. Once the demodulation is noncoherent, no carrier or timing recovery algorithms are usually used (as in most of communications systems in the marketplace).

3.3.4.2   Bipolar Phase Shift Keying (BPSK)

BPSK is another type of modulation and it is used for ISO 14443 type B. This modulation scheme modulates the carrier by changing its phases into two values as a function of the input serial data stream. Let ϕ_1 and ϕ_2 denote two distinct phases; these are usually chosen to be either 0 and 180° or 0 and 90° degrees. Then the modulated BPSK signal can be represented by

$$s(t) = A \cos(2\pi f_c t + \Phi(p(t))), \tag{3.3}$$

where
  A represents the maximum amplitude of the signal
  f_c is the carrier frequency
  $p(t)$ is the coded information bitstream

  Φ is the carrier phase function, which is defined by

$$\Phi(p(t)) = \begin{cases} \phi_1 & \text{if logic state} = 0; \\ \phi_2 & \text{if logic state} = 1. \end{cases} \tag{3.4}$$

Figure 3.10 shows the constellation for BPSK and the threshold values used by the detector on the received side for a hard decision decoder type. Decoder details are not in the scope of this chapter but the interested reader can refer to Ref. [4].

3.3.4.3   Amplitude Shift Keying (ASK)

This type of modulation can be thought of as bi-level amplitude modulation (AM). The carrier assumes two distinct values of amplitude as a function of the input signal. In contrast, the phase of the transmitted signal does not change as a function of the input serial bitstream. Figure 3.11

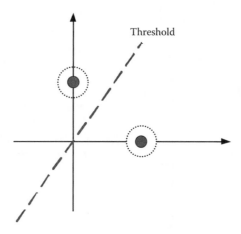

FIGURE 3.10   BPSK constellation.

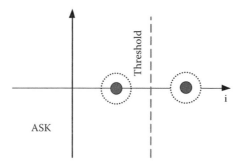

FIGURE 3.11 ASK constellation.

depicts the constellation for this type of modulation. Mathematically, the modulated ASK signal can be represented by

$$s(t) = \left[1 + kp(t)\right]\cos(2\pi f_c t + \Phi),\tag{3.5}$$

where

k is a constant, $0 < k < 1$, that determines the ASK modulation depth
f_c is the carrier frequency
Φ represents the phase (a constant)
$p(t)$ is the multiplicative coefficient function of the coded bitstream, which is assumed to take on only two values

For AM, the *modulation index* is defined as the relationship between the envelope maximum and minimum values of the carrier signal. Denoting these by A and B, respectively, the modulation index is calculated as

$$\text{Mod}_{\text{ind}} = \frac{A - B}{A + B}\tag{3.6}$$

and the modulated signal in time domain is graphically represented in Figure 3.12.

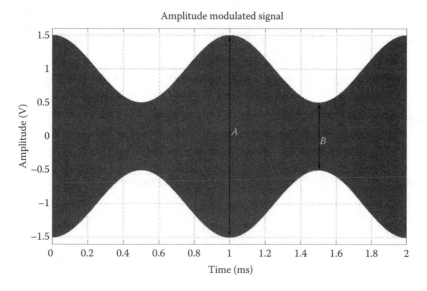

FIGURE 3.12 Amplitude modulation.

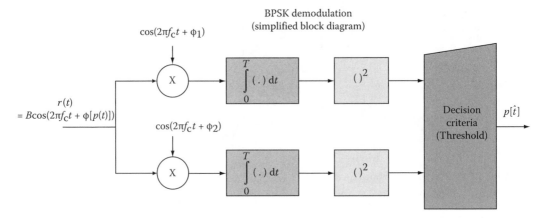

FIGURE 3.13 BPSK simplified demodulator.

3.3.5 DEMODULATOR

All three types of modulations (OOK, BPSK, and ASK) can be detected using a noncoherent demodulator. This will not affect the performance of the system due to the fact that these modulations are very robust to impairments (such as AWGN or interference).

Figure 3.13 shows a simplified demodulator for BPSK. The received signal is integrated over the period T and squared. Notice that no carrier recovery or timing recovery has been included in this demodulator block diagram. As a result, the received signal can be directly multiplied by both possible phases and then be integrated over the symbol period T and squared. A threshold value is used for the detector to estimate the symbol value that was received.

For more details about several types of demodulation schemes and performance comparison within themselves, see Ref. [4].

3.3.6 CHANNEL

The channel is modeled as a NFC channel. For simplicity it may be assumed to be instead a telecommunication channel in free space, but in practice corrections must then be applied to accommodate the differences between free-space communications channels and NFC channels.

3.4 SYSTEM ARCHITECTURE

Figure 3.2 represents a typical system architecture, although many variations on this architecture are possible. The relay attack system proposed here is composed of a reader transmitter and a reader receiver. The reader receiver has several functions: first to read the "conversation" between an legitimate reader and a legitimate tag; and second to be able to understand the protocol; and lastly to store the identity of the tag.

Note that in free space, path loss of the channel is defined as

$$P_{\text{Loss}} = 20 \log_{10}\left(\frac{4\pi d}{\lambda}\right), \tag{3.7}$$

where
  λ is the wavelength
  d is the distance between the tag and reader

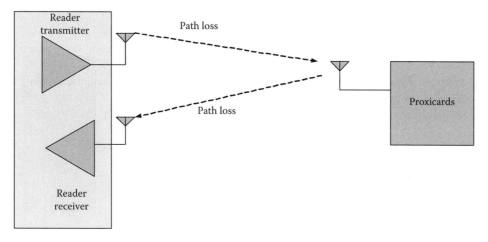

FIGURE 3.14 Relay attack system.

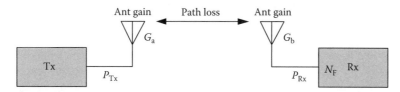

FIGURE 3.15 Link budget.

3.4.1 CHANNEL LINK BUDGET

The next step after modeling the system, described by Figure 3.1, is to determine the link budget which is based on Figure 3.14. In this section, we present the link budget of an ISO 14443 system. The link budget dictates the maximum distance over which the system will operate for a given Bit Error Rate (BER). The assumptions under which the link budget is calculated will be discussed throughout this section.

As a starting point, let us assume we have a single-input single-output communications system as depicted by Figure 3.15. The system is composed of a transmitter (Tx); a transmit antenna with a gain G_a; a receive antenna with gain G_b; and receiver circuitry that includes a low-noise amplifier (LNA), down-converter, and demodulator.

In Section 3.4.1.1, we discuss path loss and received power, total noise power, and the tag to reader link range calculation in this model.

3.4.1.1 Received Power

In the absence of any additional interference,* the receiver power P_{Rx} is the transmitter power P_{Tx} (dBi) minus the channel attenuation. For this type of communication, there is no need to consider selective fading because this is a narrow band communication protocol. The received power can be thus calculated as [5]

$$P_{Rx}(\text{dBm}) = P_{Tx}(\text{dBm}) + G_a(\text{dB}) + P_{Loss}(\text{dB}) + G_b(\text{dB}),$$

where P_{Loss} is defined in Equation 3.7. For simplicity, we use dB and dBm for power measurements rather than W or mW.

---

* Here we refer to additional interference as that created by any telecommunications equipment transmitting in a given frequency that is interfering with the given telecommunications equipment.

In its multiplicative form, one can write the received power as

$$P_{Rx}(W) = P_{Tx} G_a P_{Loss} G_b.$$

The required SNR is a function of the modulation type in use and the maximum acceptable BER. The total noise power N_t of a given receiver is calculated as

$$N_t = KTBF, \tag{3.8}$$

where
$K = 1.38 \times 10^{-23}$ is Boltzmann's constant
T is the temperature in K
B is the receiver bandwidth in Hz
F is the noise factor of the device

For a given system, the noise figure N_F is defined as the relation of input and output SNR. The lower the noise figure (or noise contribution) the better the amplifier. The noise factor F of a device is defined as the ratio of the output SNR to the input SNR, that is,

$$F = \frac{SNR_{Out}}{SNR_{In}}.$$

The noise figure, in turn, is the noise factor in dB, that is,

$$N_F = 10 \log_{10} F.$$

So, for example, at 27°C the value of noise power density per Hz resultant is

$$KT = -174 \text{ dBm/Hz}$$

so consequently by Equation 3.8, the total noise power at the receiver is given by

$$N_t(dBm) = -174 \text{ dBm/Hz} + 10 \log_{10}(B) + N_F(dB).$$

Once the SNR is known for a given receiver performance it can also be described as a function of the total noise of the receiver and the input power, via

$$SNR(dB) = P_{Rx} - N_t. \tag{3.9}$$

In other words, Equation 3.9 says that if we minimize the total noise N_t while keeping the distance constant, the performance at the receiver will improve. On the other hand, for a given performance—for example BER $= 10^{-4}$—the transmitter can be located farther away, because the total noise in the receiver is minimized and is constant.

3.4.1.1.1 Tag to Reader Link Range Calculation

Using free-space channel calculations as above, the maximum distance D_{max} theoretically achieved is presented in Table 3.3 for one set of parameters. These parameters include two which are assumed in the calculation: the received SNR and the PICC received power. The assumptions for SNR calculations are reasonable if the modulation types used for these communications are very robust for an AWGN channel model. The value of transmitted power, at -90 dBm, is also a reasonable number for PICC. The antennas gains are assumed to be 0 dBi which can also be easily improved by the usage of directive antennas.

TABLE 3.3
Link Budget for Tag to Reader Channel

Description	Value	Unit
Frequency	13.56	MHz
Wavelength	22.12	m
SNR	15	dB
Bandwidth	1000	kHz
N_F	3	dB
G_a	0	dBi
G_b	0	dBi
P_{Tx}	−90	dBm
N_t	−111	dBm
Sensitivity	−96	dBm
P_{Loss}	6	dB
D_{max}	2.36	m

3.4.2 MULTI-PROTOCOL SUPPORT

During the design of the false reader and the false tag, the designer can use a system with the capability to support existing multi-protocols and the ability to add new ones as they become available. Such an expansion will provide an attacker with a powerful tool: different single protocol tags which operate on different frequency bands.

Different Soft Defined Radios can be used in such a design to support a variety of published protocols available in the marketplace, such as ISO 15693, ISO 14443, ISO 18000-6 A/B, ICode, EPC Class 0, and EPC Class 1. The design can deal with protocol-specific characteristics such as checksums; byte-based, block or even page reading and writing; and structure and length of the command frames. The system is able to sniff RF signals in the environment within a predefined frequency bandwidth and based on the received RF signal, it is able to identify if a reader or tag is present. If so, it will successfully establish communication with the legitimate device.

3.4.2.1 Frequency Scanning

This section describes how the relay attack can be expanded to include systems with multi-protocol support. To be able to successfully establish communication in multi-protocols, this type of system has to be capable of handling several bands. Table A.1 provides more details on the frequency bands used in commonly used protocols. There are several ways to implement frequency scanning: FPGA-based NCO followed by an up-converter, fractional-N PLL, etc. Due to the broad range of bands and starting at such a low frequencies, a combination of FPGA and two or three fractional-N PLLs is needed. The FPGA will provide low frequencies up to 100 MHz, with fractional-N PLLs covering the higher range. Figure 3.16 shows a possible system design that implements frequency scanning; the system is composed by a FPGA, fractional-N synthesizer, loop filter, voltage-controlled oscillator (VCO), and up-converter. We will discuss the role of different components in this system design.

3.4.2.1.1 FPGA
The FPGA has the task of performing the digital signal processing and all other microprocessor-related functions. In this particular application, however (see Figure 3.16), its function is to control the fractional synthesizer through SPI or I2C bus type, to provide the reference clock for the synthesizer and an interface for the up-converter. It should be noted that the communication duration is only fractions of seconds and the modulation is very robust to phase noise. Consequently, the FPGA can

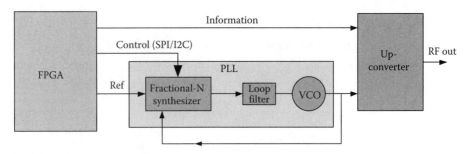

FIGURE 3.16   Frequency scanning.

be the reference source for the fractional-N synthesizer without causing degradation of the system performance. For protocols that use frequencies below 100 MHz, the synthesizer can be bypassed and the FPGA can provide the RF signals directly without the need of external up-conversion. For frequencies greater than 100 MHz, the use of one or more synthesizers is required.

3.4.2.1.2   Fractional-N Synthesizer
This component has the function of generating the correct voltage for the VCO which generates a frequency function of the input voltage. The Fractional-N has two frequencies as input: ref Clock and VCO frequency. It will divide both frequencies internally using a fractional number in such a way as to compare the phases of these signals, and will generate pulses whose widths are a function of the phase differences. The control of the fractional-N synthesizer is usually done through either an SPI or an I2C bus where the internal registers can be programmable and also some status such as PLL-locked and others can be read back.

3.4.2.1.3   Loop Filter
The pulses generated by the charge pump into the fractional-N synthesizer will pass through the low-pass filter (Loop filter), and a DC component is present in the output of the loop filter. The DC value is proportional to the width of the pulses generated in the charge pumps. Usually this is the most important and most sensitive section of a PLL. Any noise in the loop filter will make it unstable. Active components in the loop filter should also be avoided, as well as many other considerations that can be found in any PLL design books.

3.4.2.1.4   VCO
The VCO is used to generate a carrier frequency that is a function of the DC level in its input. Any noise in the DC will cause the VCO to change frequency rapidly and undesirable spurious frequencies are going to be generated. To decrease the sensitivity of VCO in the presence of noise, it is required to lower the KVCO (KVCO: VCO constant usually in MHz/V).

3.4.3   SMART ANTENNAS

3.4.3.1   Beamforming

Beamforming is a technique that uses multi-element antennas to improve the communication channel capacity. As a result, this technique increases the performance of attacks on RFID devices, but it can also be used to increase security, if implemented in the RFID systems. In general, smart antenna techniques increase the communication range, suppress interfering signals, and take advantage of scattering environments to increase the channel capacity. We illustrate these advantages using Figure 3.17. In this figure, there are three omnidirectional antennas; their transmission patterns are indicated by darkened ovals in the figure. Once the phase of each and every one of these antennas can be controlled using the weight vector $\mathbf{W} = (w_1, w_2, w_3)$ the resultant transmission pattern can be changed according to the position of the receiver.

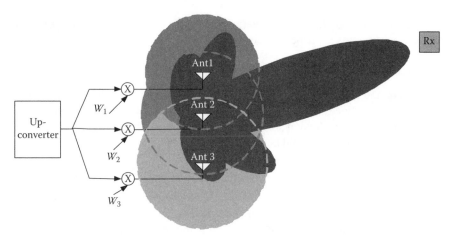

FIGURE 3.17 Antenna array.

Notice that the resultant gain in the direction of the receiver increases and if there is an undesirable signal in another direction, it can be attenuated by a certain amount of dBs. The control of the weights is usually done using digital signal processing algorithms and the implementation is usually done either using digital signal processors (DSPs) or an FPGA. The choice depends on the processing power required for implementation and also on the familiarity of the designer with one of those devices.

3.4.3.2 Beam-Selection

Beam-selection is a technique in which we assume that the antennas are directive and all of them have the same or similar characteristics. In this type of system, more than one antenna is available either in the transmitter or in the receiver. If more than one antenna is available on both sides (Tx and Rx), only one pair is selected for that particular communication link. The selection can be based on RSSI, SNR, BER, or any other performance indicator. This type of system is commonly called *space diversity* and is mostly used in the receiver only.

There are a couple of strategies for the receiver to select a particular antenna: switch and stay, and switch and compare. Figure 3.18 shows a system diversity using antenna selection. Observe that in this example, antenna 3 receives a better signal than antennas 1 and 2. Once the frequency is the

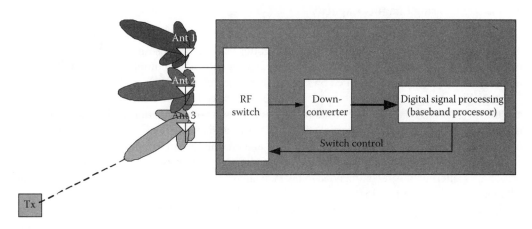

FIGURE 3.18 Antenna selection.

same no action is required for the down-converter but the baseband processor will switch to antenna 3 and stay there because this is the best received signal.

3.4.3.3   Switch and Stay

For this case when the receiver switches, it will check if the signal quality received by a certain antenna is below a predefined threshold. If yes, it will switch for the next available antenna. If the signal quality is above the threshold, it will remain receiving from that particular antenna even if the signal received from another one could be of better quality.

3.4.3.4   Switch and Compare

For this type of algorithm, the receiver will verify if the received signal quality is above or below the specified threshold. If yes, it will switch to the next antenna and then make an estimation of signal quality as well. If the new antenna is better than the previous one it will remain; if not it will switch back to the previous antenna.

3.4.4   IMPROVING THE ATTACK

The question to be answered is: How does beamforming improve the attack? The answer comes as follows. Once the gain increases in the direction of the receiver, first the attack range increases (i.e., the range between the false reader and tag as well the range between the false tag and the reader). Another factor is when the victim carrying the tag is in motion; in this case, the system can follow the victim in motion without requiring the attacker to move around, following the victim. In particular, this means that the attacker can leave the system in a position where the victim will pass by, and go to another location and get access to the desired door.

3.4.4.1   Improving the Security

As mentioned earlier, beamforming can also be used to improve security. This will happen when the attacker wants to copy the identification of the victim and replay. This type of attack usually happens when a legitimate tag is talking with a legitimate reader and the attacker tries to intercept the communications and copy the information (man in the middle attack, or replay attack). In this case if the attacker does not have this type of technology, the energy arriving at the attacker receiver is going to be much smaller in comparison with the energy received by the legitimate tag/reader, depending on the direction of the communication. In this case, the attacker system may not have good SNR and will be unable to successfully perform the attack.

3.4.4.2   Increasing Channel Capacity

Multiple antenna arrays can be used to increase the capacity of a channel exposed to a high level of scattering. This happens when both the transmitter and receiver use multi-antenna arrays and take advantage of spatial diversity. This technique is well known by multiple-input multiple-output (MIMO) and has been widely studied by industry and researchers.

A MIMO system can be depicted as Figure 3.19. Notice that the transmitter has M antennas and the receiver has N antennas. For the MIMO case presented here there is no beamforming. The antenna requirement is that they are placed in such way that spatial diversity is achieved. Notice that there are no weights that change the phase of the antennas in the transmitter but this technique in conjunction with beamforming could be used depending on the type of channel. For a channel with very large scattering, it is recommended to take advantage of the scattering and use for instance the Alamouti scheme (see Ref. [15]).

For this type of system, the channel capacity is mathematically described as

$$C = B \, \log_2 \left[\det(I_n + \text{SINR} * HH^{\text{conj}}) \right], \tag{3.10}$$

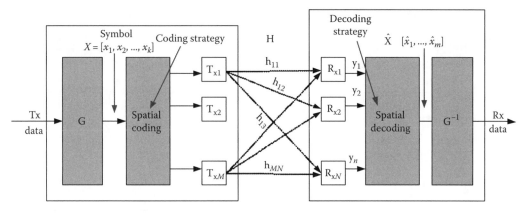

FIGURE 3.19 A MIMO system.

where

C is the channel capacity in bps/Hz

B is the bandwidth

n is the number of receiving antennas

I_n represents the $n \times n$ identity matrix

H^{conj} denotes the conjugate transpose (Hermitian transpose) of H

SINR is the signal to interference noise ratio

Observe that the capacity increases linearly with the minimum value of (M, N) and the advantages of MIMO depends on the channel matrix H. The larger the rank and the eigenvalues of $H\,H^{\mathrm{conj}}$, the larger the capacity. Degradation of MIMO systems occurs with the absence of a multipath. This may cause channel correlation and will affect the orthogonality (diversity), which will as a consequence affect the values of the eigenvalues of HH^{conj}. Another factor is unequal average branch power; this will cause bias in the eigenvalues. For further details about MIMO, the reader is referred to Ref. [16].

3.4.4.3 General Implementation Comments on Smart Antennas

The problem in using beamforming appears when the frequency range changes drastically, i.e., 13.56–900 MHz. This means that the adaptive array must work within this bandwidth. The problem is especially apparent at low frequencies where the weights of the adaptive algorithm are forced to vary considerably. For low frequencies, the wavelength is very high so this may be a showstopper in the design for low frequencies. Thus, it may not be feasible to implement beamforming at frequencies as low as 13 MHz; but for frequencies at 2.4 and 5 GHz the implementation is straightforward. These advantages are achieved at the cost of a higher complexity system, causing an increase of costs due to adding digital signal processing and multi-antenna arrays, but it is easy to envision that with new reduced IC technologies and mass productions, these costs can be lowered.

3.5 SYSTEM DELAY

For a communications system to work properly, synchronization needs to be established between the transmitter and receiver. Synchronization is one of the most important factors in making a reliable communication link work properly by providing the highest throughput the channel is capable of. A relay attack is basically the process of relaying data across to another location. Once the attacker is in the middle of an existing system, this data relaying process will introduce more delays in the overall system and it may cause loss of synchronization between the valid tag and the valid reader.

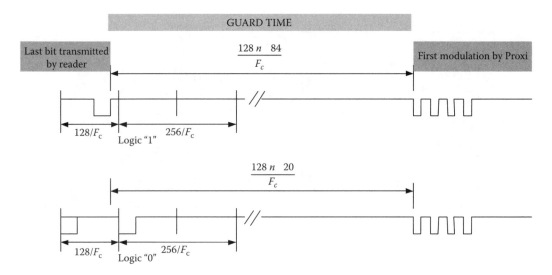

FIGURE 3.20   ISO 14443 frame delay time.

TABLE 3.4
Frame Delay Time

		Frame Delay Time	
Command Type	**n**	**Last Bit = 1**	**Last Bit = 0**
REQA, WAKE-UP, anticollision SELECT	9	$1236/f_c$	$1172/f_c$
All other commands	≥ 9	$(128n - 84)/f_c$	$(128n - 20)/f_c$

Based on this, it is very important during the design to consider the total delay of the system. This requirement comes from the ISO 14443-3 timing specifications, to maintain bit synchronization during the anticollision process. The reader can compare the systems shown in Figure. 3.1 (system in normal operation) and 3.2 (system under relay attack).

Figure 3.20 shows the frame delay timing and in the figure f_c is the carrier frequency and n is the number of bits. The bound (time-out) is calculated as per Table 3.4 provided by ISO14443-3.

Another delay that requires special attention is the frame waiting time (FWT). This is the time after which the PICC (tag) will start the response frame after the end of a PCD (reader) frame. It happens when the card is in ATS (see Figure 3.21 for details). The equation governing the FWT is

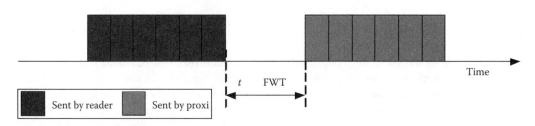

FIGURE 3.21   Frame waiting time.

TABLE 3.5
FWT Bounds

	FWI	FWT Value
FWT min	0	$302\,\mu s$
FWT typical	4	$4.8\,ms$
FWT max	14	$5\,s$

$$\text{FWT} = \frac{216 * 6}{F_c} 2^{\text{FWI}}, \tag{3.11}$$

where the frame wait time interval (FWI) value varies from 0 to 14. The default value of FWI is 4 which gives an FWT value of around 4.8 ms. Table 3.5 presents the maximum, default, and minimum values for FWT.

3.6 RELAY ATTACK IMPLEMENTATION

In theory, the relay attack is straightforward to implement. However, if the attacker wants to implement a high-tech system, using the ideas of Section 3.5, the implementation may become very expensive and time-consuming.

As mentioned in Section 3.4.2, some of the improvements need devices such as FPGA or DSPs. Using these devices, the attacker could improve the existing algorithms and using the same reconfigurable hardware, the attacker has the capability to use it to attack systems that use different protocols by a simple upgrade of the firmware and API software. The decision to use a DSP or FPGA is based on design, cost, and familiarity with a specific development platform.

3.6.1 USING THE DSP

Many DSPs are available on the market to implement the required functions. When implementing a project using DSP, it is usually necessary to estimate the Mega Instructions Per Second (MIPS) that will be required for the design before one can properly select the target device for implementation. For relay attack implementation, this requirement can be relaxed for the following two reasons: (a) the design is very simple, and (b) there will be only a couple of devices to build. Hence, a detailed MIPS calculation is not required for this type of implementation, and it suffices to select a mid-range device that can run with a clock speed greater than 600 MHz. Some typical DSPs worth mentioning are those of the TMS3xx family from Texas Instruments and Shark from Analog Devices. MATLAB and Simulink can also be used as simulation tools for the baseband processor and these tools connect directly with TI or ADI DSP families.

3.6.2 IMPLEMENTATION USING FPGA

The baseband processing implementation using FPGA can be done by partitioning the hardware in two sections: one for the fast processing implementation, which will be done in hardware, and the other for slow processing, which could be implemented using soft processors like NIOs from Altera.

Rather than going into the details of an FPGA design, let us provide, in Section 3.6.3, a brief description of the simulation and implementation environment for FPGA using different tools.

3.6.3 SIMULATION ENVIRONMENT

At the system level, the relay attack can first be implemented using MATLAB or Simulink using floating-point arithmetic. The next stage would be either to continue purely in Simulink or to blend the simulation with the DSP Builder from Altera. It is easier to use the DSP Builder, because when available the DSP Builder works in the same environment as Simulink. The migration from floating point to fixed point can be done in steps. After implementing the full system using DSP Builder components, the model can be downloaded directly to an Altera evaluation board. A designer can also generate the VHDL code automatically from the Simulink environment and then perform further simulation using, for instance, ModelSim from Mentor Graphics. Place and routing can be done using Quartus from Altera.

It should be noted that if the transmitted signal has a frequency less than 40 MHz, then the complete system can be implemented using only FPGA technology. However, as the transmitted frequency increases, it is necessary to use a proper up-converter to place the frequency in the correct spectrum location. PLL design is another factor whose complexity must be considered, as described in Section 3.4.3.

3.7   SYSTEM SECURITY

When RFID-type devices are used for logistics, access control, cashless payment systems, and even travel documents, their security becomes a very important factor to consider during the system design. We consider two categories of security mechanisms currently in use: those that are inherent to RFID systems due to their principal design, and those which can be considered as add-ons. In our discussion, we also consider different classes of security mechanisms along physical, MAC, and application layers (Figure 3.22). The reason for making such distinctions is that some of these mechanisms—such as encryption—are not robust against relay attacks, as was pointed out in Section 3.2.

3.7.1   BUILT-IN SECURITY MECHANISMS

This section is dedicated to security mechanisms which are inherent to the RFID system design, and hence do not require additional costs to employ. Two such mechanisms are the time-out and the range.

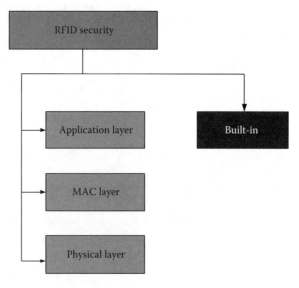

FIGURE 3.22   Security mechanisms.

3.7.1.1 Time-Out

The time-out security mechanism is closely related to the timing specifications recommended by the standards. If the added delay due to an attack is greater than what is recommended by the standard, the attack will fail due to time-out in the system. We conclude that during system design, the air interface must recommend guard bands as small as possible, even at the expense of multiple attempts by a tag to be recognized by the system. Clearly, there are some trade-offs to be considered in terms of balancing security and convenience. The use of a probabilistic threshold method may allow an appropriate balance to be struck.

3.7.1.2 Range

Passive tags usually have smaller ranges than active ones. The smaller range for a reader to recognize a passive tag makes the system more robust to replay and relay attacks, because performing such attacks requires the attacker to be able to communicate with the tags within their range. In some cases, however, such as automatic highway toll systems, restricting the range of the device is neither feasible nor desired.

3.7.1.3 Impact of Collisions during an Attack

During an attack, it is possible that collisions may occur. The probability of collision occurrence increases with the number of devices in the vicinity. In this section, we show how this can help to secure the systems against attacks.

Figure 3.4 shows the states of a typical type A system. Let us suppose that such a system supports an anticollision mechanism, as recommended by ISO 14443-3, and that the anticollision sequence is managed by the reader, working with multiple tags.

3.7.1.3.1 First Case: Multiple Readers

Consider a relay attack in which the fake reader and a second reader are both in the vicinity of the tag being attacked. In this situation, the collisions will occur between the two readers. This will happen even if the second reader is not the valid reader for the victim's tag.

The effect of reader collisions in systems such as those used in indoor environments, as well as the analysis of optimal antenna positioning to foil fake reader attacks are an active area of research (see, for example, Refs. [7,8]).

3.7.1.3.2 Second Case: Multiple Tags

Suppose now there is only one reader and the environment is populated with multiple tags. We further suppose that the attacker is trying to perform a replay attack. In this situation, it is the setup which makes the attack less likely to succeed, namely, the crowded RF environment. The attacker will need several attempts to ensure that he has correctly isolated and learned the identification of the one tag he is trying to attack.

3.7.2 OTHER SECURITY MECHANISMS

This section shall address most of the up-to-date additional security mechanisms in the literature. An important factor to consider for each of these mechanisms is that many of the proposed solutions will significantly add to the cost and the complexity of the tags and the system.

3.7.2.1 Use of Encryption

Various encryption algorithms and authentication schemes have been proposed for RFID systems in the research literature, as well as by industry. For example, TI produces one of the most popular tags, also known as digital signature transponders (DSTs). These devices have been used in a variety of applications, including vehicle immobilizer key fobs and electronic payment devices. A DST uses a proprietary stream chipper, with a 40 bit key. Bono et al. [9] claim to have mounted a successful brute-force attack against a DST tag using an FPGA array in less than two hours. They also document several instances of successful replay attacks against these tags.

Public key algorithms, and in particular elliptic curve cryptosystems, are another cryptographic solution with great potential for future use on these type of devices (see, for example, Ref. [10]). They are however quite costly in relation with the cost target for this type of application, which in most cases today is measured in cents.

3.7.2.2 Challenge–Response Protocol

Challenge–response protocols are used to verify the authenticity of both (PCD and PICC) parties in the system. During the authentication process, the reader uses these protocols to determine whether the tag is a legitimate device belonging to the system.

For example, one type of challenge–response protocol uses a pseudorandom vector $\mathbf{c} = (c_1, c_2, \ldots, c_n)$ which is generated and sent by the reader to the tag. The corresponding response of the tag is a vector $\mathbf{r} = (r_1, r_2, \ldots, r_n)$ which is produced as a function of $\mathbf{c}$. This information exchange can be signed using public key algorithms, or hash functions, if available.

In general, there are four stages in the authentication process:

1. *Initialization.* The tag sends a message to the reader, requesting to be authenticated.
2. *Challenge.* The reader generates a challenge vector and sends it to the tag.
3. *Response.* The tag signs the challenge and sends it back to the reader.
4. *Verification.* The reader verifies whether the signed version of the challenge matches the issued challenge.

This type of protocol can also be combined with distance-bound protocol applications, as mentioned in Section 3.7.2.3. The above algorithm can be modified to authenticate the reader to the tag; however, one should be mindful of the cost of and to the tags when implementing such protocols. For more information on challenge–response protocols (see Refs. [11,12]).

3.7.2.3 Distance-Bound Protocol (Security Positioning Protocol)

Unfortunately, a relay attack cannot easily be prevented by cryptographic protocols because these attacks are usually at the MAC layer. In contrast, distance bounding or security positioning is a protocol that operates at the physical layer. The main goal of this protocol is to ensure that the legitimate tag (the *prover*) is within a certain radius of the legitimate reader (the *verifier*). The prover and the verifier usually share a secret pseudorandom function that serves as a dedicated shared secret key, together with a public pseudorandom function. We also assume that the attacker has no access to the key.

One possible physical solution for implementing this protocol is to measure the round trip delay of the communication. The distance between the transmitter and the receiver can be calculated as

$$D = v_p \frac{1}{2}(\text{TOA} - \text{TOD}),$$

where

TOD and TOA represent the time of departure and the time of arrival, respectively

v_p is the propagation speed of the signal in air

The distance between the two wireless devices is usually indirectly measured by the use of a challenge–response protocol. As per Refs. [13,14], the verifier has a set of bits $\mathbf{c} = (c_1, c_2, \ldots, c_n)$

that will be sent, usually serially, to the prover. The prover must immediately respond with the response bits of the vector $\mathbf{r} = (r_1, r_2, \ldots, r_n)$, and then also send the hash function $h(\mathbf{c})$ of the original message for authentication.

Because the main goal of this protocol is to prove that the distance D is within the specified bound, for each bit c_i the verifier will expect the immediate response bit r_i with a response time in (say) nanoseconds. To establish a distance bound threshold, let d_{thresh} be a value larger than the range of the wireless capabilities for the legitimate reader and tag. Associated with this distance d_{thresh} there is the maximum allowed round trip delay τ_{max}. This means that all bits r_i received by the verifier must arrive in a time smaller than τ_{max}. Once the false reader is far away, the response sequence needs to be generated by the fake tag which needs to guess to generate $\mathbf{r}'$.

Let us assume now that the fake tag generates the bit sequence $\mathbf{r}' = (r'_1, r'_2, \ldots, r'_n)$ and furthermore these bits arrive at the reader within this time. The next step for the legitimate reader is to check the sequence received against $\mathbf{r}$. However, the probability that the fake tag has successfully generated the required sequence $\mathbf{r}$, completely at random and without additional knowledge, is

$$P(\mathbf{r} = \mathbf{r}') = 2^{-n}.$$

Furthermore, the fake tag must attempt to generate the hash of the original sequence $\mathbf{c}$. To pass as authentic, the local hash function of the fake tag, h', would have to return the correct value on $\mathbf{c}$, that is, $h'(\mathbf{c}) = h(\mathbf{c})$, which again occurs with very low probability. As a consequence, the attack will fail, thanks in part to the tight timing enforced by this protocol.

3.8 CONCLUSIONS

In this chapter, we have given a broad overview of relay attacks: their system requirements together with various hardware and software implementation issues. The susceptibility of RFID devices to relay attacks was demonstrated. This exposition illustrated the ease of implementing relay attacks on RFID systems, and discussed several different techniques that can be used to improve—or thwart—relay attacks, using techniques such as MIMO encoding and antenna arrays. Some analysis of existing and proposed security solutions was also given.

The ever-increasing reliance on RFID technologies in our society implies that greater attention must be paid to thwarting attacks to RFID systems. However, most systems in place today are completely vulnerable to the simple relay attacks discussed in this chapter. The security of this new technology is a grave and urgent concern, one which can only be successfully addressed by further research into attacks and appropriate countermeasures.

APPENDIX

STANDARDS

The RFID established standards are ISO 14443 and ISO 15693. Both use a 13.56 MHz band, and are in widespread use in the marketplace.

Different groups, such as ISO/IEC JTC/SC3/WG4/SG 3 RFID, are currently working in new standards for RFID systems. See Table A.1 for a partial list of these new standards.

ISO 14443

This section provides basic information about ISO 14443 that is one of the standards for contactless smart cards and uses 12.356 MHz as carrier. This standard uses four levels, as presented in Table A.2.

There are two different types of ISO 14443, denoted "A" and "B," whose differences are presented in Table 3.2. Notice that the transmission rate is always 106 kbps.

TABLE A.1

RFID Technical Standards

Standard	Description
ISO 18000-1	Generic Parameters for Air Interface for Global Interfaces
ISO 18000-2	Parameters for Air Interface frequency band <135 kHz
ISO 18000-3	Parameters for Air Interface frequency band 13.56 MHz
ISO 18000-4	Parameters for Air Interface frequency band 2.45 GHz
ISO 18000-5	Parameters for Air Interface frequency band 5.84 GHz
ISO 18000-6	Parameters for Air Interface frequency band 860–930 MHz
ISO 18000-7	Parameters for Air Interface frequency band 433.92 MHz

TABLE A.2

ISO 14443 Levels

Levels	Description
ISO 14443-1	Physical layer
ISO 14443-2	RF specifications
ISO 14443-3	Initialization and anticollision
ISO 14443-4	Transmission protocol

REFERENCES

[1] G. Hancke, A practical relay attack on ISO 14443 proximity cards, White paper, University of Cambridge, Computer Laboratory. http://www.cl.cam.ac.uk/~gh275/relay.pdf. 2003.

[2] ISO/IEC 14443-2, Identification cards—Contactless integrated circuit(s) cards—Proximity cards—Part 2: Radio frequency power and signal interface. 1999-06-11.

[3] ISO/IEC 14443-3, Identification cards—Contactless integrated circuit(s) cards—Proximity cards—Part 3: Initialization and anti collision. 1999-06-11.

[4] J. G. Proaks, *Digital Communications*, McGraw Hill, New York, 1989.

[5] J. Cheah, *Practical Wireless Data Modem Design*, Artech House Publishers, Reading, MA, 1999.

[6] S. Alamouti, A simple transmitter diversity scheme for wireless communications, *IEEE J. Select. Areas Commun.*, 16:1451–1458, 1998.

[7] K. Leong, M. Leng Ng, and P. Cole, The reader collision problem in RFID systems, *IEEE International Symposium on Microwave, Antenna, Propagation and EMC Technologies for Wireless Communications Proceedings*, Beijing, August 2005.

[8] K. Leong, M. Leng Ng, and P. Cole, Positioning analysis of multiple antenna in a dense RFID reader environment, *IEEE International Symposium on Application and the Internet*, SAINT Workshop, January 2006.

[9] S. Bono, M. Matthew, A. Stubblefield, A. Juels, A. Rubin, and M. Szydlo, Security analysis of a cryptographically-enabled RFID device, www.usenix.org/events/sec05/tech/bono/bono.pdf, *14th USENIX Security Symposium*, USENIX Association, 2005.

[10] D. R. Stintson, *Cryptography Theory and Practice*, Chapman & Hall/CRC, Boca Raton, FL, 2006.

[11] T. Dimitriou, A lightweight RFID protocol to protect against traceability and cloning attacks, *IEEE Securecomm.*, pp. 59–66, September 2005.

[12] M. Feldhofer, A proposal for an authentication protocol in a security layer for RFID smart tags, *IEEE MELECON 2004*, Dubrovnik, Croatia, May 12–15, 2004.

[13] G. Hancke and M. Kuhn, An RFID distance bounding protocol, *Proceedings of IEEE/Creat-Net SecureComm*, Athens, Greece, pp. 67–73, September 5–9, 2005.

[14] S. Brands and D. Chaum, Distance-bounding protocols, *Advances in Cryptology EUROCRYPT 93*, Springer-Verlag LNCS, Vol. 765, pp. 344–359, May 1993.

[15] H. Kirschenbaum and A. Wool, How to build a low-cost, extended-range RFID skimmer, http:// eprint.iacr.org/2006/054.pdf, February 2, 2006.

[16] C. Oestges and B. Clerckx, *MIMO Wireless Communications—From Real-World Propagation to Space-Time Code Design*, Academic Press, London, 2007.

4 Physical Privacy and Security in RFID Systems

Leonid Bolotnyy and Gabriel Robins

CONTENTS

User tracking and authentication with radio-frequency identification (RFID) technology have raised many privacy and security concerns that impede the widespread deployment of RFID. On the other hand, known privacy and security cryptographic defenses are too hardware-expensive to incorporate into low-cost RFID tags. To address these concerns, we propose hardware-based approaches to security and privacy in RFID that rely on physical unclonable functions (PUFs). These functions exploit the inherent variability of wire delays and parasitic gate delays in manufactured circuits, and may be implemented with an order-of-magnitude reduction in gate count as compared with traditional cryptographic functions.

We analyze privacy in RFID under the stringent assumption that adversaries can tamper with the RFID tags' hardware, and propose three different privacy models to deal with adversarial hardware-tampering attacks. We give a low-cost privacy-preserving tag identification protocol that is secure against passive adversaries. We propose several nonstandard message authentication code (MAC) algorithms that rely heavily on PUFs. We compare PUFs to digital cryptographic functions, and address other uses of PUFs to enhance RFID security. Finally, we suggest many interesting directions for future research. Our proposed protocols are efficient, and appropriate for low-cost RFID systems.

4.1 INTRODUCTION

In this chapter, we concentrate on the issue of privacy of individuals carrying radio-frequency identification (RFID)-tagged objects, and authenticity of tag-generated messages. The threat to people's privacy lies in companies, governments, and crooks-tracking people without their knowledge and consent, and then exploiting this potentially harmful information. Crooks may also try to manufacture fake products or forge identities to profit using illegal methods. In our algorithmic solutions to privacy and security in RFID, we try to keep the cost of RFID tags in check by using low-cost hardware primitives.

Much research has been done on RFID privacy and many interesting and clever solutions have been proposed. For example, Garfinkel proposes an RFID Bill of Rights [10], which is a set of laws that regulate customer tracking. However, such solutions which rely completely on legislation may not be viable in practice. Juels proposes the use of onetime pads [14] that are updated periodically. His scheme requires adversaries to not be able to eavesdrop for a number of consecutive tag identifications, which is the main limitation of the scheme. Molnar and Wagner propose a tree-based tag identification scheme [20] to considerably speed up tag identification time at a cost of significantly relaxed privacy guarantees. All these technical solutions assume that an adversary tries to break the privacy of the protocols without the power to physically tamper with the tags' memory which contains state information and secret keys, or they assume that physically compromised tags are out of the system.

Most of the literature on RFID privacy defines privacy only with respect to algorithmic attacks. In these models, an adversary is assumed to only have an algorithmic "black-box" view of a tag's computation. Tampering with a tag's secrets or utilizing side-channel information to breach RFID system's privacy is prohibited. In this chapter, we extend the RFID privacy model to include hardware-tampering attacks. To accomplish this difficult task using minimal hardware, we rely on physical unclonable functions (PUFs) [12] that exploit physical characteristics of the circuit, and are difficult to model even if adversarial physical contact with the circuit is permitted.

In this chapter, we argue that privacy is "much more difficult" to achieve than security, and propose three different RFID privacy-modeling approaches that effectively model privacy in the presence of adversarial hardware tampering. We also propose a simple PUF-based RFID identification algorithm that is privacy-preserving against passive adversaries (i.e, adversaries that can only listen to legitimate reader-to-tag communications). Privacy is not the only security concern in RFID applications. Authentication of the RFID-tagged objects and the ability to verify the authenticity of tag-generated messages are examples of other RFID security problems. We develop several PUF-based message authentication code (MAC) algorithms, and analyze their security and resistance against adversarial attacks. We compare the hardware requirements of PUF-based hash functions to those of standard hash functions, and conclude with interesting future research directions.

4.2 PHYSICAL UNCLONABLE FUNCTIONS

Existing RFID security and privacy algorithms rely on digital cryptographic primitives that require relatively high hardware cost (e.g., thousands of gates per tag to implement [9], which could be prohibitive for cheap RFID tags). We propose a physical approach to RFID security that requires only hundreds of gates to implement, and is based on PUFs. Before describing our RFID privacy and security algorithms, we introduce PUFs and their properties.

A PUF is a function that can only be evaluated by a specific instance of the underlying hardware [12]. A PUF computes its output by exploiting the inherent variability of wire delays and gate delays in manufactured circuits. These delays in turn depend on highly unpredictable factors, such as manufacturing variations, quantum mechanical fluctuations, thermal gradients, electromigration effects, parasitics, noise, etc. A good PUF is therefore not likely to be accurately modeled succinctly, nor be predicted or replicated, even using identical hardware (which will still have different manufacturing variations and associated delays, and thus yield an implemented function different from the first). Moreover, PUFs may be implemented with an order-of-magnitude reduction in gate count, as compared with traditional cryptographic functions.

The idea of a PUF originated in Ref. [25] with an emphasis on optical PUFs. A silicon PUF prototype was designed in Refs. [11,18] based on relative delay comparisons. Their design is highly reliable, can tolerate varying environmental conditions, and contains enough manufacturing variations to make each PUF circuit substantially different from the same PUFs in other chips. Although not proven, an important characteristic of the PUF proposed in Ref. [18] is that it is difficult to create an accurate model for it based on at most polynomially many known input/output pairs.

There are three important PUF parameters pivotal to our discussion. The first parameter, denoted τ, represents the probability that the PUF output for a random input challenge is the same as the PUF output of another identical tag for the same challenge. To empirically compute τ for n tags, we can determine how many of the $\binom{n}{2}$ possible tag pairs produce identical outputs for a given challenge. The probability τ is computed as the ratio of such pairs to the total number of pairs. The second parameter, denoted μ, represents the probability that the output of a PUF for a given challenge is different from the PUF's reference output (i.e., the output recorded in a stable controlled reference environment). Different operational conditions can affect μ, e.g., variations in temperature, voltage, and circuit-manufacturing variability [18]. We add a new third PUF parameter, denoted λ, representing the relative difficulty for an adversary to accurately model the PUF.

DEFINITION 1

Let s be the size of the PUF's input domain, and let t be the number of known random challenge–response pairs. We say that a PUF is $\lambda = (s, t, z)$-passively modelable if the probability of successfully determining the PUF value for an untested challenge is bounded from above by z.

DEFINITION 2

Let s be the size of the PUF's input domain, and let t be the number of adaptively learned challenge–response pairs. We say that a PUF is $\lambda = (s, t, z)$-adaptively modelable if the probability of successfully determining the PUF value for an untested challenge is bounded from above by z.

From the above definitions, an adaptively modelable PUF is also passively modelable. Even though a passively modelable PUF is weaker than an adaptively modelable one, it may be easier and cheaper to implement, while still providing sufficient privacy for some applications. All of the above factors must be considered when designing a PUF-based RFID system.

In our algorithms, we assume that an adversary cannot construct an accurate model of a PUF from a polynomial number of known input/output pairs (i.e., polynomial in the PUF's size or its input/output bit-lengths). We also assume that the pairwise PUF collision probability τ is constant for any challenge, independent of the number of identical responses that tags output to other challenges.

These assumptions are justifiable based on the practical experiments discussed in Ref. [18]. We postulate that physical tampering with a tag's PUF or probing it to measure actual wire delays will modify the PUF and thereby destroy the very function that it is designed to compute. We assume that the PUF is secure against side-channel attacks that try to recover the key (i.e., the PUF's computational behavior). Relying on these assumptions, we will sketch proofs of the privacy and security of our algorithms.

4.3 HARDWARE-TAMPERING MODELS FOR RFID PRIVACY

Existing RFID privacy models consider adversaries that can only perpetrate algorithmic attacks, and treat the cryptographic algorithm onboard a tag as a "black box" having only an input and an output. In these models, protocols have been developed that provide various levels of privacy against computationally bounded adversaries. One of the first privacy-preserving RFID schemes was designed by Weis [27], and his protocol was proved to be secure in the "Strong Privacy" model by Juels and Weis in Ref. [16]. Weis's protocol prevents an algorithmic polynomial-time-bounded adversary from distinguishing between any two adversary-chosen tags, even if an adversary knows the secret keys of all the other tags. In this chapter, we further enhance the privacy model by giving adversaries the power to tamper with tags' memory.

We start by following the idea of Gennaro et al. [13] and assume that it is possible to split a tag into two parts. The first part is read-proof (i.e., an adversary cannot read it) yet it is tamperable/modifiable, and the second one is tamper-proof (i.e., an adversary cannot tamper/modify it) yet it is readable. Having the capability of the split, we ask the question of whether RFID privacy can be achieved without requiring some part of the tag to be both read-proof and tamper-proof. (If we assume that part of a tag's memory is both read-proof and tamper-proof, then the privacy can be achieved as demonstrated by Juels and Weis [16].) We cannot allow an adversary to read the secret key used by the tag's encryption algorithm, so it has to be in a read-proof, yet tamperable part of the tag. Also, we cannot allow an adversary to modify the tag's encryption algorithm, because otherwise an adversary can just program the algorithm to output the secret key. Therefore, the algorithm should be in a tamper-proof, yet readable part of the tag.

Next, we consider what tampering capabilities to give to an adversary. To make the privacy model as general and strong as possible, we would like to only restrict an adversary to polynomial-time hardware tampering. However, this is not possible because even a simple "constant function" is sufficient to allow an adversary to track tags. Specifically, an adversary can overwrite the secret key of a tag with its own arbitrary key, and thus be able to track the tag. The above statement is simple, yet significant, because it greatly influences a possible model for a tampering adversary. To provide provable algorithmic tamper-proof (ATP) security, the authors of Ref. [13] had to augment the functionality of the device, but they did not have to limit adversarial tampering powers. Therefore, we observe that providing privacy is much more difficult than providing security. Following the impossibility of general tamper-proof privacy, we have either to significantly limit the tampering abilities of an adversary or change the privacy model substantially, thus reducing/relaxing our privacy goals.

We cannot allow an adversary to write a predetermined value in tamperable tag memory, and we must disallow/prevent an adversary from learning or even gaining considerable knowledge of a tag's secret content after adversarial tampering. These limitations leave us with only a few possible options for an allowable adversarial model. We suggest three possible directions: (1) limit the set of allowable tampering functions, (2) avoid having any readable or tamperable digital secrets onboard a tag, and (3) simplify the goals of a privacy model. Following the first direction, we can consider allowing an adversary to deterministically or probabilistically flip a tag's secret-key bits. Probabilistically flipping bits models a well-known realistic differential fault analysis attack described in Ref. [3]. We will discuss this problem formally in Section 4.3.1. It may seem really surprising that the second option is even possible. How can we guarantee privacy without having any digitally stored secrets onboard a tag? We will show that privacy can be provided without any digital secrets by extracting secrets from physical hardware components. The third direction can be taken by aiming to detect a potential privacy compromise instead of preventing it.

4.3.1 RESTRICTING MEMORY-TAMPERING FUNCTIONS

If system specification requirements requires us to have digital secret keys onboard the tags, the RFID privacy model should considerably restrict adversarial tampering capabilities. Considering the limitations of memory-tampering functions discussed above and the known realistic hardware-tampering attacks, we consider probabilistic and deterministic memory bit flips as primary candidates for adversarial tag memory-tampering functions. We leave finding other realistic memory-tampering functions and defenses against them as open problems for future research.

To defend against bit flip attacks, a tag can sign its secret key with a signature function onboard a tag. Before each secure session with the reader, the tag would check that the secret-key signature is valid by recomputing it and comparing the value to the stored result [13]. If the verification fails, a tag would output a random value. (It is important to output a random number rather than some fixed value, as suggested for security cards, because our goal is privacy, not security.) If an adversary tampers with the bits, the probability that the tag signature check succeeds is very small. However, the signature computation, verification, and storage require extra hardware, as well as extra computation power onboard the tags. We therefore investigate if a cheaper solution is possible. In other words, the question is whether bit tampering provides any real advantage to an adversary striving to compromise privacy.

To analyze this question, we extend the strong privacy definition of Juels and Weis [16], allowing an adversary to apply memory-tampering commands. We denote the set of such tampering commands by *Tamper*. Let $\text{Flip}(i, q)$ be the command that flips the ith bit of the secret key with probability q. In our analysis, we assume the strongest adversary that can flip a bit with probability 1. Note, an adversary able to make calls to $\text{Flip}(i, 1)$ can simulate $\text{Flip}(i, q)$ for any q using its own random number generator, and with probability q call $\text{Flip}(i, 1)$. Denoting $\text{Flip}(i, 1)$ by $\text{Flip}(i)$, we set $Tamper = \{\text{Flip}(i)\}$. To make our privacy model more precise for future analysis, we reproduce the privacy definition of Ref. [16] in its simplified form, and add the memory-tampering *Tamper* command(s) where appropriate. Let n be the number of tags in the system, k be the security parameter, and an adversary A be computationally limited polynomially in k. The privacy definition relies on the following experiment.

Experiment $\textbf{Exp}_A[k, n]$:

Setup:

    Generate n secret keys $(k_0, \ldots, k_n)$

    Initialize secret keys onboard the tags to $(k_0, \ldots, k_n)$, respectively

Learning:

    Adversary A communicates with the tags and the legitimate reader in any order

    Adversary A can apply *Tamper* command(s) on all tags

    Adversary A can change secret keys arbitrarily on at most $(n - 2)$ tags

Challenge:

    1. Adversary A selects two tags T_0 and T_1 whose keys it does not know

    2. One of these two tags, T_b is selected at random, where $b \in \{0, 1\}$

    3. Adversary A communicates with all the tags, except $T_{\bar{b}}$, and the legitimate reader

    4. Adversary A can apply *Tamper* command(s) to any tag

    5. Adversary A can change secret keys on any tag except T_0 and T_1

Decision:

    Adversary outputs the decision bit b'

Exp succeeds if $b = b'$

We are now ready to define privacy for a tampering-restricted adversary.

DEFINITION 3   Restricted-tampering (k,n)-privacy

The protocol is (k, n)-private if for all poly(k) limited adversaries A we have $Pr[\textbf{Exp}_A[k, n]$ succeeds in determining b] $\leq (1/2) + (1/poly(k))$.

Equipped with this expanded privacy model, we can hypothesize whether specific algorithms are privacy-preserving within this model. Several protocols (e.g., [21,22]) have been shown not private within the Strong Privacy model [16], and therefore they are also not private in our Restricted Tampering Privacy model. However, the randomized Hash algorithm of Ref. [27] that was shown private in the Strong Privacy model [16] is also private in our privacy model under the random oracle assumption [2]. In the randomized Hash protocol, the reader queries the tag, and the tag then responds with $r, f_s(r, ID)$ where r is a random number generated onboard a tag, f is the pseudorandom function, s is the tag's secret key, and ID is the tag's identifier. The reader decrypts the computation for all (s, ID) pairs and determines the identity of a tag.

THEOREM 1

For Tampering $= \{Flip(i)\}$, the randomized Hash protocol is restricted-tampering (k,n)-private in the random oracle model.

Proof: We will prove the theorem by showing that an adversary cannot distinguish with nonnegligible advantage between computations of a tag whose secret it does not know (but can tamper with), and a simulator specified as follows. This simulator will generate and send to the reader two random numbers r_1, r_2. According to the experiment, the tag will send to the reader $r, h_s(r)$ where r is a random number and s is the key to the hash function. An adversary is only allowed commands that will change the secret keys onboard a tag, but not reveal any of its bits; moreover, the hash function h is assumed to be a random function. Therefore, the adversary can only distinguish between the simulator and the tag computations if the random number generated by the tag is a duplicate used during the learning phase by the tag, or during the challenge phase either by the tag or by the simulator.

The probability that the two generated random numbers will match is $1/2^k$, and because the number of communication rounds is bounded by poly(k), the total probability is poly(k)/2^k, which is negligible for sufficiently large k. Similarly, the probability that an adversary can distinguish between the tags T_0 and T_1 during the learning phase is poly(k)/2^k. Even if an adversary unknowingly succeeds in making the secret keys on two tags the same, he will only be able to exploit this improbable event only if the two tags generate the same random number.

Although we have a proof of the security of the randomized Hash protocol, we still cannot be sure that the algorithm will be secure in practice, because a perfectly random function is a purely theoretical concept. Indeed, we need to find a hash function that will maintain its properties under the secret-key bit-flip attacks, and to be practical for RFID use, it should also require minimal hardware resources. Finding such hash functions is currently an open problem. Another interesting open problem is to find tag memory-tampering functions that would make randomized Hash protocol not privacy-preserving; yet, show that another protocol exists that is resistant to the new attacks.

4.3.2 PURELY PHYSICAL PRIVACY

We have shown that privacy in the ATP model of Ref. [13] cannot be achieved for an unrestricted adversarial hardware-tampering function. In our next purely physical privacy model, we allow an adversary to tamper with a tag's memory without limiting the tampering function. An assumption of the ATP model is that the secure content of a device is stored in tamperable memory. Therefore, to make a scheme privacy-preserving in the purely physical privacy model, the scheme should not use any secrets stored in the tamperable memory.

It is possible to construct a privacy-preserving algorithm on an RFID tag that contains *no secret data* in its onboard digital memory [4]. Instead, the secret component onboard a tag would be extracted from the physical characteristics unique to each tag, and therefore unknown to an adversary. A publicly known (tamper-proof yet readable) privacy-preserving algorithm will use this device-specific secret hardware property to provide privacy. Constructing such a device would allow us to prevent adversaries from using known memory-tampering techniques. A properly constructed PUF can provide tag-specific secrets without requiring traditional digital memory to store a secret key [4].

In our model, we rule out side-channel attacks to simplify the model and make the privacy problem manageable. Allowing side-channel attacks would make the model much more complicated. For example, following the general framework of Ref. [19] that models information leakage during algorithmic computation, we can no longer rely on the equivalence of unpredictability and indistinguishability. Intuitively, if some information is leaked, it may still be very difficult for an adversary to predict the next bit in function computation, even while knowing the previous bit sequence. However, the leaked information may suffice to distinguish observed computations from random ones.

If private tag information leaks, it is highly unlikely that an extension of the Strong Privacy model [16], similar to the one we described above, can adequately model side-channel attacks. Therefore, a more general and relaxed privacy model is required to effectively model side-channel attacks. A natural candidate model would define privacy which guarantees that an adversary can only determine a tag's identity to within a large set of tags, relative to the whole tag population. We leave it as a challenging open problem to realistically model side-channel attacks for RFID and then construct a provably privacy-preserving scheme in that model. Once such a model is developed, the next logical step would be to combine our hardware-tampering model and the side-channel model to create a universal privacy-preserving RFID Model.

4.3.3 DETECTING PRIVACY COMPROMISE

Recall that we cannot allow an adversary to arbitrarily tamper with a tag's memory, because even very simple tampering functions can break privacy. Instead of *preventing* adversarial memory tampering, we can simplify our privacy goals and instead strive to only *detect* such memory tampering. Detecting hardware tampering will allow a legitimate reader to notify the system support or a tag user that privacy may have been compromised. Reacting to such a notification, the user can either remove or substitute the compromised tag to avoid being tracked in the future. Note, if an adversary is successful in tampering with a tag's memory, he may be able to track the tag until the tag user is notified of a possible privacy violation. Also, to be able to detect adversarial hardware tampering the system requires hardware support, as purely computational (soft) tools cannot detect general hardware-tampering attacks.

Because passive RFID tags have no batteries onboard, they cannot actively notify legitimate readers (or the tags' bearer) that they have been compromised. Consequently, a user has to know identities of all the tags in his possession to verify that these tags were not tampered with. With this requirement alone (without requiring physical tag-tampering detection), a user can detect if an adversary modified the secret key, because a compromised tag is likely to be unidentifiable. However, an adversary could tamper with the state of a tag (e.g., with the counter value in the privacy-preserving "yoking-proofs" of Ref. [6]), or with the seed to a pseudorandom number generator, and thus be able to track a tag without raising suspicion.

It is possible to use the techniques discussed by Gennaro et al. [13] to protect against the leakage of secret bits; however, it was proved that a device should be able to self-destruct and store a public key (i.e., support public-key cryptography) [13]. An RFID tag can self-destruct, but only with the help of a reader that will provide enough power for it to, e.g., deliberately burn out some of its own wires. Therefore, passive RFID tags are exposed to adversarial tracking until a legitimate reader accesses them. Public-key cryptography is possible onboard an RFID tag [28]. However, it requires extra hardware resources than known low-complexity implementations of private-key cryptography and are therefore likely to require more resources than is currently acceptable for low-cost tags. In addition, the theorems proved in Ref. [13] apply to security-oriented schemes only (e.g., MACs, decryption), and provide no guarantee that an adversary cannot learn some private device information that would enable surreptitious tracking.

To detect secure content tampering onboard an RFID tag, we need a tampering-sensitive hardware component. We suggest PUFs as a promising candidate. By strategically placing the PUF circuit around the secure component (i.e., embedding the secure component within the PUF circuit), we can make it very difficult for an adversary to tamper with the secure content without also altering the PUF in a detectable way. To detect whether a PUF p was tampered with, a reader can send n challenges $c_1, \ldots, c_n$ to the tag and receive n responses $p(c_1), \ldots, p(c_n)$, and verify that at least the desired fraction of these returned values is correct. To prevent tag tracking under normal tag operation, the PUF responses can be sent encrypted with a specified encryption function f. Specifically, the tag will send to the reader $r, f(r, p(c_1)), \ldots, f(r, p(c_n))$ where r is a random number generated onboard a tag.

Particularly promising applications of detecting privacy breaches include RFID privacy protocols where secret onboard keys are shared among tags. In such protocols, the compromise of a secret key on one tag leads to the privacy degradation of other tags in the system. The tree-based authentication scheme of Molnar et al. [20] is an example of such a keys-sharing protocol. Besides sharing keys among tags, the tree-based protocol of Ref. [20] is static (i.e., the tags' secrets do not change over time), and may therefore result in a permanent privacy compromise of the system if keys are leaked to an adversary. Using PUFs onboard RFID tags, it is possible to detect potential scheme privacy compromises, and then update the secrets onboard affected tags to restore privacy.

4.3.4 SUMMARY OF RFID HARDWARE-TAMPERING PRIVACY MODELS

Assuming the existence of one-way hash functions, an algorithm can be designed that is strongly privacy-preserving against a nonhardware-tampering adversary. However, if an adversary is allowed tag hardware-tampering attacks, strong privacy is unattainable. By restricting adversarial tag memory-tampering powers, an RFID system can only tolerate a very limited set of adversarial attacks. Detecting potential privacy compromises reduces privacy goals—tags become traceable in between legitimate reader reads. Purely physical privacy relies on physical tag circuit characteristics for the extraction of a secret key. However, it requires the tag circuit to be nonreadable and nonwritable (i.e., read-proof and tamper-proof), placing considerable requirements/demands on the circuit design. The choice of the RFID privacy model and a privacy-preserving algorithm depends on the application requirements, adversarial goals, and tag resources.

4.4 PUF-BASED TAG-IDENTIFICATION ALGORITHM

We propose a simple, single-use 1-step tag identification algorithm that preserves privacy against passive adversaries [7]. An algorithm is privacy-preserving if an adversary cannot distinguish between query responses of any pair of tags. We elucidate the underlying assumptions after we describe the algorithm, because they will be more readily understood in the context of the algorithm. Our tag-identification algorithm is based on the classical cryptographic idea of using pseudonyms or onetime pads to provide security.

4.4.1 RELATED WORK

Several previous works use pseudonyms to provide RFID privacy. For example, the "minimalist-approach" of Juels [14] uses legitimate readers to update pseudonyms onboard a tag after each tag authentication. Using PUFs to generate new pseudonyms allows pseudonyms to be updated much less frequently. Another scheme builds hash-chains [1] and requires tags to implement two relatively costly digital cryptographic hash functions. In contrast, our scheme only needs a single PUF per tag (because the PUF is secret/random/unpredictable to an adversary), while still remaining unsusceptible to simple tag tracking. If the PUFs are one-way functions (but not necessarily collision resistant), we can use PUFs in place of the two hash functions [1] to protect secret tag identifiers.

4.4.2 OUR TAG-IDENTIFICATION ALGORITHM

We now describe our PUF-based tag identification protocol. Let ID be an identifier stored onboard a tag, and let p denote a tag's PUF. When a reader interrogates a tag, the tag responds with ID and updates its identifier to $p(\text{ID})$. The reader looks up a tag's ID in its database to determine the tag's identity. Note that the algorithm requires the back-end database to store the sequence $\text{ID}, p(\text{ID}), p^{(2)}(\text{ID}), \ldots, p^{(k)}(\text{ID})$ for each tag, where $p^{(i)}$ denotes the composition of the PUF p with itself i times.

Observe that it is important for the PUF responses to be reliable (i.e., return consistently identical output responses for the same inputs). Otherwise, errors will compound in long chains of PUF compositions, resulting in tag misidentification. To address this issue, we propose running PUF p multiple times for the same ID at each stage of the composition, selecting the majority answer for the new tag ID, to ensure that the value of the new ID is stable and reliable (i.e., unperturbed by random environmental factors). The reliability of the last value of k PUF compositions can be estimated as

$$R(\mu, N, k) \geq \left(1 - \sum_{m=\frac{N+1}{2}}^{N} \binom{N}{m} \mu^m (1 - \mu)^{N-m} \right)^k$$

For example, in a reference environment where the probability of an unreliable PUF value is $\mu = 0.02$, if we execute the PUF $N = 5$ times at each PUF invocation, the reliability of the last value of $k = 100$ PUF compositions is greater than ≈ 0.992268. Thus the reliability of a large number of PUF compositions can be made arbitrarily high using only a modest number of repetitions at each stage of the PUF composition chain. Note that without such iteration at each stage of the composition, the expected length of a reliable PUF composition chain is much smaller (e.g., for $\mu = 2$ percent and no iterations, i.e., $N = 1$ at each stage, the expected chain length before an erroneous value appears is $((1 - \mu)/\mu) = ((1 - 0.02)/0.02) = 49)$.

Another technique for increasing the probability of successful tag identification using PUF-computed IDs is to run several PUFs in parallel, each independently generating its own composition chain. At each query, the tag sends a tuple of PUF values to the reader (i.e., one value for every PUF onboard the tag). The reader will continue normal identification operations as long as at least one of the PUF chains is still valid (i.e., a tuple of PUF compositions is considered to be reliable if at least one of its component PUF values is reliable). Because the variance of a geometric distribution is high, such a "super chain" of tuples is expected to remain valid longer than any of its individual component PUF chains. In particular, for a tuple of size q (i.e., using q independent composition chains in parallel), the expected number of consecutive successful identifications is [4]

$$S = \sum_{x=1}^{\infty} x \cdot \left[\left(1 - (1 - \mu)^{x+1} \right)^q - \left(1 - (1 - \mu)^x \right)^q \right]$$

For example, for a tuple of size $q = 2$, we have $S \approx 73$, and for a tuple of size $q = 3$, the expected length of a valid chain rises to $S \approx 90$. This multiple-chain strategy increases the overall probability of successful tag identification. Having several PUFs onboard a tag is not an unreasonable hardware burden, because PUFs can be implemented using only a small number of gates. Alternatively, a single PUF can be used to simulate multiple PUFs by using an extra input PUF parameter to select from a family of different PUF functionalities. We can also combine the two techniques by iterating each tuple component multiple times, thus increasing the overall reliability.

In all of the above strategies, once an expected number of reliable pseudonyms is exhausted, a tag can start a new chain using a new seed identifier. Such seed identifiers can be stored onboard a tag, or provided to the tag by the reader (which implies that the back-end database must store several PUF value composition chains for each tag). In summary, given a reasonable strategy to maintain PUF computation reliability, a good PUF can extract from a single seed identifier many pseudonyms that can be used to privately identify tags.

4.4.3 ASSUMPTIONS AND REQUIREMENTS

We assume that the adversary does not carry out a denial-of-service attack. This can be enforced by considering only passive adversaries (i.e., adversaries that can only observe reader–tag communication). We assume that an adversary cannot physically overwrite identifiers on tags that it may own without damaging (or at least significantly altering) their PUF circuits. This assumption is important for the PUF of Ref. [18] that we rely upon, as it has a relatively large tag differentiation τ value (e.g., $\tau = 0.4$ [18]). If a better PUF is available (i.e., one with a smaller τ value), this assumption may become unnecessary.

We place the following requirements on secret-key construction by the back-end system. For the algorithm to ensure tag privacy, a PUF must be able to generate long chains of unique IDs (i.e., without repetitions). Moreover, different tags should not yield identical PUF outputs. This implies that the number of possible inputs/outputs should be significantly larger than the number of tags in the system. If a tag ID does repeat in the key construction predeployment phase of the system, the algorithm chooses a different ID for that tag. Because tags update their internal state after each read, the reader must supply enough power to the tags to support write operations (as tag writes require more power than reads).

4.4.4 ADVERSARIAL MODEL

The privacy of the tag identification algorithm will be determined from an experiment performed by an adversary. An adversary will observe the reader's communication with multiple tags (at most polynomially many rounds in the bit-length of PUF input), and single out two tags. The reader will then randomly select one of these two tags and run the identification algorithm once. An adversary is successful in compromising a tag's privacy if it can determine which of the two tags the reader has selected with a probability substantially greater than 1/2 (i.e., better than simply guessing).

THEOREM 2

Given a random oracle assumption [2] for PUFs, an adversary has no advantage in attempting to compromise a tag's privacy.

Proof: By observing the nonrepeating output sequences of any two tags and receiving the next output from one of them, an adversary cannot determine which one of the two PUFs computed it, because PUFs are assumed to behave as random functions.

4.5 PUF-BASED MAC ALGORITHMS

A MAC protocol is a three-tuple (K, T, V), where K is the key generation algorithm, T is the tagging algorithm, and V is the verification algorithm. The algorithm K generates a key for use by algorithms T and V. The tagging algorithm takes message m as the input, and outputs its signature σ. The verification algorithm verifies that the signature σ for a message m is authentic. A MAC protocol is secure if it is resistant against forgeries. An adversary is successful in forging a signature if it can create a valid signature for a message whose signature it has not seen before.

4.5.1 RELATED WORK

A MAC aboard a tag can be implemented using a standard cryptographic hash function (e.g., MD5, SHA-256), or be based on a block cipher such as the Advanced Encryption Standard (AES). Alternatively, for low-cost implementation, a onetime signature scheme [17] can be used, as noted in Ref. [15], where each bit position of the signature has two associated secrets—one for 0 and one for 1. Then, the signature is an ordered sequence of secrets that corresponds to 0 or 1 in each bit position. However, such a scheme requires a prohibitive amount of memory aboard a tag. A more "minimalistic" approach, where each secret is only a single bit, is suggested in Ref. [15], which in order to avoid simple forgeries, lengthens the message size and makes the message space sparser. This scheme allows the construction of a onetime MAC. We suggest a different PUF-based MAC implementation that is efficient and allows messages to be signed multiple times.

4.5.2 OUR MAC PROTOCOLS

In our PUF-based MAC protocols, tags sign messages, potentially untrusted readers collect signatures from the tags, and trusted verifiers verify the validity of the signatures. To create unforgeable signatures, tags use multiple PUF computations to sign a message [7]. Note that in a tag authentication algorithm the reader authenticates a tag, whereas in a MAC algorithm, a tag signs/authenticates a message. The message that the tag signs is used as the input to a PUF, and the PUF's output is the signature/MAC. We give two MAC protocols that rely on a PUF for security. The choice of protocol depends on the size of the message space.

We emphasize that the key used to sign a message is the PUF itself. Consequently, if the message space is small, the back-end database can make the required precomputations to learn the desired behavior of the PUF. However, if the message space is large, such exhaustive precomputations are infeasible. Therefore, the size of the message space dictates the tag design requirements and drastically affects the preferred MAC protocols used.

The PUF-based MAC protocols we describe are different from standard cryptographic MAC algorithms. First, the keys that the verifier requires to validate a signature are large, whereas standard cryptographic MACs have short keys. Second, one of our MAC algorithms cannot verify a signature without the physical presence of the tag that signed the message, and our other algorithm cannot sign arbitrary messages. These properties are unusual for classic cryptographic algorithms based on digital secrets, but are appropriate for resource-constrained RFID systems. To keep tag cost down, the computational burden is pushed to the back-end system.

Our protocols cannot be applied to all scenarios requiring MAC computations aboard tags (e.g., where the message space is large and MAC verification must be performed when a tag is not within range of a verifier). However, PUFs can still be used for general MAC constructions as part of a key generation algorithm, thus preventing physical attacks on otherwise vulnerable tags. Our MAC protocols can be used in some applications of "yoking-proofs" [6,15], where confirmation is sought that a group of tags are read simultaneously.

Our proposed MAC protocols assume a powerful adversary that can adaptively select up to polynomially many (m, σ) pairs from which it seeks to forge valid future signatures. We designed our protocols to resist forgery even if the signature verifiers are off-line (i.e., they do not participate in the signing algorithm), as is required in Refs. [6,15]. Each tag-signed message will have a time stamp associated with it. The verifier records this time stamp for each tag to ensure that a reader does not reply with a previously verified signature. We will state the scheme-specific assumptions below, during the discussions of individual protocols.

Before presenting the algorithms, we give an example of where our MAC constructions can be used. In this example, heat-sensitive objects are tagged with RFID tags containing unpowered temperature sensors [24]. As these perishable items transit from the seller/supplier to the buyer/consumer, their temperature must provably not exit a specified range. The buyer's readers in the vicinity of the objects will collect temperature readings from the tags during transport. To provide the temperature range guarantee, the buyer's readers will request that the sensing RFID tags sign the temperature values. When the shipment arrives at its destination, the buyer will verify that the temperatures/messages signed by the objects are authentic. This example is applicable to the case when the size of the message space is large and signatures need to be verified in the vicinity of the tag that signed the message. This example is also valid when the size of the message space is small (e.g., if the temperature values are discretized or rounded to the nearest degree).

Our proposed MAC protocols assume a powerful adversary that can adaptively select (m, σ) pairs from which it seeks to forge valid future signatures. An adversary can learn signatures for at most polynomially many messages. Our protocols are resistant against forgery even if the signature verifiers are off-line (i.e., they do not participate in the signing algorithm) as required in Refs. [6,15]. We assume that each tag is equipped with adaptively modelable PUF p with characteristics (τ, μ, λ) where modeling characteristic $\lambda = (s, s - 1, (1/\text{Range}(p)))$ (i.e., the PUF is a random function). Additional scheme-specific assumptions will be stated below during the discussion of individual protocols.

4.5.3 LARGE MESSAGE SPACES

When the size of the message space is large, we cannot perform all of the desired PUF computations before deploying the tags. Consequently, signature verification can only be performed when a tag in transit is within a reader's range (e.g., when it arrives at its destination). For now we disallow hardware-tampering attacks; later we will discuss how the issue of hardware tampering can be addressed.

The essential basis of a MAC signature for PUF p and message m is $p(m)$. To prevent unauthorized reuse (or replay) of such a signature, we modify the behavior of the PUF p using a unique token c to yield a more general parameterized PUF function $p_c(m)$ whose behavior depends on c. Because a passive tag does not have a clock onboard, it can instead prevent a replay attack by using a random number or a counter. Using a counter as a unique token instead of a random number has the advantage that it time stamps the message and creates a natural total ordering of the signatures (e.g., in our buyer/seller example above, the replay of past signatures will be easily detected with a counter-based scheme). However, it also has the disadvantage of revealing the state of a tag, which could leak private information onboard the tag. Note that randomness can be used in PUF computations as well, because signatures are verified in the presence of their generating tag. The signature algorithm does not accept input from the reader to allow for off-line signature verification. In online scenarios, readers can supply inputs to the signing PUF.

To reduce the probability τ of forgery using the computations of other tags, multiple PUF computations are required to create a signature. Therefore, the signature for a message m is $\{c, r_1, \ldots, r_n, p_c(r_1, m), \ldots, p_c(r_n, m)\}$, where $r_1, \ldots, r_n$ are different random numbers generated onboard the tag. This computation is similar to the one needed for tag authentication. However, in the MAC algorithm, the random input to the PUF is generated by the tag rather than the reader, and it includes a time stamp c. The verifier will store signature counter values for each tag to prevent signature replays.

The choice of n (the number of PUF computations) depends on τ, μ (Section 4.4.2), and the application requirements. To quantify the reliability of message authentication and the difficulty for an adversary to forge a signature using equivalent PUFs, we compute the probability $\mathrm{prob_v}$ that a valid signature is verified as authentic, and also the probability $\mathrm{prob_f}$ that a forged signature is incorrectly determined to be valid by the verifier. Because μ in Ref. [18] is nonnegligible, some error(s) may be allowed in PUF computations during verification. Allowing such errors, however, will make the adversary's task easier.

The probability that at most t out of n PUF responses differ from the corresponding reference responses [18] is

$$\mathrm{prob_v}(n, t, \mu) = 1 - \sum_{i=t+1}^{n} \binom{n}{i} \mu^i (1 - \mu)^{n-i}$$

On the other hand, the probability that at most t out of n responses differ from the responses of another tag is

$$\mathrm{prob_f}(n, t, \tau) = 1 - \sum_{j=t+1}^{n} \binom{n}{j} \tau^j (1 - \tau)^{n-j}$$

For example, using the empirical values reported in Ref. [18] for PUF : $\{0, 1\}^{64} \to \{0, 1\}^8$ with feed-forward arbiters, $\tau = 0.4$ and $\mu = 0.02$ (in a reference environment), and taking $n = 30$ and $t = 3$, yields a valid signature detection probability of $\mathrm{prob_v} = 0.997107$ and a forgery nonrecognition probability of $\mathrm{prob_f} = 0.000313$. These probabilities are arguably sufficient for many RFID applications. However, if greater security guarantees are necessary, n and t can be easily adjusted to increase $\mathrm{prob_v}$ and to decrease $\mathrm{prob_f}$ even further.

In general, the application will require $0 \leq \mathrm{prob_f} < \beta$ and $\alpha < \mathrm{prob_v} \leq 1$ for some reliability probability requirements α and β, and fixed n and t. Figure 4.1 shows the graphs of $\mathrm{prob_v}$ and $\mathrm{prob_f}$ from which appropriate n, t, α, and β can be determined. These graphs also show that increasing n and t causes $\mathrm{prob_v}$ and $\mathrm{prob_f}$ to rapidly converge to 1 and 0, respectively. Note that the number t of PUF responses allowed to be incorrect must be a growing function of the number of challenge–response iterations n. That is, the more challenges a tag has to respond to, the more errors it is

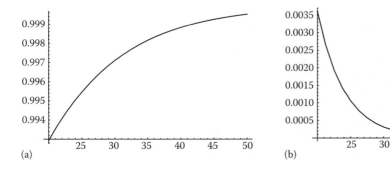

(a)   (b)

FIGURE 4.1 Graphs showing (a) the valid signature detection probability $\mathrm{prob_v}(n)$, and (b) the forgery nonrecognition probability $\mathrm{prob_f}(n)$, as functions of the number of challenges n. Following Ref. [18], we set the tag uniqueness probability $\tau = 0.4$, and the reliability probability $\mu = 0.02$. We also fix $t = 0.1 \cdot n$. The two functions are plotted on different vertical scales to better illustrate their behavior. Note that the valid signature detection probability $\mathrm{prob_v}$ quickly converges to 1, and the forgery nonrecognition probability $\mathrm{prob_f}$ quickly converges to 0, as a result of only a modest increase in the number of challenges n.

allowed to make, otherwise $prob_v$ will approach 0. In these graphs we set $t = 0.1 \cdot n$, to simplify the illustration.

THEOREM 3

Given a random oracle assumption for PUF p, the probability that an adversary can forge a signature σ for a message m is bounded from above by β (β is the upper bound of the forgery nonrecognition probability $prob_f$).

Proof: To forge a signature for a message m, an adversary must find n distinct numbers $r_1, \ldots, r_n$ as well as an unused counter value c, and compute the correct PUF values $p_c(r_i, m)$ for at least $n - t$ of them. Because p is assumed to be random and c was never input into p, an adversary can do no better than to rely on the tag(s) in its possession to create a forgery.

Because the message space is large, the key (i.e., the number of message–response pairs needed to uniquely specify a PUF's behavior in digital form, assuming a PUF is difficult to model) is too long to compute and store in the back-end database. Therefore, a tag must be within range of the verifier for MAC verification. The signature is verified by asking the tag to recompute the value of p_c on input (r, m) for each r generated by the tag during the signing algorithm. If at most t out of n responses are incorrect, the signature is assumed to be authentic, otherwise it is a forgery.

The signature verification can be authorized by a tag based on a special password sent to the tag by the verifier. The verification can proceed in the clear (i.e., be observable by an adversary) if it is a onetime operation. Otherwise, the verification should be performed securely (e.g., take place inside a radio-protected environment such as a Faraday cage). The password-based verification mechanism provides a back-door (side-channel) for an adversary to forge a signature, and thus must be protected. We assume that two passwords can be compared without leaking any side-channel information. For example, it is important to compare passwords in their entirety rather than bit-by-bit, to prevent relatively simple power analysis attacks [23].

Observe that in this algorithm, a PUF is analogous to a public key. The algorithm for a PUF computation (i.e., the PUF schematic) is known, but the private key (i.e., a PUF's input/output behavior) remains unknown. In our buyer/seller example, even though the seller possesses a tag, he cannot predict the tag's PUF computations. Private key cryptography cannot offer an equivalent solution without relying on trusted parties. This PUF property can also be leveraged to achieve private tag ownership transfer.

The MAC protocol discussions above disallowed hardware tampering. This restriction was imposed because a physical attack on the MAC protocol may allow an attacker to steal the digital password to the MAC verification algorithm, allowing him to forge a signature. We can defend against such physical tag tampering by physically locating the tag's verification password storage circuitry below the PUF's circuit/wires. This strategy will cause an invasive attempt to recover this password to physically alter (or even destroy) the PUF with high likelihood.

Similarly, the source of the message should also be protected by the PUF. However, even with such physical protection, the verifier will not know if tag tampering occurred during verification, because the verifier does not have knowledge of any secret information onboard the tag besides the verification algorithm password. Therefore, to detect forgeries, the verifier must learn some information about the PUF before the tags are deployed. Specifically, it may learn enough information to authenticate the tag using the PUF [4]. Thus, the signature is verified only after successful tag authentication.

4.5.4 SMALL MESSAGE SPACES

If the set of messages that may need to be signed is relatively small and known a priori, PUF outputs can be computed for a selected set of tokens for all of the messages. The tokens generated onboard a tag prevent adversarial replays of signatures, and the tokens cannot be random, to allow signature verification without a tag's presence. Consequently, as in the MAC protocol discussed above, counters aboard a tag can serve as tokens. Because counters are part of the signature, some private information about the tags may leak out to an adversary. Next, we discuss the key generation, tagging, and verification components of the MAC protocol.

During key generation, the verifier creates a table of PUF values for each tag and for all possible message and counter values. The tagging algorithm signs a message and the verification algorithm verifies the signature. The key generated by the back-end system is necessarily large to enable verification without the tag's presence. As before, construction of the key occurs in a secure environment before the tags are deployed. The ability to construct keys can subsequently be disabled by short-circuiting certain wires. Alternatively, a special password can be used to control access to the key learning process, and a key can be recomputed when necessary.

Let M be a small set of messages, and let k be the number of signatures a tag needs to produce. Let P be the set of tag/PUF identifiers, and let c denote a tag's counter. For each PUF $p \in P$, each message $m \in M$, and $1 \leq i \leq n$, we compute $p_c^{(i)}(m)$ for $1 \leq c \leq k \cdot q$, and $p^{(i)}$ denotes the composition of PUF with itself i times. This dataset K of PUF values serves as the key stored with the verifier. The signature for message m is $\sigma = (\{c, p_c(m), \ldots, p_c^{(n)}(m)\}, \{c + 1, p_{c+1}(m), \ldots, p_{c+1}^{(n)}(m)\}, \ldots, \{c + q - 1, p_{c+q-1}(m), \ldots, p_{c+q-1}^{(n)}(m)\})$. After signing the message, a tag will increment its counter, $c = c + q$.

Algorithm 1: Key generation for MAC

Input: Message set M; tag/PUF identifiers set P;
        # of needed signatures k; # of sub-signatures q
for *each PUF* $p \in P$ **do**
    for $i = 1$ **to** $|M|$ **do**
        for $c = 1$ **to** $k \cdot q$ **do**
            $\text{Key}[p, m_i, c] = \{c, p_c(m_i), \ldots, p_c^{(n)}(m_i)\}$

Algorithm 2: Tagging for MAC

Input: Message m; # of subsignatures q
Signature $\sigma = (\{c, p_c(m), \ldots, p_c^{(n)}(m)\},$
           $\{c + 1, p_{c+1}(m), \ldots, p_{c+1}^{(n)}(m)\}, \ldots,$
           $\{c + q - 1, p_{c+q-1}(m), \ldots, p_{c+q-1}^{(n)}(m)\})$
Side effect: $c = c + q$

To minimize an adversary's chances of mounting a successful impersonation attack, we create a composite signature consisting of a sequence of q "subsignatures," each containing n PUF computations. To avoid storing n counters onboard a tag, we instead compose the PUF with itself n times. Because the PUF's reliability is not perfect, and each repeated PUF composition depends heavily on the output of the preceding one, an invalid result computed early in the chain (see Section 4.4) may invalidate all subsequent values. To address this issue, a PUF can be run several times for each input; moreover, multiple independent PUF chains (i.e., subsignatures) can be employed. The verifier checks that at least a threshold number of subsignatures are valid. The algorithms for key generation, tagging, and verification are shown in Algorithms 1, 2, and 3, respectively.

Algorithm 3: Verification for MAC

Input: Key K; PUF p;
        # of needed signatures k; # of subsignatures q;
        allowed number t of incorrect PUF responses;
        Signature $\sigma = (\{c, p_c(m), \ldots, p_c^{(n)}(m)\},$
               $\{c+1, p_{c+1}(m), \ldots, p_{c+1}^{(n)}(m)\}, \ldots,$
               $\{c+q-1, p_{c+q-1}(m), \ldots, p_{c+q-1}^{(n)}(m)\})$
verify that $1 \leq c \leq k \cdot q$
$v = 0$
for each sub-signature σ_c **do**
     | $\sigma^* = K[p, m, c]$
     | **if** σ_c agrees with σ^* in at least $n - t$ terms **then**
     | ⌊ $v = v + 1$
    ⌊
if $v \geq$ threshold **then**
    | accept
else
    ⌊ reject

THEOREM 4

Given a random oracle assumption for a PUF p, the probability that an adversary could forge a signature σ for a message m is bounded from above by $q \cdot \beta$ (β is the upper bound of the forgery nonrecognition probability $prob_f$).

Proof: We assume that an adversary can determine (or probabilistically guess) the counter value c that a tag will use to sign its next message. However, because the PUF is assumed to be a random function, and accurate PUF modeling is not possible, an adversary can do no better than use other tags for impersonation. The success probability of forging a single subsignature is therefore bounded by β; similarly, the success probability of forging the whole signature is bounded by $q \cdot \beta$.

4.5.5 ATTACKS ON THE MAC PROTOCOLS

Relying on the properties of PUFs (described above), we consider four possible types of attacks (see Figure 4.2) against the MAC protocols described, and suggest corresponding PUF-based defenses for each one.

1. *Impersonation attacks*: An adversary can try to manufacture a duplicate of a target tag and then use it to forge signatures. Alternatively, an adversary can obtain (or steal) multiple PUF-based tags, and use their responses to impersonate the PUF of the target tag. Relying on the physical properties of PUF construction, we assume that duplicating a PUF or selecting an equivalent PUF out of a large pool, is improbable (indeed, this is exactly why such functions are called unclonable). Moreover, increasing n and t can make the valid signature detection probability $prob_v$ arbitrarily close to 1, and the forgery nonrecognition probability $prob_f$ arbitrarily close to 0, thus making impersonation improbable.

2. *Modeling attacks*: An adversary can attempt to model a PUF by observing/learning the PUF signatures for selected messages. However, the highly unpredictable factors that determine

| Impersonation | Modeling | Side-channel | Tampering |

FIGURE 4.2 Attacks on MAC protocols.

a PUF's behavior are very difficult to model [11,12,18]. Alternatively, an adversary can try to physically dissect a tag and use electrical testing probes to measure its internal wire delays. Such attacks will be foiled by the fact that the very delays thus attempted to be measured will themselves be significantly altered by the electrical coupling between the circuit and the measuring probes. Moreover, such an approach is likely to be very disruptive to the integrity of the RFID chip, e.g., it can easily damage the overlying circuit components while probing/measuring the underlying wires [12].

3. *Side-channel attacks*: Side-channel attacks such as timing and power analyses, among others, can attempt to determine the secret information stored aboard a tag. However, PUF-based secrets appear to be difficult to learn because they are difficult to represent both accurately and concisely in digital form, and thus are not easy to model. Much research is needed into side-channel attacks on PUF-based algorithm constructions.

4. *Hardware-tampering attacks*: Hardware-tampering attacks that attempt to physically probe wires run a high risk of altering (or destroying) the PUF's computational behavior. Also, attempting to physically read-off or alter the digital tag key can damage the overlying wires and alter the tag's behavior, as discussed earlier. Hardware tampering can be detected using the tag authentication protocol [4].

4.6 COMPARING PUF WITH DIGITAL HASH FUNCTIONS

It is feasible to implement PUFs with less hardware than the known cryptographic hash functions (which require thousands of gates). For example, existing designs of MD4, MD5, and SHA-256 require from approximately 7,350 to 10,868 gates to implement [9]. It is possible to use block ciphers to implement hash functions. However, even RFID-specific implementation of the AES requires 3400 gates [8]. Alternatively, a special hash function design for low-power devices can be used [29], but this scheme still requires 1701 gates for a 64-bit input size, and its security has not yet been widely accepted.

In contrast, existing PUF-based hash functions require fewer gates than the construction of [29]. Based on a suggested PUF circuit implementation of Ref. [12], we estimate that the PUF delay circuit requires about six to eight gates for each input bit, and the oscillating counter circuit that measures the delay requires about 33 gates. Therefore, a 64-bit input PUF requires only about 545 gates, considerably fewer than alternative schemes, and an order-of-magnitude improvement over standard hash functions (Table 4.1). Note, if the number of PUF output bits is not sufficient for the specific protocol, the PUF function can be executed multiple times using fresh input challenges, and outputs concatenated.

On the other hand, the low hardware complexity of PUF-based hash functions has a cost. First, the output of a PUF is only probabilistically consistent with the expected output. Second, different copies of a PUF circuit tend to have similar computational behavior (i.e., many input–output pairs are

TABLE 4.1
Comparing PUF to Standard Hash Functions

Algorithm	MD4	MD5	SHA-256	AES	Yuksel	PUF
# of gates	7350	8400	10868	3400	1701	545

identical). Third, PUF-based back-end systems must have enough memory to store all the challenge–response pairs for each chip. Fourth, discovery of challenge–response pairs for each tag consumes valuable time and resources affecting the overall cost of PUF usage. These constraints give rise to interesting algorithmic constructions.

From an adversary's viewpoint, PUFs form an attack target that is different from classical digital cryptographic systems. To break traditional keyed hash functions, adversaries attempt to determine the key of a hash function. In contrast, learning a PUF's "key" appears to be more difficult because the key is difficult to represent accurately in a concise form. Analogously to the key discovery attack on standard hash functions, an adversary could attempt to build models of PUFs. However, PUF model building seems quite difficult because PUF circuits contain numerous built-in nonlinear delay components.

Compared to their digital counterparts, PUF-based hash functions appear to be more resistant to hardware-tampering attacks and even some side-channel attacks. This is due to the apparent difficulty of creating a duplicate PUF, even when all the desired physical measurements can be made. No known digital hash function has this property. Thus, PUFs are highly desirable in otherwise vulnerable RFID systems which are too cost-constrained to implement more complicated defenses. On the negative side, PUFs rely on physical characteristics which are hard to quantify precisely, making PUF security difficult to guarantee or characterize analytically.

4.7 BUILDING PUFs

The PUF design of Ref. [12] is the first known prototype of a silicon PUF. One of the weaknesses of this design is that it employs an oscillating counter circuit to measure intrinsic delays, thus requiring a long time to sufficiently separate delay values for different challenges. Such a slower counting mechanism may not be problematic onboard an RFID tag (which is idle most of the time). However, this may slow down the manufacturing process when many challenge–response pairs need to be collected for each tag, and such manufacturing delays may translate into overall system cost increases.

The distribution of the delay values for different PUF challenges tends to be Gaussian, with many challenges producing identical (or similar) outputs even when signals take different paths through the delay circuit. Consequently, some challenges should be avoided, which requires them to be identified and filtered out during manufacturing when the database of the challenge–response pairs is created. Also, the reliability of the PUF responses is relatively low, requiring more computation rounds, while still risking producing noise. These are serious issues that may not be critical in high-end PUF systems, but are likely to be important for low-cost RFID tags.

To avoid these drawbacks, a better PUF circuit could leverage subthreshold voltage techniques [26] to compare gate polarizations, thus running quickly without using an oscillating counter. Such methods can be expected to better separate PUF values for different challenges, and thus avoid highly skewed distributions of PUF responses while still preserving PUF reliability and unpredictability. To keep the PUF modeling difficult, variable nonlinear delays can be added to the circuit [12].

4.8 FUTURE RESEARCH

More work is needed in the area of physical RFID security to help bring the ideas presented in this chapter into commercial reality. Especially important is design of novel PUF circuits with desired security properties and minimal hardware complexity. For example, subthreshold voltage-based PUFs that exploit nonlinear circuit behavior seem promising. Theoretical breakthroughs would include provable properties that characterize PUFs' security. RFID tags equipped with PUFs satisfying current RFID standards need to be fabricated and thoroughly tested under different environmental and operational conditions. The behavior of PUF-based tags should be tested under varying levels of motion, acceleration, vibration, temperature, noise, etc. In each application/scenario, the probability of output collision τ and the probability of an incorrect response μ should be characterized as functions of the operational environment. In addition, PUF's modeling characteristic λ should be determined.

New PUF-based RFID security protocols should be developed for different applications, including multi-tag regimes [5]. Utilizing PUFs onboard RFID readers (as opposed to only onboard tags) can help thwart malicious readers in the field, and make adversarial cloning of readers much more difficult. For example, when readers are off-line or the number of readers is large, it is hard to physically secure them. Therefore, the ability to ensure authenticity of a field reader is valuable. Finally, we note that although the inherent unpredictability of PUF outputs can be exploited to permit certain protocols (as done above), their predictability in some instances can be exploited to create other new protocols.

4.9 CONCLUSION

With massive RFID deployments, sensitive data transmissions may be intercepted or forged and people's privacy could be at risk. However, because low-cost RFID tags are resource-constrained, it may not be feasible to implement full-fledged cryptographic security mechanisms onboard the tags to address the security threats. We therefore proposed a PUF-based approach to RFID privacy and security. PUF-effective implementations can enable secure and minimalistic algorithms for RFID security that protect against algorithmic as well as hardware-tampering attacks. We proposed three RFID privacy models that consider tag hardware-tampering attacks and developed tag identification algorithms in these models. We also developed a simple low-cost tag identification algorithm that preserves privacy against passive adversaries.

PUFs can protect RFID tags from cloning, even if an adversary has physical access to the tags and their circuit schematics. This property makes PUF-equipped tags valuable in access control and authenticity verification applications. We developed novel protocols for MACs that require limited hardware resources, thus mitigating tag cost escalation. We compared PUFs to their digital counterparts and offered possible improvements in PUF design. We outlined numerous directions for PUF-related future research, including new PUF designs that have desired security properties, extensive PUF testing, and embedding PUFs into RFID readers.

Putting PUFs on tags is not without cost. Although PUFs are expected to require only a few hundred gates to implement, this extra hardware may increase the cost of tags over the basic tags that simply respond with a static identifier. Moreover, additional circuitry onboard a tag may necessitate readers to supply more power to the tags, or to decrease the communication distance, thus affecting the overall system design. However, at a relatively modest increase in tag cost and power consumption, the security and privacy of an RFID system can be substantially improved.

Our security and privacy algorithms have varying hardware complexity and operational requirements. Whereas our simple privacy-preserving tag identification algorithm requires a single PUF and computes it once per identification, the MAC algorithms have to perform multiple PUF computations. In addition, our MAC algorithms maintain state and require tag write operations across multiple reader accesses, thus demanding extra power to operate. The PUF complexity we discussed in Section 4.6 does not consider error correction routines that may be necessary to ensure the reliability of PUF computations. Most of our algorithms depend heavily on the reliability (correctness) of PUF computations, and thus may require tags to have PUF error correction capability, further increasing the tag cost. PUF error correction may also be required if tags operate in varying temperature and voltage environments (although there are PUF design techniques that compensate for environmental fluctuations).

Finally, note that other RFID protocols that rely on digital keys may be convertible into PUF-based variants, by following the general three-step approach: (1) learn PUF responses to select challenges, (2) run the algorithm, and (3) renew a PUF's challenge–response pairs.

ACKNOWLEDGMENT

This research was supported by grant CNS-0716635 from the National Science Foundation.

REFERENCES

[1] G. Avoine and P. Oechslin. A scalable and provably secure hash based rfid protocol. In *Proceedings of the IEEE International Workshop on Pervasive Computing and Communication Security (PerSec)*, pp. 110–114, Kauai Island, Hawaii, 2005.

[2] M. Bellare and P. Rogaway. Random oracles are practical: A paradigm for designing efficient protocols. In *Proceedings of the ACM Conference on Computer and Communications Security*, pp. 62–73, Fairfax, VA, 1993.

[3] E. Biham and A. Shamir. Differential fault analysis of secret key cryptosystems. In *Advances in Cryptology: Proceedings of CRYPTO 1997*, pp. 513–525. Springer-Verlag, Santa Barbara, CA, 1997.

[4] L. Bolotnyy. New directions in reliability, security and privacy in radio frequency identification systems. PhD thesis, University of Virginia, Charlottesville, VA, 2007.

[5] L. Bolotnyy and G. Robins. Multi-tag radio frequency identification systems. In *Proceedings of the IEEE Workshop on Automatic Identification Advanced Technologies (Auto-ID)*, pp. 83–88, Buffalo, NY, 2005.

[6] L. Bolotnyy and G. Robins. Generalized 'yoking proofs' for a group of radio frequency identification tags. In *International Conference on Mobile and Ubiquitous Systems (Mobiquitous)*, San Jose, CA, 2006.

[7] L. Bolotnyy and G. Robins. Physically unclonable function-based security and privacy in rfid systems. In *Proc. IEEE International Conference on Pervasive Computing and Communications (PerCom 2007)*, pp. 211–218, New York, 2007.

[8] M. Feldhofer, S. Dominikus, and J. Wolkerstorfer. Strong authentication for rfid systems using the AES algorithm. In B. Preneel and S. Tavares (Eds.), *Workshop on Cryptographic Hardware and Embedded Systems (CHES 2004). Lecture Notes in Computer Science*, Vol. 3156, pp. 357–370, Springer-Verlag, Cambridge, MA, 2004.

[9] M. Feldhofer and C. Rechberger. A case against currently used hash functions in rfid protocols. In *Workshop on RFID Security (RFIDSEC)*, Graz, Austria, 2006.

[10] S. Garfinkel. An rfid bill of rights. *Technology Review*, 2002, http://www.technologyreview.com/communications/12953/?a=f.

[11] B. Gassend, D. Clarke, M. van Dijk, and S. Devadas. Controlled physical random functions. In *Computer Security Applications Conference*, 2002.

[12] B. Gassend, D. Clarke, M. van Dijk, and S. Devadas. Silicon physical random functions. In *Computer and Communication Security Conference*, Washington, DC, 2002.

[13] R. Gennaro, A. Lysyanskaya, T. Malkin, S. Micali, and T. Rabin. Algorithmic tamper-proof (ATP) security: Theoretical foundations for security against hardware tampering. In *Theory of Cryptography Conference (TCC)*, pp. 258–277, 2004.

[14] A. Juels. Minimalist cryptography for low-cost rfid tags. In *International Conference on Security of Communication Networks (SCN)*, Vol. 3352, pp. 149–164, Amalfi, Italia, 2004. Springer-Verlag.

[15] A. Juels. 'yoking-proofs' for rfid tags. In R. Sandhu and R. Thomas (Eds.), *International Workshop on Pervasive Computing and Communication Security*, pp. 138–143, Orlando, FL, 2004.

[16] A. Juels and S. Weis. Defining strong privacy for rfid. Technical Report 2006/137, http://eprint.iacr.org/2006/137, *Cryptology ePrint Archive*, 2006.

[17] L. Lamport. Constructing digital signatures from a one way function. Technical Report CSL-98, SRI International, 1979.

[18] D. Lim, J. Lee, B. Gassend, G. Suh, M. Dijk, and S. Devadas. Extracting secret keys from integrated circuits. *IEEE Transactions on Very Large Scale Integration (VLSI) Systems*, 13(10):1200–1205, 2005.

[19] S. Micali and L. Reyzin. Physically observable cryptography. Report 2003/120, http://eprint.iacr.org/2003/120, *Cryptology ePrint Archive*, 2003.

[20] D. Molnar and D. Wagner. Privacy and security in library rfid issues, practices, and architecture. In *Proceedings of the ACM Conference on Computer and Communications Security*, pp. 210–219, Washington, DC, 2004.

[21] Y. Nohara, S. Inoue, K. Baba, and H. Yasuura. Quantitative evaluation of unlinkable id matching schemes. In *ACM Workshop on Privacy in Electronic Society (WPES)*, pp. 55–60, Alexandria, VA, 2005.

[22] M. Ohkubo, K. Suzuki, and S. Kinoshita. Efficient hash-chain based rfid privacy protection scheme. In *International Conference on Ubiquitous Computing (UBICOMP)*, Nottingham, England, 2004.

[23] Y. Oren and A. Shamir. Power analysis of rfid tags, 2006, http://www.wisdom.weizmann.ac.il/yossio/rfid/.

[24] M. Philipose, J. Smith, B. Jiang, A. Mamishev, S. Roy, and K. Sundara-Rajan. Battery-free wireless identification and sensing. In *Pervasive Computing*, 2005.

[25] P. Ravinkanth. Physical one-way functions. PhD thesis, MIT, Cambridge, MA, 2001.

[26] A. Wang, B. Calhoun, and A. Chandrakasan. *Sub-threshold Design for Ultra-Low Power Systems*, 2006. Springer.

[27] S. Weis. Security and privacy in radio-frequency identification devices. Master's thesis, MIT, Cambridge, MA, 2003.

[28] J. Wolkerstorfer. Scaling ECC hardware to a minimum. In *ECRYPT Workshop—Cryptographic Advances in Secure Hardware (CRASH)*, 2005.

[29] K. Yuksel. Universal hashing for ultra-low-power cryptographic hardware applications. Master's thesis, Worcester Polytechnic Institute, Worcester, MA, 2004.

5 Authentication Protocols in RFID Systems

Goran Pantelić, Slobodan Bojanić,
and Violeta Tomašević

CONTENTS

The authentication protocols for radio-frequency identification (RFID) systems are vital security procedures aimed to protect user privacy and remove other security vulnerabilities. Various authentication protocols have been developed to enable reliable authentication of RFID system components and to provide object identification. The major challenge in the design of these protocols is to achieve the balance between required security level and the limited computational capabilities of RFID devices.

As there are different types of RFID tags that have to bear in mind when designing the authentication protocols, the care also has to be taken about the requirements of specific RFID application. Most of the RFID tags are low-cost devices with limited resources that cannot afford standard cryptographic operations but there are also some RFID tags with certain level of embedded cryptographic capabilities. Having in mind these different types of tags, the most relevant RFID authentication protocols are presented in this chapter.

5.1 SECURITY PREMISES

Radio-frequency identification (RFID) systems are widely applied for the identification of physical objects. They have been used for product authentication in supply chain, stock control, and also in other fields such as authentication of the books in the libraries [1], or Euro banknotes [2], etc. Being one of the most pervasive computing technologies, it has to enable secure identification and other security issues to be solved reliably. They are vulnerable to various security threats and implying privacy problems mainly due to the restricted computational power and the memory size of low-cost tags.

The main security threats in RFID systems are generally classified into two categories. The first one relates to the attacks that disrupt the system functionality where denial-of-service (DoS) is the most illustrative example. The second category refers to the privacy issue that is a primary subject of this chapter. Two main problems in RFID privacy are data leakage from RFID tag, and malicious tracking and personal identification by tracing the person's RFID tag with corresponding unique ID.

The leakage of sensitive information about the user (e.g., banking data, medical data, etc.) in communication between the reader and the tag in most cases can occur without the user's knowledge. Therefore, various measures have been proposed to ensure safe data transmission between the tag and the reader. The malicious tag tracking is based on the ability to distinguish the tag that has already been identified and assuming the relation between the tag and the holder, the tag tracing converts to the person tracing. The usual contra measure is to provide variable tag output but that induces other security problems and additional measure's have to be involved.

Apart from the two above-mentioned problems that are the most relevant in security of RFID technology, there are also the other ones: physical attacks, DoS, counterfeiting, spoofing, eavesdropping, traffic analysis, etc. The physical attacks manipulate tags physically that is in general carried out in a laboratory. The examples are the probe attacks, material removal through shaped charges or water etching, radiation imprinting, circuit disruption, etc. DoS may be very expensive and disruptive attack. A common example of DoS attack to RFID systems is the signal jamming of radio frequency (RF) channels. Counterfeiting attacks consist in modifying the identity of an object and it is generally achieved by means of tag manipulation. Spoofing is a type of attack when an attacker successfully impersonates a legitimate tag as, for example, in the case of a man-in-the-middle attack. In eavesdropping type of attacks, unintended recipients intercept and read messages. Traffic analysis is the process of interception and study of messages to extract information from communication patterns. It can be performed even with encrypted messages that cannot be decrypted. In general, the greater the number of messages observed, the more information can be inferred from the traffic.

According to the type of RFID applications and applied tags, many security methods have been developed. Most of them are based on protocols that allow authorized persons to identify the tags without an adversary being able to trace them.

In Sections 5.2 and 5.3, the RFID systems are described and numerous authentication protocols are reviewed ranging from those for low-cost RFID tags that are used in most cases in practice to those for the tags with greater cryptographic capabilities. In Section 5.4, the hash-lock authentication protocols, the lightweight protocols (e.g., HB, efficient mutual authentication protocol (EMAP), etc.), the challenge–response protocol either with symmetric cipher or public-key (PK) primitive are presented among the others. Basic cryptographic functionalities that are applied for securing RFID tags like hash functions, random generators, and certain encryption algorithms are also explained.

5.2   RFID SYSTEM COMPONENTS

The typical RFID system consists of three main components: the transponder or RFID tag, the transceiver or RFID reader, and the back-end server (Figure 5.1). RFID tags or transponders are small devices that contain a microchip and an antenna coil as a coupling element. The microchip has memory and some computation capabilities depend on its purpose and complexity. The memory of the tag may be read-only, write-once read-many, or fully rewritable.

The tags can also be classified according to their power source as passive, semipassive, and active tags. Passive RFID tags do not have their own power supply. To be operational they receive power from the electrical field generated by the reader. They cannot initiate any communication, and they have the shortest read range [3]. They are quite small and the cheapest. The semipassive tags do have an internal battery but they cannot initiate communication. They may only respond to incoming transmissions. The active tags have a battery, they can initiate their own communications and have significantly greater read range and cost. Thus, each type of the tags could be applied for a different set of applications.

According to physical characteristics and the applications, tags are categorized into several types. The Auto-ID Center has proposed one of the first categorization and defined the five classes of RFID tags, where each successive class has been more sophisticated then the one bellow it. Class 1 refers to simple passive, read-only identity tags with onetime field-programmable nonvolatile memory. Class 2 encompasses passive tags with up to 65 kB read–write memory. Class 3 is a type of semipassive tags with up to 65 kB, like a Class 2 tags with a built-in battery to support increased read range. Class 4 relates to active tags that may broadcast a signal to a reader. Class 5 is a type of active tags that can communicate with other Class 5 tags or other devices. Later, a new Class 0 has been adopted as a type of passive, read-only tags like Class 1 but programmed in factory at the time the microchip was made. The classes of the tags have changed over time but the main principles have stayed similar to the originally proposed ones.

RFID readers or transceivers are composed of an RF module, control unit, and a coupling element. They provide a number of different functions on transponders, like interrogation, powering, identification, reading, and writing, and may have additional storage and processing capabilities. They communicate to the back-end server and may perform some complex cryptographic operations instead of the tags. The communication channel between tags and readers is considered insecure because it is based on the air interface while the communication channel between readers and back-end servers is considered secure. Readers might be handheld devices and also mobile ones based on wireless network.

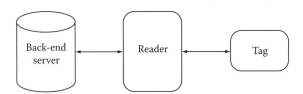

FIGURE 5.1   Typical RFID system.

Back-end servers are usually composed of a database and processing logic. They receive data from readers, store data into a database, and provide access to the data. The database contains data for each tag it managed like unique identifier, key, random identifier, or other primitives. Also, it stores transaction logs, product information, or key management data. It is assumed that the communication channel between back-end severs and a reader is considered secure as the common solution like the communications protocol secure sockets layer (SSL) can be used to protect it.

5.3 OVERVIEW OF RFID AUTHENTICATION PROTOCOLS

To be massively used, RFID tags have upper cost limits that restrain their resources. Therefore they are not able to offer significant processing capabilities that are necessary for resource-demanding cryptographic functions. To compensate these hardware restrictions, a special attention has to be paid in the designing of the security protocols for tags authentication.

5.3.1 Basic Features of the Authentication Protocols

In general, an authentication should assure the identity of a participant at the other end of a communication channel. In that sense, several concepts can be distinguished: identification, unilateral authentication, and mutual authentication. An identification protocol enables the RFID reader to obtain the identity of a tag without proofing it. A unilateral authentication protocol enables, the one entity (e.g., reader) to be convinced of the identity of the other entity (tag). A mutual authentication is a case when the both sides, the reader and the tag, are convinced about the identity of the other side.

In the basic identification protocol used in RFID applications (Figure 5.2), the reader sends a request to the tag, and the tag replies by sending an answer to the reader. This answer usually contains a unique identifier ID of the tag. The system database checks if the ID exists in it, and if it is the case, it considers it as a legitimate ID.

The authentication protocols can be viewed as challenge–response protocols. The first step in challenge–response protocol is that the verifier A sends a challenge request to the other entity B. This challenge can be only starting signal or any data (e.g., a randomly chosen number or nonce). To prevent any eavesdropper to repeat this data, the challenge should be different for the subsequent requests. The entity B processes the challenge and sends a response to the A. The response should be changed in every new round. It can be the tag's identifier or its processed value. When the verifier A receives the response, it validates the response to be sure that the B is legitimate. To assure the response be always different, it is necessary that the tag information is refreshed (e.g., the tag identifier can be renewed). If the tag is not capable to refresh this information itself then it can be performed with the help of reader.

The basic authentication protocol in RFID applications is realized as two-way challenge–response protocol (Figure 5.3). The reader sends a nonce a to the tag that prepares an answer and reply a response ID and $F(ID, a)$, where F is some function (e.g., any encryption function). When

System (ID)   Request   →   Tag (ID)
                   ← ID

FIGURE 5.2   Basic identification scheme.

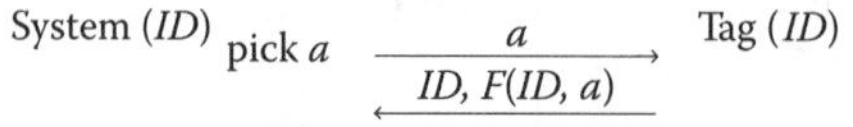

System (ID) pick a a →   Tag (ID)
                   ← $ID, F(ID, a)$

FIGURE 5.3   Basic challenge–response authentication scheme.

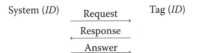

FIGURE 5.4   Three-way challenge–response protocol.

the reader receives the response, it checks if *ID* exists in the database and processes *F(ID, a)* to check if it can obtain the same nonce *a*.

The three-way challenge–response protocols (Figure 5.4) use the third step in which the reader sends new answer to the tag. This answer can be used for the authentication of the reader (in the similar way as the tag is authenticated), or can contain data that is used for tag refreshment.

There exist several methods that carry out these authentication protocols. Although most of them are designed to use basic cryptographic primitives, some of them use the stronger cryptography when strong authentication is needed. The main difference is about whether symmetric or asymmetric (PK) cryptography is used. In symmetric cryptography, the signer and the verifier share a secret key to encrypt and decrypt response. These algorithms are faster but there is a problem of the key exchange. The public-key cryptography (PKC) enables that the response is encrypted by the signer using the private key, and the verifier decrypt it with the PK. There the key exchange problem is resolved because the private key is kept secret in the signer's environment whereas the PK is available to any verifier.

5.3.2   CRYPTOGRAPHIC PRIMITIVES IN RFID TAGS

To ensure a reliable identity of the entities, the tags with more hardware resources enable the authentication protocols that use basic cryptographic functions like hash functions, or random generator, up to some cryptographic algorithms as well. In that sense, the basic cryptographic features that can be used in RFID tag having in mind their limited hardware possibilities are presented in this section.

5.3.2.1   Hash Function

The basic operation of hash functions is to map a given input x of arbitrary finite length, to an output y of fixed length n. The purpose of hash functions in cryptographic sense is to provide data integrity and message authentication. So, they should satisfy a one-way requirement that mean that for output y it is computationally infeasible to compute input x. Also, for given x, it is computationally infeasible to find $x_1 \neq x$ such that $H(x) = H(x_1)$.

Hash chain is a variant of hash functions performed in several rounds, where an output of the round $n - 1$ is input to the round n:

Seed: s_0
1st round: $H_1 = H(s_0)$
2nd round: $H_2 = H(H(s_0))$
. . .
nth round: $H_n = H(H_{n-1})$.

The use of hash chain values is in reverse order, i.e., from H_n to H_1, to provide that no one can predict the next value from the current value (without the knowledge of the seed).

5.3.2.2   Random Number Generator

The random numbers are necessary in a wide variety of operations, such as key generation or challenge–response protocols. A basic purpose of the random number generator is to produce a sequence of 0s and 1s such that the next bit cannot be predicted based on the previous bits. However, it is very difficult to build true random number generation, thus a computer is usually used only for pseudorandom number generator (PRNG) that produces a sequence of bits that has a random looking distribution. With each different seed (a typically random stream of bits used to generate a usually longer pseudorandom stream), the PRNG generates a different pseudorandom sequence. The quality of the PRNG is a significant criterion for the safety of the whole cryptographic system.

5.3.2.3 Cryptographic Algorithms

The purpose of the cryptographic algorithms is to provide a confidentiality of data exchanged between entities. A basic classification of the cryptographic algorithms is on symmetric and asymmetric ones. Whereas the symmetric algorithms use the same secret key for the both entities, the asymmetric ones use a private key to encrypt data and appropriate PK to decrypt data.

The mostly used symmetric algorithms that can be implemented on the RFID tags are the Data Encryption Standard (DES) [4] and Advance Encryption Standard (AES) [5]. They are block cipher algorithms. The DES uses 56-bit keys, whereas the AES as the successor supports key sizes of 128, 192, and 256 bits. For the authentication protocol, the AES with minimum key size of 128 bits is proposed, having in mind a real possibility of its realization on the reduced chip area of the tags.

Although the most of authors have agreed that no real possibilities to asymmetric algorithms could be soon used on the RFID tags, there were recent works that demonstrated this probability. The elliptic curve cryptography (ECC) [6] is the first asymmetric algorithm that could be really designed for restricted capabilities of RFID tags.

5.3.3 Tag Requirements for Authentication

The important condition for massive usage of RFID tags is, beside reliable functionality, low cost that proportionally limits the complexity of the embedded chip. The complexity of the tag's chip can be described by several parameters where the gate count is the most significant one. The gate count of current low-cost tags is 5,000–10,000 [7]. A considerable fraction of them is necessary to implement the basic tag functions, thus the number of gates that are available for security operations is estimated to be below 5000 [8].

The basic hardware requirements for tag's authentication are an accessible memory and logical operations that can be performed. The elementary logical bit-wise operations (AND, XOR) can be realized with a small number of gates. The hash function is more hardware demanding. It is estimated that the standard hash function like SHA-1 needs about 20,000 gates but some low-cost designs with less than 2000 gates have been already presented [9]. The simple PRNG can be implemented using the currently tag's resource like by keying a hash function. The stronger cryptographic requirements that are related to encryption purposes demand substantially more gates. It is estimated that implementation of standard AES algorithm needs about 20,000–30,000 gates [10]. However, some implementations require less gates bringing cost-efficient strong authentication closer to reality for RFID tags. One variant of DES algorithm needs about 1900 gates [11], and an implementation of 128-bit AES requires only 3600 gates [12]. The PK algorithms are still considered too expansive for tags. Some ECC algorithms designs require about 14,000 gates [13], and it seems that it would be feasible to use them in future tag's authentication protocols.

5.4 SECURITY SCHEMES

There are several different methods and authentication protocols that are aimed to protect privacy in RFID systems as mentioned above. In general, they can be classified as palliative techniques, hash-lock-based authentication protocols, lightweight protocols, and protocols with classical cryptographic symmetric-key or PK primitive. This section provides a review of existing and proposed RFID authentication approaches.

5.4.1 PALLIATIVE TECHNIQUES

The palliative techniques are simple methods that do not use cryptographic primitives as the main measure to prevent a misuse data from the tags, but they are based on the other approaches. The widespread solution particularly in supply chains is to kill the tag. Each tag has a unique password (e.g., 24 bits) and upon receiving the valid password (kill command), the tag will be deactivated. The major inconvenience of this method is that the tag cannot be used again.

The second way is to isolate the tags from any kind of electromagnetic field by enclosing them in a Faraday cage. However, this solution is suitable for few applications, like e-passport or money wallets.

The third technique is based on preventing the reader from understanding the reply from the tag. The blocker tag [14] is an example where the tag simulates the full spectrum of possible serial numbers for tags, and prevents a reader from determining which tags are presented in its environment.

5.4.2 HASH FUNCTION-BASED AUTHENTICATION PROTOCOLS

The hash function as a basic cryptographic primitive is used in the most of the authentication protocols. In this section, some hash-based protocols are described that can also serve as the base for other protocols, like hash lock, randomized hash lock, hash-based varying identifier, hash chain, and modification of hash chain.

5.4.2.1 Hash-Lock Scheme

The hash-lock scheme presented by Weis et al. [15] is based on the feature that the tag can be locked without storing the access key, but only a hash of the key on the tag. Each tag has implemented a low-cost hash function h, and a portion of memory to store a unique identifier ID and a $metaID$ as a hash value of its key k, $metaID = h(k)$. The back-end database stores the key k and $metaID$ for each of tags.

The protocol is illustrated in Figure 5.5. When a reader sends request to a tag, the tag replies $metaID$ to the reader. The reader forwards a received $metaID$ to a database to find appropriate key k for it, and sends the key k to the tag. When the tag receives key k, it calculates the hash value for this key and compares it with the stored $metaID$. If those two values are matched, the tag sends its own ID to the reader.

The primary disadvantage of this scheme is that a $metaID$ is constant so that attackers can eavesdrop it, identify each tag, and trace the tag compromising the location privacy of tag holders.

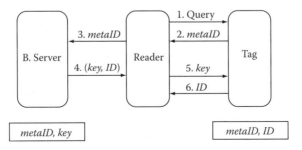

FIGURE 5.5 Hash-lock scheme.

5.4.2.2 Randomized Hash-Lock Scheme

To improve a hash-locked scheme, Weis et al. [15] proposed an extension that requires the tag has a PRNG beside a hash function and a unique identifier *ID*.

Figure 5.6 presents this protocol. After the reader queries the tag, the tag picks pseudorandom number r and calculates hash value c, $c = hash(ID||r)$ that is different for every session. Then the tag transmits its c and r to the reader that forward them to a back-end server. The server finds the unique identifier *ID* of the tag comparing c with the analog construction of r and all *ID*s that are stored in database. If the appropriate *ID* is found, it is sent back to the tag.

In this way, the tag's output is always random, but the tag can be traced if the tag's *ID* is exposed.

5.4.2.3 Hash-Based Varying Identifier

The hash-based varying identifier proposed by Henrici and Müller [16] extends the randomized version of the original hash-lock scheme [15]. Each tag contains a unique identifier *ID*, the current session number *TID*, and the last successful session number *LID*. Also, the tag has a hash function whereas a PRNG is realized only on back-end server. When the system is started, the database contains the entries the one for each tag it manages. Each entry contains the same data as is stored in the tag, an identifier *ID*, a hash value of *ID*, *h(ID)*, *TID* that are set up with random values and *LID* equals *TID*.

As shown in Figure 5.7, after the reader sends a request to the tag, the tag increases its current session number by 1. It then sends back $h(ID)$, $h(TID \oplus ID)$, and $\triangle TID = TID - LID$ to the reader which forward the values to the database. The $h(ID)$ allows the database to recover the identity of the tag in its data; whereas the $\triangle TID$ is used by the database to recover *TID* and therefore to compute $h(TID \oplus ID)$.

The database checks the validity of these values according to its recorded data. Then, it sends a random number R and the value $h(R \oplus TID \oplus ID)$ to the tag.

Because the tag knows *ID* and *TID* and receives R, it can check received value $h(R \oplus TID \oplus ID)$. If it is correct, the tag calculates its new identifier by $R \oplus ID$, and replaces *LID* by *TID*, that will be used in the next authentication.

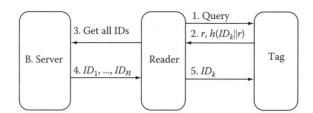

FIGURE 5.6 Randomized hash-lock scheme.

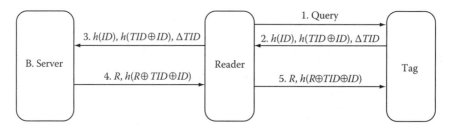

FIGURE 5.7 Hash-based varying identifier.

To be able to recover a system, the database entry is not erased when the database has replied to the tag, but a copy is kept until the next correct session.

This scheme protects location privacy making a tag's *ID* randomized in every interrogation. However, location privacy of tag bearers is compromised because the response of tag is constant until the next authentication session. Adversaries can track tag bearers whose tags are long distance from readers and scarcely have chance to be queried. Using *TIDs* the replay attacks cannot compromise the scheme because tags and back-end servers are mutually authenticated in every interrogation. Errors in message transfer can be detected and the scheme is reliable for data loss because it can provide the data from the previous record. Although authors claim that the scheme protects the man-in-the-middle attack, it can be compromised. The attacker can query any tag before the tag is interrogated by the legitimate reader, and he can be authenticated with the obtained data.

5.4.2.4 Hash-Chain Scheme

Ohkubo et al. [17] proposed a scheme (OSK) in which two hash functions G and H are embedded in the tag. The identifier of the tag is modified each time it is queried by a reader such that it can be recognized by authorized parties only. The hash-chain technique is used to refresh the tag identifier using two hash functions G and H (Figure 5.8).

The tag has a unique identifier *ID* and an initial secret information s_1. In the ith transaction with the reader, the tag sends an answer $a_i = G(s_i)$ to the reader, and renews secret $s_{i+1} = H(s_i)$ as determined from previous secret s_i, as in Figure 5.9. The reader sends a_i to the back-end database. The back-end database maintains a list of pairs (*ID*; s_1) for all tags it manages. To check received value a_i, the beck-end server calculates hash chain $ar_i = G(Hi(s_i))$ from each of the initial values until it finds the expected a_i or until it reaches a given maximum limit m on the chain length.

In this way, the tag's output a_i can neither be used to know secret s_i, nor output a_{i+1}. Also, the secret s_i cannot be derived from s_{i+1}. This protocol neither supports authentication of the reader nor prevents replay attacks.

5.4.2.5 Modification of Hash-Chain Scheme

Avoine et al. [18] proposed a modification of the OSK hash-chain protocol [17]. They introduce a fixed public string w, and nonce r to query the tag, as described in Figure 5.10. The tag sends an answer $a_i = G(s_i \oplus r)$ to the reader, and renews secret $s_{i+1} = H(s_i)$. The back-end system knows

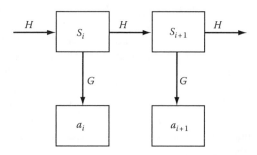

FIGURE 5.8 Hash-chain scheme.

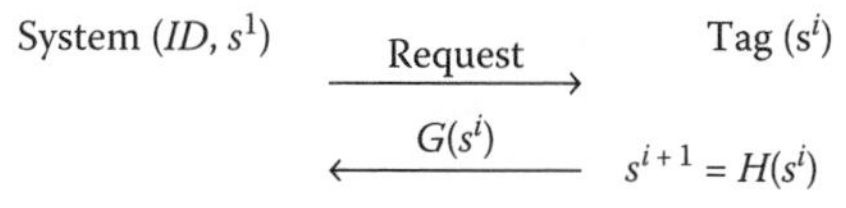

System (ID, s^1) $\xrightarrow{\text{Request}}$ Tag (s^i)

$\xleftarrow{\quad G(s^i) \quad}$ $s^{i+1} = H(s^i)$

FIGURE 5.9 OSK hash-chain protocol.

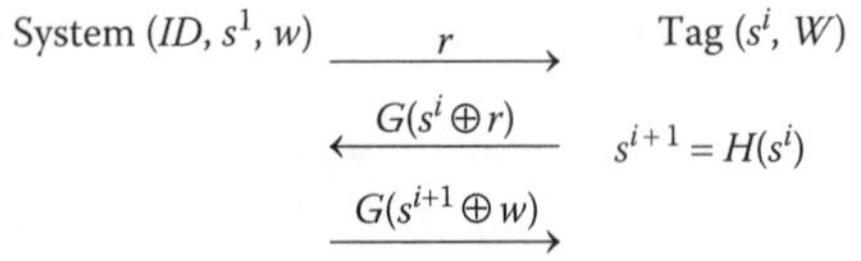

$$\text{System } (ID, s^1, w) \xrightarrow{\quad r \quad} \text{Tag } (s^i, W)$$

$$\xleftarrow{\quad G(s^i \oplus r) \quad} \quad s^{i+1} = H(s^i)$$

$$\xrightarrow{\quad G(s^{i+1} \oplus w) \quad}$$

FIGURE 5.10   Modification of OSK hash-chain protocol.

r, and checks a_i on the similar way as described above. If it finds this value, then it calculates and sends value $G(s_{i+1} \oplus w)$ to the tag to provide an authentication of the reader.

To reduce the complexity of the OSK protocol [17], Avoine et al. [18,19] presented a technique that introduces a specific time-memory trade-off based on Hellman's [20] and Oechslin [21] works. They suggest using a specific time-memory trade-off that reduces the amount of work T needed to invert any given value in a set of N outputs of a one-way function F with help of M units of memory.

5.4.3   LIGHTWEIGHT AUTHENTICATION PROTOCOLS

The basic idea in the realization of privacy protecting in RFID is to provide efficient mechanisms that can offer acceptable security level according to the purpose of RFID application, having in mind a limited hardware capabilities of the tags. In that sense the lightweight authentication protocols are designed with intention to exploit the existing capabilities of the tags in the efficient way. In this section, we shall describe some of the protocols that use less or more of cryptographic functionalities, although some of them have stronger requirements and needs: Juels's minimalist cryptographic protocol, Yet Another TRrivial RFID Authentication Protocol (YA-TRAP), HB+, EMAP, banknote protection protocol, and Molnar and Wagner (MW) private authentication protocol.

5.4.3.1   Juels's Minimalist Cryptographic Protocol

A mutual authentication protocol designed on the principle of minimalist cryptography is presented by Jules [22]. It is based on random identifier or pseudonyms stored in the tag and back-end server.

Each tag has a list of pseudonyms α_1, α_2, ..., α_n. Each pseudonym α_i is associated with two unique random values β_i and γ_i also stored inside the tag. Thus, after the initialization both tag and system contain k 3-tuples $(\alpha_i, \beta_i, \gamma_i)$. This list is refreshed by the reader after successful mutual authentication between tag and verifier. To refresh these values, a vector of m random values is associated with each of these values. The updating of these values is performed using onetime pads that have been transmitted across multiple authentication protocols. Thus an adversary that only eavesdrops periodically is unlikely to learn the updated α_i, β_i, and γ_i values.

As described in Figure 5.11, when the tag is queried by a reader, in the $(i + 1)$th identification it sends the next pseudonym $\alpha_{(i \bmod k)+1}$ from the list (in Figure 5.11, i is stored into the counter c, which is initially equal to 0). The back-end server checks if this pseudonyms exist in database and sends appropriate key $\beta_{(i \bmod k)+1}$. The tag checks this value and sends the third element of 3-tuple $\gamma_{(i \bmod k)+1}$ to the system which checks whether it is the expected value. In this way, the tag and system are authenticated to each other, and then data for tag's refreshing is sent.

The updating process uses a composition of onetime pads across multiple verifier tag sessions (e.g., that pads from two different sessions are XORed with a given tag value k to update it.). Then even if the adversary intercepts the pad used in one session, it may be seen that it will learn no information about the updated value of k. The system sends $3k$ vectors of m fresh random values that are used to update the α_i, β_i, γ_i. Let κ be such a value, $\Delta\kappa = (\delta_\kappa^{(1)}, \ldots, \delta_\kappa^{(m)})$ the vector associated to κ, and $\tilde{\Delta}\kappa = (\tilde{\delta}_\kappa^{(1)}, \ldots, \tilde{\delta}_\kappa^{(m)})$ the vector sent by the system to update $\Delta\kappa$. The update procedure works as follows:

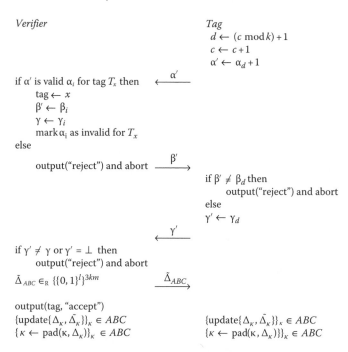

FIGURE 5.11   Juels's minimalist cryptography protocol.

1. $\delta_\kappa^{(i)} \leftarrow \delta_\kappa^{(i+1)} \oplus \tilde{\delta}_\kappa^{(i)}$ where $1 \leq i < m$
2. $\delta_\kappa^{(m)} \leftarrow \tilde{\delta}_\kappa^{(m)}$
3. $\kappa \leftarrow \kappa \oplus \delta_\kappa^{(1)}$

This protocol really uses elementary operation like memory management, string comparisons, and a basic XOR, but the drawback to this approach is that the transmission cost to maintain pads is *3km* bits per session, i.e., it is linear in the length of individual tag values and in the number of consecutive authentication sessions relative to which we wish to achieve the safe protocol.

5.4.3.2   HB+ Protocol

Juels et al. [23] proposed a low-cost HB+ authentication protocol as a modification of human authentication protocol of Hopper and Blum (HB) [24], based on the usage of "Learning Parity in the Presence of Noise" (LPN) problem.

First we shall describe HB authentication protocol. Suppose that a participant *PA* and a device *C* share the same k-bit secret key x, and *PA* like to authenticate herself to *C*. *C* selects a random challenge a and sends it to *PA*. *PA* computes the binary inner product $a \cdot x$ and sends back to *C*, which computes $a \cdot x$, and accepts if *PA*'s parity bit is correct. The probability that someone will guess the correct parity bit in r rounds is 2^{-r}. An eavesdropper can use the GAUSSIAN elimination to calculate the value of x. If *PA* injects noise into her response, it can prevent the passive eavesdroppers to reveal x. *PA* can intentionally send the wrong response with constant probability $\eta \in (0, 1/2)$. In that case *C* authenticates *PA* if less than ηr of her responses is incorrect.

One round of the HB protocol in RFID environment is described in Figure 5.12 (the tag plays the role of the participant *PA* and the reader of the *C*). Each authentication consists of r rounds. As it is shown, this protocol requires only AND and XOR logical operations that can be easy implemented

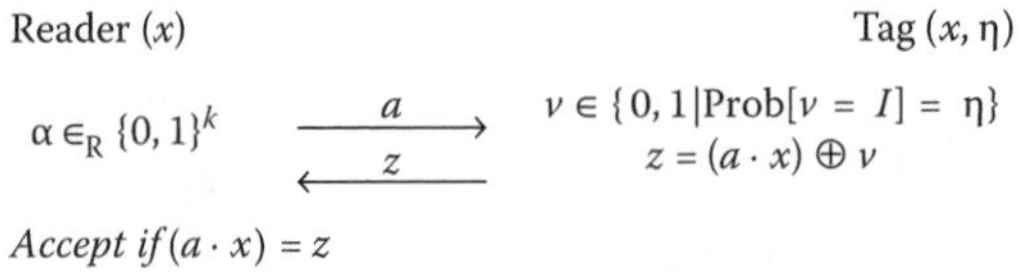

Reader (x) Tag (x, η)

$\alpha \in_R \{0, 1\}^k$ $\quad\xrightarrow{\quad a \quad}\quad$ $v \in \{0, 1 | \text{Prob}[v = I] = \eta\}$
$\qquad\qquad\quad\xleftarrow{\quad z \quad}\qquad$ $z = (a \cdot x) \oplus v$

Accept if $(a \cdot x) = z$

FIGURE 5.12   One round of the HB protocol.

Reader (x, y) Tag (x, y, η)

$\alpha \in_R \{0, 1\}^k$ $b \in_R \{0, 1\}^k$
$\qquad\qquad\qquad\qquad\qquad$ $v \in \{0, 1 | \text{Prob}[v = 1] = \eta\}$

$\qquad\qquad\quad\xleftarrow{\quad b \quad}$
$\qquad\qquad\quad\xrightarrow{\quad a \quad}$ $z = (a \cdot x) \oplus (b \cdot y) \oplus v$
$\qquad\qquad\quad\xleftarrow{\quad Z \quad}$

Accept if $(a \cdot x) \oplus (b \cdot y) \oplus v = z$

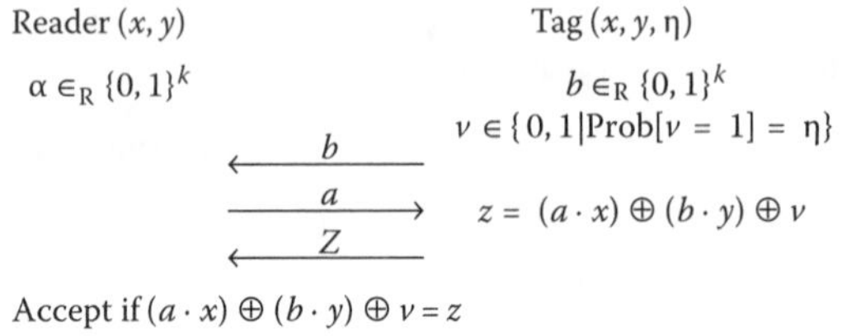

FIGURE 5.13   One round of the HB+ protocol.

in hardware. Also, a noise bit v can be simply generated from physical properties like thermal noise, diode breakdown noise, or any other methods.

The HB protocol can prevent the passive attackers but cannot prevent the active attackers. The HB+ protocol introduces an additional random secret key, so the tag and the system share two k-bit secret key x and y. The tag first generates as random k-bit vector b and sends it to the reader (Figure 5.13). The reader also generates a k-bit random vector a and sends to the tag. The tag then computes response $z = (a \cdot x) \oplus (b \cdot y) \oplus v$, and replies to the reader. The reader computes a value $(a \cdot x) \oplus (b \cdot y)$ and if it equals to z it accepts, otherwise ejects. The reader will authenticate the tag after r rounds if the tag's response is incorrect in less than ηr rounds.

The HB+ protocol is secure against an active and passive attacker, and could be implemented in low-cost tags. However, its security is based on the LPN problem, whose hardness over random instances remains an open question.

5.4.3.3   YA-TRAP

The YA-TRAP [25] proposes a monotonically increasing time stamp on the tag to provide tracking resistant authentication. Each tag contains three values: K, T_t, and T_{max}, where K is a tag-specific value that serves as tag identifier and cryptographic key. The size of K is usually 160 bits. T_t is the internal time stamp that designates time at which it was last interrogated by a reader (it can be initialized with, e.g., the time stamp of manufacture). T_{max} is the top value for the time stamp. Also, the PRNG is required. In practice, it can be realized as an iterated keyed hash, e.g., hash message authentication code (HMAC) started with a random secret seed and keyed on K. The back-end server also contains a unique value K for each of the tags.

As described in Figure 5.14, a reader sends the current time T_r to the tag. The tag compares T_r with T_t. If T_r predates T_t, i.e., $T_r < T_t$, then the tag outputs a random response. Otherwise, the tag outputs $H_r = HMAC_k(T_r)$, where $HMAC_k$ represents an HMAC computed with secret key K. Simultaneously, the tag updates its time stamp T_t with current time T_r. The back-end server checks whether value H_r exists for any secret key in its database. If so, it accepts the tag, otherwise rejects it.

YA-TRAP is susceptible to a trivial DoS attack, when the adversary can send an inaccurate time stamp and incapacitate a tag either fully or temporarily.

1 Tag ← Reader: T_r

2 Tag:
   [2.1] if $((T_r - T_t) \leq 0)$ OR $(T_r > T_{max})$
    $H_r = PRNG_i^j$
   [2.2] else $T_t = T_r$, $H_r = HMAC_{Ki}(T_t)$

3 Tag → Reader: H_r
 In real-time mode, the following steps take place immediately
 following step 3. In batch mode, they are performed later.

4 Reader → Server: T_r, H_r

5 Server:
  [5.1] Let $s = LOOKUP(HASH\_TABLE_{Tr}, H_r)$
  [5.2] If $(s = -1)$ MSG = TAG-ERROR
  [5.3] Else $MSG = G(K_s)$ (or MSG = "VALID")

6 Server → Reader: MSG

FIGURE 5.14 YA-TRAP.

5.4.3.4 Banknote Protection Protocol

To improve banknotes forgery resistance and tracing by law enforcement agency, Juels et al. [2] presented an authentication protocol based on a re-encryption of a digital signature of the banknote's serial number. This approach uses optical data printed on the banknote and electronic data stored on embedded RFID tag.

The participants in this system are a central bank, a law enforcement agency, the merchants, and the consumers. The central bank creates banknotes with unique serial number, and signs them using its private key. The law enforcement encrypts banknotes by its PK, tracks their flow and detects fake ones. The merchants handle the banknotes and may inform the law enforcement agency about some irregularities. The consumers hold the banknotes and want to protect their privacy.

As we already mentioned the banknotes contain optical and electronic data. Optical data are banknote serial number, denomination, digital signature, etc., and can be encoded in a human-readable form or in a machine-readable form. Electronic data are digital signed by the central bank and encrypted with the law enforcement agency's PK and a random number. An RFID tag consists of two cells γ and δ whose access is key-protected, and can be derived from the banknote optical data. The cell γ is always readable but keyed-writable. The other cell δ is both keyed-readable and keyed-writable. The basic idea is the serial number of the banknote signed by the central bank, and encrypted with the law enforcement agency PK and a random value r, is stored into the cell γ. The random value r is stored into the cell δ. The digital signature of the serial number is re-encrypted by merchants as often as possible. After the re-encryption is performed, the new encrypted value is put into γ and the used random value r is put into δ. Because both cells γ and δ are keyed-writable, an optical contact with the banknote is necessary for deriving the access-key and thereby re-encrypting the banknote.

The operations that should be performed in process of banknote creation are following. Let $Sign(k; m)$ be the signature on a message m with a key k, and $Enc(k; m; r)$ the encryption of m under the key k with the random number r. We note $\|$ the concatenation of two bit-strings. The central bank B and law enforcement agency L, respectively, own a pair of public/private keys $(PK_B; SK_B)$ and $(PK_L; SK_L)$.

For every banknote i, B selects a unique serial number S_i and computes its digital signature $\Sigma_i = Sign(SK_B, S_i \| den_i)$ where den_i is the banknote denomination. B then computes an access key D_i such that $D_i = h(\Sigma_i)$, prints S_i and Σ_i on the banknote, and computes $C_i = Enc(PK_L, \Sigma_i \| S_i; r_i)$ where r_i is a random number. C_i is written into cell γ and r_i is written into cell δ (Figure 5.15).

RFID	
Cell γ universally readable/keyed-writable	Cell δ keyed-readable/keyed-writable
$C = Enc(PK_L, \Sigma \| S, r)$	r

Optical	
S	$\Sigma = Sign(SK_B, S \| den)$

FIGURE 5.15 Banknote data.

The merchant M verifies and re-encrypts a received banknote i with the following steps:

1. M reads the optical data S_i and Σ_i and computes access key $D_i = h(\Sigma_i)$.
2. M reads C_i, stored in γ, and keyed-reads r_i which is stored in δ.
3. M checks that $C_i = Enc(PK_L; \Sigma_i \| S_i; r_i)$.
4. M chooses randomly r_{in} and keyed-writes it into δ.
5. M computes $C_{in} = Enc(PK_L; \Sigma_i \| S_i; r_{in})$ and keyed-writes it into γ.

If one of these steps fails then the merchant should inform the law enforcement agency.

When the law enforcement agency L wants to check or track any banknote, L should read cell γ to obtain a encrypted value C, and decrypts C to calculate the plaintext $\Sigma \| S = Dec(SK_L; C)$. Then L verifies digital signature Σ and finally obtains the banknote serial number S.

The authors suggest using an ElGamal-based encryption scheme and the Boneh–Shacham–Lynn signature scheme, both using elliptic curves.

This solution does not require any capabilities beyond the limited ones of the current generations of RFID tags. The intensive cryptographic operations take place in devices for handling banknotes, rather than in the banknotes themselves. Some analyzes of this protocols warn to exist the attacks that compromise the privacy of the banknotes' holders [26].

5.4.3.5 MW Private Authentication Protocol

Molnar and Wagner [1] proposed private mutual authentication protocols for library RFID using a shared secret and pseudorandom function (PRF) residing on the tag. They identify private authentication as an approach where a reader and tag that share a secret efficiently authenticate each other without revealing their identities to an adversary. In addition, they introduce a tree-based technique to reduce identification complexity.

Each tag contains an identifier ID and a secret s. Also, the system database stores a pair (ID, s) for each tag it manages.

The protocol is explained in Figure 5.16. The system sends a nonce a to the tag. When the tags receives a, it picks a nonce b, computes $\sigma = ID \oplus f_s(0, a, b)$ where f_s is a PRF, and sends b and σ to the system. The system finds a pair (ID, s) in the database such that $ID = \sigma \oplus f_s(0, a, b)$. If this pair is found, the system computes $\tau = ID \oplus f_s(1, a, b)$ and sends to the tag. Then the tag authorizes the reader checking whether $ID = \tau \oplus f_s(1, a, b)$.

To decrease the system workload $O(n)$ for n secret keys, i.e., tags, the tree-based technique has been proposed to reduce it to $O(\log_n)$. Instead of searching a whole range of n keys, it uses a

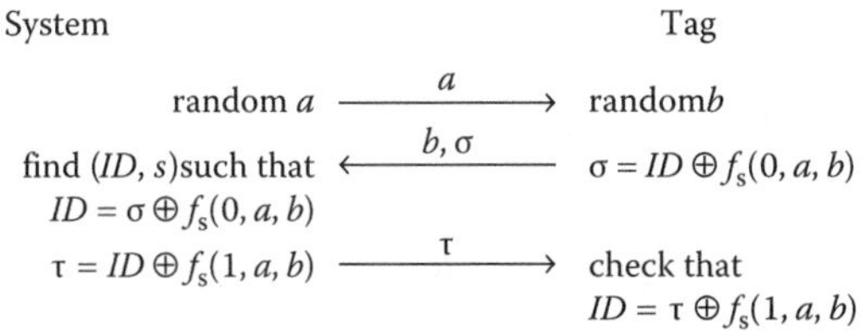

System   Tag

$$\text{random } a \xrightarrow{\quad a \quad} \text{random } b$$

$$\text{find } (ID, s) \text{ such that} \xleftarrow{\quad b, \sigma \quad} \sigma = ID \oplus f_s(0, a, b)$$
$$ID = \sigma \oplus f_s(0, a, b)$$
$$\tau = ID \oplus f_s(1, a, b) \xrightarrow{\quad \tau \quad} \text{check that}$$
$$ID = \tau \oplus f_s(1, a, b)$$

FIGURE 5.16   MW authentication protocol.

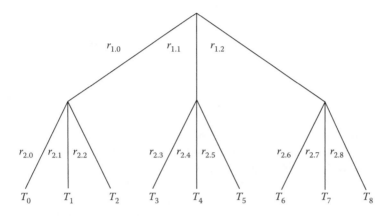

FIGURE 5.17   Tree-based tag secret.

hierarchical tree structure with branching factor δ. Let us imagine that the tags are the leaves of this tree and each edge is associated with a value. Each tag has to store the values along the path from the root of the tree to itself. On the other side, the reader knows all the keys. Let n be the number of tags managed by the system and $l := [\log_\delta n]$ be the depth of the tree with a branching factor δ. Each edge in the tree is valued with a randomly chosen secret $r_{i,j}$ where i is the level in the tree and j is the branch index. Figure 5.17 represents such a tree with parameters $n = 9$ and $\delta = 3$. The secret keys of a given tag is the list of the values $r_{i,j}$ from the root to the leaf. For example, the secret of T_5 in Figure 5.17 is $[r_{1,1}, r_{2,5}]$.

The reader has to query the tag by level from the root to the leaves. It means that the protocol described above is performed for each secret of the explored subtree to find out on which one the tag is. If it fails for all secrets of current level, the protocol is stopped. If the reader has been successfully authenticated at each level, the protocol is succeeded. In this way, the time needed for authentication can be significantly reduced, but it has to be kept in mind that the small branching factor can enable the tag tracing.

5.4.3.6 EMAP

EMAP [27] is a mutual authentication protocol that uses pseudonyms and elementary logical operations on the tag. Each tag contains 96-bit unique identifier (ID) and requires 480 bits of rewritable memory to store index-pseudonym (IDS) and key K. An index-pseudonym of 96-bit length is used as the index to a server's database table where all information about the tag is stored. The key K is divided into four parts each with 96 bits, $K = K1\|K2\|K3\|K4$.

The EMAP has three stages: tag identification, mutual authentication, index-pseudonym and key updating. The nth protocol run of EMAP is shown in Figure 5.18.

Tag identification

      Reader $\rightarrow$ Tag   *hello*

      Tag $\rightarrow$ Reader: $IDS_{\text{tag}(i)}^{(n)}$

EMAP mutual authentication.

      Reader $\rightarrow$ Tag: $A\|B\|C$

      Tag $\rightarrow$ Rader $D\|E$

where

$$A = IDS_{\text{tag}(i)}^{(n)} \;\oplus\; K1_{\text{tag}(i)}^{(n)} \;\oplus\; n1$$

$$B = (IDS_{\text{tag}(i)}^{(n)} \;\vee\; K2_{\text{tag}(i)}^{(n)}) \;\oplus\; n1$$

$$C = IDS_{\text{tag}(i)}^{(n)} \;\oplus\; K3_{\text{tag}(i)}^{(n)} \;\oplus\; n2$$

$$D = (IDS_{\text{tag}(i)}^{(n)} \;\wedge\; K4_{\text{tag}(i)}^{(n)}) \;\oplus\; n2$$

$$E = (IDS_{\text{tag}(i)}^{(n)} \;\wedge\; n1 \vee n2) \;\oplus\; ID_{\text{tag}(i)} \oplus_{I=1}^{4} KI_{\text{tag}(i)}^{(n)}$$

FIGURE 5.18   EMAP.

The reader starts the tag identification by sending a hello message to the tag, which answers by sending its current index-pseudonym (*IDS*). The server uses this *IDS* to access the tag's corresponding secret key ($K = K1\|K2\|K3\|K4$) from the database.

When the server finds the key K, it generates two random numbers $n1$ and $n2$. With $n1$, $n2$, and the subkeys $K1$, $K2$, and $K3$, the reader generates the messages A, B, and C (see Figure 5.18), and then sends them to the tag. With the messages A and B, the tag can authenticate the reader and obtain $n1$. Once the reader is authenticated, the tag obtains the random number $n2$ from the message C, and generates messages D and E. The tag's identifier ID is transmitted securely to the reader in E. If the reader computes a valid ID from E, the tag is successfully authenticated.

After the reader and the tag have authenticated each other, they carry out the index-pseudonym and key updating operations based on the following:

$$IDS_{\text{tag}(i)}^{(n+1)} = IDS_{\text{tag}(i)}^{(n)} \oplus n2 \oplus K1_{\text{tag}(i)}^{(n)}$$

$$K1_{\text{tag}(i)}^{(n+1)} = K1_{\text{tag}(i)}^{(n)} \oplus n2 \oplus \left([ID_{\text{tag}(i)}]_{[1:48]}\|F_{\text{p}}\left(K4_{\text{tag}(i)}^{(n)}\right)\|F_{\text{p}}\left(K3_{\text{tag}(i)}^{(n)}\right)\right)$$

$$K2_{\text{tag}(i)}^{(n+1)} = K2_{\text{tag}(i)}^{(n)} \oplus n2 \oplus \left(F_{\text{p}}\left(K1_{\text{tag}(i)}^{(n)}\right)\|F_{\text{p}}\left(K4_{\text{tag}(i)}^{(n)}\right)\|[ID_{\text{tag}(i)}]_{[49:96]}\right)$$

$$K3_{\text{tag}(i)}^{(n+1)} = K3_{\text{tag}(i)}^{(n)} \oplus n1 \oplus \left([ID_{\text{tag}(i)}]_{[1:48]}\|F_{\text{p}}\left(K4_{\text{tag}(i)}^{(n)}\right)\|F_{\text{p}}\left(K2_{\text{tag}(i)}^{(n)}\right)\right)$$

$$K4_{\text{tag}(i)}^{(n+1)} = K4_{\text{tag}(i)}^{(n)} \oplus n1 \oplus \left(F_{\text{p}}\left(K3_{\text{tag}(i)}^{(n)}\right)\|F_{\text{p}}\left(K1_{\text{tag}(i)}^{(n)}\right)\|[ID_{\text{tag}(i)}]_{[49:96]}\right)$$

where $F_{\text{p}}(X)$ is a parity function where the 96-bit number X is divided into twenty-four 4-bit blocks. Parity is generated for each block, with a total of 24 parity bits. Also, the expression $[ID_{\text{tag}(i)}]_{[j:k]}$ denotes bit sequence from the jth to the kth positions of $ID_{\text{tag}(i)}$.

This protocol does not have intensive computational requirements. It uses simple functions, bitwise XOR, bitwise AND, bitwise OR, and logic needed for the parity function. As these functions can be easy implemented it could be suitable for low-cost RFID tags.

The EMAP has a disadvantage because the tag does not know if the reader verified messages D and E, exchanged during the last step of the authentication. If they are not verified successfully, the reader will not update the tag's entry in the database, while the tag would have updated its memory. It is a potential risk because it might leave both the sides in an unsynchronized state. Thus, the desynchronization attack can cause the database and the tag permanently out of synchronization.

5.4.4 CLASSICAL CRYPTOGRAPHIC AUTHENTICATION PROTOCOLS

The classical cryptographic authentication protocols are based on cryptographic algorithms adapted to RFID tags' possibilities. According to the applied algorithm, they are divided into two general categories, symmetric and asymmetric. Most of them use the symmetric algorithms, but some combine the both. Also, to support the limited hardware resources of the tags, some protocols use a computational power of the back-end server for complex cryptographic operation. However, there are some solutions where cryptographic functionality is carefully designed and implemented on the tags. We will present the approaches that use AES and DES algorithms and the possibilities of usage of an ECC PK algorithm. We have already described in Section 5.4.3.4 the banknote protection protocol [2], based on re-encryption and combined usage of symmetric and PK cryptographic, with the help of an external device unit.

5.4.4.1 Protocols Based on AES

Feldhofer et al. [12] proposed an authentication protocol based on AES algorithm realized on RFID tag. Also, they presented an effective low-cost 128-bit AES implementation with less then 4000 gates, which is suitable for this environment and enables stronger authentication in RFID systems.

Each tag has a unique identifier ID and a randomly chosen secret key s, which are also stored by the system's database. The system generates a 128-bit random number a and sends it to the tag (Figure 5.19). Upon receiving this message, the tag generates a random number b, encrypts these values by computing $\sigma = AES_s(a, b)$, and sends them back to the reader. The reader decrypts the received data $AES_s^{-1}(\sigma)$ and compares it for each ID and, to find the known random number a. If they are equal, the reader can believe in the authenticity of the tag.

Figure 5.20 illustrates the improvement of the above protocol introducing an additional step to provide mutual authentication. When the system has found the value s and identified the tag, it computes $\tau = AES_s(b, a)$ and sends to the tag. Then the tag checks whether τ is a valid encryption of b and a, to be sure that the reader is a legitimate.

5.4.4.2 Protocol Based on DES

The DES algorithm implemented on the RFID tags could be equally used in the protocols described above if it justifies a required security level of the applications. Although DES demands less hardware resources and thus is cost effective, the AES low-cost implementation causes that AES is preferred

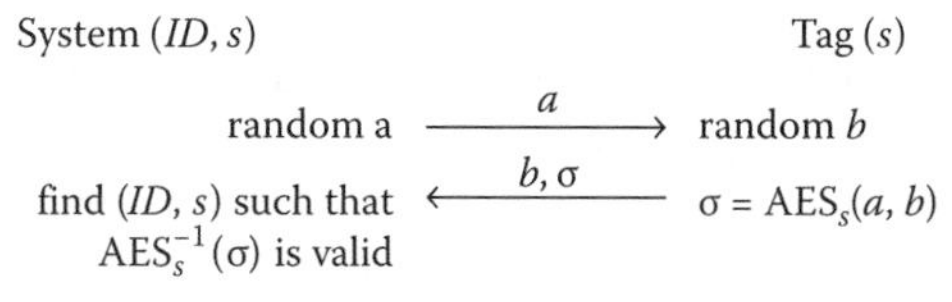

FIGURE 5.19 AES-based authentication protocol.

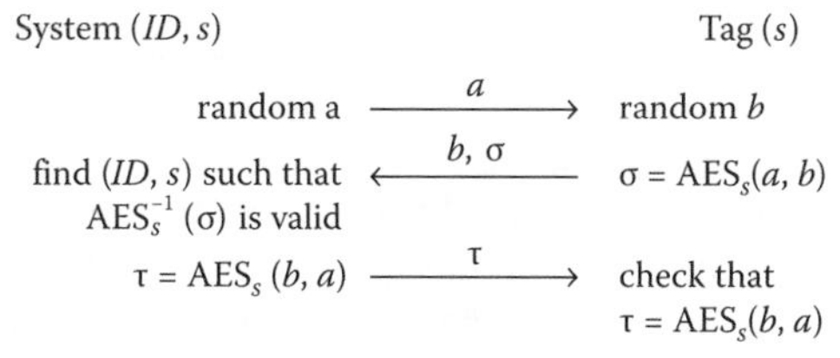

FIGURE 5.20 AES-based mutual authentication protocol.

solution. However, new designs with reduced complexity and cost could be applied for large range of the applications.

Poschmann et al. [11] propose a new block cipher, DES Lightweight (DESL) extension, which is a modification on the standard DES algorithm adapted and especially suited for RFID tags where low power consumption and small chip size are required. It uses only 1900 gates compared with 3700 gate for well-known AES implementation [12].

DESL is based on standard DES but to maximal reduce the hardware requirements, it replaces the eight original S-Boxes by a single new one. Its security features are limited, thus it is relevant for application where short-term security is needed, or where the values protected are relatively low. However, we could agree there are applications where such a security level is adequate, and where DESL implemented tags are appropriate solution.

5.4.4.3 Protocol Based on PKC

PKC offers more sophisticated solutions and facilitates problems that exist in currently authentication protocols (e.g., key management). However, the suitability of PK algorithms for RFID is an open issue because important research problems like limitations in costs, area, or power should be solved to enable its wider usage.

Battina et al. [13,28] discussed the feasibility of PK-based secure identification protocols for RFID tags. They investigate an Okamoto protocol implemented over ECC.

Wolkerstorfer [29] showed that ECC-based PKC is feasible on RFID-tags by implementing the ECDSA on a small IC. In Battina et al. [13,28], it is shown that realization of ECC on the tag would require between 8,500 and 14,000 gates depending on the implementation characteristics.

These works encourage the new investigation that would lead to minimizing the area, reducing the operating frequency, lower power, and provide evidence that ECC on RFID tags might be a feasible solution in the near future.

5.5 OPEN ISSUES

The authentication protocols described above represent some of the typical examples for different kinds of protocols, having in mind a requested security level of applications and hardware resource capabilities of the tags. We may say that there is not any generally applicable protocol but it should be adapted from case to case. On the other hand, many analyses [30–32] have shown some vulnerabilities and disadvantages of these protocols in regard to the different attacks. Thus, further development and researches could be deployed in several directions.

One of the major challenges in design of these protocols is to compose the right balance between needed security and the limited computational capabilities of RFID tags. Most of applications use low-cost tags with very few resources, and thus cannot offer sufficiently strong security level. The second group of protocols, able to guarantee a stronger security, uses the tags with more cryptographic capabilities. It leads to higher cost and fewer numbers of users. It would be necessary to introduce a formalization of security of RFID protocols, and a formalization of adversary model also, what could make progress and improvement in protocol analysis.

The complexity of the protocols is an issue that should be discussed to reduce time needed for authentication. In most of the cases, the complexity depends on the number of tags managed by the system because the verifier does not know in advance the identity of the entity it communicates with. Thus, it cannot assume which key should be used and therefore has to check all keys into the database until the right one is found. Some proposals like memory trade-off and tree-based structure are presented, but more efficient solution could be obtained using asymmetric cryptography. However, asymmetric cryptography is not really applicable yet, especially for widespread usage due to its high hardware demands.

The majority of protocols use an online back-end server as the database and for complex computation. This network approach introduces some additional questions where key distribution is one of the most important. The key distribution and exchange of secrets between the participants could be regarded independent on network services. The efficient way to resolve this issue would be symmetric and particularly asymmetric encryption. However, as we have already seen, it would add a new complexity and cost to the system, whereby asymmetric solution is not real yet. On the other hand, an off-line approach that is not based on the online server could be of interest for some applications.

The design of hardware resource of the RFID tags leads to additional security capabilities and opens possibilities to make classical cryptography closer to the usage in RFID world. The symmetric algorithms (DES, AES) are already really applicable into the tags, while asymmetric ones are not ready yet. However, the new research results have shown that PKC could be feasible into the processors of the tags. Thus, further researches focused on the reduction of power consumption and chip area, design of functional efficiency and protocols based on PKC for RFID are needed to enable asymmetric solution into RFID systems in the near future.

Besides the technical problems mentioned above, many practical problems related to RFID employment should also be resolved. For example, because RFID systems use the electromagnetic spectrum, they are relatively easy to jam using energy at the right frequency. Also, RFID reader or tag collision problems are likely to occur. Another problem appears because an RFID tag cannot make a difference between the readers. Sometimes, the content of a tag can be read from a distance after the item leaves the supplier chain. Also, RFID tag can be read without owner's knowledge.

In the scope of ubiquitous computing, there is an increasing trend in integrating RFID and Wireless Sensor Networks (WSN), which consists of a large number of sensors with sensing, data processing, communication, and network capabilities. It is expected that this integration can exponentially increase visibility and monitoring capabilities, and provide more diversity of security and privacy challenges.

Even considering that the technological problems could be solved, widespread implementation of RFID systems will not become a reality until some nontechnological issues stay open. The main problem is the lack of standards for the data structures to be used with RFID. Different manufactures implement RFID in different ways. RFID devices rarely leave their networks. On the social aspect, laws concerning the intrusion on RFID in user's privacy are expected. For full adoption of the technology and its future, the education of people about RFID's potential benefits is necessary.

5.6 CONCLUSION

This chapter presents the authentication protocols in RFID environment used as security measures aimed to enable reliable authentication of the participants and to protect user privacy. The basic components and several potential security threats in RFID systems are also described. To provide a satisfactory security level of the protocols, the cryptographic functionalities like hash functions, random generator, and some encryption algorithms are implemented in RFID tags in accordance with their capabilities. In this chapter, we have presented several authentication protocols for low-cost RFID tags used in many practical applications, as well as the most relevant authentication protocols with cryptographic capabilities. Some representatives of the hash-lock authentication protocols, lightweight protocols, and challenge–response protocols either with symmetric cipher (AES algorithm) or PK primitive (ECC) are explained.

Having in mind the limited hardware resources of RFID tags, one of the major challenges in design of these protocols is the trade-off between needed security and the computational capabilities of the tags. Although the symmetric cryptographic algorithms have become a reality in RFID, the new research results have shown that PKC could also be feasible in this context. This could lead to more sophisticated security solutions.

REFERENCES

[1] D. Molnar and D. Wagner, Privacy and security in library RFID: Issues, practices, and architectures, *Conference on Computer and Communications Security—CCS'04*, ACM Press, Washington, DC, October 2004.

[2] A. Juels and R. Pappu, Squealing Euros: Privacy protection in RFID-enabled banknotes, *Financial Cryptography '03, Lecture Notes in Computer Science*, Springer-Verlag, Vol. 2742, Le Gosier, Guadeloupe, French West Indies, 2003.

[3] S. Shepard, *RFID Radio Frequency Identification*, McGraw-Hill, New York, 2005.

[4] National Institute of Standards and Technology (NIST), Data Encryption Standard, FIPS PUB 46-2, http://www.itl.nist.gov/fipspubs/fip46-2.htm, 1993.

[5] National Institute of Standards and Technology (NIST), Advanced Encryption Standard, FIPS PUB 197, http://www.csrc.nist.gov/publications/fips/fips197/fips-197.pdf, 2001.

[6] Standards for Efficient Cryptography Group, Elliptic Curve Cryptography, http://www.secg.org/download/aid-385/sec1_final.pdf, Retrieved September 20, 2000.

[7] S. Sarma, S. Weis, and D. Engels, Radio-frequency identification: Security risks and challenges, *RSA Laboratories Cryptobytes*, Vol. 6, No. 1, 2003.

[8] J. Yang, J. Park, H. Lee, K. Ren, and K. Kim, Mutual authentication protocol for low-cost RFID, *Ecrypt Workshop on RFID and Lightweight Crypto*, Graz, Austria, July 2005.

[9] K. Yüksel, Universal hashing for ultra-low-power cryptographic hardware applications, Master's thesis, Worcester Polytechnic Institute, 2004.

[10] CAST Inc., AES and SHA-1 cryptoprocessor cores, http://www.cast-inc.com. visited in September 2007.

[11] A. Poschmann, G. Leander, K. Schramm, and C. Paar, A family of light-weight block ciphers based on DES suited for RFID applications, *Workshop on RFID security*, Graz, Austria, July 2006.

[12] M. Feldhofer, S. Dominikus, and J. Wolkerstorfer, Strong authentication for RFID systems using the AES algorithm, *Workshop on Cryptographic Hardware and Embedded Systems—CHES 2004, Lecture Notes in Computer Science*, Springer-Verlag, Vol. 3156, Boston, MA, August 2004.

[13] L. Batina, J. Guajardo, T. Kerins, N. Mentens, P. Tuyls, and I. Verbauwhede, An elliptic curve processor suitable for RFID-Tags, *IACR eprint*, July 2006.

[14] A. Juels and J. Brainard, Soft blocking: Flexible blocker tags on the cheap, *Workshop on Privacy in the Electronic Society WPES'04*, ACM Press, Washington, DC, October 2004.

[15] S. Weis, S. Sarma, R. Rivest, and D. Engels, Security and privacy aspects of low-cost radio frequency identification systems, *Proceedings of the First Security in Pervasive Computing, Lecture Notes in Computer Science*, Springer-Verlag, Boppard, Germany, Vol. 2802, 2004.

[16] D. Henrici and P. Müller, Hash-based enhancement of location privacy for radio-frequency identification devices using varying identifiers, *Proceedings of PerSec'04 at IEEE PerCom*, Orlando, FL, March 2004.

[17] M. Ohkubo, K. Suzuki, and S. Kinoshita, Cryptographic approach to privacy-friendly tags, *RFID Privacy Workshop*, MIT, Cambridge, MA, 2003.

[18] G. Avoine and P. Oechslin, A scalable and provably secure hash based RFID protocol, *International Workshop on Pervasive Computing and Communication Security—PerSec 2005*, IEEE Computer Society Press, Kauai Island, Hawaii, March 2005.

[19] G. Avoine, E. Dysli, and P. Oechslin, Reducing time complexity in RFID systems, *Selected Areas in Cryptography—SAC 2005, Lecture Notes in Computer Science*, Springer-Verlag, Kingston, Canada, August 2005.

[20] M. Hellman, A cryptanalytic time-memory trade off, *IEEE Transactions on Information Theory*, IT-26(4), 1980.

[21] P. Oechslin, Making a faster cryptanalytic timememory trade-off, *Advances in Cryptology CRYPTO'03, Lecture Notes in Computer Science*, Springer, Santa Barbara, CA, 2003.

[22] A. Juels, Minimalist cryptography for low-cost RFID tag, *Conference on Security in Communication Networks—SCN'04, Lecture Notes in Computer Science*, Springer-Verlag, Amalfi, Italia, September 2004.

[23] A. Juels and S. Weis, Authenticating pervasive devices with human protocols, *Advances in Cryptology—CRYPTO'05, Lecture Notes in Computer Science*, Springer-Verlag, Vol. 3126, Santa Barbara, CA, August 2005.

[24] N. Hopper and M. Blum, A secure human–computer authentication scheme, Technical Report CMU-CS-00-139, Carnegie Mellon University, Pittsburg, PA, 2000.

[25] G. Tsudik, YA-TRAP: Yet another trivial RFID authentication protocol, *International Conference on Pervasive Computing and Communications—PerCom 2006*, IEEE Computer Society Press, Pisa, Italy, March 2006.

[26] G. Avoine, Privacy issues in RFID banknote protection schemes, *International Conference on Smart Card Research and Advanced Applications—CARDIS*, Kluwer Academic Publishers, Toulouse, France, August 2004.

[27] P. Peris-Lopez, J. C. Hernandez-Castro, J. M. Estevez-Tapiador, and A. Ribagorda, EMAP: An efficient mutual authentication protocol for low-cost RFID tags, *OTM Federated Conferences and Workshop: IS Workshop*, Montpellier, France, November 2006.

[28] L. Batina, J. Guajardo, T. Kerins, N. Mentens, P. Tuyls, and I. Verbauwhede, Public-key cryptography for RFID-tags, *International Workshop on Pervasive Capacity and Communications Security PerSec 07*, New York, 2007.

[29] J. Wolkerstorfer, Scaling ECC hardware to a minimum, *ECRYPT Workshop—Cryptographic Advances in Secure Hardware—CRASH 2005*, Leuven, Belgium, September 6–7, 2005.

[30] A. Juels and S. Weis, Defining strong privacy for RFID, *IACR ePrint*, April 2006.

[31] G. Avoine, Radio frequency identification: Adversary model and attacks on existing protocols, Technical Report LASEC-REPORT-2005-001, September 2005.

[32] T. Li and R. Deng, Vulnerability analysis of EMAP—An efficient RFID mutual authentication protocol, *Proceedings of the Availability, Reliability and Security AReS 07*, April 2007.

6 Lightweight Cryptography for Low-Cost RFID Tags

Pedro Peris-Lopez, Julio Cesar Hernandez-Castro, Juan M. Estevez-Tapiador, and Arturo Ribagorda

CONTENTS

The design of lightweight cryptography to conform with real tag requirements is an imperative. However, tags vary enormously in terms of the resources they are equipped with (memory, circuitry, power consumption, etc.). Tags can be classified according to memory type, power source, price, etc. Chien recently proposed a system of tag classification according to the operations supported on-board [7]. In this chapter, we present the main advances that have been made in radio-frequency identification (RFID) security. First, lightweight cryptographic primitives are analyzed. The analysis reveals that while great strides have been made in lightweight stream/block cipher design, much work remains to be done in the case of lightweight hash functions and pseudorandom number generators. Second, we examine the prominent lightweight protocols. These schemes fall into three main categories: (1) ultralightweight protocols based on simple bit-wise operations, (2) those based on a hard problem, such as the human protocols based on the learning parity with noise problem (HB and its variants), and (3) the EPC+ family of protocols, which aim to provide an appropriate security level upon the framework of the EPC-C1G2 standard. Finally, we conclude this chapter by pointing out the most common weaknesses of the proposals put forward to date.

6.1   INTRODUCTION

The major challenge faced when trying to provide security for low-cost radio-frequency identification (RFID) tags is their very limited resources—for example, tag memory will be restricted to several hundred bits, and approximately 250–4000 logic gates out of the total tag space can be devoted to security-related tasks. So readers should bear in mind that these constrained devices are unable to support on-chip standard cryptographic primitives.

6.1.1   TAG SPECIFICATIONS

In an RFID system, each object will be labeled with a tag. Each tag contains a microchip with some computational and storage capabilities, and a coupling element, such as an antenna coil for communication. Tags can be classified according to two main criteria:

1. **Memory type**: The memory element serves as writable and nonwritable data storage. Tags can be programmed to be read-only, write-once read-many, or fully rewritable. Depending on the kind of tag, tag programming can take place at the manufacturing level or at the application level.
2. **Power source**: A tag can obtain power from the signal received from the reader, or it can have its own internal power source. The way the tag gets its power generally defines the category of the tag.
   Passive RFID tags: Passive tags do not have an internal source of power. They harvest their power from the reader that sends out electromagnetic waves. They are restricted in their read/write range as they rely on RF electromagnetic energy from the reader for both power and communication.
   Semipassive RFID tags: Semipassive tags use a battery to run the microchip's circuitry but communicate by harvesting power from the reader signal.
   Active RFID tags: Active tags possess a power source that is used to run the microchip's circuitry and to broadcast a signal to the reader.

Another relevant parameter is tag price,* which creates a broad distinction between high-cost and low-cost RFID tags. Each time a new protocol is defined, the class of tag for which the proposed protocol is appropriate should also be specified. We note that depending on the class of tag, the security level that can be supported will also be different. For example, the security level of a high-cost tag used in e-passports should not equal that of a low-cost tag employed in the supply chain (i.e., tags compliant to EPC Class-1 Generation-2 specification). To clarify the kind of systems we refer to as low-cost/high-cost RFID tags, Table 6.1 summarizes their specifications, these being relevant to current-commercial RFID tags. Additionally, we encourage interested readers to study Ref. [2] which quantifies the technical requirements necessary to evaluate tag costs.

---

* The rule of thumb of gate cost says that every extra 1K gates increases chip price by 1¢ [1].

TABLE 6.1

Specifications for Low-Cost and High-Cost RFID Tags

	Low-Cost RFID Tag	High-Cost RFID Tag
Standards	EPC Class-1 Generation-2 ISO/IEC 18000-6C	ISO/IEC 14443 A/B
Power source	Passively powered	Passively powered
Storage	32–1K bits	32–70 KB
Circuitry (security processing)	250–4K gates Standard cryptographic primitives cannot be supported	Microprocessor Implement 3DES, SHA-1, RSA
Reading distance (commercial devices)	Up to 3 m	Around 10 cm
Price	0.05–0.1 €	Several euros
Physical attacks	Not resistant	Tamper resistance EAL 5+ security level
Resistance to passive attacks	Yes	Yes
Resistance to active attacks [3–6]	No	Yes

In Ref. [7], Chien proposed a tag classification mainly based on the operations supported on chip. High-cost tags divide into two classes: "full-fledged" and "simple." Full-fledged tags support on-board conventional cryptography-like symmetric encryption, cryptographic one-way functions or even public cryptography. Simple tags support random number generation and one-way hash function. Likewise, two classes are distinguished for low-cost RFID tags. "Lightweight" tags are those whose chip supports a random number generation and simple functions like cyclic redundancy code (CRC) checksum, but not hash function. "Ultralightweight" tags can only compute simple bit-wise operations like XOR, AND, OR, etc.

6.1.2 SECURITY CONCERNS

RFID technology is employed today in a great number of applications. However, security aspects do not play an important role in the introduction of this promising technology. We should have learned from past errors such as, for example, those related to bluetooth or WiFi technology. However, the security level offered by commercial solutions is still very low (i.e., Texas Instruments DST tags [8] and Philips Mifare cards [9]).

The two main problems related to RFID technology are privacy and tracking:

1. **Privacy**: Tag content, which may include sensitive information, is revealed when insecure tags are interrogated by readers. Tags and readers should be authenticated to correct this problem. However, readers are frequently not authenticated, and tags usually answer in a completely transparent way.
2. **Tracking**: A problem closely related to privacy is tracking, or violations of location privacy. Even if access to tag content were only allowed to authorized readers, tracking still might not be guaranteed. The answer provided by tags is usually a constant value (i.e., a static identifier). In this situation, an attacker is able to establish an association between tags and their owners. Additionally, we can relax our conditions and assume that tags only contain product codes, rather than a unique identifier. In spite of this, Weis et al. claim that tracking is still possible using an assembly of tags (a constellation) [10].

In addition to the previous threats, there are certain other aspects that should be considered. For depth in all these matters we recommend reading of Refs. [11–13] which survey the most important advances in RFID technology. RFID technology operates by radio, so messages can be easily eavesdropped. Even if communications are encrypted, traffic analysis is feasible, allowing information to be extracted from the communications patterns.

Furthermore, a tag's memory is insecure and susceptible to physical attacks, completely revealing its contents. Counterfeiting (modification of the identity of an item) is also generally based on tag manipulation.

Additionally, active attacks are possible, in which an attacker is able to transmit data to one or both of the involved parties (reader and tag), or stop the data stream in one or both directions. This sort of attack can usually be ruled out when a protocol design conforms to low-cost tags requirements (see Table 6.1). Denial-of-service is another important threat. For example, attackers could prevent the normal operation of tags and readers by jamming the RF channel.

6.2 CRYPTOGRAPHIC PRIMITIVES

A great number of proposals rely on the use of cryptographic primitives: ciphers, hash functions, and pseudorandom number generators (PRNGs). However, the severe tag requirements for low-cost RFID tags prevent the use of standard primitives. Progress in the design of primitives has been different. Many lightweight stream/block ciphers have been designed recently. However, the design of lightweight hash functions and PRNGs remains a pending task. In this section, we present some of the main proposals appropriate to low-cost RFID tags.

6.2.1 Symmetric Ciphers

Block and stream ciphers are two categories of ciphers used in classic cryptography.

Block cipher: A block cipher is a type of symmetric-key encryption algorithm that transforms a fixed-length block of plaintext data into a block of ciphertext data of the same length. The transformation is controlled using a second input—the secret key. The fixed-length is usually designed as the block size, 64-bits being a value normally found in many ciphers. However, block size has nowadays been increased to 128-bits as processors are more sophisticated.

Techniques known as modes of operation have to be used when we encrypt messages longer than the block size. To be useful, a mode must be at least as secure and as efficient as the underlying cipher.

Stream cipher: A stream cipher is a type of symmetric encryption, much faster than block ciphers. In contrast to block ciphers that operate on block of data, stream ciphers operate on smaller units of plaintext, usually bits. If we encrypt, several times any particular plaintext with a block cipher using the same key, the same ciphertext is obtained. However, with a stream cipher, the transformation of these smaller plaintext units will vary, depending on when they are encountered during the encryption process.

A stream cipher generates a keystream—a sequence of bits used as a key. Encryption is performed by combining the keystream with the plaintext, usually the bitwise XOR operation is employed. When the keystream is independent of the plaintext and ciphertext, the stream cipher is said to be synchronous. Alternatively, it can depend on the data and its encryption, in which case the stream cipher is said to be self-synchronizing.

6.2.1.1 AES

In 2005, Feldhofer et al. proposed a hardware implementation of the Advanced Encryption Standard (AES) which is optimized for low-resource requirements [14]. The authors claim that the proposed implementation will serve for considerable time as a reference for AES-128 implementations that support encryption and decryption including key setup. The two main objectives of this implementation are optimization of the silicon area (smallest possible footprint) and power consumption. High data throughput, on other hand, is of minor importance.

Most AES operations are byte oriented, executing efficiently on 8-bit processors. As 8-bit operations can be combined to form 32-bit operations, AES can be efficiently implemented on 32-bit processors too. However, the most common implementation found is a 128-bit architecture. With this architecture, a higher degree of parallelism is obtained, so permitting higher throughput. Feldhofer et al. decided to implement AES with encryption and decryption using a fixed key size of 128 bits. The low-power requirements do not permit use of 128-bits operations and even a 32-bit architecture still exceeds the restrictions. An 8-bit architecture, as seen in Figure 6.1, was finally proposed. For more details, reader is referred to the original paper. We summarize now the most relevant aspects from their implementation:

Die size: The core occupies a silicon area of 0.25 mm^2 on a $0.35 \, \mu\text{m}$ CMOS. This compares roughly to 3.4K gate equivalents, or to the size of a grain of sand.

Performance: The functionality of cheap devices, even at very low supply voltages, has been tested. The chip works correctly with a supply voltage higher than 0.65 V. The encryption of one 128-bit block requires 1032 including loading data and data reading. The maximum throughput for encryption is 9.9 Mbps. Similar values are obtained for decryption.

Power consumption: A charge transfer method has been employed to measure the power consumption on the AES chip. The mean current consumption of the chip measured is $3.0 \, \mu\text{A}$ at the target clock frequency of 100 kHz and a supply voltage of 1.5 V.

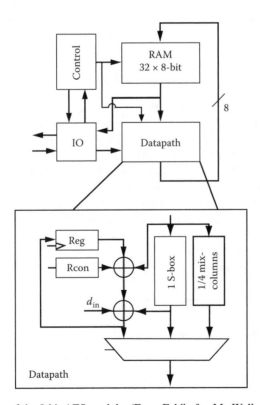

FIGURE 6.1   Architecture of the 8-bit AES module. (From Feldhofer, M., Wolkerstorfer, J., and Rijmen, V., AES implementation on a grain of sand. In *Proc. on Information Security*, Vol. 152, pp. 13–20. IEEE Computer Society, 2005. With permission.)

TABLE 6.2

AES-128: Performance Comparison

AES-128 Version	Technology (μm)	Area (GEs)	Throughput (Mbps)	Max. Frequency (MHz)	Power (μW)
Feldhofer [14]	0.35	3400	9.9	80	4.5
Satoh [15]	0.11	5400	311	130	—
Mangard [16]	0.6	7000	70	50	—

Finally, we can compare the Feldhofer et al. proposal with other efficient AES implementations. Table 6.2 summarizes the specifications of these proposals. The Feldhofer proposal requires the lower chip area for its implementation, with a reduction of at least 40 percent. From a throughput perspective, Satoh [15] and Mangard [16] are superior. However, 10 Mbps may be enough for the intended applications. Finally, the Satoh and Mangard proposals do not include power consumption results because they did not manufacture their design in silicon.

6.2.1.2 DES and Its Variants

Poschmann et al. [17] proposed a serialized version of data encryption standard (DES) which processes 4- and 6-bit data words instead of 32 or 48 bits (see the original paper for details). This implementation requires 2310 equivalent gates, and 144 clock cycles are consumed to encrypt a plaintext. The security of this cipher is limited by the use of a 56-bit key. A brute-force attack using software takes a few months and hundreds of personal computers (PCs), but only few days with a special purpose machine such as COPACOBANA [18]. So the above implementation is intended for applications demanding short-term security or when the protected contents have a relatively low value. A key-whitening technique can be employed when a higher security level is required, yielding DESX:

$$\text{DESX}_{k.k1.k2}(x) = k2 \oplus \text{DES}_k(k1 \oplus x)$$

The key space increases from 56 to 184 bits. However, the security level of DESX is bounded by 118 bits due to the time-memory trade-offs [19]. This scheme demands around 14 percent extra gates because of the use of the bank of XOR gates and the additional registers.

The same authors proposed a compact version called DES lightweight extension (DESL). DESL is based on DES design, but the eight original S-boxes are replaced by a single new one (7 S-boxes and a multiplexer are eliminated). The S-box has been optimized allowing the cipher to be inmune against common attacks (i.e., linear and differential cryptanalysis, Davies–Murphy attack). This DES variant demands 1850 equivalent gates, consuming 144 clock cycles to encrypt a plaintext. Finally, authors proposed the use of key whitening when a higher security level is required. This version, called DESXL, requires 2170 equivalent gates, consuming the same number of clock cycles. Table 6.3 summarizes the specifications of the different DES variants.

TABLE 6.3

DES Variants: Performance Comparison

DES Version	Technology (μm)	Area (GEs)	Clock Cycles	Current Consumption (μA)
DESL	0.18	1,848	144	0.89
DES	0.18	2,309	144	1.19
DESX	0.18	2,629	144	—
DESXL	0.18	2,168	144	—

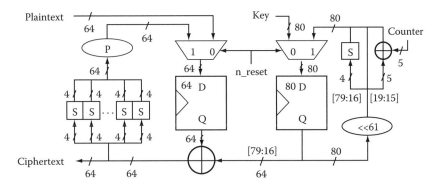

FIGURE 6.2 Present cipher. (From Bogdanov, A., Knudsen, L.R., Leander, G., Paar, C., Poschmann, A., Robshaw, M.J.B., Seurin, Y., and Vikkelsoe, C., PRESENT: An ultra-lightweight block cipher. In *Proc. CHES07. Lecture Notes in Computer Science*, 4727, pp. 450–466. Springer-Verlag, Vienna, Austria, 2007. With permission.)

6.2.1.3 Present

Recently, Bogdanov et al. proposed an ultralightweight block cipher named Present [20]. This cipher is an example of a substitution–permutation network (SPN) and consists of 32 rounds. The block length is 64 bits and two key lengths of 80 and 128 bits are supported. Figure 6.2 depicts the data path of an area-optimized version of Present-80. As any other SPN, the cipher comprises three stages: a key mixing step, a substitution layer, and a permutation layer. For the key mixing, authors choose a simple XOR operation. The key schedule consists basically of a 61-bit rotation together with an S-box and a round counter (Present-80 uses a single S-box, whereas Present-128 demands two S-boxes). The substitution layer comprises 16 S-boxes with 4-bit inputs and outputs (4 × 4). The authors recommend the version with the 80-bit key for constrained devices (i.e., low-cost RFID tags or sensor networks). The implementation of Present-80 requires around 1.6K equivalent logic gates, and 32 clock cycles are needed to encrypt a 64-bit plaintext (200 kbps). Present-80 is therefore more efficient than AES-128 [14] from a hardware perspective. However, from a security perspective, AES has been studied for many years and a deeper security analysis of Present is required [21].

6.2.1.4 Other Block Ciphers

Recently, many block ciphers (i.e., HIGHT, Clefia, SEA, TEA, etc.) have been proposed. HIGHT is block cipher with 64-bit block length and 128-bit key length [22]. This cipher requires 3048 gates on 0.25 μm technology and consumes 34 clock cycles for encryption. Clefia is a new 128-bit block cipher supporting key lengths of 128, 192, and 256 bits [23]. The authors claim this cipher achieves enough resistance against known attacks and performs well both in hardware and software. The circuit of CLEFIA with 128-bit key by area optimization requires around 6K gates. The Scalable Encryption Algorithm (SEA$_{n,b}$) is a block cipher targeted for small embedded applications [24,25]. This cipher can be parameterized according to the plaintext size n, key size n, and the processor (or word) size b. Although it was initially designed for software implementations, its implementation in a field programmable gate array (FPGA) device has been also examined [26]. Finally, the Tiny Encryption Algorithm (TEA) and its variants (XTEA and XXTEA) are optimized for software architectures [27–29]. The TEA family uses only addition, XOR, and shift, so that those can be implemented efficiently on 8-bit platforms.

6.2.1.5 Grain

Grain [30,31] is a stream cipher designed for restricted hardware environments and submitted to eSTREAM in 2004 by Martin Hell, Thomas Johansson, and Willi Meier. It has been selected as

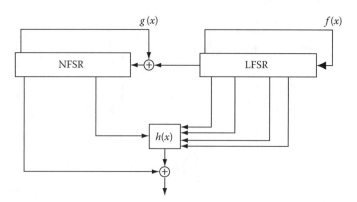

FIGURE 6.3 Grain cipher. (From Hell, M., Johansson, T., and Meier, W., Grain: A stream cipher for constrained environments, 2005. http://www.ecrypt.eu.org/stream/. With permission.)

Phase 3 Focus Candidate for Profile 2 by the eSTREAM project. This cipher is suitable for devices in which gate count, power consumption, and memory are very limited.

We shall give a brief description of the Grain cipher (the reader is referred to the original paper for more details). Grain cipher is a bit-oriented synchronous stream cipher. So the key stream is generated independently from the plaintext. Two shift registers are employed, one with linear feedback (LFSR) and one with nonlinear feedback (NFSR). The NFSR is used in combination with a nonlinear output function to introduce nonlinearity in the cipher. To balance the state of the NSFR, the input to the NFSR is masked with the output of the LFSR. Both shift registers are 80 bit in size. The key size is 80 bits and the IV size is specified to be 64 bits. Figure 6.3 shows the architecture of the cipher.

In the initial proposal, both shift registers are regularly clocked so the cipher will output 1 bit/clock. Johansson and Meier looked at how to increase the speed of the cipher at the expense of more hardware. Specifically, this can be done by just implementing the feedback function and the output function several times. To achieve the objective, in registers each bit is shifted t steps instead of one when the speed is increased by a factor t. An example of the architecture proposed when the speed is doubled is shown in Figure 6.4.

The authors studied the hardware complexity for its implementation. They elaborated a design based on standard FPGA architectures. The design has been implemented in three different FPGA families. Table 6.4 summarizes the results obtained.

Finally, the authors claim that the Grain design provides much better security than both E0 and A5/1 while maintaining a low gate count.

6.2.1.6 Trivium

Trivium is a synchronous stream cipher designed to provide a flexible trade-off between speed and gate count in hardware, and reasonably efficient software implementation [32]. It has one of the simplest architecture of the eSTREAM candidates and is consequently particularly easy to implement. Trivium generates up to 2^{64} bits of output from an 80-bit key and an 80-bit IV. As for most stream ciphers, this process consists of two phases: first the internal state of the cipher is initialized using the key and the IV, then the state is repeatedly updated and used to generate keystream bits.

Trivium's 288-bit internal state consists of three shift registers of different lengths. The key stream generation consists of an iterative process which extracts the values of 15 specific state bits and uses them both to update 3 bits of the state and to compute 1 bit of keystream. The state bits are then rotated and the process is repeated until the requested $N < 2^{64}$ bits of keystream have been

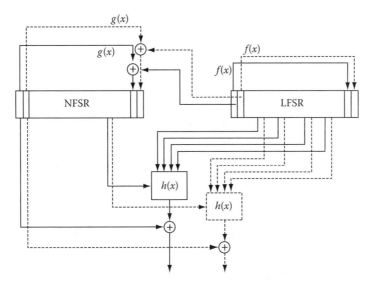

FIGURE 6.4 Grain cipher when the speed is double. (From Hell, M., Johansson, T., and Meier, W., Grain: A stream cipher for constrained environments, 2005. http://www.ecrypt.eu.org/stream/. With permission.)

TABLE 6.4
Gate Count and Throughput of Grain

t	Gate Count	Throughput (Mbit/s)		
		MAX 3000A	**MAX II**	**Cyclone**
1	1450	49	200	282
2	1637	98.4	422	576
4	2010	196	632	872
8	2756	—	1184	1736
16	4248	—	2128	3136

generated. Figure 6.5 shows a graphical representation of the keystream generation. To initialize the cipher, the key and IV are written into two of the shift registers and the remaining bits are set to a fixed pattern of zeros and ones. The cipher state is then updated $4 \times 288 = 1152$ times in the same way as explained above, so that every bit of the internal state depends on every bit of the key and of the IV in a complex nonlinear way.

Authors suggest a bit-oriented architecture for compact implementation. Additionally, the parallelization of operations allows power-efficient and fast implementations. As any state bit is not used for at least 64 iterations after its modification, up to 64 iterations can be computed at once. So the 3 AND gates and 11 XOR gates in the original scheme are duplicated a corresponding number of times. Table 6.5 summarizes the results obtained.

6.2.1.7 Other Stream Ciphers

The ECRYPT Stream Cipher Project is a multiyear effort to identify new stream ciphers that might become suitable for widespread adoption. Profile 2 is oriented to stream ciphers for hardware

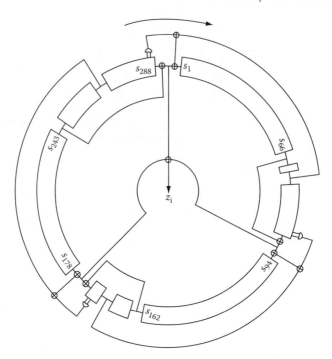

FIGURE 6.5 Trivium cipher. (From De Cannire, C. and Preneel, B., Trivium specification, 2005. http://www.ecrypt.eu.org/stream/. With permission.)

TABLE 6.5
Performance of Trivium

Version	Gates	Bits/Cycle
Trivium	3488 (288 flip-flops + 3 AND + 11 XOR)	1
Trivium-8	3712 (288 flip-flops + 24 AND + 88 XOR)	8
Trivium-16	3968 (288 flip-flops + 48 AND + 176 XOR)	16
Trivium-32	4480 (288 flip-flops + 96 AND + 352 XOR)	32
Trivium-64	5504 (288 flip-flops + 192 AND + 704 XOR)	64

applications with restricted resources. Besides Grain and Trivium, other candidates for Profile-2 are the following:

Cipher	Authors
DECIM	Côme Berbain, Olivier Billet, Anne Canteaut, Nicolas Courtois, Blandine Debraize, Henri Gilbert, Louis Goubin, Aline Gouget, Louis Granboulan, Cédric Lauradoux, Marine Minier, Thomas Pornin, and Hervé Sibert
Edon80	Danilo Gligoroski, Smile Markovski, Ljupco Kocarev, and Marjan Gusev
F-FCSR	Thierry Berger, François Arnault, and Cédric Lauradoux
MICKEY	Steve Babbage and Matthew Dodd
Moustique	Joan Daemen and Paris Kitsos
Pomaranch	Cees Jansen, Tor Helleseth, and Alexander Kolosha

The reader is referred to `http://www.ecrypt.eu.org/stream/` page for more details. Additionally, hardware implementation and performance metrics of these algorithms have been assessed in detail [33–35].

6.2.2 ASYMMETRIC CIPHERS

The major vulnerability related to the use of symmetric key systems is the need to share a secret key, this being required for both encryption and decryption. In asymmetric (or public-key) cryptography, this is not an issue in the same way. Two keys, mathematically related, are used and work together. For example, a plaintext encrypted with one of the keys can only be decrypted with the other associated key. Of these two keys one will be typically be kept private by its owner and the other will be made as widely known as possible. So there is almost no need to share keys, avoiding the risk of compromising security.

Whether a public-key cryptosystem can be implemented on an RFID tag or not remains an open problem. Elliptic curve cryptography (ECC) is emerging as an attractive public-key cryptosystem for this kind of device. Compared to traditional cryptosystems like RSA or discrete logarithms, ECC offers equivalent security with smaller operand length, which results not only in lower computational requirements but also power consumption and storage. ECC has been commercially accepted and recently endorsed by the U.S. government.

In Ref. [37], the most recent works in this research area are described. Batina et al. investigated several options considering ECC over F_{2^p}, p a prime, operands ranging between 130 and 140 bits of length, and composite fields. Instead, Kumar et al. proposed ECC over binary fields ($F_{2^{131}}$ for short-term security—$F_{2^{131}}$ for medium-term security). The authors claim that the use of binary fields rather than prime fields offers two main advantages: carry-free arithmetic and simplified squaring arithmetic. A performance comparison of the main contributions of ECC implementations for constrained devices is summarized in Table 6.6.

6.2.3 HASH FUNCTIONS

Basically, a cryptographic hash function is a transformation that takes an input and returns a fixed-size string, which is called the hash value. Specifically, it is commonly assumed that a cryptographic hash function should meet the following prerequisites:

- Preimage resistant: given h, it should be hard to find any m such that $h = \text{hash}(m)$.
- Second preimage resistant: given an input $m1$, it should be hard to find another input, $m2$ (not equal to $m1$) such that $\text{hash}(m1) = \text{hash}(m2)$.
- Collision resistant: it should be hard to find two different messages $m1$ and $m2$ such that $\text{hash}(m1) = \text{hash}(m2)$.

TABLE 6.6
ECC Performance Comparison

Source	Field	Total Area (GEs)	Technology (μm)	Frequency (MHz)
Kumar et al. [36]	$GF(2^{113})$	10,113	0.35	13.560
	$GF(2^{131})$	11,970	0.35	13.560
	$GF(2^{163})$	15,064	0.35	13.560
	$GF(2^{193})$	17,723	0.35	13.560
Batina et al. [37,38]	$GF(2^{67})^2$	12,944	0.25	0.175
	$GF(2^{131})$	14,735	0.25	0.175
Gaubatz et al. [39]	$GF(p_{100})$	18,720	0.13	0.5
Wolkerstorfer [40]	$GF(p^{191})$	23,000	0.35	68.500
Ötztürk et al. [41]	$GF(p_{166})$	30,333	0.13	20.000

For a hash output of n bits, compromising these should require 2^n, 2^n, and $2^{n/2}$. Additionally, some precautions should be taken when a new protocol is designed. Because most hash functions are built using the Merkle–Damgard construction, these are vulnerable to length-extension attacks: given $h(m)$ and length(m) but not m, by choosing a suitable m', an attacker can calculate $h(m\|m')$, where $\|$ denotes concatenation.

Hash functions are considered a better choice within the RFID security community from the point of view of implementation. As a result, most of the proposed protocols are based on the use of hash functions. Specifically, since the work of Sarma [42] in 2002, there has been a huge number of solutions based on this idea [43–46].

Although the above proposals apparently constitute good and secure approaches, engineers face the nontrivial problem of implementing cryptographic hash functions with only 250-4K gates [47]. The best implementation of SHA-256 requires around 11K gates and 1120 clock cycles to perform a hash calculation on a 512-bit data block [48]. As far greater resources are necessary compared with a low-cost RFID tag, it may seen sensible to propose the use of other smaller hash functions. However, neither functions such as SHA-1 (8.1K gates, 1228 clock cycles) nor MD5 (8.4K gates, 612 clock cycles) fit in a tag [48]. Recently, some authors have suggested the use of a "universal hash function" [49]. Although this solution only needs around 1.7K gates, deeper security analysis is needed and this has not yet been done. Furthermore, this function only has a 64-bit output, which does not guarantee an appropriate security level because finding collisions is a relatively easy task because of the birthday paradox (around 2^{32} operations).

6.2.3.1 Tav-128 Bits

Recently, Peris et al. proposed a 128-bit hash function, named Tav-128, that can be fitted in low-cost RFID tags and provides a suitable security level for most applications [50]. The code of the proposed hash function is included in Appendix A. Tav-128 is composed of two main parts:

Expansion algorithm: Some of the recent cryptanalytic attacks on many of the most important hash functions [51,52] rely on the fact that these constructions generally use a very linear (LFSR-based) expansion algorithm. To avoid this, the authors decided to make the expansion of the Tav-128 hash function (corresponding to algorithms C and D) highly nonlinear.

Filter phase: A filter phase (corresponding to algorithms A and B) was included to avoid giving the attacker direct access to any bit of the internal state. The authors claim that without this possibility, some attacks that have been made on other cryptographic primitives in the past are precluded. In fact, decreasing the control that the attacker has over the hash function inputs complicates his task significantly.

The statistical properties of Tav-128 output over a very low entropy input was examined with several suites of randomness tests. The authors claim that although passing these tests does not prove security, it does point out the nonexistence of trivial weaknesses.

One of the most relevant aspects in the design of Tav-128 is its hardware complexity. Only simple operations were employed to respect tag requirements. Specifically, the following operators were used: right shifts, bitwise xor, and addition mod 2^{32}. Although the proposed primitive was not implemented in hardware, an overestimation of its gate count was accomplished. The authors reckoned that 2578 equivalent logic gates are needed for implementing Tav-128. Another key aspect is its throughput. The authors estimated that 1568 clock cycles (8.2 kbps at 100 kHz) are consumed in executing one Tav-128 hash. Finally, the authors studied how the throughput can be increased by 25 percent by reducing the number of rounds in the $r2$ loop to six.

6.2.4 PSEUDORANDOM NUMBER GENERATORS

In many cryptographic applications, PRNGs are necessary. In the RFID context, PRNG use has been proposed almost because the first appearance of publications concerned with security. In 2003,

Weis et al. proposed the randomized hash-locking scheme, based on a hash function and a random number generator (RNG) to prevent tracking [53]. This proposal presents scalability problems, its applicability being limited only to small tag populations. Molnar et al. proposed a simple protocol for enhancing passwords in RFID tags [54]. There are also others papers in which the use of a PRNG has been proposed [55–59].

Recently, EPCGlobal (EPC-C1G2) and ISO (ISO/IEC 18000-6C) ratified the use of a PRNG for low-cost RFID tags. A generator conforming with these specifications [60,61] should meet the following randomness criteria:

Probability of a single $RN16$: The probability that any $RN16$ drawn from the RNG has value $RN16 = j$ for any j, shall be bounded by

$$\frac{0.8}{2^{16}} < P(RN16 = j) < \frac{1.25}{2^{16}} \tag{6.1}$$

Probability of simultaneously identical sequences: For a tag population of up to 10,000 tags, the probability that any of two or more tags simultaneously generate the same sequence of $RN16$s shall be less than 0.1 percent, regardless of when the tags are energized.

Probability of predicting an $RN16$: An $RN16$ drawn from a tag's RNG 10 ms after the end of Tr, shall not be predictable with a probability greater than 0.025 percent if the outcomes of prior draws from RNG, performed under identical conditions, are known.

Furthermore, when designing a PRNG conforming to the EPC-G1C2, we should take into account the severe hardware limitations of these systems. Another relevant aspect is the matter of temporal requirements, which demands that a given number of tags should be read in a given amount of time. Compared to previous Class-1 EPC tags (150 tags/second), Generation-2 readers should be able to read 450 tags/second [60,62]. This puts a severe limitation on the maximum number of cycles ($1/f$) a tag can spend to generate a random number.

6.2.4.1 LAMED

So far, no public algorithm conforming to EPC-C1G2 has been published. On other hand, one can find commercial tags that satisfy this specification [63,64]. However, the algorithms of the PRNGs supported are not public, and if you try to obtain them you receive a negative response. Manufacturers should learn from past disasters, such as Texas DST or Philips Mifare cards, whose security resided in their obscurity and was quickly breached [8,9].

Recently, Peris et al. proposed a new PRNG (LAMED) conforming to low-cost tag requirements and EPC-C1G2 specification [65]. The methodology to obtain the core of the proposed PRNG was based on the use of genetic programming. We now briefly describe the most important aspects for their experimentation (we refer the reader to the original paper for further details):

1. Function set: As the building block of the individual, only very efficient operations easy to implement in hardware were employed: vrotd (one-bit right rotation), xor (addition mod 2), and (bitwise and), or (bitwise or), and not (bitwise not). The sum (sum mod 2^{32}) operator was also necessary, to avoid linearity.
2. Terminal set: The terminals were represented by two 32-bit unsigned integers (a_0, a_1).
3. Fitness function: The nonlinearity of the generator was evaluated by means of the avalanche effect. In fact, they proposed the use of the strict avalanche criterion [66], an even more demanding property.
4. The number of nodes was limited to 65 to try to ensure a high degree of avalanche effect.

The code for the implementation of LAMED is included in Appendix B, where $^\wedge$ is the **xor** operator and $\texttt{vrotdk}(v,k)$ means rotations of v, k times.

The seed of the PRNG consists of an initialization vector (iv) and a key (s). The iv may be public, but it is very important that it is never reused together with the same key. It can also be kept secret, effectively extending the key length up to 64 bits, depending on the security needs of the specific application. The key is a secret only known by an authorized reader and the tag. Usually, the secret (s) is set at manufacture time, and is stored in the relevant row of the back-end database. In Equations 6.2, and 6.3, the proposed update function for the internal state of LAMED is shown.

$$a_0^{n+1} = \begin{cases} a_1^n + iv & \text{if } n \text{ is odd} \\ a_1^n \oplus iv & \text{if } n \text{ is even} \end{cases} \tag{6.2}$$

$$a_1^{n+1} = \begin{cases} 0(a_0^n, a_1^n) \oplus s & \text{if } n \text{ is odd} \\ 0(a_0^n, a_1^n) + s & \text{if } n \text{ is even} \end{cases} \tag{6.3}$$

The output length is 32 bits. As the specification EPC-C1G2 proposes the use of a 16-bit PRNG, Peris et al. designed a 16-bit version of our PRNG, named LAMED-EPC, with an additional xor operation before the final output. The 32-bit output is divided in two halves, $MSB_{31:16}$ and $LSB_{15:0}$. These two halves will then be xored to obtain a 16-bit output with higher entropy. The authors maintain that their proposal is EPC-C1G2 compliant and has the additional advantage that a 32-bit PRNG is also supported, which could be relevant for particular applications and also increases its flexibility and, probably, its longevity, as mentioned in Refs. [6,67]. Furthermore, the access and kill PIN are 32-bit values. The use of 32-bit random numbers would avoid the complex multistep procedure for using the access and kill command proposed in the standard.

The security margin of a protocol using a 16-bit PRNG is usually bounded by $1/2^{16}$. Moreover, a generic time-memory-data trade-off attack costs $O(2^{\frac{n}{2}})$, see Ref. [68], where n is the number of inner state variables in the PRNG. In LAMED-EPC, with a public iv, which is the weakest security configuration possible, the total of state variables is 32. Thus the expected complexity of a time-memory is lower limited by $O(2^{16})$.

An extensive security analysis of LAMED, consisting of examining the statistical properties of the output over a random initialization of the iv and the key (s), were performed by the authors. Specifically, ENT [69], Diehard [70,71], NIST [72], and David Sexton [73] batteries were employed. From the obtained results, Peris et al. concluded that LAMED's output successfully passed all the randomness tests. Additionally, the authors performed an extensive analysis of LAMED-EPC by evaluating its conformity with the EPC-C1G2 specification. From this last analysis, the authors obtained two main conclusions. First, the probabilities associated with LAMED-EPC output are well within the limits set by the specification. Second, the prediction analysis gives no indication that the output could be predicted significantly better by knowledge of prior outputs without knowing the secret key and the IV, than just by chance.

Finally, Peris et al. proposed a logic design to implement the proposed PRNG. They estimated that fewer than 1.6K equivalent logic gates are needed to implement LAMED (1566 LG) and LAMED-EPC (1585 LG). Additionally, LAMED and LAMED-EPC throughput were also derived, obtaining 17.2 and 8.2 kbps, respectively.

6.3 LIGHTWEIGHT PROTOCOLS

6.3.1 Naïve Proposals

In 2003, Vajda and Buttyán proposed a set of lightweight challenge and response authentication protocols [74]. They started analyzing the following simple protocol:

$$R \to T : \quad x \oplus k = a$$
$$T \to R : \quad f(x) \oplus k = b \tag{6.4}$$

where

    R and T symbolize the reader and the tag entity, respectively
    k is the secret key shared between both entities
    x is an n-bit random challenge
    f is an n-bits to n-bits mapping function

The mutual information $(I(h, k) = H(x \oplus f(x)))$ between the key and the exchanged messages $(h = (a, b))$ is the entropy of $x \oplus f(x)$. Bearing this in mind, there are several ways to strengthen the security of the above protocol:

1. **Nonlinearity**: The set of preimages will be more hard to compute by an attacker if a nonlinear function f is used.
2. **Mixed operations**: XOR operation (which is linear over binary vectors) could be replaced by modular addition or modular integer powering. In this case, the analysis and combination of messages would not be so easy.
3. **Compression**: Authentication does not need to be based on invertible transformations.
4. **Keys**: Different keys may be used in the two directions.

Protocol 1: XOR: A protocol based on the xor operation and with different keys used in both directions is proposed below:

$$R \to T : \quad x \oplus k_1$$
$$T \to R : \quad x \oplus k_2 \tag{6.5}$$

The above protocol is probably secure if the keys are randomly selected in each run of the protocol. As low-cost RFID tags are very restricted, a probably secure key update algorithm is necessary. The authors then proposed a new protocol, where a lightweight block stream generator with secret seed value k^0 is employed:

$$R \to T : \quad x \oplus k_i$$
$$T \to R : \quad x \oplus k_0 \tag{6.6}$$

where $k^i = \prod(k^{(i-1)})$, and $\prod : \{0, 1\}^n \to \{0, 1\}^n$ is a permutation, a special stream generator that expands seed k^0.

    Protocol 2: Subset: The following protocol was proposed, where a tag sends back a m-bit portion of the challenge as a replay:

$$R \to T : \hspace{4cm} x \oplus k$$
$$T \to R : \quad f(x) = \left(x_{L, x_{R[0..7]}}, x_{L, x_{R[8..15]}}, \ldots, x_{L, x_{R[8m..8m+7]}} \right) \tag{6.7}$$

The challenge is divided into two parts: $x = (x_L, x_R)$. The jth byte of x_R, denoted by $x_{R,[8j..8j+7]}$, addresses a bit of x_L, denoted by $x_{L, x_{R,[8j..8j+7]}}$, which is considered to be the jth bit of the output vector. The following parameters are assumed: $n = 384(= 256 + 128)$, $|x_L| = 256$, $|x_R| = 128$ bits, and $m = 16$. Under these conditions, the probability that an attacker will successfully impersonate a tag using a random response is bounded by $2^{-m} = 2^{-16}$. This value is unacceptable for standard cryptographic applications but it may suffice for some RFID applications.

Protocol 3: Squaring: A protocol based on the squaring of a $2n$ bit portion of the key shared between the tag and reader was proposed:

$$
\begin{aligned}
R \to T : &\qquad\qquad\qquad\qquad x \\
T \to R : &\quad k_L \oplus ((k_R + x)^2 \bmod 2^n)
\end{aligned}
\tag{6.8}
$$

where
k_L and k_R are two halves of a $2n$ bit secret key $k = (k_L, k_R)$
"+" symbol represents integer addition

Protocol 4: Knapsack:

$$
\begin{aligned}
R \to T : &\quad d \oplus k, \kappa(x, d) \\
T \to R : &\qquad\quad x \oplus k'
\end{aligned}
\tag{6.9}
$$

where
k is an m-bit secret key
k' is an n-bit secret key
x is an n-bit challenge
d is an m-bit trapdoor
κ is a punctured multiplicative knapsack; in other words, a public set of n s-bit prime numbers, stored both by the reader and the tag

A brief description of the protocol can be the following. R selects randomly $n/2$ elements from this set (knapsack) and multiplies together the selected primes. The n-bit challenge x contains 1 at those bit positions which correspond to the primes selected (an order is assumed among the primes) and 0 at the remaining positions. R chooses t integers randomly from the range of 1, 2, ..., $s \cdot n/2$, and marks bits of binary representation of the product at bit positions corresponding to the selected integers. The marked bits are deleted and the binary string is shrunk in size accordingly. The resultant punctured string is the output of mapping κ. Trapdoor d consists of the integers used in puncturing, by appending these integers in order. It follows that the output of κ has length $s \cdot n/2 - t$ bits; furthermore the trapdoor is $m = t \cdot \log(s \cdot n/2)$ bits long. When the tag receives the reader's message, it knows the punctured position according to the trapdoor, and the punctured bits will be found by exhaustive search.

All of these protocols assume that each tag shares a secret with the reader, so it is quite difficult for the reader to find which secret corresponds to which tag. To limit the impact of a potential discovery by a cryptographic adversary, they should be changed regularly and kept secure during distribution and in service. The process of selecting, distributing, and storing keys is known as key management; this is difficult to achieve reliably and securely.

The aim of the above schemes is tag authentication, but the security level is very low and can be breached by a powerful adversary. Additionally, they do not solve important problems such as reader-to-tag authentication or tracking, to name just a few issues.

6.3.2 LIST OF IDENTIFIERS

Juels [75] proposed a solution based on the use of pseudonyms, without using hash functions at all. The RFID tag stores a short list of random identifiers or pseudonyms (α_1, α_2, α_3, ..., α_k). When the tag is queried, it emits the next pseudonym in the list. An adversary can, however, gather all the names on the list by querying a tag multiple times. Then the fraudulent tag could impersonate a honest tag. This is the sort of cloning attack to which standard RFID tags with static identifiers are vulnerable (i.e., EPC-C1G2 specification).

To prevent such an attack, some solutions are proposed: tags could release their name only at a certain prescribed rate, or pseudonyms could be refreshed only by authorized readers. If the second solution is employed, a mutual authentication between the reader and the tag is required. Juels proposed a lightweight mutual authentication protocol (MAP) based on the release of keys shared between both parties. The verifier authenticates to a tag by releasing a key β_i, which is unique to a pseudonym α_i. Once the verifier has authenticated to the tag, the tag authenticates itself to the reader by releasing an authentication key γ_i. Like β_i, this authentication key γ_i is unique to a pseudonym α_i. After mutual authentication, key (β_i, γ_i) and pseudonym (α_i) updating is accomplished. The reader transmits onetime padding data that the tag uses in the updating stage. Although encryption is not explicitly involved by means of onetime pads, it is equivalent to encryption. Pads can be considered keys used to "encrypt" and thereby update the α_i, β_i, and γ_i values. Indeed, each tag stores a series of pads. The stored pads are updated with new material on each authentication. This new pad material is sent in clear on the channel, but the updating procedure ensures that it will be used only after a certain number (m) of updates. This number should be chosen such that an adversary cannot observe m consecutive authentications.

As it has been shown, Juels's protocol does not require the use of any cryptographic primitive. However, it involves the exchange of four messages and needs key updating, which may be costly and difficult to perform securely. Moreover, the assumption that an attacker cannot observe m consecutive authentications does not hold for many real scenarios.

6.3.3 Ultralightweight Authentication Protocol

In 2006, Peris et al. proposed some lightweight protocols, arousing interest in lightweight cryptography for low-cost RFID tags. Chronologically, M^2AP [76] was the first proposal, followed by EMAP [77] and LMAP [78]. In spite of the fact that some authors have pointed out certain weaknesses in the aforementioned protocols [79–81], they have their own merits due to its lightweight nature. LMAP might be considered an early version of ULAP [82], the protocol presented in this section.

6.3.3.1 Suppositions of the Model

The proposed protocol is based on the usage of pseudonyms, specifically *index-pseudonyms (IDSs)*, and the related *session index-pseudonyms (SessionIDSs)*. A (L-bit length) *IDS* is a unique index of a table row where all the information on a tag is stored. The *IDS* is never sent through the channel—instead an L-bit *SessionIDS* is transmitted. Each tag has an associated key, divided in four L bits components ($K = K1 \parallel K2 \parallel K3 \parallel K4$).

All costly operations such as random-number generation are performed by the reader. On the other hand, as tags are very computationally limited devices, only the simplest operations are available: bitwise xor ($\oplus$), bitwise or ($\vee$), and sum mod 2^m ($+$). The authors claim that although it could seem that this seriously limits the strength of the resulting protocol, there are other proposals in cryptography that give adequate security levels by composing very simple and efficient operations, such as Salsa20 [83] and TEA family [27–29].

As most low-cost tags are passive, the communication must be initiated by the reader. The authors also suppose that both the backward and forward channel can be passively eavesdropped by an attacker. Note that this assumption does not imply that the air channel can be actively manipulated by an adversary, ruling out, therefore, active attacks. Finally, they also assume that the communication channel between the reader and the database is secure.

6.3.3.2 Proposed Protocol

The protocol can be split into four main stages: tag identification, mutual authentication, index-pseudonym updating, and key updating.

1. **Tag identification**:

$$Reader \rightarrow Tag : hello||challenge \qquad (6.10)$$

$$Tag \rightarrow Reader : SessionIDS \qquad (6.11)$$

The reader sends a *hello* message concatenated with a random number (challenge) to the tag, and the tag answer consists of a *SessionIDS*. The *SessionIDS* will be computed in the following way (for more details see the original paper):

```
SessionIDS = IDS
for(i=0;i<32;i++){
SessionIDS = (SessionIDS>>1) + SessionIDS + SessionIDS
             + challenge;}
```

2. **Mutual authentication**:

$$Reader \rightarrow Tag : A||B||C \qquad (6.12)$$

$$Tag \rightarrow Reader : D \qquad (6.13)$$

$$A = SessionIDS_{tag(i)}^{(n)} \oplus K1_{tag(i)}^{(n)} \oplus n1 \qquad (6.14)$$

$$B = (SessionIDS_{tag(i)}^{(n)} \vee K2_{tag(i)}^{(n)}) + n1 \qquad (6.15)$$

$$C = SessionIDS_{tag(i)}^{(n)} + K3_{tag(i)}^{(n)} + n2 \qquad (6.16)$$

$$D = (SessionIDS_{tag(i)}^{(n)} + ID_{tag(i)}) \oplus n1 \oplus n2 \qquad (6.17)$$

  a. **Reader authentication**: The reader generates two nonces $n1$ and $n2$. With $n1$ and subkeys $K1$ and $K2$, the reader generates submessages A and B. With $n2$ and $K3$, it generates submessage C.
  b. **Tag authentication**: With submessages A and B, the tag authenticates the reader and obtains $n1$. From submessage C, the tag obtains the random number $n2$. Nonces $n1$ and $n2$ are used in the *IDS* and key ($K = K1 \parallel K2 \parallel K3 \parallel K4$) updating. Once these verifications are performed, the tag generates the answer message D to authenticate itself and send its static identifier securely.
3. **Index-pseudonym updating and key updating**.[c.] After the mutual authentication phase, the protocol prescribes an index-pseudonym and key updating stage. The equations for the index-pseudonym and key update phase are as follows:

$$IDS_{tag(i)}^{(n+1)} = \left(IDS_{tag(i)}^{(n)} + \left(n2 \oplus K4_{tag(i)}^{(n)}\right)\right) \oplus ID_{tag(i)} \qquad (6.18)$$

$$K1_{tag(i)}^{(n+1)} = K1_{tag(i)}^{(n)} \oplus n2 \oplus \left(K3_{tag(i)}^{(n)} + ID_{tag(i)}\right) \qquad (6.19)$$

$$K2_{tag(i)}^{(n+1)} = K2_{tag(i)}^{(n)} \oplus n2 \oplus ID_{tag(i)} \qquad (6.20)$$

$$K3_{tag(i)}^{(n+1)} = \left(K3_{tag(i)}^{(n)} \oplus n1\right) + \left(K1_{tag(i)}^{(n)} \oplus ID_{tag(i)}\right) \qquad (6.21)$$

$$K4_{tag(i)}^{(n+1)} = \left(K4_{tag(i)}^{(n)} \oplus n1\right) + ID_{tag(i)} \qquad (6.22)$$

The authors performed a statistical analysis of the exchanged messages. Additionally, they analyzed the security of the protocol against the most relevant attacks. Finally, the performance of the scheme was assessed to show that it can be implemented even in very low-cost tags. From the above analysis, it resulted that around 300 equivalent logic gates (8-bit architecture) and $7L$ bits of memory are required to implement the protocol. L may be set to 96 bits, which is a compatible length with all the encoding schemes (GTIN, SSCC, GLN, GRAI, GIAI, and GID) defined by the EPC [84].

6.3.4 HUMAN PROTOCOLS

In Ref. [85], Weis introduced the concept of human computer authentication protocol due to Hopper and Blum, adapted to low-cost RFIDs (HB protocol). The security of protocols based on this concept is based on the *Learning Parity with Noise Problem* (LPN), whose hardness over random instances still remains an open question.

Figure 6.6a illustrates a single round of the HB authentication protocol [86]. Suppose that the reader and the tag share a k-bit secret x, and the tag would like to authenticate itself to the reader. The

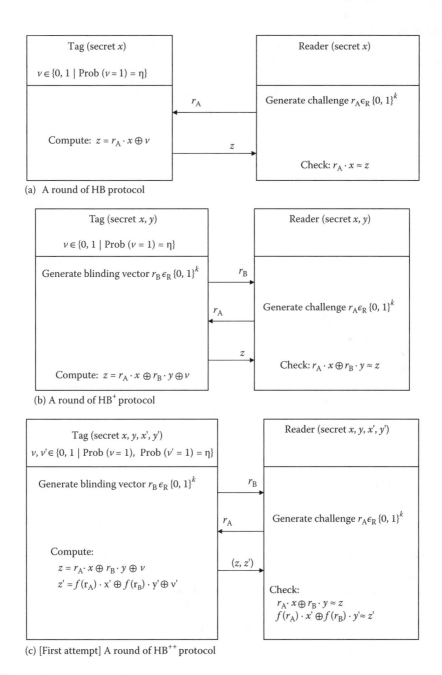

(a) A round of HB protocol

(b) A round of HB$^+$ protocol

(c) [First attempt] A round of HB^{++} protocol

FIGURE 6.6 Human protocols (I).

reader selects a random challenge $a \in (0, 1)^k$ and sends it to the tag. The tag responds to the reader challenge by computing the binary inner-product $a \cdot x$ and injecting noise into the result. The tag intentionally sends the wrong response with probably $\eta \in (0, \frac{1}{2})$. This interaction must be repeated q rounds and the reader will authenticate the tag's identity if fewer than $q\eta$ of his responses are incorrect.

The above protocol is resistant to passive attacks, but not to active attacks. Weis et al. proposed a new version of its protocol (HB$^+$) to offer protection against active attacks [86]. The main differences with respect to the HB protocol are the following. They introduce another k-bit secret key (y) shared between the reader and the tag. The tag and not the reader initiates the protocol, transmitting a k-bit blinding vector. Finally, z is computed as the scalar product of the newly introduced secret key (y) and the blinding vector transmitted by the tag, xored with the z in HB. A round of HB$^+$ is shown in Figure 6.6b. Although Juels et. al claimed that HB$^+$ is resistant to active attacks, Gilbert et al. showed how a man-in-the-middle attack can be accomplished [87].

To avoid Gilbert et al.'s attack on HB$^+$, Bringer et al. [88] proposed two protocols (HB^{++}[first attempt] and HB^{++} (Figures 6.6c and 6.7a), attempting to secure against such man-in-the-middle

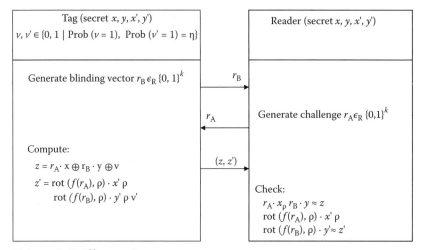

(a) A round of HB^{++} protocol

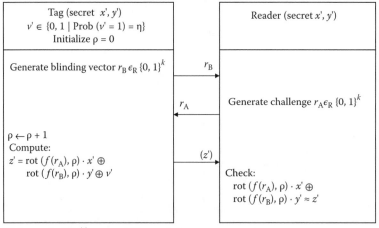

(b) A round of HB^{++} protocol (Piramuthu)

FIGURE 6.7 Human protocols (II).

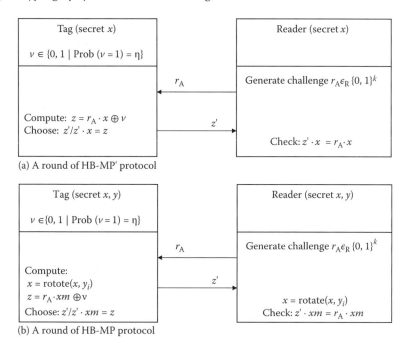

(a) A round of HB-MP' protocol

(b) A round of HB-MP protocol

FIGURE 6.8   Human protocols (III).

attacks. However, these protocols are vulnerable to attacks from an adversary that pretends to be a genuine reader [89]. Piramuthu proposed a new protocol (see Figure 6.7b) inspired by the HB^{++} protocol. The main changes introduced are as follows:

1. To thwart the attacker when an adversary pretends to be a valid reader, z, and the related vectors (x, y) and v were omitted. Additionally, the protocol is kept more lightweight.
2. To prevent the use of the same ρ until protocol completion, updating of ρ is accomplished every time z is computed.

Recently, Munilla and Peinado proposed another protocol [90]. First a protocol inspired by HB and named HB-MP' was proposed (see Figure 6.8a). The authors acknowledge that the above protocol resulted vulnerable to a simple man-in-the-middle attack, just like the HB^+ protocol. To avoid this weakness, a new protocol named HB-MP was worked out (see Figure 6.8b). We briefly describe a round of the HB-MP' protocol. Suppose that the reader and the tag share a k-bit secret x, and the tag would like to authenticate itself to the reader. The reader selects a random k-bit binary vector a and sends it to the tag. The tag computes the binary inner-product $a \cdot x$ and injects noise into this result. Then, the tag looks for a k-bit binary vector b such that $b \cdot x = z$. The tag sends back b to the reader. The reader checks the equality of $b \cdot x$ and $a \cdot x$. If it is correct, the tag is authenticated. This protocol differs from the protocols based on the LPN problem. However, the authors maintain that the problem of finding x, knowing the vectors a and b, is at least at difficult as solving the LPN problem.

6.3.5   ENHANCING SECURITY OF EPCGLOBAL GEN-2

The EPC Class-1 Generation-2 specification (named in the future as EPC-C1G2) [60] can be considered the "universal standard" for low-cost RFID tags. Tags are passive and operate on the UHF frequency band (860–960 MHz). Their very constrained resources and storage capabilities dictate that EPC-C1G2 tags cannot afford traditional cryptographic primitives. Tags include a 16-bit PRNG and 16-bit CRC checksum. Additionally, simple operations such as bitwise operations are supported. Tags have two 32-bit PINS: kill PIN and access PIN. After a tag receives the kill password, it renders it silent thereafter. The access password is employed to access data store in the reserved memory.

The two main operations for managing tag populations are inventory and access. A detailed description of the inventory operation is as follows (note that RN16 represents a 16-bit random or pseudorandom number):

1. Reader sends a request message (*query*) to a tag, initiating an inventory round.
2. Each tag which receives the query picks a 16-bit random value using the PRNG, and loads this value into a slot counter. When the slot counter becomes zero, the tag backscatters the random value (*RN*16) to the reader.
3. Reader acknowledges the tag with an *ACK* containing this same RN16.
4. Tag compares the random number in the *ACK* with the *RN*16 it sends. If it is right, the tag backscatters its *PC*, Electronic Product Code (*EPC*), and *CRC*-16.

After acknowledging a tag, a reader may choose to access it. A reader will access a tag as follows:

1. Reader issues a *ReqRN*, containing the previous RN16, to the acknowledged tag.
2. Tag compares the random number in the *ReqRN* with the *RN*16 in the tag. If the values coincide, the tag generates and stores a new *RN*16 (denoted handle), and backscatters the handle.
3. Write, Kill, and Access commands send 16-bit words (either data or half-passwords) from reader to tag. These commands use one-time-pad-based link cover-coding to obscure the word being transmitted, as follows:
   a. Reader issues a *ReqRN*, to which the tag responds by backscattering a new *RN*16.
   b. Reader then generates a 16-bit ciphertext string comprising a bitwise XOR of the 16-bit word to be transmitted with this new RN16.
   c. Tag decrypts the received ciphertext string by performing a bitwise XOR of the received 16-bit ciphertext string with the original RN16.

As many authors have mentioned [6,67,91,92], EPC-G1C2 specification poses important security flaws. Possible threats include the following: in the inventory command, the EPC is transmitted as plaintext, something that generates privacy and spoofing problems. Tracking could be done in a very straightforward way, because the EPC is fixed. Moreover, the security of the access command is weak, so performing a passive attack is very simple. An attacker eavesdropping on the backward and forward channel can pick up the random number sent by the tag. Next, the attacker can decrypt the ciphertext sent by the reader by performing an XOR with the previous collected random number. So the plaintext or PIN can be disclosed by this quite simple mechanism, which constitutes an important security pitfall.

6.3.5.1 Mutual Authentication Scheme Utilizing Tag's Access Password

In 2007, Konidola and Kim proposed an interesting paper, attempting to correct some of the security shortcomings of the EPC-C1G2 specification [93]. The authors hold that the proposed scheme protects the tag access password from the XOR operation. A brief description of the mutual authentication protocol is as follows:

1. First, the tag is singulated and backscatters its *EPC* number. Then the reader sends a *ReqRN* to the tag. The tag backscatters two generated random numbers: RN_1^{Tag} and RN_2^{tag}.

2. Reader also generates two 16-bit random numbers: RN_1^{Reader} and RN_2^{Reader}. These four random numbers and the access password are used to construct R_1 and R_2 responses.

$$R_1 = PWD_M \oplus PadGen(RN_1^{Tag}, RN_1^{Reader}) \tag{6.23}$$

$$R_2 = PWD_L \oplus PadGen(RN_2^{Tag}, RN_2^{Reader}) \tag{6.24}$$

where PWD_M and PWD_L are the 16 most significant and 16 least significant bits of the access password, respectively.

Next, two 16-bit random numbers (RN_3^{Reader}, RN_4^{Reader}), which will be used in tag authentication, are generated and transmitted to the tag.

3. Tag checks $R1$ and $R2$. If these are correct, the process continues but aborts if they fail. The tag also generates two new random numbers (RN_3^{Tag}, RN_4^{Tag}), and builds $R3$ and $R4$ responses.

$$R_3 = PWD_M \oplus PadGen(RN_3^{Tag}, RN_3^{Reader}) \tag{6.25}$$

$$R_4 = PWD_L \oplus PadGen(RN_4^{Tag}, RN_4^{Reader}) \tag{6.26}$$

These new random numbers and responses are sent to the reader.

4. Reader verifies $R3$ and $R4$. If these are correct, the tag is authenticated. Otherwise an alarm is raised.

The *PadGen* is a pad generation function that produces a 16-bit output from its inputs and access password. The inputs values are used as location numbers to retrieve the individual access passwords bits stored in those locations. For example $PadGen(7E2B_h, 2B5F_h)$ with $PWD_M = 1110\ 0101\ 0100\ 1000_2$ and $PWD_L = 1110\ 1000\ 1100\ 1010_2$.

- $7E2B_h$ = 7th 14th 2th 11th location of $PWD_M = 1010_2$
- $7E2B_h$ = 7th 14th 2th 11th location of $PWD_L = 0110_2$
- $2B5F_h$ = 2th 11th 5th 15th location of $PWD_M = 1010_2$
- $2B5F_h$ = 2th 11th 5th 15th location of $PWD_L = 1000_2$
- Combining the above four results we have a 16-bit pad value $PadGen(7E2B_h, 2B5F_h) = 1010\ 0110\ 1010\ 1000 = A6A8_h$

In Ref. [94], some weaknesses of the Konidola et al. scheme are shown. Specifically, Lim and Li show how a passive attacker can recover the password of the tag by eavesdropping a single run of the protocol and performing some correlation analysis on the captured information.

6.3.6 ABSTRACTIONS OF INTEGER ARITHMETICS

Lemieux and Tang proposed a mutual authentication scheme based on infinite, nonassociative, and usually nonabelian structures that authors have termed Abstractions of Integer Arithmetic (AIA) and Partial Abstractions of Integer Arithmetic (PAIA) [95]. To give an easily understandable explanation of the proposed protocol, the scheme is introduced to the reader using integer arithmetic. However, infinite nonabelian groupoids or quasigroups should be used to achieve a secure solution (see the original paper for more details).

The proposed authentication protocol is based on the multiplication of two integers, as displayed in Figure 6.9. We assume that Alice and Bob share a secret number $K = k_n \cdots k_2 k_1$ and a common secret digit d. In each round, a randomly digit is concatenated on the left hand side of the number M. M is initialized with the digit $m1 = d$, which is only shared between Alice and Bob. In each round:

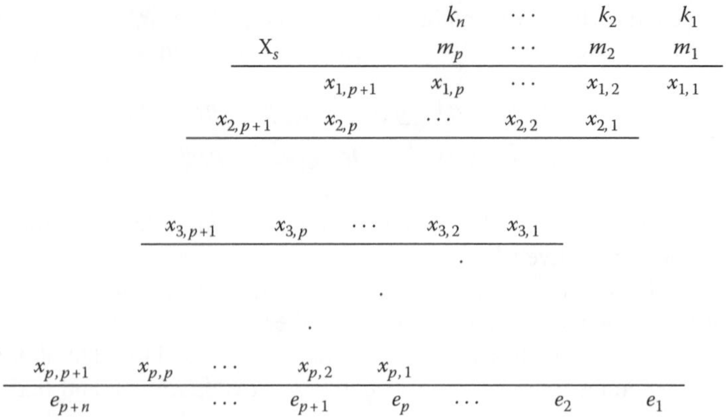

			k_n	$\cdots$		k_2	k_1
X_s			m_p	$\cdots$		m_2	m_1
		$x_{1,p+1}$	$x_{1,p}$	$\cdots$		$x_{1,2}$	$x_{1,1}$
	$x_{2,p+1}$	$x_{2,p}$	$\cdots$			$x_{2,2}$	$x_{2,1}$
$x_{3,p+1}$		$x_{3,p}$	$\cdots$		$x_{3,2}$	$x_{3,1}$	
$x_{p,p+1}$	$x_{p,p}$	$\cdots$		$x_{p,2}$	$x_{p,1}$		
e_{p+n}		$\cdots$		e_{p+1}	e_p	$\cdots$	e_2	e_1

FIGURE 6.9 Multiplication of two integers. (From Lemieux, S. and Tang, A., Clone resistant mutual authentication for low-cost RFID technology. In *Cryptology ePrint Archive*, Report 2007/170. IACR, 2007. With permission.)

1. Alice randomly generates a new digit m_i, concatenates this value to $M = m_i m_{i-1} \cdots m_1$ and computes $E = K \times M$. Then the pair (e_i, m_i) is transmitted to Bob, where e_i is the ith digit of E.
2. Bob computes the same product and checks the received e_i. If it is correct, he randomly generates m_{i+1} and repeats the process.

After r rounds of consecutive success, both parties are convinced that they share the same secret number K and secret digit d. Encryption and decryption of a message can be computed in a similar manner as authentication. In this case, m_i represents the next digit in the message that Alice wishes to send Bob. She encrypts it and the ciphertext e_i is sent to Bob. Bob decrypts e_i to recover m_i and the process continues with the next digit m_{i+1}.

The authentication and encryption/decryption schemes, based on integer arithmetic, are completely nonsecure. Two main assumptions are necessary to guarantee an appropriate security level. First, given numbers E and M, one cannot divide E and M even if it is known that M is a factor of E. Second, multiplication is not commutative. Therefore, AIAs or PAIAs are required to implement the protocol.

6.4 CONCLUSIONS

Whenever a new technology appears (i.e., bluetooth and wireless technology), the main concerns are price and operativity, the security aspect aside. We believe that past errors should be avoided and the use of secure solutions should be generalized. Since 2003, there has been a great number of research publications centered on security. However, the majority of proposals to secure RFID tags make the same two errors:

- Class of tag for which the proposed protocol is intended is not specified in the proposals. This is a very important point, as the number of available resources (memory, circuitry, and power) will depend on this decision. So, not all tags will support the same kind of

operations. For example, public cryptography is applicable for high-cost RFID tags, but it exceeds the capabilities of low-cost RFID tags. Additionally, each RFID class should have a different security level. It is not sensible for a low-cost RFID tag (i.e., a tagged biscuit packet) to have the same security level as that of a high-cost RFID tag (i.e., an e-passport).
• Proposed protocols are not realistic as regards tag resources. Although it is quite clear that there has been a great advance in the design of lightweight stream/block ciphers, the design of lightweight hash functions and PRNGs remains as a pending task. As we have mentioned, many proposals are based on hash-functions or PRNGs. In spite of this, many authors claim that its schemes are appropriate for low-cost RFID tags. However, a maximum of 4K gates can be devoted to security functions in this class of tag. As seen in Section 6.2.3, great resources are needed to implement traditional cryptographic primitives. On other hand, lightweight cryptographic primitives are not proposed.

To summarize, lightweight cryptography is imperative. In this chapter, the main proposals in this research area were presented. First, lightweight cryptographic primitives such as Present (block) or Grain (stream) ciphers and LAMED PRNG were examined. Then protocols based on noncryptographic primitives were analyzed. Finally, an extensive bibliography has been included, which the reader can consult should he wish to research specific aspects further.

APPENDIX A: TAV-128

```
/*******************************************************************************/
Process the input a1 modifying the accumulated hash a0 and the state
/*******************************************************************************/
void tav(unsigned long *state, unsigned long *a0, unsigned long*a1)
{ unsigned long h0,h1; int i,j,r1,r2,nstate; /* Initialization */
 r1=32; r2=8; nstate=4;
 h0=*a0; h1=*a0;
/* A - Function */ for(i=0;i<r1;i++){h0=(h0<<1)+((h0+(*a1))>>1);} /*
B - Function */ for(i=0;i<r1;i++){h1=(h1>>1)+(h1<<1)+h1+(*a1);}
 for(j=0;j<nstate;j++) {
   for(i=0;i<r2;i++)
      {
      /* C - Function */ h0^=(h1+h0)>>3; h0=((((h0>>2)+h0)>>2)+(h0<<3) +(h0<<1))^0x736B83DC;
      /* D - Function */ h1^=(h1^h0)>>1; h1=(h1>>4)+(h1>>3)+(h1<<3)+h1;
      } // round-r2
      state[j]+=h0; state[j]^=h1;
 } // state
/* a0 updating */
 *a0=h1+h0;
}


/*******************************************************************************/
Initialization of the state and a0 with random values obtained from
www.random.org
/*******************************************************************************/
{void init_state(unsigned long *state, unsigned long *a0)}
 {
     state[0]=0xa92be51d; state[1]=0xba9b1ef0;
     state[2]=0xc234d75a; state[3]=0x845c2e03;
       a0[0]=0x768c7e74;
}
```

APPENDIX B: LAMED

```
---------------------------------------------------------------
                        #1 If n is odd
                        #2     a0=a1+iv
                        #3     a1=out^s
                        #4 If n is even
                        #5     a0=a1^iv
                        #6     a1=out+s
---------------------------------------------------------------
#1  aux1 = a0 + a1;              #12 aux3 = aux3 ^ aux1;
#2  aux2 = a0 ^ a1;              #13 aux3 = vrotdk(aux3,3);
#3  aux3 = vrotdk(aux1,5);       #14 aux3 = aux3 + a1;
#4  aux3 = aux3 + aux2;          #15 aux3 = vrotdk(aux3,2);
#5  aux3 = vrotdk(aux3,3);       #16 aux3 = aux3 + aux1;
#6  aux3 = aux3 ^ aux1;          #17 aux3 = vrotdk(aux3,4);
#7  aux3 = vrotdk(aux3,4);       #18 aux3 = aux3 ^ a1;
#8  aux3 = a1 + aux3;            #19 aux3 = vrotd(aux3);
#9  aux3 = vrotdk(aux3,2);       #20 aux3 = aux3 + aux2;
#10 aux3 = aux3 + aux1;          #21 aux3 = vrotdk(aux3,2);
#11 aux3 = vrotdk(aux3,2);       #22  out = aux1 ^ aux3;
---------------------------------------------------------------
```

REFERENCES

[1] S. Weis. Security and privacy in radio-frequency identification devices. In Master's Thesis, MIT, Cambridge, MA, 2003.

[2] M. Lehtonen, T. Staake, F. Michahelles, and E. Fleisch. From identification to authentication—A review of RFID product authentication techniques. In *Networked RFID Systems and Lightweight Cryptography*, pp. 169–187. Springer, 2007, Chapter 9.

[3] Y. Cui, K. Kobara, K. Matsuura, and H. Imai. Lightweight asymmetric privacy-preserving authentication protocols secure against active attack. In *Proc. PerSec'07*. IEEE Computer Society, New York, 2007.

[4] A. Juels, D. Molnar, and D. Wagner. Security and privacy issues in e-passports. In *Proc. of SecureComm'05*. IEEE Computer Society, Athens, Greece, 2005.

[5] S. Karthikeyan and M. Nesterenko. RFID security without extensive cryptography. In *Proc. SASN'05*, Alexandria, VA, 2005.

[6] D. Nguyen Duc, J. Park, H. Lee, and K. Kwangjo. Enhancing security of EPCGlobal Gen-2 RFID tag against traceability and cloning. In *Proc. Symposium on Cryptography and Information Security*, 2006.

[7] H.-Y. Chien. SASI: A new ultra-lightweight RFID authentication protocol providing strong authentication an strong integrity. *IEEE Transactions on Dependable and Secure Computing*, 4(4), 337–340, 2007.

[8] S. Bono, M. Greem, A. Stubblefield, A. Juels, A. Rubin, and M. Syzdlo. Security analysis of a cryptographically-enabled device. In *Proc. SSYM'05*. Usenix Association, Berkeley, CA, 2005.

[9] N. Karten and H. Plotz. Mifare little security, despite obscurity. http://events.ccc.de/congress/2007/Fahrplan/events/2378.en.html, Berlin, Germany, 2007.

[10] S. Weis, S. Sarma, R. Rivest, and D. Engels. Security and privacy aspects of low-cost radio frequency identification systems. In *Proc. SPC'03. Lecture Notes in Computer Science*, 2802, pp. 454–469. Springer-Verlag, 2003.

[11] A. Juels. RFID security and privacy: A research survey. Manuscript, 2005.

[12] P. Peris-Lopez, J.C. Hernandez-Castro, J.M. Estevez-Tapiador, and A. Ribagorda. RFID systems: A survey on security threats and proposed solutions. In *Proc. PWC06. Lecture Notes in Computer Science*, 4217, pp. 159–170. Springer-Verlag, 2006.

[13] S. Piramuthu. Protocols for RFID tag/reader authentication. *Decision Support Systems*, doi:10.1016/j.dss.2007.01.003, 2007.

[14] M. Feldhofer, J. Wolkerstorfer, and V. Rijmen. AES implementation on a grain of sand. In *Proc. Information Security*, Vol. 152, pp. 13–20. IEEE Computer Society, 2005.

[15] A. Satoh, S. Morioka, K. Takano, and S. Munetoh. A compact Rijndael hardware architecture with S-Box optimization. In *Proc. ASIACRYPT'01. Lecture Notes in Computer Science*, 2248, pp. 239–254. Springer-Verlag, Gold Coast, Australia, 2001.

[16] N. Pramstaller, S. Mangard, S. Dominikus, and J. Wolkerstorfer. Efficient AES implementations on ASICs and FPGAs. In *Proc. Fourth Workshop on the Advanced Encryption Standard "AES—State of the Crypto Analysis". Lecture Notes in Computer Science*, 3373, pp. 98–112. Springer-Verlag, Bonn, Germany, 2004.

[17] A. Poschmann, G. Leander, K. Schramm, and C. Paar. New light-weight crypto algorithms for RFID. In *IEEE International Symposium on Circuits and Systems 2007, ISCAS 2007*, pp. 1843–1846, New Orleans, LA, 2007.

[18] S. Kumar, C. Paar, J. Pelzl, G. Pfeiffer, and M. Schimmler. Breaking ciphers with COPACOBANA–A cost-optimized parallel code breaker. In *Proc. of CHES'06. Lecture Notes in Computer Science*, 4249, pp. 101–118. Springer-Verlag, Yokohoma, Japan, 2006.

[19] G. Leander, C. Paar, A. Poschmann, and K. Schramm. New lightweight DES variants. In *Proc. FSE'07. Lecture Notes in Computer Science*, 4593, pp. 196–120. Springer-Verlag, Luxembourg City, Luxembourg, 2007.

[20] A. Bogdanov, L.R. Knudsen, G. Leander, C. Paar, A. Poschmann, M.J.B. Robshaw, Y. Seurin, and C. Vikkelsoe. PRESENT: An ultra-lightweight block cipher. In *Proc. CHES'07. Lecture Notes in Computer Science*, 4727, pp. 450–466. Springer-Verlag, Vienna, Austria, 2007.

[21] M. Wan. Differential cryptanalysis of PRESENT. In *Cryptology ePrint Archive*, Report 2007/408, 2007.

[22] D. Hong, J. Sung, S. Hong, J. Lim, S. Lee, B.-S. Koo, C. Lee, D. Chang, J. Lee, K. Jeong, H. Kim, J. Kim, and S. Chee. HIGHT: A new block cipher suitable for low-resource device. In *Proc. CHES'06. Lecture Notes in Computer Science*, 4249, pp. 46–59, Springer-Verlag, Yokohoma, Japan, 2006.

[23] T. Shirai, K. Shibutani, T. Akishita, S. Moriai, and T. Iwata. The 128-bit blockcipher CLEFIA (extended abstract). In *Proc. FSE'07. Lecture Notes in Computer Science*, 4593, pp. 181–195. Springer-Verlag, Luxembourgh City, Luxembourgh, 2007.

[24] F.-X. Standaert, G. Piret, N. Gershenfeld, and J.-J. Quisquater. SEA: A scalable encryption algorithm for small embedded applications. In *Proc. CARDIS'06*, pp. 222–236. Springer-Verlag, Tarragona, Spain, 2006.

[25] F. Mace, F.-X. Standaert, and J.-J. Quisquater. ASIC implementations of the block cipher SEA for constrained applications. In *Proc. of RFIDSec'07*, pp. 103–114, Malaga, Spain, 2007.

[26] F. Mace, F.-X. Standaert, and J.-J. Quisquater. FPGA implementation(s) of a scalable encryption algorithm. In *IEEE Transactions on Very Large Scale Integration (VLSI) Systems*, 16, pp. 212–216, 2008.

[27] D. Wheeler and R. Needham. TEA: A tiny encryption algorithm. In *Proc. of FSE'04*, pp. 363–366, New Delhi, India, 1994.

[28] R. Needham and D. Wheeler. TEA extensions. Technical report, Computer Laboratory, University of Cambridge, 1997.

[29] R. Needham and D. Wheeler. Correction to XTEA. Technical report, Computer Laboratory, University of Cambridge, 1998.

[30] M. Hell, T. Johansson, and W. Meier. Grain: A stream cipher for constrained environments. http://www.ecrypt.eu.org/stream/, 2005.

[31] M. Hell, T. Johansson, and W. Meier. A stream cipher proposal: Grain-128. http://www.ecrypt.eu.org/stream/, 2006.

[32] C. De Cannire and B. Preneel Trivium specification. http://www.ecrypt.eu.org/stream/, 2005.

[33] M. Rogawski. Hardware evaluation of eSTREAM candidates: Grain, Lex, Mickey128, Salsa20 and Trivium. http://www.ecrypt.eu.org/stream/, 2007.

[34] T. Good and M. Benaissa, Hardware results for selected stream cipher candidates. http://www.ecrypt.eu.org/stream/, 2007.

[35] P. Bulens, K. Kalach, F.-X. Standaert, and J.-J. Quisquater. FPGA implementations of eSTREAM phase-2 focus candidates with hardware profile. http://www.ecrypt.eu.org/stream/, 2007.

[36] S. Kumar and C. Paar. Are standards compliant elliptic curve cryptosystems feasible on RFID? In *Proc. RFIDSec'06*, Graz, Austria, 2006.

[37] L. Batina, J. Guajardo, T. Kerins, N. Mentens, P. Tuyls, and I. Verbauwhede. Public-key cryptography for RFID-tags. In *Proc. PerCom'7*, pp. 217–222, New York, 2007.

[38] L. Batina, N. Mentens, K. Sakiyama, B. Preneel, and I. Verbauwhede. Low-cost elliptic curve cryptography for wireless sensor networks. In *Proc. ESAS'06. Lecture Notes in Computer Science*, 4357, pp. 6–17. Springer-Verlag, Hamburg, Germany, 2006.

[39] G. Gaubatz, J.-P. Kaps, and B. Sunar. Public key cryptography in sensor networks—Revisited. In *ESAS'04*, Heidelberg, Germany, 2004.

[40] J. Wolkerstorfer. Scaling ECC hardware to a minimum. In *ECRYPT Workshop—Cryptographic Advances in Secure Hardware—CRASH'05*, Leuven, Belgium, 2005 (invited talk).

[41] E. Öztürk, B. Sunar, and E. Savascedil. Low-power elliptic curve cryptography using scaled modular arithmetic. In *Proc. of CHES'04. Lecture Notes in Computer Science*, 3156, pp. 92–106. Springer-Verlag, Cambridge, MA, 2004.

[42] S.E. Sarma, S.A. Weis, and D.W. Engels. RFID systems and security and privacy implications. In *Proc. of CHES'02. Lecture Notes in Computer Science*, 2523, pp. 454–470. Springer-Verlag, Redwood Shores, CA, 2002.

[43] E.Y. Choi, S.M. Lee, and D.H. Lee. Efficient RFID authentication protocol for ubiquitous computing environment. In *Proc. of SECUBIQ'05. Lecture Notes in Computer Science*, 3823, pp. 445–954. Springer-Verlag, Nagasaki, Japan, 2005.

[44] D. Henrici and P. Müller. Hash-based enhancement of location privacy for radio-frequency identification devices using varying identifiers. In *Proc. PerSec'04*, pp. 149–153. IEEE Computer Society, Orlando, FL, 2004.

[45] M. Ohkubo, K. Suzuki, and S. Kinoshita. Cryptographic approach to "privacy-friendly" tags. In *Proc. RFID Privacy Workshop*, MIT, Cambridge, MA, 2003.

[46] J. Yang, J. Park, H. Lee, K. Ren, and K. Kim. Mutual authentication protocol for low-cost RFID. In *Proc. of Workshop on RFID and Lightweight Cryptography*, Graz, Austria, 2005.

[47] D. Ranasinghe, D. Engels, and P. Cole. Low-cost RFID systems: Confronting security and privacy. In *Auto-ID Labs Research Workshop*, Zurich, Switzerland, 2004.

[48] M. Feldhofer and C. Rechberger. A case against currently used hash functions in RFID protocols. *Proc. of Workshop on RFID and Lightweight Cryptography*, Graz, Austria, 2006.

[49] K. Yksel, J.P. Kaps, and B. Sunar. Universal hash functions for emerging ultra-low-power networks. In *Proc. CNDS'04*, Athens, Greece, 2004.

[50] P. Peris-Lopez, J.C. Hernandez-Castro, J. Estevez-Tapiador, and A. Ribagorda. An efficient authentication protocol for RFID systems resistant to active attacks In *Proc. SecUbiq'07. Lecture Notes in Computer Science*, 4809, pp. 781–794. Springer-Verlag, Taipei, Taiwan, 2006.

[51] X. Wang, D. Feng, X. Lai, and H. Yu. Collisions for hash functions MD4, MD5, HAVAL-128 and RIPEMD. *Cryptology ePrint Archive*, Report 2004/199, 2004.

[52] X. Wang, Y. Lisa Yin, and H. Yu. Finding collisions in the full SHA-1. In *Proc. CRYPTO'05*, Santa Barbara, CA, pp. 17–36, 2005.

[53] S.A. Weis, S.E. Sarma, R.L. Rivest, and D.W. Engels. Security and privacy aspects of low-cost radio frequency identification systems. In *Proc. Security in Pervasive Comp. Lecture Notes in Computer Science*, 2802, pp. 201–212. Springer-Verlag, Boppard, Germany, 2004.

[54] D. Molnar and D. Wagner. Privacy and security in library RFID: Issues, practices, and architectures. In *Proc. ACM CCS'04*, pp. 210–219. ACM Press, Washington, DC, 2004.

[55] C. Chatmon, T. Van Le, and M. Burmester. Secure anonymous RFID authentication protocols. Technical report TR-060112, 2006.

[56] T. Dimitriou. A lightweight RFID protocol to protect against traceability and cloning attacks. In *Proc. of SecureComm'05*. IEEE Computer Society, Athens, Greece, 2005.

[57] S. Lee, T. Asano, and K. Kim. RFID mutual authentication scheme based on synchronized secret information. In *Symposium on Cryptography and Information Security*, Hiroshima, Japan, 2006.

[58] K. Rhee, J. Kwak, S. Kim, and D. Won. Challenge–response based RFID authentication protocol for distributed database environment. In *Proc. SPC'05. Lecture Notes in Computer Science*, 3450, pp. 70–84. Springer-Verlag, Boppard, Germany, 2005.

[59] G. Tsudik. YA-TRAP: Yet another trivial RFID authentication protocol. In *Proc. PERCOM'06*. IEEE Computer Society, Pisa, Italy, 2006.

[60] Class-1 Generation-2 UHF air interface protocol standard version 1.0.9: "Gen-2". http://www.epcglobalinc.org/standards/, 2005.

[61] ISO/IEC 18000-6:2004/Amd:2006. http://www.iso.org/, 2006.

[62] M. Roberti, K.A. Vice, Y. Marguire, M. Reynolds, and R. Dunn. EPC Generation-2: Everything you need to know. In *RFID Journal Webinars*. Technical report, 2005.

[63] Texas Instruments. TI-RFid. http://www.ti.com/rfid/shtml/rfid.shtml, 2006.

[64] Symbol. RFID: A revolution in asset management. http://www.symbol.com/products/rfid-readers, 2006.

[65] P. Peris-Lopez, J.C. Hernandez-Castro, J.M. Estevez-Tapiador, and A. Ribagorda. LAMED—A PRNG for EPC Class-1 Generation-2 RFID specification. *Journal of Computer Standards & Interfaces*, 31(1), 88–97, January 2009.

[66] R. Forré. The strict avalanche criterion: Spectral properties of boolean functions and an extended definition. In *Proc. CRYPTO'88, Lecture Notes in Computer Science*, 403, pp. 450–468. Springer-Verlag, Santa Barbara, CA, 1990.

[67] K. Hyun Kim, E. Young Choi, S. Mi Lee, and D. Hoon Lee. Secure EPCglobal Class-1 Gen-2 RFID system against security and privacy problems. In *Proc. OTM-IS'06. Lecture Notes in Computer Science*, 4277, pp. 362–371. Springer-Verlag, Montpellier, France, 2006.

[68] A. Biryukov and A. Shamir. Cryptanalytic time-memory-data tradeoffs for stream ciphers. In *Proc. Advances of Cryptology-ASIACRYPT*, Vol. 1976, pp. 1–13, Kyoto, Japan, 2000.

[69] J. Walker. *Randomness battery*. http://www.fourmilab.ch/random/, 1998.

[70] G. Marsaglia. The Marsaglia random number CDROM including the DIEHARD battery of tests of randomness. http://stat.fsu.edu/pub/diehard, 1996.

[71] G. Marsaglia and W.W. Tsang. Some difficult-to-pass tests of randomness. *Journal of Statistical Software*, 7(3), pp. 1–8, 2002.

[72] A. Rukhin, J. Soto, J. Nechvatal, M. Smid, E. Barker, S. Leigh, M. Levenson, M. Vangel, D. Banks, A. Heckert, J. Dray, and S. Vo. A statistical test suite for random and pseudorandom number generators for cryptographic applications. NIST special publication 800-22, http://csrc.nist.gov/rng/, 2001.

[73] David Sexton's battery. http://www.geocities.com/da5id65536, 2005.

[74] I. Vajda and L. Buttyán. Lightweight authentication protocols for low-cost RFID tags. In *Proc. UBICOMP'03*, Seattle, WA, 2003.

[75] A. Juels. Minimalist cryptography for low-cost RFID tags. In *Proc. SCN'04. Lecture Notes in Computer Science*, 3352, pp. 149–164. Springer-Verlag, Amalfi, Italy, 2004.

[76] P. Peris-Lopez, J.C. Hernandez-Castro, J. Estevez-Tapiador, and A. Ribagorda. M2AP: A minimalist mutual-authentication protocol for low-cost RFID tags. In *Proc. UIC'06. Lecture Notes in Computer Science*, 4159, pp. 912–923. Springer-Verlag, Wuhan and Three Gorges, China, 2006.

[77] P. Peris-Lopez, J.C. Hernandez-Castro, J. Estevez-Tapiador, and A. Ribagorda. LMAP: A real lightweight mutual authentication protocol for low-cost RFID tags. *Proc. of Workshop on RFID and Lightweight Cryptography*, Graz, Austria, 2006.

[78] P. Peris-Lopez, J.C. Hernandez-Castro, Juan M. Estevez-Tapiador, and A. Ribagorda. EMAP: An efficient mutual authentication protocol for low-cost RFID tags. In *Proc. IS'06. Lecture Notes in Computer Science*, 4277, pp. 352–361. Springer-Verlag, Montpellier, France, 2006.

[79] T. Li and R. Deng. Vulnerability analysis of EMAP—An efficient RFID mutual authentication protocol. In *Proc. ARES'07*. IEEE Computer Society, Vienna, Austria, 2007.

[80] M. Bàràsz, B. Boros, P. Ligeti, K. Lòja, and D.A. Nagy. Breaking EMAP. In *Proc. SecureComm'07*, Nice, France, 2007.

[81] M. Bàràsz, B. Boros, P. Ligeti, K. Lòja, and D.A. Nagy. Passive attack against the M2AP mutual authentication protocol for RFID Tags. In *Proc. of RFID'07—The First International EURASIP Workshop on RFID Technology*, Vienna, Austria, 2007.

[82] P. Peris-Lopez, J.C. Hernandez-Castro, J.M. Estevez-Tapiador, and A. Ribagorda. An ultra light authentication protocol suitable for resource-limited Gen-2 RFID tags. *Journal of Information Science and Engineering*, 25(1), January 2009.

[83] D.J. Bernstein. Salsa20 specifications. http://www.ecrypt.eu.org/stream/, 2005.

[84] EPCglobal. http://www.epcglobalinc.org/, 2007.

[85] S. Weis. Security parallels between people and pervasive devices. In *Proc. PerSec'05*, pp. 105–109. IEEE Computer Society Press, 2005.

[86] A. Juels and S. Weis. Authenticating pervasive devices with human protocols. In *Proc. CRYPTO'05, Lecture Notes in Computer Science*, 3126, pp. 293–308. Springer-Verlag, Santa Barbara, CA, 2005.

[87] H. Gilbert, M. Robshaw, and H. Sibert. An active attack against HB$^+$—A probably secure lightweight authentication protocol. Manuscript, 2005.

[88] J. Bringer, H. Chabanne, and E. Dottax. HB^{++}: A lightweight authentication protocol secure against some attacks. In *Proc. SecPerU'06*. IEEE Computer Society, Lyon, France, 2006.

[89] Selwyn Piramuthu. HB and related lightweight authentication protocols for secure RFID tag/reader authentication. In *Proc. CollECTeR'06*, Basel, Switzerland, 2006.

[90] J. Munilla and A. Peinado. HB-MP: A further step in the HB-family of lightweight authentication protocols. *Computer Networks*, 51(9), 2262–2267, 2007.

[91] D. Bailey and A. Juels. Shoehorning security into the EPC standard. *Proc. SCN'06. Lecture Notes in Computer Science*, 4116, pp. 303–320, Springer-Verlag, Maiori, Italy, 2006.

[92] A. Juels. Strengthening EPC tags against cloning. Manuscript, 2005.

[93] D.M. Konidala, Z. Kim, and K. Kim, A Simple and cost-effective RFID tag-reader mutual authentication scheme. In *Proc. of Workshop on RFID and Lightweight Cryptography*, pp. 141–152, Graz, Austria, 2007.

[94] T.L. Lim and T. Li. Addressing the weakness in a lightweight RFID tag-reader mutual authentication scheme. *Proc. Global Communications Conference (GLOBECOM), IEEE*, pp. 59–63, Washington, DC, 2007.

[95] S. Lemieux, and A. Tang Clone Resistant Mutual Authentication for Low-Cost RFID Technology. In *Cryptology ePrint Archive*, Report 2007/170. IACR, 2007.

7 Distance-Bounding Protocols for RFID

Jorge Munilla Fajardo and Alberto Peinado Domínguez

CONTENTS

Radio-frequency identification (RFID) technology is widely being deployed today in applications which require security such as payment and access-control applications. Although many solutions have been proposed to secure these RFID systems, most of them are still vulnerable to different attacks related to location: distance fraud attacks, relay attacks (also known as mafia fraud attacks), and terrorist attacks. All of these attacks share a wrong assumption about the distance between the verifier (reader) and the prover (transponder, tag, or card). In the distance fraud attack, a card operates from out of the range where it is supposed to be. Relay attack is a kind of man-in-the-middle attack, where a rogue card circumvents the security mechanisms by getting the right answers from the legitimate card via a rogue reader. Both genuine parties, reader and card, remain unaware. In the terrorist attack, a legitimate card colludes with the adversary, giving him the necessary information to access the system by impersonating it a limited number of times.

The described attacks require simpler technical resources than tampering or cryptanalysis, and they cannot be prevented by ordinary security protocols that operate in the high layers of the protocol stack. The main countermeasure against these attacks is the use of so-called distance-bounding protocols, which verify not only that the card knows a cryptographic secret but also that it is within a certain distance. To achieve this aim, distance-bounding protocols must be tightly integrated into the physical layer. This chapter describes the different proposed solutions to implement distance-bounding protocols in RFID. Benefits, problems, and open issues of such protocols are presented through the chapter.

7.1 INTRODUCTION

Radio-frequency identification (RFID) is nowadays more and more used in secured applications. These applications require implementation of security mechanisms to make them resistant to the different risks that they face, e.g., pseudonyms to avoid malicious traceability, cover-coding or encipherment to hide information from eavesdroppers, and passwords or symmetric keys to authenticate legitimate readers and cards. Section 7.2 is an overview of the change process from simple identification systems to secured systems. Section 7.3 explains, however, that most of the implemented secured solutions are still vulnerable to the different attacks related to location: distance fraud, relay, and terrorist attacks. The main solution against such attacks is the use of so-called distance-bounding protocols. Section 7.4 reviews the state of the art of the distance-bounding protocols for RFID, and Section 7.5 focuses on those which measure the round-trip delay of electromagnetical signal to bound the distance. Hancke and Kuhn's protocol is comprehensively described and compared to others. Finally, Section 7.6 concludes by summarizing the key points of the chapter and by pointing out some of the open research issues in this field.

7.2 FROM IDENTIFICATION TO AUTHENTICATION

RFID devices were originally thought as a means of identifying people or objects using a radio-frequency (RF) transmission [1]. Communication takes place between a reader (interrogator) and a transponder (silicon chip connected to an antenna), often called a tag. Tags send their identification numbers (IDs) when they receive the query from a reader. This communication process presents different security problems [2]—lack of privacy or confidentiality, malicious traceability, loss of data integrity, etc., which must be treated in different ways depending on the application [3]. Thus, there are applications where these security problems are not considered important, in other cases the main problem is the traceability, as occurs with medical products, but also more and more applications require tag authentication. Examples of this last case are the payment and access control applications, e.g., RFID devices used in the automotive industry to start the ignition system [4], credit cards [5], public transports [6], or the new digital passports [7]. This section reviews the security mechanisms generally implemented to authenticate the tags, and introduces the types of tags usually used for these applications.

7.2.1 AUTHENTICATION MECHANISMS

Authentication implies to corroborate the identity, it is not enough with the claimed or stated identity. This means that it needs the design of techniques that allow one party, called verifier (reader), to gain assurances that the identity of another, called prover or claimant (tag), is as declared, thereby preventing impersonation. Authentication can be based on

- Something known: a secret or private key whose knowledge must be demonstrated.
- Something possessed: the possession of typically a physical accessory.
- Something inherent: using biometric characteristics for instance.

The authentication method in RFID is usually a combination of the first and the second case; i.e., the tag is authenticated based on the knowledge of a private key, and people or objects are authenticated based on the tag they bear or is attached to them. On the other hand, we call the reader attention to the fact that when biometric systems are simultaneously employed, RFID devices sometimes contribute nothing to the security, and they provide only the identity of the person that presents himself; i.e., RFID devices are used for convenience as a simple identification mechanism to reduce the search time for the user in the database.

The use of time-invariant passwords can be viewed as the simplest scheme of authentication (weak authentication). This scheme is the same as used when we introduce the personal identification number (PIN) code in our mobile phone or the password in our personal computer. This scheme based on time-invariant password is however not suitable for RFID, as the messages are transmitted over the air channel and therefore they are very vulnerable to interception. An adversary could access the system with a simple replay attack. Time-variant password schemes are the next step from fixed password schemes. This way, time-variant IDs are used in RFID to avoid traceability. Nevertheless, they are still not very convenient to authenticate tags/readers because they remain vulnerable to an active adversary who get an unused password, for the purpose of subsequent impersonation; i.e., the adversary interrogate a victim tag in advance to get a new (unused) password, and then replays it to the reader at the control point. To prevent this attack, the tag must transmit the cryptographic evidences in a way that it does not provide an adversary with useful information for a subsequent authentication (useless except for the genuine reader), or reveal it only to a party which itself is known to be authentic. Challenge–response techniques with unilateral (only one party is authenticated) or mutual authentication (both parties are authenticated) address this problem. Unilateral authentication is more vulnerable to active attacks by using chosen challenges than mutual authentication.

The basic idea of cryptographic challenge–response protocols is to demonstrate the knowledge of a secret without revealing the secret itself to the verifier during the protocol. This is done by providing a response to a time-variant challenge, which depends on the secret and the challenge. Challenge–response protocols can be further based on symmetric or asymmetric techniques [8]. Up to recently, asymmetric techniques were considered too complex to be implemented in RFID devices. They have a much higher computational cost than symmetric techniques, which means higher power consumption and larger chip areas. These, as explained in the following sections, are two resources very restrained in RFID tags. So, although lately elliptic curves cryptography (ECC) and other public-key alternatives ([9] or [10]) are being pointed out as possible, symmetric techniques are definitely much more popular than asymmetric techniques to be implemented in RFID applications ([11], [12], or [13]). On the other hand, there has also been a common assumption in the literature that hash functions are implementable in a manner more suitable than block ciphers in RFID tags. However, some works have recently questioned this [14].

7.2.2 CONTACTLESS SMART CARDS

To conclude this section, the main characteristics of the tags usually used in applications that require authentication (payment and access control) are presented here.

The idea of using RF to identify things/people is not new; it was already used by the Royal Air Force in the Second World War to distinguish allies from enemy aircraft, and radar systems or even mobile phones can be used for it. It is the ability to develop very small and cheap transponders what has actually caused the recent interest for RFID technology. Thus, although there exist transponders with battery, most of RFID devices do not have their own battery, and they take the energy that they need from the reader's electromagnetic field. It means extraordinary lifetimes but shorter ranges. Power available to carry out any sort of calculation decreases with increasing the distance between the tag and the reader. It implies that the operation ranges for security applications are relatively short; i.e., a few meters in the best case and few centimeters in the normal case (depending on the computational complexity, size of the antennas, used frequency, reader's power, etc.). So, it can be said that for security applications, the power consumption along with the chip area is the most restrictive resource.

Maybe the best example of secured RFID devices are the contactless smart cards [15]. Smart cards like mobile phone or bank cards are electronic data-carrying devices with additional computing capacity. Contactless smart cards substitute communication and power supply via galvanic contact, used in the traditional smart cards, for wireless communication and power supply by using instead electromagnetic fields [16]. This avoids their tendency to malfunction due to the vulnerability of the contacts to wear, corrosion, and dirt, and speed up the processes because it removes the time-consuming insertion.

The frequency of 13.56 MHz (HF) is preferred for most contactless smart cards, as the frequency range of 125–135 kHz (LF) only allows low bit rates, and the frequency ranges of 860–960 MHz and 2.45 GHz (UHF) do not usually transfer enough power for cryptographic security and privacy techniques (however, Ref. [17] described a secured RFID system in UHF). On the other hand, the capability of HF to penetrate substances such as metals or liquids, or the possibility of longer operation ranges in UHF sometimes represent, talking about security, disadvantages more than advantages. Contactless smart cards operating at 13.56 MHz (and LF) use inductive coupling between two coils, reader antenna and card antenna, to supply energy to the card from the reader and send information (whereas backscattering is used in UHF). Contactless smart cards are generally activated automatically when they are close to a reader. From the card to reader, the procedure for sending data is the use of load modulation (ohmic or capacitive). In the load modulation, the reader's field is influenced by the card by varying the parameters of the resonant circuit in time with the data stream.

The HF standards ISO/IEC 15693 [18], for "vicinity cards" with operating ranges of up to 1 m, and especially ISO/IEC 14443 [19], for "proximity cards" with operating ranges of up to 10 cm, are the most widely used standards for secured applications with contactless smart cards [20]; or used as starting points to add proprietary features. ISO 14443 specifies A and B operation modes that use slightly different communication procedures. ISO 14443A uses 100 percent amplitude shift keying (ASK) modulation with modified Miller coding (106 kbps, 9.4 µs per bit) in reader-to-card communication. The length of the blanking intervals of the Miller coding is just 2–3 µs to guarantee the power supply. For data transfer from the card to the reader, load modulation with a 847 kHz subcarrier is used. The modulation is performed by on/off keying of the subcarrier using a Manchester coded data stream (also 106 kbps). In Type B cards, 10 percent ASK modulation is used as the modulation procedure for the data transfer from the reader to the card. A simple non return to zero (NRZ) coding is used for bit coding. From the card to the reader, load modulation with a 847 kHz subcarrier is also used here, but in this case it is modulated by 180° phase shift keying of the subcarrier using the NRZ coded data stream.

The standard also defines two different anticollision protocols (14443-3) for Type A and Type B cards. The reader periodically sends a REQ (Request) command, which places all cards into the range in the READY state (cards are passive, the communication is always initiated by the reader). These cards send back an Answer to Request (ATQ) command so that the reader knows that there is at least one card within range. If there are more than one card, the reader has to perform the corresponding anticollision protocol. Having completed the anticollision procedure, card communication will be under control of the reader, allowing only one card to talk at a time. For Type A cards, a binary search tree algorithm is used to select the card; the reader transmits a number of bits which each card has to compare to its ID. If it matches, the card responds with the remainder numbers of its ID. This process is repeated until only one card with a specific ID is selected (the card enters the ACTIVE state). To maintain the bit synchronization during the anticollision process, the response time is specified. Type B cards use the Aloha anticollision protocol; the reader dynamically sets a number of time slots, and the cards send their ATQ command in one of these slots. The strategy of the card to choose the slot is under the control of the application designer. To guarantee the synchronization of the cards with the slots, the reader transmits its own slot marker at the beginning of each slot. A short time after the transmission of a slot marker, the reader can determine whether a smart card has begun to transmit an ATQ within the current slot. If not, the current slot can simply be interrupted by the transmission of the next slot marker to save time.

Because many of RFID secured applications, as just explained, are designed to be implemented using contactless smart cards, the word "card" will be used throughout the rest of the chapter to refer to RFID tags with cryptographic computational capabilities.

7.3 ATTACKS RELATED TO LOCATION

Section 7.2 explains the change from simple identification to authentication, and how challenge–response protocols can be used to authenticate the cards (also the reader in case of mutual authentication). However, in spite of these added security capabilities, RFID systems are still vulnerable to the following attacks related to location verification: distance fraud attacks, relay attacks or aka mafia fraud attacks, and terrorist attacks. All of these attacks share a wrong assumption about the distance between the verifier (reader) and the prover (card); i.e., as cards are only activated when they are close to a reader, when a reader detects a card, it assumes that the card, and by implication the user, is nearby. Such attacks work on the physical layer and therefore cannot be prevented by any kind of security protocols that operate in the higher layers of the protocol stack.

7.3.1 DISTANCE FRAUD ATTACK

A RFID system is vulnerable to distance fraud attack if the reader is not able to verify whether the card is within a certain distance. An example of this attack is shown in Figure 7.1. Thus, this type of attack is not carried out by an external adversary but by a swindling genuine card, which operates from out of the range where it is supposed to be. The swindling card can use a modified (option A) or an additional transmitter to deceive the reader (option B). This attack, although could initially seem not very dangerous, must be particularly taken into account for those systems where the access rights change according to the physical location [21].

7.3.2 MAFIA FRAUD ATTACK OR RELAY ATTACK

Relay attack, also known as mafia fraud attack, is conceptually depicted in Figure 7.2. The term "mafia fraud" was coined in Ref. [22] as response to the article [23] where Shamir said (in relation to the protection of credit card with his protocol [24]), "I can go to a Mafia-owned store a million successive times and they still will not be able to misrepresent themselves as me". The attack is a man-in-the-middle attack where the genuine reader interacts with a rogue card that manages to fool the reader into thinking that it is directly communicating with the genuine card. Reader's challenges are relayed to the rogue reader that gets the right responses to the challenges from the genuine card. For example, to open a vehicle, an adversary with a rogue card placed near the vehicle establishes contact with the legitimate reader, while an accomplice with a rogue reader, placed near the owner,

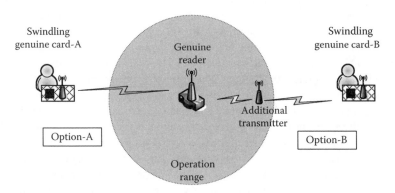

FIGURE 7.1   Sketch of distance fraud attacks.

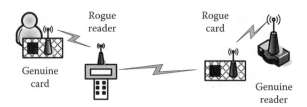

FIGURE 7.2   Sketch of a relay attack.

powers up his card. Then both rogue parties relay all the messages. The electronic protection is thus breached, and both genuine parties, reader and card, remain unaware.

The most significant adversary's limitations, which could be the difference between a theoretical attack and a realistic one, are the delay introduced by the adversary's devices and the distances between the rogue devices and the genuine ones. A longer distance means less exposure to eye contact or security camera.

7.3.3   TERRORIST ATTACK

In the terrorist fraud attack, a genuine card conspires with an adversary giving him the necessary information to impersonate it, but only once (or a limited number of times). It means that the adversary is given by the card the necessary information to pass a single run of the protocol (the responses to the challenges), but not the secret information to pass the protocol whenever he wishes. Clearly, if the adversary knew the long-term private key of the card, the adversary would be trivially successful, as the legitimate card and the adversary would be the same party from a cryptographic point of view.

7.3.4   EXAMPLES OF PRACTICAL ATTACKS

The relay attack is the only one of the attacks described that does not need intentional misbehavior of a genuine card. Thus, this attack is considered the most dangerous of the attacks related to location verification, and most of the research activity has focused on it. The other two attacks are usually marginally treated along with relay attack. Below, some examples of what have been just mentioned are given by looking through some published attacks concerning location verification in RFID.

Kfir and Wool [25] showed in 2005 that contactless smart card technology was vulnerable to relay attacks. They define a "low-tech" adversary capable of building a pickpocket system. This system essentially comprises two devices, which they call "the ghost" (rogue card) and "the leech" (rogue reader). The application used as example is a point-of-sale system, where the reader is connected to the merchant's cash register. To pay, the contactless card is waved near the reader, the reader powers up the card, executes an authentication protocol, and upon successful authentication, the customer's electronic wallet or the credit card is charged. They state that timing constraints imposed by the standard do not represent any problem; time-outs of 5 ms in the anticollision specification (ISO 14443-3) and 5 s for the frames (ISO 14443-4) are long enough to be met by almost any modern digital communication channel. They do not have any problem with the bit synchronization, in contrast with the following examples that will be explained next, because they focus on Type B cards, where the anticollision protocol is Aloha. Type A cards, as explained in Section 7.2.2, use a binary search tree algorithm and therefore do require bit synchronization. In relation to the distance limitations, they consider, instead a passive card, an active rogue card to increase the distance from which it can receive the reader's signal. This way the rogue card does not need to be within "activation range" because it is not longer powered by the reader's transmitted energy. The authors find by simulation that this rogue card, ignoring any regulatory limitations and not using the standard passive load modulation but active modulated transmission synchronized with the reader's carrier, could be up to 50 m away from the genuine reader. On the other hand, the rogue reader can be up to 40–50 cm away from the genuine cards, again calculated by simulation. The latter range is much more constrained than the previous one because the reader has to power up the genuine card and simultaneously detect the slight variations that the load modulation produces in its carrier. Such variations are dependent on the transmitted power, sensitivity of the devices, signal to noise relationship, antenna diameter, and the quality factor of the coupled antennas [26]. These parameters are linked, depend on others, and therefore have to be chosen properly. Specifically, the authors use a relatively big antenna, maintaining still certain mobility, whose dimensions are $40 \times 40 \, cm^2$ with a current between 1 and 4 A. On the other hand, they take advantages of ISO-14443 standard, which allows the reader to request an unlimited number of retransmissions, to request several (between 5 and 16) retransmissions for each frame. Clearly, more retransmissions mean longer exposure time and higher probability of being uncovered. Finally, the authors explain that a victim's near-field communication (NFC) device, like a personal digital assistant (PDA) or a mobile phone, could be hacked to act as the rogue reader. NFC devices (NFCIP standard) offer a gateway between RFID devices and larger devices like cell phones, PDAs, and laptops. An NFC device can act both as a reader and as a card, and can be easily programmed. Thus, according to the authors, they become a convenient platform from which to attack contactless smart card systems.

Also in 2005, Hancke [27] went further from simulation, and demonstrated the effectiveness of relay attack by developing a low-cost hardware capable of carrying out a relay attack against Type A contactless smart cards. The rogue card's operating range is not a problem, and the limitations, as in the previous example, are given by the communications between the rogue reader and the genuine card. Communication between the adversary's devices is achieved by using two half-duplex RF channels. These channels are implemented using integrated transmitter and receiver with bandwidths of up to 115 kbps, and operation ranges of up to 50 m. The delay introduced by the adversary's system is about 15–20 μs. The most restrictive specification is the timing requirements specified by ISO 14443-3 to maintain bit synchronization during the anticollision process (the card has to answer when is due, after 86 or 91 μs). Furthermore, the genuine reader transmits REQ commands periodically and as a result the rogue card has to request information from the rogue reader with the same regularity until the targeted card is within range.

Carluccio et al. [28] present a very detailed implementation of a developed system to carry out relay attacks against ISO-14443A. They also point out the possibility of active attacks by modifying the data before relaying them. According to the text, relay attacks using this hardware have been carried out against:

- Digital passport (e-passport), issued in January 2006 by the Federal Republic Germany
- Student identity chip card used at the Ruhr-University in Bochum, used for paying in the cafeteria
- Philips classic Mifare and DESFire cryptographically enabled smartcards
- Atmel AT88SC153 smartcard
- Ticket for the FIFA World Cup 2006 in Germany

On the other hand, they explain as the timing requirements specified in the ISO 14443, which could hinder the attack, as explained in Hancke's attack, are not met by the used commercial RFID reader.

In spite of that, the authors of the above examples focus on relay attacks, it must be noted that those systems could be easily adapted to carry out distance fraud and terrorist attacks.

7.4 DISTANCE-BOUNDING PROTOCOLS

Distance-bounding protocols are the main technical countermeasure against the attacks described in Section 7.3 [29]. These protocols are tightly integrated into the physical layer of the communication protocol stack, as attacks related to location are orthogonal to any security mechanism implemented in the higher layers. When such protocols are used, not only the verifier confirms that the prover knows a cryptographic secret but also the verifier checks that the prover is within a certain distance (or at least to avoid relay attacks that the information is not being relayed). So, two different parts are merged into a distance-bounding protocol to prevent unauthorized permissions: an identification part and a distance-bounding mechanism. The former works as a "conventional" authentication protocol and provides assurance that the prover knows a cryptographic secret, while the latter allows the verifier to upper bound the distance between them to guarantee that it is the same party that executes the identification part.

Identification part is, as explained in Section 7.2.1, usually based on challenge–response protocols with symmetric keys. Whereas, distance-bounding mechanisms fall into two main types of categories: measuring the signal strength and measuring the round-trip delay.

The first one analyzes the received signal strength to correlate it to card distance [30]. Besides practical problems like tag orientation or the presence of obstacles (e.g., metal or water), such mechanisms are not suitable for security-critical applications due to the following reasons. First, they assume that the prover has a standard device, and thus they are vulnerable when modified devices are used instead (which can change the broadcast power or directional characteristics of devices to spoof locations). In fact, it is not difficult to build directional antennas to increase largely the sending or receiving range. And the second and foremost, they are completely useless against relay attacks; i.e., the received signal power from the rogue card or rogue reader is exactly the same as if they were genuine. As a result, protocols based on the round-trip delay measurement are preferred.

Protocols based on the round-trip delay measurement try to calculate the propagation time as accurately as possible to determine the distance between the verifier and the prover. The variations of the processing time are the main difficulty to extract the propagation time from the overall round-trip delay. So, some authors (an example can be found in Ref. [21]) propose the use of ultrasound. Sound travels far slower than light and therefore greater spatial resolution can be achieved with simpler hardware. The processing delay can be neglected compared to the propagation time, and the accuracy of the measurements is not so critical. Greater processing delay can be tolerated, what is crucial when low-cost devices like RFID are used. However, it presents some serious security problems. The use of ultrasound is effective against distance fraud attack but it is very vulnerable against relay attacks which use wormholes of electromagnetic waves. Even its supposed security against distance fraud attack can be circumvented if the adversary could, in any way, modify the medium, and thus change the speed of sound. Due to these problems, we will focus on those protocols which measure the round-trip time (or time of flight) of electromagnetic signals, and therefore Section 7.5 presents the most relevant distance-bounding protocols of this type proposed for RFID.

An alternative to distance-bounding protocols to avoid relay attack (not distance fraud or terrorist attack) is the use of metal shielding (Faraday cage) to prevent unauthorized usage of a card. The card does not become active until the owner has performed an action, e.g., press a button or open a cover. Other options are the use of biometric measures or type additional passwords. Unfortunately such solutions eliminate some of the conveniences of contactless applications, i.e., handy and fast systems.

7.5   TIMING-BASED PROTOCOLS

This section is a detailed review of the distance-bounding protocols proposed for RFID which upper bounds the distance between the reader and the card by measuring the propagation time of electromagnetic signals. So, the distance is calculated as

$$d = c \cdot \frac{t_{\mathrm{m}} - t_{\mathrm{d}}}{2} \tag{7.1}$$

$$t_{\mathrm{m}} = 2 \cdot t_{\mathrm{p}} + t_{\mathrm{d}} \tag{7.2}$$

where

    c is the propagation speed of light

    t_p is the one-way propagation time

    t_m is the measured total elapsed round-trip time

    t_d is the processing delay at the card

Determining as accurately as possible the processing time is the key to reliably isolate the propagation component from the measured total time, and thus variable processing times constitute a serious problem. Besides being invariable, it is also convenient that this processing time is as short as possible, because an adversary could overclock the card to absorb the delay introduced by his devices. The amount of introduced delay that can be absorbed is largely dependent on the clock frequency and the number of clock cycles required to compute the response. The processing time includes the time to detect the challenge, compute and transmit the response, and it will be in general very long with respect to the propagation time; i.e., the propagation time is at most 2/3 ns for an operation range of 10 cm; whereas a single clock cycle takes 74 ns at the HF standard (ISO-14443) reader supplied clock rate of 13.56 MHz.

Before going on to describe the protocols, some notations that are used in the rest of this section are introduced:

$\{\}^n$ denotes the set of all strings of bit-length n

A|B denotes the concatenation of the bitstrings A and B

s_i given a string s, denotes the ith less significant bit of s

$=$ is used to indicate assignment to a variable

$\oplus$ this symbol denotes the OR exclusive operation (XOR)

Start clock and *Stop clock* are two instructions used to measure the time; the verifier computes
    the time between these instructions (Δt)

ID_U is the identity string of the user U

7.5.1 BRANDS AND CHAUM'S PROTOCOL

In 1993, when S. Brands and D. Chaum described their protocol, the RFID technology was not as fashionable as nowadays [31]. So, although they proposed the use of their protocol for electronic payment and access control, they did not mention anything about RFID technology. However, Brands and Chaum's protocol has been included here because it was the first distance-bounding protocol based on timing measurements of single bits (others measure the round-trip time of entire data packets [32]), and it has been used as starting point by different authors to propose their protocols for RFID—even the term "distance-bounding protocol" comes from here.

The essential idea of this protocol is quite simple. The verifier sends out a challenge bit and starts a timer. The prover receives the challenge, computes the response bit, and sends it back to the verifier, which stops the timer. The verifier uses the round-trip time to extract the propagation time and determine the distance between them. If the distance is under a certain threshold and the response is right, then the verifier considers with probability 1/2 that no attacks occurs in this round. This is repeated n times until the probability of detecting an attacker is high enough.

The protocol is comprised of three stages: the commitment phase, the fast bit exchange, and finally a signing phase where the parties acknowledge the challenges and the responses which have been used. Figure 7.3 shows a sketch of the protocol:

Step 1 The verifier V generates uniformly at random n bits C_i, and the prover P generates uniformly
    at random n bits R_i. This step could be done beforehand.

Step 2 Now the low-level distance-bounding exchanges take place. V sends bit C_i to P, and P sends
    bit R_i to V immediately after receiving C_i. It is repeated n times, for $i = 1, \ldots, n$.

Step 3 P concatenates the $2n$ bits C_i and R_i, and signs the resulting message m with his secret key.
    P sends this signature to V.

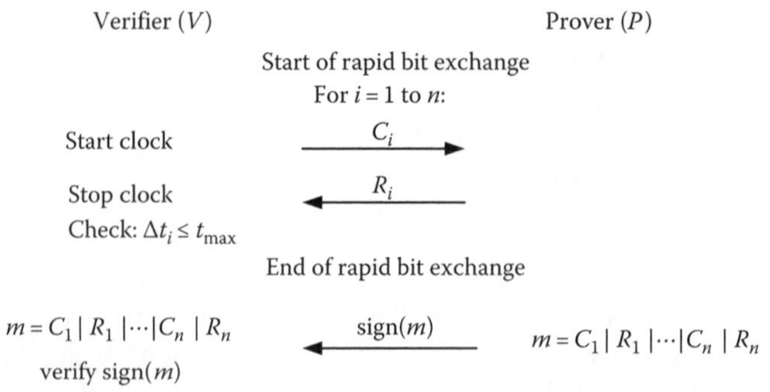

Verifier (V)   Prover (P)

Start of rapid bit exchange
For $i = 1$ to n:

Start clock $\xrightarrow{\quad C_i \quad}$

Stop clock $\xleftarrow{\quad R_i \quad}$
Check: $\Delta t_i \le t_{max}$

End of rapid bit exchange

$m = C_1 \mid R_1 \mid \cdots \mid C_n \mid R_n$ $\xleftarrow{\quad \text{sign}(m) \quad}$ $m = C_1 \mid R_1 \mid \cdots \mid C_n \mid R_n$

verify sign(m)

FIGURE 7.3 Brands and Chaum's protocol.

Verifier (V)   Prover (P)

$C_i \in_R \{0,1\}$ $R_i \in_R \{0,1\}$

$\xleftarrow{\quad ...,M_i^2,..., \text{commit}(..., R_i, ...) \quad}$ $M_i \in_R Z_k^*$

Start of rapid bit exchange
$\xrightarrow{\quad C_i \quad}$
$\xleftarrow{\quad R_i \quad}$

End of rapid bit exchange

$\xleftarrow{\quad ...,X^{C_i \oplus R_i} M_i, ...,(\text{open commit}) \quad}$

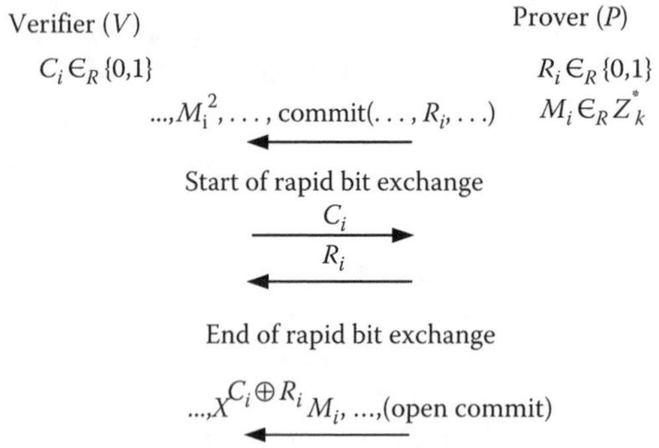

FIGURE 7.4 Brands and Chaum's protocol with the Fiat–Shamir identification scheme.

The verifier will only accept the prover as valid if the maximum of the delay times between sending out bit C_i and receiving bit R_i back, for $i = 1, \ldots, n$, is less than a certain value (t_{max}), in accordance with the maximum allowed distance, and the received signature is a correct signature of P on $m = C_1 \mid R_1 \cdots \mid C_n \mid R_n$.

Assuming that the signature scheme is secure, the number of exchanges, n, is the main security parameter because the average success probability of a relay attack equals $(1/2)^n$; i.e., an adversary who sends back at random n bits during the rapid bit exchange, will be successful only if those bits are equal to the n bits sent back by the genuine prover.

This protocol, in the form as described above, is not effective against distance fraud attack. If the swindling prover knows at what time V will send out bits, can send R_i out at the correct time before receiving C_i. Brands and Chaum describe two protocol variants to overcome this problem. One modification can be that V sends bits out with randomly chosen delay times. The adversary cannot send out bits before he has received bit because V will not accept a response bit R_i before he has sent out bit C_i. Another solution is that R_i is dependent on C_i. In this way, the authors propose to form the responses by XOR-ing the received challenge bits with v, $R_i = C_i \oplus v_i$, where v is a bitstring known by both parties. The choice of the bitstring v is not described by the authors.

Brands and Chaum also show how to integrate this distance-bounding mechanism with a public-key identification scheme. Specifically, a parallel version of the basic Fiat–Shamir identification (zero-knowledge) protocol [24] is presented. This protocol, where both parties share a secret information X, is depicted in Figure 7.4 and it is comprised of the following steps:

Step 1 P generates uniformly at random n numbers $M_i \in Z_k^*$, and sends their squares $M_i^2 \text{mod}(k)$ to V. P also generates uniformly at random n bits R_i and commits to these bits (and their order) by sending a commitment on them to V.

Step 2 V generates uniformly at random n bits C_i.

Step 3 Now the low-level distance-bounding exchanges can take place. The following steps are repeated n times, for $i = 1, \ldots, n$.

V sends bit C_i to P.

P sends bit R_i to V immediately after he receives C_i from V.

Step 4 P opens the commitment on the bits R_i made in Step 1 by sending the appropriate information to V. Furthermore, P determines the n responses $X^{v_i} M_i$ corresponding to the challenges $v_i = C_i \oplus R_i$, for $1 \le i \le n$, and sends them to V.

V determines the n challenges v_i in the same way as P did, and verifies that the n responses are correct. Then V verifies whether the opening of the commitments by P is correct. If this holds and the maximum of the delay times between sending C_i and receiving R_i is under the threshold, V accepts P as valid. Again to be resistant against distance fraud attack, the protocol has to be modified to avoid that an adversary can send R_i before receiving C_i. To reach this objective, in Step 1 P commits to n bits h_i, and in Step 3 P will reply with response bits $R_i = C_i \oplus h_i$. Finally, the responses of P in Step 4 must be computed with respect to the multibit challenge $C_1|R_1| \cdots |C_n|R_n$.

The authors also show how to incorporate the distance-bounding mechanism to other schemes with public-key cryptographic operations. Nevertheless, we will not give more details about them because, as aforementioned, public-key cryptography is in most cases too computationally complex to be implemented in the highly resource-constrained RFID devices.

Before concluding this section, we would like to draw the reader's attention to the fact that an important part of the security of Brand and Chaum's protocol rely on the final signed message of the challenges and responses exchanged. With this final message, the prover provides a value that is cryptographically derived from the challenges, thereby confirming to the verifier that the real prover was indeed involved in the protocol. It avoids that an adversary could carry out a relay attack asking the prover the responses in advance. An adversary who pretends to get the R_i in advance from the genuine prover, by supplying it with guessed challenges C_i^*, will succeed in doing so only with probability 2^{-n} without being detected; i.e., when $C_i^* = C_i$ for $i = 1, \ldots, n$.

Finally, some real characteristics (especially noise) of the communication channel have not been taken into account, and as a result this protocol is not able to cope with bit errors occurring during the fast bit exchange; a single bit error occuring causes the protocol to fail.

7.5.2 HANCKE AND KUHN'S PROTOCOL

Hancke and Kuhn's protocol was the first distance-bounding protocol designed specifically for RFID devices [33]. The authors used as reference point Brands and Chaum's protocol, described in Section 7.5.1, and modified it according to the particular characteristics of the RFID technology: it is taken into account that RFID devices are computationally weak, and the protocol is able to deal with bit errors caused by noise in the channel.

A sketch of this protocol is depicted in Figure 7.5. The protocol is based on the timing measurement of single bit round trips combined with a symmetric-key identification mechanism, where card and reader share a common secret value K:

Step 1 It starts by having reader and card exchange random nonces, N_r and N_c, which will never be used again. With these nonces and the key K, the parties use a hash function to compute an unpredictable pseudorandom string H of length $2n$ bits. Then the parties split it into two n-bit strings, $v0$ and $v1$.

Step 2 A rapid n-round challenge–response phase begins. For the ith round, the ith bit of $v0$ is answered if the ith challenge is zero ($C_i = 0$), and the ith bit of $v1$ otherwise ($C_i = 1$). The processing time is reduced because a small number of gate delays are required to compute the response (simple lookup memory). The reader checks that the received response is correct and also that it has been received within a period.

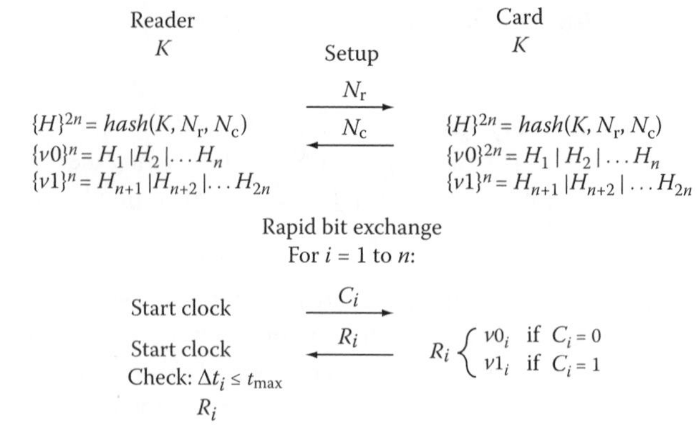

Reader   Card
K Setup K

$$\xrightarrow{\quad N_r \quad}$$
$$\xleftarrow{\quad N_c \quad}$$

$\{H\}^{2n} = hash(K, N_r, N_c)$ $\{H\}^{2n} = hash(K, N_r, N_c)$
$\{v0\}^n = H_1 |H_2|...H_n$ $\{v0\}^{2n} = H_1 | H_2 |...H_n$
$\{v1\}^n = H_{n+1} |H_{n+2}|...H_{2n}$ $\{v1\}^n = H_{n+1} |H_{n+2}|...H_{2n}$

Rapid bit exchange
For $i = 1$ to n:

Start clock $\xrightarrow{\quad C_i \quad}$
Start clock $\xleftarrow{\quad R_i \quad}$ $R_i \begin{cases} v0_i & \text{if } C_i = 0 \\ v1_i & \text{if } C_i = 1 \end{cases}$
Check: $\Delta t_i \le t_{max}$
R_i

FIGURE 7.5 Hancke and Kuhn's protocol.

The communication method used for the rapid exchanges is different from that used for the ordinary communication for two reasons:

First, the ISO 14443 card to reader modulation scheme is inappropriate for this time-critical phase because of its low bit rate. Let us assume that the distance resolution is the propagation distance in one bit period, an ultrawide band (UWB) channel is needed to achieve a resolution of 10 cm.

Second, detector and corrector mechanisms cannot be used for this communication, as it would mean additional and variable cycles of processing.

The latter and the high sensitivity to the background noise of the UWB link made this channel very unreliable. Hence, unlike Brands and Chaum's protocol, we don't find any signing phase here. Due to the high BER (bit error rate) of the channel, the challenges or the responses may be easily corrupted, and therefore, it is not possible to send a final signed message of the actual challenges and responses exchanged. It could be substituted for a signed message after each round, but it would overload and make slower the authentication protocol. The carrier of 13.56 MHz, which does not provide enough distance resolution, is used to provide a time base for synchronizing the UWB communication. Biphase modulation, where bits are represented by pulses of opposite polarity, has been considered the best suited to the distance-bounding purpose.

The nonce N_c is not necessary, as Avoine explains [2], if the card has a simple trusted time reference that keeps it from running at twice the normal clock frequency, preventing an adversary to query the card $2n$ times (two consecutive executions of the protocol) with the same N_a but complementary challenges in each run. A simple band-pass filter could be used to detect these large deviations (at least twice its normal speed) from its nominal clock frequency.

A secure hash function (i.e., one-way, collision resistant, and pseudorandom) has been chosen as keyed public pseudorandom function to generate the pseudorandom bitstrings. Therefore, it will be necessary a key establishment process to set up the shared key K.

The main disadvantages that can be pointed out in the protocol of Hancke and Kuhn are

- It is vulnerable to terrorist attack. A conspirator genuine card could hand over the bitstrings $v0$ and $v1$ to an adversary. With this information the adversary is able to pass the protocol without knowing the long-term private key of the genuine card.
- Success probability that an adversary impersonates a genuine card, as the authors themselves mention, is not $(1/2)^n$ but $(3/4)^n$. In fact, the adversary can query the card with any value (1 or 0) before receiving the challenge, obtaining in this way the right response for this value. When the actual challenge is sent by the reader, the adversary already knows

the response whether this challenge coincides with the value that was previously queried. In case it does not coincide, the adversary randomly answers one of two possibilities. It must be reminded that Brands and Chaum's protocol avoided this problem (or attack) with the signed message at the end of the protocol.

- Possible corruption of challenges or responses implies that a genuine card could be falsely rejected (false-reject). To avoid this, the protocol handles communication errors simply by tolerating some bit errors during the rapid bit exchanges; i.e., the reader will accept a card as valid even if some of its responses are not correct. For instance, a reader can accept a card as valid even if at most C of the n received responses are incorrect. Unfortunately, this increases the success probability for an adversary, who now can fail up to C responses:

$$p_{\text{adv-succ}} = \sum_{k=0}^{k=C} \binom{n}{k} \cdot \left(\frac{3}{4}\right)^{n-k} \cdot \left(\frac{1}{4}\right)^{k} \tag{7.3}$$

So, the parameter C has to be chosen properly, according to BER, to ensure two requirements: being resistant to impersonation (low $p_{\text{adv-succ}}$) along with completeness (the reader accepts a genuine card with overwhelming probability). In general, this problem compels to increase the number of rounds n.

- It adds quite a lot of complexity (with its corresponding cost). Let us assume the distance resolution is the propagation distance in one bit period, the spatial resolution of a RF channel of bandwidth B is roughly c/B, where c is the speed of light. This means that to achieve a resolution of 10 cm is necessary to add an expensive and complex 3 GHz bandwidth radio transceiver to the reader and to each card.

7.5.3 REID ET AL.'S PROTOCOL

Reid et al. [34] propose a new protocol which overcomes some disadvantages of Hancke and Kuhn's protocol. So, the authors state that their protocol is the first symmetric-key-based distance-bounding protocol that is resistant to terrorist attacks. They specify that it is based on symmetric key because Bussard [35] had already described a distance-bounding protocol which used asymmetric techniques and which was capable of protecting against terrorist attacks. The basic idea of Bussard's protocol is to force the conspirator prover to give its long-term key away to the adversary if it wants him to pass the security protocol. This protocol works as follows. The prover computes $v = E_{Ks}(K)$, where K is its long-term private key and Ks a generated session key. The verifier then sends challenge bits, C_i, to the prover. If $C_i = 1$, the prover must respond with v_i. If $C_i = 0$, the prover sends Ks_i back to the verifier. Thus, to respond correctly and timely the adversary must be in possession of v and Ks. This is enough information for the adversary to recover the long-term key K. This same basic idea, as it will be shown, is used by Reid et al. to make their protocol resistant to terrorist attacks. On the other hand, the distance-bounding mechanism is in essence the same as used by Hancke and Kuhn's protocol but the UWB channel is substituted for another based on side-channel effects.

The protocol of Reid et al. is shown in Figure 7.6, and it consists of the following steps:

Step 1 There is an initial exchange of nonces and identities (sound protocol). Both parties use a key derivation function, KDF, to derive a session symmetric encryption key, Ks. This session key is used to encrypt the long-term shared secret K, $v = Ks \oplus K$. KDF is a pseudorandom function, e.g., a MAC algorithm such as HMAC [36].

Step 2 The fast challenge–response phase begins. It is similar to Hancke and Kuhn's one, except that now the ith bit of v is returned when $C_i = 0$, and the ith bit of the key Ks otherwise.

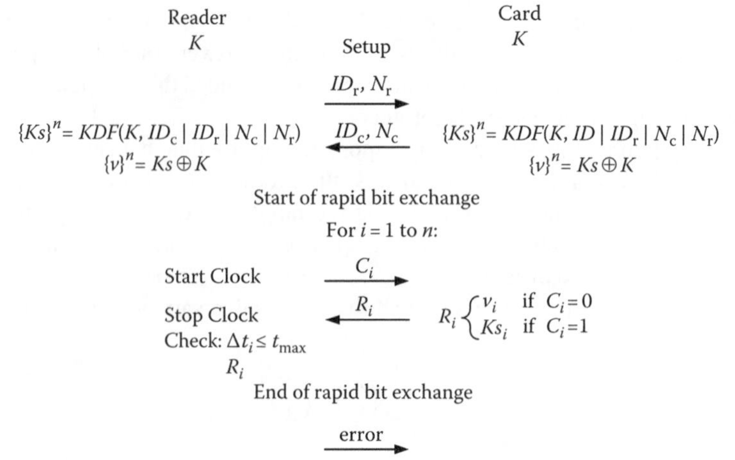

Reader   Card
K Setup K

$$\xrightarrow{\quad ID_r, N_r \quad}$$

$\{Ks\}^n = KDF(K, ID_c \mid ID_r \mid N_c \mid N_r)$ $\xleftarrow{\quad ID_c, N_c \quad}$ $\{Ks\}^n = KDF(K, ID \mid ID_r \mid N_c \mid N_r)$
$\{v\}^n = Ks \oplus K$ $\{v\}^n = Ks \oplus K$

Start of rapid bit exchange
For $i = 1$ to n:

Start Clock $\xrightarrow{\quad C_i \quad}$

Stop Clock $\xleftarrow{\quad R_i \quad}$ $R_i \begin{cases} v_i & \text{if } C_i = 0 \\ Ks_i & \text{if } C_i = 1 \end{cases}$
Check: $\Delta t_i \leq t_{max}$
R_i

End of rapid bit exchange

$$\xrightarrow{\quad \text{error} \quad}$$

FIGURE 7.6   Reid et al.'s protocol.

Step 3 If any response bit is incorrect, the reader will send an extra error message to the card, that points out which round these errors occurred in. Note that this message has nothing to do with the final signed message of the Brand and Chaum's protocol. It has been included, as discussed below, to protect against other more refined terrorist attacks.

This protocol solves some of the problems found in Hancke and Kuhn's protocol:

- It is resistant to terrorist attack. An adversary needs to have Ks and v to respond timely and correctly to the challenges. Because knowing Ks and v implies to know K ($K = Ks \oplus v$), no card should hand this information over to the adversary. Furthermore, the protocol is also resistant to a more refined terrorist attack where the conspirator genuine card does not provide the adversary with Ks and v, but two bitstring, Ks^* and v^*, with m instances for which $Ks_i^* \neq Ks_i$ or $v_i^* \neq v_i$ but not both; i.e., if the adversary computes $K^* = Ks^* \oplus v^*$, then m bits of the computed secret will be flipped with respect to the real secret K. The adversary can guess the position of these bits with probability, $\binom{n}{m}^{-1}$, much smaller than the probability of passing the security protocol (2^{-m}). If for instance, $n = 128$ and $m = 10$, the adversary's probability of being accepted is 2^{-10}, whereas the probability of guessing K is less than 2^{-40}. So, in this more refined terrorist attack the conspirator genuine card does not ensure that the adversary will be accepted, but increases his chances without risking its long-term key too much. The last message with the errors is specifically sent to avoid this, because with this information the adversary can guess K with the same probability of being accepted (2^{-m}).
- It does not use the complex and expensive UWB radio link for the time-critical phase. It is substituted for a novel low-cost communication approach that leverages the phenomena of side-channel leakage to deliver a low-latency channel.

As mentioned earlier, this protocol does not resort to the expense and complexity of a UWB radio. The authors propose to take advantage of the generally undesirable side-channel effects. Side-channel effects are generally considered as serious security vulnerabilities because they are physical leakages (e.g., power consumption, timing information, electromagnetic emanations, and radiated heat) about the internal computing, which could be used by an adversary to infer secret cryptographic information [37]. In contrast, this protocol looks for optimizing it so that the reader can measure directly the physical side effect of calculation process to deduce the result. This can reduce the latency because the reader can detect the response bits as the prover calculates them. Detecting the response as soon as possible increases the timing resolution and as a result the distance resolution. Because the reader provides the cards with power, this physical parameter results especially suitable for being monitored (simple power analysis). This way, the reader monitors the amplitude modulation on f_c rather than accessing the data via processing a side band. Timing resolution will depend on how quickly the card can change the load to produce a detectable carrier amplitude change on the reader's antenna circuit. A special XOR gate is proposed to be implemented in the cards which have deliberately pronounced (increasing or reducing power consumption), output-dependent leakage characteristics which can be easily detected by the reader. The required circuitry can be implemented at modest cost in both card (special XOR gate) and reader (peak or fading detector).

Although this communication method has not been implemented, the authors estimate that at closer distances the timing resolution could be as low as a half of cycle, $0.5f_c = 37$ ns. The authors argue that this resolution is enough to detect highly sophisticated relay attacks. However, experimental results show that as the quality of the inductive coupling between the card and reader antenna loops degrades the carrier amplitude changes decrease and it takes more cycles to produce a detectable modulation change. As the coupling quality degrades with increasing the distance, the operation ranges must be reduced to keep low the latency. Thus, a timing resolution of 300 ns is estimated at higher distances, which is still 50 times smaller than the delay introduced by the devices in the low-cost Hancke's relay attack (see Section 7.3.4).

However, this protocol still presents some disadvantages:

- Timing resolution reduces with increasing the operation range. Thus, to keep a fine timing resolution the operation range must be small (300 ns for 4–5 cm).
- Even the theoretical finest timing resolution of 37 ns is not enough to detect mafia fraud attacks. Light travels more than 11 m in that time.
- In the analysis of the channel latency carried out by the authors, only the rate of change in carrier amplitude that cards can effect via load modulation is estimated. The time that cards would need to detect the challenges and compute the responses is not included.
- Viability of the communication channel has not been proved, and further investigation is needed into the impact of different RF noise environments on modulation detection.
- Unreliability of the channel causes the challenges or the responses to be easily corrupted. Due to this, a signing phase at the end of the protocol is not possible and thus the probability that an adversary succeeds is again $(3/4)^n$. Moreover, the reader will have to accept a card as valid even if some, according to the expected error rate, of its responses are not correct. As a result, the number of rounds, n, has to be increased, and more time is needed to complete the protocol.
- Protocol is vulnerable to traceability. Cards send their static identities and therefore they can be tracked by an adversary.

7.5.4 PROTOCOLS ENHANCED WITH VOID CHALLENGES

Distance-bounding protocols for RFID enhanced by using void challenges are described in Ref. [38]. Munilla and Peinado propose a modification of Hancke and Kuhn's protocol to decrease the adversary's success probability, and thus reduce the average time to complete the protocol (fewer rounds are necessary for the same probability). Such modification is based on what the authors call "void challenges". A void challenge is a challenge which the reader intentionally leaves without sending and which is used to detect if an adversary is trying to get the responses in advance from the card.

The modified protocol defines a crude time base with intervals, T_c, and the exchanges occur within such intervals. The carrier wave used to power the card and carry out the nondistance-bounding part can be used to define such time base (HKP defines a time base for the UWB link in the same way). In addition to the bitstrings $v0$ and $v1$, a new random n-bit string P is generated by the parties in

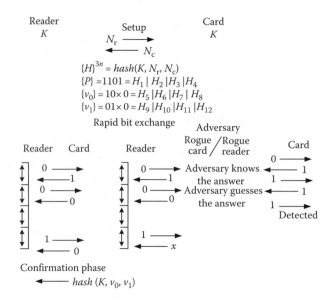

FIGURE 7.7   Hancke and Kuhn's protocol enhanced with void challenges.

each execution of the protocol. The bitstring P indicates the void challenges; i.e., the intervals when the reader does not send the challenge (e.g., $P_i = 1$ reader sends and $P_i = 0$ it does not). Hence, two types of intervals or rounds can be distinguished: full challenges (the challenge is sent) and void challenges (the challenge is not sent). These void challenges, as aforementioned, will allow the card to detect if an adversary is trying to get the responses in advance. The modified protocol is depicted in Figure 7.7; in the left part a normal run is shown and in the right part is shown how an adversary is detected. When the card detects an adversary, it stops sending responses. The protocol ends with a confirmation message to verify that no adversary has been detected. Note that this final message does not change if any challenge or response is corrupted (it is different from the signed message in Brand and Chaum's protocol).

The parameter p_f is defined as the probability of an interval being full; i.e., each bit of the bitstring P has a probability p_f of being 1. If this parameter is chosen properly the success probability for an adversary decreases. Specifically, for $p_f = 4/5$ the minimum adversary's success probability is achieved: $(3/5)^n$; where n is the number of rounds. Thus, fewer rounds are needed for the same adversary's success probability. This reduction in the number of rounds compared to Hancke and Kunh's protocol is especially significant when noise in the channel is taken into account, and the number of rounds has to be increased to tolerate some wrong responses.

The drawbacks of this protocol are the followings:

- It is not easy to generate a bitstring P with $p_f = 4/5$. So, for practical implementations a value of $p_f = 3/4$ is recommended. With this value, the adversary's success probability is not $(3/5)^n$, but a bit higher, $(5/8)^n$.
- Length of the pseudorandom bitstring v generated by the hash function needs to be greater to compute P.
- Although it is not a serious problem because the required signal to activate the card is usually very strong compared to background noise, false alarms must be taken into account because cards have to distinguish between three physical states: 0, 1, and voids. A false alarm occurs in a void-challenge interval when the card receives a signal and it believes that an adversary is trying to get the responses in advance when it is actually noise. A genuine reader is rejected as result of it.

- Computing and the sending of the final message takes an extra time. Reduction in the number of rounds should make up for this time. Hence, this protocol is really effective only when the number of rounds is high (a low adversary's success probability is wished or the channel is especially noisy).

7.5.5 OTHER PROTOCOLS

In this section, other timing-based protocols are briefly described. These works have not been, in our opinion, sufficiently discussed and we strongly recommend the reader being observant of new research works related to them that could appear:

- Protocol described in Ref. [39] is a modification of Hancke and Kuhn's protocol that seeks to reduce the number of rounds. As already explained, a final signed message on the actual challenges and responses exchanged avoids that an adversary can ask the card in advance. However, this message cannot be used if the channel is noisy, and sending a signed message after each round is not efficient. To overcome this problem, this paper proposes to use an Error Correcting Code [8].
- Another proposal that purport to reduce the adversary's success probability is described in Ref. [40]. The basic idea is to authenticate the reader by sending several temporal keys, again to avoid that an adversary can ask the card in advance.
- Taking advantage of the special characteristics of the RFID inductive communication in combination with the use of void challenges, described in Section 7.5.4, in Ref. [41] is proposed a low-cost distance-bounding protocol with zero-processing delay, capable of detecting relay attacks.

7.6 CONCLUSIONS

RFID devices, and particularly contactless smart cards, are more and more used in payment and access control applications. These applications require the implementation of security mechanisms. However, in spite of these added security capabilities, most of RFID devices are still vulnerable to the different attacks related to location: distance fraud attacks, relay attacks or mafia fraud attacks, and terrorist attacks. This chapter describes comprehensively such attacks and presents the main countermeasure against them: distance-bounding protocols.

Distance-bounding protocols, as shown, combine physical and cryptographic properties not only to verify that the card knows a cryptographic secret, but also to determine an upper bound on the distance between the reader and the card. Distance-bounding techniques, because of security reasons, usually rely on the measurement of round-trip time of electromagnetic signals. These measurements are performed during a n-round rapid challenge–response phase. Some important details must be taken into account:

- It must be impossible for the card to send the response before receiving the challenge. This implies that the response must be dependent on the challenge. The time that the card takes to detect the challenge, compute, and transmit the response is the processing time. Propagation time is estimated by subtracting the processing time from the measured overall round-trip time.
- Communication channel must allow to exchange bits with very low latency.
- Protocol must cope with bit errors taking place during the rapid bit exchanges.
- To avoid terrorist attacks, the long-term key and the responses must be intermingled in a cryptographic way.

Thus, the future of distance-bounding protocols for RFID depends on the ability to overcome the following issues:

- Processing time must be as short and invariant as possible (nought preferably).
- New low-latency channels that can be implemented at modest cost are necessary. The length of the symbols used to represent the bits must be minimized.
- Reducing the number of rounds as much as possible, i.e., reducing the time needed to complete the protocol.
- Besides relay attack, to be secure, distance-bounding protocols have to prevent distance fraud and terrorist attacks.

REFERENCES

[1] LARAN RFID, A basic introduction to RFID technology and its use in the supply chain, *RFID Journal*, http://www.rfidjorunal.com/whitepapers/download/46.
[2] G. Avoine, Cryptography in radio frequency and fair exchanges protocols, PhD thesis, École Polytechnique Féderale de Lausanne, Lausanne, Switzerland, 2005.
[3] T. Karygiannis, B. Eydt, G. Barber et al., *Guidelines for Securing Radio Frequency Identification (RFID) Systems*, NIST (National Institute of Standards and Technology), Gaithersburg, MD, April 2007.
[4] P. Schmitt, R. Ulrich, and J. Canvin, RFID in the automotive industry, 2006. http://www.odette.org/Newsletter07/RFID@Automotive.pdf.
[5] Mastercard and Visa agree to a common contacless communication protocol, http://www.corporate.visa.com/md/nr/press252.jsp.
[6] Philips, Philips' MIFARE Identification Chips just the ticket for London's Oyster Smart Card, http://www.semiconductors.philips.com/news/content/file_910.html, 2002.
[7] J.H. Hoepman, E. Hubbers, B.Jacobs, M.Oostdijk, and R. Wichers Scchreur, Crossing borders: Security and privacy issues of the European e-passport, *Advances in Information and Computer Security, First International Workshop on Security (IWSEC'06)*, Kyoto, Japan, *Lecture Notes in Computer Science*, Springer-Verlag, Vol. 4266, pp. 152–167, 2006.
[8] A. Menezes, P.C. Van Oorschot, and S.A. Vanstone, *Handbook of Applied Cryptography* (fifth printing), CRC Press, Boca Raton, FL, August 2001.
[9] J. Wolkerstorfer, Is elliptic-curve cryptography suitable to secure RFID tags? *Workshop on RFID and Light-Weight Crypto*, Graz, Austria, July, 2005.
[10] M. Girault, L. Juniot, and M.J.B. Robshaw, The feasibility of on-the-tag public key cryptography, *Conference on RFID Security*, Málaga, Spain, July 2007.
[11] M. Aigner and M. Feldhofer, Secure symmetric authentication for RFID tags, *Telecommunication and Mobile Computing—TCMC 2005*, Graz, Austria, March 2005.
[12] S. Piramuthu, HB and related lightweight authentication protocols for secure RFID tag/reader authentication, *CollECTeR Europe Conference*, Basel, Switzerland, June 9–10, 2006.
[13] M. Feldhofer, J. Wolkerstorfer, and V. Rijmen, AES implementation on a grain of sand, *IEE Proceedings on Information Security*, Vol. 152, pp. 13–20, October 2005.
[14] M. Feldhofer and C. Rechberger, A case against currently used hash functions in RFID protocols, *RFIDSec 06*, Graz, Austria.
[15] K. Finkenzeller, *RFID Handbook: Fundamentals and Applications in Contactless Smart Card and Identification*, 2nd edn, Wiley, New York, 1993.
[16] D. Paret, *RFID and Contactless Smart Card Applications*, John Wiley & Sons, New York, 2005.
[17] H.J. Chae, D.J. Yeager, J.R. Smith, and K. Fu, Maximalist cryptography and computation on the WISP UHF RFID tag, *Conference on RFID Security*, Málaga, Spain, July 2007.
[18] ISO 15693, *Identification Cards–Contactless Integrated Circuit Cards–Vicinity Cards*, International Organization for Standardization, Geneva, http://www.iso.org.
[19] ISO 14443, *Identification Cards–Contactless Integrated Circuit Cards–Proximity Cards*, International Organization for Standardization, Geneva, http://ww.iso.org.
[20] T. Phillips, T. Karygiannis, and R. Kuhn, Security standards for the RFID market, *IEEE Security and Privacy*, Vol. 3, No. 6, pp. 85–89, November/December 2005.

[21] N. Sastry, U. Shankar, and D. Wagner, Secure verification of location claims, *Proceedings of the 2003, ACM Workshop on Wireless Security*, San Diego, CA, pp. 1–10, 2003.

[22] Y. Desmedt, C. Goutier, and S. Bengio, Special uses and abuses of the Fiat–Shamir passport protocol, *Advances in Cryptology—CRYPTO'87: Proceedings*, Springer, Berlin/Heidelberg, Vol. 293, p. 21, 1988.

[23] J. Gleick, A new approach to protecting secrets is discovered, *New York Times*, pp. C1 and C3, February 17, 1987.

[24] A. Fiat and A. Shamir, How to prove yourself: Practical solutions to identification and signature problems, *CRYPTO '86*, Santa Barbara, CA. *Lecture Notes in Computer Science*, Springer-Verlag, Vol. 263, pp. 186–199, 1987.

[25] Z. Kfir and A. Wool, Picking virtual pockets using relay attacks on contactless smartcard systems, *Cryptology ePrint Archive*, Report 2005/052, http://eprint.iacr.org, 2005.

[26] J. Sorrels, Optimizing read range in RFID systems, EDN, December 2000. http://www.edn.com/article/CA84480.html.

[27] G. Hancke, A practical relay attack on ISO 14443 proximity cards, http://www.cl.cam.ac.uk/~gh275/relay.pdf, 2005.

[28] D. Carluccio, T. Kasper, and C. Paar, Implementation details of a multi purpose ISO 14443 RFID-tool, *Workshop on RFID Security*, Graz, Austria, 2006.

[29] D. Singelee and B. Preneel, Location verification using secure distance bounding protocols. *2nd IEEE International Conference on Mobile, Ad Hoc and Sensor Systems, MASS-2005*, Washington, DC, 2005.

[30] K. Fishkin and S. Roy, Enhancing RFID privacy via antenna energy analysis, *RFID Privacy Workshop*, Cambridge, MA, November 2003.

[31] S. Brands and D. Chaum, Distance bounding protocols, *Advances in Cryptology EUROCRYPT'93, Lecture Notes in Computer Science*, Springer-Verlag, Berlin, Vol. 765, pp. 344–359, 1994.

[32] B. Walters and E. Felten, Proving the location of tamper resistant devices, http://www.cs.princeton.edu/ bwaters/research/, February 2003.

[33] G. Hancke and M. Kuhn, An RFID distance bounding protocol, *Proceedings of the IEEE, SecureComm*, 2005.

[34] J. Reid, J.M.G. Nieto, T. Tang, and B. Senadji, Detecting relay attacks with timing-based protocols, *Proceedings of the 2nd ACM Symposium on Information, Computer, and Communication Security*, pp. 204–213, Singapore, 2007.

[35] L. Bussard, Trust establishment protocols for communicating devices, PhD thesis, Institut Eurécom, Télécom, Paris, 2004.

[36] M. Bellare, R. Canetti, and H. Krawczyk, HMAC: Keyed-hashing for message authentication, Internet Request for Comment RFC 2104, Internet Engineering Task Force, February 1997.

[37] S. Mangard, E. Oswald, and T. Popp, *Power Analysis Attacks: Revealing the Secrets of Smart Cards*, Springer Science + Business Media, New York, 2007.

[38] J. Munilla and A. Peinado, Distance bounding protocols for RFID enhanced by using void-challenges and analysis in noisy channels, *Wireless Communications and Mobile Computing*, Vol. 8, pp. 1227–1232, John Wiley & Sons, January 2008.

[39] D. Singelée and B. Preneel, *Distance Bounding in Noisy Environments, Lecture Notes in Computer Science*, Springer, Berlin/Heidelberg, Vol. 4572, pp. 101–105, 2007.

[40] Y.-J. Tu and S. Piramuthu, RFID distance bounding protocols, *First International Workshop*, Vienna, Austria, 2007.

[41] J. Munilla, A. Ortiz, and A. Peinado, Distance bounding protocols for RFID, *Workshop on RFID Security*, Graz, Austria, 2007.

8 Secure Proximity Identification for RFID

Gerhard P. Hancke and Saar Drimer

CONTENTS

Physical distance serves as a measure of trust and in some systems users are granted privileges based on their perceived location. Radio-frequency identification (RFID) tokens are often used for proximity identification due to their limited range of communication. RFID systems are, however, particularly vulnerable to relay attacks, which allow an attacker to pretend that a valid token is present by relaying the communication between a reader and a remote token. The increasing availability of RFID equipment for development makes relay attacks relatively easy to implement. Distance-bounding protocols allow a device to cryptographically determine an upper bound on the distance to another device and are an effective countermeasure against relay attacks. RFID systems would therefore benefit from distance-bounding protocols even though the practical implementation of suitable cryptographic mechanisms and communication channel requires careful consideration because of resource constraints. This chapter describes the implementation of relay attacks against RFID tokens and discusses several protocols and communication channels, which have been proposed for cryptographically bounding the distance to low-resource tokens.

8.1   INTRODUCTION

Many systems use location information as part of an authentication process that grants privileges and services. This is so, even though location is quite easy to fake if the system was not specifically designed to detect location spoofing attacks, commonly referred to as "distance fraud." The increasing availability and ubiquity of RFID tokens and readers, the forthcoming wide deployment of near-field communication (NFC) devices, and the assumptions made about location, all make distance fraud a critical concern to the integrity of these devices and systems.

RFID systems use the physical constraints of the communication channel to judge the proximity of a token. These systems assume that the token is within a certain distance as the operational range of NFC is seen as limited. Based on this assumption, RFID devices are often used to associate a token with a specific location, with a typical RFID system having trusted readers placed at known locations for this purpose. When a token is within its communication range, the reader identifies it and reports the required information to the back-end system. Because the location of the reader is fixed and known, the system determines that the token and the object or person with which it is associated are within a close range of the reader. If an attacker can fool the system into believing that a token is at one location by relaying communications from a token at another location, the communication distance assumptions of the system are violated.

Cryptographic "distance-bounding" protocols allow one or more participating principals in an authentication process to put an upper bound on the distance between them. In this chapter, we discuss relay attacks followed by a review of distance protocols along with their merits with respect to their security properties, their communication overhead, and the resources they require for implementation. We also discuss practical implementations of both relay attacks and distance-bounding protocols.

8.2   RELAY ATTACKS

In 1976 Conway [12, p. 73] described a scenario where someone who does not know the rules of chess can beat a Grandmaster. This is done by challenging two Grandmasters for a postal game of chess and playing moves from one against the other by "relay," simply by changing the "sender" information. This results in the chess novice "in the middle" either winning against one or drawing against both. Desmedt et al. [15] demonstrated that this type of attack can also be applied to challenge–response security protocols and called it "mafia fraud." The relay attack allows the attacker to fool an authentication protocol by simply passing along challenges and responses between legitimate participants. Thus, the only difficulties the attacker must overcome are timing restrictions and the practical engineering challenges of relaying signals between the participants. Relay attacks are a special case of "man-in-the-middle" (MITM) attacks, where the attacker is not required to modify the data in transit. As a result, attackers do not need to find any exploitable weaknesses in the algorithms or protocols to successfully attack the system.

RFID systems can be susceptible to relay attacks through temporarily placing a "clone" of a legitimate token close to a reader to gain the privileges associated with this token's proximity. We present an access control system using "proximity cards," shown in Figure 8.1a, as an example. As their name suggests, these cards are tokens that are read at close range, and once authenticated the system grants entry to whoever is in possession of the card. Because the authentication process receives no input from the holder of the card, an attacker can covertly place his own reader close to a card of a legitimate owner, the fake token next to the door's reader, and be granted entry, as shown in Figure 8.1b. Unless the system imposes strict timing constraints, the legitimate owner (and card) can be situated anywhere during the attack. The attacker needs two devices:

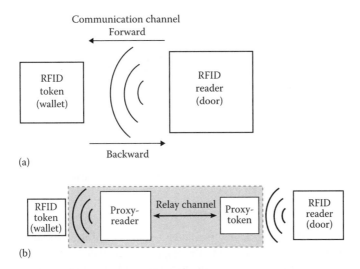

FIGURE 8.1 Example of a relay attack on a proximity access control system; using proxy devices, the attacker can violate location assumptions made by the system. (a) Ordinary communication between an RFID reader and a token. (b) An attacker uses a proxy-reader and token to mount a relay attack.

a "proxy-reader" that communicates with the legitimate card, and a "proxy-token" that communicates with the door's reader. Both proxy devices establish communication with their respective legitimate counterparts at the same time and the authentication protocol executes as normal. These proxy devices can communicate using any means necessary to meet timing restrictions, so the attacker is not restricted to any particular communication medium for the relay channel.

Because the door's reader has no way of knowing that it is actually communicating with a card that is far away, the access is granted to the attacker with the legitimate card owner unaware of the attack. Although some RFID tokens perform mutual authentication and encrypt the subsequent communication, the attacker does not need to know the plain-text data or cryptographic keys as long as he can continue relaying data back and forth. It is therefore irrelevant whether the reader authenticates the token cryptographically, because the relay attack cannot be prevented by application-layer mechanisms.

Hancke [18] practically demonstrated this attack on an ISO 14443 RFID system, commonly used for access control and ticketing applications, with a specially constructed proxy-token and proxy-reader linked by a short-range radio channel. The attacker can implement his own custom hardware for the proxy-token and reader or alternatively use existing hardware, such as NFC devices [30].

Relay attacks are a general class of attacks, and are not only applicable to RFID systems. Drimer and Murdoch [16] demonstrated such attacks on a live contact-based payment system in the United Kingdom called "Chip and PIN." The attack is shown in Figure 8.2, where fraudsters cause an unwitting cardholder to authorize a $2000 transaction by entering her PIN while thinking that she is only paying $20. The cardholder is fooled into this by being presented with a fake terminal for PIN entry, one that does not communicate with the bank, but rather, with a laptop behind the counter. This laptop is communicating wirelessly with another laptop inside of a backpack on the back of another fraudster in a shop elsewhere. This second laptop is connected to a fake card that is inserted into a real terminal; once both real and fake cards are inserted, the $2000 transaction data is relayed between the real terminal and card, and authorized by the cardholder through entry of the correct PIN. The cardholder will only find out the fraud had occurred when she examines her financial records, perhaps weeks after the fact. The authors have shown that the timing margins at the protocol level are so wide that even if fraudsters are on opposite sides of the world, or are experiencing network latency, the attack will still succeed.

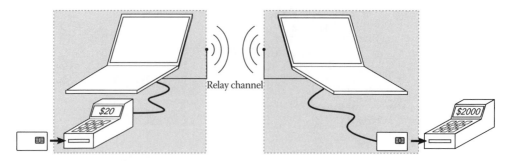

FIGURE 8.2   Drimer and Murdoch's relay attack on a contact-based payment system; shaded areas mark the equipment that is under the control of the attackers.

One of the drawbacks of the above attack is that it requires careful coordination by the fraudsters to make sure that both cards are inserted into the respective terminals at about the same time. With RFID, this is made much easier. RFID transactions seldom require user interaction, such as PIN entry, in addition to presenting the card to a reader at a close proximity. This potentially allows the attacker to activate the card several times, or skim it, without the knowledge of the cardholder. People often scan their wallet, purse, or bag containing the token, which means that an attacker never needs to reveal his hardware. This allows attackers to hide their proxy-token from the reader/merchant, and conversely, allows a corrupt reader to not only read the payment card but also read other ones within the bag or wallet. People may hold several contactless cards together, so while someone is paying for a subway fare, someone else is getting unauthorized access to their office. With contact cards, there is a fair chance that the merchant may notice that something is not right with the card, or even be required to handle it. With RFID payment this is discouraged as the whole point is to streamline the transaction such that there is no interaction with a merchant. The contact card attack remains relevant to our discussion because it demonstrates that current systems, even those used in security-sensitive environments such as banking, do not have the necessary mechanisms in place to prevent relay attacks. The only challenge left for an attacker is handling the RF communication channel between reader and token. A practical relay attack on an RFID token is described in Section 8.2.2.

8.2.1   Security Implications

There are several ways in which an attacker can benefit from a relay attack. We describe worst-case scenarios to illustrate how relay attacks may be used to circumvent security measures, even though these may not necessarily be practical or be able to bypass current fraud detection mechanisms. We predominantly use payment systems as examples so we will begin with defining a few terms. The *acquirer bank* issues *merchants* with card reader (terminals) so that they can accept card payments, while the *issuer bank* provides cardholders with the payment cards, or tokens. Cardholders initiate a transaction by inserting a token into (for contact cards), or waving it in front of (for contactless), a payment terminal at the merchant. This transaction is verified through an *operator network*, which communicates with the respective banks to authorize the transaction and informs the merchant that the goods and services can be supplied to the cardholder. This process ensures that money will eventually be transferred from the cardholder's issuer bank account to the merchant's acquirer bank account. In some of the cases, "off-line" transactions are made where the authentication process occurs in bulk at the end of the day and after the merchant already provided the goods or services.

A relay attack should not only be seen as an attack by a fraudulent third party against an honest merchant and holder. This attack can also be used by a fraudulent merchant who can set up a proxy-token at the reader supplied by the acquirer while his accomplice wanders around a busy train station sapping money from tokens using a proxy-reader. This attack could go unnoticed if the merchant conducts transactions of small value with several victims. The victims, however, may not notice a single fraudulent transaction when they check their statements because it is quite unlikely that they remember the specific stall from which they purchased a sandwich or newspaper a few weeks back. Similarly, the issuer bank or operator cannot easily distinguish this fraudulent transaction activity from regular purchasing patterns of the merchant's reader. The enterprising merchant can even use several proxy-readers sending information to a single proxy-token to allow him to have multiple "readers" without purchasing additional hardware from the acquirer, possibly circumventing expensive licensing agreements.

There have been proposals for the use of multi-application RFID tokens. A token, for example, might be required to act both as a credit/debit and a transport card with readers located in stores, or at public rail stations. In such systems, a fraudulent merchant could possibly set up a fake top-up reader to act as a proxy-reader, which then selectively relays communication to the transport authority and the debit card readers. Alternatively, it may be possible to covertly attach a small loop antenna onto the transport authority's reader, which acts as the antenna of the proxy-reader relaying information to the debit card reader. A person wishing to top-up his travel credit first enters the amount he wishes to add, and after payment he briefly touches his card to the reader for the credit to be loaded. Using the relay setup, it may be possible for the merchant to also charge the debit card during the time that the card is near the reader. The holder may not notice the extra time taken for the debit card transaction, because both transactions can be conducted before he moves the card away. To implement a suitable proxy-reader for this scenario is difficult, as it would need to modulate the forward data onto the RF carrier of the real reader. It would not be able to modulate only its own carrier because the real reader's carrier will cause interference, for example, the token will always receive a carrier even if the proxy-reader stops transmitting. As a result the proxy-reader would need to cancel the genuine carrier by transmitting a 180° phase shifted version to achieve 100 percent amplitude-shift keying (ASK) modulation. This attack could possibly be detected if the travel and payment systems look for simultaneous transactions.

A fraudulent holder can also benefit from a relay attack by setting up the attack using a proxy-reader close to his own token. This is done by creating several proxy-tokens, all of which communicate with the proxy-reader, such that each act as a virtual clone of the original. Theoretically, this allows several "cardholders" to share the same valuable token. For example, if one owner is issued with a yearly public transport pass he can issue proxy-tokens to some of his friends who would now be able to use the same transport token assuming that they do not trigger back-end fraud detection. Another advantage the owner can gain by implementing an "attack" against his own token is the ability to control the communication. The owner can therefore implement an active relay attack (MITM) and selectively modify the communication. This can possibly allow the attacker to exploit further vulnerabilities in the security protocols of the RFID system if data is not properly signed, for example.

The relay attack is not without limitations. Unless there are vulnerabilities in the security protocol, an attacker cannot modify the data he relays without being detected. Therefore, it has limited success against systems that require additional verification of the holder, or implement "two factor" authentication, such as presenting a card and providing a secret PIN. An attacker, for example, would struggle to execute the attack against RFID-enabled passports if the photo read from his "passport" does not resemble him. The attacker must also have access to the token for the full duration of his interaction with the reader. Some additional synchronization is therefore needed between the attackers to present the proxy-token to a reader at the time when the proxy-reader is within range of a suitable token.

8.2.2 PRACTICAL RELAY ATTACKS

The theory of a relay attack is straightforward, though implementing it in practice can be an engineering challenge. As a practical example, we discuss the hardware required to execute a such an attack against an RFID system adhering to ISO 14443A. As already described, a relay attack system

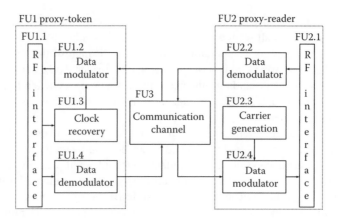

FIGURE 8.3   Functional diagram of the relay system.

consists of three main parts: a proxy-token, a proxy-reader, and a communication channel. A block diagram outlining the tasks and interactions of each "functional unit" (FU) is shown in Figure 8.3 (full details of the attack and hardware designs for ISO 14443A proxy-tokens and readers are in Hancke [19] and Kasper [29]).

8.2.2.1   Proxy-Token (Functional Unit 1)

The proxy-token communicates with the reader and behaves as a normal token. There is no need for the attacker to extend the token's operating range as the device can be held in close proximity to the reader. The token does not need to be powered by the reader and can have its own power supply. The RF interface (FU1.1) is coupled with the reader's antenna from which it receives a 13.56 MHz carrier signal. The carrier is modulated by the reader for reader-to-token communication (forward channel) and by the token for token-to-reader communication (backward channel). The antenna's resonant circuit should be tuned to 13.56 MHz and have a sufficient Q-value to allow both the forward and backward channel data to be transmitted. Q, or the quality factor, determines the peak amplitude and bandwidth of the tuning circuit's frequency response. The data modulator (FU1.2) modulates the backward channel data onto the carrier using load modulation. The modulator needs to switch the impedance of the token, thereby changing the amplitude of the carrier according to the 106 kbps Manchester-encoded data it is provided with. The modulator also requires an 847 kHz sub-carrier, which is provided by the clock recovery section (FU1.3). The data demodulator (FU1.4) recovers the forward channel data. It should therefore amplitude demodulate the incoming signal and output 106 kbps Modified Miller-encoded data.

8.2.2.2   Proxy-Reader (Functional Unit 2)

The proxy-reader communicates with the token and acts like a normal reader. The proxy-reader's operating range is determined by the distance over which it can power the token, and its ability to receive the token's answer. This range is dependent on the transmitted power in addition to the diameter and the Q-factor of the antenna [40,41]. Ideally, the attacker would try to extend the operating range to avoid detection, so the attacker might implement an extended range skimming attack [18,31]. The RF interface (FU2.1) should therefore be designed for the required operating range. The only part of the unit that needs to be covert is the antenna, as it needs to be close to the victim for a short period without being noticed. The creativeness of the implementation is left to the attacker, but an antenna can be built into a briefcase, clothes, a fake racquet, etc. The data demodulator (FU2.2) recovers the load modulated side-band data located at 13.56 MHz ± 847 kHz and outputs 106 kps Manchester-encoded data. The data modulator (FU2.4) amplitude modulates the 106 kbps Modified Miller-encoded data onto the generated 13.56 MHz carrier (FU2.3) with a modulation index of 100 percent.

8.2.2.3 Communication Channel (Functional Unit 3)

The communication channel relays data between the proxy-reader and proxy-token. The channel receives Modified Miller-encoded data as input from the proxy-token and outputs Modified Miller-encoded data to the proxy-reader. Similarly, the channel receives Manchester-encoded data as input from the proxy-reader and outputs Manchester-encoded data to the proxy-token. Additional signal processing might need to be implemented to allow for communication channel constraints between the proxy-token and proxy-reader, i.e., data buffering, adding signal delay or altering modulation and encoding schemes. The communication channel potentially causes the largest time delay in the system. It is therefore important that the channel is designed to relay the data over the required distance while staying within any time limits imposed by the RFID system.

8.2.2.4 Timing Requirements

We now discuss the timing requirements during communication as specified by the ISO 14443A standard.

ISO 14443, Part 3: The reader periodically polls for new tokens using the Type A Request command (*REQA*). The minimum time between the start bits of two consecutive *REQA* commands is specified as $7000/f_{\text{carrier}} \approx 500\,\mu\text{s}$. The token must be able to respond to the *REQA* command with a Type A Answer To Request response (*ATQA*) within 5 ms after first receiving an unmodulated carrier. These requirements do not impose an upper bound on the attack delay, because there is nothing linking a specific *ATQA* to a *REQA*. An attacker can therefore answer any of the subsequent *REQA* commands once he has determined the token's response. An attacker's response would, however, need to adhere to the frame delay time (FDT) used to ensure bit synchronization. FDT is specified as $(n \cdot 128 + 84)/f_{\text{carrier}}$ if the last data bit sent by the reader was "1" and $(n \cdot 128 + 20)/f_{\text{carrier}}$ if the last data bit sent was "0". FDT is calculated using $n = 9$ for *REQA* and *SELECT* commands, and $n \geq 9$ for all other commands. The proxy-token must therefore ensure that the start bit of the response is aligned to a valid FDT value. For $n = 9$ the reader will expect the token's response to start after 91 or 86 μs, depending on the last data bit sent by the reader. The token will only respond at those times, because it thinks that it is speaking to a real reader, which means that the relay process will have to be very quick or that the attacker will have to get some information, such as the values of the token's *ATQA* and unique identifier (UID) in advance.

ISO 14443, Part 4: The frame waiting time (FWT) specifies the time within which a token shall start its response after the end of the reader's data. FWT is defined as $(256 \cdot 16/f_{\text{carrier}}) \times 2^{\text{FWI}}$, where frame waiting integer (FWI) is a value from 0 (FWT = 300 μs) to 14 (FWT = 5 s) with a default of 4 (FWT = 4.8 ms). The value of the FWI is defined by the token in the *ATS* response. If implemented, the FWT defines an upper bound on the relay delay, so in the default case an attacker would need to relay the required data in 4.8 ms, which is a very long time when considering the capabilities of current communication systems.

Both Hancke [19] and Kasper [29] note that the systems they tested did not strictly enforce the low-level timing constraints. The exact value of n does not seem to matter, although the reader does expect the token's response to adhere to the general FDT bit-grid. The total response time of the proxy-token, which is the relay delay plus the time taken by the token to respond therefore has to be set to a multiple of 9.44 μs using an adjustable delay. A possible explanation is that the sampling clock generated by the receiver corresponds to the bit-grid defined by FDT. A response that is phase shifted relative to the sampling clock will be sampled at the incorrect time intervals, causing the data to be evaluated incorrectly. An example of the additional delay introduced by the attacker is shown in Figure 8.4.

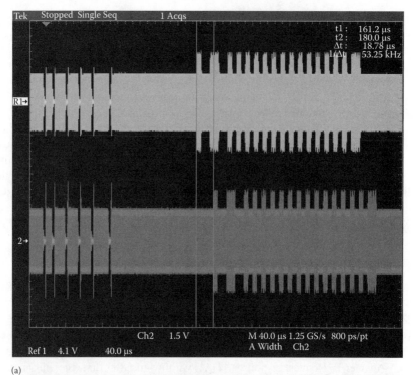

(a)

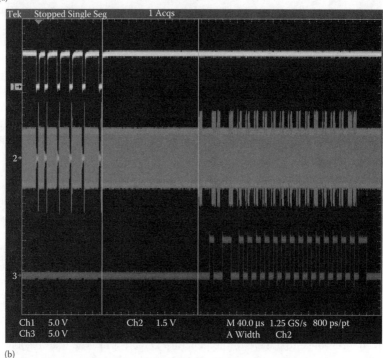

(b)

FIGURE 8.4 Delay introduced by the relay hardware. (a) Timing comparison between a token (top) and proxy-token (bottom). The tokens send an *ATQA* in response to a *REQA* command. (b) Example of a relayed data sequence at the proxy-reader. Modified Miller *REQA* command (top), carrier modulation (middle), and Manchester *ATQA* response (bottom). (From Hancke, G.P., A practical relay attack on ISO 14443 proximity cards. http://www.cl.cam.ac.uk/~gh275/relay.pdf. With permission.)

The attacker only encounters a possible problem when he participates in the anticollision procedure together with a real token. Because the real token adheres to the $n = 9$ condition, a delayed response would be detected, as the total length of the response would be greater than expected. For example, in the case where the attack delay results in $n = 11$, the reader will receive two extra bit periods of data because the relayed response starts two bit periods after the real token's response. Alternatively, the misaligned bits will be interpreted by the reader as collisions and it will be unable to select a token. Within the context of this attack, this scenario should not occur as the attacker's proxy should be the only 'token' interacting with the reader.

Even though low-level timing constraints were not enforced, the attacker is still potentially limited by higher layer time-outs. Reader configuration software in some cases allow for a time-out condition to be set for communication between the reader and the token. For the reader tested in Ref. [19], the time-out could be set from 300 μs to 76.2 ms for the *REQA* and *SELECT* commands. The default value for the time-out is set at 4.8 ms. For any further communication, the time-out could be set from 300 μs to 19.7 s, with a default value of 230 ms. Although the implemented time-outs complicate the relay attack by placing constraints on the hardware it does not prevent the attack. The additional delay introduced by the attacker in Refs. [19,29] was under 300 μs for both cases. It should therefore be feasible for an attacker with the necessary resources to implement a relay attack that is completed within 4.8 ms. The attacker can also circumvent the time-outs for the *REQA* and *SELECT* commands by getting the token's *ATQA* and *SAK* responses along with its UID. These values can then be stored in the proxy-token and sent to the reader when required without any delay [29]. In this attack scenario, the attacker would be able to execute the relay attack even if the reader enforced the $n = 9$ FDT condition. We can conclude, then, that the specified time-outs and timing constraints defined in the ISO standard do not provide adequate protection against relay attacks.

8.3 DISTANCE BOUNDING

Location provides a measure of trust with regards to security and some systems grant users privileges, or services, based on their perceived proximity. Trusted proximity measurements can enhance traditional authentication mechanisms [5] and provide additional assurance, such as a metric for secure routing in ad hoc networks [7]. Verifying the location of a device, through the use of secure protocols, has therefore become important in pervasive environments [9].

We now consider the scenario where two devices establish trust by verifying the proximity between themselves cryptographically, without the help of a trusted third party, such as a network of trusted devices at known locations. For this we define two principals: *verifier* and *prover*. The verifier is the principal that grants privileges and requires the knowledge of the distance between it and the prover who needs to prove that it is within an allowed range.

Brands and Chaum [4] were the first to address the vulnerability of cryptographic protocols to distance fraud by introducing distance-bounding protocols. These protocols allow a verifier to conclusively determine an upper bound for the physical distance between it and the prover. Secure distance-bounding protocols are integrated into the underlying communication channel and are meant to detect any extra delay in the prover's expected response. Distance-bounding protocols, if implemented correctly, can be an effective way to prevent relay attacks and distance fraud. Device proximity is also useful for mapping the topology of the network and for geographically-aware routing algorithms [28]. Distance bounding has therefore also been proposed as a protective measure for wireless networks, where relay attacks (known in this context as "wormholes attacks" [7,24]) may be used to circumvent and subvert key establishment and routing protocols [24–26].

As shown in Figure 8.5, distance bounding only involves two parties, the prover and the verifier, and allows the verifier to place an upper bound on the physical distance to the prover. In contrast, secure location services provide relative or absolute location of devices within a specific network [2,42]. A device within the network not only estimates the distance to another device but also

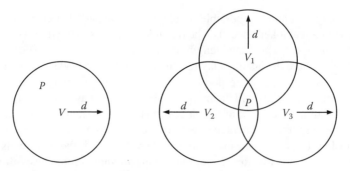

FIGURE 8.5 Location-based services provide absolute or network-relative position in relation to anchor nodes or stations; distance bounding provided relative distance information between two devices.

collaborates with additional devices, including "anchor" or base stations that provide trusted reference locations [27]. As a result, the verifier can be assisted by other devices to cross reference, repeat, and verify measurements to defend against malicious behavior [8,9,34,35], thereby protecting against relay attacks and fraudulent senders. Unlike these secure location services, the distance-bounding verifier relies exclusively on information gained from executing the protocol with the prover. Distance-bounding protocols can, however, be used as building blocks in secure localization or positioning systems. Secure distance-bounding protocols require accurate and reliable distance measurements from underlying communication channels, so the security of the protocol does not only depend on the cryptographic mechanisms, but also on how the physical attributes of the communication channel are used to measure proximity. Conventional location-finding techniques generally used for distance measurement are

- **Received-signal-strength (RSS)**: Uses the inverse relationship between signal strength and distance to estimate the distance to other nodes.
- **Angle-of-arrival (AoA)**: Examines the direction of received signals to determine the locations of transmitters or receivers.
- **Time-of-flight (ToF)**: Measures elapsed time for a message exchange to estimate the distance based on the communication medium's propagation speed.

Systems have been demonstrated that can estimate location with typical errors as small as 1.5 m, by processing RSS information from multiple base stations [2,10,32]. A trust system for RFID that is based on signal strength has also been proposed [17]. However, RSS and AoA methods are not ideal because attackers can easily alter RSS, by either amplifying or attenuating a signal, and AoA, by reflecting or retransmitting signals from a different direction.

This leaves only ToF as a possible mechanism for securely determining proximity and the method most often described in distance-bounding protocols. Both RF [42] and ultrasound channels [23,36] have been used in indoor ToF location systems. The propagation speed of sound is much slower than that of light, so it is easier to obtain high spatial resolution using simple hardware. This property, however, makes ultrasound vulnerable to a relay attack where messages are forwarded over a faster communication medium. In contrast, the propagation speed of radio waves in air approaches the speed of light and because information cannot exceed this speed, relay attacks can be resisted if this speed is assumed. In this case, an attacker will only be able to make a device appear further away by blocking a legitimate device's communication and sending a delayed version to the intended receiver.

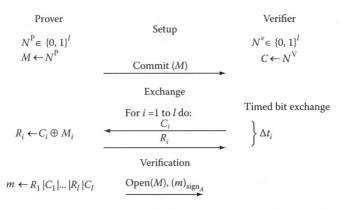

FIGURE 8.6　Brands–Chaum distance-bounding protocol.

8.3.1　ToF Distance-Bounding Protocols

In 1993, Brands and Chaum [4] described the first distance-bounding protocol based on timing the single-bit round-trip time in a cryptographic challenge–response exchange as shown in Figure 8.6. Since then several protocols have been published that allow a verifier to determine an upper-bound on the physical distance to a specific prover. The Brands–Chaum protocol, however, forms the basis for many of the newer proposals so it remains a good example with which to illustrate the general architecture of distance-bounding protocols.

Distance-bounding protocols should take into account two fundamental attacks:

- **Distance fraud**: A prover convinces a verifier that the distance between them is shorter than it really is. If a legitimate prover colludes with a third party for this purpose, this is sometimes called a "terrorist attack" [14].
- **Relay attack**: A prover convinces a verifier that it is a different prover, which is actually situated further away. This is done by relaying information from a legitimate prover and verifier pair using a fraudulent pair between them. This attack was also described as "mafia fraud" in Ref. [15] and is also known as a "wormhole" attack in the sensor network context.

If the prover gives up its response data to a third party, near the verifier, the distance bound will obviously fail. The verifier will not be able to detect that the real prover is not within the bounded distance because it cannot distinguish between the prover and the proxy-"prover". Some protocols, such as [6], try to discourage the prover from colluding by making it reveal a valuable secret if it reveals the response data. In theory, this sounds feasible but is there really such an entity as a half-fraudulent prover that selectively reveals key material? And if the prover controls the third party he should have no problem revealing a valuable secret. It is a general condition of security that the participants do not reveal their key material because doing so would compromise most protocols. Preventing collusion altogether is therefore nearly impossible, which is probably why many distance-bounding proposals do not include this attack within the threat model.

The Brands–Chaum protocol protects against both distance fraud and relay attacks. A fraudulent prover or a third party attacker that attempts to preemptively guess all the response bits R_i will succeed with probability 2^{-l}, where l is the number of bit exchanges performed. Similarly, a third party who prematurely requests R_i from the prover by supplying it with guessed challenges C_i', will succeed with probability 2^{-l} without being detected. The protocol fails if a single bit error occurs during the exchange stage because the verification signature will be incorrect. The ability to cope with bit errors during the exchange stage was addressed by later proposals, as discussed in

Section 8.3.2. The protocol does not protect against a prover colluding with an attacker because the prover suffers no penalty for releasing M to a collaborating attacker.

Similarly to the Brands–Chaum proposal, other distance-bounding protocols also consist of three primary stages:

- **Setup**: The verifier and prover prepare for the time-critical exchange stage.
- **Exchange**: Timed exchange of challenges and responses.
- **Verification**: The verifier checks the correctness of the prover's responses and uses the round-trip time to calculate the distance between them.

Each protocol can be classified by how it implements the different stages of the distance-bounding process. *Timed authentication* protocols, as first proposed by Beth and Desmedt [3], are the simplest form of ToF-based distance bounding, with the verifier timing normal data exchanges. The basic idea is to execute a challenge–response authentication protocol under a very tight time-out constraint. For example, a verifier V transmits a random n-bit nonce $N_V \in_R \{0, 1\}^n$ to the prover P, who replies with a message-authentication code $h(K, N_V)$, where h is a keyed pseudorandom function and K is a shared secret key. Numerous protocols have been proposed using different constructions for pseudorandom functions keyed with shared secrets, public-key mechanisms, or trusted third parties. The main weakness of the protocols in this category is that they cannot ensure accurate round-trip time measurements. To determine the response, the prover needs to perform calculations during the timed exchange stage, which introduces processing delays and can lead to errors in the distance estimate. This processing delay is significant when considering that a $1\,\mu s$ change in the measurement results in additional $166\,m$ in the distance estimate. To minimize these delays, secure distance-bounding protocols should use simple operations with predictable processing delays such as XOR or single-bit lookups.

In protocols using *precommitment with a bitwise XOR exchange*, like the Brands–Chaum proposal, the verifier generates a random challenge bit string, $C = (C_1, C_2, \ldots, C_l)$, although the prover generates a random response mix, $M = (M_1, M_2, \ldots, M_l)$. The prover commits to M, for example, by transmitting a collision-resistant message authentication code $h(K, M)$. The verifier then sends one C_i after another, which the prover receives as C_i'. It then instantly replies with a bit $R_i = C_i' \oplus M_i$, which is calculated by XOR-ing each received challenge bit with the corresponding bit of M. Finally, the prover reveals M and authenticates C'. The commitment on M is needed to prevent the prover from sending a random bit R_i early and then setting $M_i = C_i' \oplus R_i$ after receiving C_i'. Authenticating C' keeps attackers from sending fake C_i bits prematurely to the prover to learn bits of M_i for responding early to the verifier.

Within *precomputed table lookup* protocols, the verifier generates a random challenge bit string $C_1, C_2, \ldots, C_l$ and a nonce N_V, which is sent to the prover. The prover responds with its nonce N_P. Both the prover and the verifier then use a pseudorandom function h and a shared key K to calculate two n-bit sequences M^1 and M^2:

$$\left(M_1^1, M_2^1, M_3^1, \ldots, M_l^1, M_1^2, M_2^2, M_3^2, \ldots, M_l^2\right) := h_K(N_V, N_P)$$

The prover's reply bit $R_i = M_i^{C_i'}$ to each C_i' received from the verifier is the result of a 1-bit table lookup in M^1 or M^2, selected by the received challenge bit C_i' (for $1 \leq i \leq l$). The verifier checks whether at least k of the l R_i' bits that it receives match its locally calculated $R_i^{C_i}$ values. The values k and l are security parameters.

8.3.2 DISTANCE-BOUNDING PROTOCOLS FOR RFID

In this section, we discuss three protocols in detail that were proposed for use in the RFID environment. For an overview of further distance-bounding protocol proposals in literature, please see Ref. [20].

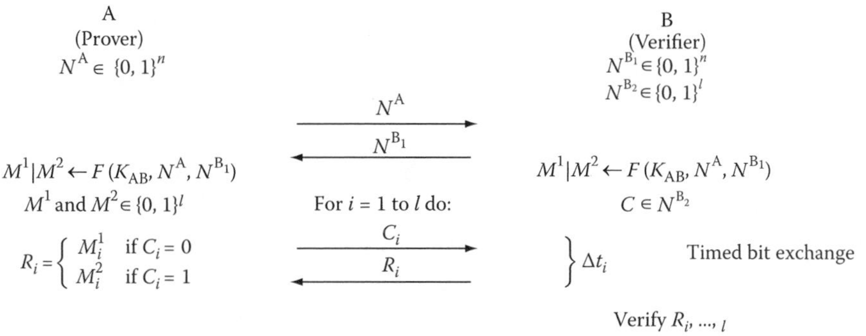

A
(Prover)
$N^A \in \{0, 1\}^n$

B
(Verifier)
$N^{B_1} \in \{0, 1\}^n$
$N^{B_2} \in \{0, 1\}^l$

$$N^A \longrightarrow$$

$$N^{B_1} \longleftarrow$$

$M^1 | M^2 \leftarrow F(K_{AB}, N^A, N^{B_1})$
M^1 and $M^2 \in \{0, 1\}^l$

$M^1 | M^2 \leftarrow F(K_{AB}, N^A, N^{B_1})$
$C \in N^{B_2}$

For $i = 1$ to l do:

$$R_i = \begin{cases} M_i^1 & \text{if } C_i = 0 \\ M_i^2 & \text{if } C_i = 1 \end{cases}$$

$$C_i \longrightarrow$$
$$R_i \longleftarrow$$

$\Big\} \Delta t_i$ Timed bit exchange

Verify $R_i, ..., _l$

FIGURE 8.7   Hancke–Kuhn distance-bounding protocol.

8.3.2.1 Hancke and Kuhn

Hancke and Kuhn [22] proposed a protocol, shown in Figure 8.7, which minimizes data transmission and is resistant to bit errors in the exchange stage. This protocol uses a 1-bit lookup table, a concept also proposed by Bussard and Bagga [6], and precomputation instead of commitment. The authors assume that the exchange stage is performed using a "fast" channel and that other communication is sent over a "slow" error-corrected channel. The "fast" channel provides adequate timing resolution for distance bounding and uses modulation techniques resistant to relay attacks, but at the same time it is susceptible to bit errors.

The verifier generates a random nonce N^{B_1} of length n and a challenge bit string N^{B_2} of length l. The nonce is then transmitted to the prover and the bit string kept as the challenge. The prover and verifier calculate $F(K_{AB}, N^{B_1})$, where F is a pseudorandom function, and split the result into two response strings M^1 and M^2, both of length l. If the prover has the ability to generate a nonce of length n, the prover can send N^A to the verifier and the response strings can be determined by calculating $F(K_{AB}, N^{B_1}, N^A)$. This is also required if it is possible for a proxy-verifier to run the protocol with the prover twice, thereby recovering both M^1 and M^2, before the verifier starts its exchange stage. Because the verifier now knows M^1, M^2, and C, it can calculate the expected R_i values. The verifier then transmits one challenge bit C_i at a time (for all $i = 1, \ldots, l$), to which the prover responds immediately with $R_i = M_i^1$ if $C_i = 0$ and $R_i = M_i^2$ if $C_i = 1$. The verifier times the round-trip delay between sending each bit C_i and receiving the corresponding response bit R_i. There is no data transmitted during the verification stage. The verifier only checks whether at least k of the l response bits that it received match the expected R it calculated earlier. By specifying threshold k the protocol still works even if $l - k$ bit errors occur. The values k and l are security parameters and should be chosen to ensure acceptable false acceptance and false rejection probabilities. F should be chosen such that it can generate an output of length $2l$. This protocol protects against distance fraud and relay attacks. The main weakness is that an attacker can send his own challenge before the verifier and recover half of the response strings' content. This means that the attacker can guess the correct reply R_i with probability of 3/4.

8.3.2.2 Munilla et al.

Munilla et al. [37] proposed a protocol, similar to Hancke–Kuhn, that aims to provide additional security during the exchange stage by randomizing the time at which the verifier sends the challenge bit. The protocol is shown in Figure 8.8. The exchange stage is split into l time slots and the prover knows in which slot the verifier will issue a challenge. This discourages an attacker from gathering response string data, by sending a challenge before the verifier, because the prover could detect the attack if it receives an unexpected challenge. This means that the attacker can only guess the correct

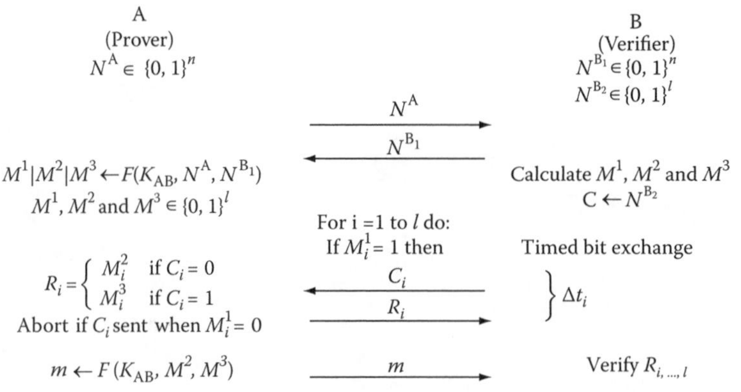

A
(Prover)
$N^A \in \{0, 1\}^n$

B
(Verifier)
$N^{B_1} \in \{0, 1\}^n$
$N^{B_2} \in \{0, 1\}^l$

$\xrightarrow{\quad N^A \quad}$

$\xleftarrow{\quad N^{B_1} \quad}$

$M^1 | M^2 | M^3 \leftarrow F(K_{AB}, N^A, N^{B_1})$
M^1, M^2 and $M^3 \in \{0, 1\}^l$

Calculate M^1, M^2 and M^3
$C \leftarrow N^{B_2}$

For i =1 to l do:
If $M^1_i = 1$ then

Timed bit exchange

$R_i = \begin{cases} M^2_i & \text{if } C_i = 0 \\ M^3_i & \text{if } C_i = 1 \end{cases}$

$\xleftarrow{\quad C_i \quad}$

$\left. \begin{array}{c} \\ \\ \end{array} \right\} \Delta t_i$

Abort if C_i sent when $M^1_i = 0$

$\xrightarrow{\quad R_i \quad}$

$m \leftarrow F(K_{AB}, M^2, M^3)$

$\xrightarrow{\quad m \quad}$

Verify $R_{i, ..., l}$

FIGURE 8.8   Void-challenge distance-bounding protocol.

reply R_i with a probability of 1/2. The prover does not know the value of the challenge, only the time slot, so it cannot commit distance fraud by preemptively sending a response.

The verifier generates a random nonce N^{B_1} of length n and a random bit string N^{B_2} of length l. The nonce is then transmitted to the prover and the bit string kept as the challenge. The prover generates a random nonce N^A of length n and transmits it to the verifier. Both parties then calculate $F(K_{AB}, N^{B_1}, N^A)$, where F is a pseudorandom function, to determine three bit strings M^1, M^2, and M^3, each with length l. M_1 indicates the time slots in which the challenges will be sent while M_2 and M_3 are the response strings. The exchange stage is broken up into l time-slots and the verifier will transmit a challenge bit C_i only during time slot numbered i if $M^1_i = 1$. The prover responds immediately with $R_i = M^2_i$ if $C_i = 0$ and $R_i = M^3_i$ if $C_i = 1$. The verifier times the round-trip delay between sending each bit C_i and receiving the corresponding response bit R_i. During verification the prover confirms the values it calculated for M^2 and M^3. The pseudorandom function should be chosen such that it generates an output of length $3l$.

For this protocol to work, the communication channel would need three symbols, which would allow the prover to distinguish between challenge "0," challenge "1," and no challenge. Only aborting if the prover receives $C_i = 1$ when $M^1_i = 0$, in other words when the symbol for no challenge is the same as for challenge "0," is not sufficient. In this case, a prover will not be able to detect an attacker's challenge if he sent a "0" in a time slot where a challenge is not expected. This means that the attacker would be able to gather all of M^2 without being detected, by preemptively challenging the prover with only "0" challenges during every time slot. This protocol protects against distance fraud and relay attacks. This protocol can be made resistant to communication errors in a similar way to the previous protocol by specifying acceptance thresholds in the prover and verifier.

8.3.2.3   Reid et al.

Reid et al. [39] proposed a protocol that discourages collusion, also known as the "terrorist attack," assuming both parties have the ability to perform symmetric encryption. The protocol is shown in Figure 8.9. This protocol is based on the protocol proposed by Bussard–Bagga [6], although in this proposal no public-key cryptography is required. As is the case in the Hancke–Kuhn protocol there is no data is exchanged during the verification stage.

The verifier generates a random nonce N^{B_1} of length n and a random bit string N^{B_2} of length l. The nonce is then transmitted to the prover and the bit string kept as the challenge. The prover generates a random nonce N^A of length n and transmits it to the verifier. Both parties then calculate $F(K_{AB}, N^{B_1}, N^A)$, where F is a pseudorandom function, to determine M^1 with length l. M^2 is determined by encrypting K_{AB} with M^1, which means that if a prover reveals the response strings he

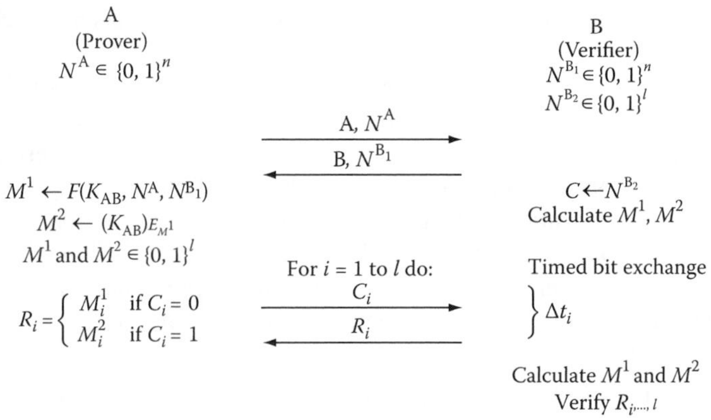

$$\begin{array}{ll}
\text{A} & \text{B} \\
\text{(Prover)} & \text{(Verifier)} \\
N^A \in \{0, 1\}^n & N^{B_1} \in \{0, 1\}^n \\
& N^{B_2} \in \{0, 1\}^l
\end{array}$$

$$\xrightarrow{\quad A, N^A \quad}$$

$$\xleftarrow{\quad B, N^{B_1} \quad}$$

$$M^1 \leftarrow F(K_{AB}, N^A, N^{B_1}) \qquad\qquad C \leftarrow N^{B_2}$$
$$M^2 \leftarrow (K_{AB})_{E_{M^1}} \qquad\qquad \text{Calculate } M^1, M^2$$
$$M^1 \text{ and } M^2 \in \{0, 1\}^l$$

For $i = 1$ to l do: Timed bit exchange

$$R_i = \begin{cases} M_i^1 & \text{if } C_i = 0 \\ M_i^2 & \text{if } C_i = 1 \end{cases} \qquad \xrightarrow{\quad C_i \quad} \quad \Big\} \Delta t_i$$

$$\xleftarrow{\quad R_i \quad}$$

Calculate M^1 and M^2
Verify $R_{i\cdots l}$

FIGURE 8.9 "Terrorist"-resistant distance-bounding protocol.

also reveals his valuable shared key K_{AB}. The verifier then transmits one challenge bit C_i at a time (for all $i = 1, \ldots, l$), to which the prover responds immediately with $R_i = M_i^1$ if $C_i = 0$ and $R_i = M_i^2$ if $C_i = 1$. The verifier times the round-trip delay between sending each bit C_i and receiving the corresponding response bit R_i. During the verification stage the verifier checks whether the prover's responses matches the expected values it calculated locally. The protocol also allows an attacker to send his own challenge and recover half of the response strings' content before the verifier starts the exchange stage. This protocol protects against distance fraud and relay attacks while discouraging the prover to participate in a "terrorist attack." This means that the attacker can guess the correct reply R_i with a probability of 3/4.

The authors state that their protocol can be made resistant to bit errors by using the same method used for the Hancke–Kuhn protocol. If this is the case, the verifier checks whether at least k of the l response bits that it received matches the expected R it calculated, so the protocol still works even if $l - k$ bit errors occur. The values k and l should be chosen to ensure acceptable false acceptance and false rejection probabilities. The verifier only checks whether at least k of the l response bits that it received matches the expected R it calculated. By specifying threshold k the protocol still works even if $l - k$ bit errors occur. If the verifier allows for bit errors, the prover might be able to collaborate in a "terrorist attack" without revealing K_{AB}. In this case, the prover deliberately changes bits in M^1 and M^2, which will complicate an attempt to recover K_{AB} by decrypting M^2 with M^1. The prover changes enough bits to make a brute-force search on K_{AB} difficult without causing more than $l - k$ bit errors in the third-party attacker's response.

8.4 DISTANCE-BOUNDING CHANNELS

Time-of-flight distance-bounding protocols must be integrated into the physical layer of the communication channel to accurately determine the distance between the prover and verifier. This means that the security of the distance bound depends not only on the cryptographic protocol itself but also on the practical implementation and the physical attributes of the communication channel. The communication channel used for the exchange must therefore not introduce any latency that the attacker can exploit to circumvent the physical distance bound. Any latency added for signal processing, or signals traveling at velocities slower than in other channels, provides an attacker an opportunity to seem closer than he really is. For example, using ultrasound as a channel is not effective because the propagation speed is much slower than that of radio waves, leaving the system vulnerable to relay attacks if the attacker manages to transmit the signals using that "faster" medium. Despite the importance of the implementation of distance bounding on a suitable exchange channel, we found that it is not often considered in the literature.

Latency at the packet level (i.e., data format), the physical layer (i.e., coding and modulation), or the communication channel can be exploited by the attacker to gain a time advantage during the distance bound [11,21]. These communication channel vulnerabilities undermine the security of distance-bounding protocols. Conventional RF channels, often found in RFID and sensor networks, are therefore not suitable for implementing secure distance bounding. For this reason, special consideration must be given to the communication channel used for distance bounding, and the designer must include any potential vulnerabilities into the final distance-bound estimate. Ideally, distance bounding should be implemented using a specially designed channel.

8.4.1 RELAY-RESISTANT COMMUNICATION CHANNELS

Conventional communication channels are designed for reliable data transfer. Channels allow for redundancy and timing tolerances to prevent bit errors, but unfortunately, this also introduces latency that an attacker can exploit. System designers planning to use distance-bounding protocols must therefore implement special low-latency channels. Proposals for the implementation of distance-bounding channels are currently confined to the HF RFID environment [22,37,39], although there are also a few proposals for creating unforgeable RF channels that could prevent relay attacks [25,33,38].

8.4.1.1 Unforgeable Channels

Several proposals attempt to construct unforgeable channels to prevent relay attacks. The basic principle in this case is that the proxy-prover will not be able to impersonate the real prover if he cannot exactly replicate the communication channel. Alkassar et al. [1], for example, suggest that channel-hopping radio is difficult to track and thus difficult to relay. The verifier can also try to uniquely identify the prover by using the physical characteristics of the channel. Rasmussen and Čapkun [38] propose that a verifier can construct a unique "fingerprint" for each prover by using the attributes of the received RF signal. In another proposal, DeJean and Kirovski [13] suggest that a prover can identify itself by intentionally making its channel characteristics unique. This is achieved by placing a random constellation of conductive or dielectric objects within the token, which would alter the near-field response of a token when exposed to RF signals. Neither of these methods provides any accurate proximity information apart from "in communication range", nor do they protect against a fraudulent prover.

Further proposals hide additional information within the transmitted data. Hu et al. [25] suggest adding geographical information, referred to as "packet leashes," to transmitted packet data. This method, however, requires the verifier to know its location, which disqualifies it for two-party distance bounding as it requires collaboration with additional parties. Kuhn [33] proposes that the prover transmits a hidden "watermark" along with the data, which is subsequently revealed so that the verifier can retroactively check whether the data it received was transmitted by the prover. This method was suggested for global positioning system where the sender is trusted, and therefore it does not protect against a fraudulent prover.

8.4.1.2 Direct Carrier Sampling

There are two proposals for a distance-bounding channel where the verifier directly samples the modulated carrier. This means that the verifier could determine the prover's response without performing demodulation and decoding, thus reducing communication channel latency. Both proposals are tailored to the HF RFID environment and depend on the load modulation process, which allows the token to amplitude modulate the carrier transmitted by the reader.

In the proposal by Munilla et al. [37], the reader transmits a periodic sequence of pulses that are 100 percent ASK modulated onto the carrier. The pulses act as synchronization bits with the periods in between, when the carrier is off, referred to as slots. In some slots, the reader will switch on the carrier for a short period of time to indicate that it wants a response. The token knows when to expect these requests and preemptively switches its impedance to indicate the answer. Then, when the reader switches on the carrier, the envelope of the signal rises immediately to a level that indicates the token's answer state. An example of how the channel works is shown in Figure 8.10a. The reader measures the time from the point when it switches on the carrier until the token's response can be determined. To determine the token's response, the reader continuously samples the envelope of the carrier until it finishes rising and becomes stable. Once the envelope reaches this steady state, the verifier checks the amplitude level to see whether load modulation is on or off. The timing process is shown in Figure 8.10a.

The time it takes for the two levels to be reliably distinguished, and the difference between the envelope amplitude for the two states, depends on the distance between the token and the reader. The authors state that the timing resolution of the channel is less than 1 μs. Because the token knows when the reader will issue a challenge, and is in fact expected to respond preemptively, this implementation does not allow for the prevention of distance fraud. The token would also need to be protected against a proxy-reader transmitting a weak carrier, which appears to the token to be "off," to probe the state of the load early. Another practical drawback is that the carrier is switched off regularly, which means that the token has no source of power for long periods of time.

The proposal by Reid et al. [39] assumes that the token will reply after a fixed time t_{wait}. In practice the token waits for a predetermined number of cycles of the 13.56 MHz carrier, which would synchronize its response to an accuracy of $1/13.56\,\text{MHz} = 75\,\text{ns}$. The reader measures the time from the end of its command to the moment that the response is detected, with the distance-bounding time measurement then taken as $t_{\text{m}} - t_{\text{wait}}$. An example of a challenge–response sequence is shown in Figure 8.11a. The time at which the response is received is measured using a special detector that tries to determine the exact moment that the amplitude of the carrier is first modulated. This involves sampling the peaks of the HF carrier and comparing the latest sample to a threshold calculated from the eight previous samples. The resolution of the system is once again dependent on the distance between the token and the reader, with the authors stating that a 300 ns resolution was obtained when the token and the reader were 4–5 cm apart.

This channel could be vulnerable to distance fraud if the prover does not wait t_{wait} and transmits its response preemptively. The authors also state their assumption that the token is protected against overclocking and that the RF carrier operates within the ±7 kHz tolerance specified by the relevant standard. This does not seem to be a valid assumption for tokens currently available as it was practically demonstrated that these tokens can be overclocked [21]. If the token is overclocked it

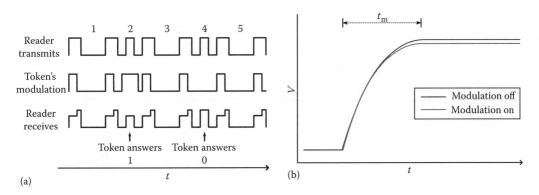

FIGURE 8.10 The void-challenge distance-bounding channel. (a) Example of a bit exchange sequence. (b) Timing of a single bit exchange.

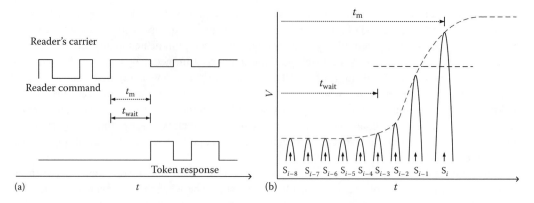

FIGURE 8.11 Accurately timing the token's response by early modulation detection. (a) Example of a challenge–response sequence. (b) Timing the start of the token's response.

will respond with the correct answer earlier, which will give the attacker additional time to relay the response back to the prover.

8.4.1.3 Wideband Pulses

Hancke and Kuhn proposed a crude ultrawideband channel for near-field systems [22] that adheres to the principals of secure distance bounding defined in Ref. [11]. Making the bit period as short as possible would limit the attacks described in this chapter, although this requirement might compromise the reliability of the channel and result in bit errors. If this is the case, the distance-bounding protocol would need to allow for these errors during the timed exchange stage. The reader and the token use the 13.56 MHz carrier for loose synchronization and the response is sent immediately after receiving the challenge using an asynchronous circuit which limits the effect of overclocking attacks. A challenge and response bit exchange occurs on each rising edge of the carrier, as shown in Figure 8.12a.

The reader starts timing on the zero-crossing of the carrier, waits for t_t, and then transmits the challenge bit C_i. The token also waits for the zero-crossing of the carrier before it starts the sampling process. The sampling time t_s is fixed and dependent on the token's hardware implementation. The reader tries to ensure that the token samples C_i' correctly by adjusting delay $t_t \approx t_s$, essentially aligning

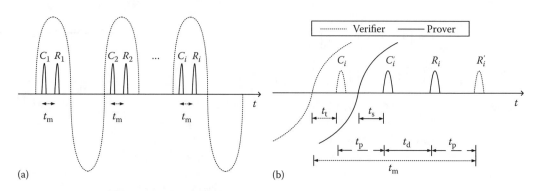

FIGURE 8.12 Ultrawideband pulse exchange using carrier synchronization. (a) Example of a bit-exchange sequence. (b) Timing of a single-bit exchange.

the challenge bit period with the time the token samples. By varying the delay t_t during the first few values of i until a delay has been found that results in the correct response bits, the reader can adjust itself automatically to any component tolerances and instabilities that may affect the exact sampling time in the token. After a brief processing delay t_d, the prover transmits a response bit R_i. After time t_m, the reader samples the channel to determine R_i'.

In a very simple implementation, the reader is equipped with two adjustable delay circuits. The first delay t_t is used to position the challenge bit such that the token has the best chance to sample the challenge bit correctly. The second delay element t_m is used to time the moment after the zero-crossing when the reader has the best chance to sample the incoming R_i bit correctly. It is up to the reader to repeat the protocol and try different values for t_t and t_m until R_i' matches the expected result well. The total number of bits exchanged n should be chosen large enough such that enough bits remain to satisfy the security requirement of the challenge–response phase after the delay-element adjustment phase. In a more sophisticated implementation, the verifier samples a response for multiple delays that are of interest, and then searches in the recorded results for the lowest value t_m with an acceptable response.

In neither case are high clock-frequency circuits nor precise reference frequencies, needed in the token. The timing of a single bit exchange is shown in Figure 8.12b. The propagation time t_p can then be calculated by the reader as follows: $t_p = (t_m - t_t - t_d)/2$. As with all the other channels presented here, a prover could commit distance fraud if it managed to decrease the expected t_d. Minimizing t_d limits the amount of time the attacker could gain. Drimer and Murdoch [16] used a similar technique to implement distance bounding to prevent relay attacks against contact smart cards, as explained in Section 8.4.2.

8.4.2 EXAMPLE IMPLEMENTATION OF DISTANCE BOUNDING

In Sections 8.3.1 and 8.3.2, we described several distance-bounding protocols for RFID; to the best of our knowledge, however, none have been securely implemented in practice. We now describe an implementation of the Hancke–Kuhn protocol adapted to contact-based smartcards and ISO 7816 by Drimer and Murdoch [16]. This implementation was proposed as a secure solution to the relay attack on contact-based payment cards we outlined in Section 8.2.

Smartcards (provers) are low resource devices so additions to their circuitry should be minimized to keep their cost low. Payment terminals (verifiers), on the other hand, are costly devices and can accommodate moderate changes and additions without adversely affecting adoption or market appeal. With that in mind, the frequency of the smartcard's operation is maintained at 1–5 MHz per ISO 7816, while the terminal is required to operate rapid bit exchange circuitry at higher frequencies to obtain a high distance resolution.

Table 8.1 lists the signals of the design followed by the circuit diagram shown in Figure 8.13 and the signal waveforms in Figure 8.14. The goal is to allow the verifier to securely determine the maximal distance the prover is away and to account for any timing advantages the attacker may be able to obtain. There is an implicit assumption that potential attackers do not have access to the internal operation of the terminal and that extracting secret material out of the smartcard itself, or interfering with its internal operation, is not economical or possible. Otherwise, we must assume that the attacker is very capable, being able to transmit signals at the speed of light and overcome signal integrity issues associated with transmitting signals over the relay channel.

The verifier's clock frequency, f_V, determines the distance resolution it can obtain, and because signals cannot travel faster than the speed of light, c, the upper-bound distance resolution is c/f_V. Thus, for example, 200 MHz would provide a 1.5 m resolution, meaning that we can determine distance at multiples of this unit. For the smartcard, ISO 7816 specifies an operational frequency of 1–5 MHz; this frequency is maintained in the implementation by using delay lines that create copies of the clock received by the prover, which is shifted by t_p. This allows creating two consecutive rising edges, effectively simulating a very fast clock without actually needing one on the device.

TABLE 8.1

Signals and Their Associated Timing Parameters

Symbol	Description
CLK_V, f_V	Verifier's clock and frequency; determines the distance resolution
$CLK_{V \to P}$, f_P	Prover's clock and frequency; received from verifier
DRV_C	While asserted the challenge is transmitted
t_n	Length of time verifier drives the challenge on to the I/O
$SMPL_C$	Prover samples challenge on rising edge
t_m	Length of time between assertion of DRV_C to assertion of $CLK_{V \to P}$
DRV_R	Prover transmits response
t_p	Amount of delay applied to $SMPL_C$
$SMPL_R$	Verifier samples response on rising edge
t_q	Time from assertion of $CLK_{V \to P}$ to rising edge of $SMPL_R$; determines upper bound of prover's distance
t_d	Propagation delay through distance d

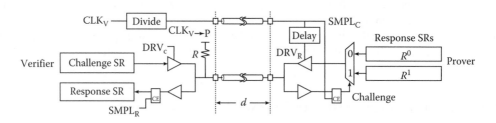

FIGURE 8.13 Simplified diagram of the distance-bounding circuit. DRV_C controls when the challenge is put on the I/O line. CLK_V controls the verifier's circuit; it is divided and is received as $SMPL_C$ at the prover where it is used to sample the challenge. A delay element produces DRV_R, which controls when the response is put the I/O, while at the verifier $SMPL_R$ samples it. The pull-up resistor R is present to pull the I/O line to a stable state when it is not actively driven by either side.

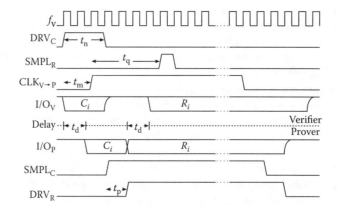

FIGURE 8.14 Waveforms of a single bit-exchange of the distance-bounding protocol. f_V is the verifier's clock; DRV_C drives the challenge on to I/O; $SMPL_R$ samples the response; $CLK_{V \to P}$ is the prover's clock; I/O_V and I/O_P are versions of the I/O on each side accounting for the propagation delay t_d; $SMPL_C$ is the received clock that is used to sample the challenge; and DRV_R drives the response on to the I/O.

The design has four shift registers (SRs): the verifier's challenge and received response registers and the prover's two response registers. The challenge SR is clocked by CLK_V and is shifted one clock cycle before it is driven on to the I/O line by DRV_C. The verifier's response register is also clocked by CLK_V and data is shifted in on the rising edge of $SMPL_R$. On the prover side, the registers are clocked and shifted by $SMPL_C$.

The verifier and prover communicate using a bidirectional I/O with tristate buffers at each end. These buffers are controlled by the signals DRV_C and DRV_R and are implemented such that only one side drives the I/O line at any given time to prevent contention (the situation when both ends try to simultaneously assert complementary signals onto the I/O). This is a consequence of adapting the Hancke–Kuhn protocol to a wired medium, and implies that the duration of the challenge must be no longer than necessary, so as to obtain the most accurate distance bound. A pull-up resistor is also present, as with the ISO 7816 specification, to maintain a high logic state when the line is not driven by either side. As a side note, if the constraints imposed by ISO 7816 were not crucial for the application, two unidirectional wires for the challenge and response could be used for easier implementation with tighter distance bounds.

8.4.2.1   Operation

The first operation is driving the challenge on to the I/O line by DRV_C for t_n, which is a multiple of f_V period. The verifier then sends a clock edge $CLK_{V \rightarrow P}$ to the prover t_m after the rising edge of DRV_C. Both $CLK_{V \rightarrow P}$ and the I/O line have the same propagation delay t_d due to the distance d between the verifier and prover. When the clock edge arrives to the prover (now the signal is called $SMPL_C$), it is used to sample the challenge into a register. The same clock edge also shifts the two response registers, one of which is chosen by a 2:1 multiplexer that is controlled by the sampled challenge. DRV_R is a delayed replica of $SMPL_C$, which is created using a delay-line element of t_p allowing the response SR signals to shift and propagate through the multiplexer, preventing the intermediate state of the multiplexer from being leaked through the I/O. Without this delay, the attacker could discover both responses to the previous challenge in the case where $C_i \neq C_{i-1}$. When DRV_R is asserted, the response is being driven on to the I/O line.

At the verifier, the response is sampled by $SMPL_R$ after t_q from the assertion of $CLK_{V \rightarrow P}$. The value of t_q determines the distance measured and should be long enough to account for the propagation delay that the system was designed for (including on-chip and package delays), and short enough to not allow an attacker to be further away than desired, with the minimum value being $t_p + 2t_d$. With a single iteration, the verifier can discover the prover's maximum distance away according to its time measurement setting. With multiple iterations, however, more accurate measurement can be made by changing this setting from high to low until the exchange fails, yielding the minimum measurable distance with the resolution governed by the operating frequency.

The protocol was implemented on a Xilinx Virtex II Pro field programmable gate array (FPGA) development board, and Figure 8.15 shows oscilloscope traces of a single bit challenge–response exchange over a $50\,\Omega$, 30 cm printed circuit board transmission line. In this case, the challenge is 1 and the response is 0 with indicators where $SMPL_R$ has sampled the response. The first, after $t_{qfail} = 15$ ns has sampled too early while the second, $t_{qpass} = 20$ ns, which is a single period of f_V later, has correctly sampled the 0 as the response. The delay $t_d = 2.16$ ns, can also be seen and is, of course, due to the length of the transmission line. If the attacker exploited all possible attacks previously discussed and was able to transmit signals at c, he would need to be within approximately 6 m, although the actual distance would be much shorter for a less than ideal attacker, roughly confined to a space of a room.

An attacker can initiate the rapid bit exchange after the setup phase, randomly guessing the challenge bits and, on average, be correct for half of them. For challenge bits guessed incorrectly, the attacker can randomly guess the response and, again, get half of those correctly. Therefore, we can expect an average success rate of 3/4 per bit. Assuming no errors are tolerated in the response, for

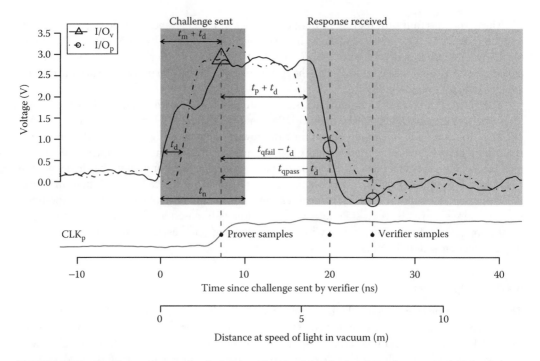

FIGURE 8.15 Oscilloscope trace of a single-bit exchange of the distance-bounding protocol. Delay is introduced by a 30 cm transmission line between the verifier and prover. Timing parameters are $t_n = 10$ ns, $t_m = 5$ ns, and $t_p = 8$ ns. Two values of t_q are shown, one where the bit was correctly received $t_{q_{pass}} = 20$ ns and one where it was not, $t_{q_{fail}} = 15$ ns. t_d was measured to be 2.16 ns which over a 30 cm wire corresponds to propagation velocity of 1.4×10^8 m/s. Note that before the challenge is sent, the trace is slowly rising above ground level; this is the effect of the pull-up resistor as also seen in (a) after the protocol completes. The shown signals were probed at the I/Os and do not precisely represent when they actually appear inside the FPGA. For example, the I/O introduces 3–5 ns delay to the signal so in actuality the logic will "see" the falling edge slightly after what is represented in the figure.

a 64-bit exchange the probability of success is then $(3/4)^{64} \approx 1$ in 2^{26}, and of course, the size of the registers can be adjusted to reduce this probability. A replay attack is resisted by a mutual exchange of nonces so the attacker cannot run the protocol twice with complementary challenge vectors. If the attacker can invest in the required equipment, it would be possible for him to sample the verifier's challenge bit earlier than the prover would gain a slight timing advantage. Similarly, superior equipment may allow the attacker to get the response just before the sampling time, gaining an advantage. These timing advantages should be taken into account according to potential adversaries, with the maximum timing budget for the attacker being t_{m+q}.

8.5   RESEARCH DIRECTIONS AND FUTURE WORK

We have seen that several deployed systems are vulnerable to relay attacks and location fraud. The main solution for secure proximity identification we focused on was cryptographic distance bounding, which allows the principle involved in an authentication process to establish an upper bound on the physical distance between them. Currently there are a number of distance-bounding protocol proposals, of which we briefly discussed those specifically tailored to RFID systems. However, none of these protocols have been implemented and demonstrated to be secure for RFID systems.

The main technical challenge in implementation is the integration of the cryptographic protocol with the physical layer of the underlying communication channel, which would allow for accurate, and secure, distance measurements. Conventional communication channels have been shown to be vulnerable to distance fraud. As a result, designers must give additional consideration to the channels used for distance bounding and allow for any potential vulnerabilities in the final distance-bound estimate. Ideally, distance bounding should therefore be implemented using a specially designed channel. Distance-bounding channels for RFID systems have been proposed, but not yet to be properly implemented. We believe that the next logical research task is to follow the successful implementation of distance bounding for contact tokens to a secure implementation for RFID devices.

REFERENCES

[1] A. Alkassar, C. Stuble, and A. Sadeghi. Secure object identification: Or solving the chess grandmaster problem. *Proceedings of New Security Paradigms Workshop*, pp. 77–85, 2003.

[2] P. Bahl and V.N. Padmanabhan. RADAR: An in-building RF-based user location and tracking system. *Proceedings of Nineteenth Annual Joint Conference of the IEEE Computer and Communications Societies*, pp. 775–784, 2000.

[3] T. Beth and Y. Desmedt. Identification tokens—or: Solving the chess grandmaster problem. *Proceedings of Advances in Cryptology (CRYPTO), LNCS*, Springer-Verlag, 537, pp. 169–177, 1990.

[4] S. Brands and D. Chaum. Distance bounding protocols. *Advances in Cryptology, EUROCYPT '93, LNCS*, Springer-Verlag, 765, pp. 344–359, 1993.

[5] L. Bussard and Y. Roudier. Embedding distance-bounding protocols within intuitive interactions. *Security in Pervasive Computing: First International Conference, LNCS*, Springer-Verlag, 2802, pp. 143–156, 2004.

[6] L. Bussard and W. Bagga. Distance-bounding proof of knowledge to avoid real-time attacks. *Proceedings of IFIP International Information Security Conference*, pp. 223–238, 2005.

[7] S. Čapkun, L. Buttyán, and J. Hubaux. SECTOR: Secure tracking of node encounter in multi-hop wireless networks. *Proceedings of ACM Workshop on Security in Ad Hoc and Sensor Networks (SASN)*, ACM Press, 2003.

[8] S. Čapkun, M.C. Čagalj, and M. Srivastava. Securing localization with hidden and mobile base stations. Technical Report, NESL, UCLA, 2005. http://www.syssec.ethz.ch/research/TechrepCovert.pdf

[9] S. Čapkun and J.P. Hubaux. Secure positioning in wireless networks. *IEEE Journal on Selected Areas in Communications: Special Issue on Security in Wireless Ad Hoc Networks*, 24(2), 221–232, 2006.

[10] S. Čapkun, M. Srivastava, M. Čagalj, and J. Hubaux. Securing positioning with covert base stations. NESL, UCLA Technical Report TR-UCLA-NESL-200503-01, 2005. http://lcawww.epfl.ch/capkun/spot/

[11] J. Clulow, G.P. Hancke, M.G. Kuhn, and T. Moore. So near and yet so far: Distance-bounding attacks in wireless networks. *European Workshop on Security and Privacy in Ad-Hoc and Sensor Networks (ESAS), LNCS*, Springer-Verlag, 4357, pp. 83–97, 2006.

[12] J.H. Conway. *On Numbers and Games*. Academic Press, 1976.

[13] G. DeJean and D. Kirovski. RF-DNA: Radio-frequency certificates of authenticity. *9th International Workshop Cryptographic Hardware and Embedded Systems (CHES 2007), LNCS*, Springer-Verlag, 4727, pp. 346–363, 2007.

[14] Y. Desmedt. Major security problems with the 'unforgeable' (Feige)-Fiat-Shamir proofs of identity and how to overcome them. *Proceedings of SecuriCom*, pp. 15–17, 1988.

[15] Y. Desmedt, C. Goutier, and S. Bengio. Special uses and abuses of the Fiat-Shamir passport protocol. *Advances in Cryptology (CRYPTO), LNCS*, Springer-Verlag, 293, pp. 21–39, 1987.

[16] S. Drimer and S.J. Murdoch. Keep your enemies close: Distance bounding against smartcard relay attacks. *Proceedings of USENIX Security Symposium*, pp. 87–102, 2007.

[17] K.P. Fishkin and S. Roy. Enhancing RFID privacy via antenna energy analysis. Presented at the *RFID Privacy Workshop*, 2003.

[18] G.P. Hancke. Practical attacks on proximity identification systems (short paper). *Proceedings of IEEE Symposium on Security and Privacy*, pp. 328–333, 2006.

[19] G.P. Hancke. A practical relay attack on ISO 14443 proximity cards. http://www.cl.cam.ac.uk/~gh275/relay.pdf

[20] G.P. Hancke. Security of proximity identification systems. PhD dissertation, February 2008.

[21] G.P. Hancke and M.G. Kuhn. Attacks on time-of-flight distance bounding channels. *Proceedings of First ACM Conference on Wireless Network Security (WISEC'08)*, pp. 194–202, March 2008.

[22] G.P. Hancke and M.G. Kuhn. An RFID distance bounding protocol. *Proceedings of IEEE/CreateNet SecureComm*, pp. 67–73, 2005.

[23] A. Harter, A. Hopper, P. Steggles, A. Ward, and P. Webster. The anatomy of a context-aware application. *Proceedings of Fifth Annual ACM/IEEE International Conference on Mobile Computing and Networking, MOBICOM'99*, pp. 59–68, 1999.

[24] Y.C. Hu, A. Perrig, and D.B. Johnson. Rushing attacks and defense in wireless ad hoc network routing protocols. *Proceedings of ACM Workshop on Wireless Security*, pp. 30–40, 2003.

[25] Y.C. Hu, A. Perrig, and D.B. Johnson. Packet leashes: A defense against wormhole attacks in wireless networks. *Proceedings of INFOCOM*, pp. 1976–1986, 2003.

[26] C. Karlof and D. Wagner. Secure routing in wireless sensor networks: Attacks and countermeasures. *Elsevier Ad Hoc Networks*, 1, 293–315, 2003.

[27] H. Karl and A. Willig. *Protocols and Architectures for Wireless Sensor Networks*. Wiley, 2005.

[28] B. Karp and H.T. Kung. GPSR: Greedy perimeter stateless routing for wireless networks. *Proceedings of MOBICOM 2000*, pp. 243–254, 2000.

[29] T. Kasper. Embedded security analysis of RFID devices. Diploma thesis, Ruhr-University Bochum, Germany, 2006.

[30] Z. Kfir and A. Wool. Picking virtual pockets using relay attacks on contactless smartcard systems. *Proceedings of IEEE/CreateNet SecureComm*, pp. 47–58, 2005.

[31] I. Kirschenbaum and A. Wool. How to build a low-cost, extended-range RFID skimmer. *Proceedings of 15th USENIX Security Symposium*, pp. 43–57, 2006.

[32] J. Krumm and E. Horvitz. LOCADIO: Inferring motion and location from Wi-Fi signal strengths. Presented at the *First Annual International Conference on Mobile and Ubiquitous Systems: Networking and Services Mobiquitous 2004*, 2004. http://research.microsoft.com/~horvitz/locadio.pdf

[33] M.G. Kuhn. An asymmetric security mechanism for navigation signals. *6th Information Hiding Workshop, LNCS*, Springer-Verlag, 3200, pp. 239–252, 2004.

[34] D. Liu, P. Ning, and W. Du. Attack-resistant location estimation in sensor networks. *Proceedings of IEEE Information Processing in Sensor Networks*, pp. 99–106, 2005.

[35] D. Liu, P. Ning, and W. Du. Detecting malicious beacon nodes for secure location discovery in wireless sensor networks. *Proceedings of International Conference on Distributed Computing Systems*, pp. 609–619, 2005.

[36] R. Mayrhofer, M. Hazas, and H. Gellersen. An authentication protocol using ultrasonic ranging. Lancaster University Technical Report COMP-002-2006, 2006. http://eis.comp.lancs.ac.uk/index.php?id=361

[37] J. Munilla, A. Ortiz, and A. Peinado. Distance bounding protocols with void challenges for RFID. *Proceedings of Workshop on RFID Security (RFIDSec)*, pp. 15–26, 2006.

[38] K.B. Rasmussen and S. Capkun. Implications of radio fingerprinting on the security of sensor networks. *Proceedings of IEEE SecureComm*, 2007.

[39] J. Reid, J.M.G. Nieto, T. Tang, and B. Senadji. Detecting relay attacks with timing-based protocols. *Proceedings of the 2nd ACM Symposium on Information, Computer and Communications Security*, pp. 204–213, 2007.

[40] J. Sorrels. Optimizing read range in RFID systems, EDN. http://www.edn.com/article/CA84480.html

[41] Texas Instruments. HF antenna design notes, Technical Application Report. http://www.ti.com/rfid/docs/manuals/appNotes/HFAntennaDesignNotes.pdf

[42] J. Werb and C. Lanzl. Designing a positioning system for finding things and people indoors. *IEEE Spectrum*, 35(9), 71–78, 1998.

9 Public Key in RFIDs: Appeal for Asymmetry

Erwing R. Sanchez, Filippo Gandino,
Bartolomeo Montrucchio, and Maurizio Rebaudengo

CONTENTS

Radio-frequency identification (RFID) devices represent a promising new technology that, in the near future, could be applied in a wide range of applications. The increasing success presents a side effect due to the necessity to protect data stored in RFIDs guaranteeing security and privacy. This chapter presents, on one side, the motivations for addressing security and privacy in RFIDs and, on the other side, it analyzes different approaches proposed in the literature that aim at solving these issues. Such different approaches, based on asymmetric encryption, are described and compared by analyzing their properties and their applicability to RFID devices, taking into account their limitations in terms of computation performance and power consumption.

9.1   INTRODUCTION

The increasing relevance of computer and communication networks drives to security and data protection. To address security and privacy constraints, complex networks have historically relied in encryption, which denotes the action of encoding or transforming information to hide it to nontrusted individuals. There are several methods for encrypting information. *Symmetric* and *asymmetric* encryptions are typical examples of encrypting algorithms. Although symmetric encryption utilizes the same secret *key* to encrypt and decrypt information, asymmetric encryption exploits different keys for each operation. Now, in many countries, a great attention is paid on privacy risks [34,35,43]. In classical scenarios where digital security is essential, e.g., e-business or bank wire transfers over Internet, asymmetric encryption offers a secure method for authenticating and protecting information.

Radio-frequency identification (RFID) systems constituted by passive, low-cost transponders are currently being used in a variety of applications and environments [16]. Born as an identification technology, RFIDs are expected to deluge every ordinary activity in an ubiquitous way. Contactless cards, warehouse labels, and generic product identifiers are currently exploiting RFID's capabilities. In recent years, well-defined difficulties have been arisen related to privacy and security in different RFID-based schemes. Theoretically, any security problem may be solved with the same algorithms employed in classical network security; however, generally this is not possible because of inherent RFID resource constraints.

Symmetric encryption and challenge–response techniques have been successfully implemented in RFID systems, and satisfactory results in terms of security have been obtained for specific applications. Asymmetric encryption provides advantages that symmetric encryption does not; e.g., because it does not require key distribution, benefits such as off-line authentication might be obtained. Nevertheless, asymmetric encryption is less diffused in RFIDs mainly because its hardware requirements are larger than in other security schemes.

Different asymmetric encryption approaches have been studied and proposed. Each proposal differs from others in the kind of encryption algorithm adopted. Relevant approaches are based on the Rivest, Shamir, and Adleman (RSA) algorithm [8], elliptic curve cryptography (ECC) [3], ElGamal scheme [15], or NTRU scheme [33].

RSA-like encryption engines are based on the complexity of computing logarithms in finite fields. Good and proven security is obtained by using it; however, its key length is the main drawback that affects its smooth implementation in RFID systems. An encryption core based on this kind of algorithm requires key lengths of, at least, several hundreds of bits which combined with the kind of operations that must be performed provides an unhealthy environment for RFID deployment. Finally, RSA-like algorithms require a huge quantity of memory in the RFID transponder.

ECC-based algorithms are based on the same mathematical principles as RSA except by the fact that operations are carried out within an elliptic curve in the finite field. As a result, key lengths are reduced in a significant way maintaining many of the security advantages of RSA.

ElGamal scheme is based on the intractability of the discrete logarithm problem and the Diffie–Hellman problem. Because two modular exponential operations are required, the efficiency is lower than in RSA, and the size of the keys is larger. However, this approach retains several interesting properties, such as semantic security and homomorphism.

NTRU is an encryption algorithm based on the complexity of finding a very short vector in a high-dimension lattice which has been proposed recently. This approach allows reducing execution time because keys may be managed in short chunks, thus taking advantage of general purpose processors.

Many solutions have been proposed in the literature aiming at providing RFID with asymmetric cryptographic capabilities. The remaining of this chapter will be organized as follows. First, notions on public-key cryptography are presented. Second, some state-of-the-art approaches are considered; they are classified and compared according to the type of cryptography they used regardless of their application or implementation. Approaches implemented in hardware are presented and compared in terms of their dimension, technology, and performance. Finally, conclusions are presented based on the compared approaches.

9.2 PUBLIC-KEY FUNDAMENTALS

Cryptographic algorithms have been used for decades to guarantee communication privacy. Cryptography deals with authentication systems with the aim of ensuring the genuineness of a message to the receiver. We will focus on cryptographic systems based on public-key cryptosystems firstly presented in Ref. [14]. Many other applicable algorithms based on public-key cryptosystems have been proposed in the literature like RSA [36], ElGamal's scheme [15], and Knapsack scheme [13].

In a public-key cryptosystem, given a pair of families $\{E_K\}_{K \in \{K\}}$ and $\{D_K\}_{K \in \{K\}}$ of algorithms representing inverting transformations,

$$E_K : \{M\} \to \{M\}$$

and

$$D_K : \{M\} \to \{M\},$$

on a finite message space $\{M\}$, the following statements must be true:

1. For every $K \in \{K\}$, E_K is the inverse of D_K.
2. For every $K \in \{K\}$ and $M \in \{M\}$, algorithms E_K and D_K are easy to compute.
3. For almost every $K \in \{K\}$, each algorithm equivalent to D_K is computationally infeasible to derive from E_K.
4. For every $K \in \{K\}$, it is feasible to compute inverse pairs E_K and D_K from K.

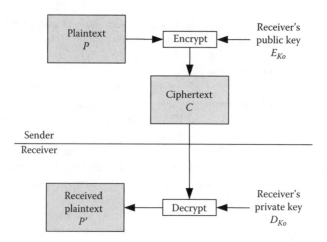

FIGURE 9.1 Public-key encryption.

Therefore, by making $K = Ko$, a pair of ciphering functions D_{Ko} and E_{Ko} are fixed. The third property allows making public the key E_{Ko} without compromising the security of the secret key D_{Ko}. There are two main branches of public-key cryptography depending on the utilization of the secret and public keys: public-key encryption and digital signature.

9.2.1 PUBLIC-KEY ENCRYPTION

The basic idea behind public-key encryption is to ensure confidentiality in a communication channel. A message is encrypted in a way that cannot be decrypted by other person than the intended recipient. Formally, a plaintext message $P \in \{M\}$, is ciphered by means of the public key E_{Ko}. The result is a ciphertext message $C \in \{M\}$ that can be deciphered using the secret key D_{Ko}. Thus, the following relations are true:

$$C = E_{Ko}(P),$$
$$P' = D_{Ko}(C).$$

Where $P' = P$. Figure 9.1 presents the concept of public-key encryption as a means to obtain confidentiality between two communicating parties. The receiver's public key is available to anyone who wants to communicate with him or her.

9.2.2 DIGITAL SIGNATURE

When digitally signing, messages are authenticated. A message is signed by means of the sender's private key; in that way anyone who has access to the public key is able to verify the authenticity of the message. Hence, a plaintext message $P \in \{M\}$, is signed by means of the secret key D_{Ko}. The result is a ciphertext, signed message $S \in \{M\}$ that can be deciphered and verified using the public key E_{Ko}. Accordingly, the following relations remain true:

$$S = D_{Ko}(P),$$
$$P' = E_{Ko}(S).$$

Where, if performed correctly, $P' = P$. Figure 9.2 shows how a signature operation is performed. The sender uses his or her own private key to sign a message, and everyone is able to verify its authenticity by exploiting the sender's public key.

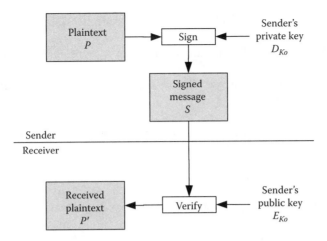

FIGURE 9.2   Digital signature.

9.2.3   EXAMPLE

Consider a simplified version of the RSA algorithm where the ciphering functions are defined as follows:

$$E_{Ko}(x) = x^e \bmod n,$$
$$D_{Ko}(x) = x^d \bmod n.$$

By selecting the parameters in the following way:

$$e = 17,$$
$$d = 2753,$$
$$n = 3233,$$

it is possible to have both a public key (17, 3233) and a private key (2753, 3233). For example, suppose that a message $P = 123$ is encrypted with the public key. Hence,

$$E_{Ko}(123) = 123^{17} \bmod 3233 = 855.$$

Similarly, the corresponding decryption is performed as follows:

$$D_{Ko}(855) = 855^{2753} \bmod 3233 = 123.$$

According to RSA Laboratories [37], as of 2007, valid key dimensions for a long-term security should be at least 1024-bit long. They show that 663-bit keys have been factored with an effort equivalent to 55 years on a single 2.2 GHz Opteron CPU; however, with a cluster of 80 2.2 GHz CPUs the key was broken in about three months.

9.2.4   RELEVANT ATTACKS

In the following, a short description of some attacks on public-key encryption will be presented:

- *Chosen-plaintext* attack: The adversary chooses a set of plaintexts and obtains the corresponding ciphertexts, to deduce information to recover the plaintext of an specific ciphertext.

- *Adaptive chosen-plaintext* attack: The adversary can choose a plaintext based on the previously obtained ciphertexts.
- *Chosen-ciphertext* attack: The adversary can obtain a set of plaintexts corresponding on the chosen set of ciphertexts.
- *Adaptive chosen-ciphertext* attack: The adversary can choose a ciphertext based on the previously obtained plaintexts.

Although chosen-ciphertext attacks are very dangerous for public-key cryptography, in the RFID context they are normally not possible.

9.2.5 PUBLIC-KEY PROPERTIES

In the following, some properties of cryptographic schemes are presented.

9.2.5.1 Homomorphism [17]

Let $\mathcal{M}$ denote the set of the plaintexts and $\mathcal{C}$ the set of ciphertexts. An encryption scheme is said to be homomorphic if for any given encryption key k the encryption function E satisfies:

$$\forall m_1, m_2 \in \mathcal{M}, \quad E(m_1 \odot_{\mathcal{M}} m_2) \longleftrightarrow E(m_1) \odot_{\mathcal{C}} E(m_2).$$

With the multiplication operators a scheme is multiplicatively homomorphic, so the multiplication of two encrypted values is equal to the encrypted multiplication of two values.

9.2.5.2 Semantic Security [21]

This property requires that no information about the plaintext can be learned from the ciphertext. More strictly, an encryption scheme (G, E, D) is semantically secure if for all probabilistic polynomial time algorithms M and A, functions h, polynomials Q there is a probabilistic polynomial time B such that for sufficiently large k,

$$\Pr(A(1^k, c, e) = h(m)|(e, d) \leftarrow G(1^k); \; m \leftarrow M(1^k); \; c \leftarrow E(e, m))$$

$$\leq \Pr(B(1^k) = h(m)|m \leftarrow M(1^k)) + \frac{1}{Q(k)}.$$

Semantic security does not consider the case of the chosen ciphertext attack.

9.2.5.3 Key Privacy [5]

This property requires that, from a ciphertext encrypted by a public key selected in a set of known keys, an eavesdropper is not able to identify the employed key.

9.3 RFID SCENARIO

To face the RFID public encryption question proposed in this chapter, some notions and models should be considered. Within this section tags properties are depicted, spotlighting different nomenclature and current organization. In addition, a communication model suitable for reader-tag data transfer is presented.

9.3.1 RFID TAG PROPERTIES AND ORGANIZATION

Basic architecture of an RFID tag is presented in Figure 9.3. A *tag* (also known as a *transponder*) is composed by a *radio* frequency interface block, a *memory* component and a *logic* element.

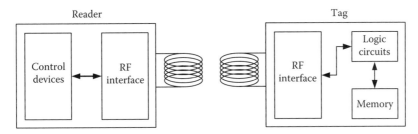

FIGURE 9.3   Tag architecture and reader.

Tags have usually no battery called *passive* ones, so they must acquire the power from the external radio-frequency communication. Only *active* tags have their own power supply. Both active and passive tags have limited computational capacities due to energy constraints, but these limits are much harder for passive ones. The major concern of an RFID reader consists in accessing the tag memory. Memory, which plays an important role in the RFID architecture, may be a read-only memory (ROM) or an electrically erasable memory (EEPROM). It contains the *unique* identification number and may have storage capacity for other writable information.

Classification is generally done by means of tag physical characteristics such as computational resources, storage capacity, operational frequency, modulation technique, etc. RFID transponders may contain computational resources from just a few-gate state machine to robust microcontrollers. Their memory capacity may oscillate from a single bit to about 128 kb. Operational frequency used in an RFID system varies: it ranges from low frequencies—several kilohertz—to ultrahigh frequencies—a couple of gigahertz. Modulation techniques found in RFID systems are frequency-shift keying (FSK), amplitude-shift keying (ASK), and phase-shift keying (PSK) [16].

From the point of view of cryptography, RFID tags can be classified according to their

- Computational performance: whether a tag has cryptographic capabilities or not
- Memory features: whether a tag has usable storage capacity or not

Some RFID tags are able to perform cryptographic operations because of their internal logic circuitry. The majority of RFID devices, however, have not real capabilities for cryptanalysis functions in part due to their power constraints. Most of RFID tags are passive ones, with limited processor performance and, hence, computational resources. As a result, according to the tag computational performance, it is feasible to conceive two categories of RFID devices: *basic* tags, meaning those that cannot execute standard cryptographic operations (e.g., encryption, pseudorandom number generation and hashing), and *crypto* tags, meaning those that can.

Current versions of RFID tags may have read-only or read and write memory. The size of the user memory may reach several kilobytes. Writable user memory allows RFIDs to be utilized to store dynamic data that enhance application fields; however, it also requires data protection to keep information safe.

9.3.2   RFID COMMUNICATION CHANNEL

Almost all communication channels are devised using a layered approach. RFID communication among readers and tags is typically described with the model shown in Figure 9.4. This model is often presented with some limited formal differences [2,16,29]. It is compatible with the ISO standard 18000-1 [25], and may be easily expanded to RFID devices complying with any other directive.

The RFID communication model is constituted by three layers: application, communication, and physical layer. Briefly, RFID layers have the following characteristics:

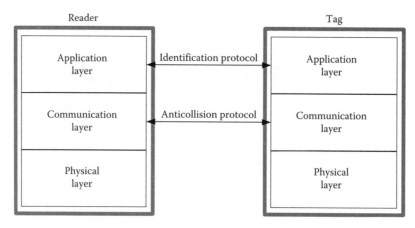

FIGURE 9.4 RFID communication model.

- Application layer processes the information defined by the user. That information includes data about the tagged object or an identifier.
- Communication layer outlines the way in which readers and tags can communicate. Anticollision protocols may be found in this layer.
- Physical layer defines the physical air interface, i.e., the frequency, modulation of transmission, data encoding, timings, and so on.

9.3.3 RFID CRYPTOGRAPHY AIMS

The main aims of using cryptography in RFID context are the authentication, the data protection, and the privacy protection.

The authentication aims at guaranteeing the originality of a commodity by using a signature system. Two malicious counterfeiting actions are

- Cloning that represents the creation of new RFID tags equal to the first one
- Forgery that represents the creation of RFID tags with new data and that can pass rightly an authenticity check

The data protection aims at avoiding the access to some information in the tag memory by encoding the data, or managing a challenge–response-like protocol. Two threats that data protection must contrast are

- Spying that represent the reading of reserved data
- Tampering that consists in the malicious changing of data recorded in the tag memory; a scheme that avoids tapering is called *tamper-resistant*, a scheme that only detects tampering actions is called *tamper-evident*

The privacy protection based on public-key cryptography aims at avoiding privacy threats by ciphering, hiding, and changing the information transmitted by an RFID tag. There are many privacy threats connected to RFIDs [27,28], that require different solutions:

- Serial number of a tag can be associated with the customer's identity, so it could be possible to monitor the customer.
- Knowing the object identified by a serial number, it is possible to get information for profiling. Besides knowing which object a person buys, it is possible to know how often a person uses it as well.

- Even without associating a tag number with a person identity, a set of tags can track an unidentified person, violating the "location privacy" [6].
- The transfer of a tag from a set to another set means that an object passes from one person to another, so it is possible to know that there is a relation between those persons.
- When the memory tag holds information about the tagged commodity, by reading the memory it could be possible to know which commodities a person possesses.

In the following sections of this chapter, the target of different public key approaches, and how they face the concerning problems, will be discussed.

9.4 RSA-LIKE CRYPTOGRAPHY

RSA cryptosystem was invented by Rivest et al. [36] and is the best known of the integer factorization family of cryptosystems where its strength lies in the mathematical complexity of factoring large integers. In this scheme, large enough integers are selected to make brute-force factorization infeasible.

RSA is, probably, the most widely used public-key cryptosystem. Because multiplication of integers modulo n is a relatively cumbersome procedure to implement, and because an exponentiation operation requires repeated multiplication, the RSA system cannot achieve the speed of private systems. This is, of course, true for other public-key cryptosystems as well. RSA encryption and signature verification can be speeded up significantly by selecting a small exponent (typical values are either 3 or $2^{16} + 1$).

9.4.1 MATHEMATICAL DESCRIPTION

Based on the integer factorization problem, RSA cryptosystems are created from two large prime numbers p and q. In RSA, the product $n = pq$ generates the group $G = \mathbb{Z}_n^*$, which is a multiplicative group of units in the integers module n. The order of G is $\phi(n) = (p-1)(q-1)$, where ϕ denotes the *Euler Phi function*. Because no efficient algorithms are known for taking bth roots in $\mathbb{Z}_n^*$ without the knowledge of p and q, hence, breaking the RSA cryptosystem is believed to be equivalent to factoring n.

9.4.1.1 Keys Generation

Public and private generation is performed in the following steps:

1. Select two large prime numbers p and q.
2. Compute $n = pq$.
3. Compute $\phi(n) = (p-1)(q-1)$.
4. Find two numbers a and b that satisfy $ab \equiv 1 \pmod{(\phi(n))}$.
5. The public key is the pair (n, b) and the secret key is (p, q, a).

9.4.1.2 Encryption

In an RSA cryptosystem, message x is translated into a ciphertext $c(x)$, i.e., encrypted, as follows:

$$c(x) = x^b \bmod n$$

9.4.1.3 Decryption

On the other hand, decryption in an RSA cryptosystem is performed in the following way:

$$p(y) = y^a \bmod n$$

9.4.2 High-Level Approach

Some approaches utilize RSA-like encryption for dealing with security in RFID transponders. Most of them are part of a broader protocol with a specific aim that takes advantage of RSA encryption in an indirect way, avoiding encryption requirements within the RFID tag.

A common use of encryption in RFID systems is to reduce counterfeit. In Ref. [7], authors come up with a solution for providing a level of security against counterfeit tags based on RSA-like encryption. The basic idea behind their approach is to use a key-couple K_P and K_S, the public and secret key, respectively, to encode the identifier of the tag I_T. In particular, they encrypt I_T with K_S that gives as a result C_T, a ciphered version of the identifier. C_T is then memorized in the tag's memory allowing final users to verify the authenticity of the tag by decrypting it with K_P and comparing the result with I_T; after the memorization, the writing action for the involved area of memory is disabled.

This approach is a simple authentication protocol that bypasses the need of encryption capabilities inside RFID transponders.

The cloning action is possible, but it clearly requires the production of new tags. The forgery action needs to break the RSA cryptosystem system. A spying action can only get the identifier. The ciphertext is unchangeable, so the system is tamper-resistant. The system does not provide protection against tracking or personal belongings spying.

9.4.3 Hardware Approach

An authentication protocol for RFIDs that has been implemented in hardware and adopts RSA-like cryptography is introduced in Ref. [8]. In that approach, authors present an RFID tag with encryption capabilities. The tag encrypts the identifier of the tag I_T, providing a ciphertext C_T. C_T can be decrypted and compared to I_T. According to authors, choosing the public key K_P or the secret key K_S for encryption depends on the kind of application that is intended. For instance, if encryption is performed with K_S then decryption must be performed with K_P, and thus the application should be an authentication or anticounterfeit protocol. Authors claim that, when using a technology of 90 nm and 1.9 MHz as operational frequency, they reach about 1600 μW of power consumption and an encryption average time of 540 ms for a 1024 bit key. Even though timing and power results are not fully satisfactory, this approach provides a good starting point for online encryption in RFID transponders.

9.5 ELLIPTIC CURVE CRYPTOGRAPHY

ECC is a term used to mean a multitude of different cryptographic key exchange and agreement protocols. The building block of all these protocols is the scalar point multiplication which also represents the computationally most expensive operation. The construction of elliptic curve groups may be performed by exploiting different types of finite fields, the most common one being the Galois field with prime characteristics or binary extension fields, e.g., GF(p) and GF(2^k). Low power implementations of ECC in hardware may be achieved by means of efficient arithmetic that can lead to feasible approaches for public-key cryptography within the RFID domain.

Several elliptic curve cryptographic schemes are related to schemes based on the discrete logarithm problem. To show theoretical constructs of ECC and provide a possible implementation architecture, the particular approach of Menezes–Vanstone [39] is considered throughout this section. For the description of other schemes quoted in this section, we will refer the readers to the original papers.

9.5.1 Mathematical Description

ECC relies on a group structure induced on an elliptic curve. The set of points on an elliptic curve (with one special point added, known as the point at infinity $\mathcal{O}$) together with point addition as a binary operation has the structure of an abelian group, i.e., a group where the point addition is commutative. A finite field of characteristic 2 is considered, i.e., GF(2^n). A nonsupersingular elliptic curve E over $GF(2^n)$ is defined as the set of solutions $(x, y) \in GF(2^n) \times GF(2^n)$ of the generic equation

$$y^2 + xy = x^3 + ax^2 + b,$$

where $a, b \in GF(2^n)$, $b \neq 0$, together with $\mathcal{O}$.

9.5.1.1 Keys Generation

Public and private keys are generated as follows:

1. Choose a specific elliptic curve, for instance: $E : y^2 \equiv x^3 + \alpha \cdot x + b \bmod p$.
2. Choose a primitive element $\alpha = (x_\alpha, y_\alpha) \in E$.
3. Pick a random integer $a \in \{2, 3, \ldots, \#E - 1\}$.
4. Compute $a \cdot \alpha = \beta = (x_\beta, y_\beta)$.
5. The public key is then (E, p, α, β), and the private key is (a).

9.5.1.2 Encryption

The encryption algorithm executes the following steps:

1. Pick a random $k \in \{2, 3, \ldots, E - 1\}$.
2. Compute $k \cdot \beta = (c_1, c_2)$.
3. Encrypt $e_{E,p,\alpha,\beta}(x, k) = (Y_0, Y_1, Y_2)$ where $Y_0 = k \cdot \alpha$, $Y_1 = (c_1 \cdot x_1) \bmod p$, and $Y_2 = (c_2 \cdot x_2) \bmod p$.

9.5.1.3 Decryption

To decrypt, the following steps must be performed:

1. Compute $a \cdot Y_0 = (c_1, c_2)$.
2. Decrypt as follows: $d_a(Y_0, Y_1, Y_2) = (Y_1 \cdot c_1^{-1} \bmod p, \ Y_2 \cdot c_2^{-1} \bmod p) = (x_1, x_2)$.

9.5.2 HIGH-LEVEL APPROACH

Work proposed in literature related to this kind of encryption relies on the different mathematical operations within the elliptic curve to introduce the required levels of difficulties for public-key algorithms. Several approaches utilize this encryption for off-line authentication of RFID tags.

One of them is presented in Ref. [40], where authors construct an elaborated algorithm to be used for authentication based on ECC. Their approach exploits, other than the ECC-based public-key algorithm, a physical unclonable function (PUF) that is embedded within the transponder.

A PUF, as defined by the authors, is a function that has the property of being easy to evaluate but hard to characterize. That is, the amount of knowledge that an attacker can obtain from studying the responses to randomly distributed challenges (i.e., queries) to the PUF is negligible. In addition, PUFs should be unique, i.e., responses measured on different PUFs should be, with high probability, far apart. Besides, the PUF should be inseparably linked to the chip, meaning that any attempt to remove the PUF from the chip leads to the destruction of the chip and the PUF.

As a valid implementation for the PUF, authors propose the silicon PUF (SPUF) [19]. SPUFs have advantages compared to other PUF techniques because they can be constructed in silicon base materials, and thus common CMOS manufacturing processes can be exploited. SPUFs are circuits designed to be sensitive to time delays which vary across integrated circuits due to process variations in transistors and wires. Therefore, even a person who has the detailed information of the SPUF circuit cannot physically clone it because circuit delays depend on process variations that are, even, beyond the manufacturer's control.

The main elements in the off-line identification approach are

1. A reader and a transponder with identification I_D and a PUF
2. A standard identification scheme composed by K_g which is a generation-key algorithm; P, an interactive protocol used by the *prover* (the transponder); and V, an interactive protocol used by the *verifier* (the reader)
3. A secure signature mechanism (which is typically ECC-based) constituted by SK_g, a generation-key algorithm; S_e which is the signature algorithm; and V_f which is the verification algorithm

The overall identification scheme is generated in two stages, namely *enrollment* and *authentication*.

In the enrollment stage, SK_g is conceived as master key generation algorithm MK_g to generate a secret key *msk* and a public key *mpk*, used to sign and verify, respectively. K_g is used as a UK_g which is an algorithm that creates a public key pair (pk, sk) for each tag. A certificate, that functions as a signature, is then stored in the transponder based on pk and the signature mechanism, i.e., *msk*. In the authentication stage, the tag sends the certificate to the reader. If it is valid, the reader and the tag start the standard identification protocol. If the tag finishes the protocol, the reader validates the tag correctly.

This anticounterfeiting algorithm protects effectively against cloning and forgery actions.

9.5.3   HARDWARE APPROACHES

In the following, several approaches implemented in hardware are presented.

9.5.3.1   Gaubatz et al.

In Ref. [20], authors offer an ECC hardware implementation addressed to highly constrained devices that may be utilized in RFIDs. Their core operates at a frequency of 500 kHz. Their architecture occupies a chip area equivalent to 18,720 gates using a 0.13 μm CMOS technology. They state that consumption is under 400 μW in signing and encrypting messages. Furthermore, authors claim that, with their scheme, a message of less than 200 bits is encrypted in approximately 817 ms.

9.5.3.2   Batina et al.

Batina et al. [4] improved the implementation proposed by Gaubatz et al. [20]. Their processor included a modular arithmetic logic unit capable of computing additions and multiplications using the same cells without having a full-length array of multiplexers. Results are better compared to other implementations. The total amount of gates required is around 12,000 in a 0.13 μm CMOS technology; consumption is less than 30 μW and message encryption is performed in 115 ms operating at 500 kHz.

9.5.3.3   Wolkerstorfer

Wolkerstorfer developed an integrated chip that can provide good performance for signing messages [42]. The chip has an equivalent area of 23,000 gates implemented in 0.35 μm CMOS technology. It reaches the operating frequency of 68.5 MHz.

9.5.3.4   Kumar–Paar

The hardware implementation presented in Ref. [30] is a valid approach of an ECC processor. It is able to operate at 13.56 MHz, which is a standard frequency in RFIDs. The area of the chip is about 12,000 gates according to the authors. They state that operating at that frequency, with a technology of 0.35 μm, their processor has a timing performance of about 18 ms.

9.6 ElGamal CRYPTOGRAPHY

In Ref. [15], ElGamal presented the ElGamal encryption system and the ElGamal signature scheme. The ElGamal encryption is based on the intractability of the discrete logarithm problem and the Diffie–Hellman problem.

The generalized version of this encryption scheme can work in any finite cyclic group G, but the group determines the security and efficiency of the scheme. Two right groups are

- Multiplicative group Z_p^* of integers modulo p
- Group of points on an elliptic curve over finite field

The efficiency of ElGamal encryption is lower than RSA, because it requires two modular exponential operations, and because the resulting ciphertext size of an encryption is twice as the plaintext one.

ElGamal scheme employs randomization. This is a reason of the large size utilized, but it is a protection against statistical and chosen-plaintext attacks.

9.6.1 MATHEMATICAL DESCRIPTION

In this section, the generalized version of ElGamal is described. The problem to solve to break ElGamal encryption is the Diffie–Hellman problem, which is strictly related to the widely studied discrete logarithm problem, so the security of ElGamal encryption is often considered based on the second one.

The generalized discrete logarithm problem is given a finite cyclic group G of order n, a generator α of G, and an element $\beta \in G$, find the integer x, $0 \leq x \leq n - 1$, such $\alpha^x = \beta$.

The generalized Diffie–Hellman problem is given a finite cyclic group G, a generator α of G, and group elements α^a and α^b, find α^{ab}.

The group G satisfies the computational Diffie–Hellman assumption if no efficient algorithm can compute α^{ab}. A stronger assumption, useful for the demonstration of security properties, is the decisional Diffie–Hellman assumption [10], which is based on the decisional Diffie–Hellman problem.

The generalized decisional Diffie–Hellman problem is given a finite cyclic group G, a generator α of G, and group elements α^a and α^b, distinguish α^{ab} from α^z.

The group G satisfies the decisional Diffie–Hellman assumption if no efficient algorithm can distinguish α^{ab} from α^z.

9.6.1.1 Keys Generation

Public and private generation is performed in the following steps:

1. Select an appropriate cyclic group G of order n, with generator α.
2. Choose a random integer a, $1 \leq a \leq n - 1$.
3. Compute the group element α^a.
4. The public key is (α, α^a), together with a description of how multiply elements in G.
5. The private key is a.

9.6.1.2 Encryption

In ElGamal encryption, message m is translated into a ciphertext c, i.e., encrypted, as follows:

1. Represent m as an element of the group G.
2. Select a random integer k, $1 \leq k \leq n - 1$.
3. Calculate $\gamma = \alpha^k$ and $\delta = m \cdot (\alpha^a)^k$.
4. The ciphertext is $c = (\gamma, \delta)$.

9.6.1.3 Decryption

Decryption in ElGamal encryption is performed in the following way:

Calculate $m = \gamma^{-a} \cdot \delta$.

9.6.1.4 Properties

Some properties of ElGamal are:

- Semantic security: Tsiounis and Yung [41] demonstrated that the semantic security of the ElGamal encryption and the decision Diffie–Hellman assumption are equivalent, so if the decision Diffie–Hellman problem is hard over G, ElGamal encryption possesses the property of semantic security.
- Homomorphism: ElGamal is a homomorphism cryptosystem, if the encryption of a message is

$$\varepsilon(m) = (\alpha^k, m \cdot (\alpha^a)^k),$$

then

$$\varepsilon(m_1) \cdot \varepsilon(m_2) = (\alpha^{k_1}, m_1 \cdot (\alpha^a)^{k_1})(\alpha^{k_2}, m_2 \cdot (\alpha^a)^{k_2})$$
$$= (\alpha^{k_1+k_2}, (m_1 \cdot m_2)(\alpha^a)^{k_1+k_2}) = \varepsilon(m_1 \cdot m_2 \bmod n).$$

- Key privacy: Bellare et al. [5] proved that the ElGamal scheme provides anonymity under chosen-plaintext attack.

About the key length, in 1996, Menezes et al. [32] recommended 1024 bit or larger moduli for long-term security.

9.6.2 HIGH-LEVEL PROTOCOLS AND APPROACHES

Works proposed in literature related to this kind of encryption relies on the different mathematical operations within the ElGamal approach.

9.6.2.1 Re-Encryption

In 2001, European Central Bank proposed to embed RFID tags in Euro banknotes; RFID tags in banknotes can be used like a money flow tracking mechanism and an anticounterfeiting systems. In Ref. [26], authors consider the special problem of consumer privacy protection for RFID-powered banknotes, but their approach is potentially applicable to a generic RFID-based authentication system.

Their broader protocol can employ: a generic public-key cryptosystem, which may be based on ElGamal cryptosystem [15], thanks to its readiness to encoding over elliptic curves; and a digital signature of any type, which may be based on elliptic curves, thanks to its relative good security with compact size. In this particular case, a public key PK_L and a secret key SK_L are generated by an appropriate law entity or agency. That pair is used to manage the privacy protection. An RFID tag, within this system, possesses a unique identification S_i which is the same as the serial number assigned to the banknote. A central bank generates a signing key pair PK_B and SK_B, which are used to guarantee the authenticity of the information. The central bank generates the signature Σ_i on S_i by using SK_B, then the bank encrypts Σ_i and S_i by means of PK_L and a randomly value called the encryption factor r_i, and thus obtaining a ciphertext C_{Si}. Now the bank generates an access key D_i by using the public hash function h. Finally the bank writes S_i and Σ_i on the banknote, it inserts C_{Si} in a cell of the tag memory that is protected from the writing by the access key D_i, and it inserts r_i in a cell that is protected from both writing and reading by D_i. Everyone can read and write the information in the memory by using the data written on the banknote and h.

Every merchant with an optic access to a banknote can verify its authenticity by encrypting Σ_i and S_i by PK_L and r_i, and then checking the resulting C_{Si}; decrypting the signature Σ_i by using PK_B and checking the resulting S_i.

By using this system, the agency in charge of inspecting banknotes transactions can decrypt C_{Si}, which is readable without an optic access to the banknote, because it has SK_L, and compute back the original serial number C_{Si}.

To avoid privacy risks, authors proposed a re-encryption mechanism to be performed periodically; hence, by using again PK_L, selected users re-encrypt Σ_i and S_i by PK_L and a new r'_i, creating a new C'_{Si}. Therefore, after every re-encryption the tag will emit different data, which cannot be linked to the previous ones.

Authors presents a numerical sample that uses the signature scheme of Boneh et al. [9] and the Fujisaki–Okamoto [18] variant on ElGamal. With a serial number size of 40 bits, the relative signature size is 154 bits, the plaintext size is 194; considering an elliptic-curve-based group of 195 bit order, the ciphertext size is 585 bits, so the required memory is 780 bits.

The described scheme is based on mixnets [12]; the main difference is that in mixnet the entities performing re-encryption do not know the plaintext, whereas in Ref. [12] plaintext, which is the serial number, is known.

The described scheme allows the money tracking, and it provides a partial defense against counterfeiting and privacy violation.

The cloning action is possible on every banknote in the optic and radio contact. The forgery action needs to break the signature system. A problem is that, also if it is secure, there is no possibility to change the key couple, so forgers have many years to break it. The signature is written on banknotes, so, after the secret key has been unveiled, it is unchangeable and there is no more proof of the authenticity of the information on all the printed banknotes.

The only information to hide is the serial number; all the information in the tag memory, if maliciously changed, can invalid the money tracking system. The spying and tapering actions are possible only by using an optic reading, or by breaking the cryptosystem. Moreover, if the secret key has been unveiled, the couple of keys can be easily changed.

To avoid people tracking, the ciphertext emitted by the tag is changed by the re-encryption. However, it is possible to suppose that the re-encryption is executed only when there is a banknote transfer between a customer and an authorized entity, so the banknotes emit the same numbers for the whole time they are hold by the same person. Furthermore, a person can bring a set of banknotes, so he or she is identifiable by a set of ciphertexts with periodical input and output of elements.

The exchange of money between persons does not involve the re-encryption, so it could be easy detecting relation between private persons. The presence of the tag could show the presence and the approximate quantity of money. This information can be very useful for merchants and thieves. Authors suggest that some additional RFID privacy systems can reduce the exposition to these threats.

9.6.2.2 Universal Re-Encryption

In Ref. [22], a modified version of the re-encryption approach exploited in mixnet [12] is presented. The main difference from re-encryption is that universal re-encryption does not need the knowledge of the public key under which a ciphertext was computed. The properties of universal re-encryption are demonstrated exploiting ElGamal cryptosystem [15], due to its security properties and its homomorphism. The required storage and computation resources are twice as standard ElGamal ones.

The authors also propose a possible application of universal re-encryption to generic RFID tags. Like for the scheme described in the re-encryption section, some entities perform re-encryption of the public information of tags to avoid people tracking. Differently than the previous scheme, here various kind of entities with their couple of keys, which do not have information about the other entities and about the key previously used on a tag, can re-encrypt the output of the tag by using their own public key.

The universal re-encryption offers the opportunity to extend the RFID re-encryption protection privacy system to all RFID tags, but authors do not go in deep about practical implementation problems.

Saito et al. [38] examine the special problem of the malicious modification of the information in the tag memory; they found two dangerous attacks and they propose two protection schemes. Both the attacks exploit an encryption with special parameters, which avoid a correct privacy re-encryption, to make the tag traceable.

The first proposed scheme is "the re-encryption protocol with check." In this protocol, when a reader writes a ciphertext on a tag, the tag must check if the ciphertext is correctly re-encryptable, and so it can avoid malicious ciphertext. The second proposed scheme is the "re-encryption protocol using a onetime pad." In this protocol the tag re-encrypts the ciphertext by using a onetime pad, which is written and updated by an authorized reader and is stored in the tag memory. The re-encryption computational work is devised between the generation and the application of the onetime pad. Authorized readers need a key, stored in a database, to update the onetime pad, and so to determine the next emitted ciphertexts.

The first protocol protects only against two specific attacks, instead the second brings a higher protection, by allowing only authorized access, but it limits strictly the universality of the updating operation, that is the core of the universal re-encryption. Both the protocols require tags with computation capacity and storage memory, but they are not quantified. In the absence of an efficient physical implementation, these protocols look too expensive for low-cost RFID tags.

The universal re-encryption owns some good properties, but it also presents some gaps that must be still managed. The scheme does not provide protection against cloning, forgery, and tampering actions. It provides protection only against spying and tracking actions, but a malicious re-encryption can make the tracking possible.

9.6.2.3 Insubvertible Encryption

The scope of the insubvertible encryption [1] is to solve the security problems of the universal re-encryption, which is based on ElGamal encryption.

As explained before in this section, with universal re-encryption a malicious re-encryption can make the RFID tag traceable. The insubvertible encryption consists of a certificate attached to an ElGamal encryption. A valid encryption requires the use of the correct certificate. The cyphertext can be randomized by everyone, but the entity that performs the re-encryption can identify if the cyphertext is safe. A cyphertext cannot became back untraceable, but the re-encryption entity can delete the cyphertext and write a new untraceable cyphertext that is without meaning.

The privacy protection of the presented cryptographic primitive is based on three properties:

* This scheme inherits semantic security from ElGamal.
* The certificate is an extension of the signature proposed by Camenisch an Lysyanskaya [11], which depends on LRSW assumption [31], so the certificates are unforgeable.
* The scheme provides key privacy.

The described scheme is a good evolution of universal re-encryption, in fact some security problems are solved.

The cloning action is possible; to avoid it, additional protection systems are required. The forgery action needs to break the certificate system.

The spying is not possible, because to get the encrypted information needs to break the cryptosystem, which is based on ElGamal. The scheme is tamper-evident, but not tamper-resistant; at the first re-encryption the tampering will be detected, but the previous data are lost. However, the tamper-evidence is limited, because the insertion of safe data, e.g., copied from another tag, is not detectable.

The re-encryption, which can be performed by everyone, can potentially limit the tracking to short path, but it requires the presence of several re-encryption entities. Perhaps the evaluation of mobile re-encryption tools could be interesting.

The extension of the system to many kind of commodities, and the spying resistance, protect people against the malicious identification of their personal belongings.

9.7 NTRU CRYPTOGRAPHY

NTRU is a cryptosystem that is, apparently, highly efficient and appropriate for embedded applications such as smart cards or RFID tags. Although it has not been profoundly tested to establish its resistance to cryptographic attacks, there is theoretical evidence of its efficiency that claims that its provided level of security is comparable to RSA scheme's [23,24].

9.7.1 MATHEMATICAL DESCRIPTION

NTRU is based on arithmetic in a polynomial ring

$$R = \mathbb{Z}(x)/((x^N - 1), q)$$

defined by the parameters set (N, p, q) that presents the following properties:

- All elements of the ring are polynomials of degree at most $N - 1$, where N is prime.
- Polynomial coefficients are reduced either mod p or mod q, where p and q are relatively prime integers or polynomials.
- p is considerably smaller than q, which lies between $N/2$ and N.
- All polynomials are univariable over the variable x.

The main operation inside the ring is the multiplication which is commonly represented with the asterisk ⊛. It can be best described as the discrete convolution product of two vectors, where the coefficients of the polynomials form vectors as follows:

$$a(x) = a_0 + a_1 x + a_2 x^2 + \cdots + a_{N-1} x^{N-1} = (a_0, a_1, \ldots, a_{N-1})$$
$$b(x) = (b_0, b_1, b_2, \ldots, b_{N-1})$$
$$c(x) = (c_0, c_1, c_2, \ldots, c_{N-1})$$

Then the coefficients c_k of $c(x) = a(x) \circledast b(x) \bmod q$, p are each computed as

$$c_k = \sum_{i+j=k \bmod N} a_i b_j.$$

The modulus of reduction of each coefficient c_k of the resulting polynomial is either q for key generation and encryption (Sections 9.7.1.1 and 9.7.1.2) or p for decryption (Section 9.7.1.3).

9.7.1.1 Keys Generation

To generate the private key $f(x)$, the following steps should be executed:

1. Choose a random polynomial $F(x)$ from the ring R. $F(x)$ should have binary or ternary coefficients.
2. Construct $f(x) = 1 + pF(x)$.

On the other hand, the public key $h(x)$ should be derived from $f(x)$ in the following way:

1. Choose a random polynomial $g(x)$ from the ring R.
2. Calculate the inverse $f^{-1}(x)$ mod q.
3. Calculate $h(x) = g(x) \circledast f^{-1}$ mod q.

9.7.1.2 Encryption

In a NTRU cryptosystem, encryption is performed as follows:

1. Encode plaintext message into a polynomial $m(x)$ with coefficients from either 0, 1 (binary) or $-1, 0, 1$ (ternary).
2. Choose a random polynomial $\phi(x)$ from ring R.
3. Compute ciphertext polynomial $c(x) = p\phi(x) \circledast h(x) + m(x)$ mod q.

9.7.1.3 Decryption

In opposition to the encryption, the decryption is performed in the following way:

1. Use the private key $f(x)$ to compute the message polynomial $m'(x) = c(x) \circledast f(x)$ mod q.
2. Map the coefficients of the message polynomial to plaintext bits.

9.7.2 HARDWARE APPROACH

A good hardware implementation of NTRU cryptosystem is found in Ref. [20]. Authors employed a small number of gates that may be suitable for RFID-constrained schemes. Their approach is contained in just 3000 gates with a consumption of about 20 μW. The operating frequency is 500 kHz, which is enough for most of RFID applications. According to the authors, operations such as encryption or verification are performed in about 58 ms, while decryption is calculated to be executed in about 117 ms and messages are signed in 234 ms.

9.8 APPROACHES DISCUSSION AND CONCLUSIONS

Most of the algorithms covered so far have been implemented in hardware for RFIDs or other constrained devices. In this section, these solutions are summarized and compared according to their performance and feasibility.

Tables 9.1 and 9.2 summarize all hardware approaches reviewed in this chapter. Although it is not easy to have concrete conclusions by analyzing these hardware approaches, they provide a good understanding of state-of-the-art solutions to public-key problems in RFIDs.

TABLE 9.1

ECC Hardware Approaches

	ECC			
	Gaubatz et al. [20]	Batina et al. [4]	Wolkerstorfer [42]	Kumar–Paar [30]
Gates	18,720	12,000	23,000	12,000
CMOS technology	0.13 μm	0.13 μm	0.35 μm	0.35 μm
Frequency	500 kHz	500 kHz	68.5 MHz	13.5 MHz
Power (avg.)	~394 μW	~30 μW	Not available	Not available
Time performance	~817 ms	115 ms	Not available	~18 ms

TABLE 9.2
RSA and NTRU Hardware Approaches

	RSA	NTRU
	Bernardi et al. [8]	Gaubatz et al. [20]
Gates	Not available	3000
CMOS technology	90 nm	Not available
Frequency	1.9 MHz	500 kHz
Power (avg.)	$\sim 1600\,\mu W$	$\sim 20\,\mu W$
Time performance	~ 540 ms	~ 234 ms

TABLE 9.3
High-Level Protocols (1)

	Anticounterfeiting Encryption [7]	Re-Encryption [26]
Cloning	Possible	Possible
Forgery	Not possible	Not possible
Spying	Possible	Only with optic reading
Tampering	Tamper-resistant	Only with optic reading
Tracking	Possible	Limited
Set tracking	Possible	Possible
Set relation	Possible	Possible
Belongings monitoring	Possible	Only money presence

As stated before, asymmetric encryption may provide the advantages to RFID security that symmetric encryption lacks. However, as in classical asymmetric approaches, those advantages have costs which are commonly related to higher computational requirements. This seems evident by observing, for instance, the expected dimension or the power consumption of the presented approaches.

Nevertheless, according to the estimates of the algorithms and hardware approaches shown, it is feasible to conceive public-key encryption as a real possibility in RFID systems because they have acceptable limitations and are sufficiently fast.

In Tables 9.3 and 9.4, a comparison of high-level protocols is shown.

Anticounterfeiting with off-line encryption [7] and universal re-encryption [22] seem to be weak protocols, because they have no protection against many threats. They are specific protocols that try to solve only some problems and that need to work together with other schemes. From the privacy point of view, the best analyzed scheme is the insubvertible encryption [1], which provides protection against all the privacy threats.

Public-key cryptography represents a great opportunity of security improvement, both through hardware implementations and high-level protocols. In this chapter, relevant novelties about the application of public-key cryptography to RFID context were described; the strength and weakness points of the described approaches were highlighted.

TABLE 9.4
High-Level Protocols (2)

	Universal Re-Encryption [22]	Insubvertible Encryption [1]
Cloning	Possible	Possible
Forgery	Possible	Not possible
Spying	Not possible	Not possible
Tampering	Possible	Partly tamper-evident
Tracking	Possible through tampering	Short paths
Set tracking	Possible through tampering	Short paths
Set relation	Possible through tampering	Not possible
Belongings monitoring	Not possible	Not possible

REFERENCES

[1] G. Ateniese, J. Camenisch, and B. de Madeiros, Untraceable RFID tags via insubvertible encryption, in *Proceedings of the 12th ACM Conference on Computer and Communications Security*, Alexandria, VA, 2005.

[2] G. Avoine and P. Oechslin, RFID traceability: A multilayer problem, in *Financial Cryptography and Data Security, FC 05, Lecture Notes in Computer Science*, pp. 125–140, Springer, 2005.

[3] L. Batina, J. Guajardo, T. Kerina, N. Mentena, P. Tuyls, and I. Verbauwhede, An elliptic curve processor suitable for RFID-tags, *Cryptology ePrint Archive*, Report 2006/2007, 2006.

[4] L. Batina, N. Mentens, K. Sakiyama, B. Preneel, and I. Verbauwhede, Low-cost elliptic curve cryptography for wireless sensor networks, in L. Buttyan, V. Gligor, and D. Westhoff (eds.), *Security and Privacy in Ad-Hoc and Sensor Networks, Lecture Notes in Computer Science*, Vol. 4357, pp. 6–17, Springer, Hamburg, Germany, 2006.

[5] M. Bellare, A. Boldreva, A. Desai, and D. Pointcheval, Key-privacy in public-key encryption, in *ASIACRYPT '01, Lecture Notes in Computer Science*, Vol. 2248, pp. 566–582, Gold Coast, Australia, 2001.

[6] A. Beresford and F. Stajano, Location privacy in pervasive computing, *IEEE Pervasive Computing*, 2(1), 46–55, 2003.

[7] P. Bernardi, F. Gandino, F. Lamberti, B. Montrucchio, M. Rebaudengo, and E. R. Sanchez, An anti-counterfeit mechanism for the application layer in low-cost RFID devices, in *Proceedings of the IEEE International Conference on Circuits and Systems for Communications*, Bucharest, Romania, 2006.

[8] P. Bernardi, F. Gandino, B. Montrucchio, M. Rebaudengo, and E. R. Sanchez, Design of an UHF RFID transponder for secure authentication, in *Proceedings of the ACM Great Lakes Symposium on VLSI*, pp. 387–392, Stresa, Italy, 2007.

[9] D. Boneh, B. Lynn, and H. Shacham, Short signatures from the weil pairing, *Journal of Cryptology*, 17(4), 297–319, 2004.

[10] D. Boneh, The decision Diffie–Hellman problem, in *Proceedings of the Third Algorithmic Number Theory Symposium, Lecture Notes in Computer Science*, Vol. 1423, pp. 48–63, 1998.

[11] J. Camenisch and A. Lysyanskaya, Signature schemes and anonymous credentials from bilinear maps, in *Proceedings of the 24th Annual International Cryptology Conference, CRYPTO 2004*, Santa Barbara, CA, 2004.

[12] D. Chaum, Untraceable electronic mail, return addresses, and digital pseudonyms, *Communications of the ACM*, 24(2), 84–88, 1981.

[13] B. Chor and R. Rivest, A Knapsack-type public-key cryptosystem based on arithmetic in finite fields, *IEEE Transactions on Information Theory*, IT-34(5), 901–909, 1988.

[14] W. Diffie and M. E. Hellman, New directions in cryptography, *IEEE Transactions on Information Theory*, IT-22, 644–654, 1976.

[15] T. ElGamal, A public-key cryptosystem and a signature scheme based on discrete logarithms, *IEEE Transactions on Information Theory*, IT-31, 469–472, 1985.

[16] K. Finkenzeller, *RFID Handbook*, John Wiley & Sons, New York, 2003.

[17] C. Fontaine and F. Galand, A survey of homomorphic encryption for nonspecialists, *EURASIP Journal on Information Security*, 2007.

[18] E. Fujisaki and T. Okamoto, Secure integration of asymmetric and symmetric encryption schemes, in *Proceedings of CRYPTO '99, Lecture Notes in Computer Science*, Vol. 1666, pp. 537–554, London, U.K., 1999.

[19] B. Gassend, D. Clarke, M. van Dijk, and S. Devadas, Silicon physical random functions, in *Proceedings of the 9th ACM Conference on Computer and Communications Security*, Washington, DC, 2002.

[20] G. Gaubatz, J. P. Kaps, E. Ozturk, and B. Sunar, State of the art in ultra-low power public key cryptography for wireless sensor networks, in *Proceedings of the IEEE International Conference on Pervasive Computing and Communication Workshops*, pp. 146–150, Kauai Island, HI, 2005.

[21] S. Goldwasser and S. Micali, Probabilistic encryption, *Journal of Computer and System Sciences*, 2, 270–299, 1984.

[22] P. Golle, M. Jakobsson, A. Juels, and P. Syverson, Universal re-encryption for mixnets, in *CT-RSA04*, Lecture Notes in Computer Science, Vol. 2964, pp. 163–178. Springer-Verlag, San Francisco, CA, 2004.

[23] J. Hoffstein, J. Pipher, and J. H. Silverman, NTRU: A ring-based public key cryptosystem, in J. P. Buhler (Ed.), *Algorithmic Number Theory (ANTS III), Lecture Notes in Computer Science*, Vol. 1423, pp. 267–288, Springer-Verlag, Portland, OR, 1998.

[24] J. Hoffstein and J. H. Silverman, Optimizations for NTRU, in *Proceedings of Public Key Cryptography and Computational Number Theory*, Warsaw, Poland, 2000.

[25] ISO/IEC 18000-1, Information technology AIDC techniques—RFID for item management—air interface, part 1—generic parameters for air interface communication for globally accepted frequencies. http://www.iso.org.

[26] A. Juels and R. Pappu, Squealing euros: Privacy protection in RFID-enabled banknotes, in *Financial Cryptography 03*, Vol. 2742, pp. 103–121. Springer-Verlag, Gosier, Guadeloupe, FWI, 2003.

[27] A. Juels, S. Garfinkel, and R. Pappu, RFID privacy: An overview of problems and proposed solutions, *IEEE Security and Privacy*, 3(3), 34–43, 2005.

[28] A. Juels, RFID security and privacy: A research survey, *IEEE Journal on Selected Areas in Communications*, 24(2), 381–394, 2006.

[29] H. Knospe and H. Pohl, RFID security, *Information Security Technical Report*, pp. 39–50, Elsevier, 2004.

[30] S. Kumar and C. Paar, Are standards compliant elliptic curve cryptosystems feasible on RFID?, in *Proceedings of Workshop on RFID Security*, 2006, Graz, Austria.

[31] A. Lysyanskaya, R. Rivest, A. Sahai, and S. Wolf, Pseudonym systems, in *Selected Areas in Cryptography '99, Lecture Notes in Computer Science*, Vol. 1758, pp. 184–199, Springer Verlag, Kingston, Ontario, Canada, 2000.

[32] A. J. Menezes, P. C. Van Oorschot, and S. A. Vanstone. *Handbook of Applied Cryptography*. CRC Press, Boca Raton, FL, 1996.

[33] NTRU, GenuID, http://www.ntru.com/products/genuid.htm.

[34] Privacy Commissioner of Canada, Annual Report to Parliament 2005—Report on the Personal Information Protection and Electronic Documents Act, pp. 39–42.

[35] Privacy Rights Clearinghouse, RFID Position Statement of Consumer Privacy and Civil Liberties Organizations, 2003.

[36] R. L. Rivest, A. Shamir, and L. Adelman, A method for obtaining digital signatures and public-key cryptosystems, *Communications of the ACM*, 21(2), 120–126, 1978.

[37] RSA LABS, http://www.rsa.com/rsalabs/

[38] J. Saito, J. C. Ryou, and K. Sakurai, Enhancing privacy of universal re-encryption scheme for RFID tags, in *EUC'04, Lecture Notes in Computer Science*, Vol. 3207, pp. 879–890. Springer-Verlag, Aizu-Wakamatsu City, Japan, 2004.

[39] D. R. Stinson, *Cryptography: Theory and Practice*, CRC Press, 1995.

[40] P. Tuyls and L. Batina, Rfid-tags for anti-counterfeiting, in *Topics in Cryptology-CT-RSA '06*, Springer Berlin/Heidelberg, 2006.

[41] Y. Tsiounis and M. Yung, On the security of ElGamal-based encryption, in *Public Key Cryptography*, *Lecture Notes in Computer Science*, Vol. 1431, pp. 117–134, Springer-Verlag, Pacifico Yokohama, Japan, 1998.

[42] J. Wolkerstorfer, Scaling ECC hardware to a minimum, in *PECRYPT Workshop—Cryptographic Advances in Secure Hardware—CRASH 2005*, 2005, Leuven, Belgium.

[43] Working party On the protection Of individuals with regard to the processing of personal data, *Working Document on Data Protection Issues Related to RFID Technology*, Article 29 Data Protection Working Party, 2005.

10 Scalable RFID Privacy Protecting Schemes

Sepideh Fouladgar and Hossam Afifi

CONTENTS

The privacy issues raised by radio-frequency identification (RFID) technology have been widely dealt within the state of the art and numerous solutions have been developed. The main drawback of proposed security solutions is often the fact that they entail nonnegligible costs, first in terms of memory and computation capacity for the RFID tag, and second in terms of computations amount for the back-end database to identify the tag. These costs restrict the potential of large-scale deployment of secure RFID tags.

This chapter examines the privacy protecting schemes that aspire to limit these constraints for the different entities of the RFID system. It first introduces the scalability problems that privacy protecting schemes may lead to. Then, it presents the different categories of solutions with their strengths and weaknesses, such as protocols based on cryptanalytical techniques, privacy protecting schemes with synchronization between the RFID tags and the back-end database, and finally solutions based on temporary or even permanent delegation to trusted readers.

10.1   INTRODUCTION

To be widely deployed, radio-frequency identification (RFID) tags need to remain low cost and adhere to a minimalist design. This low price should not, however, affect the security and the privacy of the communications between tags and readers. Designing scalable RFID privacy protecting schemes is then a trade-off between security requirements and the cost limitation of the whole RFID system [1].

The price for a widespread deployment of RFID tags is estimated to 5¢. According to Refs. [2,3], to construct such a low-price RFID tag, the integrated circuit cost should not exceed 2¢. This condition limits the number of gates to 7.5 to 15K gates. As a result, the number of gates available for security cannot exceed 2.5 to 5K gates. To solve the privacy and security issues raised by RFID technology, each security scheme should take into account the RFID tags weak computational capabilities. This chapter presents various privacy protecting schemes designed for RFID technology that aim to reduce the cost and enhance scalability of RFID tags private identification.

10.2 PRIVACY PROTECTING SCHEMES AND SCALABILITY ISSUES

When a reader emits a query, the tags located in its read range respond without alerting their owners. As a consequence, if a tag replies with a constant bit-string (static identifier or even cryptographically protected identifier), the person bearing the tag broadcasts this value along its way, enabling clandestine readers to track him. Likewise, if a tag replies with a value that can be related to a particular item, thanks for example to an object naming service, clandestine readers will be able to harvest information about the person carrying the tag [4–6].

To overcome the privacy issues raised by RFID tags, researchers proposed various privacy protection schemes. These solutions mainly aim to achieve private identification, which means enabling authorized readers to identify and authenticate an RFID tag, while excluding unauthorized readers from determining any information about the identity of tags they interact with.

To manage private identification, Weis et al. [2] first proposed to build an RFID pseudonym protocol. In their approach, the tag replies with a different pseudonym each time it is queried by readers, so that it cannot be identified by illegitimate parties. The pseudonym is either based on a freshly generated random value encoded with a tag fixed secret or a tag secret encoded and updated at each query [7,8]. These values are stored on the tag and are known to authorized parties. As the pseudonym changes at each query, the tag cannot be tracked. In addition, only authorized entities possessing the tag secret have the ability to tell if two pseudonyms came from the same tag and link these pseudonyms to the tag's identity. Thus, clandestine information collection is no more possible.

Pseudonym generation often requires that RFID tags embed low-cost practical implementations of cryptographic functions that are sufficiently secure. Several recent research works [9–11] propose efficient hardware implementations specifically designed for ultra-low-power devices and open the way to the manufacturing of low-cost cryptographically enabled tags.

The tag–reader protocol remains very simple as shown in Figure 10.1: a reader R interrogates a tag T, and the tag responds with its current pseudonym p. Usually, the tag shares its secret key used to compute pseudonyms with a trusted online database D. This latter can thus identify the tag by means of key search, i.e., the database searches the space of all tag-specific keys it possesses, for

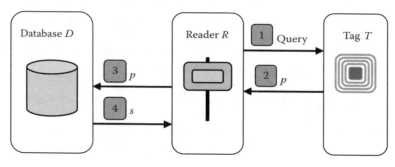

FIGURE 10.1 Pseudonym-based privacy protecting scheme.

the one which permits to obtain the pseudonym broadcasted by the tag. The reader R connected to the online database, simply acts as a relay passing pseudonyms to D. If R possesses the appropriate rights, the online database identifies the tag from the pseudonym it broadcasts, and replies to the reader with the tag's identity s.

The major drawback of the key search approach is that the computational cost for the identifying entity is of $O(N)$ where N is the size of the pseudonym set. If N is large, the proposed scheme becomes poorly scalable. To reduce the cost of brute-force key search and enhance scalability of such privacy protecting schemes, various solutions have been proposed in the literature. These solutions can be classified in two categories:

- Protocol-based solutions in which one tries to minimize the amount of time required for tags and database elementary calculations and communications. Reducing the tag's computational and communication load can be done by designing simple pseudonym generation algorithms and minimalist communication protocols. To reduce the time complexity of the database key search, one can use cryptanalytical techniques such as time-memory trade-off or synchronization methods.
- Architectural-based solutions in which the database asks the readers to contribute to key search calculations and tag identifications. This can be done by delegating temporarily or permanently the tag identifications to authenticated readers.

  It is worth to note that protocol and architectural approaches can be conveniently combined.

The following sections describe various solutions to reduce tag identification period in pseudonym-based approaches.

10.3 PROTOCOL APPROACH

A tag can generate pseudonyms either in a deterministic fashion or thanks to an encoded random variable. In the first case, pseudonyms can be precomputed hash values of an initially embedded secret, which are stored in a onetime lookup table. In the second case, the tag can generate and encode a nonce thanks to a fixed secret key. In this section, we describe some of the main pseudonym-based protocols proposed in RFID literature.

10.3.1 Time-Memory Trade-Off Protocol

The time-memory trade-off technique is based on the fact that the problem of key search is nearly identical with that of breaking keys. This section first introduces a protocol with tags emitting a bounded number of pseudonyms. It then presents the application of time-memory trade-off on this protocol.

10.3.1.1 Bounded Number of Pseudonyms

Ohkubo et al. [12,13] first introduced a privacy protecting scheme with a bounded number of pseudonyms. In fact, in their proposed solution, they assume that a tag never emits more than m pseudonym values over its lifetime. Similar to other proposed solutions, the main idea of this scheme is to modify a tag's reply each time it is queried by readers, so that it cannot be identified by illegitimate parties. In this scheme, readers are not trusted and a back-end database D is in charge of the identification of a set of n tags.

To create fresh pseudonyms at each query, the tags embed two one-way hash functions H and G. At setup, a given tag T_i (where $1 \le i \le n$) stores an initial secret information or identifier s_i^1 which is different for each tag. This preliminary value is generated randomly. The tag uses its embedded

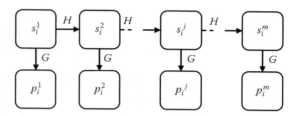

FIGURE 10.2   Ohkubo et al.'s hash chain.

hash function H to update the value of the current identifier s_i^j (where $1 \leq j \leq m$) while function G is used to produce the tag's pseudonyms from the previously generated s_i^j values.

As shown in Figure 10.2, when the jth identification request is sent to a given tag T_i, the querying reader receives back the pseudonym $p_i^j = G(s_i^j)$ where s_i^j is the current value of the tag identifier. Then, the tag renews its identifier s_i^j by computing $s_i^{j+1} = H(s_i^j)$. As the reader is unable to identify the tag from the broadcasted pseudonym, it forwards p_i^j to database D which records the set of the tags setup identifier values $\{s_i^1 | 1 \leq i \leq n\}$.

Database D needs to identify the tag from the forwarded pseudonym p_i^j. For this purpose, the database can construct the hash chain for each tag $\{T_i | 1 \leq i \leq n\}$, i.e., compute $\{G(H^j(s_i^1)) | 1 \leq j \leq m\}$, until it finds a match with p_i^j or it reaches the limit m of the chain length. As it is assumed that a tag never emits more than m values over its lifetime, the database can also construct a onetime lookup table on the first m outputs for all tags.

This solution achieves three important requirements to protect RFID privacy. First of all, the tag's output is indistinguishable from truly random values. Then, this output cannot be linked to the tag's identifier. Finally, even if the adversary obtains the secret key stored in the tag, he cannot associate the current output with old ones. However, this solution can have a complexity in terms of hash computations of $N = mn$ which, when the number of tags n and the maximum number of read operation m are important, is poorly sclalable.

10.3.1.2   Time-Space Trade-Off

To reduce the complexity of key search in the protocol presented above, Avoine et al. [14,15] suggest a time-memory trade-off using Hellman research works [16]. They base their protocol on the fact that the database key search in an RFID system can be treated as a cryptanalytic problem. More precisely, in Ohkubo et al.'s protocol, the problem becomes how to find out the tag's identity from pseudonym p_i^j among all potential tag pseudonyms.

In 1980, Hellman described a cryptanalytic time-memory traded-off which reduces the time of cryptanalysis by using precomputed data stored in memory. Hellman mainly states that, if there are N possible solutions to search over, the time-memory trade-off allows the solution to be found in T time with M words of memory, provided that the time-memory product TM is of $O(N^\alpha)$ with α close to 1. In fact, cryptanalysis is a searching problem of complexity N that allows two extreme solutions:

- Exhaustive search where $T = N$ and $M = 1$
- Table lookup where $T = 1$ and $M = N$

Between these two extreme solutions, Hellman proposes a time-memory trade-off with $M = O(N^{\frac{2}{3}})$ and $T = O(N^{\frac{2}{3}})$ which is much more cost effective than exhaustive search and table lookup.

To find s_i^1 from a given p_i^j, Hellman introduces a function f such that $f(i) = R(p_i^j)$, where R is an arbitrary reduction function that enables to obtain an output of length l_r from a bit-string of length

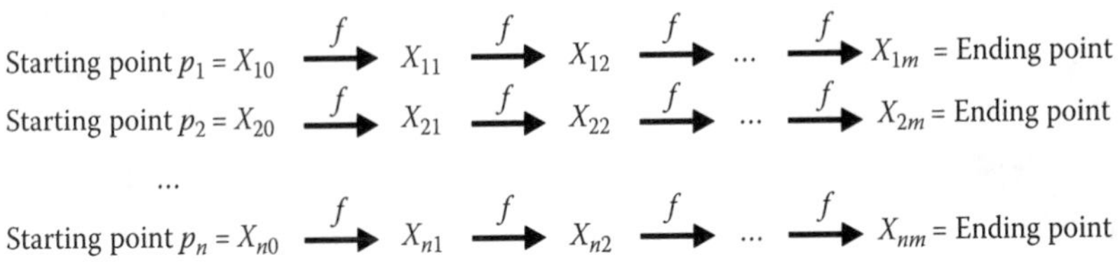

Starting point $p_1 = X_{10}$ $\xrightarrow{f}$ X_{11} $\xrightarrow{f}$ X_{12} $\xrightarrow{f}$ $\cdots$ $\xrightarrow{f}$ X_{1m} = Ending point

Starting point $p_2 = X_{20}$ $\xrightarrow{f}$ X_{21} $\xrightarrow{f}$ X_{22} $\xrightarrow{f}$ $\cdots$ $\xrightarrow{f}$ X_{2m} = Ending point

$\cdots$

Starting point $p_n = X_{n0}$ $\xrightarrow{f}$ X_{n1} $\xrightarrow{f}$ X_{n2} $\xrightarrow{f}$ $\cdots$ $\xrightarrow{f}$ X_{nm} = Ending point

FIGURE 10.3 Hellman chains.

l bits where $l_r < l$. As part of precomputation, the database chooses n starting points $p_1, p_2, \ldots, p_n$ from the key space $\{s_i^j\}$ and computes for each starting point:

$$X_{i0} = p_i \quad \text{for } 1 \leq i \leq n$$
$$X_{ij} = f(X_{i,j-1}) \quad \text{for } 1 \leq j \leq m$$

X_{i0} corresponds to the starting point of the ith chain of Hellman table (see Figure 10.3) and X_{im} to its ending point. The parameter m is chosen by the database to trade-off time against memory. To reduce memory requirement, only X_{i0} and X_{im} are sorted by ending points and stored in a table. Now, suppose a given pseudonym p_i^j from tag T_i at step j. The database applies the reduction function R to obtain

$$y_1 = R(p_i^j)$$

It then checks if y_1 is one of the endpoints stored in the table. If this is the case, the tag T_i has been identified. If y_1 is not an endpoint, the database computes

$$y_2 = R(y_1)$$

and checks if it is an endpoint. It continues this process until an endpoint is found or the y calculation reaches X_{i0}. The chance of finding a tag by using a table of n tags and m steps is

$$P \geq \frac{1}{N} \sum_{i=1}^{n} \sum_{j=0}^{m-1} \left(1 - \frac{im}{N}\right)^{j+1}$$

The calculation time and the memory size for this method are $T = m^2$, $M = nm$ with $nm^2 = N$. For a single table of a large size, there is a chance that chains X_{ij} starting at different pseudonyms collide and merge. This reduces the efficiency of a single table. To obtain a high probability of success, it is better to generate multiple tables using different reduction functions for each table. The probability of success using t tables is then given by Oechslin [17]

$$P \geq 1 - \left(1 - \frac{1}{N} \sum_{i=1}^{n} \sum_{j=0}^{m-1} \left(1 - \frac{im}{N}\right)^{j+1}\right)^t$$

To reduce the collisions and merging of X_{ij} chains, Oecshlin suggests a rainbow table of size $n \times m$ which means that different successive reduction functions are used at each point of the chain. Each chain of Figure 10.3 starts with reduction function 1 and ends with reduction function $m - 1$. The probability of success for this table is given by Oechslin [17]

$$P \geq 1 - \prod_{j=1}^{m} \left(1 - \frac{n_i}{N}\right) \quad \text{where } n_1 = n \quad \text{and} \quad n_{k+1} = N\left(1 - \exp^{-\frac{n_k}{N}}\right)$$

TABLE 10.1

Comparison between Exhaustive Time and Optimal Trade-Off Time

Memory (Megabytes)	32	64	128	256	512	1024
Optimal operation time (s)	54	6.8	0.84	0.11	0.013	0.0016
Exhaustive time/optimal time	1.2	9.5	76	606	4854	38836

In Refs. [14,15], Avoine et al. optimize and apply this method to the protocol proposed in Ref. [12], to obtain s_i^1 from $p_i^j = G(H^{j-1}(s_i^1))$. They not only store the starting point and the ending point of each chain but also add some of its intermediate elements. Authors illustrate the time-space trade-off with the case of a library which is able to serve up to 1 million items ($n = 2^{20}$). They suppose that the number of read operations on a single tag between two recalculations of the tables is 1024 ($m = 2^{10}$). The chosen key size is 128 bits. Finally, they assume that the system is capable of carrying out 2^{24} hash operations per second. With these assumptions, the advantage of time-memory trade-off over the exhaustive search is given in the Table 10.1 [15]. (Assumptions: $n = 2^{20}$; $m = 2^{10}$; probability $= 0.999$; key size $= 128$ bits; exhaustive search time $= 64$ s.)

10.3.2 SYNCHRONIZATION TECHNIQUES

To avoid brute-force key search, the entity which needs to identify the tag can maintain synchronized state with the tags. In fact, if the identifying entity knows the approximate state of a given tag, it can construct a onetime lookup table of tag output values and reduce the computational cost of key search. Synchronization can be achieved for example through the use of a counter incremented by the tag at each query.

10.3.2.1 Synchronization Thanks to a Transaction Counter

In Ref. [18], Henrici et al. propose to synchronize the back-end database and the tags it manages, through a transaction counter Δ_c which is emitted by each tag and that gives the number of transactions since the last successfull authentication with the database. Each tag T embeds a hash function H and a conjunction function, that can be a simple exclusive-OR function, which are used to change the tag's identifier on every read attempt.

Each RFID tag has to store various data among which a back-end database identifier ID_D that indicates the database which is in charge of the tag, the tag's current secret identifier s, the number of the current transaction c, and the last successful transaction number c_{lst}. The corresponding database records for each tag it is responsible for, the tag's current identifier s and its hashed value $H(s)$ (used as primary index of the table), the transactions numbers (c and c_{lst}) along with an associated database entry (AE) that aims to prevent replay attacks (Figure 10.4).

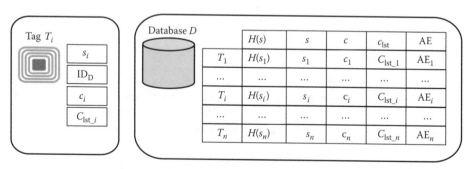

FIGURE 10.4   Data recorded by the tag and the database.

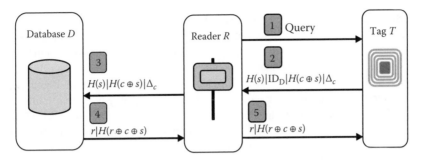

FIGURE 10.5   Diagram of exchanged messages in Henrici et al.'s privacy protecting scheme.

At setup, a given tag's fields s and c are set to random values and c_{lst} is initialized with c's value. The database stores a row of data for each tag where s, c_{lst}, and c are set to the same values than those stored in the tag. The value of AE is not set because there is no associated entry at setup.

When a tag is queried (Figure 10.5), it first increments its transaction counter c and replies to the reader with the value $H(s), \mathrm{ID_D}, H(c \oplus s)$ along with the difference between its current transaction number and the number of the last successful transaction $\Delta_c = c - c_{lst}$. As the reader cannot identify the tag from this information, it forwards the value $H(s), H(c \oplus s), \Delta_c$ to the back-end database D depicted by $\mathrm{ID_D}$. The database searches for the row indexed by $H(s)$. Once found, it adds the stored value of c_{lst} and the Δ_c forwarded by the reader to obtain the value of the tag's current transaction counter c. To authenticate the tag to the database and prevent replay attacks, D computes $H(c \oplus s)$ and checks if this value is identical to the one received from the reader. Similarly, it compares the received c and the computed one, which should be higher. If these conditions are respected, the value of c stored in the database is updated: otherwise the message is discarded.

To avoid tracking, the tag has to change its identifier. For this purpose, D generates a random number r and creates a new identifer s_{new} by performing $r \oplus s$. If AE does not exist for s (this is the case for the first update of the tag identifier), the database creates a new row for s_{new}. The AE-field is updated in both rows (old and new tag identifier rows) so that they reference each other. If an AE exists, the s field of this record row is updated to s_{new} and its $H(s)$ field is updated to $H(s_{new})$. The transaction counter of the newly selected row c along with the corresponding c_{lst} are updated to the current transaction counter value.

Finally, the database replies with $r, H(r \oplus c \oplus s)$ to the reader which forwards this value to the tag. On reception, the tag checks if the transmitted hash matches its own computation. If not, the message is discarded and the identification session stops. Otherwise, the tag updates the value of its identifier thanks to the received r and set c_{lst} to the value c. In order to get information about the tag, the reader needs to query the database another time by using the contents of its first message $(H(s), \mathrm{ID_D}, H(c \oplus s))$ as identifier and session key.

There are always two table rows of the database devoted to each tag, except when the tag is queried for the first time. One corresponds to the tag's last identifier and the other to the identifier newly generated by the database. Thus, if the reply message from the database to the tag is lost or intercepted, the row more recently updated by the database becomes invalid. Then, in the following tag identification process, the tag can be identified thanks to the old table row and the invalid new row gets overwritten.

This solution reduces the brute-force key search time; however, it exposes the tags to potential adversarial tracking and denial-of-service (DoS) attacks as detailed in Ref. [19].

10.3.2.2   Synchronization with Mutual Authentication

Juels [20] proposes a privacy protecting scheme where pseudonyms are updated thanks to onetime pads. This solution aims to prevent tracking and man-in-the-middle attacks for tags unable to perform cryptographic operations like hashing.

In this scheme, when a reader queries an RFID tag, the latter responds alternatively with various pseudonyms α_1, α_2, α_3, ..., α_k, to avoid tracking. A verifier (e.g., a back-end database or an authorized reader) can use these pseudonyms to identify the tag. However, this basic scheme of pseudonym generation is not safe. In fact, an adversary can lead a man-in-the-middle attack and clone the tag very easily. For this purpose, the attacker queries the tag to obtain its pseudonym α_i and then transmits this pseudonym to the verifier to receive the tag's identity (there may be more sophisticated attacks like relay attacks). To avoid man-in-the-middle attacks, the verifier authenticates itself to the tag by releasing a key β_i which is unique for a given pseudonym α_i. Similarly, the tag authenticates itself to the verifier with a key γ_i also unique for a given α_i.

To have a system that lasts and that can face various probing attacks from an adversary, the values of α_i, β_i, γ_i are updated after each successful tag-verifier mutual authentication. To prevent the eventual corruption of secrets used in this update process, authors renew the α_i, β_i, γ_i values thanks to onetime pads transmitted through multiple authentication sessions. A onetime pad is a random bit-string known by two parties that can use it to communicate secretly. For example, if node A shares a onetime pad δ of l bits with node B, it can safely transmit to B an l-bit message M by performing $M \oplus \delta$.

Let κ be a pseudonym or an authentication key stored on the tag, i.e., $\kappa \in \{\alpha_i\} \bigcup \{\beta_i\} \bigcup \{\gamma_i\}$ and m be a parameter that governs the eavesdropping resistance of this system. For each value κ, the RFID tag and the verifier both store a vector of one-time pads:

$$\Delta_\kappa = \{\delta_\kappa^{(1)}, \delta_\kappa^{(2)}, \delta_\kappa^{(3)}, \ldots, \delta_\kappa^{(m)}\}, \quad \text{where} \quad \kappa \in \{0, 1\}^l \quad \text{and} \quad \delta_\kappa^{(i)} \in \{0, 1\}^l.$$

After each successfull authentication, the Δ_κ vectors are updated. For this purpose, $\delta_\kappa^{(1)}$ is discarded, the other pads of the vector are shifted so that $\delta_\kappa^{(j)} = \delta_\kappa^{(j+1)}$ and $\delta_\kappa^{(m)}$ is filled with a bit-string of zeros. Therefore, Δ_κ becomes

$$\Delta_\kappa = \{\delta_\kappa^{(1)}, \delta_\kappa^{(2)}, \delta_\kappa^{(3)}, \ldots, \delta_\kappa^{(m-1)}, 0^l\}.$$

The tag also receives from the verifier, the vectors Δ_κ':

$$\Delta_\kappa' = \{\delta_\kappa'^{(1)}, \delta_\kappa'^{(2)}, \delta_\kappa'^{(3)}, \ldots, \delta_\kappa'^{(m)}\}.$$

Finally, Δ_κ is updated by performing an exclusive OR between elements of Δ_κ and Δ_κ', i.e., $\delta_\kappa^{(j)} = \delta_\kappa^{(j)} \oplus \delta_\kappa'^{(j)}$. Once Δ_κ is updated, the values of κ are updated in their turn. For this purpose, both tag and verifier perform $\kappa_{new} = \kappa \oplus \delta_\kappa^{(1)}$.

Figure 10.6 presents the messages exchanged for tag-verifier authentication and key and pseudonym update. At setup, a given RFID tag embeds a counter c which is initialized to zero and incremented at each identification query. It also stores the indice of the current pseudonym, $d = (c \mod k) + 1$, where k denotes the number of pseudonyms stored in this tag. When the tag T is queried by a reader, it increments its counter c and replies with pseudonym $\alpha = \alpha_d$. If α is a valid α_i value (i.e., $\alpha \in \{\alpha_i\}$ where $1 \leq i \leq k$), the verifier identifies the tag and deduces the corresponding authentication keys $\beta = \beta_d$ and $\gamma = \gamma_d$. The value of α is also withdrawn from the set of potential tag T's pseudonyms. Finally, the verifier replies to the tag with its authentication key β. If the value received from the verifier corresponds to the expected β_d, the tag sends to the verifier, the authentication key $\gamma = \gamma_d$. If the received γ matches the value expected by the verifier, this latter generates three onetime pad matrix which are transmitted to the tag:

$$\Delta'A = \{\delta_{\alpha_i}'^{(1)}, \delta_{\alpha_i}'^{(2)}, \delta_{\alpha_i}'^{(3)}, \ldots, \delta_{\alpha_i}'^{(m)} | 1 \leq i \leq k\},$$

$$\Delta'B = \{\delta_{\beta_i}'^{(1)}, \delta_{\beta_i}'^{(2)}, \delta_{\beta_i}'^{(3)}, \ldots, \delta_{\beta_i}'^{(m)} | 1 \leq i \leq k\},$$

$$\Delta'\Gamma = \{\delta_{\gamma_i}'^{(1)}, \delta_{\gamma_i}'^{(2)}, \delta_{\gamma_i}'^{(3)}, \ldots, \delta_{\gamma_i}'^{(m)} | 1 \leq i \leq k\}.$$

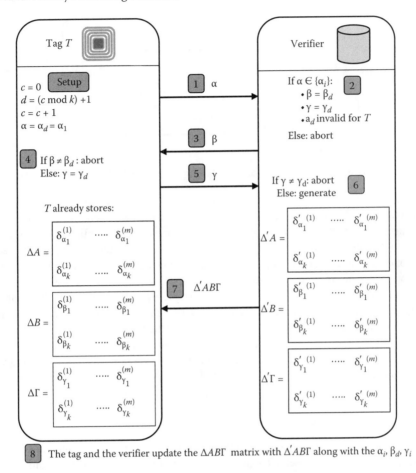

FIGURE 10.6 Diagram of exchanged messages in Juel's minimalist protocol.

Finally, both the tag and verifier update their onetime pad matrix $\Delta AB\Gamma$, their pseudonyms $\{\alpha_i\}$ along with their authentication keys $\{\beta_i\}$ and $\{\gamma_i\}$ using the aforementioned method.

In this scheme, the tag loops through a bounded sequence of d output values. It passes to the next sequence of d outputs only after mutual authentication with the verifier. To identify the tag, the verifier maintains a dynamic lookup table of size $O(dn)$. This solution minimizes the tag computations but necessitates an important amount of memory to store the different pseudonyms, authentication keys, and onetime pad matrix. Another drawback of this solution is that the verifier needs to transfer a maximum of $3klm$ bits to the tag for each identification.

10.3.2.3 Perfect Syncronization

In Ref. [21], Dimitriou proposes a scheme where the tag emits a new pseudonym only after successfull mutual authentication with the back-end database. Indeed, at setup a given tag T embeds an initial secret identifier s_0 set to a random value. A back-end database D also records this value along with $H(s_0)$ that is used as the main key for the access to any T-related information.

When a reader R queries the tag, it joins to its query a freshly generated nonce N_R. In response, the tag generates a new nonce N_T and replies with $H(s_i)|N_T|H_{s_i}(N_T|N_R)$ where s_i is the tag's identifier at the ith query, H is a one-way hash function, and H_{s_i}, a keyed hash function where s_i is the key. When receiving these values, the reader forwards them to the back-end database D along with nonce

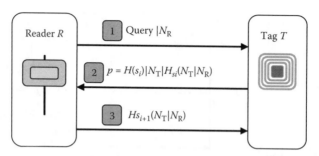

FIGURE 10.7 Diagram of exchanged messages in Dimitriou's privacy protecting scheme.

N_R. The database uses H_{s_i} to search and recover the tag's identity s_i. Then, to check the authenticity of the message, it computes $H_{s_i}(N_T|N_R)$ that acts a s a message authentication code. Finally the database generates the message $H_{s_{i+1}}(N_T|N_R)$ where s_{i+1} is the new tag's identity obtained from s_i thanks to a one-way function computation. This message is sent to the reader which forwards it to the tag. Upon reception, the tag generates s_{i+1} from s_i using the same one-way function used by the database and computes the message authentication code $H_{s_{i+1}}(N_T|N_R)$ sent by the database. If the received and the computed values match, the received message is considered to be authentic and the tag deletes s_i and accepts s_{i+1} as its new identity. Otherwise the tag keeps its old identity and discards the message. This third message is used to prevent desynchronization between the tag and database as the tag only updates its secret after reception and authentication of this message (Figure 10.7).

This approach greatly facilitates key search, however, between identification sessions the output of a tag is static. Consequently, tags can be tracked during such intervals of time.

10.3.2.4 Syncronization with Universal Time Stamps

Tsudik [22] proposes a protocol based on the work of Herzberg et al. [23] that considered anonymous authentication of mobile users. The main approach of this protocol is to obtain an ephemeral user identifier or pseudonym, computed with a one-way hash function of current time and a secret permanent key.

At setup, each tag is initialized with a key s_i which is unique to this tag. s_i is both a tag identifier and a cryptographic key and the secret key shared between the tag and the remote database. The tag also embeds an initial and a maximum time stamp value, respectively, t_0 and t_{max}, which may not be unique to the tag. t_0 is the initial time value assigned to the tag. t_{max} can be the lifetime of the tag or the maximum time for the use of the protocol. Each tag should also possess a uniquely seeded pseudorandom number generator (PRNG). This can be achieved through an iterated keyed hash started with a random seed and keyed on s_i. Thus, for a given tag T_i, $PRNG_i^j$ is the jth invocation of the tag's unique PRNG.

At the beginning of the protocol, readers give the tag a valid time stamp t_{stored} which is recorded by the tag. Each time a reader queries a given tag, it also gives the tag the current time t. If the time stamp t received from the reader is higher than the time already registered t_{stored} and smaller than t_{max}, the tag stores the new time stamp value and hashes t with its secret key s_i. Otherwise, the tag replies with a random value.

When the reader receives back the pseudonym $H(s_i, t)$ or a given random value, it is unable to deduce the tag's identity. For this purpose, it transmits this value to the database.

The database should already have computed the values of

$$H(s_i, t_0), H(s_i, t_0 + \Delta t), H(s_i, t_0 + 2\Delta t), \ldots, H(s_i, t_{max})$$

where $(1 \leq i \leq n)$ and n is the number of tags which the database is in charge of. On reception of $H(s_i, t)$, the database searches for this value in his or her precomputed table. If $H(s_i, t)$ is not found, the reader recieves back a *KO* message. Otherwise, the database's response to the reader depends on the application. If authenticated readers have the right to identify or track the tags, the database returns $G(s_i)$ where G is a one-way hash function. If the reader only needs to know if the tag is valid, the database replies to the reader with an *OK* message.

In this protocol, the tag only makes a lightweight cryptographic operation and needs little amount of memory. Moreover, the database does not need to perform any other real-time operation than a simple table lookup. However, as pointed out by the authors, the main drawback of this protocol is that if an attacker sends a time stamp value much higher than the current time or than t_{max}, the tag is either fully or temporarily incapacitated.

To mitigate this kind of attacks, Tsudik proposes in Ref. [24] an enhanced version of the protocol described in Ref. [22] where at each query, readers present to the tag an epoch token. This token enables to limit to a predefined interval, the ability of an attacker to give a time stamp value higher than previously received time stamp. Thus, the degree of vulnerability to DoS attacks depends of the frequency of epoch tokens changes. If the epoch is too long, a rogue reader will be able to incapacitate the tag for a long period. If the epoch is too short, the tag can only be incapacitated for a short time but as the reader needs to ask the database for epoch tokens, interactions between the database and readers can become important.

In this protocol, each tag is initialized with a root of the hash chain $ET_0 = H^z(X)$ of length $z = t_{max}/INT$ where *INT* is the duration between two changes of an epoch token, e.g., 12 hours. At a given time, the tag also embeds the value of t_{stored} corresponding to the time of the last successful query and the corresponding last epoch token ET_{stored}.

When a reader queries a tag, it joins to its query the current epoch token ET_r. The tag then calculates $\delta = t - t_{stored}$ and $\nu = \frac{t}{INT} - \frac{t_{stored}}{INT}$. If the reader is genuine, ν is the number of epochs between the last time the tag was successfully queried and now. If $\delta > 0$ and $t < t_{max}$, the tag computes ν iterations of the hash function $H()$ over the previously stored epoch token ET_{stored}. If this value matches ET_r, the tag can tell that t is at most *INT* time units later than the actual current time and it changes t_{stored} to t and ET_{stored} to ET_r.

10.4 ARCHITECTURE APPROACH

The protocols presented in the Section 10.3 do not request the readers contribution. In fact, in these protocols, the readers act as simple transmission relays between the tags and the trusted database. The limitations of this online approach are clear. A centralized database is often in charge of a large number of tags and must compute all the possible tag outputs until it finds a match. This can make scalability difficult. Moreover, each time a reader needs to identify a tag, it has to interact with the centralized database. In many applications, this reading latency can be disqualifying. Finally, if the database becomes unavailable for some reasons such as network connectivity failure, etc., all the reading operations of the tags relying on that database will be stopped.

Delegation is a solution to these drawbacks. The idea of delegation is to enable readers to decode pseudonyms without referring to the online database. In fact, if a reader is authenticated and has delegation rights, the database not only gives the reader the tag identity but also the tag-specific secret used to create its pseudonyms, providing the ability to identify the tag to the reader. Delegation may be permanent or temporarily. In the first case, the readers have permanent delegation for the tags in their read range and the database is solicited only when a new tag arrives or an old tag leaves the system. In the second case, the database delegates temporarily a set of tags to a reader for a limited number of queries and updates or retire the delegation according to a delegation policy.

10.4.1 DISTRIBUTED ARCHITECTURE AND COOPERATING READERS

Solanas et al. [25] propose a tag identification architecture where readers are static devices with permanent delegation rights. They are intelligently distributed to cover the area in which the tags are roaming. An area ψ is covered through the use of a number of readers. Tags enter and leave ψ through designated points called system access points (SAP) and system exit points (SEP), respectively. A reader R_i covers a square cell A_i. For the sake of simplicity authors assume that readers have the same coverage range; therefore, all cells A_i have the same size. Cells are disjoint and cover entirely the area ψ.

Let RAD be the radius of the smallest circle containing a square cell and rad the greatest circle contained in a square cell. If the distance between the tag and the reader R_i is less than rad, the tag is located in the square area covered by R_i. If the distance between the tag and the reader R_i is less than RAD and greater than rad, to determine the location of the tag, R_i needs the help of an adjacent reader $R_j \in R_i^{\text{adj}}$ where R_i^{adj} is the set of readers adjacent to R_i. If the distance between the tag and the reader R_i is greater than RAD, the tag is off-range of R_i (Figure 10.8).

Authors consider a class of RFID tags capable of simple cryptographic operations like one-way hash functions. This protocol is based on the improved randomized hash-locks proposed in Ref. [26], where a reader queries a tag and sends a random number N_R along with its query. Upon reception, T generates its own nonce N_T and replies with N_T, $H(N_R|N_T|s)$. To determine the tag identity s, the reader computes N_T, $H(N_R|N_T|s)$ for all tag identifiers it possesses until it finds a match with the received pseudonym (Figure 10.9).

Readers share information like the identifier of the tag and that of the reader which covers the tag. They use three kind of messages to share information: tag arrival, tag roaming, and tag departure. This information is stored in the local cache of each reader involved in the message exchange. To avoid unnecessary replication, each reader removes from its cache the information related to tags that are no longer in its cell or in adjacent cells. There are three protocols corresponding to the tag lifecycle: arrival protocol, roaming protocol, and departure protocol.

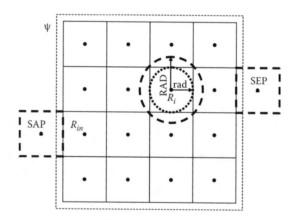

FIGURE 10.8   Readers coverage.

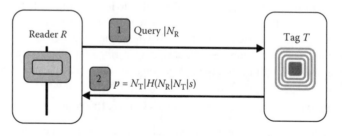

FIGURE 10.9   Improved randomized hash-locks protocol.

- Arrival protocol: It starts when a new tag T enters the system and is detected by a SAP. The SAP is a reader connected to a computer that can efficiently access a database of tag identifiers. If the identity of T is not found in the SAP's database, then the SAP raises an alarm signaling an unidentified tag. If the tag can be identified, the SAP sends to the reader receiving the entering tag R_{in}, a message containing the tag's identity s. R_{in} adds s to its cache and is therefore able to identify T. Then, R_{in} informs its adjacent readers R_{in}^{adj} that T is in its cell by forwarding them the tag's identity s. Therefore, the adjacent readers add T to their caches and record R_{in}'s identity.
- Roaming protocol: The roaming protocol is launched when any tag T moves from its current cell to another adjacent cell. When tag T is detected by a new reader R_i, this latter has in its cache the identity of the tag because it is adjacent to the previous tag location. Therefore, it is able to identify the tag. Then, R_i informs its adjacent readers R_i^{adj} that T is in its cell by forwarding them the tag's identity s. If an adjacent reader has no information about T in its cache, it adds T to its cache and records R_i's identity. If an adjacent reader has information about T in its cache but R_i was not the tag's last location, it updates the reader's identifier information with R_i's identity. The reader of the tag's last location forwards R_i's message to its adjacent readers. If these readers are also adjacent to R_i, they do nothing. Otherwise, they remove any T-related information from their caches. At the end, only readers adjacent to the current tag location keep information on T in their cache.
- Departure protocol: It starts when a tag T is detected by a SEP. The SEP informs its adjacent readers that T must be removed from their caches and all SEP adjacent readers remove the information they have stored about T. The reader of the tag's last location forwards the SEP's message to its own adjacent readers which remove any information on T from their caches.

This scheme enables scalable identification as the readers do not need to refer to a back-end database to identify tags. However, a delegated reader can be compromised. Moreover, one may not want to put unlimited trust on readers. Therefore, to ensure better security, delegation should not be permanent.

10.4.2 TREE-BASED DELEGATION

To reduce the burden on the back-end database, Molnar et al. [5,27,28] propose a delegation protocol based on a tree of secrets of depth $d = d_1 + d_2$ (Figure 10.10). In this scheme, each tree node, except the root, is associated with a k bit secret. Secrets until depth d_1 are chosen by random; those from depth d_1 to depth d can be derived from their ascendant node. To each node at depth d_1 is associated a tag. Thus, tags are roots of subtrees of depth d_2. A tag receives all the keys of the path from the tree root to its node. Therefore, each tag needs to store d_1 secrets. The next d_2 levels of the tree contain secrets that can be derived from ascendant nodes by applying PRNG. For a binary tree, a 0 bit is concatenated to the ascendant node secret and the PRNG is used to generate the left leave secret from this value. The same procedure is used for the right leave by concatenating a bit of value 1 to the ascendant node secret.

The tag holds a counter in which minimum value stands for the leave of the tag subtree which is most to the left while its maximum stands for the leave most to the right. Therefore, to each counter value is associated a different path of secrets going from the tree root to the leave corresponding to the tag's current counter value. For example in Figure 10.10, when T_1's counter value is "01," the path of secrets is $\{k_{root}, k_0, k_{00}, k_{001}\}$. When queried by a reader, the tag responds by encoding a random value with successively all of the d secret keys of the path. The tag will use each time a different path of secrets because its counter is incremented at each query.

If a reader is not delegated, it forwards the pseudonym received from the tag to the back-end database D. This latter possesses the whole tree of secrets. The first d_1-encoded random values of

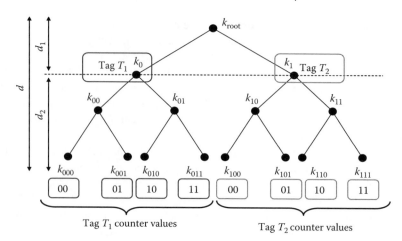

FIGURE 10.10   Binary tree of secrets.

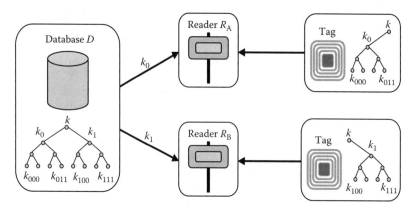

FIGURE 10.11   Back-end database gives delegation to readers R_A and R_B.

the pseudonym allow the database to identify the tag. If the readers credentials are valid, D transmits to the reader the decoded identity.

If a reader has the associated credentials, the back-end database can decide to delegate the ability to decode pseudonyms to the reader and thus delivers to this latter, a set of nodes under depth d_1, corresponding to a given interval of leaves (Figure 10.11). Thanks to the last d_2-encoded random values of the tags pseudonyms and the subtrees of secrets arising from these nodes, the delegated reader will be able to recognize tags which counter values are situated in this interval of leaves, while they are situated in this interval. However, the reader will have to make a brute-force search among keys it possesses to identify the tags.

Thanks to the tree architecture of secrets, the back-end database can make a depth first search and cut the branches it does not need to explore. Compared to brute force, tree architecture decreases greatly tag identification complexity for D, reducing it to a maximum of $O(b \log_b N)$ where b is the tree branching factor. The main inconvenient of this protocol comes from the fact that the tree structure implies intersections between the key sets, thus compromise of secrets in one tag can lead to compromise of secrets in other tags. Avoine et al. analyze the consequences of this drawback in Ref. [14] and show that for some acceptable security requirements, the branching factor b of the tree of secrets should be important.

10.4.3 DELEGATION WITH TWO SECRET KEYS

In this protocol [29,30], authors assume that a tag T embeds two secret keys. One of the keys, K_p is used to compute pseudonyms. The other, K_u is used to update both keys K_p and K_u. T also embeds a counter incremented at each query. A given T is passive and possesses a small rewritable memory to store K_p, K_u and the counter's value. T also needs to embed low-cost hash function H an XOR gate and a random number generator in order to create its pseudonyms and update its secrets. Similarly to previously presented designs interactions between the database D and each reader R are performed over a suitable secure communications protocol.

At setup, each tag T shares two secret keys K_p and K_u with the online database D. For each tag it manages, D stores these keys along with the tag's identifier s. The tags counter c is initialized to zero and will be incremented at each readers query.

When a reader R first meets a tag T, it needs to forward the pseudonym received from the tag to the online database D to identify the tag. As D shares with the tag the pseudonym key K_p, it can decode the tag's pseudonym. If the reader R possesses suitable credentials for tag T's identification, D gives R the tag's identifier s.

To prevent limitations of permanently online systems, the idea is to delegate the ability to decode pseudonyms to selected readers by giving them key K_p. "Delegation request" subprotocol is initiated when R asks D for the ability to decode tags pseudonyms on its own. If R has delegation rights, database D joins in its reply along with s, the tags pseudonym key K_p. Once R is delegated for tag T, it is able to identify T, without referring to the database. The sequence of messages exchanged for delegation request is illustrated in Figure 10.12. We describe the detailed procedure for each step:

- **Step 1**: The reader R queries tag T joining to its demand, a freshly generated nonce N_R. This nonce enables R to prevent replay attacks from a fake tag.
- **Step 2**: In response to this query, T increments its counter c, generates a nonce N_T, and computes its pseudonym $p = N_R|N_T|H(N_T|K_p)$. N_T ensures that the tag creates a fresh pseudonym at each query and protects the tag bearer against tracking. The reader is unable to find out the tag's identity s or its secret keys from p because it does not know the tag's pseudonym key K_p.
- **Step 3**: R forwards p along with its credentials Cred$_R$.
- **Step 4**: If Cred$_R$ is not valid, the protocol ends. If the reader has the rights for tag identification, database D decodes the tag's pseudonym p by searching the space of all tag K_p keys it possesses and computing p until the calculated value matches the received pseudonym and returns s. If the reader has delegation rights, the database joins K_p to its reply.

  Once the reader R is authenticated and granted delegation for a given tag T, it is able to decode T's pseudonyms by itself. Consequently, it does not forward the tag pseudonym to

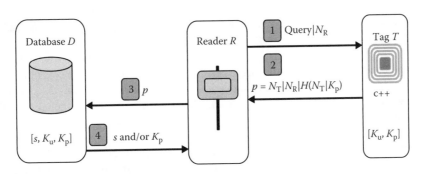

FIGURE 10.12 Delegation request.

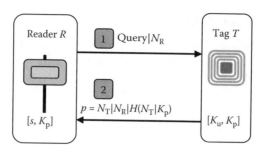

FIGURE 10.13   Delegation.

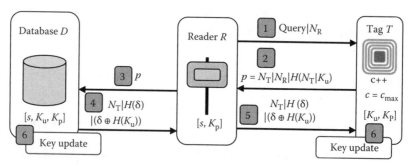

FIGURE 10.14   Delegation update.

D in step 3, but computes p for all the keys it possesses until it finds the matching key and the corresponding s. Only two messages are exchanged to identify the tag (Figure 10.13).

As delegation should not be permanent, it is necessary to regularly update key K_p to end or update the reader's delegation status. For this purpose, T embeds a counter which is incremented at each query. When the counter reaches its maximum value c_{max}, keys are updated thanks to key K_u. This mechanism permits to limit readers delegation to a number of c_{max} queries for a given tag. The sequence of messages exchanged for key update is illustrated in Figure 10.14. We describe the detailed procedure for each step:

- **Step 1**: R generates a fresh nonce N_R and queries its surrounding tags.
- **Step 2**: T increments its counter c which reaches the maximum value c_{max}. Then, T generates a fresh nonce N_T and computes a new pseudonym $p = N_R|N_T|H(N_T|K_u)$ using key K_u instead of K_p.
- **Step 3**: When the reader receives this pseudonym, it is unable to decode it because R only knows key K_p. As a consequence, R forwards the tags pseudonym along with its credentials $Cred_R$ to database D.
- **Step 4**: If $Cred_R$ is not valid, the protocol ends. If the reader has the rights for tag identification, database D identifies the tag thanks to key K_u. Then D generates a random value δ and updates K_u and K_p while keeping their old values. K_p and K_u are updated as follows: $K_{Pnew} = H(K_p \oplus \delta)$ and $K_{unew} = H(K_u \oplus \delta)$. Then, D replies to R with the tag identity and an acknowledgment of key update which is $N_T|H(\delta)|\delta \oplus H(K_u)$.
- **Step 5**: R forwards the received message to the tag.
- **Step 6**: T decodes the received message with K_u and gets δ. Like database D, T updates K_u and K_p. Once the tag has updated its keys, previously delegated readers have to refer to the database to decode tags pseudonyms or to extend their delegation rights.

In this solution, tags thwart malicious traceability by replying with a fresh pseudonym $H(N_T|K)$ (where K is K_p for delegation) at each reader's query. N_T which is a random number value generated at each query, guarantees the freshness of the pseudonym while shared keys restrict the ability to identify the tag to authorized principals. To ensure confidentiality and integrity, sensitive data is hashed because illegitimate readers should not be able to determine secret keys or s from exchanged messages and consequently attempts to alternate communications can be detected. The back-end database authenticates the tag thanks to the shared secrets K_p and K_u. K_p also enables authentication between the tag and a delegated reader. The back-end database and the readers authenticate each other thanks to specific credentials (e.g., certificates). To detect replay attacks from a fake tag broadcasting a pseudonym previously generated by a legitimate tag, a random value N_R is generated by the reader, each time it queries the tag.

The loss or blocking of the readers request and tag's reply messages is a DoS attack preventing tag identification. This kind of attack is not inherent to the proposed scheme but an issue in any wireless system. However, these attacks cannot remain undetected for a long time. In the case of delegation update subprotocol, messages of step 2 or 5 may be lost, intercepted, or blocked. Consequently, tag T does not change its keys. Although T has not received δ from the back-end database D, it keeps on sending pseudonym $H(N_T|K_u)$ without incrementing its counter until a reader transmits successfully step 2 or 5 messages. As D keeps the old key values, T can still be identified. This procedure ensures the synchronization between D and T. Another feasible attack is that of an illegitimate reader incrementing the tags counter through rapid-fire interrogation, until it reaches c_{max}. This kind of DoS attack can be detected if delegation update is frequently requested for a given tag.

In terms of computations, T requires a nonce generation, a hash calculation, and an XOR computation to create a new pseudonym. T also needs to increment its counter. When readers are delegated, the back-end database D makes no computations. On the other hand, when a reader first asks for delegation, it requires an average of n calculations (n times one hash and one XOR) to identify the tag from the pseudonym forwarded by the reader, where n is the number of tags relying on D. The maximum number of operation is $2n$. In fact, when D receives a delegation request from a familiar reader R, it first considers that R asks for delegation update as it may occur more frequently than delegation request. It then searches the space of all K_u keys to identify the tag, before searching the space of all K_p keys. A solution to reduce the maximum complexity of this search (from $2n$ to n) can be to add a flag to the tags response to distinguish delegation request and delegation update. This complexity can also be globally reduced if D keeps for each reader an updated table of the tags which are in its corresponding read range. If a reader is delegated, it takes an average $m/2$ calculations for a reader to decode the pseudonym, where m is the number of tags the reader has delegation for. We assume that m is much smaller than n.

To check D's acknowledgment and update its keys in delegation update subprotocol, T executes two hash calculations and two XOR operations. For the reader, when the tag delegation comes to an end and the pseudonym is encoded with key K_u, it takes an average of $m/2$ calculations for R to find out that it cannot decode the tags pseudonym. D requires a random number generation, two hash and two XOR operations, to update keys K_s and K_u.

In terms of memory, the tag needs to store two 64 bit secret keys K_p and K_u and hold a 20 bit counter. It needs also some buffer memory for cryptographic operations.

Authors assume that most of the time readers are delegated, i.e., delegation subprotocol is the usual and commonly employed procedure. Consequently, D makes few computations, the tag identification complexity is of $O(m/2)$, and communications are mainly done between tags and readers where only two messages are exchanged.

10.5   CONCLUSION

To hold the promising possibilities of RFID tags, one should understand their risks. The security and privacy issues raised by the indiscriminate nature of RFID tags demand an effective technique to ensure users privacy. The RFID pseudonym protocol presented in this chapter aims to achieve private identification, i.e., to enable authorized readers to identify RFID tags while preventing unauthorized ones from extracting any information about the tags they interact with.

Although pseudonym protocols can bring strong privacy to an RFID system, they also raise complex scalability problems highlighted in this chapter. In fact, scalability of RFID privacy protecting scheme is major challenge that may hinder the pervasive deployment of RFID tags.

This chapter has introduced various protocols that intend to advance in the direction of more scalable privacy protecting designs. In fact, the aforementioned solutions try to minimize the amount of memory and cryptographic computations on the tag, or the size of exchanged messages on the tag–reader interface or again the number of real-time computations and the amount of storage on the back-end database.

We presented some protocol-based solutions in which reducing the tag or the back-end database computational, communication, or storage load is done thanks to simplified pseudonym generation algorithms, synchronization methods, or cryptanalytical techniques.

We also presented some architectural-based solutions where readers are totally or partially trusted and asked to contribute to key search calculations and tag identifications.

However, proposals often involve some kind of trade-off either the addition of storage to reduce computational load or the suppression of a security or privacy property.

We believe that these protocol and architectural-based solutions combined with an efficient design and implementation of cryptographic primitives will achieve scalable security in RFID systems.

10.6   FURTHER INFORMATION

Some recent papers study the security issues of low-cost RFID tags [31–33]. The Internet site held by Avoine [34], presents an exhaustive list of research works on RFID security and privacy.

REFERENCES

[1] EPC Global Inc. EPC Radio-Frequency Identity protocols Class-1 Generation-2 UHF RFID. Referenced 2007 at http://www.epcglobalinc.org.

[2] S.A. Weis, S. Sarma, R.L. Rivest, and D.W. Engels, Security and privacy aspects of low-cost radio frequency identification systems, in *Proceedings of IEEE PerCom'03*, March 2003.

[3] S.A. Weis, Security and privacy in radio-frequency identification devices, Master's thesis, MIT, Cambridge, MA, May, 2003.

[4] S.A. Weis, Security parallels between people and pervasive devices, in *Proceedings of IEEE PerSec'05*, 2005.

[5] D. Molnar and D. Wagner, Privacy and security in library RFID: Issues, practices, and architectures, in *Proceedings of ACM Conference on Communications an Computer Security*, 2004.

[6] G. Avoine and P. Oeschlin, RFID traceability: A multilayer problem, *Financial Cryptography 2005*, *LNCS* 3570, Springer-Verlag, pp. 125–140, 2005.

[7] A. Juels, S. Garfinkel, and R. Pappu, RFID privacy: An overview of problems and proposed solutions, in *Proceedings of IEEE Security and Privacy*, 2005.

[8] A. Juels, RFID Security and privacy: A research survey, *IEEE JSAC*, 24(2), 381–394, February 2006.

[9] K. Yüksel, J.P. Kaps, and B. Sunar, Universal hashing for ultra-low-power cryptographic hardware applications, in *Proceedings of Communication Networks and Distributed Systems Modeling and Simulation Conference*, January 2004.

[10] M. Feldhofer, J. Wolkerstorfer, and V. Rijmen, AES implementation on a grain of sand, *Information Security, IEE Proceedings*, 152(1), 13–20, October 2005.

[11] P. Kaps and B. Sunar, Energy comparison of AES and SHA-1 for ubiquitous computing, in *Proceedings of EUC'06*, December 2006.

[12] M. Ohkubo, K. Suzuki, and S. Kinoshita, Cryptographic approach to privacy-friendly tags, in *Proceedings of RFID Privacy Workshop*, 2003.

[13] M. Ohkubo, K. Suzuki, and S. Kinoshita, Efficient hash-chain based RFID privacy protecting scheme, in *Proceedings of Ubicomp, Workshop Privacy*, 2004.

[14] G. Avoine, E. Dysli, and P. Oeschlin, Reducing time complexity in RFID systems, in *Proceedings of SAC'05*, August 2005.

[15] G. Avoine and P. Oeschlin, A scalable and provably secure hash based protocol, in *Proceedings of IEEE PerSec'05*, 2005.

[16] M.E. Hellman, A cryptanalytic time-memory trade-off, in *Proceedings of IEEE Symposium of Information Theory*, 1980.

[17] Ph. Oechslin, Making a faster cryptanalytic time-memory trade-off, in *Advances in Cryptography—CRYPTO'03, LNCS*, Springer, 2003.

[18] D. Henrici and P. Muller, Hash-based enhancement of location privacy for radio frequency identification devices using varying identifiers, in *Proceedings of IEEE PerSec'04*, March 2004.

[19] G. Avoine, Adversarial model for radio frequency identification, *Cryptography ePrint Archive*, Report 2005/049. Referenced 2005 at http//eprint.iacr.org, 2005.

[20] A. Juels, Minimalist cryptography for low cost RFID tags, in *Proceedings of SCN'04*, 2004.

[21] T. Dimitriou, A lightweight RFID protocol to protect against traceability and cloning attacks, in *Proceedings of IEEE SecureComm'05*, 2005.

[22] G. Tsudik, YA-TRAP: Yet another trivial RFID authentication protocol, in *Proceedings of IEEE PerCom'06*, March 2006.

[23] A. Herzberg, H. Krawczyk, and G. Tsudik, On traveling incognito, *IEEE Workshop on Mobile Systems and Applications*, December 1994.

[24] G. Tsudik, A family of dunces, trivial RFID identification and authentication protocols, in *IACR Eprint*, September 2007.

[25] A. Solanas, J. Domingo-Ferrer, A. Martinez-Balleste, and V. Daza, A distributed architecture for scalable private RFID tag identification, *Elsevier Computer Networks*, 51(9), 2268–2279, June 2007.

[26] A. Juels and S.A. Weis, Defining strong privacy for RFID, in *Proceedings of RFID Privacy Workshop*, 2003.

[27] D. Molnar, A. Soppera, and D. Wagner, A scalable, delegatable pseudonym protocol enabling ownership transfer of RFID tags, in *Proceedings of SAC'05*, August 2005.

[28] A. Soppera and T. Burbridge, Secure by default: The RFID acceptor tag (RAT), in *Proceedings of RFIDSec'06*, July 2006.

[29] S. Fouladgar and H. Afifi, A simple delegation scheme for RFID systems (SiDeS), in *Proceedings of IEEE RFID'07*, March 2007.

[30] S. Fouladgar and H. Afifi, An efficient delegation and transfer of ownership protocol for RFID tags, in *Proceedings of EURASIP Workshop RFID'07*, September 2007.

[31] H. Chien and C. Chen, Mutual authentication protocol for RFID conforming to EPC Class 1 Generation 2 Standards, *Computers Standards and Interfaces*, February 2007.

[32] S. Lemieux and A. Tang, Clone resistant mutual authentication for low-cost RFID technology, *IACR Eprint*, May 2007.

[33] B. Defend, K. Fu, and A. Juels, Cryptanalysis of two lightweight RFID authentication schemes, in *Proceedings of IEEE PerSec'07*, March 2007.

[34] http://www.avoine.net/rfid/

11 A Secure RFID Access Control Mechanism

Dijiang Huang and Zhibin Zhou

CONTENTS

Extremely limited in computational and energy capability, radio-frequency identification (RFID) tags, especially passive tags, can hardly authenticate the scanning readers. Thus, information leakage of RFID tags is one of the most challenging problems holding back users' confidence in adopting RFID technologies. In this chapter, we propose Smart RFID Keeper (SRK)—an off-tag RFID data access control mechanism, which is installed in RFID-enabled environments to regulate RFID scanning. To solve the information leakage threat, SRK is designed to (1) authenticate the user, (2) detect and counter unauthorized accesses to RFID tags, and (3) enforce fine-grained access policy.

11.1   INTRODUCTION

Radio-frequency identification (RFID) tags are small, wireless devices that help identify objects and people, as the successor of widely used optical bar code. It can be expected that RFID tags will proliferate into billions and eventually into trillions in the near future. RFID technologies has been widely deployed in Medicare, manufacturing, supply chain management, and many other areas. New security threats have been identified. People who carry RFID tags may be tracked and profiled without being noticed; intruders may use counterfeited RFID tags to acquire physical access to the protected buildings [1].

Among other security threats, the leakage of information on RFID tags to unauthorized users has been identified as a major challenge to RFID technologies. The information leakage threat stems from one of the fundamental design principles of RFID system: low cost. It is reported that electronic product code (EPC) [2] tags costed several U.S. cents apiece [1,3] in 2006. In the quest for minimal cost, the RFID tags, especially passive tags, possess only a couple of thousand gates and no battery. The passive RFID tags, such as EPC, energize themselves by absorbing radio wave emitted from readers and reply to interrogations of any readers without authenticating the identities of readers. Moreover, robust cryptographic algorithms are too complicated to be implemented economically on the RFID tags.

Much work has been proposed to solve the information leakage problem by integrating light-weighted cryptographic algorithms into the tag architecture and using tag–reader authentication protocols to control the access to RFID tags. We call these solutions on-tag access control mechanism, and we argue that on-tag solutions have the following drawbacks: (1) complicating the architecture of RFID tags violates the minimalist design principle and increases the cost of individual tags, (2) with the presence of tag–reader authentication protocols, readers are able to identify less tags in each second, because extra clock cycles need to be allocated to the reader–tag protocols, (3) lightweight authentication protocols are vulnerable to sophisticated attacks, and (4) scalability of complicateds RFID architectures is still an open problem.

The off-tag solutions [4,5] to information leakage take advantage of auxiliary devices, which are much powerful than RFID tags, to regulate the access of readers to RFID tags and require little change to the architecture of RFID tags. Thus, universally used passive RFID tags such as EPC tags can be protected by these solutions. In this chapter, we propose an off-tag access control mechanism: Smart RFID Keeper (SRK), trying to solve the information leakage problem. Our proposed off-tag solutions avoid the above-mentioned drawbacks of on-tag solutions: (1) the architecture of tags remains minimalist, (2) RFID singulation protocols are simple and no tag–reader authentication process exits, (3) with enough resource, auxiliary devices can authenticate the readers with robust cryptographic algorithms and protocols, and (4) the system is easy to extent.

The research of countermeasures for information leakage of RFID systems is largely motivated by the immense market. In the coming years, RFID will be intensively used in various circumstances, especially (among others) in Medicare environments. A number of hospitals in the United States and Europe have deployed RFID and even more Medicare organizations are going to follow the trend and enjoy the benefits in facilitating the data collection, instrument locating, and people tracing. In Medicare domain, information collection policies associated with RFID applications are frequently identified as a major problem holding back users' confidence in adoption of these technologies. To protect the privacy of the patients, Health Insurance Portability and Accountability Act was set forth in 2002, which imposes liability on covered entities for failing to protect privacy of the patients and insured records. Thus, data collection control for RFID system is quite critical for the Medicare organizations to protect the privacy of patients and to avoid liability. Similarly, as RFID is widely utilized in the supply chain and manufacturing environments, information leakage of RFID systems may be taken advantage by competitive commercial opponents for espionage purpose.

To counter the information leakage threat, we propose SRK, whose architecture is shown in Figure 11.1. A room watch dog (RWD) is a device installed in every protected room and consists of an authentication module and an access control module. The authentication module authenticates the users and the access control module is to detect and block unauthorized tag-reading operations. If users get authenticated by the authentication module, they are authorized to read the tags; if not, access control module can detect and block the unauthorized reading. An authentication server (AS) supports the authentication module of RWD to authenticate the identities of users. In this

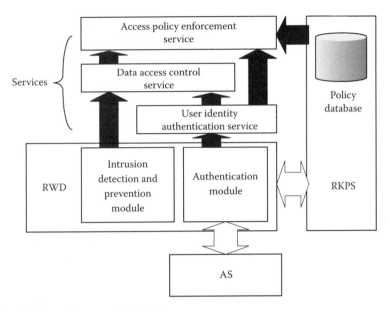

FIGURE 11.1 System architecture of the SRK.

work, we propose one architecture with online AS and one architecture with offline AS (please see Sections 11.3 and 11.4 for detailed information). Furthermore, with an RFID keeper policy server (RKPS), fine-grained access control can be enforced. SRK provides three security services: (1) user identity authentication service: a user should be authenticated by RKPS and authenticated users are authorized to perform *read* operations; (2) data access control service: unauthorized read operations to RFID tags are blocked by RWD, and (3) access policy enforcement service: fine-grained policy can be enforced to the access of RFID tags. In this work, we also provide performance and security analysis of SRK.

Chapter organization: The rest of this chapter is organized as follows. In Section 11.2, we present a brief literature survey related to our research. In Section 11.3, we describe the system models in our design. UIA service, data access control service, and access policy enforcement service are presented in Sections 11.4, 11.5, and 11.6, respectively. We then follow up with some evaluation and analysis results in Section 11.7. Finally, we conclude our work and provide future research in Section 11.8.

11.2 RELATED WORKS

Researchers have recognized privacy problems with RFID systems for a few years [6–8]. In particular, information leakage of passive RFID is a critical but unsolved issue. Much research has been done on the countermeasures for information leakage issues. A very comprehensive survey can be found in Ref. [1], in which the authors presented a survey on privacy and security research issues for RFID systems and many tag–reader authentication protocols. Due to the page limitations, we just mention some of the newest works to the best of our knowledge. We can classify these solutions into on-tag solutions [9–17] and off-tag solutions [4,5]. Meanwhile, due to the lightweight feature of on-tag solutions, many security vulnerabilities have been identified by cryptanalysis researchers in Refs. [18–22].

In Ref. [23], the authors presented various collision resolution protocols for multi-access communications between RFID tags and readers. Access control through blocking techniques has been proposed in Ref. [4] based on multi-access control protocols. The authors claimed that their approach can be also used to block the reader-driven time division multiple access (TDMA) singulation algorithms. However, the research in Ref. [24] pointed out that the straightforward implementation of RFID blocking is vulnerable to the differential signal analysis (DSA). In Ref. [5], we proposed RFID keeper to address the access control problems in RFID, which blocks the unauthorized reading based on various RFID singulation protocols.

11.3 SYSTEM MODELS

In this section, we present the network model and the attack model that support the proposed SRK architecture.

11.3.1 NETWORK MODEL

In this work, we describe two network models: one with online AS and one with offline AS. The network models are illustrated in Figures 11.2 and 11.3, respectively.

The RWD is installed in each protected room to authenticate the identity of users and control the data access to the tags. The authentication module of RWD authenticates the identities of users (refer to Section 11.4 for more details). The access control module can detect and block unwanted scanning of RFIDs (Please refer to Section 11.5 for more details).

The RKPS provides fine-grained access control policy along with RWD. We assume that the communication between the RWD and RKPS cannot be compromised by attackers.

In Figure 11.2, an online AS stores users' credentials and connects to every RWD through a secure channel.

In Figure 11.3, an offline AS determines some system parameters prior to the setup of network. These parameters, which are required in the authentication process, are preinstalled into each RWD and readers.

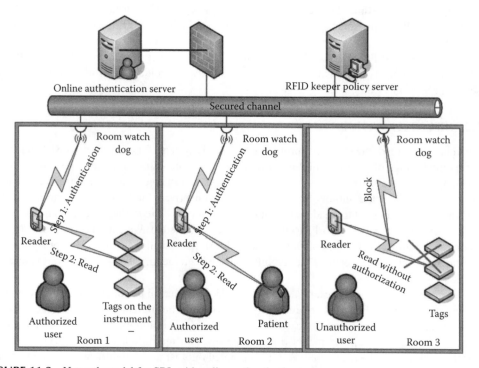

FIGURE 11.2 Network model for SRL with online authentication server.

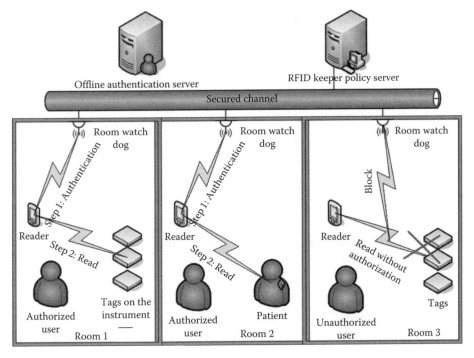

FIGURE 11.3   Network model for SRL with offline authentication server.

11.3.2   ATTACK MODEL

To present our proposal, we need to define the assumptions on the setting of environments and possible attackers. We assume that minimalist design of RFID tags, i.e., the passive RFID tags without cryptographic functions, are used in our environment. The attackers, i.e., attack source, can be either insiders or outsiders who are not authorized to collect data from RFID tags. The attack method includes (1) unintentional mis-operations of insiders, e.g., a nurse fails to follow the required authentication process before his or her scanning RFID tags attached to patients; (2) intentional, unauthorized reading RFID tags from insiders or outsiders, e.g., a malicious attacker tries to harvest patients' healthcare records held in the RFID tags. We assume that the attackers cannot compromise the AS and RKPS. We also assume that the attackers do not have unbounded computational capability to (1) break the ciphers encrypted using symmetric key encryption algorithms; (2) break discrete logarithm problems (DLP) [25].

It should be noted that the communication between passive tags and readers is in plain text and thus vulnerable to passive eavesdropping attacks. How to prevent passive eavesdropping attacks is still an open problem in this work. In Ref. [1], the authors argue that, due to limitation of RFID tags response distance, the eavesdropping attackers need to get physical proximity to RFID tags.

11.4   USER IDENTITY AUTHENTICATION SERVICE

UIA service authenticates the user before the user (reader) can read the data on the RFID tags. This service is provided by authentication module of RWD and AS, which authenticate the user with authentication protocol. In this work, we propose two authentication protocols, namely *Protocol*$_1$ and *Protocol*$_2$. *Protocol*$_1$ is constructed using one-way hash functions and the AS needs to be online. *Protocol₂* is constructed using asymmetric cryptographic functions and AS can be offline. We compare the two protocols in the following table:

	Protocol₁	*Protocol₂*
Cryptographic functions	Symmetric key functions	Asymmetric key functions
Computational overhead	Low	High
AS	Online	Offline

11.4.1 PROTOCOL₁

Notations used in *Protocol₁* are presented in the following table.

Symbols	Descriptions
RWD	Room Watch Dog
RKS	RFID Keeper Server
Uid	User Identification
Rid	Room Identification
Pw	Symmetric Password shared by the user and RKS
Req	Scan Purpose and Tag Selection Request
T	Security Token
Sq	Sequence Number
Rp	Room Policy
Ps	Authentication Passed
Dn	Authentication Denied
Cr	Challenge Request
Bye	Logout Request

The *Protocol₁* consists of three processes that are shown in Figure 11.4. A user is first required to pass the authentication by AS before he or she can use RFID readers to interact with RFID tags. Otherwise, the user will be blocked by the RWD (see descriptions in Section 11.5). During the login process, a user needs to provide his or her *Uid*, *Pw*, and *Req* to pass authentication. To do this, the reader generates a security $Token = Hash_{Pw}(Uid, Sq)$ and then sends message $<Uid, T, Req, Sq>$ to RWD. RWD forwards the message together with the room ID to the online AS through a secure channel. AS needs to process the following validations in sequence:

1. It looks for the $<Uid, Pw>$ in the user-information database.
2. It verifies the *Token* by computing $Hash_{Pw}(Uid, Sq)$.
    (a) If the verification is passed, AS replies the pass message (*Ps*), the room policy (*Rp*), and the increased sequence number ($Sq + 1$) to the RWD and handhold reader.
    (b) Otherwise, AS returns a denial message (*dn*).

After the authentication procedure, RWD needs to challenge the users periodically to prevent session hijacking attacks (Section 11.7). The challenge process is similar to the login process except that RWD first sends the challenge of $<Cr, Uid>$ to the reader.

After the user extracts data from the tags, she or he needs to go through logout process that gracefully terminates the communication session.

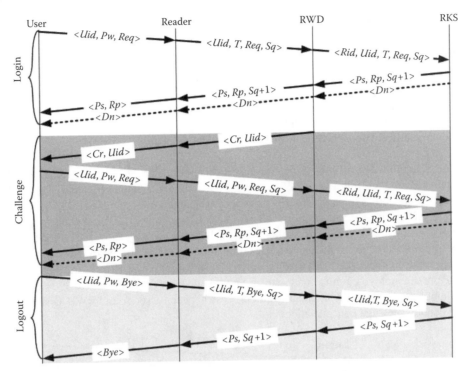

FIGURE 11.4 Information flow of authentication protocol. Three phases: login, challenge, and logout. Note that the dashed arrows stand for the information flow when the authentication failed.

11.4.2 PROTOCOL₂

Notations used in *Protocol₂* are presented in the following table.

Symbols	Descriptions
ID	Identification of user of RWD
Ph	Secured hardware card storing users private key
Pw	Password only known to user
Req	Scan Purpose and Tag Selection Request
N	Nonce Number
V	Validation of Identity
K	Shared Key between User and RWD
Ps	Authentication Passed
Dn	Authentication Denied
Cr	Challenge Request
Bye	Logout Request

Before proceeding to the details of our protocol, we give the definition of pairing, building block for *Protocol₂*. Pairing is a bilinear map function $\hat{e}$ in the following form:

$$\hat{e} : \mathbb{G}_1 \times \mathbb{G}_1 \rightarrow \mathbb{G}_2, \tag{11.1}$$

where $\mathbb{G}_1$ and $\mathbb{G}_2$ are groups with large prime order q. DLP is hard for $\mathbb{G}_1$ and $\mathbb{G}_2$.*

---

* It is computationally infeasible to extract the integer $x \in \mathbb{Z}_q^* = \{i | 1 \leq i \leq q-1\}$, given $P, Q \in \mathbb{G}_1$ (respectively, $P, Q \in \mathbb{G}_2$) such that $Q = xP$ (respectively, $Q = P^x$).

Pairing has the following properties:

- Bilinearity:

$$\hat{e}(aP, bQ) = \hat{e}(P, Q)^{ab}, \quad \forall\, P, Q \in \mathbb{G}_1, \quad \forall\, a, b \in \mathbb{Z}_q^*. \tag{11.2}$$

- Nondegeneracy: If $\hat{e}(P, Q) = 1$ for all the $Q \in \mathbb{G}_1$, then $P = \mathcal{O}$. Alternatively, for each $P \neq Q$ there exists $Q \in \mathbb{G}_1$ such that $\hat{e}(P, Q) \neq 1$.
- Computability: There exists an efficient algorithm to compute the pairing.

Before setup of SRK, offline AS determines pairing parameters $\{\hat{e},\ \mathbb{G}_1,\ \mathbb{G}_2,\ H_1,\ H_2\}$ and these parameters are installed into every RWD, prior to the setup of network. H_1 and H_2 are collision-resistant hash functions: H_1, mapping strings to nonzero elements in $\mathbb{G}_1$ and H_2, e.g., SHA-1, mapping arbitrary inputs to fixed-length outputs. AS also selects a system-wise master key $g \in \mathbb{Z}_q^*$. The parameters $\{\hat{e},\ \mathbb{G}_1,\ \mathbb{G}_2,\ H_1,\ H_2\}$ are preloaded to each device, while g is well guarded by AS. The AS is responsible to generate the private key corresponding to specific ID of users and RWD. For simplicity, we denote user (i.e., staff reading the RFID) with a capitalized letter. An ID of user A, ID_A, is a structure of employment information of A, and private key s_A is computed as $s_A = g \cdot H_1(ID_A)$.

Without loss of generality, authentication process between User A and RWD is shown below, where $\parallel$ denotes message concatenation.

Login

$A \rightarrow Reader$	:	ID_A, Ph, Pw, Req
$Reader \rightarrow RWD$	:	ID_A, Req, N_1
$Reader \leftarrow RWD$	:	ID_{RWD}, N_2
$Reader \rightarrow RWD$	:	$V = H_2(N_1 \| N_2 \| 0 \| K)$
$Reader \leftarrow RWD$	:	$Ps(Dn)$

User A starts the protocol by entering his ID_A, password Pw, and requested operation Req to the RFID reader. Also, A may need to insert a secure hardware key, which stores A's private key s_A, protected by password of A. Then, the reader sends ID_A, Req, and a random nonce number N_1 to RWD. Upon receiving the request, RWD replies its ID ID_{RWD} and another random nonce number N_2. RWD also computes a pair-wise master key $K' = \hat{e}(H_1(ID_A), s_{RWD})$ as well as a expected validation of identity $V' = H_2(N_1 \| N_2 \| 0 \| K')$. Upon reception of reply from RWD, the reader calculates the master $K = \hat{e}(H_1(ID_{RWD}), s_A)$ and the validation of identity $V = H_2(N_1 \| N_2 \| 0 \| K)$. According to Equation 11.2, if and only if user input the right password, we can have

$$K' = \hat{e}(H_1(ID_A), s_{RWD})$$
$$= \hat{e}(H_1(ID_A), gH_1(ID_{RWD}))$$
$$= \hat{e}(H_1(ID_A), H_1(ID_{RWD}))^g$$
$$= \hat{e}(gH_1(ID_A), H_1(ID_{RWD}))$$
$$= \hat{e}(s_A, gH_1(ID_{RWD})) = K$$

and hence

$$V' = H_2(N_1 \| N_2 \| 0 \| K')$$
$$= H_2(N_1 \| N_2 \| 0 \| K) = V$$

After a successful three-way handshake, A and RWD authenticated each other and shared the same pair-wise master key K. Using K, A and RWD can derive Γ pairs of shared session key (*Skey*) and key index (*Kid*) as

$$\begin{cases} Skey_\gamma = H_2(N_1||N_2||2 \cdot \gamma||K) \\ Kid_\gamma = H_2(N_1||N_2||2 \cdot \gamma + 1||K) \end{cases} \tag{11.3}$$

After the authentication procedure, RWD needs to challenge the users periodically to prevent session hijacking attacks. The challenge process is as follows:

Challenge

$RWD \leftarrow Reader \quad : \quad Cr, Kid_i, N_3 \quad 0 \le i \le \gamma$

$Reader \rightarrow RWD \quad : \quad H_2(Skey_i||N_3)$

After the user extracts data from the tags, he or she needs to go through logout process that gracefully closes the communication session.

Logout

$A \rightarrow Reader \quad : \quad Bye$

$Reader \rightarrow RWD \quad : \quad Bye, Kid_i, N_4, H_2(Skey_i||N_4)$

$Reader \leftarrow RWD \quad : \quad Bye$

We briefly analyze the security of this protocol here.

- If the attacker does not know the private key $gH_1(ID_A)$, the probability that attacker can derive K is $1/p$, where p is the group size of $\mathbb{G}_1$ and $\mathbb{G}_2$. In practice, p is supposed to be large prime numbers (larger than 160 bits), so we can see that $1/p$ is negligible.
- If the attacker compromises one or more private keys $gH_1(ID_A)$, the probability that the attacker can compromise the system-wise master key is equivalent to the probability that the attacker can break DLP in $\mathbb{G}_1{}^*$ or $\mathbb{G}_2.^\dagger$ According to our assumption in attack model, this probability is negligible.

11.5 DATA ACCESS CONTROL SERVICE

The data access control service protects information stored in tags from unauthorized accesses by detecting unauthorized access and injecting interference into wireless communication channels. This service is provided by access control module of RWD. The injected inference signals will prevent unauthorized users from correctly receiving the messages sent by RFID tags. To achieve the functionality, RWD must include both intrusion (unauthorized access) detection and access control capabilities. The basic components and functions are presented in the following table:

Components	Functions
Intrusion Detector	No_Reading_Command(); No_Power_Field()
RFID Simulator Group	Randomly_Choose_Simulator_Subset(); Send_Noise()
Signal Modulation Randomizer	Randomly_Modulate_Signal()

11.5.1 TAG ACCESS DETECTION

In our presented authentication protocols, the readers have to be authenticated before accessing the information. Otherwise, any reading activity without being authenticated is considered as an *intrusion*. The data access control contains an intrusion detector that can detect read operations

---

* Extract g given $gH_1(ID_A)$ and $H_1(ID_A)$.

$\dagger$ Extract g given $\hat{e}(A, B)^g$ and $\hat{e}(A, B)$ where A and B are members in $\mathbb{G}_1$.

specified in reader–tag protocols. For example, most reader–tag protocols require the reader to send a read command to a tag first. If an RWD detects an unauthorized read command, it will send out blocking signals, i.e., a random noise in the same frequency band. The response time of RWD is very critical for the access control. This is because lagged blocking signals cannot prevent adversaries from reading the information replied by tags.

11.5.2 DATA ACCESS CONTROL

Most of multi-access protocols are based on TDMA. In this work, we only consider how to block TDMA-based protocols. In TDMA-based access control protocols, a reader can only communicate with one tag at a time. Thus, if multiple tags response to a reader simultaneously, the access fails. Intensive research [23] has been done to identify each participating tag from one another. To block the data access (i.e., the *read* function), RWD needs to make it impossible for unauthorized readers to differentiate tags' responses. To this end, we propose a set of blocking methods to disable RFID multi-access schemes [23]. In addition, we need to randomize the injected blocking signals to prevent attackers from using DSA [24] to filter out the injected blocking signals.

To perform blocking, RWD randomly chooses a group of *RFID Simulators*. The selected RFID Simulators simulate the RFID tags' signal and perform blocking functions based on specific multi-access control protocols (see later this section for detailed descriptions of blocking functions). In addition, the signal modulation randomizer (SMR) changes the strength of simulators' signals and introduces randomly generated noises to the signal sent by the RFID Simulators. The signal strength keeps on changing, and thus, the DSA attackers cannot filter out blocking signals.

In TDMA, the available channel capacity is divided for participants chronologically. There are two classes of TDMA technologies for RFID multi-access control: (1) tag-driven TDMA and (2) reader-driven TDMA. Tag-driven TDMA functions asynchronously without relying on the control of readers. For example, using ALOHA [26], a tag starts to transmit its ID once it is in the power field and has data to transmit. The tag will continuously retransmit data in random time intervals. Once there is a collision, the tag can either mute or reduce the retransmission frequency.

To block ALOHA TDMA approaches, RWD is required to send randomly generated IDs in very high frequency. Because the blocking signal is randomly generated, it is impossible for the attackers to differentiate the blocking signal from the real signal. This blocking method can either cause collision with the real tags' signals or overwhelm unauthorized readers in a short time by filling their memory space.

The pseudo codes for blocking the tag-driven TDMA are given as in the following algorithm:

```
        Normal_Condition
While(No_Power_Field( ));
Jump to Blocking_Condition;
        Blocking_Condition
While(block_time_interval){
    Randomly_Choose_Simulator_Subset( );
    Randomize_Modulate_Signal( );
    Send(ID);
    Decrease(block_time_interval);}
Jump to Normal_Condition;
```

Using reader-driven TDMA, the readers and RFID tags work in a synchronous manner, i.e., the readers and tags all work in constant time intervals (aka time splits). In Figure 11.5, we present the taxonomy of reader-driven TDMA presented in Ref. [23]. Here, we explain each of them based on the proposed blocking techniques.

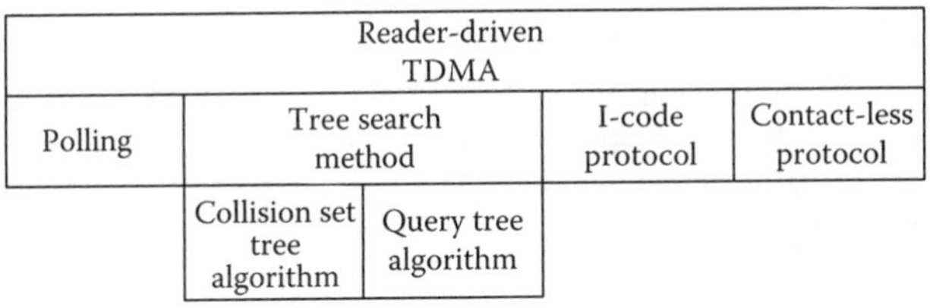

Reader-driven TDMA			
Polling	Tree search method	I-code protocol	Contact-less protocol
	Collision set tree algorithm / Query tree algorithm		

FIGURE 11.5 Taxonomy of reader-driven TDMA.

Using *polling approach*, the reader first identifies each tag in the range by differentiating the prefix of its serial number. For example, the reader first asks: "Does anyone have prefix A?" If some tags reply with "yes!," the reader will ask: "Is there anyone with the prefix Aa?" where a is either 0 or 1. The reader interrogates each tag one-by-one until the reader repeats this process recursively to identify each tag. The proposed blocking technique should prevent an unauthorized reader from identifying each tag. To this end, RWD can randomly select a group of RFID Simulators and reply the reader with randomly modulated signal "yes." As a consequence, the multiple random replies force the reader to try every possible serial number prefix. In RFID GEN2 standard, each RFID tag contains a unique 92-bit serial number and the size of entire possible tag serial number will be 2^{92}. This set is very large, and it is impractical in terms of time and memory for the readers to query every prefix or store the identified serial numbers. Therefore, the reader may be expected to stall after querying thousands of prefix.

The *tree search method* [27] can be divided into two classes: (1) collision set tree algorithm [28] and (2) query tree algorithm [29]. The collision set tree algorithm splits the group of colliding tags into A disjoint subsets, where $|A| > 1$. The subsets get smaller and smaller until the number of tags within a subset equals to 1, and each tag will be uniquely identified. Once a subset is completely resolved, the waiting subsets are resolved in a first-in last-out order. When several tags collide in the same time split, each tag randomly selects to which subgroup it should belong. Thus, it is the tag's responsibility to track the current position in the stack so that the individual tag can decide the time to retransmit the signal. To block this algorithm in each time split, RWD can randomly select a group of the RFID Simulators, which transmits randomly modulated response to the reader. Thus, the subset is never decreased to 1 and the reader will never identify the real RFID tags. The following algorithm illustrates how procedure of blocking algorithm for collision set tree algorithm.

```
        Normal_Condition
While(No_Reading_Command( ));
Jump to Blocking_Condition;
        Blocking_Condition
While(block_time_interval){
    Randomly_Choose_Simulator_Subset( );
    Randomize_Modulate_Signal( );
    Send(ID);
    Decrease(block_time_interval);
    Wait_for_Next_Time_Slot( );}
Jump to Normal_Condition;
```

The query tree algorithm consists of several rounds of queries and responses. In each round, the reader asks tags whether their serial numbers contain a certain prefix B. Then, the tags compare the prefix B with their serial numbers. If a tag has the queried prefix, it replies with the $B1$ or $B0$, where 1 or 0 is the bit in its serial number next to the prefix. If only one tag replies, this tag is identified because no collision occurred; otherwise, the reader will add a bit to the prefix and continue its queries. In Ref. [4], the authors proposed a blocking approach by replying to both *B*1 and *B*0 to block the reader. Similarly, in our approach, RWD randomly chooses a group of RFID Simulators and transmits randomly modulated responses, i.e., *B*1 and *B*0 to the reader. Similar to the polling approach, RKS will overwhelm the reader by forcing him/her to try every possible prefix. The pseudo codes for blocking the query tree algorithm are given as follows:

```
            Normal_Condition
        While(No_Reading_Command( ));
        Jump to Blocking_Condition;
            Blocking_Condition
        While(block_time_interval){
            Randomly_Choose_Simulator_SubsetA( );
            Randomly_Choose_Simulator_SubsetB( );
            Randomize_Modulate_Signal( );
            SubsetA_Send("1");
            SubsetB_Send("0");
            Decrease(block_time_interval);
            Wait_for_Next_Time_Slot( ); }
        Jump to Normal_Condition;
```

I-code protocol [30,31] uses a random seed to randomize each tag's response time split and uses an estimated flexible size of frame to minimize the probability of collisions. To block the I-code protocol, RWD can use a similar strategy of blocking the query tree algorithm. *Contact-less protocol* [32] works similar to the combination of collision subset tree algorithm and query tree algorithm. The reader queries the specific bit position one-by-one and separates the colliding tags into subsets. To block contact-less protocols, RWD can act similarly as blocking the query tree algorithm.

11.6 ACCESS POLICY ENFORCEMENT SERVICE

We propose in this section, with the help of RKPS, fine-grained policy can be enforced to the access control of RFID readers to RFID tags. Up to now in this chapter, SRK can authenticate the ID of users-prevent unauthorized users from reading the tags in the protected rooms. However, fine-grained access control is still not available, that is, SRK cannot differentiate authenticated users from each other and grant different access privileges to users in different roles. However, fine-grained access control to RFID tags is not just interesting add-ons in some environments. Consider the following example:

Example 1

In a hospital, patients' health information may include highly private data such as prescriptions for antidepressants, cancer, long-ago abortions, AIDS or HIV, testing for the Alzheimer gene, children's autism or ADD, sexual impotency prescriptions. These private information may be stored in some RFID tags to facilitate data collecting performed by responding nurses or doctors. Suppose this hospital uses same RFID tags to manage and track the properties (medical equipments) and another group of staff are responsible to scanning the tags on properties. However, possible information leakage can occur when one corrupted staff scans the tags of a patient. Without fine-grained access control, the hospital can hardly differentiate the staff from doctors, neither to protect the patients' privacy.

Now, we are ready to present the design details of this service. Remember in the user authentication protocol, users are required to provide his or her ID to RWD. Here, we define the structure of ID to be

$$\{Department,\ EmployeeID\}$$

Immediately after a user submits the ID, the RWD will follow the authentication protocol and, simultaneously, forward the ID to RKPS. In RKPS, policies may be structure of the following items:

RoomNumber(s)	One or multiple numbers of target rooms
Department	Department regulated by this policy item
EmploymentID(s)	(optional) One or more employee ID(s) regulated by this policy item
Prefix[a . . .]	RFID binary prefix allowed to be scanned

If RKPS finds an item in the policy database, allowing a staff from property control department to scan the RFID tags with prefix [000]. RFKS will return "allow to scan prefix [000]" to RWD. In Ref. [4], the authors propose a reader friendly selective blocking protocol, which is based on the query tree algorithm mentioned in Section 11.5. In this protocol, the so-called blocker tag announces the predefined privacy zones (subtree of RFID serial numbers with certain prefix) to the RFID readers, which should avoid reading tags in these zones. In SRK, we modify this protocol to let the RWD announce the privacy zone to readers. After RFKS returns "allow to scan prefix [000]" to RWD, RWD will send Ps and announce the allowed prefix [000] to the reader. Afterward, RWD can watch over the scanning activity of user and enforce the access policy; that is, if the reader follows the policy, RWD will not block it and, if not, RWD will block the reader.

We give a possible example of application of access policy enforcement service:

Example 2

A hospital makes use of RFID tags to tag the property whose prefix is [000] and each patient bears an RFID tag whose prefix is [111]. Property control staffs can only read the tags starting with [000] and the doctors can read the tags starting with [111]. If, property control staff's read tries to read tags starting with [111] (which can only happen in mistake operations or malicious attack), RWD can detect the protocol-dependent reading command (see Section 11.5) and block the readers.

11.7 EVALUATION AND ANALYSIS

11.7.1 Computation Performance Evaluation

In this section, we evaluate the computation performance of UIA protocols described in Section 11.4. In both protocols, RWD and RFID readers are required to compute some cryptographic functions. In $Protocol_1$, an one-way hash function is used; in $Protocol_2$, cryptographic pairing and two hash functions are used. RWD and RFID readers are lightweighted embedded devices, with limited computing capability. We need to make sure that such cryptographic functions can be computed efficiently in such lightweight devices. We evaluate the timing of computing pairing and hash functions in personal computer (PC) and personal digital assistant (PDA). We assume the hardware configuration is similar to the tested PDA and, thus, we can assess the computation performance of RWD and readers in our protocols by evaluating the timing performance of PDA. The hardware configuration of the tested PDA is provided in the following table:

Components	Specifications
Processor	I32 bit-624 MHz Intel Bulverde technology-based RISC processor
OS	Microsoft Windows Mobile 2003 Second Edition software for Pocket PC
Memory	92 MB total memory (128 MB ROM and 64 MB SDRAM) up to 135 MB user available memory that includes 80 MB iPAQ File Store

In the following table, we show the timing results [33] of computing Tate pairing [34,35] in PC and PDA.

Devices	Processor	Timing (ms)
PC	1GHz Pentium III	20
Pocket PC	32 bit-624 MHz Intel Bulverde technology-based RISC processor	550

Figure 11.6 shows the timing results [33] of various hash functions in PDA. From the data above we can see that Tate pairing and hash functions can be efficiently computed in RWD and RFID readers, if they are equipped with similar computing capability with PDAs. Moreover, because cryptographic pairing function is significantly expensive than hash functions and $Protocol_2$ computes pairing once, we can expect RWD and reader to complete $Protocol_2$ in less than 600 ms.

11.7.2 SECURITY PERFORMANCE ANALYSIS

In this section, we analyze the security of SRK under the session hijacking attack, eavesdropping attacks, and denial-of-service (DoS) attacks.

11.7.3 SESSION HIJACKING ATTACKS

In the session hijacking attacks, the attackers read the tags after a user was authenticated. One attacking scenario happens when the attacker reads the tags while the authenticated users are reading

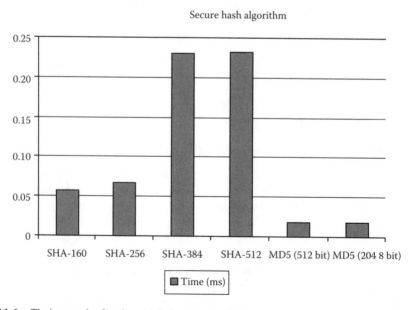

FIGURE 11.6 Timing result of various hash functions in PDA.

the tags in the same room. To counter this attack, RWD is required to detect the number of performing readers in the protected room. If the number of readers is larger than the authenticated readers in this room, a session hijacking attack can be identified.

Another attacking scenario happens when an authorized user leaves the protected room without logging out and the attacker masquerades as the authorized user to read the tags. The periodical challenge initialized by RWD can effectively reduce the amount of information leaked. The attacker, who is unaware of the password of previous logged in user, will fail the challenge and RWD will send blocking signals. Moreover, RKS will log this event and the user who forgot to logout will be notified.

11.7.4 EAVESDROPPING ATTACKS

Communications between RFID tags and readers are vulnerable to the eavesdropping because very few passive tags use cryptographic protections. However, due to the short reading range of passive tags, the eavesdroppers need to be the physical proximity of RFID tags [1], which is a sporadic activity.

11.7.5 DENIAL-OF-SERVICE ATTACKS

We analyze two types of DoS attacks to the proposed approaches: (1) *denial of reading*: the attackers prevent authorized users from reading RFID tags; and (2) *denial of authentication*: the attackers prevent RWD from authenticating users. These DoS attacks are illustrated in Figure 11.7.

11.7.5.1 Denial of Reading

To prevent authorized users from reading RFID tags, the attackers can keep on sending read commands without being authenticated. As a result, RWD will send blocking signals, which will also block authorized readers. RWD should report to system administrators after several consecutive read commands without authentications.

11.7.5.2 Denial of Authentication

To perform the denial of authentication attacks, attackers can block the RWD from detecting reader's signals by covering the RWD with a metal shield (as illustrated in Figure 11.7). Then, attackers can read RFID tags without being detected by the RWD. To counter this attack, we propose three approaches: (1) passive listening, (2) active scanning, and (3) interactive challenging.

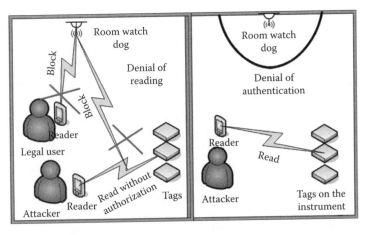

FIGURE 11.7   Denial of reading and denial of authentication.

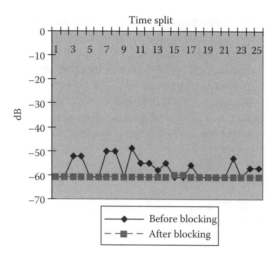

FIGURE 11.8 The strength of noise before and after blocking the RWD.

By deploying a metal cover over an RWD, attackers can block read commands from reaching the RWD. However, this cover will also block the random signal and noise in background. Thus, RWD can detect the presence of metal cover by monitoring the background noise level in its room. If the RWD is blocked, the noise level will decrease, or vice versa. Figure 11.8 illustrates the change of background noise in 2.4 GHz frequency band before and after blocking. The data are collected for every 0.01 s and we collect 25 samples for analysis.

We can differentiate two data series by computing their standard deviations:

$$\sigma = \sqrt{\frac{1}{N} \sum_{k=1}^{n} (x_i - \overline{x})^2}$$

The following table shows the results of standard deviations of the noise level before and after blocking the RWD.

Data series	Standard deviation
Before blocking the RWD	4.2454
After blocking the RWD	0.2769

In the experiment, we collected 25 samplings and the sampling frequency is 100 Hz. It takes 0.25 s to collect the data series.

In active scanning approach, RWD periodically sends cosine signals and collects the reflected signals. In the normal condition, RWD keeps the fingerprint of reflected signal. If the reflected signal changes significantly, it is quite possible that this RWD is subject to denial of authentication attack.

In interactive challenging approach, two RWDs, sharing a pair-wise key, challenge each other periodically by sending a random number. After receiving a challenge, the responding RWD uses an one-way hash function H to derive the response $Hkey(random\ number)$.

We note that all presented three approaches are vulnerable to active and sophisticated attacks. For example, attackers can deploy man-in-the-middle attacks to the interactive challenging approach. However, we consider these activities are difficult because the attacks are required to set up sophistic devices and the activity is rather sporadic.

11.8 CONCLUSION AND FUTURE WORK

In this work, we describe the SRK as an RFID data access control infrastructure. In our design, SRK can authenticate the identity of user, control the access to RFID tags and enforce fine-grained access control policy to the environment. We also evaluate the performance and analyze security of the SRK. We postulate that SRK is very fit for many RFID deployed areas, such as hospitals, and corporation warehouses. We also find that the SRK is vulnerable to passive eavesdropping attacks and some sophisticated active attacks. Our future work should be concentrated on countering these attacks. One possible research direction is combining the off-tag mechanism with minimal on-tags operations, like XOR operation. We believe that the combined design may achieve more satisfactory results than the previously proposed solutions, in terms of security and cost-effectiveness.

REFERENCES

[1] A. Juels and R. Pappu. Squealing euros: Privacy protection in RFID-enabled banknotes. In R. N. Wright, editor, *Financial Cryptography—FC'03*, of *Lecture Notes in Computer Science*, vol. 2742, pp. 103–121, Le Gosier, Guadeloupe, French West Indies, January 2003. IFCA, Springer-Verlag.

[2] Epcglobal inc. In *www.epcglobalinc.org*.

[3] S. Sarma. Towards the five-cent tag. Technical report, MIT-AUTOID-WH-006, MIT Auto-ID Center, Cambridge, MA, 2001.

[4] A. Juels, R. Rivest, and M. Szydlo. The blocker tag: Selective blocking of RFID tags for consumer privacy. In *Proceedings of ACM Conference on Computer and Communications Security—CCS*, pp. 103–111, Washington, DC, October 2003.

[5] Z. Zhou and D. Huang. RFID Keeper: An RFID data access control mechanism. In *Proceedings of IEEE Globalcom Conference*, pp. 4570–4574, 2007.

[6] A. Juels. RFID security and privacy: A research survey. *IEEE Journal on Selected Areas in Communications*, 24(2):381–394, 2006.

[7] D. McCullagh. RFID tags: Big Brother in small packages. *CNET News. com*, 13:2010–1069, 2003.

[8] S. Sarma, S. Weis, and D. Engels. RFID systems and security and privacy implications. In B. Kaliski, C. Kayaço, and C. Paar, editors, *Cryptographic Hardware and Embedded Systems—CHES 2002*, of *Lecture Notes in Computer Science*, vol. 2523, pp. 454–469, Redwood Shores, CA, August 2002. Springer-Verlag.

[9] S. Weis, S. Sarma, R. Rivest, and D. Engels. Security and privacy aspects of low-cost radio frequency identification systems. In D. Hutter, G. Müller, W. Stephan, and M. Ullmann, editors, *International Conference on Security in Pervasive Computing—SPC 2003*, of *Lecture Notes in Computer Science*, vol. 2802, pp. 454–469, Boppard, Germany, March 2003. Springer-Verlag.

[10] M. Burmester, B. de Medeiros, and R. Motta. Robust, anonymous RFID authentication with constant key-lookup. *Cryptology ePrint Archive*, Report 2007/402, 2007.

[11] G. Tsudik. A family of dunces: Trivial RFID identification and authentication protocols. *Cryptology ePrint Archive*, Report 2006/015, 2007.

[12] S. Lemieux and A. Tang. Clone resistant mutual authentication for low-cost RFID technology. *Cryptology ePrint Archive*, Report 2007/170, 2007.

[13] C. C. Tan, B. Sheng, and Q. Li. Severless search and authentication protocols for RFID. In *International Conference on Pervasive Computing and Communications—PerCom 2007*, New York, March 2007. IEEE, IEEE Computer Society Press.

[14] Y. Cui, K. Kobara, K. Matsuura, and H. Imai. Lightweight asymmetric privacy-preserving authentication protocols secure against active attack. In *International Workshop on Pervasive Computing and Communication Security—PerSec 2007*, pp. 223–228, New York, USA, March 2007. IEEE, IEEE Computer Society Press.

[15] M. McLoone and M. Robshaw. Public key cryptography and RFID tags. In M. Abe, editor, *The Cryptographers' Track at the RSA Conference—CT-RSA, Lecture Notes in Computer Science*, San Francisco, CA, February 2007. Springer-Verlag.

[16] M. Conti, R. D. Pietro, L. V. Mancini, and A. Spognardi. RIPP-FS: An RFID identification, privacy preserving protocol with forward secrecy. In *International Workshop on Pervasive Computing and Communication Security—PerSec 2007*, pp. 229–234, New York, March 2007. IEEE, IEEE Computer Society Press.

[17] J.-S. Chou, G.-C. Lee, and C.-J. Chan. A novel mutual authentication scheme based on quadratic residues for RFID systems. *Cryptology ePrint Archive*, Report 2007/224, 2007.

[18] B. Defend, K. Fu, and A. Juels. Cryptanalysis of two lightweight RFID authentication schemes. In *International Workshop on Pervasive Computing and Communication Security—PerSec 2007*, pp. 211–216, New York, March 2007. IEEE, IEEE Computer Society Press.

[19] T. Li and R. H. Deng. Vulnerability analysis of EMAP - an efficient RFID mutual authentication protocol. In *Second International Conference on Availability, Reliability and Security—AReS 2007*, Vienna, Austria, April 2007.

[20] T. Li and G. Wang. Security analysis of two ultra-lightweight RFID authentication protocols. In *IFIP SEC 2007*, Sandton, Gauteng, South Africa, May 2007. IFIP.

[21] P. Peris-Lopez, J. Hernandez-Castro, J. Estevez-Tapiador, and A. Ribagorda. Cryptanalysis of a novel authentication protocol conforming to EPC-C1G2 standard. *Computer Standards amd Interfaces*, 31(2), pp. 261–526, 2009.

[22] A. Bogdanov. Attacks on the KeeLoq Block Cipher and Authentication Systems. In *Proceedings of the 3rd Conference on RFID Security*, 2007.

[23] D.-H. Shiha, P.-L. Suna, D. C. Yenb, and S.-M. Huangc. Taxonomy and survey of RFID anti-collision protocols. *Computer Communications*, 29:2150–2166, 2006.

[24] M. Rieback, B. Crispo, and A. Tanenbaum. Keep on blockin' in the free world: Personal access control for low-cost RFID tags. In *Proceedings of the International Workshop on Security Protocols (IWSP), Lecture Notes in Computer Science*, Cambridge, U.K., April 2005. Springer-Verlag.

[25] A. Menezes, P. C. van Oorschot, and S. A. Vanstone. *Handbook of Applied Cryptography*. CRC Press, Boca Raton, FL, 1997.

[26] M. Liard and V. D. Corporation. *The Global Markets and Applications for Radio Frequency Identification and Contactless Smartcard Systems*. Venture Development Corp., Natick, MA, 2003.

[27] J. Capetanakis. Tree algorithms for packet broadcast channels. *IEEE Transactions on Information Theory*, 25:505–515, 1979.

[28] D. Hush and C. Wood. Analysis of tree algorithms for RFID arbitration. In *Proceedings of IEEE International Symposium on Information Theory*, p. 107, August 1998.

[29] C. Law, K. Lee, and K. Siu. Efficient memory-less protocol for tag identification. In *Proceedings of the 4th International Workshop on Discrete Algorithms and Methods for Mobile Computing and Communications*, pp. 75–84, August 2000.

[30] H. Vogt. Multiple object identification with passive RFID tags. In *Proceedings of IEEE International Conference on Systems, Man and Cybernetics*, vol. 3, p. 6, October 2002.

[31] H. Vogt. Efficient object identification with passive RFID tags. In *Proceedings of International Conference on Pervasive Computing*, pp. 98–113, April 2002.

[32] M. Jacomet, A. Ehrsam, and U. Gehrig. Contactless identification device with anti-collision algorithm. In *Proceedings of IEEE Computer Society, Conference on Circuits, Systems, Computers and Communications, Athens (CSCC)*, July 1999.

[33] A. Ramachandran, Z. Zhou, and D. Huang. Computing cryptographic algorithms in portable and embedded devices. In *Proceedings of the IEEE International Conference on Portable Information Devices (PORTABLE)*, pp. 1–7, 2007.

[34] P. Barreto, H. Kim, B. Lynn, and M. Scott. Efficient algorithms for pairing-based cryptosystems. *Advances in Cryptology–Crypto*, 2442:354–368, 2002.

[35] D. Boneh and M. Franklin. Identity-based encryption from the Weil pairing. *SIAM Journal of Computing*, 32(2):586–615, 2003.

12 Threat Modeling in EPC-Based Information Sharing Networks

Alexander Ilic, Trevor Burbridge, Andrea Soppera,
Florian Michahelles, and Elgar Fleisch

CONTENTS

EPC-based information sharing networks are a global effort to standardize the supply chainwide exchange of operational trace data. Due to the complex nature of supply chains and different information sharing relationships, security is a critical issue. Prior research and end-user feedback suggest that there is currently a limited understanding of how to assess and address security threats that could affect multiple parties. In this chapter, we describe a threat model that can help to compensate this shortcoming. Our model helps to assess current as well as future risks. Our findings suggest that designers, operators, and users of EPC-based information sharing networks should focus on providing accountability as a key aspect of improving collective security.

12.1   INTRODUCTION

Today's global market places face information uncertainty. The demands in almost every industrial sector are volatile, and product and technology life-cycle times have shortened dramatically. Many companies have experienced difficulties to predict the effect of market changes and the effect of understocking or overstocking increases. In this dynamic context, we see the need for supply chains that are able to cope with high level of heterogeneity and customization. RFID technology is a cost-efficient way of gathering trace data about logistic objects. Amongst other benefits, RFID is said to optimize supply chain operations [1], reduce theft [2], and prevent counterfeiting [3]. The EPCglobal Architectural Framework [4] offers standards for gathering, filtering, and sharing trace data with other partners in a supply chain through the EPC information service (EPCIS). Sharing trace data through information sharing network beyond a single organization enables a radical new degree of supply chain visibility and traceability. Capturing and sharing supply chain information is valuable for many trading partners. This information can be used to improve and customize services and processes, to provide statistical and marketing information and could, in certain situations, be sold to third parties. On the other hand, we have to be careful that misuse and unauthorized access to this information could violate service agreements, cause fraud, and, in certain cases, disrupt critical supply chain processes. The risk is that a great deal of dependency on external processes and information could lead to a loss of control and expose a company to greater supply chain vulnerability.

Organizations perceive and address security issues in different ways, ranging from completely ignoring them and losing control of confidential information (mainly due to lack of awareness), to being so cautious as to prevent new technology being deployed because of the lack of expertise in recognizing and dealing with potential threats. The prevailing, dominant, strategy is to consider these threats as an internal risk, and to manage them locally (within the bounds of the enterprize). The wider supply chain context is only rarely considered, and there is minimal support for those needing to optimize large-scale, global-level, supply chains. This is paradoxical and most likely contrary to the real source of the greatest threat. It can be argued that the biggest risk to an enterprise may in fact be in the wider supply chain network, and the data control mechanisms applied within an organization itself is just a small part of the security it really needs. The result of this is often an exposure to higher levels of risk as a result of miscommunication and lack of tools to express authorization to electronically manage information. Consequently, EPC-based information sharing networks suffer from so-called interdependent security problems (as described by Kunreuther and Heal [5], e.g.).

It is important for senior managers to identify the most relevant and critical threats and to concentrate on sharing this information across the supply chain partners so that an appropriate supply chainwide security strategy can be put in place. Overall we seek to provide means for simplifying security management experience so that organizations can feel they are in control of their confidential data and that this data is managed in an accountable way. The goal of this chapter is to provide a tool and a method for reasoning so that the most relevant and critical threats can be identified. As the suspected interdependent nature of security problems increases the problem domain complexity, a structured approach is needed. Hence, we employ a threat modelling approach that can serve as a basis for existing enterprise risk management frameworks. The purpose of our threat model is therefore to establish where potential weak areas lie and what impact threats for internal processes and for the wider network have. Once an organization understands its potential threats, it can then start to put in place an appropriate security strategy (counter measures), and will have a clear picture of how security breaches could compromise their own processes, as well as potentially damaging their customers or partners.

To achieve the goal of providing a structured understanding for IT security threats associated with EPC-based information sharing networks, this chapter is structured as follows. First, we start with related work and present that, to our best knowledge, no threat model for an EPCglobal-based information sharing network exists. In Section 12.3, we develop our threat model based on the information life cycle of an EPCglobal-based network. Section 12.4 then completes the threat model by providing an attacker perspective on the different life-cycle phases. We use a qualitative analysis approach, where we introduce the potential attacker types together with their motivations and capabilities. Different attacks are enumerated, described, and categorized against their threat to the classical security goals of confidentiality, integrity, and availability (CIA). Section 12.5 shows the practical relevance of our threat model for improved security risk management. We provide an application guideline that is concluded with a fictive example. In Section 12.6, we discuss our learnings and findings. Finally, Section 12.6 concludes and summarizes the key results.

12.2   RELATED WORK

The goal of this chapter is to provide a threat model to better understand the nature of security problems in the domain of EPC-based information sharing networks. In general, security is a topic that is largely discussed in the area of RFID. A recent research survey of Juels [6] shows, the academic community is currently mainly focused on securing RFID tags or the tag to reader link. A reason for this may be the current hype on privacy issues [7] due to insecure tag implementations and the amplifications of public perception in the media. Avoine and Oechslin [8] recognize that RFID technology imposes a multilayer privacy problem. Their perspective focuses on a physical, a communication, and a simplified application layer. Garfinkel et al. [9] look at the RFID privacy problems not only from a multilayer perspective but also beyond the scope of a single organization. Also, general security RFID security documents such as the National Institute of Standards and Technology (NIST) [10], boot-sector infector (BSI) [11], and BRIDGE [12] report confirm that there are security issues beyond the protection ability of a single entity. Explicitly focused on the security of the EPCglobal network specification is the work of Konidala et al. [13], which assesses the security of individual interfaces and elicits a broad range of security threats. However, although all of the previously mentioned works state one or more solutions concerning the hardware and software levels, they rarely discuss the interorganizational and network aspects of security investments. On a general perspective for interorganizational security problems, Kunreuther and Heal [5] discuss the class of so-called interdependent security problems. They use a game-theoretical approach to prove that organizations are better off if they cooperate in different scenarios. Yet, they confirm that each party may have the incentive to "cheat" and save on investment, at the same time increasing the risk of a potential loss to itself and other partners through security vulnerabilities. A threat analysis should therefore always consider the risk of contagion from other organizations that have not yet implemented the same level of security. Moreover, as Anderson [14,15] indicates, a solution to the interdependent security problems requires properly aligned incentives for each participating organization to cooperate for higher collective security. In contrast to the cited papers above, we look at the RFID security from an interorganizational and economically motivated perspective to demonstrate that EPC-based information sharing networks suffer from interdependent security problems. We use a structured threat modeling approach to identify potential threats and weak areas in EPC-based information sharing networks. The threat model is hereby a suitable representation to identify threats in a certain domain. The idea is that this domain knowledge of security threats can feed into existing risk management processes or frameworks of organizations and therefore improve the overall security management process.

12.3 THREAT MODEL OVERVIEW

A system may be exposed to many different kinds of threats. For the remainder of this chapter, we will focus on threats that could emerge from a previously unknown vulnerability. As the probability of such an event is not predictable outside a specific context, we will focus on understanding within which areas threats can theoretically occur. Our threat model is based on a simple information life cycle for RFID read-event data, which will be introduced in the following section. The threat model comprises the following three components:

- The system model, which offers a suitable perspective on the system that should be protected
- The attack sources, which describe the characteristics of likely attacker types
- A threat list, which contains some attacks against the classical security goals of CIA

12.3.1 INFORMATION LIFE CYCLE

Instead of focusing only on technology aspects, this chapter is concerned with the security problems associated with the exchange of item-level event information in EPC-based networks. We assume that the information of interest is generated through reads of RFID tags. As these RFID tags are attached to logistic objects, the supply chainwide sharing of these read events may be of significant business value. Generally, if data is generated at one organization and should be shared with another one, the following steps occur. First, the event data is created by an organization. Second, the organization prepares and approves the data for sharing with other selected parties. Finally, interested and authorized parties can search and retrieve the data from the offering party. The described flow resembles an information life cycle (Figure 12.1), which is a suitable baseline for analyzing information security risks [5].

The advantage is that the life-cycle model helps to structure the weak areas of a system by decomposing it into functional phases critical for the information handling. Like other academic papers (e.g., [17]), we map the life-cycle phases to specific architectural system components. Figure 12.2 shows one loop of the resulting information life-cycle model for EPCglobal-based information sharing networks. The loop consists of five distinct life-cycle phases, in which organizations can take one or more of the following roles suitable for the exchange of RFID data traces [18]: data supplier*, data consumer, or metadata operator.

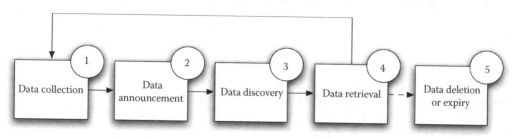

FIGURE 12.1   Generic trace data information life cycle with its five phases.

---

* As we focus on the information sharing aspects, we will use the term trace data supplier instead of distinguishing between trace data creator and trace data publisher.

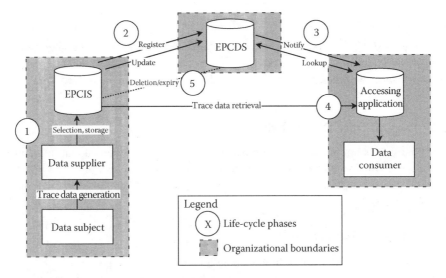

FIGURE 12.2 Information life cycle of the EPCglobal-based network.

12.3.2 System Model Based on Life-Cycle Phases

The following sections describe the functionality, components, and interactions in detail. The description is relevant, as each of the components constitutes potential entry points or vulnerable assets for the overall system.

12.3.2.1 Phase 1: Trace Data Creation and Storage

In this phase, trace data is generated and prepared for sharing throughout the network. First, a trace data supplier uses RFID readers to interrogate tags and create trace data events for specific trace data subjects. The subjects can be any type of tagged object such as container, pallets, cases, or even individual items. Moreover, the RFID tags contain unique identifiers that act as proxy for identifying trace data subjects. By means of these unique identifiers, trace data can be collated to establish trace histories of the data subjects across multiple organizations. Second, the trace data supplier selects the trace data information he or she would like to share with others and stores this information in a database, the so-called EPCIS repository. The EPCIS repository contains access policies that determine which data can be seen by whom and makes sure that an organization can control access to its data.

12.3.2.2 Phase 2: Trace Data Announcement

As information about a certain trace data subject is distributed over a supply chain (due to its logistic flow), the EPCIS discovery service (EPCDS) provides a service to determine which EPCIS repository might have information about a particular object. For this concept to work, each partner offering data traces regarding a particular EPC must announce to the EPCDS that they have related trace data by sending an update announcement to the EPCDS.

12.3.2.3 Phase 3: Trace Data Lookup and Notification

In this phase, a trace data consumer submits a query to the EPCDS to find out which EPCIS repositories have data about specific objects. The query may be a onetime call or a standing query. In the case of a standing query, the EPCDS sends notifications to a trace data consumer whenever a trace data announcement matches their expressed interests.

12.3.2.4   Phase 4: Trace Data Retrieval

After querying the EPCDS for potential information sources (phase 3), phase 4 is now concerned with the actual trace data information retrieval. Depending on the credentials of the trace data consumer and the security policy of the trace data suppliers, the trace data can be retrieved from the different EPCIS systems.

12.3.2.5   Phase 5: Trace Data Deletion

Although this issue is not discussed on a broader scope, it can be assumed that in an information life cycle all of the trace data will not be retained forever. Therefore, we foresee a phase 5, which is typical for almost every life cycle, where trace data is purged through an explicit operation or reaching an expiry time.

12.4   ATTACKER PERSPECTIVE

To complete our threat model, we will now apply an attacker-centric perspective against our life-cycle model established in the previous section to explain the extent and nature of threats. The attacker-perspective was chosen over a system-centric view, as actual implementations of EPC-based information sharing networks may differ in their security strengths and vulnerabilities [19].

12.4.1   ATTACKER TYPES AND CAPABILITIES

In the following section, we describe and characterize the most important attacker types. The three types were chosen due to their access abilities (internal/external) and their main motivation (benefits/damage).

Being aware of the attacker characteristics helps to conduct better risk assessment. The attractiveness to a certain attacker type and the characteristics of the attacker types (as summarized in Table 12.1) largely determines the probability and damage potential of attacks during the life-cycle phases.

12.4.1.1   Competitors

Malicious organizations may want to attack the trace data network to either strengthen their position, to harm their competitor, or a combination of both. Typically the goal is to steal confidential information to gain competitive advantage or to disrupt the information integrity and thereby affect business processes. Process failure can result in direct and indirect financial damage, while subversion of a

TABLE 12.1
Summary of the Characteristics of Investigated Attacker Types

	Competitor	Insider	Saboteur
Main motivation	Competitive benefit	Personal benefit	Damage
System knowledge	Limited knowledge	Full knowledge	Limited knowledge
Trust level	Untrusted	Trusted	Untrusted
Probable entry points	External interfaces	From within	External interfaces
Critical phases	Data announcement	Data creation	Data announcement
	Data retrieval	Data announcement	Data lookup
		Data retrieval	Data deletion
		Data deletion	
Attack scale	Single target	Single target	Single–multiple targets

process can result in benefits to the attacker such as the availability of private assets. One example of this subversion is the use of regular distribution channels for the sale of counterfeit goods. What makes a competitor an attractive target is that the damage and losses caused can directly translate into the other organization's benefits. The access and knowledge to the network's security weaknesses is, however, fairly limited. Competitors need to find vulnerabilities in a very cautious way. They will therefore likely target the vulnerabilities where the attack is easy to perform and hard to trace, which usually lie in system configuration and interaction [20]. Potential entry points may therefore focus on the public network interfaces of a trace data supplier (in phases 2 and 4). Moreover, the manipulation of physical items or tags in phase 1 is also possible, as is observation of network traffic during phases 2, 3, and 4.

12.4.1.2 Insiders

Insiders are employees of network participants that have malicious intentions of disrupting the network or stealing information for their personal benefit. Personal motivational reasons often include low wages and working environment, affiliation with a competitor or terrorist organization, or personal benefits, for example due to predictive stock market reactions. Insiders are particularly dangerous, as they can have the full knowledge of the internal system of an organization and the resources at their disposal to run an extensive and well-prepared attack. They have a trusted status within one organization and can exploit this to harm either the whole system (including the organization they work for) or specific targets. Unlike other attacker types, insiders do not need to rely on finding vulnerabilities. Instead, they can abuse their privileges or attack the network via hidden attacks. Attack situations may become particularly attractive if observation and therefore punishment is difficult or unlikely [21]. Entry points for attacks usually come from within an organization and can consist of both remote and local proximity attacks. Life-cycle phases 1, 2, 4, and 5 are particularly vulnerable to an attacker within the data supplier organizations.

12.4.1.3 Saboteurs

In contrast to other attacker types, the motivation of saboteurs is not primarily to get personal benefits from attacks but rather to cause as much damage as possible in as little time as possible [17]. Like competitors, they need to invest in finding vulnerabilities or to use an insider (e.g., social engineering) before being able to mount an attack. Once they find a vulnerability that is applicable to more than one particular target, they will likely aim at exploiting the vulnerability and attack multiple targets. Potential entry points include especially centralized or shared network elements such as the EPCDS, attacks on which would affect phases 2, 3, and 5. An attack affecting these phases could cause damage to all participants by disrupting the service availability or metadata integrity. If an attack is targeted more specifically at individual targets, potential entry points can be found in the EPCDS interface (phase 2) and the EPCIS interface (phase 4). Also, saboteurs are able to mount attacks on phase 1, by using either insiders or specially prepared tagged objects equipped with malicious software (e.g., RFID-virus [22]) or other hardware (e.g., blocker tags [6], radio jamming).

12.4.2 THREATS

In this section, we discuss threats against components of the trace data systems. To structure the discussion, we categorize potential attacks against the information life-cycle phases and refer them to the (CIA) security goals. The following list briefly explains each of the CIA goals:

- Confidentiality: Only authorized parties should have access to the trace data at specified times and in a specified manner. This applies to data in storage (tag, EPCIS), during processing (ALE) or in transit (over a network).

- Integrity: The trace data should remain accurate and complete. In addition, system components should retain their integrity and operate as intended.
- Availability: Data, networks, and information systems must be available in a timely manner to meet the requirement of business operations.

Attacks that compromise the CIA goals can result in numerous threats to the business. Each business must analyze the severity of the business threat that can result from the attack on the trace data system. Such threats may include the stalling or subversion of a business operation. For example, a shipment may be stopped or delayed, or sent to the wrong location. Attacks on the trace data may also be used for activities such as theft or the introduction of counterfeit goods into the existing supply chain. Compromising the confidentiality of any trace data activity may also be used to infer business activity and implement competitive strategies, resulting in a loss of market or suppliers.

12.4.2.1 Attacks during Trace Data Creation and Storage

Trace data is generated by reads of RFID tags and the resultant processing. Attacks are possible on the tags themselves, along with the collection and processing networks and the trace data storage systems. The communication networks, such as the wireless tag–reader protocol and the trace data supplier's internal networks should also be considered open to attack.

- Confidentiality: Such attacks comprise of both unauthorized access to trace data and eavesdropping on legitimate communications. For example, tags may be read by unauthorized readers for competitive intelligence, identifying opportunities for theft, or the cloning of the tag or other communications. Network traffic may also be observed and unauthorized access attempts made to trace data collection components or storage systems.
- Integrity: Attacks on the integrity of the system may impact on the CIA and accountability of the trace data. Attackers may seek to compromise the integrity of the trace data by attacking the elements or networks within the trace data supplier. The attacker may also target the tag or reader devices that may be physically accessible at certain points in their lifetime. The trace data integrity may be compromised by modification or removal of the data on the tag, EPCIS, or as it passes through any network or intermediate systems. Cloning and replay of trace data should also be considered. The tag itself may be cloned for later presentation to a tag reader, for example on a counterfeit good. Communications may also be replayed to the original trace data supplier's systems, or systems within a different organization. Other injection attacks may use falsified information, delivering this into the system where sufficient checks are not performed on the data integrity or the identity of the injecting system.
- Availability: Attackers may seek to remove the availability of system components (and hence trace data) from dependent systems and processes. Access may be disrupted by attacking the system components or communications capabilities. External attackers may attack external interfaces and components. This will include physical attacks on tags and readers and disruption of the tag communication, for example through radio or protocol jamming [6]. Wireless networks used for the tag communication and wireless reader devices are particularly vulnerable.

12.4.2.2 Attacks during Trace Data Announcement

Attackers can target the systems involved in the announcement of trace data, including the originating trace data supplier, the EPCDS, or the intervening network such as the Internet.

- Confidentiality: Attackers may seek to gain access to the announcement of trace data. They can do this by eavesdropping on the network used to communicate with the EPCDS from the trace data supplier. Attackers may also impersonate an authorized recipient of the announcement, for example subscribing to trace data announcements with false credentials at either the EPCDS or the trace data supplier systems.
- Integrity: Attackers may attack the integrity of trace data announcements by modifying or removing announcements, or injecting false or replayed announcements. This may cause trace data consumers to miss the announcement of trace data, be mislead about the existence of trace data, or be diverted to incorrect trace data suppliers.
- Availability: Attackers may attack the availability of the EPCDS update interface, along with the network carrying such updates and the systems in the trace data supplier producing updates. Such availability attacks will affect the integrity of the trace data held in the EPCDS or the timely availability of the trace data for use within business processes. Because the EPCDS update interface is likely to be available to other entities over the Internet, it is particularly vulnerable to large-scale denial-of-service (DoS) attacks from external entities such as saboteurs.

12.4.2.3 Attacks during Trace Data Search

Attackers can target the search activity between the trace data consumer and the ECPDS. This can involve attacks on the communication network, the EPCDS, or the trace data consumer systems.

- Confidentiality: Attackers will attempt to compromise the confidentiality of the trace data announcements held in the EPCDS, and may also eavesdrop on the trace data searches and responses from other parties. The availability and interest of parties in EPC identifiers may constitute sensitive business information. The patterns of EPCs announced and accessed may be mined to infer business information. Such patterns may include the parties and EPCs involved along with the timing of the announcements/searches, and any other information that may be available such as geographic location. Even if such communications are securely encrypted, the network traffic may still be mined to infer business activity. For example, attacker may learn that a certain pattern of announcements and searches occurs when Company A receives a palette of a specific type of goods.
- Integrity: Attackers may mount man-in-the-middle attacks to affect the trace data consumer, along with attacking the integrity of the trace data held by the EPCDS or the operation of the EPCDS itself. This can subvert operations relying solely on the trace data announcements, cause trace data to remain unnoticed (stalling business operations), or lead trace data consumers to perform trace data retrieval on the incorrect systems.
- Availability: Attackers may launch DoS attacks to exhaust the trace data search capabilities. This will prevent trace data consumers from being able to search and retrieve trace data announcements. Processes will fail to act on new trace data in a timely manner, producing delays in business operations.

12.4.2.4 Attacks during Trace Data Retrieval

Attackers may target the trace data supplier and consumer systems or the communications network used to transfer trace data.

- Confidentiality: Attackers may seek to compromise the confidentiality of the trace data maintained in the EPCIS or eavesdrop on communications between the trace data consumer and trace data supplier. Along with the confidentiality of the trace data, the confidentiality of the trace data consumer should also be considered. The trace data requests will reveal detailed information about the trace data consumer's operations.

- Integrity: Attackers may compromise the integrity of the trace data in the EPCIS, or the integrity of the networked communications between the trace data consumer and supplier. The trace data consumer may be misled by removing trace data from the retrieval response, or by modifying or fabricating additional trace data.
- Availability: Attackers may target the trace data consumer and supplier external interfaces or communication networks to remove their ability to perform trace data retrieval. Because the EPCIS is a widely reachable service, it is vulnerable to DoS attacks. Although the EPCIS may restrict service to only trusted trace data consumers (under normal operation or during times of service overload), attacks to deny network bandwidth will remain possible. Solutions to availability threats should consider solutions that address both the network and system availability.

12.4.2.5 Attacks during Trace Data Deletion

Attacker may target the trace data and announcement storage systems or the operations to remove or renew trace data and announcements.

- Confidentiality: Attackers may seek access to the trace data deletion information. Such messages may inform the attacker that the trace data was present, along with revealing information about the lifetime and usefulness of the trace data. Depending on the system implementation, the attacker may listen to expiry and refresh messages, or explicit deletion instructions.
- Integrity: Attackers may attack the integrity of the trace data deletion communications or seek to delete trace data (announcements). Deletion of the announcement information from the EPCDS will mean that trace data is not found by trace data consumers. Removal of trace data from the trace data supplier will cause confusion as trace data consumers attempt to retrieve data that no longer exists, particularly if this breaks service level agreements (SLAs) for the retention of data and incurs financial or other penalties.
- Availability: Attackers may attack the availability of the systems and networks during the trace data deletion phase. This may result in the data being retained unintentionally, or may actually lead to the premature removal of data (e.g., if a refresh instruction is disrupted).

12.5 APPLICATION GUIDELINES

Organizations rely on risk management to select cost-effective countermeasures for mitigating potential threats. Risks are usually assessed in the dimensions of negative impact (potential damage, unfavorable adverse effects, and consequences) and probability (at which a risk is likely to occur) [23]. To evaluate risks according to these factors, a comprehensive understanding of the situation is required. In IT security, threat modeling is regarded as an enabling step for effective security risk management [24]. Organizations can use our threat model as a tool to better understand and estimate security risks associated with trace data sharing networks. A security process based on threat modeling, as described in Ref. [25], is depicted in Figure 12.3.

12.5.1 GENERAL GUIDELINES

To be able to actually use the proposed threat model as a basis for risk management, the following steps need to be applied to put the threat model into an organization's context (Figure 12.3). The context allows for determining the individual threat exposure depending on the roles and phases of the life-cycle model. The threat model supports the identification of the risks based on contextual factors as shown in Table 12.2. After the risk identification phase, risks can be evaluated by using our context factors together as input failure mode and effect analysis (FMEA) [26] model or to a proven

Risk identification

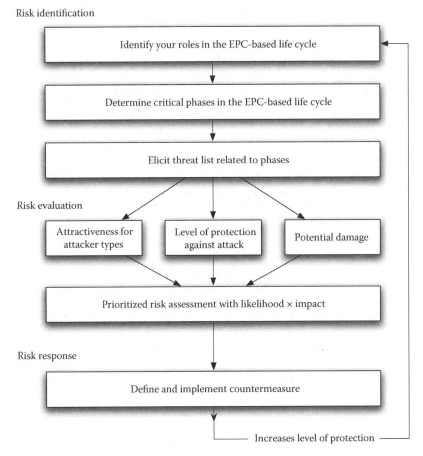

FIGURE 12.3 Threat modeling as basis for security risk management. (Adapted from Hasan, R. et al. Toward a threat model for storage systems, *Proceedings of the 2005 ACM Workshop on Storage Security and Survivability*, ACM Press, New York, 2005, pp. 94–102.)

TABLE 12.2
Summary of the Relevant Context Factors for Risk Identification and Risk Evaluation

Threat Model Elements	Subjective Context Factors	Influence on
Roles	Relevance, goals	Threat exposure, number of risks
Phases	Dependency on others	Threat exposure, number of risks
Attacker characteristics	Attractiveness for attacker	Impact, probability
Attack	Own protection strength	Impact, probability

framework as COSO risk management framework [27]. The objective is to estimate the dimensions of probability and negative impact for each identified threat. The resulting threat list can now be prioritized and visualized with a likelihood/impact diagram [27]. Depending on an organization's risk appetite [27], appropriate risk responses and countermeasures may now be specified. Finally, the implementation of these actions mitigates or prevents the assessed risks and improves the overall risk profile for a given organization.

What is an FMEA?

An FMEA is an risk assessment technique for systematically identifying potential failures in a system or a process. FMEA is normally used within the design phase with the aim to avoid future failures. The objective is to prioritize our security threats according to four criteria: how serious the consequences are, how frequently they occur, how easily the attack can be detected, and how attractive a successful attack is for the attacker. The four criteria, as discussed in Table 12.2, are impact (I), probability (P), threat exposure (T), and ability to control the attack (D).

   The four criteria are then associated with a range from 1 (lowest risk) to 3 (highest risk) as discussed in the table below. The overall risk for each threat is then called risk priority number and it is obtained by multiplying the four scores together. In the table below we describe an example of a rating system, the analysis performed at points 3 and 4 will help to identify the correct value for each index.

12.5.2 THREAT ANALYSIS STEP-BY-STEP

In the following, we describe each step of Figure 12.3 in detail and relate to the existing frameworks and proven methodologies where possible. In line with the general remarks above, we note that the application of the threat model builds the foundation for the risk identification step. The steps of risk evaluation and risk response are captured here only to provide a sound application example. Actual implementations of risk evaluation and risk response may depend on an organizations practice. The threat model's output, a customized list of threats, is, however, vital for their success. It adds the domain specific threat knowledge required for determining the right actions.

12.5.2.1 Risk Identification

To identify the risk profile of the EPC-based network system is important to analyze the critical operations together with the critical sources of risk within the specific organization's context.

- Identify key roles: The objective is to identify which role an organization takes up for a specific EPC-based information sharing application. This could be one or more of the roles stated in Section 12.3, namely, trace data consumer, trace data supplier, or metadata operator. For example, in an e-pedigree application, a manufacturer could take up only the role of a trace data supplier, whereas a retailer would take up only the role of a trace data consumer. All parties in between might take up both, the roles of a trace data supplier and a trace data consumer.
- Identify critical phases: The objective of this point is to identify which phases of our life-cycle information model if compromised or sabotaged could affect internal and external supply chain operations of the organization. It is likely that components of the EPC-based information sharing networks which are involved in the critical phases are those, where we want to focus our future security investments. A critical phase could be identified by fulfilling one of the following characteristics:
  1. An element of the system on which many others could depend, for example the "trace data announcement" phase where information contained in an EPCIS is needed to enable a timely track and trace for other trading partners.
  2. An element of the system with limited amount of alternatives, for example the phase "trace data retrieval" where the only source of information is a single EPCIS repository. If this repository is compromised, no other way of retrieving the required data is possible.

3. An element of the system that is associated with a high risk environment, for example the phase of "trace data search" where a publicly running web service (e.g., EPCDS) could expose confidential supply chain information without a secure access control mechanism.

12.5.2.2   Risk Evaluation

Rather than evaluating in depth all the possible security risks that a company might face, the threat model analysis helps to isolate the most relevant threats based on the previous steps, the attacker types, and relevant supply chain scenarios. EPC-based networks can be seen as a complex web of interconnected nodes and relationships. The nodes represent components, EPCIS, discovery service, and the links are the means by which information is exchanged—network connection. The security threats represent the risk of failure of these nodes and links and our goal is to identify which combination of these nodes and links are critical.

- Attractiveness for attacker types: How likely is for a certain element of the system to attract a certain type of attack? Where are the protection mechanisms? How much additional capacity is available if the system fall under a DoS attack that consumes system's resources? Traditionally, we could expect that if a component transports valuable information then it represents a high risk element. However, for an EPC-based network the risk of failures for most services does not depend on a single component, for example an e-pedigree service relies on the integrity of a set of supply chain record and an attacker could just decide to perform an action against the weakest link to bring the whole system offline.
- Own protection strength: What are the security mechanisms already in place? Are standard monitoring tools available to warn about security vulnerabilities? Do I have good communication with suppliers and customers to develop a greater understanding of potential vulnerabilities and attacker strategy? Ideally organization needs to be able to react quickly, and protection mechanisms should be reviewed regularly as part of the risk assessment process.
- Internal and external potential damage: Threats identified in step 2 could lead to various damages to internal supply chain processes and logistic operations (roles). The challenge is to isolate the impact of these threats for a specific scenario.
- Risk assessment with likelihood per impact: The purpose of this step is to define where the greatest threats lie. Generally accepted risk management frameworks such as COSO [27] or FMEA [26] can help to quantify the dimensions of individual risks by evaluating the combination of probability and impact. Note that the previously gathered threat domain knowledge with the list of several potential risks is used as an input to them. The output is a prioritized list that reflects the risk estimation of a particular context. A brief description of FMEA is provided in Table 3. A curious reader may refer to Ref. [27] for more details.

12.5.2.3   Risk Response

Once the major threats of the system have been identified and prioritized, we can develop specific countermeasures to mitigate the potential damage of an attack or to prevent an attack nearly completely. At this stage we could also consider to redesign some processes if the probability of occurrence and severity of the attacks are too high. Again it is essential that security issues receive attention on an ongoing basis, the risk identification, and the evaluation task needs to be performed on a regular basis to ensure an appropriate mitigation strategy. Standardization could also play

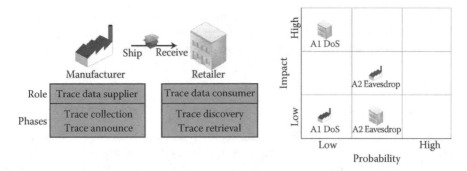

FIGURE 12.4 Example of a fictive two-tier supply chain with the associated threat model roles, phases, and the probability/impact matrix for the selected attacks.

a fundamental role. The EPC global standard will drive for standardization of platforms and components that should reduce the complexity to manage this process across multiple organizations and increase the visibility of potential threats across the chain. However, we should not forget to diversify our technology suppliers, if all components come from the same suppliers it is likely that a single vulnerability could have major effect on our internal system. Access to threat analysis and attack reports from other organization is also another major component that should be considered to mitigate the risk and increase the resilience of our systems. Within a supply chain we should create a collaborative working environment that enables to share relevant information about upstream and downstream threats and that motivate commitment to mitigate and address these security vulnerabilities. The EPCglobal network is already built on these principle and a proposal could then be to provide an extended management of these risks. We will detail this discussion in Section 12.6.

12.5.3 EXAMPLE OF AN INTERDEPENDENT SECURITY PROBLEM

Consider a two-tier supply chain with a manufacturer and a retailer who implement an e-Pedigree application to ensure food traceability. Figure 12.4 shows the scenario put into the threat model context with the corresponding roles and phases. The traceability scenario requires the retailer to verify the pedigree of all incoming objects. Therefore, the retailer is dependent on the availability of the manufacturer's EPCIS database. In contrast, the manufacturer requires confidentiality of the trace data, as the data might be misused by a competitor to reveal shipment quantities between manufacturer and retailer. Figure 12.4 shows the perceived risks for two selected attacks against the life-cycle phase trace retrieval. The attack A1 represents a DoS attack against the EPCIS of the manufacturer and the attack A2 denotes an eavesdropping attack on the data exchanged during trace data retrieval. Although both parties might perceive the probability of the attack realization equally, the impact of the damage can be considerably different (as depicted on the matrix in Figure 12.4). For example, the manufacturer might perceive the potential damage of A1 as low while the retailer would suffer from a high damage potential. The manufacturer is not strongly dependent on the EPCIS availability and therefore such an attack might just consume more bandwidth and traffic costs, but not threaten the business at all. In contrast, the retailer faces process holdups or delays that could cause high costs. The interdependent security problems become apparent when looking at the ability of each party to reduce the imposed risks. For example, the manufacturer can employ encryption and access control to prevent eavesdropping attack A2. However, if the retailer treats security for this aspect loosely and leaks the encryption key to a malicious party, the whole security of the encrypted data traffic is compromised. In conclusion, the security of one party is strongly dependent on the other parties interacting in a certain life-cycle phase.

12.6 DISCUSSION

As illustrated in Section 12.5.3, EPC-based information sharing networks suffer from interdependent security problems. Taking into account that supply networks are rarely as simple as two symmetric partners, this section discusses how to improve the overall collective security of the multipartner supply chain community. Even though the threat model does not claim completeness, it provides a structured and practical way of assessing risk exposure for individual parties, along with an assessment of risk that they place on other parties and the recompense that they can expect for removing those threats. Different risk perceptions are the source for unbalanced motivations for investing in security. If, for example, the costs of security are higher than the risks against that partner in isolation then clearly no partner will ever invest in security, regardless of the behavior of its partners. However, if the costs of security are less than the internal risks combined with the external benefit to other partners, then there exists another equilibrium where all partners can benefit from the combined investment in security. The problem is therefore to convince all parties in the system to move to this beneficial collaborative equilibrium. There are several options to achieve this goal. In the following, we discuss cooperative, noncooperative, and externally motivated solutions. In a cooperative approach, organizations would share their views on threat probabilities and especially threat impacts. The threat model would be used for a joint risk assessment with a bilateral understanding of the risks and attractiveness for certain attacker types. The result could be a joint action plan for protecting identified critical points. In a noncooperative approach, organizations would assess and implement security measures based on their own risk perceptions. Each party would be held accountable for the losses of other parties resulting in the failure of its security measures against previously set critical points. These points and penalties are usually coordinated through contracts such as SLAs. Simpler market mechanisms may include the choice of whether to do business with a trading partner, knowing that our organization will be exposed to uncontrolled risks. A business may chose only to do business with partners who can show compliance to a security accreditation, technical standards, and business practices. With supply chain, wide contracts, the benefits of proper security investments to external parties can be internalized, making a decision to implement security straightforward for every party. Therefore, we reason that a secure EPC-based information sharing system must include clear accountability. Such accountability can include records of who submitted trace data, along with who accessed data, and for what purpose. Data may be signed as proof-of-origin, and systems provided to ensure nonrepudiation of trace data, only when implementing proper accountability, incentives, or penalties can be applied effectively. In externally motivated solutions, coordinating bodies, such as an industry consortium or government agency, can be used to encourage the implementation of security across all partners. This can be achieved through different means such as subsidies for implementing security, fines for failure to adopt industry standards, and even regulation. In such circumstances, regulation can be in the interests of all the parties because it forces a multilateral move toward security. Again, accountability is a key property of the technical solution to allow for implementing this approach. Based on the threat model, we reason that the shared motivation to implement secure trace data systems is not sufficient without the tools to implement security, and the ability to gain assurance that supply chain partners have also done so. Security does not stop at the product selection and integration, but continues with the business practice. Regular audits from external trusted agencies can ensure that trace data partners continue to operate their business to manage the risks that can be introduced to their partners' supply chain processes. Technology can assist with the accountability of trace data operations, preventing many attacks and ensuring that others can be traced and corrective action taken to reduce future threats. The above discussion has largely been around the motivations of the trace data supplier and trace data consumer relationships to implement security; however, there are other parties within a trace data network that must also be considered. Parties such as the trace data operator (implementing the EPCDS) must consider both the trace data suppliers and trace data consumers that it works with. In this case, however, it is expected that security failures will result in internalized losses through the breach of SLAs, and the loss of business to other trace data operators.

12.7 CONCLUSION

The objective of this chapter has been to provide a tool to identify and prioritize potential risks associated with EPC-based information sharing networks. We developed a general trace data information life-cycle model that allows further tailoring to specific organizations. We introduced threat modeling as a basis for individual risk management and outlined factors that should be considered within each life-cycle stage to analyze the threat. These factors include the role performed during the life-cycle phase, attractiveness for certain attacker types and the protection strength of the implemented system against specific attacks. We then discussed some guidelines to be able to actually use the proposed threat model as a basis for risk management on a specific context. The context allows for determining the individual threat exposure depending on the roles and phases of the life-cycle model. We focused on the fact that when tailoring the threat model to a specific context, the interdependent nature of the security risks become apparent. For the fictive example of a retailer and manufacturer, we show the magnitude of one's risks is strongly dependent on the actions of the other party. With increasing complexity of supply chains, the interdependent security risks become pervasive and require a supply chainwide solution. Therefore, we discussed the potential to mitigate the interdependent security problems by cooperative risk assessment, market mechanisms such as contractual incentive design, and external enforcement.

Our findings suggest that designers, operators, and users of EPC-based information sharing networks should focus on providing accountability as a key to improve collective security. Technical accountability mechanisms within standardized security frameworks are essential to the enforcement of service contacts or regulatory practices and are also essential to identify the root of any attack and remove future threats. In addition, because security incidents are not completely preventable, the issue of recovery has major practical relevance. For example, how long does it take until a network can recover from a compromised digital signature key? As EPCglobal-based information sharing networks support business processes, they not only need to focus on how to manage the security risks, but also how quickly they can recover and restore operations.

We highlight that future research is needed to investigate the role of security frameworks and contractual design for making interdependent security problems explicit and their resolution more efficient.

REFERENCES

[1] H. L. Lee and O. Ozer, Unlocking the value of RFID, Graduate School of Business, Stanford University, Working paper, 2005.
[2] A. D. Smith, Exploring the inherent benefits of RFID and automated self-serve checkouts in a B2C environment, *International Journal of Business Information Systems*, 1, 2005, 149–181.
[3] T. Staake, F. Thiesse, and E. Fleisch, *Extending the EPC Network: The Potential of RFID in Anti-Counterfeiting*, ACM Press, New York, 2005, pp. 1607–1612.
[4] K. Traub et al., The EPCglobal architecture framework, EPCglobal Final Version, 2005.
[5] H. Kunreuther and G. Heal, Interdependent security, *Journal of Risk and Uncertainty*, 26, 2003, 231–249.
[6] A. Juels, RFID security and privacy: A research survey, *IEEE Journal on Selected Areas in Communications*, 24, 2006, 381–394.
[7] F. Thiesse, RFID, privacy and the perception of risk: A strategic framework, *Journal of Strategic Information Systems*, 2007.
[8] G. Avoine and P. Oechslin, RFID traceability: A multilayer problem, *Financial Cryptography and Data Security*, 2005, 125–140.
[9] S. L. Garfinkel, A. Juels, and R. Pappu, RFID privacy: An overview of problems and proposed solutions, *IEEE Security & Privacy Magazine*, 3, 2005, 34–43.

[10] T. Karygiannis et al., *Guidelines for Securing Radio Frequency Identification (RFID) Systems*, Special publication, National Institute of Standards and Technology, 2007, pp. 800–898.

[11] Security Aspects and Prospective Applications of RFID Systems, Federal Office for Information Security, Bonn, Germany, 2004.

[12] M. Aigner et al., D-4.1.1: Security analysis, A. Ilic, Ed., *Building Radio Frequency IDentification for the Global Environment (BRIDGE)*, 2007.

[13] D. M. Konidala, W.-S. Kim, and K. Kim, Security assessment of EPCglobal architecture framework, Auto-ID Labs White Paper Series, WP-SWNET-017, 2006, Auto-ID Labs, Available from http://www.autoidlabs.org.

[14] R. Anderson, Why information security is hard—An economic perspective, *Computer Security Applications Conference*, *Proceedings of the 17th ACSAC*, 2001, pp. 358–365.

[15] R. Anderson and T. Moore, The economics of information security, *Science*, 314, 2006, 610–613.

[16] R. Bernard, Information lifecycle security risk assessment: A tool for closing security gaps, *Computers & Security*, 26, 2007, 26–30.

[17] R. Hasan et al., Toward a threat model for storage systems, *Proceedings of the 2005 ACM Workshop on Storage Security and Survivability*, ACM Press, New York, 2005 pp. 94–102.

[18] M. Bauer et al., Emerging markets for RFID traces, Arxiv preprint cs.CY/0606018, 2006.

[19] D. M. Nicol, Modeling and simulation in security evaluation, *IEEE Security and Privacy*, 3, 2005, 71–74.

[20] S. E. Schechter and M. D. Smith, How much security is enough to stop a thief. *Proceedings of the Financial Cryptography Conference*, Guadeloupe, Springer, January, 2003.

[21] T. Moore, Countering hidden-action attacks on networked systems, *Proceedings of the Fourth Workshop on the Economics of Information Security*, 2005.

[22] M. R. Rieback, B. Crispo, and A. S. Tanenbaum, Is your cat infected with a computer virus? *IEEE Computer Society*, 2006, 169–179.

[23] Y. Y. Haimes, *Risk Modeling, Assessment, and Management*, John Wiley & Sons, 2004.

[24] S. Evans et al., Risk-based systems security engineering: Stopping attacks with intention, *IEEE Security & Privacy Magazine*, 2, 2004, 59–62.

[25] S. Myagmar, A. J. Lee, and W. Yurcik, Threat modeling as a basis for security requirements, *Symposium on Requirements Engineering for Information Security (SREIS)*, 2005.

[26] D. Bell, L. Cox, S. Jackson, and P. Schaefer, Using causal reasoning for automated failure modes & effects analysis (FMEA). *IEEE Annual Reliability and Maintainability Symposium*, 1992, pp. 343–353.

[27] Enterprise risk management—Integrated framework, American Institute of Certified Public Accountants, 2004.

13 RFID-Based Secure DVD Content Distribution

Shiguo Lian and Zhongxuan Liu

CONTENTS

As one of the main multimedia distribution channels, digital versatile disc (DVD) has been one of the main objectives attacked by pirates. Since DVD has the feature that once distributed, it is difficult to be controlled, finding the secure way to protect DVD contents is a challenging task. Until now, there have been some solutions, among which, content-scrambling system (CSS) and advanced access content system (AACS) are two important ones. Although AACS is much more powerful than CSS, such as adopting longer key, and using broadcasting encryption to block certain user, audio watermark to resist recording, and video fingerprint to trace traitors, some recent attacks have broken it. In this chapter, after reviewing the existing techniques and protocols of DVD content distribution techniques including CSS, AACS, and their following progresses, we emphasize on the usage of radio-frequency identification (RFID) for protecting DVD copyright and also indicate that introducing RFID for DVD content distribution will bring some other priorities. After that, we propose a method using RFID in key management for video decryption and fingerprint embedding. This chapter concludes by identifying some open research issues in secure DVD distribution based on RFID.

13.1   INTRODUCTION

Due to the success of Internet technology and the broadband network, more and more people are sharing digital audio and video (AV) contents through P2P network, although lots of those contents are copyright-protected. To reduce the risk of piracy, content owners and publishers want their contents to be protected throughout the producing, distribution, and consumption process. Digital rights management (DRM) system [1–3] can satisfy the Content Providers by encrypting digital AV content and limiting the access right to only those people who have acquired a proper license. Although total security is not possible, DRM can prevent elementary attackers, make it difficult and costly for skilled attackers, and diminish commercial opportunities for professional attackers. However, although DRM adds persistent security to the digital AV contents that it protects, it also adds extra steps to the process of acquiring those contents. There is always a tradeoff between convenience and security, which depends on the content owners to establish a balance.

Radio frequency identification (RFID) [4–6] is an automatic identification technology that can be used to provide electronic identity to an item/object. RFID tags today are small computer chips with or without self-contained energy, each connected to a miniature antenna, which can be attached to physical objects. A typical RFID system consists of transponders (tags), reader(s), antennas, and a host (computer to process the data). Communication in RFID occurs through radio waves, where information from a tag to a reader or vice versa is passed via an antenna. Unique identification or electronic data is stored in RFID tags, which consists of serial numbers, security codes, product codes, and other object specific data. The data on the tag can be read wirelessly by an RFID reader.

Digital versatile disc (DVD) [7], as one of the main multimedia distribution channel, is being widely used. And the attacks on DVD content arise as frequently as the means for protection DVD content [8–10]. This condition origins from DVD's property that it is difficult to control the DVD content once the DVD is distributed. By combining DVD with RFID, identifying DVD becomes easier and copying content becomes more difficult [11,12]. Thus, the DRM system for DVD can be improved by making use of RFID's identification capability or computing capability.

In this chapter, we first review the existing DRM systems for DVD content, and then investigate the secure DVD distribution scheme based on RFID. Furthermore, we propose a secure DVD content distribution scheme using RFID to control the key management and fingerprint embedding. Additionally, some open issues in DVD distribution based on RFID are presented.

The rest of the chapter is arranged as follows. In Section 13.2, the existing DRM systems for DVD content are reviewed, and some DVD protection schemes based on RFID are investigated in Section 13.3. In Section 13.4, the secure video distribution scheme based on DVD and RFID is proposed and the example for MPEG2 video is presented and evaluated in Section 13.5. Finally, some open issues are given in Section 13.6, and conclusions are drawn in Section 13.7.

13.2 DRM SYSTEMS FOR DVD CONTENT

In this section, the existing DRM systems for DVD content are first reviewed to indicate the difference between them and the RFID-based schemes and the priorities of the RFID-based ones.

13.2.1 CONTENT-SCRAMBLING SYSTEM

As the former of DVD, video compact disk (VCD) has no security scheme, which makes pirate very easy. Then, for DVD standard, a scheme named content-scrambling system (CSS) [8] is used as the security scheme. The CSS was defined by the DVD Copy Control Association (CCA) and released in 1996. In this system, the movies put on DVD-video disk are in an encrypted form and a key is required to decode and watch the content. Thus, a DVD player is able to play the content only if it has such a key. To provide the DVD player with the playback key, the manufacturer must first register with the DVD CCA to gain access to the decryption key. Different manufacturers will have different keys and the key can be used to identify the manufacture once a key is published. After the user gets the DVD disk distributed via stores or through Internet purchases, the DVD can be played on a registered player by an unlimited number of times. Thus, the physical disc can be given to anyone else so that the usage is not limited to particular user or playback device.

Recently, CSS has been broken, and the programs to decrypt DVD can be found all over the Internet. Till now, more than six million people have used the hacked CSS decryption algorithm [13]. It is estimated by Hollywood studios that the annual revenue loss caused by the piracy is no smaller than three billion dollars.

13.2.2 ADVANCED ACCESS CONTENT SYSTEM

The Advanced Access Content System (AACS) [9,14] is a standard for content distribution and digital rights management, which is particularly intended to protect DVD content. The specification was publicly released in April 2005, and developed by AACS Licensing Administrator, a consortium that includes Disney, Intel, Microsoft, Panasonic, Warner Brothers, IBM, Toshiba, Sony, etc.

AACS uses cryptography to control the usage of DVD content, which encrypts DVD content with advanced encryption standard (AES) [15] under the control of title keys. The title keys are derived from the combination of media key, Volume ID, and the cryptographic hash of the usage rules. The main difference between AACS and CSS is the decryption key. In CSS, all players are equipped with the same decryption key. DVD content is encrypted under the title key, which is itself encrypted under a certain key. It is difficult to revoke the illegal players. Differently, in AACS, each player uses a unique set of decryption keys. This approach allows licensors to "revoke" individual players and make them out of work when the players compromise or publish their decryption keys.

Additionally, the audio watermarking technology [16] is recommended by AACS. The inaudible mark can be embedded in the sound tracks of media content by modifying the sound slightly. For example, the mark is embedded by varying the wave of speech or music in a regular pattern to convey a digital code. The variation is too subtle to be heard by human ears. If an AACS-compliant player does not detect the inaudible mark, the disc is regarded as a pirate copy, and the player will refuse to play it.

Till now, AACS has not been finalized because it is operated under "interim agreement." Since 2006, some AACS systems have been constructed, although several of them have been broken by extracting the decryption keys from the weakly protected software players [17].

13.2.3   BD-ROM MARK

The ROM Mark [18] works in physical layer and aims to prevent bit-by-bit data copy. The ROM Mark is a unique identifier embedded in prerecorded ROM discs. The data can be read by the disc player but not transmitted out, and therefore they are not available to any readers or players. The ROM Mark can be operated only by the equipment licensed by BD-ROM manufacturers, which confirms the security. Additionally, the ROM Mark can be used to control the playback of media content. For example, before playback, the ROM Mark's existence is decided, and only the disc containing the mark is played, while the other discs are forbidden. However, because the special machinery is required to embed the ROM Mark in disc making, it costs much and is not easy to change the ROM Mark.

13.3   DVD CONTENT PROTECTION COMBINED WITH RFID

RFID can be used to improve the performance of secure DVD content distribution. In the following content, some distribution schemes combining DVD disc with RFID will be investigated.

13.3.1   DVD PROTECTION BASED ON RFID AND USER FINGERPRINT

13.3.1.1   Application Scenarios

In Ref. [11], a secure DVD distribution method based on RFID and user fingerprint is presented. It aims to confirm that users can only play back the DVD bought by themselves in the DVD player registered by themselves, and thus to prevent illegal redistribution. In the application scenario, as shown in Figure 13.1, there are four partners, i.e., license server, DVD reader with fingerprint scanner, DVD shop, and user. When the user buys a DVD from the DVD shop, the DVD shop scans user's fingerprint, and stores the user's fingerprint and access right into the RFID after communicating with the license server. The license server stores the user's fingerprint and access right into a database. When the user watches the DVD with the DVD reader, the DVD reader scans user's fingerprint, matches it with the one in RFID, and decrypts the media content.

Thus, the DVD shop is equipped with an RFID reader and a secured Internet connection to the license server. The user has a DVD player with RFID reader and fingerprint scanner. The whole

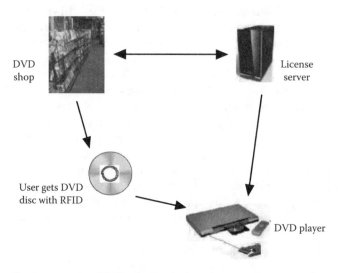

FIGURE 13.1   Application scenario of DVD distribution based on user fingerprint.

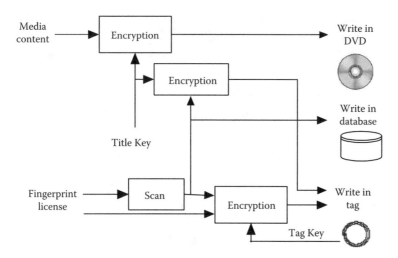

FIGURE 13.2 The process of DVD content making.

application process can be partitioned into two parts, i.e., DVD content making and DVD content reading, which are described as follows.

13.3.1.2 DVD Content Making

As shown in Figure 13.2, when the user buys a DVD, the DVD shop does the following operations:

First, it encrypts the media content with Title Key and writes the encrypted content into DVD disc.

Second, it encrypts Title Key with User Key and writes the encrypted Title Key in the RFID tag. Here, User Key is generated using some secured algorithms and based on the information of Fingerprint in Tag, unique Tag ID, Movie Title, Content Owner, etc.

Third, it scans user's fingerprint, encrypts the fingerprint with Tag Key, and writes the encrypted Tag Key in the tag. Here, Tag Key is generated based on the unique ID of RFID tag, Movie Title, Content Owner, etc., according to some secured algorithms [6].

Fourth, it encrypts the License information with Tag Key and writes the information in the tag.

Fifth, it transmits and stores some important information, such as Tag ID, License, Movie Title, Fingerprint, DVD ID, etc., in the license servers database.

13.3.1.3 DVD Content Reading

Consumer needs to register his fingerprint to his DVD player. Once his fingerprint is successfully protected by a master password and stored in his DVD player, it could be used all along for playback of any DVD. When such a DVD is played back, the DVD player does some operations as shown in Figure 13.3.

First, it generates the Tag Key, decrypts the Fingerprint and License in the tag, and matches the decrypted fingerprint with the one stored in the DVD player.

Second, if the fingerprint is successfully matched, DVD player regenerates the User Key and decrypts the Title Key.

Third, with the Title Key, the player decrypts the media content and plays the content.

13.3.1.4 Advantages and Disadvantages

This method has some advantages. For example, the fingerprint is combined with the DVD. Thus, any copyright pirate can be traced back. Additionally, the playback is controlled by the user fingerprint. Thus, the DVD bought by a user can only be played back on the DVD player registered by the

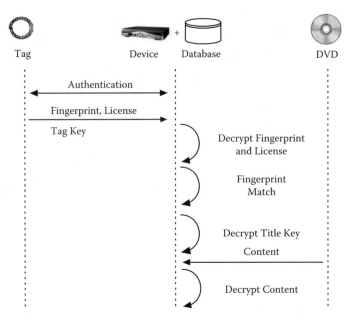

FIGURE 13.3 The process of DVD reading.

corresponding user, which prevents the illegal distribution of DVD copy. The shortcomings of the method lie in two aspects. First, using human's fingerprint may make some user uncomfortable, which limits the user's population. Second, communication with license server (registration and authentication) for every time of buying will give too much burden to the server.

13.3.2 DVD PROTECTION BASED ON RFID AND MEDIA FINGERPRINT

13.3.2.1 Application Scenarios

In Ref. [12], the secure DVD distribution scheme based on RFID and content signature is presented. It aims to confirm that the illegal-copied DVD discs are not able to be played back by the DVD players. In the application scenario shown in Figure 13.4, there are four partners, i.e., Content Provider, License Administrator, User and DVD player. Content Provider computes the media fingerprint with

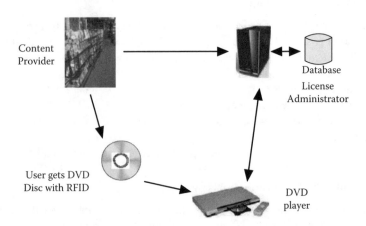

FIGURE 13.4 Application scenario of DVD distribution based on media fingerprint.

the signature function, writes the media fingerprint in the RFID tag and media content in DVD disc, and transmits the signature function to License Administrator. License Administrator stores the signature function in a database. DVD player verifies the consistence of the RFID tag and DVD disc, and plays only the legal discs.

13.3.2.2 DVD Content Making

In DVD content making, the Content Provider does some operations, as shown in Figure 13.5.

First, it computes the media fingerprint with a signature function. As described in the related works [19,20], the media fingerprint refers to content-based identification. It is the feature or signal of media content, which can represent or characterize the content. The signature function is the algorithm used to generate the fingerprint. Actually, there have been already developed many robust ways for fingerprint [21].

Second, it writes the fingerprint in the tag and media content in the DVD disc.

Third, it transmits the signature function to License Administrator and stores the function in a database.

13.3.2.3 DVD Content Reading

In DVD content reading, the DVD player does some operations, as shown in Figure 13.6.

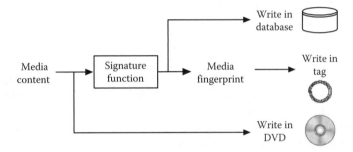

FIGURE 13.5 The process of DVD making.

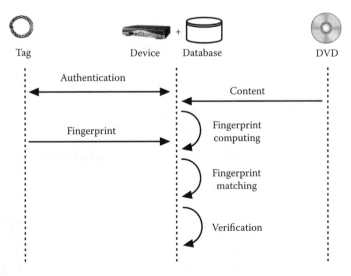

FIGURE 13.6 The process of DVD playing.

First, it reads media content from the DVD disc, gets the signature function from the database in License Administrator, and computes the media fingerprint from media content with the signature function.

Second, it gets the original fingerprint from the tag and compares the original one with the computed one.

Third, if the two fingerprints are same, it plays the media content. Otherwise, it regards the content as an illegal copy and refuses to play the content.

13.3.2.4 Advantages and Disadvantages

In this scheme, the authentication method is effective in a sense that it guarantees the digital contents originality in a more precise way. Thus, the scheme prevents the DVD player from playing illegal DVD content. Here, the illegality includes three cases, i.e., only the tag is copied, only the media content is copied, or both the tag and content are copied. Since the player calculates a fingerprint on the basis of current digital content in the disc, and compares it to the fingerprint of the original digital content stored in the RFID tag, it can guarantee the media content's originality in a precise way. However, the signature function is often of high computational complexity, which may limit the suitability for real-time applications. Additionally, the DVD player needs to communicate with License Administrator frequently, which requires the DVD player connected with networks and limits the application regions.

13.4   SECURE DVD DISTRIBUTION BASED ON RFID AND DIGITAL FINGERPRINTING

In this section, we propose a secure DVD content distribution scheme based on RFID and digital fingerprinting. Digital fingerprinting is the technique to embed identity information into media content. Different from the previous schemes, this scheme uses RFID to generate the media keys and determine the embedded fingerprint bits. The media copy can be traced by the embedded fingerprint code when it is illegally distributed.

To make the scheme easy to be understood, we first introduce some basic techniques, including watermarking, fingerprinting, and partial encryption.

13.4.1   SOME BASIC TECHNIQUES

13.4.1.1   Digital Watermarking

Digital watermarking [22,23] is the technique that embeds some information into media content by modifying media pixels slightly. The embedded information can be extracted or detected from the marked media content and used for authentication. Generally, according to embedding strength, it can be classified into two types, i.e., visible watermarking and invisible watermarking. In visible watermarking, the embedded watermark is visible on media content, while it is invisible in invisible watermarking. Because invisible watermarking does not affect media content's commercial value, it is used more widely than visible watermarking.

As a good invisible watermarking algorithm, some performances are required, including imperceptibility, robustness, embedding capacity, etc. The imperceptibility means that there is no perceptual difference between the marked media content and the original one. Generally, peak signal-to-noise ratio (PSNR) is used as the metric for the imperceptibility. The robustness denotes the capability for the watermark to survive some acceptable operations, such as recompression, filtering, adding noise, etc. Generally, the correct detection rate (the ratio between the detected bits and the total bits) measures the robustness of the watermarking scheme. The embedding capacity is represented by the number of bits that can be embedded in the unit area of media content. Generally, there is a triangle relation between the imperceptibility, robustness, and embedding capacity. For example, the imperceptibility contradicts with the robustness, and the embedding capacity contradicts with both the imperceptibility and robustness.

In implementation, various embedding/detection algorithms can be implemented. According to the operation domain, the algorithms can be classified into spatial-domain method, frequency-domain method, or temporal-domain method. In spatial-domain method, the watermark bit is embedded into the media pixels directly by replacing or modulation. For example, the least significant bit (LSB) algorithm [24] embeds the watermark by modifying the LBSs of a media pixel. In frequency-domain method [25–27], media content is transformed from spatial domain to frequency domain, and watermarked in frequency domain. For example, the transformation may be discrete cosine transformation (DCT), discrete wavelet transformation (DWT), fast Fourier transformation, etc. In temporal-domain method, the watermark bits are embedded into the temporal information of video or audio sequences [28,29].

13.4.1.2 Digital Fingerprinting

Digital fingerprinting is the technique to embed identity information (also named fingerprint code) into media content with watermarking algorithms. Thus, when the media copy is leaked out, the illegal distributor can be traced by detecting the embedded fingerprint code. There are two apparent differences between watermarking and fingerprinting. First, digital fingerprinting embeds different information into the same media content. Thus, given the same media content, it produces different copies for different users. Second, digital fingerprinting faces the important threat, named collusion attack, which combines several copies together to make a new copy without fingerprint code. The considered combination operations [30] include linear collusion (averaging and cut-and-paste), nonlinear collusion (minimum/maximum/median/minmax/modified negative/randomized negative attacks) and linear combination collusion attack (LCCA) [31].

In digital fingerprinting, the most important research topic is collusion-resistant fingerprint encoding that forms the fingerprint code with high ability to resist collusion attacks. The fingerprint encoding methods can be classified into three types, i.e., orthogonal fingerprinting, coded fingerprinting, and desynchronized fingerprinting.

Orthogonal fingerprinting [32,33] makes the fingerprint code orthogonal to each other utilizing orthogonal random signals of uniform or Gaussian distribution and ensures the colluded copy still has detectable fingerprint. The shortcomings of orthogonal fingerprint are the high cost on fingerprint detection and the limitation in customer and colluder population [30] (more customers and colluders will reduce the correct detection rate).

The coded fingerprinting carefully designs the fingerprint in combinatorial codeword that can detect the colluders partially or completely [34–36]. These methods have some problems, i.e., it is hard to form the code supporting large number of users and colluders, and the fingerprint often suffers LCCA attack [31].

The desynchronized fingerprinting produces a media copy by modifying media content via shifting, rotation, or translation and embeds fingerprint code into the desynchronized media content. Since each media copy is different from each other, the collusion between different copies causes great quality degradation to the colluded copy. Thus, it avoids the collusion attacks. This first method for images is proposed by Celik et al. [37] and is extended to videos by Mao and Miheak [38]. Additionally, some methods working in compression domain are presented in [39,40]. The problem to be solved is how to evaluate and keep the quality of the desynchronized media content.

13.4.1.3 Partial Encryption

Partial encryption encrypts only some significant parts of media content with ciphers [41]. It is suitable for multimedia encryption because of several reasons. First, for multimedia content, the encryption focuses on content protection. That is, the encrypted multimedia content is too chaotic to be understood. Encrypting only some significant parts can affect media content's intelligibility and reduces its commercial value. Second, encrypting only parts of media content can decrease the encrypted data volumes, improve the encryption efficiency and meet real time applications. Additionally, some format information in media content can be left unchanged, which increases the encrypted media content's error-resilience.

Till now, various partial encryption schemes have been proposed. For example, for the raw image [42], only the most significant bit-planes of image pixels are encrypted, while the other ones are left unchanged. For the image encoded with JPEG2000 [43], only the significant data layers are encrypted, while the other data layers are left unchanged. For the image encoded with H.264 [44], only some significant parameters are encrypted, while the other parameters are left unchanged. For the audio encoded with mp3 [45], only some variable-length codes of the coefficients in low frequency are encrypted, while the others are left unchanged. These partial encryption schemes reduce the computational complexity and are able to meet real-time applications.

13.4.2 APPLICATION SCENARIOS

This scheme is suitable for the application scenarios similar to the ones in the previous two schemes. The User buys the DVD from Content Provider. Content Provider writes the encrypted media content together with the encrypted parameters in the DVD disc, and writes the keys for content encryption and parameter encryption in the RFID tag. When the DVD is played back, the DVD player reads the parameters from the DVD disc and transmits them to the RFID tag, and the tag decrypts the parameters and then generates the media keys. Based on the parameters and media keys, the DVD player plays the media content. The decrypted and displayed media content contains the unique code corresponding to the DVD player, which can be used to trace the illegal leakage through "analog hole." That is, if the media content is captured from the displayer or leaked out from the gap between decryption and display, the DVD player can be identified.

13.4.3 DVD CONTENT MAKING

In DVD content making, the Content Provider does some operations, as shown in Figure 13.7.

First, it generates some Content Keys from the Key Seed. Some existing key generation methods can be used here.

Second, it encodes media content with such codec as MPEG2, H.264, AVS, etc., and operates the content with secure processing under the control of the Content Keys. The secure processing, as shown in Figure 13.8, produces some parameters. In secure processing, the media content is first lengthened by repeating some segments, e.g., I-frame, then marked by the watermarking technique [23], and finally, encrypted by ciphers [15]. The processed media content is written in the DVD disc.

Third, it encrypts the generated parameters with Parameter Key, and writes the encrypted parameters in the DVD disc.

Fourth, it writes the Key Seed, Parameter Key, and Tag ID in the RFID tag.

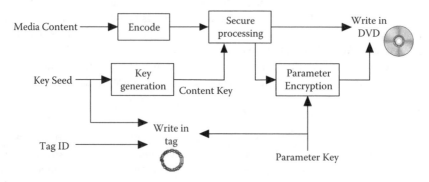

FIGURE 13.7 DVD making based on watermarking.

FIGURE 13.8   Architecture of the secure processing.

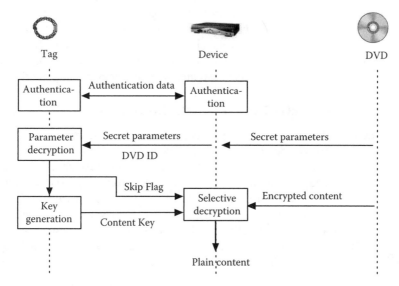

FIGURE 13.9   DVD reading based on selective decryption.

13.4.4   DVD CONTENT READING

In DVD content reading, the DVD player and RFID tag do some operations, as shown in Figure 13.9. First, the DVD player and RFID tag authenticate each other according to the methods in the previous schemes. Then, they act according to the following steps.

First, the DVD Player reads the secret parameters from DVD disc and sends them together with DVD ID to RFID Tag.

Second, the RFID Tag decrypts the parameters, and together with the DVD ID, produces the Skip Flag. Skip Flag tells whether to skip the segment or decrypt the segment (e.g., I-frame).

Third, the RFID Tag generates the Content Key for each segment with the Key Seed.

Fourth, the RFID Tag sends the Content Key or Skip Flag to the DVD Player.

Fifth, the DVD Player reads each media segment from DVD disc, and skips or decrypts the segment.

13.4.5   DVD CONTENT TRACING

In this scheme, the content leakage from "analog hole" or hardware fragileness is considered. If a media copy is redistributed publicly, e.g., over Internet, the original DVD player that leaks out the content can be traced according to the following steps.

First, the detector gets each segment, e.g., I-frame, from the media content, and transforms the media segment into frequency domain according to the encoding process.

Second, the detector detects the fingerprint bit from the media segment with the fingerprint detection method.

Third, the detector detects the remaining fingerprint bits according to Step 1 and Step 2.

Fourth, the detector searches the fingerprint code composed of the fingerprint bits in the database to get the corresponding DVD player.

13.4.6 ADVANTAGES AND DISADVANTAGES

Compared with the existing DVD distribution schemes combined with the RFID tag, the proposed scheme has two priorities. First, the Content Key is generated by the RFID tag, which obtains higher security than the one generated by the DVD player. Second, the media content is fingerprinted by the DVD ID during DVD reading, which can trace the DVD player that leaks the media content out. The disadvantage is that the RFID tag needs to compute the Content Key frequently. According to this case, some means should be taken to meet real-time applications. Additionally, the fingerprint's robustness against camera capture is a challenge. Some means should be taken to make the embedded fingerprint survive such operations as camera capture, recompression, edition, filtering, noise, etc.

13.5 EXAMPLE FOR MPEG2 CONTENT DISTRIBUTION

Until now, MPEG-2 [46] is still the most widely used video coding standard applied in DVD, DVB, etc. MPEG-2 video includes multiple GOP (Group Of Pictures) which includes multiple frames. There are three classes of frames in MPEG-2 video: I-frame (use intra-frame encoding and can be independently decoded), P-frame (decoded referencing former frames), B-frame (decoded referencing forward and backward I- or P-frames). Here I- and P-frames are called referenced frames because they can be used for compensating P- and B-frames, B-frames are called unreferenced frames because they cannot be used for compensating. For I-frame, it is partitioned into blocks, each block is transformed by DCT, and the DCT coefficients, including DC and ACs, are encoded by quantization and entropy coding. In the following content, we propose the scheme for secure MPEG-2 video distribution.

13.5.1 CONTENT MAKING

In this scheme, we encrypt the video content frame by frame. Thus, the Content Key corresponding to each frame is generated one by one. Here, the AES cipher is used to encrypt the Key Seed iteratively, and the result of each iteration is used as the Content Key. The parameters include frame type (I-frame, B-frame, or P-frame) and repetition flag (repeated or nonrepeated). The parameters are also encrypted by AES cipher. In the following content, the Secure Processing composed of three steps is focused.

13.5.1.1 I-Frame Repetition

The I-frame is repeated to contain different watermark bits. Since P-frame or B-frame needs to be decoded by referencing to the adjacent I-frame or P-frame, repetition of P-frame or B-frame is not suitable for decoding. Table 13.1 gives a simple example for I-frame repetition. As can be seen, in the repeated GOP, only the I-frame is repeated for few times, e.g., once here, and the GOP's length increases accordingly.

13.5.1.2 Watermark Embedding

The watermark bits are embedded into I-frames (original I-frame and repeated I-frame). For example, "0"-bit is embedded into the first I-frame, while "1"-bit is embedded into the second I-frame. The

TABLE 13.1

Example of Frame Repetition

Original GOP: I		B	B	P	B	B	I		B	B	P...
Repeated GOP: I	I	B	B	P	B	B	I	I	B	B	P...
Watermark Bit: 0	1						0	1	...		

watermark embedding method in Ref. [25] is adopted. Taking the watermark bit "0" for example, first, the ACs coefficients in the middle frequency of each DCT block are extracted and form a coefficient sequence. Then, the random sequence corresponding to "0"-bit is embedded into the coefficient sequence with additive embedding. Finally, the marked coefficients in the coefficient sequence are returned to the DCT blocks.

13.5.1.3 Content Encryption

To improve the computing efficiency, the media content is encrypted with a partial encryption scheme [47]. That is, for each frame, only the DCs and ACs' signs in DCT blocks are encrypted, while other parameters are left unchanged. In detail, the encryption process is composed of three steps. First, the DCs and ACs' signs are extracted and form a data sequence. Then, the data sequence is encrypted by AES cipher under the control of Content Key. Finally, the encrypted DCs and ACs' signs are returned to the DCT blocks.

13.5.2 CONTENT READING

In content reading, the parameters are read from the DVD disc and transmitted to the RFID tag. The tag decrypts the parameters with AES cipher under the control of Parameter Key. Additionally, the tag produces the fingerprint code from the DVD ID with collusion-resistant fingerprint encoding method, e.g., Wu et al.'s method [35]. Then, the tag will determine the Skip Flag based on the parameters and fingerprint code, and the DVD player decrypts media content with Selective Decryption. They are described as follows.

13.5.2.1 Generating Skip Flag

The Skip Flag is generated according to the process shown in Figure 13.10. Firstly, the frame type is decided. If it is not an I-frame, the Skip Flag is 0 (not skip). Otherwise, the watermark bit is decided. If the watermark bit in the parameter is same to the fingerprint bit corresponding to the DVD ID, then Skip Flag is 0 (not skip). Otherwise, the Skip Flag is 1 (to be skipped).

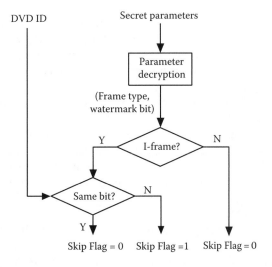

FIGURE 13.10 Generation of Skip Flag.

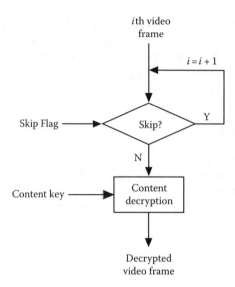

FIGURE 13.11   The selective decryption process.

13.5.2.2 Selective Decryption

After receiving the Skip Flag and Content Key from the RFID tag, the DVD player do Selective Decryption according to the steps shown in Figure 13.11. First, it decides whether the frame is to be decrypted or not. If Skip Flag is 1, the frame is skipped. Otherwise, the frame will be decrypted. Second, it uses the Content Key to decrypt the frame with the method symmetric to the encryption process in content making. That is, only the DC and ACs' signs in each DCT block are decrypted with AES cipher under the control of the Content Key. As can be seen, the selective decryption process skips some I-frames according to the fingerprint code derived from the DVD ID, and produces the content that contains the fingerprint code.

13.5.3 Experimental Results

In this scheme, the DC and ACs' signs of each DCT block in all the frames are encrypted with AES cipher. The AES cipher itself is proved as a secure cipher. Then, the perceptual security determines the scheme's security. Since the DC and ACs' signs are in close relation with media contents' intelligibility, the encrypted media content is often too chaotic to be understood. Some examples are shown in Figure 13.12.

The partial encryption is adopted, which obtains higher efficiency compared with the MPEG2 encoding/decoding process. Table 13.2 shows the time ratio between encryption and MPEG2 encoding. As can be seen, the encryption operation is quite efficient compared with encoding operation.

In this scheme, the fingerprint is embedded by modulating ACs in each DCT block. Figure 13.13 shows the original and marked copies. As can be seen, the fingerprint is invisible. Additionally, the increased data volumes caused by I-frame repetition are no more than 5 percent of the original file, which is acceptable by DVD discs.

The scheme's robustness against recompression is tested. Generally, the fingerprint can still be detected when the bit-rate of recompression is no smaller than 512 kbps. The robustness against some other attacks, such as camera capture, can be improved by introducing some other embedding means [23].

Original   Encrypted

Original   Encrypted

FIGURE 13.12   Perceptual security of the encrypted video frames.

TABLE 13.2
Time Ratio between Encryption and MPEG2 Encoding

Video Sequence	Frame Size	Time Ratio (Percent)
Football	CIF	1.3
Foreman	CIF	1.5
Tempete	CIF	0.9
Akiyo	QCIF	3.4
Salesman	QCIF	4.6
Mobile	QCIF	5.3

13.6   OPEN ISSUES

Using RFID for secure DVD distribution is still a new topic, and there are some open issues in the existing schemes based on RFID.

First, RFID tag's computing capability is more and more required by secure schemes. In previous schemes, RFID tag is only used as a storage device containing the key or fingerprint. Considering that more and more RFID tags have the capability of complex computing, they can do some secure operations that have been done by the DVD player, such as decryption and decision. Thus, the

FIGURE 13.13 Imperceptibility of the marked video frames.

scheme's security can be improved. However, since different tags may have different computing capability, the tags should be selected according to the required decryption or decision operations.

Second, the suitable RFID tag reader should be equipped with the DVD player. During DVD reading, the information exchange between RFID tag and DVD player may be frequent. For example, if each video frame's Content Key is generated by the tag and transmitted to the player, the tag reader needs to have high access speed to meet the real-time frame rate, 25 fps. This depends on the tag's manufacture.

Third, to provide a secure system, the Content Provider and DVD player's manufacturer should work consistently. The provider writes media content and keys in DVD disc and RFID tag. Secure and frequent communication between DVD player and DVD disc, and RFID tag makes the player work fluently. Thus, Content Provider and manufacturer need to obey the same interface or protocol. This is still a barrier to the wide application of such secure systems.

Fourth, in home network environment, the communication between the DVD player and other devices needs to be defined. First, the consistence should be constructed between multiple devices, which include the input/output interfaces or access protocols. Second, since different devices may use different DRM systems, the interoperation between DRM systems should be defined to make the media content played on different devices.

Fifth, media content tracing is still an open issue. In existing DVD distribution schemes, good means have been proposed to protect media content from content making, content distribution to content playing. However, there are no suitable means for the content leakage during or after playing.

For example, if the media content is captured by camera and redistributed on Internet, it can be viewed freely. The digital fingerprinting has been introduced to trace media content, but whether fingerprint code can survive various attacks, e.g., editing or collusion attacks, is not confirmed. Thus, better means are expected to improve fingerprinting schemes' robustness, and new means need to be proposed for countering the content leakage.

13.7 CONCLUSIONS

In this chapter, the DRM systems for DVD content are reviewed, and the secure DVD content distribution schemes based on RFID are investigated. Additionally, a secure DVD distribution scheme based on RFID and digital fingerprinting is proposed and evaluated. The RFID-based schemes obtain some good advantages compared with traditional schemes, such as the ability to limit the user to view his own DVD disc on his own DVD player, to refuse playing the copied media content, or to trace the illegal redistribution. However, since it is just at the beginning of using RFID for secure DVD distribution, some open issues need to be solved, which are listed in the chapter. These interesting topics are expected to attract more and more researchers and engineers.

REFERENCES

[1] Open Mobile Alliance, Digital Rights Management 2.0 (OMA DRM 2.0), 03 Mar 2006.
[2] HDCP (High-bandwidth Digital Content Protection System), http://en.wikipedia.org/wiki/HDCP.
[3] Digital Video Broadcasting Content Protection & Copy Management (DVB-CPCM), DVB Document A094 Rev. 1, July 2007.
[4] K. Finkenzeller, *RFID Handbook: Fundamentals and Applications in Contactless Smart Cards and Identification*, 2nd edition, John Wiley & Sons, Munich, Germany, 2003.
[5] The History of RFID Technology, RFID J., 20 December 2005; www.rfidjournal.com/article/articleview/1338/1/129.
[6] S. Bono et al., Security analysis of a cryptographically-enabled RFID device, *Proceedings of 14th USENIX Security Symposium*, USENIX, 2005, pp. 1–5; http://spar.isi.jhu.edu/ mgreen/DSTbreak.pdf.
[7] Digital Versatile Disc (DVD). http://en.wikipedia.org/wiki/DVD.
[8] DVD Copy Control Association. http://www.dvdcca.org/css/.
[9] Advanced Access Content System. http://en.wikipedia.org/wiki/Advanced_Access_Content_System.
[10] J.A. Bloom et al., Copy protection for DVD video, *Proceedings of IEEE*, 87(7), 1267–1276, July 1999.
[11] B. Lan and T. Tan, A DRM system implementing RFID to protect AV content, In *Proceedings of 2006 IEEE 10th International Symposium on Consumer Electronics (ISCE'06)*, pp. 1–3, 2006.
[12] J.H. Suh and S.C. Park, u-DRM: A unified framework of digital rights management based on RFID and application of its usage data, *International Journal of Computer Science and Network Security*, 6(8B), 151–155, 2006.
[13] Decrypted CSS (DeCSS). http://en.wikipedia.org/wiki/DeCSS.
[14] http://en.wikipedia.org/wiki/Advanced-Access-Content-System.
[15] R.A. Mollin, *An Introduction to Cryptography*, 2nd edition, CRC Press, Boca Raton, FL, 2006.
[16] P. Bassia, I. Pitas, and N. Nikolaidis, Robust audio watermarking in the time domain, *IEEE Transactions on Multimedia*, 3, 232–240, June 2001.
[17] E. Felten, AACS Decryption Code Released, http://www.freedom-to-tinker.com/?p=1104.
[18] http://en.wikipedia.org/wiki/ROM-Mark.
[19] E. Allanmanche, J. Herre, O. Helmuth, B. Frba, T. Kasten, and M. Cremer, Content-based identification of audio material using MPEG-7 low level description, *Proceedings of the International Symposium of Music Information Retrieval*, pp. 197–204, 2001.
[20] P. Cano, E. Battle, T. Kalker, and J. Haistsma, A review of algorithms for audio fingerprinting, *Proceedings of International Workshop on Multimedia Signal Processing*, pp. 169–173, 2002.
[21] K. Mihcak and R. Venkatesan, New iterative geometric methods for robust perpetual image hashing, security and privacy in digital rights management, *Proceedings of ACM CCS-8 Workshop DRM 2001*, pp. 13–24, 2001.

[22] M. Barni and F. Bartolini, Data hiding for fighting piracy, *IEEE Signal Processing Magazine*, 21(2), 28–39, 2004.

[23] I.J. Cox, M.L. Miller, and J.A. Bloom, *Digital Watermarking*, Morgan-Kaufmann, San Francisco, CA, 2002.

[24] M.U. Celik, G. Sharma, A.M. Tekalp, and E. Saber, Lossless generalized-LSB data embedding, *IEEE Transactions on Image Processing*, 14(2), 253–266, February 2005.

[25] F. Hartung and B. Girod, Watermarking of uncompressed and compressed video, *Signal Processing*, 66(3), 283–301, May 1998.

[26] H. Liu, N. Chen, J. Huang, et al., A robust DWT-based video watermarking algorithm, *IEEE International Symposium on Circuits and Systems*, pp. 631–634, 2002.

[27] C.Y. Lin, M. Wu, J.A. Bloom, et al., Rotation, scale, and translation resilient watermarking for images, *IEEE Transactions on Image Processing*, 10(5), 767–782, May 2001.

[28] B. Mobasseri, Direct sequence watermarking of digital video using m-frames, In *International Conference on Image Processing, ICIP'98*, pp. 399–403, 1998.

[29] M.D. Swanson, B. Zhu, A.H. Tewfik, and L. Boney, Robust audio watermarking using perceptual masking, *Signal Processing*, 66(3), 337–355, 1998.

[30] W. Trappe, M. Wu, Z.J. Wang, and K.J.R. Liu, Anti-collusion fingerprinting for multimedia, *IEEE Transactions on Signal Processing*, 51, 1069–1087, 2003.

[31] Y. Wu, Linear combination collusion attack and its application on an anti-collusion fingerprinting, *IEEE International Conference on Acoustics, Speech, and Signal Processing*, 2, 13–16, 2005.

[32] Z.J. Wang, M. Wu, H.V. Zhao, W. Trappe, and K.J.R. Liu, Anti-collusion forensics of multimedia fingerprinting using orthogonal modulation, *IEEE Transactions on Image Processing*, 14(6), 804–821, 2005.

[33] G. Tardos, Optimal probabilistic fingerprint codes, In *STOC'03: Proceedings of the Thirty-Fifth Annual ACM Symposium on Theory of Computing*, ACM Press, New York, pp. 116–125, 2003.

[34] D. Boneh and J. Shaw, Collusion-secure fingerprinting for digital data, *IEEE Transactions on Information Theory*, 44(5), 1897–1905, September 1998.

[35] M. Wu, W. Trappe, Z.J. Wang, and K.J.R. Liu, Collusion-resistant fingerprinting for multimedia, *IEEE Signal Processing Magazine*, pp. 15–27, March 2004.

[36] W. Kim and Y. Suh, Short N-secure fingerprinting code for image, *2004 International Conference on Image Processing*, pp. 2167–2170, 2004.

[37] M.U. Celik, G. Sharma, and A.M. Tekalp, Collusion-resilient fingerprinting by random pre-warping, *IEEE Signal Processing Letters*, 11(10), 831–835, 2004.

[38] Y.N. Mao and K. Mihcak, Collusion-resistant intentional de-synchronization for digital video fingerprinting, *IEEE International Conference Image Processing '05*, Vol. 1, pp. 237–240, 2005.

[39] Z. Liu, S. Lian, and Z. Ren, Image desynchronization for secure collusion-resilient fingerprint in compression domain, *PCM 2006*, pp. 56–63, 2006.

[40] Z. Liu, S. Lian, R. Wang, and Z. Ren, Desynchronization in compression process for collusion resilient video fingerprint, *IWDW 2006*, pp. 308–322, 2006.

[41] S. Lian, Z. Liu, Z. Ren, and H. Wang, Secure advanced video coding based on selective encryption algorithms, *IEEE Transactions on Consumer Electronics*, 52(2), 621–629, 2006.

[42] M. Podesser, H.P. Schmidt, and A. Uhl, Selective bitplane encryption for secure transmission of image data in mobile environments, In *CD-ROM Proceedings of the 5th IEEE Nordic Signal Processing Symposium (NORSIG 2002)*, Tromso-Trondheim, Norway, pp. 1037–1042, October 2002.

[43] S. Lian, J. Sun, D. Zhang, and Z. Wang, A Selective image encryption scheme based on JPEG2000 codec, *2004 Pacific-Rim Conference on Multimedia (PCM2004)*, Springer, *LNCS*, Vol. 3332, Tokyo, Japan, pp. 65–72, November 30–December 3, 2004.

[44] S. Lian, Z. Liu, Z. Ren, and H. Wang, Commutative encryption and watermarking in compressed video data, *IEEE Circuits and Systems for Video Technology*, 17(6), 774–778, June 2007.

[45] A. Torrubia and F. Mora, Perceptual cryptography on MPEG Layer III bit-streams, *IEEE Transactions on Consumer Electronics*, 48(4), 1046–1050, November 2002.

[46] ISO/MPEG-2. ISO 13818-2: Coding of moving pictures and associated audio, 1994.

[47] S. Lian, Z. Liu, Z. Ren, and Z. Wang, Selective video encryption based on advanced video coding, In *Proceedings of 2005 Pacific-Rim Conference on Multimedia (PCM2005)*, Part II, *LNCS*, Vol. 3768, Jeju Island, Korea, pp. 281–290, November 13–16, 2005.

Part II

Security in Wireless Sensor Networks

14 A Survey on Security in Wireless Sensor Networks

Qinghua Wang and Tingting Zhang

CONTENTS

Recent advances in electronics and wireless network technologies have offered us access to a new era where wireless sensor networks formed by interconnected small intelligent sensing devices provide us the possibility to form smart environments. Considering the specialty of wireless sensor network, the security threats and possible countermeasures are quite different from those in Internet and Mobile Ad Hoc Networks (MANETs). On the one hand, the wireless communication, large-scale and possibly human unattended deployment make attacks in wireless sensor networks relatively easier to perform. Furthermore, all features that make sensor nodes cheap and thus sensor network application affordable, such as limited energy resource, limited bandwidth, and limited memory, also make many well-developed security mechanisms inappropriate in sensor networks. On the other hand, the user unfriendly interface makes the physical compromise of a sensor node difficult, the relatively simple communication profile makes the intrusion detection easy to perform, and also the redundant deployment makes the new type of network more fault-tolerant. Thus, we need a complete redesign of sensor network security mechanisms from technique to management.

14.1 INTRODUCTION

Wireless sensor networks (WSNs) are making their way from research to real-world deployment. Body and personal-area networks, intelligent homes, environmental monitoring, or inter-vehicle communications: there is almost nothing left that is not going to be "smart" and "networked" [1]. Although a great amount of research has been devoted to the pure networking aspects, WSNs will not be successfully deployed if security, dependability, and privacy issues are not addressed adequately. These issues become more important because WSNs are usually used for very critical applications. Furthermore, WSNs are very vulnerable and thus attractive to attacks because of their limited prices and human-unattended deployment.

The goal of this chapter is to present a framework for implementing security in WSNs. The security threats faced by WSNs are first identified. The state of the art for WSN security is then further elaborated by detailing exisiting countermeasures.

14.2 SECURITY THREATS IN WSNs

WSNs are becoming popular in more and more applications, because of their sensing ability in the physical world, large scale human-unattended deployment, and the most important: simple and cheap devices. A typical WSN consists of hundreds or even thousands of tiny and resource-constraint sensor nodes. These sensor nodes are distributed and deployed in uncontrollable environment for the collection of security-sensitive information. Individual sensor node relies on multihop wireless communication to deliver the sensed data to a remote base station. In a basic WSN scenario, resource constraint, wireless communication, security-sensitive data, uncontrollable environment, and even distributed deployment are all vulnerabilities. These vulnerabilities make WSNs suffer from an amazing number of security threats. WSNs can only be used in the critical applications after the potential security threats are eliminated.

In the following of this section, we classify WSN threats according to the OSI model, which is also followed by WSN design.

14.2.1 PHYSICAL LAYER THREATS

Comparing WSNs with traditional networks, there are more threats to WSNs in the physical layer, due to the nontamper-resistant WSN nodes and the broadcasting nature of wireless transmission [2]. Typical types of attacks in the physical layer include physical layer jamming and the subversion of a node.

Physical layer jamming: It is a type of attack which interferes with the radio frequencies that network's nodes are using. A jamming source may randomly disturb a part of or the whole network.

Subversion of a node: If a sensor node is captured, an attacker can extract sensitive information such as cryptographic keys or other data on the node. The node may also be altered or replaced to create a compromised node which the attacker controls.

14.2.2 LINK LAYER THREATS

The data link layer is responsible for the multiplexing of data streams, data frame detection, medium access, and error control. The following attacks can happen in the link layer of WSNs.

Link layer jamming: The link layer jamming attacks focus on disturbing the communication between sensor nodes around the jammer. This kind of jamming attack utilizes the weaknesses of some link layer protocols [3].

Eavesdropping: An adversary can gain access to private information by monitoring transmissions between nodes.

Collisions: An adversary may strategically cause collisions in specific packets such as ACK control messages. A possible result of such collisions is the costly exponential back-off in certain media access control protocols.

Resource exhaustion: Denial-of-service (DoS) attack because of resource exhaustion can be caused by purposely introduced repeated collisions. For example, a naive link-layer implementation may continuously attempt to retransmit the corrupted packets. Unless these hopeless retransmissions are discovered or prevented, the energy reserves of the transmitting node and those surrounding it will be quickly depleted.

Traffic analysis: The basic idea of traffic analysis attack [4] is that the nodes near to the sink forward a significantly greater volume of packets than nodes further away from the sink. By listening to the network traffic at various locations in a sensor network, an adversary is able to locate the important nodes, such as the base station.

Packet-tracing: The packet-tracing attack [5] is a case of that an equipped adversary can tell the location of the immediate transmitter of an overheard packet. The adversary is thus able to perform hop-by-hop trace toward the original data source, causing the disclosure of the source privacy.

14.2.3 NETWORK LAYER THREATS

Threats in the network layer mostly aim at disturbing data-centric and energy efficient multihop routing, which is the main design principle in WSNs. The following threats and attacks in this layer are identified in Refs. [2,6].

Spoofed, altered, or replayed routing information: The most direct attack against a routing protocol in any network is to target the routing information itself although it is being exchanged between nodes. An attacker may spoof, alter, or replay routing information to disrupt traffic in the network. These disruptions include the creation of routing loops, attracting or repelling network traffic from selected nodes, extending and shortening source routes, generating fake error messages, partitioning the network, and increasing end-to-end latency.

Sybil: The Sybil attack is the forging of multiple identities of a compromised node. This attack can affect fault-tolerant schemes, distributed storage, and network-topology maintenance. When the local entity selects a subset of identities to save replicated copies, it can be duped into selecting a single remote entity multiple times, thereby defeating the distributed storage. This attack can be used against routing algorithms in sensor networks [7]. One vulnerable mechanism is multipath or dispersity routing where seemingly disjoint paths could in fact go through a single malicious node presenting several Sybil identities.

Selective forwarding: A significant assumption made in multihop networks is that all nodes in the network will accurately forward received messages. An attacker may create malicious nodes which selectively forward only certain messages and simply drop others. A specific form of this attack is the black hole attack in which a node drops all messages it receives.

Sinkhole: In a sinkhole attack, an attacker makes a compromised node look more attractive to surrounding nodes by forging routing information, creating a metaphorical sinkhole with the adversary at the center. This type of attack makes selective forwarding very simple, as all traffic from a large area in the network will flow through the adversary's node.

Wormholes: A wormhole is an attack that tunnels messages received in one part of the network over a low-latency link and replays them in a different part. This link may be established either by a single node forwarding messages between two adjacent but otherwise nonneighboring nodes or by a pair of nodes in different parts of the network communicating with each other. The latter case is closely related to the sinkhole attack, as an attacking node near the base station can provide a one-hop link to that base station via the other attacking node in a distant part of the network.

Hello flood attacks: Many protocols which use HELLO packets to announce new nodes to their neighbors, and a node receiving such a packet may assume that it is within (normal) the radio range of the sender and is therefore a neighbor. An attacker may use a high-powered transmitter to trick a large area of nodes into believing they are neighbors of that transmitting node. If the attacker falsely broadcasts a superior route to the base station, all of these nodes will attempt transmission to the attacking node, despite many being out of radio range in reality. The location-based protocols are also subject to this attack.

Acknowledgment spoofing: Routing algorithms used in sensor networks sometimes rely on implicit or explicit link layer acknowledgements. An attacking node can spoof the acknowledgments of overheard packets destined for neighboring nodes to provide false information to those neighboring nodes. The false acknowledgments can convince the sender that a weak link is strong or that a dead or disabled node is alive. By encouraging the target nodes sending information through the false strong links will result selective forwarding attacks.

Flooding: Whenever a protocol is required to maintain state at either end of a connection it becomes vulnerable to memory exhaustion through flooding. An attacker may repeatedly make new connection requests until the resources required by each connection are exhausted or reach a maximum limit. In either case, further legitimate requests will be ignored.

Desynchronization: It is a case of disruption of an existing connection. An attacker may cause an end host to request the retransmission of missed frames. If timed correctly, an attacker may degrade or even prevent the ability of the end hosts to successfully exchange data, thus causing them instead to waste energy by attempting to recover from errors which never really existed.

14.2.4 APPLICATION LAYER THREATS

Many WSNs' applications heavily rely on coordinated services such as localization, time synchronization, and in-network data processing to collaboratively process data [8]. Unfortunately, these services represent unique vulnerabilities such as false data filtering, clock unsynchronization, and false data injections.

False data filtering: The energy limited WSN usually use in-network data aggregation. The need for aggregation makes end-to-end cryptography infeasible [8]. An attack on an aggregation point allows an adversary to corrupt not only all the data from the down stream nodes but also the overall data aggregation result observed at the base station. Thus, an attack can seriously hamper sensing applications by manipulating data even without having to disrupt other fundamental components in an in-network data aggregation WSN.

Clock unsynchronization: Time synchronization is a critical building block in distributed WSN. Time unsynchronization can disrupt sleep scheduling. An attacker node can send a falsified synchronization message to its neighbor during this time exchange period. This will make other nodes calculate an incorrect phase offset and skew.

False data injections: The nature of in-network aggregation is vulnerable to false data injection. Attackers can launch an outsider attack by sending their own packets to inject data. An insider attack can also be launched by compromising several sensor nodes, and then use the compromised nodes to inject false data into the network.

14.3 COUNTERMEASURES

WSN threats presented in Section 14.2 either violate network secrecy and authentication, such as packet spoofing, or violate network availability, such as jamming attack, or violate some other network functionalities. Generally, countermeasures to the threats in WSNs should fulfill the following security requirements:

- Availability, which ensures that the desired network services are available whenever required
- Authentication, which ensures that the communication from one node to another node is genuine
- Confidentiality, which provides the privacy of the wireless communication channels
- Integrity, which ensures that message or the entity under consideration is not altered
- Nonreputation, which prevents malicious nodes to hide or deny their activities
- Freshness, which implies that the data is recent and ensures that no adversary can replay old messages
- Survivability, which ensures the acceptable level of network services even in the presence of node failures and malicious attacks
- Self-security, countermeasures may introduce additional hardware and software infrastructures into the network, which must themselves be secure enough to withstand attacks

Depending on the applied applications, countermeasures should also fulfill appropriate performance requirements. Because of the resource-constraint nature of WSNs, an inevitable performance requirement for countermeasures is low-overhead. Other applicable performance requirements could be low-cost, easy deployment, real-time requirement, etc. In the real implementation, there is usually a trade-off between the security provided and the overhead introduced by the applied countermeasure.

In the following, we will elaborate available countermeasures covering all aspects, including passive and active defense technologies, and solutions from a social engineering point of view. To be noticed, most proposed countermeasures in the literature either explicitly or impliedly assume the countermeasure proposed works under some kind of attack model. An attack model specifies how much information and how many resources the potential attacker has and the attacker is supposed to be capable of launching arbitrary attacks limited only by the information and resources available to him. However, because an attacker is rarely assumed to be all-powerful, a single countermeasure may not provide sufficient security if there is an all-powerful attacker.

14.3.1 KEY MANAGEMENT

When setting up a sensor network, one of the first security requirements is to establish cryptographic keys for later secure communication. The established keys should be resilient to attacks and flexible to dynamic update. The task that supports the establishment and maintenance of key relationships between valid parties according to a security policy is called key management. Desired features of key management in sensor networks include energy awareness, localized impact of attacks, and scaling to a large number of nodes [9].

Recently, numerous key management schemes have been proposed for sensor networks. Many schemes, referred to as static schemes, have adopted the principle of key predistribution with the underlying assumption of a relatively static short-lived network. An emerging class of schemes, dynamic key management schemes, assumes long-lived networks requiring network rekeying for sustained security and survivability. Also, there are some special kind of key management schemes supporting in-network processing, which is an important energy-saving mechanism in many proposed WSNs.

14.3.1.1 Static Key Management Schemes

These schemes assume that once administrative keys are predeployed in the nodes, they will not be changed. Administrative keys are generated prior to deployment, assigned to nodes either randomly or based on some deployment information, and then distributed to nodes. For communication key management, static schemes use the overlapping of administrative keys to determine the eligibility of neighboring nodes to generate a direct pairwise communication key [9].

Most existing schemes in this category are built on the seminal random key predistribution scheme introduced by Eschenauer and Gligor [10]. Subsequent extensions to that scheme include using key polynomials [11] and deployment knowledge [12] to enhance scalability and resilience to attacks. Generally, these static key predistribution schemes consist of three phases: (1) key setup prior to deployment, (2) shared-key discovery after deployment, and (3) path-key establishment if two sensor nodes do not share a key.

Random pairwise key predistribution scheme: Random key predistribution scheme was proposed first by Eschenauer and Gligor [10]. Given an n sensor WSN, the basic random key predeployment strategy proposed is composed of the following steps in the key predistribution phase:

1. A large pool of P keys (2^{17}–2^{20} keys) is generated
2. Each sensor is assigned k random distinguished keys from the pool

At the shared-key discovery phase, each node broadcasts its set of key identifiers and receives one message from each node within its radio range. Nodes which discover that they contain a shared key can then verify that their neighbor actually holds the key through a challenge–response protocol. The shared key then becomes the key for that pair. The probability of key share among two sensor nodes is $((P - k)!)^2/(P - 2k)!P!$.

After the shared-key discovery phase is finished there will be a number of unused keys left in each sensor's key ring and these keys can be put to work by each sensor node for path-key establishment. If two sensor nodes do not share a common key, they can find an intermediate node that has shared pairwise keys with both of them. The intermediate node can act as a key distribution center to set up a pairwise key between them.

Chan et al. [13] developed the q-composite key predistribution and the random pairwise keys schemes. The q-composite key predistribution requires two sensors share at least q predistributed keys to establish a pairwise key. The random pairwise keys scheme has the property that compromised sensors do not lead to the compromise of pairwise keys shared between noncompromised sensors.

However, these approaches still have some limitations. For the basic probabilistic and the q-composite key predistribution, a small number of compromised sensors may reveal a large fraction of pairwise keys shared between noncompromised sensors. For random pairwise keys scheme, it cannot support a big size network.

Polynomial pool-based pairwise key predistribution: Polynomial pool-based pairwise key predistribution was developed in Ref. [11]. In the proposal, the following bivariate t-degree polynomial is used.

$$f(x,y) = \sum_{i,j=0}^{t} a_{ij} x^i y^j$$

The polynomial is over a finite field F_q, where q is a prime number that is large enough to accommodate a cryptographic key, such that it has the property of $f(x,y) = f(y,x)$. It is assumed that each sensor has a unique ID. For any two sensor nodes u and v, they share a common key $f(u,v)$.

To predistribute pairwise key, the setup server computes a random polynomial share of $f(u,y)$ for each sensor u. To establish a pairwise key for sensors u and v, node u needs to compute the common key $f(u,v)$ by evaluating $f(u,y)$ at point v, and node v needs to compute the same key $f(v,u) = f(u,v)$ by evaluating $f(v,y)$ at point u.

The security proof in Ref. [11] ensures that this scheme is unconditionally secure and t-collusion resistant. That is, the collusion of no more than t compromised sensor nodes knows nothing about the pairwise key between any two noncompromised nodes.

The basic polynomial pool-based key predistribution has a limit. It can only tolerate no more than t compromised nodes, where the value of t is limited by the memory available in sensor nodes. In a large sensor network it is more likely that there are more than t sensor nodes are compromised.

An extension method [14] is to use a random strategy for subset assignment during the setup phase. That is, for each sensor, the setup server selects a random subset of generated polynomials and assigns the polynomial shares of these polynomials to the sensor. If no more than t shares on the same polynomial are disclosed, no pairwise keys constructed using this polynomial between any two noncompromised sensor nodes will be disclosed.

If two sensors fail to establish a pairwise key directly, they must start path-key establishment phase. During this phase, a source sensor node tries to find another node that has direct pairwise keys with both nodes. The common node acts as a KDC. It generates a random key for and sends to the pair nodes in a secure channel. In practice, a sensor may be restricted to only contact its neighbors within a certain range for creating a path key.

Location based pairwise key predistribution scheme: The location-based pairwise key predistribution scheme [12] uses the sensor location information to predistribute pairwise keys. The idea is to have each sensor share pairwise keys with its c closest neighbors. For each sensor u, the setup server first discovers a set S of c sensors whose expected locations are closest to the expected location of u. For each sensor v in S, the setup server randomly generates a unique pairwise key $K_{u,v}$. $(u, K_{u,v})$ and $(v, K_{u,v})$ are sent to nodes v and u, respectively.

If two sensors u and v want to set up a pairwise key to secure the communication between them, they only need to check whether they have a predeployed pairwise key with the other party. The algorithm to identify such a common key is trivial, because each pairwise key in a particular sensor was associated with a sensor ID.

To add a new sensor after deploying the sensor network, the setup server has to inform a number of existing sensors in the network about the addition of the new sensor. It may introduce a lot of communication overhead. This can be improved by a technique based on a pseudorandom function (PRF) [15] and a master key shared between each sensor and the setup server. For each pair neighbor sensor nodes u and v, node u saves a master key K_u and node v saves a pairwise key $K_{u,v}$, where $K_{u,v} = \text{PRF}(K_u, v)$. The direct key establishment stage is similar to the basic scheme. The only difference is that one of the two sensors has a predistributed pairwise key and the other only needs to compute the key using its master key and the ID of the other party.

Group-wise key distribution: The former introduced key management schemes are focused on pairwise key distribution. For groupwise key distribution, a straightforward approach is to use existing pairwise keys to establish groupwise keys. For example, lightweight key management system [16] considers a WSN where group of sensor nodes are deployed in different phases. It proposes to distribute groupwise keys through the links which are secured with pairwise keys. Yet another approach is to predistribute polynomial shares to sensor nodes by using which group members can generate a common group key [11].

14.3.1.2   Dynamic Key Management Schemes

Dynamic key management schemes may change administrative keys periodically, on demand or on detection of a node capture. The major advantage of dynamic keying is the enhanced network survivability, because any captured key(s) is replaced in a timely manner in a process known as rekeying. Another advantage of dynamic keying is providing better support for network scalability.

The major challenge in dynamic keying is to design a secure yet efficient rekeying mechanism. A proposed solution to this problem is by using exclusion-based system (EBS), a combinatorial formulation of the group key management problem developed in Ref. [17]. The EBS assigns each node k keys from a key pool of size $k + m$. If a node capture is detected, rekeying occurs. In the rekeying process, replacement keys are generated, and then encrypted with all the m keys unknown to the captured node, and finally distributed to other nodes that collectively know the m keys. A disadvantage to this EBS scheme is that a small number of nodes may collude and collectively reveal all the secret keys.

LOCK [9] is an EBS-based dynamic key management scheme for clustered sensor network. The physical network model is a three-tier WSN with the base station at the top, followed by cluster heads, then sensor nodes. There is no assumption about location knowledge in LOCK. When the nodes are initially released into the environment, they create a set of backup keys. These sets of backup keys are only shared with the base station, not the local cluster heads. If a node is captured, other nodes are rekeyed locally so that the compromised node is unable to communicate with others. If a cluster head is compromised, the base station initiates a rekeying at the cluster head level. Also, nodes within the cluster governed by a compromised cluster head directly rekey with the base station.

14.3.1.3   Key Management Schemes Supporting In-Network Processing

There are some key management schemes specially designed for supporting in-network processing, which is one of the important mechanisms in sensor networks. LEAP [18] is a key management protocol for sensor networks that is designed to support in-network processing, while at the same time restricting the security impact of a node compromise to the immediate network neighborhood of the compromised node. LEAP includes efficient protocols for supporting four types of key schemes for different types of messages broadcasted in WSNs and includes an efficient scheme for local broadcast authentication. LEAP is an efficient scheme for key establishment that resists many types of attacks in the network, including the Sybil, sinkhole, wormhole, and so on. LEAP also provides efficient schemes for node revocation and key updating in WSNs.

The concealed data aggregation (CDA) approach [19] proposes to use symmetric additively homomorphic encryption transformations for end-to-end encryption of sensed data and for reverse multicast traffic between the monitoring sensor nodes and the sink node. CDA enables intermediate aggregator nodes to aggregate ciphers without the cost of decrypting and reencrypting these messages. These aggregator nodes are not required to store sensitive keys. Aggregation supported by CDA is based on homomorphic encryption transformations, and the supported aggregation functions cover averaging, movement detection, and variance functions.

14.3.1.4   Open Issues

Due to resource constraints of WSNs, there is usually a trade-off between the network performance and the security provided by key management systems. More schemes should be developed to make efficient use of sensor nodes' limited resources. Also, the rekeying in dynamic key management schemes relies on efficient node failure detection and compromised node discovery. However, node failure detection and intrusion detection (ID) in WSNs remain as open research areas at the time of this writing.

14.3.2 Authentication

As sensor networks are mostly deployed in human-unattended environments for critical sensing measurements, the authentication of the data source as well as the data are critical concerns. Proper authentication mechanisms can provide WSNs with both sensor and user identification ability, protect the integrity and freshness of critical data, and prohibit and identify impersonating attack. Traditionally, authentication can be provided by public-key schemes as digital signature and by symmetric-key schemes as message authentication code (MAC). Besides, key-chain schemes using symmetric keys determined by asymmetric key-exchange protocols are also popular for broadcast authentication in WSNs.

14.3.2.1 Authentication Based on Asymmetric Public-Key System

In a public-key scenario, each entity has a certificate $<PK>$ issued by a certificate authority (CA) and an assigned public/private key pair. A unilateral message authentication works as follows: Let $SIG(m, SK)$ be a signature of the message m by the private key SK, and $VER(S, m, PK)$ be the verification of the signature S to the message m by the public key PK. $VER(S, m, PK)$ is valid if S is the signature of m by the corresponding secret key of PK, i.e., if $S = SIG(m, SK)$, and it is invalid otherwise. Here, a receiver can actively send a random challenge nonce, and ask the sender to include this nonce in the replying authenticated message. By doing this challenge–response operation, the receiver can guarantee the freshness of the authenticated message by verifying the received nonce. This public-key based unilateral message authentication scheme can be easily extended to a mutual message authentication scheme, and it can be freely scaled to a broadcast authentication scenario.

However, public-key schemes require much longer keys and are much heavier to execute. Because WSNs face memory limitations, slower CPUs that limit the speed of execution, limited communication bandwidth, and energy constraints, it is usually assumed that public-key cryptography cannot be used in WSNs. There are ongoing works [20] to customize public-key cryptography and elliptic key cryptography for low-power devices, such approaches are still considered as costly due to high processing requirements.

14.3.2.2 Authentication Based on Symmetric-Key System

In a symmetric-key scheme, the two communicating entities must first agree on a shared symmetric key in a secure and trusted way. Then, this agreed symmetric key can be used to create MAC, which is a cryptographic checksum that is appended to the message. During verification phase, the receiver generates another MAC from the received message using the agreed symmetric key and compare it to the MAC appended in the received message. If the two MACs are the same and the receiver trusts that the symmetric key used is shared with the correct sender, the message authentication has been verified. It is obvious the MAC also protects the integrity of a message. Similar to that in public-key-based authentication, the freshness of authenticated message can also be guaranteed by a challenge–response method.

It can be seen that a symmetric-key-based end-to-end authentication can be easily realized, as well as the two participating entities agree on a pairwise key by some way. Traditionally this is done by encrypting the symmetric key using public-key cryptography or by using a trusted third party called KDC. In WSNs, the way to build a pairwise key between entities can be found in the well-discussed key management schemes (see Section 14.3.1). Once a pairwise key is established between two entities, they can easily agree on more pairwise keys by encrypting these keys using the first established pairwise key. The symmetric keys for authentication can be agreed similarly.

However, a MAC only provides mutual (pairwise) authentication. It is difficult to provide broadcast authentication with MACs. Performing a unilateral message authentication with each receiver is inefficient. Also, the simplest approach of distributing a global secret key mechanism cannot be applied to broadcast authentication, because any compromised receiver can easily forge messages from the sender. A considerable broadcast authentication approach based on MACs is the one proposed by Canetti et al. [21]. They propose an asymmetric MAC scheme which is based on multiple symmetric keys and MACs for n receiving parties. The goal is to achieve the same capabilities as a digital signature with symmetric MACs only. Let S be a set of keys. Each entity then gets a subset of the keys $S_i := k_1, \ldots, k_n \in S$ at initialization time. The main set and the subsets are chosen in such a way that the probability each two subsets have at least one key in common is high. If Alice wants to authenticate a message m, she computes the MAC over the message by each key of her subset $M_i := MAC(m, k_i)$ for all $k_i \in S_i$, and broadcasts $m, M_1, \ldots, M_n$. Due to the design of subsets S_i, each receiver is able to find at least one key that she shares with Alice to verify m. This approach provides a computationally efficient broadcast authentication scheme by using only symmetric keys. However, there is a large message overhead to carry n MACs for each broadcast, and is not easily applicable to resource-constrained WSNs.

14.3.2.3 Broadcast Authentication Based on Key-Chains

As discussed before, asymmetric public-key-based authentication can easily perform broadcast authentication, but is not suitable for WSNs because of its high computation, communication, and storage overhead. Symmetric-key-based authentication performs great in the two-party communication case, but not in the broadcast authentication case. Key-chain-based schemes are however found to be promising for broadcast authentication in WSNs.

A key-chain is defined by a sequence of keys generated based on a one-way hash function. First, there is a randomly choosed secret w. Then, a one-way hash function h is used repeatedly on this initial secret w to generate a sequence of keys: $w, h(w), h(h(w)), \ldots, h^t(w)$. Here, $h^i(w)$ is the repeated ith iteration of h. In a basic broadcast authentication scheme, the sender is assumed to have generated such a key-chain. First, each receiver in the broadcast group should be delivered with the last key (i.e., $h^t(w)$) in the key-chain by an authenticated initial key-exchange. Then for the ith authentication, the sender broadcast key k_i together with the authenticated message and each receiver verifies whether $h(k_i) = k_{i-1}$, if we define $k_i = h^{t-i}(w)$. If the verification succeeds, the receiver will store k_i for the next authentication. Because the verification by hash calculating has low computation overhead, this scheme can be used by WSNs. The one-way feature of hash function also ensures the difficulty to impersonate the sender. The authenticated initial key-exchange can be achieved by a public-key-based signature scheme, or by preloaded pairwise keys between the sender and each receiver. However, this basic key-chain-based scheme suffers from malicious attacks in real implementation, because a powerful adversary can eavesdrop the authenticated message and forge its own message using the captured key.

TESLA [22] reinforces the basic scheme by introducing a clock and by delayed release of keys by the sender. In TESLA, the sender sends message m_i together with a MAC authenticated by k_i in time interval t_i. Such a message is only accepted during the time interval t_i but not later. In the next time interval, the sender releases k_i and the receivers can verify the buffered m_i. μTESLA [23,24] proposed by Perrig et al. is a broadcast authentication protocol based on TESLA. Instead of using a digital signature scheme in TESLA, μTESLA authenticates initial key-exchange message using a symmetric node-to-base-station authenticated channel, in the case that the base station is the sender. In the case that a sensor node is the sender, the sensor node always broadcasts the data through the base station. Thus, there is no expensive node-to-node key agreement and it is not necessary for a sensor node to store the key-chain and disclose keys. Also, time interval is usually small for security reason, thus there may be not always broacast messages within each time interval. Thus, the key disclosure time delay in μTESLA is on the order of a few time intervals, as long as it is greater than any reasonable round-trip time between the sender and the receivers. An μTESLA broadcast example is shown in Figure 14.1, where the key disclosure delay is two time intervals. Obviously, μTESLA is more suitable for WSNs than TESLA.

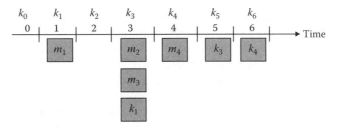

FIGURE 14.1 μTESLA broadcasting authenticated packets with disclosure delay being two time intervals.

μTESLA is still considered to have a limited scalability because of its unicast-based initial key-exchange. Liu and Ning [25] propose an enhancement to the μTESLA system that uses broadcasting of the key chain commitments rather than μTESLA's unicasting technique. They present a series of schemes starting with a simple predetermination of key chains and finally settling on a multilevel key chain technique. The multilevel key chain scheme uses predetermination and broadcasting to achieve a scalable key distribution technique that is designed to be resistant to DoS attacks, including jamming.

LEAP [18] also defines a local broadcast authentication scheme using a one-way key chain like μTESLA, but without the delayed key disclosure. The first key in the chain is sent to each neighbor encrypted with their pairwise keys. Then for each message the node has to send, it attaches the next key in the key chain. This is similar to the basic scheme proposed at the beginning of this section. To prevent outsider attacks, the keys in the key chain can be XORed with the cluster key. Insider attacks are still possible but depressed.

14.3.2.4 Open Issues

Although symmetric-key and a few public-key-based authentication can provide acceptable security with moderate communication, computation, and storage overheads, it is still valuable to develop more economic authentication ways in resource-constrained WSNs.

The authentication delay introduced in key-chain-based broadcast authentication cannot satisfy real-time applications. Also, each receiver has to buffer the received packets before the key disclosure. This is quite memory consuming and may be vulnerable to inside DoS attack. The requirement in TESLA for loose time synchronization may not be true in some applications, and also there are many ways to disrupt the time synchronization. So, it is desirable to have alternative broadcast authentication approaches which should be independent of time synchronization.

14.3.3 INTRUSION DETECTION

Security technologies, such as authentication and cryptography, can enhance the security of sensor networks. Nevertheless, these preventive mechanisms alone cannot deter all possible attacks (e.g., insider attackers possessing the key). Intrusion detection (ID), which has been successfully used in Internet, can provide a second line of defense.

ID involves the runtime gathering of data from system operation, and the subsequent analysis of the data. ID systems can be classified according to the detection techniques they use: signature-based detection, specification-based detection, and anomaly detection. Signature-based detection needs knowledge to build attack signatures and suffers from the inability to detect unknown attacks. At current stage of sensor network development, most of known possible attacks are only imagined or copied from other mature networks like Internet. Whether these known attacks would be serious problems and whether any unknown serious attack could happen in sensor networks still remain unclear. Unlike those unclear attack signatures, people have exact knowledge about what each designed protocol functions like. If a sensor node does not act according to the protocol specification, people have high confidence to declare that node to be malicious. Refs. [26,27] are examples that use finite state machines for specifying correct routing protocol behaviors and then detect runtime violation of the routing specifications. Such specification-based detection has an advantage of low false alarm. However, specification-based detection cannot detect malicious behaviors which do not violate protocol specifications. In that case, anomaly detection which not only detects incorrect behaviors (which violate specifications), but also detects abnormal behaviors (which do not violate specifications) can serve as a complement to specification-based detection strategy. In anomaly detection, profiles of normal behaviors of systems, usually established through automated training, are compared with the actual activities of systems to flag any significant deviation. Although anomaly detection has the advantage in detecting attacks other technologies cannot do, it usually suffers from a high false alarm rate. Besides the classification according to the detection techniques, ID systems can also be classified according to the place it is located. ID systems installed and run on a single node are called host-based ID system (IDS), and this kind of ID system usually use the information (e.g. system logs) acquired from the host node to detect an attack or misbehavior, and is usually only responsible for the security of the host node. ID systems which are installed on gateway nodes or separate monitors usually take network traffic as data source and are responsible for the security of a part or the whole network. This kind of ID system is called network-based IDS. Currently, most of the proposed ID systems for WSNs are network-based and use either specification-based or anomaly-based detection techniques.

14.3.3.1   Detecting Techniques of ID Systems

Although attackers in WSNs are supposed to behave in an arbitrary way and there are difficulties to statistically model attacks, it does not mean attacks inside WSNs cannot be detected. Researchers believe the behavior of an attacker will violate the normal behavior profile of the network [26,28]. Ref. [29] proposes that a successful attack in a WSN will violate one of the following rules: (1) Interval rule: A failure is raised if the time past between the reception of two consecutive messages is larger or smaller than the allowed limits. (2) Retransmission rule: The monitor listens to a message, pertaining to one of its neighbors as its next hop, and expects that this node will forward the received message, which does not happen. (3) Integrity rule: The message payload must be the same along the path from its origin to a destination, considering that in the retransmission process there is no data aggregation by other sensor nodes. (4) Delay rule: The retransmission of a message by a monitor's neighbor must occur before a defined timeout. (5) Repetition rule: The same message can be retransmitted by the same neighbor only by a limited number of times. (6) Radio transmission range: All messages listened to by the monitor must be originated (previous hop) from one of its neighbors. (7) Jamming rule: The number of collisions associated with a message sent by the monitor must be lower than the expected number in the network.

We can see the violation of the above listed rules can be detected by checking traffic features, and network traffic is actually the most used data source by ID systems for WSNs. The following introduced ID techniques are all based on analyzing traffic features. Instead of directly implementing the listed rules above, most of the proposed ID techniques identify anomalies based on the preknown specification knowledge and the knowledge learned based on some kind of artificial intelligence.

Ref. [26] proposes a specification-based ID to detect attacks on the Ad hoc On-Demand Distance Vector (AODV) routing [30], which is a classic routing protocol designed for mobile ad hoc and sensor networks. In this approach, the authors use finite state machines to specify correct AODV routing behavior and use distributed network monitors to detect runtime violation of the AODV specifications. The rationale behind is that AODV protocol has specified the sequence relations among different kinds of routing messages, and such sequence relations can be depicted by finite state machines. Any observation of the violation of the protocol specifications triggers an alert. Ref. [27] also proposes a specification-based ID to detect routing attacks. In their approach, they use finite state machines to specify dynamic source routing (DSR) [31] behavior, which is another classic routing protocol for ad hoc and sensor networks.

Besides that the sequence relations among some special kinds of packets (e.g., routing messages) can be specified according to protocol specifications, Ref. [28] suggests that the sequence relations among general packets can also be learned automatically by online training, thus attacks can be detected by comparing runtime traffic patterns with learned historic traffic patterns.

Ref. [32] uses another traffic feature instead of packet sequence relations. It records the arrival time of each observed packet and check the mean and the standard deviation of interarrival times of the packets in a long term receive buffer and a short term intrusion buffer. An arrival is considered anomalous if the statistics in these two buffers deviate a lot.

Refs. [33,34] introduce a new data mining method that uses *cross-feature analysis* to capture the interfeature correlation patterns in normal traffic, thus it can make decisions based on multiple traffic features. These patterns can be used as normal profiles to detect deviations (or anomalies) caused by attacks. Concretely, this approach computes a classifier C_i for each feature f_i using $\{f_1, f_2, \ldots, f_{i-1}, f_{i+1}, \ldots, f_L\}$, where $\{f_1, f_2, \ldots, f_L\}$ is the feature set. C_i can be learned from a set of training data. It predicts the most likely value of f_i based on the values of other features. Based on a set of rules presented, this approach can identify the attack type of several well-known attacks. In some cases the rules can also identify the attacking or misbehaving nodes.

14.3.3.2 Deployment Strategies of ID Systems

Typically, there are different strategies when an ID system comes to deployment. A straight forward way is to install an ID system for every single node in the network, and the ID system is only responsible for the security of the host node. The goodness of this method is that there is almost no communication overhead. However, this method only makes sense when the ID system itself is more secure and thus more trustable than the host node. An alternative is to make ID system monitor neighboring nodes, based on the fact a node can also oversee the packets transmitted by neighboring nodes in the wireless world. Several systems proposed [35–37] do use neighbor monitoring to detect network anomalies. Due to the fact that a normal sensor node is usually resource-constrained and easy to be compromised, some researchers [26] also suggest to use special monitoring nodes for ID usage. Because most of network traffic will finally pass through cluster heads in a cluster-based network, Ref. [34] only deploys ID systems on cluster heads to save costs. The drawback of this deployment strategy is that the deployment loses monitoring coverage. Because ID systems at the places of cluster heads have no way to know details inside the network, it is difficult to detect and effectively respond to some kind of attacks, such as jamming attack, impersonating attack. Ref. [38] proposes an optimal placement that the detection modules are only placed on all the nodes that belong to the minimum cut-set. Here a cut-set is a set of sensor nodes such that all paths from all cluster heads to the base station traverse this set of node. This deployment is optimal under the assumption that the objective of an intruder is to cause harm to or failure of the base station computer system by having malicious packets delivered to it. All the above strategies make the ID systems distributed deployed among a few nodes and cooperation could exist to make the distributed system as efficient as a central-based system. Ref. [39] assumes that a central-based ID system can exist but this system cannot protect all sensor nodes simultaneously due to system limitations. Thus, the ID system only chooses the most vulnerable node to protect at any given time based on a game approach, which is formulated as a two-player attack–defense problem.

14.3.3.3 Alert and Response

Once an attack is detected, the attack information should be alerted to both uplink nodes and downlink nodes. The uplink nodes are supposed to finally forward the alert to the base station, where the administrator can be notified with all security related information through a client system. The awared administrator then determines the possible responses, such as manually revoke the malicious node. The downlink nodes are the places the attack comes from. After receiving the alert, each downlink node will check its neighborhood situation and will rebuild routing table to bypass the possible malicious node if necessary. Finally, the malicious node will be revoked from the network.

14.3.3.4   ID System Evaluation

The two common used measurements for evaluating the performance of ID systems are false positive rate (FP) and false negative rate (FN). FP is defined as the proportion of normal events that are erroneously classified as abnormal. FN is defined as the proportion of abnormal events that are erroneously classified as normal. Obviously, a good ID system should have both low FP and low FN. However, trade-off is usually to be made between FP and FN, given these two measurements is usually influenced in the opposite way by adjusting system parameters. Besides FP and FN, the overhead caused by ID system is also concerned. Considering the extreme resource-constrained specialties of WSNs, a good ID system should introduce as little overhead as possible. Although WSNs are designed for low rate communication, a broad range of real-time applications, such as medical care, highway traffic coordination, and even video transmission, have also been introduced. When an ID system is designed for such real-time application, it should also fulfill the real-time requirement thus will not cause performance degradation of the underlying application.

14.3.3.5   Open Issues

Although ID is a well-implemented technology in Internet security, it is still among the least explored security related technologies for WSNs. The reason behind is simple: It is very difficult to implement an ID system that is effective for WSNs. First, an ID system is usually installed on a powerful gateway computer in the Internet structure, but there is no such kind of gateway in the WSN structure. Although a cluster head can watch most of traffic in its corresponding cluster and is supposed to be more powerful than ordinary sensor nodes, it cannot be compared with a powerful gateway in Internet. Second, there are only two places in WSNs for an ID system to be installed on: either normal sensor nodes, or specialized network monitors. If an ID system is installed on normal sensor nodes how can the security of the ID system itself be protected and thus the alerts by the ID system can be trusted, because a normal sensor node is easy to be tampered. If an ID system is installed on specialized monitors, how to get these cheap but still powerful monitors is the problem, because a lot of applications can only be possible given that the sensor network deployment is cheap enough. Third, many proposed ID technologies have a nonzero FP. A striking fact is that an ID system with a low 0.1 percent FP and a zero FN still generates 10 false alarms in every 11 alerts, if the probability for an observed event (e.g., packet arriving) being caused by an attack is 0.01 percent (and many people believe the actual attack probability is even below this number). Obviously, this seemingly low FP is not tolerable for WSNs where attacks will not happen as often as that in Internet. But for many ID technologies proposed, they even cannot guarantee an FP as low as this.

14.3.4   FAULT AND INTRUSION TOLERANCE

WSNs consist of a large number of tiny sensor devices that have limited power and limited sensing, computation, and wireless communication capabilities. Sensor nodes usually operate in unattended and even harsh environments, and as a result, sensor nodes are prone to failures and are vulnerable to malicious attacks. Therefore, for reliable and secure computation and communication in WSNs, fault tolerance and intrusion tolerance become two essential attributes that should be designed into WSNs [40]. Concretely, the goal to obtain a fault and intrusion tolerant WSN can be depicted as the following problem in the design stage: minimize the total cost of a WSN, given the constraint that the expected network operation time should still be longer than the desired network lifetime even after one or several faults and intrusions happen.

Fault tolerance and intrusion tolerance are related and thus we put them together to elaborate. The common point between faults and intrusions is that they both cause errors inside the system. Therefore, the system can malfunction due to the errors caused. The difference is that faults cause errors randomly, but intrusions are usually done deliberately and will preferentially target the most important component in the system. Further, faults can exist everywhere in the system and can happen anytime, but the scopes of intrusions are subject to the abilities of attackers. In terms of available techniques, there are similarities for fault tolerance and intrusion tolerance. For example, redundancy is efficient for both fault and intrusion tolerance. However, encryption and authentication technologies are only useful for intrusion tolerance. In the following, the analysis of errors in WSNs together with available tolerance technologies will be elaborated.

14.3.4.1   Causes of Errors in WSNs

Errors in WSNs can happen if there are faults or intrusions. An intuitionistic description about all kinds of errors in WSNs can be found in Figure 14.2 [40,41].

Generally, faults can be the result of a variety of things that occur within WSNs, external to WSNs, or during the component or network design process. At the highest level is the possibility of specification mistakes such as incorrect algorithms, architectures, hardware/software design, and vulnerabilities/flaws in security policies. The next cause of faults is implementation mistakes. The implementation can introduce errors due to poor design, poor component selection, poor construction, software coding mistakes, and vulnerabilities/flaws in security mechanisms. The third cause of errors is component defects. Typical examples include manufacturing imperfection, and component wear-out. The fourth cause of faults is external disturbances such as channel noises, radiation and electromagnetic interference, operator mistakes, battle damages, and environmental extremes (e.g., earthquakes, floods, fire, hurricanes). The fifth cause of faults is resource constraints. Due to the limited power, memory, computation capability, and communication bandwidth, WSNs can fail because of the exhaust of their resources.

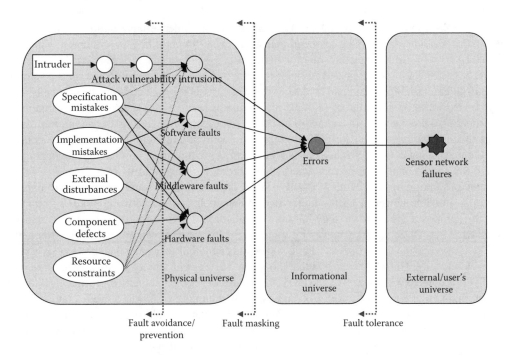

FIGURE 14.2   Three-universe model for WSN.

Besides inherent faults, successful intrusions can also cause errors in WSNs. Typical examples include sensor node capture or compromise, traffic jamming, and routing attacks.

14.3.4.2 Error Tolerance Techniques

Although error tolerance has been studied for several decades in computer and VLSI systems, tremendous intrinsic reliability of VLSI integrated circuits technology and operation in well-conditioned environments restrict the importance of error tolerance in great majority of computing systems [42]. Currently, the researches in error tolerance of WSNs are mainly focused on improving network robustness with the presence of errors in measurements [42–44], unreliable radio channel [45,46], failed nodes [40,47–49], and malicious attacks [40,50–53], because these four observations are either different from or more obvious in WSNs than in traditional systems and have significant higher possibilities to breakdown the system than other possible causes (e.g., failures in specification, operating system, and processor).

Tolerance to measurement faults: Ref. [42] explicitly takes into account the possibility of sensor measurement faults and develops a distributed Bayesian algorithm for detecting and correcting such faults. The approach proposed exploits the notion that measurement errors due to faulty equipment are likely to be uncorrelated, although environmental conditions are spatially correlated. Thus, each node can estimate the correctness of its own reading according to the readings of its neighbors. Simulations show this method can correct most of the error readings while introducing a small number of new errors. However, this approach does not guarantee an estimation consensus because each sensor has a different set of neighbors.

Ref. [40] envisions that each of five primary types of resources computing, storage, communication, sensing, and actuating can replace each other with suitable change in system and application software. Thus, the authors propose a heterogeneous backup scheme, where a single type of resource backs up different types of resources. The approach is in particularly well suited for addressing transient errors and errors in measurements as the authors say.

Tolerance to unreliable channel and failed nodes: Unreliable channel and node failure propose a big challenge to reliable communication in WSNs. Ref. [44] found packet loss was a severe problem in their GlacsWeb project. Wet and windy locations often hinder reliable radio communications. Theoretical calculations of radio losses in glacier ice were found to be a poor guide to actual performance. Thus, a redundancy transmission policy was used in their implementation. That is, every packet was sent multiple times at the sender to get a lucky arrival. A more common used strategy by fault-tolerant routing techniques is to send the same packet through multiple disjoint paths. Usually, the existence of such kind of multiple disjoint paths between any pair of nodes is guaranteed by providing the network with a property called k-connectivity, where k is the number of disjoint paths that can be guaranteed. Ref. [52] suggests to deploy a minimum number of relay nodes so that the underlying network achieves certain connectivity requirement. Sometimes the deployment of a WSN can be unnecessarily redundant when all the nodes work at its full capability. To save the precious energy resources in WSNs, Ref. [43] presents a distributed algorithm for assigning minimum possible power to all the nodes in a WSN, such that the network is still k-connected. Ref. [46] proposes a rotation strategy for an overredundant WSN. During each round, a Connected Dominated Set (CDS) is first achieved by some available protocol such as Span [53]. After getting a CDS, each node in CDS will select some nodes in its neighborhood such that this node is k vertex connected locally. This chapter shows that the global vertex connectivity is guaranteed by required local vertex connectivity as well as the CDS. The nodes unselected in one round will be set to sleep in that round for energy saving. Ref. [47] argues that maintaining a specific degree of fault tolerance between any two sensors is not critical in WSNs, where data is only transmitted from sensors to sink(s). Thus, the authors introduce the k-degree anycast topology control (k-ATC) problem with the objective of selecting each sensor's transmission range such that each sensor is k-vertex sink(s) connected and the maximum sensor transmission power is minimized. However, the way to increase reliability by sending copies of the same packet over multiple paths is not efficient [54]. Ref. [54] proposes to distribute the fragments of the same packet among multiple paths to achieve both load balance and energy saving. In their approach, the sender node uses forward error correction (FEC) erasure coding to encode each packet into multiple fragments and distributes these fragments among multiple disjoint paths. The reliability of this approach is provided by the property that the receiver only needs to receive one part of these encoded fragments to rebuild the original data packet. Some earlier works also focuses on providing fault tolerance for single path routing by providing a way to quickly recover from a broken route path. For example, Ref. [45] gives a way for the base station to efficiently trace failed nodes and quickly rebuild route paths according to the knowledge of network topology. Except for routing, the communication problems caused by link and node failures also have a negative influence on in-network processing. In Ref. [55], error correcting codes and soft-decision decoding schemes are used in the data fusion process to achieve fault tolerance.

Tolerance to malicious attacks: Due to the unique features of WSNs, they face a combination of threats that are not normally faced by traditional wired and wireless networks, and are vulnerable to a variety of security attacks. Thus, it is important for a WSN to possess the capability in keeping systems working correctly, despite the occurrence of malicious attacks. Usually, this capability is called intrusion tolerance. Encryption and authentication technologies either prevent the happening of an intrusion (e.g., by hindering unauthorized packets) or mitigate the consequences of an intrusion (e.g., the privacy of an encrypted packet can be kept after an eavesdropping attack) and thus they are currently the mostly used intrusion tolerance techniques. It is shown [38] that the conventional public-key cryptography, such as RSA, ECC, can be deployed on the current resource-constrained sensor nodes. Among all available cryptographic algorithms, ECC, AES, and NtruEncrypt are the most promising candidates for low-power implementations for WSNs. Ref. [48] provides intrusion tolerance against isolation of a base station by introducing multipath routing redundancy and by ensuring that each path is routed toward a different base station. Similar idea is seen in Ref. [54], where information arrives at the sink via multiple proxy nodes. The difference is that each original packet in Ref. [54] is first split into N fragments and then encoded into $N + K$ fragments using FEC coding before distributedly transmitting them through multiple paths. To alleviate the consequences caused by key compromising, Ma, etc. propose to encrypt different fragments of a packet using different cryptographic algorithm in Ref. [38]. After that, the encrypted fragments can be further encoded using FEC and transmitted to the destination through multiple disjoint paths like that in Ref. [54]. INSENS [49,50] is an intrusion tolerant routing protocol that limits the effects of a single compromised node to a localized portion of the network. In INSENS, only the loosely authenticated base station is allowed to broadcast to the entire network and initiates a route discovery process. Thus, DoS-style flooding attacks are prevented. Rate control is also applied in INSENS to further restrict attacks. Finally, a heuristic multipath discovery algorithm is used to improve routing redundancy.

The Byzantine faulty behavior [56] is a specialty that can only happen in a malicious environment. In a typical Byzantine attack, a malicious sensor node can send different values to other sensor nodes. This will greatly influence applications which rely on collaboration among local nodes to produce a result global to the region. Ref. [51] proposes a fault-tolerant clock synchronization scheme and Refs. [57,58] propose a fault-tolerant collaborative target detection scheme with the assumption that a Byzantine faulty behavior can exist.

14.3.4.3 Error Tolerance Modeling and Evaluation

In Ref. [59], a Bernoulli node model, sensor networks are modeled by the unit disc graph, random point process and Bernoulli nodes with binary states, i.e., operation and failure. Fault tolerance can be investigated through the probability of node failures. Ref. [42] assumes the errors due to sensor faults and the fluctuations in the environment can be modeled by Gaussian distributions with a zero mean and a standard deviation. In Ref. [60], the reliability of a set of sensors, which is defined as the probability that no sensor in this set fails during a given time interval, is modeled using Poisson distributions.

Redundant connection is usually considered to be the key in providing network stability. Refs. [61,62] show that error tolerance is not shared by all redundant systems: heterogeneous scale-free networks with power-law node degree distribution exhibit high fault tolerance. However, fault tolerance comes at a high price in that these networks are extremely vulnerable to attacks. In Refs. [61,62], the node failure is simulated by the removal of randomly selected nodes. Instead, the attack is simulated by the targeted removal of the most important nodes (e.g., the most connected nodes). In Ref. [61], error tolerance is studied by measuring the change of network diameter, defined as the average length of the shortest paths between any two nodes in the network. In Ref. [62], error tolerance is evaluated by a measure called global efficiency, defined as the average of the efficiency over all couple of nodes, and the efficiency for a couple of nodes is defined by the reciprocal of the time it takes to send a unit packet of information through the fastest path.

Ref. [63] addresses the problem of fault-tolerant estimation and the design of fault-tolerant sensor networks from a control system point of view. Fault tolerance is defined with respect to a given estimation objective, namely a given functional of the system state should remain observable when sensor failures occur. Three criterias are used to evaluate the system's fault tolerance: (strong and weak) redundancy degrees (RD), sensor network reliability (R), and mean time to nonobservability (MTTNO). Sensor networks are designed by finding redundant sensor sets whose RD or R or MTTNO are larger than some specified values.

Ref. [39] considers the reliability and security modeling of WSNs in an integrated manner. The approach proposed unifies the attributes of fault tolerance and intrusion tolerance through the development of a WSN three universe model (WSN-TUM) (see Figure 14.2). In Ref. [39], security failures due to the occurrence of malicious intrusions are represented in a dynamic fault tree (DFT), and traditional failures due to the malfunctions of the system's constituent components are represented in a combination of probabilistic graph model and DFT model. The resulting solution technique is applicable to Markov analysis and combinatorial methods such as binary decision diagrams for the analysis of both reliability and security.

14.3.4.4 Open Issues

Sensor networks suffer from faults and attacks due to the low-price nodes and the deployment in complicated environments. Thus, fault and intrusion tolerance attract attention from the beginning of sensor network research. However, fault and intrusion tolerant techniques usually introduce extra redundancy, and the introduced extra redundancy either costs more money, or expends more precious energy and bandwidth resources. This is contrary to the usual sensor network design goal, such as minimum cost and minimum energy consumption. Further, the introduced extra redundancy also proposes some new challenges in network operation. For example, a sleep/wake-up scheduling algorithm with synchronization will be necessary in a redundant sensor network. The extra communication also increases the risk to reveal a cryptographic key. So, how to make a good balance among system cost, system reliability and system security is the challenge for future work.

14.3.5 Privacy Protection

As WSN applications expand to include increasingly sensitive measurements in both military tasks and everyday life, privacy protection becomes an increasingly important concern. For example, few people may enjoy the benefits of a body area WSN, if they know that their personal data such as heart rate, blood pressure, etc., is regularly transmitted without proper privacy protection. Also, the important data sink in a battlefield surveillance WSN may be first destroyed, if its location can be traced by analyzing the volume of radio activities.

Generally, privacy in WSNs can be classified into two categories [5]: content privacy and contextual privacy. Threats against content privacy arise due to eavesdropping and tampering. This type of threats is partially countered by encryption and authentication (see Sections 14.3.1 and 14.3.2). The following introduced anonymity protection and aggregation privacy can be classified into this category of content privacy. However, even after strong encryption and authentication mechanisms are applied, wireless communication media still exposes contextual information about the traffic carried in the network. For example, an adversary can deduce the direction of wireless communications by eavesdropping and analyzing the patterns of network traffic. In particular, the location information about senders/receivers may be derived based on the direction of wireless communications.

14.3.5.1 Anonymity

Anonymity in sensor networks includes sender anonymity, receiver anonymity, and message transmission anonymity between the sender and the receiver. A third party cannot determine the sender and the receiver's identities through reading a message. The adversary also cannot determine whether two communication segments belong to the same communication between a sender and a receiver.

Pseudonyms can be used for each sensor node in WSNs to protect its real ID. However, using fixed pseudonym cannot prevent leaking identity information of sensor nodes in the long term. Thus, random dynamic pseudonyms are used to protect a sensor's real ID. Randomly selecting a pseudonym from a pool and hashing-based ID random pseudonym are two methods that can protect sensor ID under the assumption that the secret keys shared by sensor nodes and the base station cannot be compromised.

Misra and Xue [64] proposed a cryptographic anonymity scheme (CAS) in which the pseudonym of a sensor node is generated from keyed hash functions. Authors in Ref. [65] propose a hashing-based ID randomization (HIR) and a reverse hashing ID randomization (RHIR)-based that use a one-way keyed hash chain for producing a sequence of hash values as sensor's IDs. In the HIR methods, each sensor i will have a dynamic pseudo-ID for every neighbor j. The dynamic pseudonym of i for neighbor j is a hash value of real ID i keyed by the shared key K_{ij} of node i and j. This pseudo-ID of node i for j, ID_{i4j}, is used only once. The next one will be $H_{K_{ij}}(ID_{i4j})$, which node i will use when it sends another message to node j. Both the immediate neighbor receiver and the final base station will have no problem to find the real ID of i, if they know how many times the hash execution has been used on the real ID. RHIR uses the same operations as the HIR method, except that it uses the one-way keyed hash chain in reverse. In this method, a sensor node needs to compute the hash chain first and store it locally. However, RHIR offers better security.

14.3.5.2 Aggregation Privacy

Due to resource limitation and especially due to power limitation, many sensor networks need collaboration on in-network processing to reduce the amount of raw data sent. In the in-network processing, networked sensor nodes process data aggregation statistic functions in collaboration with one another. The aggregation function can be SUM, AVERAGE, etc. Because the readings of sensors could reveal sensitive information of an individual person or household, we need a way to preserve data privacy in the process of aggregating sensed data.

Authors in paper [66] present cluster-based private data aggregation (CPDA) and slice-mix aggregate privacy-preserving data aggregation (SMART) schemes for additive aggregation functions.

Cluster-based private data aggregation [66]: The first step in CPDA scheme is to construct clusters with size bigger than 3 to perform intermediate aggregations. The second step of CPDA is to calculate the desired aggregate value within clusters. Given a cluster consisting of member nodes $A_1, A_2, \ldots, A_n$ with sensor reading data $a_1, a_2, \ldots, a_n$ respectively. Let the cluster leader to be A_1 without loss of generality. Nodes within this cluster share a common knowledge of nonzero seeds, $x_1, x_2, \ldots, x_n$, which are distinct with each other. Each node A_i first calculates $v_{A_j}^{A_i} = a_i + \sum_{t=1}^{n} r_t^{A_i} x_j^t$ for $j = 1, \ldots, n$, where $r_1^{A_i}, \ldots, r_n^{A_i}$ are n random numbers generated by node A_i, and known only to node A_i. Then node A_i encrypts $v_{A_j}^{A_i}$ and sends to every other node A_j, using the shared key between A_i and A_j. When node A_i receives all $v_{A_i}^{A_j}$ from all other nodes, it calculates assembled value $FA_i = \sum_{j=1}^{n} v_{A_i}^{A_j} = \sum_{j=1}^{n} a_j + \sum_{t=1}^{n} r_t x_i^t$, where $r_t = \sum_{t=1}^{n} r_t^{A_j}$. After that, nodes $A_2, \ldots, A_n$ broadcast

$F A_2, \ldots, F A_n$ to the cluster leader A_1. Because A_1 knows all the seeds x_i and $F A_i$, the cluster leader A_1 can deduce the aggregate value $f(\overline{a}) = \sum_{j=1}^{n} a_j$. The intermediate aggregate values in each cluster will be further aggregated (along an aggregation tree) on their way to the data sink.

Slice-mix aggregate privacy-preserving data aggregation [66]: In SMART, each node hides its private data by slicing the data and sending encrypted data slices to different aggregators which are selected randomly. When an aggregator receives an encrypted slice, it decrypts the data using its shared key with the sender. Upon receiving the first slice, the node waits for a certain time, which guarantees that all slices of this round of aggregation are received. Then, it sums up all the received slices and forward data to a query server. When the server receives the aggregated data, it calculates the final aggregation result.

14.3.5.3 Location Privacy

It is very important to protect the privacy of a node's location in WSNs. In many applications, sensors are attached to persons, vehicles, etc. Location information has the potential to allow an adversary to physically locate a person and critical nodes, and therefore most wireless users have legitimate concerns about their personal safety if such information should fall into the wrong hands.

Sink location privacy: In many sensor networks, the sink is the most critical node of the whole network as to collect data from all sensors. A sensor network can be rendered useless by taking down its sink. In some scenarios, the sink itself can be highly sensitive. Imagining a sensor network deployed in a battlefield where the receiver is carried by a soldier, the soldier will be in great danger if the location of the sink is exposed to adversaries [67].

In WSNs, the location of a sink can be traced by eavesdropping, traffic analysis [4], and packet-tracing attack [5] (see Section 14.2.2). Different approaches have been designed to protect sink location privacy. These approaches can be classified as destination ID protection, randomized traffic volumes, fake path injection, and random delay transmission time.

- Destination ID protection: MASK [68] deals with passive eavesdropping attacks in mobile ad hoc networks. To achieve anonymity in communications, it conceals the nodes' network/MAC addresses by dynamic pseudonyms rather than static MAC and network addresses.
- Randomized traffic volumes: In a typical sensor network, the area close to the base station bears high traffic volume. This feature can be used by rate monitoring attacks to determine the locations of important nodes, such as the cluster heads and the base station, in a WSN. The goal of this protection method is to introduce randomized traffic volumes throughout the sensor network away from the base station, to deceive and misdirect an adversary so that the true path toward the base station cannot be easily found. In Ref. [4], a controlled random walk is introduced into the multihop path traversed by a packet through the WSN toward the base station. This distributes packet traffic, thereby rendering less effective rate monitoring attacks. Also, multiple random areas of high communication activity are created to deceive an adversary as to the true location of the base station, which further increases the difficulty of rate monitoring attacks.
- Fake path injection: The object is to inject fake paths to make paths become a completely random for the adversary who follows the overheard packets. Multipath routing is a way to achieve this goal [4]. Also, fake packets can be sent such that the probability of forwarding a packet to any neighbor is equalized, thus the adversary is confused which neighbor is the real parent [67].
- Random delay transmission [48]: The transmissions of the packets are randomly delayed to hide the traffic pattern and the parent–child relationship under a certain traffic rate model. This packet-tracing attack protection approach introduces extra delay for delivering packets in a sensor network.

Source location privacy: It is an important security issue. The loss of source location privacy can enable the exposure of identity information, because location information enables the binding between cyberspace information and physical world entities. In a WSN, location information often means the physical location of the event, which is crucial for some applications. For example, the location of a soldier should not be exposed in a battlefield.

For source location privacy, random walk [69], random delay transmission, and flooding are methods of projection. Flooding is the worst method for protecting source location privacy in terms of the time it takes for an adversary to get a good enough estimate of the source location [69]. Random delay transmission cannot protect the source location privacy for a patient adversary who uses backtracking strategy. The use of random walk can protect source location privacy well. In a random walk, the forwarding decision is made locally and independent of the source location. An eavesdropper cannot distinguish two random walks from two different sources. Also, the random walk randomizes the routing path, thus making backtracking strategy difficult to use. However, a pure random walk tends to stay around the source and the average delivery time for a packet hitting sink eventually goes to infinite. This drawback in a pure random walk can be eliminated by filtering out the nodes visited in early random walk path so that the random walk is always trying to cover an unvisited area.

14.3.5.4   Open Issues

One of the most notable challenges threatening the successful deployment of WSNs is privacy. Although many threats against content privacy can be countered by encryption and authentication. The threats against contextual privacy such as location privacy cannot be easily treated. Currently, the location information of source and sink is mostly hidden by disrupting the normal routing path or by introducing background noise traffic. Besides the acquired privacy, the results by doing this also include the degraded routing performance, the missed real-time requirement, and the shortened network lifetime. The development of privacy protection methods that minimize their negative effects to network performance never comes easy.

14.3.6   SECURITY MANAGEMENT

Security management is the process of managing, monitoring, and controlling the security related behavior of a network, and it plays an important role in network management. The primary function of security management is controlling access points to critical or sensitive data that is stored on devices attached to the network. Security management also includes the seamless integration of different security function modules, like encryption, authentication, ID, etc. Besides these, security management in WSNs should not incur too much communication, computation and storage overheads, and should be compatible with other network management functionalities.

Until now, security management research in WSNs is still in the germinal stage and there are only a few works on that. In the following, we first summarize the important access control in WSNs, and then provide a new policy-based sensor network security management (Sec-SNMP) architecture.

14.3.6.1   Access Control

Access control is the ability to permit or deny the use of something by someone. The access control in WSNs include the authorization to network devices and user authentication.

Device authorization: It defines the operations which a network device permitted to do. It is good to always limit the operation ability of a network device to those really necessary. Hierarchical network structure is found not only providing better scalability, but also better security management support. In a hierarchical sensor network, the device authorization can be defined according to node role and node location. For example, the cluster heads are allowed to talk with the base station and issue a global broadcast, while common nodes are only allowed to broadcast within the range of a cluster. By doing this, the influence of compromised common nodes can be limited inside the corresponding clusters.

User authentication: The user authentication in WSNs can be divided into two categories: base-station level user authentication and in-field user authentication.

The base station acts as the access point between users and the underlying mesh network. Users (including administrator and clients) of WSNs usually sit before a remote monitor and connect to the base station or a backup data server via Internet or satellites. Because both the client side and the server side are powerful, the user authentication and access control problem is the same as that in traditional computer network area, and there are a bunch of solutions on this.

Considering the scenario that users walk around a deployed sensing field and inject queries into the network through a PDA like mobile device is reasonable. Efficient in-field user authentication in such scenario is necessary to protect sensible information from being acquired by unauthorized users. Ref. [70] proposes an authenticated querying scheme that provides sensor nodes the ability of identifying an in-field user based on public-key cryptography. In this scheme, the user standing somewhere in the sensor field first broadcasts his or her identity and certificate to the sensors in his or her communication range. Each receiver within range will then send a challenge "nonce" to the user. The user answers the challenge of each sensor by signing the received nonce. Then each sensor independently extracts the nonce using the user's public key included in the first received certificate message. If the nonce is the same as the one sent, the user is verified. Intrusion tolerance is provided by only permitting the user's access to WSN data when up to a threshold of sensors in his or her communication range have verified his or her identification. Ref. [71] leverages the underlying pair-wise key predistribution scheme for the WSNs and modifies it slightly to accommodate authenticated querying. In this scheme, the users and the nodes are preloaded with information ensuring both sides can compute a shared symmetric key and then use it to compute and verify the MAC. Ref. [72] presents a distributed in-field user authentication scheme based on self-certified keys cryptosystem (SCK). In this scheme, a KDC is used to preload or dynamically establish pairwise keys between queried sensors and the in-field user. To conquer adversaries, sensor nodes choose to accept or reject query request by a local voting. Ref. [73] assumes the sensor network is deployed in a confined area and there is a powerful gateway node. Before issuing any queries into the system, a user must first register at the gateway node through one-hop communication. Upon successful registration, the user can submit a query to the sensor network system within a predefined time period. The query however has to be processed with the nearest login-node in the field, which has access to the user's registration information and can bridge the authentication process between the in-field user and the gateway node where the user is registered. The authentication approach is based on a strong-password.

14.3.6.2 Policy-Based Sensor Network Security Management

In this section, we propose a security management approach called Sec-SNMP-[74], which organizes and manages security related behaviors in WSNs based on security policies. As for security policy, it is a high-level definition of what it means to be secure for a system. In Sec-SNMP, the top security policy is that the sensor network should fulfill security requirements addressed at the beginning of Section 14.3. When it comes to implementation, the top security policy is divided into many policy items which specify actions when the predefined conditions satisfy. Figure 14.3 shows the Sec-SNMP architecture. The considered WSN system consists of two parts: infrastructure servers and mesh network. Infrastructure servers provide remote data acquisition and query service, strong authentication service, and network management service. Mesh network provides data collection, event detection, and authenticated in-field query service. In Sec-SNMP, the security management service is collaboratedly provided by the Sec-SNMP manager standing on the infrastructure server side and the Sec-SNMP agents distributedly installed on the sensor side.

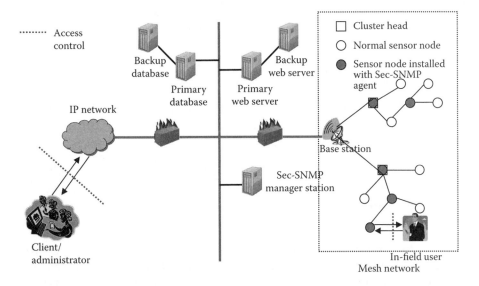

FIGURE 14.3 Sec-SNMP architecture.

Sec-SNMP manager: It provides the interface between the human network administrator and the managed WSN system. It consists of a security policy base, a security management information base (Security MIB), a security state base, and a security event processing module.

- Security policy base stores all available security policy rules configured by the network administrator.
- Security MIB provides static security management information, including both network and network component security configuration. In a policy-based security management, a security MIB tells which object is imposed with which security policy.
- Security State Base keeps the up-to-date network and component dynamic states and are stored according to WSN security models. Examples of WSN security models include network-topology map, network routing path map, network behavioral history, etc. The security state base is dynamically updated on receiving alerts and query results from Sec-SNMP agents, and it provides necessary information when the administrator decides to update security configuration.
- Security event processing module consists of key management engine, authentication engine, intrusion detection engine, fault detection engine, etc. It is responsible for security event detecting, identification, and processing. It also provides the data source for security policy matching and enforcement.

The Sec-SNMP manager runs an application that provides an interface for the administrator to edit the Security MIB and the security policy base. It also runs a policy control and deployment protocol (PCDP) [74] for its communication with Sec-SNMP agents.

Sec-SNMP agent: It is a piece of software located in the managed device and provides the interface between the Sec-SNMP manager and the physical devices or software applications being managed. Each Sec-SNMP agent is a mini Sec-SNMP manager, and is responsible for accepting the security policy configuration from the Sec-SNMP manager, for enforcing configured local security policies, and for reporting local security states to the Sec-SNMP manager. To fulfill these functions, it needs to keep a local Security MIB, a local Security Policy Base, a local Security State Base, and a local Security Event Processing Module.

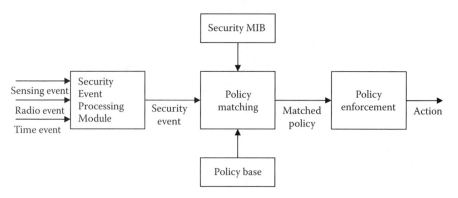

FIGURE 14.4 Security policy enforcement.

As for the implementation of Sec-SNMP agent, we use the way that Almajali and Elrad proposed in their Remote Dynamic Policy Deployment Framework (RDPD) [74]. Simply speaking, Sec-SNMP agent is run as an agent service application, which relies on a filter driver called network driver component to control the traffic that flows in and out through sensors. Sec-SNMP agent stays in contact with Sec-SNMP manager using the PCDP defined in [74]. Sec-SNMP agent communicates with the Network Driver Component to enforce the policies configured by Sec-SNMP manager.

Security policy enforcement: In Sec-SNMP, a security policy tells which action should be executed when a security event happens. Examples of security events include authentication failed/succeeded, failed node detected, malicious node detected, etc. Examples of adoptable actions include alerting to manager, accounting, reestablishing routing, going to sleep, droping packets, etc.

Security policies need to be enforced after configuration. Figure 14.4 shows the flow of policy enforcement. When a general event (e.g., sensing event, radio event, time event, etc.) is captured by the system, it is forwarded to the security event processing module for security analysis. The security event processing module consists of key management engine, authentication engine, ID engine, fault detection engine, etc. Thus, it has the ability to translate a general input event into a security event that is defined by security management manager. The identified security event is further forwarded to the policy matching engine for extracting applicable policy on applicable object. Finally, the extracted applicable policy is enforced according to the action defined by this special policy.

14.3.6.3 Open Issues

There are open issues to implement security management for WSNs. Although some existing security mechanism for WSNs already consider management in their solutions, incorporating these independently developed security mechanisms to put up one security solution is supposed to be challenging. Another problem could be the development of expressive languages or metadata for representing management policies and for representing the MIB among security agents, security management server, and security administrator.

Generally, security management in WSNs is vital important, but there are still too few works on this area. There have been some works [75,76] on general network management and fault management on WSNs. Incorporating security management and other network management services using one framework is a favorite.

14.4   CONCLUSION

Due to low-price requirement and human-unattended deployment, WSNs with unusual vulnerabilities are facing unusual security challenges. This chapter identified the vulnerabilities associated with the operational paradigms currently employed by WSNs. Following that, security solutions covering both passive and active network defense technologies, and social engineering considerations were elaborated. For each potential solution, the challenges on practical implementation were also discussed. The future of WSN security relies on how well the proposed challenges are solved.

REFERENCES

[1] C. Castelluccia, H. Hartenstein, C. Paar, and D. Westhoff, Security in ad-hoc and sensor networks, in *Proc. 1st European Workshop on Security in Ad-Hoc and Sensor Networks (ESAS 2004)*, LNCS 3313, 2004.

[2] Y. Wang, G. Attebury, and B. Ramamurthy, A survey of security issues in wireless sensor networks, *IEEE Communications Surveys & Tutorials*, 8(2), 2nd quarter 2006.

[3] Y. Law, P. Hartel, J. Hartog, and P. Havinga, Link-layer jamming attacks on S-MAC, Technical Paper, University of Twente, the Netherlands, 2005.

[4] J. Deng, R. Han, and S. Mishra, Countermeasures against traffic analysis attacks in wireless sensor networks, in *Proc. 1st International Conference on Security and Privacy in Communication Networks*, 2005, pp. 113–126.

[5] P. Kamat, Y. Zhang, W. Trappe, and C. Ozturk, Enhancing source-location privacy in sensor network routing, in *Proc. 25th IEEE International Conference on Distributed Computing Systems (ICDCS)*, June 2005, pp. 599–608.

[6] C. Karlof and D. Wagner, Secure routing in wireless sensor networks: Attacks and countermeasures, in *Proc. 1st IEEE International Workshop on Sensor Network Protocols and Applications*, May 2003, pp. 113–127.

[7] J. Newsome, E. Shi, D. Song, and A. Perrig, The sybil attack in sensor networks: Analysis & defenses, in *Proc. 3rd International Symposium on Information Processing in Sensor Networks (ISPN'04)*, April 2004, pp. 259–268.

[8] E. Sabbah, A. Majeed, K. Kang, K. Liu, and N. AbuGhazaleh, An application driven perspective on wireless sensor network security, in *Proc. 2nd ACM Workshop on Quality of Service and Security for Wireless and Mobile Networks*, Torremolinos, Spain, October 2006, pp. 1–8.

[9] M. Eltoweissy, M. Moharrum, and R. Mukkamala, Dynamic key management in sensor networks, *IEEE Communications Magazine*, 44(4), April 2006, pp. 122–130.

[10] L. Eschenauer and V.D. Gligor, A key-management scheme for distributed sensor networks, in *Proc. 9th ACM Conference on Computer and Communications Security*, November 2002, pp. 41–47.

[11] C. Blundo, A. De Santis, A. Herzberg, S. Kutten, U. Vaccaro, and M. Yung, Perfectly-secure key distribution for dynamic conferences, in *Advances in Cryptology—CRYPTO '92*, LNCS 740, 1993, pp. 471–486.

[12] D. Liu and P. Ning, Location based pairwise key establishments for static sensor networks, in *1st ACM Workshop on Security of Ad Hoc and Sensor Networks*, 2003, pp. 72–82.

[13] H. Chan, A. Perrig, and D. Song, Random key predistribution schemes for sensor networks, in *IEEE Symposium on Research in Security and Privacy*, May 2003, pp. 197–213.

[14] D. Liu and P. Ning, Establishing pair-wise keys in distributed sensor networks, in *Proc. 10th ACM Conference on Computer and Communications Security*, October 2003, pp. 52–61.

[15] O. Goldreich, S. Goldwasser, and S. Micali, How to construct random functions, *Journal of the ACM*, 33(4), October 1986, pp. 792–807.

[16] B. Dutertre, S. Cheung, and J. Levy, Lightweight key management in wireless sensor networks by leveraging initial trust, Technical Report, SRI-SDL-04-02, System Design Laboratory, April 2004.

[17] L. Morales, I.H. Sudborough, M. Eltoweissy, and M.H. Heydari, Combinatorial optimization of multicast key management, in *Proc. 36th Annual Hawaii International Conference on System Sciences*, January 2003.

[18] S. Zhu, S. Setia, and S. Jajodia, LEAP: Efficient security mechanisms for large-scale distributed sensor networks, in *Proc. 10th ACM Conference on Computer and Communications Security (CCS'03)*, October 2003, pp. 62–72.

[19] D. Westhoff, J. Girao, and M. Acharya, Concealed data aggregation for reverse multicast traffic in sensor networks: Encryption, key distribution, and routing adaption, *IEEE Transactions on Mobile Computing*, 5(10), October 2006, pp. 1417–1431.

[20] D. Malan, M. Welsh, and M. Smith, A public-key infrastructure for key distribution in tinyos based on elliptic curve cryptography, in *Proc. 1st IEEE International Conference on Sensor and Ad Hoc Communications and Networks (SECON'04)*, October 2004, pp. 71–80.

[21] R. Canetti, J. Garay, G. Itkis, D. Miccianicio, M. Naor, and B. Pinkas, Multicast security: A taxonomie and some efficient constructions, in *Proc. 18th IEEE Conference on Computer Communications (INFOCOM'99)*, March 1999.

[22] A. Perrig, R. Canetti, D. Song, and D. Tygar, Efficient authentication and signing of multicast streams over lossy channels, in *Proc. The 2000 IEEE Symposium on Security and Privacy*, May 2000.

[23] A. Perrig, R. Szewczyk, V. Wen, D. Culler, and J.D. Tygar, SPINS: Security protocols for sensor networks, in *Proc. 7th Annual International Conference on Mobile Computing and Networks*, July 2001.

[24] A. Perrig, R. Szewczyk, J.D. Tygar, V. Wen, and D. Culler, SPINS: Security protocols for sensor networks, *Wireless Networks*, 8(5), September 2002, pp. 521–534.

[25] D. Liu and P. Ning, Multi-Level μTESLA: A broadcast authentication system for distributed sensor networks, *ACM Transactions on Embedded Computing System*, 3(4), November 2004, 800–836.

[26] C. Tseng, P. Balasubramanyam, C. Ko, R. Limprasittiporn, J. Rowe, and K. Levitt, A specification-based intrusion detection system for AODV, in *Proc. 1st ACM Workshop on Security of Ad Hoc and Sensor Networks*, 2003, pp. 125–134.

[27] P. Yi, Y. Jiang, Y. Zhong, and S. Zhang, Distributed intrusion detection for mobile ad hoc networks, in *Proc. The 2005 Symposium on Applications and the Internet Workshops (SAINT-W'05)*, 2005, pp. 94–97.

[28] Q. Wang and T. Zhang, Detecting anomaly node behavior in wireless sensor networks, in *Proc. 21st International Conference on Advanced Information Networking and Applications Workshops*, Vol. 1, May 2007, pp. 451–456.

[29] A.P. da Silva, M. Martins, B. Rocha, A. Loureiro, L. Ruiz, and H. Wong, Decentralized intrusion detection in wireless sensor networks, in *Proc. 1st ACM International Workshop on Quality of Service & Security in Wireless and Mobile Networks (Q2SWinet'05)*, ACM Press, October 2005, pp. 16–23.

[30] C.E. Perkins and E.M. Royer, Ad hoc on-demand distance vector routing, in *Proc. 2nd IEEE Workshop on Mobile Computing Systems and Applications*, New Orleans, LA, February 1999, pp. 90–100.

[31] D.B. Johnson, D.A. Malty, and J. Broch, DSR: the dynamic source routing protocol for multi-hop wireless ad hoc networks, in *Ad Hoc Networking*, Chapter 5, edited by C.E. Perkins, Addison–Wesley, 2001, pp. 139–172.

[32] Y. Huang, W. Fan, W. Lee, and P.S. Yu, Cross-feature analysis for detecting ad-hoc routing anomalies, in *Proc. 23rd International Conference on Distributed Computing Systems*, May 2003, pp. 478–487.

[33] I. Onat and A. Miri, A real-time node-based traffic anomaly detection algorithm for wireless sensor networks, in *Proc. of Systems Communications*, August 2005, pp. 422–427.

[34] Y. Huang and W. Lee, A cooperative intrusion detection system for ad hoc networks, in *Proc. 1st ACM Workshop on Security of Ad Hoc and Sensor Networks*, 2003, pp. 135–147.

[35] K. Ioannis, T. Dimitriou, and F. C. Freiling, Towards intrusion detection in wireless sensor networks, *13th EuropeanWireless Conference*, April 2007, Paris.

[36] I. Khalil, S. Bagchi, and C. Nina-Rotaru, DICAS: Detection, diagnosis, and isolation of control attacks in sensor networks, in *Proc. 1st International Conference on Security and Privacy for Emerging Areas in Communications Networks (SecureComm 2005)*, September 2005, pp. 89–100.

[37] I. Onat and A. Miri, An intrusion detection system for wireless sensor networks, in *Proc. IEEE International Conference on Wireless and Mobile Computing, Networking and Communications (WiMob'2005)*, Vol. 3, August 2005, pp. 253–259.

[38] F. Anjum, D. Subhadrabandhu, S. Sarkar, and R. Shetty, On optimal placement of intrusion detection modules in sensor networks, in *Proc. 1st International Conference on Broadband Networks*, 2004, pp. 690–699.

[39] A. Agah, S.K. Das, K. Basu, and M. Asadi, Intrusion detection in sensor networks: A non-cooperative game approach, in *Proc. 3rd IEEE International Symposium on Network Computing and Applications*, 2004, pp. 343–346.

[40] R. Ma, L. Xing, and H.E. Michel, Fault-intrusion tolerant techniques in wireless sensor networks, in *Proc. 2nd IEEE International Symposium on Dependable, Autonomic and Secure Computing*, September 2006, pp. 85–94.

[41] L. Xing and H.E. Michel, Integrated modeling for wireless sensor networks reliability and security, *Annual Symposium on Reliability and Maintainability (RAMS'06)*, January 2006, pp. 594–600.

[42] F. Koushanfar, M. Potkonjak, and A. Sangiovanni-Vincentelli, Fault tolerance techniques for wireless ad hoc sensor networks, in *Proc. IEEE Sensors 2002*, Vol. 2, June 2002, pp. 1491–1496.

[43] B. Krishnamachari and S. Iyengar, Efficient and fault-tolerant feature extraction in wireless sensor networks, in *Proc. IPSN 2003*, LNCS 2634, 2003, pp. 488–501.

[44] B. Krishnamachari and S. Iyengar, Self-organized fault-tolerant feature extraction in distributed wireless sensor networks, in *Proc. of Information Processing in Sensor Networks*, April 2003.

[45] I. Saha, L.K. Sambasivan, S.K. Ghosh, and P.K. Patro, Distributed fault tolerant topology control in wireless ad-hoc sensor networks, in *Proc. IFIP International Conference on Wireless and Optical Communications Networks*, April 2006.

[46] K. Martinez, J.K. Hart, and R. Ong, Environmental sensor networks, *Computer*, 37(8), August 2004, 50–56.

[47] J. Staddon, D. Balfanz, and G. Durfee, Efficient tracing of failed nodes in sensor networks, in *Proc. 1st ACM International Workshop on Wireless Sensor Networks and Applications*, ACM Press, September 2002, pp. 122–130.

[48] Y. Chen and S.H. Son, A fault tolerant topology control in wireless sensor networks, in *Proc. 3rd ACS/IEEE International Conference on Computer Systems and Applications*, 2005.

[49] M. Cardei, S. Yang, and J. Wu, Fault-tolerant topology control for heterogeneous wireless sensor networks, in *Proc. IEEE International Conference on Mobile Adhoc and Sensor Systems (MASS 2007)*, October 2007, pp. 1–9.

[50] J. Deng, R. Han, and S. Mishra, Intrusion tolerance and anti-traffic analysis strategies for wireless sensor networks, in *Proc. IEEE International Conference on Dependable Systems and Networks (DSN)*, 2004.

[51] J. Deng, R. Han, and S. Mishra, INSENS: Intrusion-tolerant routing in wireless sensor networks, in *Proc. 23rd IEEE International Conference on Distributed Computing Systems (ICDCS 2003)*, 2003.

[52] J. Deng, R. Han, and S. Mishra, A performance evaluation of intrusion-tolerant routing in wireless sensor networks, in *Proc. 2nd IEEE International Workshop on Information Processing in Sensor Networks (IPSN'03)*, 2003.

[53] K. Sun, P. Ning, and C. Wang, Fault-tolerant cluster-wise clock synchronization for wireless sensor networks, *IEEE Transactions on Dependable and Secure Computing*, 2(3), July–September 2005, 177–189.

[54] W. Zhang, G. Xue, and S. Misra, Fault-tolerant relay node placement in wireless sensor networks: Problems and algorithms, in *Proc. 26th IEEE International Conference on Computer Communications (INFOCOM 2007)*, May 2007, pp. 1649–1657.

[55] B. Chen, K. Jamieson, H. Balakrishnam, and R. Morris. Span: An energy-efficient coordination algorithm for topology maintenance in ad hoc wireless networks, *ACM Wireless Networks Journal*, 8(5), September 2002, 481–494.

[56] P. Djukic and S. Valaee, Minimum energy fault tolerant sensor networks, in *Proc. IEEE Global Telecommunications Conference Workshops (GlobeCom'04)*, November–December 2004, pp. 22–26.

[57] T. Wang, Y.S. Han, C. Biao, and P.K. Varshney, A combined decision fusion and channel coding scheme for distributed fault-tolerant classification in wireless sensor networks, *IEEE Transactions on Wireless Communications*, 5(7), July 2006, 1695–1705.

[58] L. Lamport, R. Shostak, and M. Pease, The Byzantine generals problem, *ACM Transactions on Programming Languages and Systems*, 4(3), July 1982, 382–401.

[59] T. Clouqueur, P. Ramanathan, K.K. Saluja, and K. Wang, Value-fusion versus decision-fusion for fault-tolerance in collaborative target detection in sensor networks, in *Proc. 4th Annual Conference on Information Fusion*, August 2001, pp. TuC2/25-TuC2/30.

[60] T. Clouqueur, K.K. Saluja, and P. Ramanathan, Fault tolerance in collaborative sensor networks for target detection, *IEEE Transactions on Computers*, 53(3), March 2004, 320–333.

[61] C. Yi, P. Wan, X. Li, and O. Frieder, Fault tolerant sensor networks with Bernoulli nodes, in *Proc. IEEE Wireless Communication and Networking Conference (WCNC'03)*, March 2003.

[62] G. Hoblos, M. Staroswiechi, and A. Aitouche, Optimal design of fault tolerant sensor networks, in *Proc. The 2000 IEEE International Conference on Control Applications*, September 2000, pp. 467–472.

[63] R. Albert, H. Jeong, and A. Barabasi, Error and attack tolerance of complex networks, *Nature* 406, July 2000, 378–382.

[64] P. Crucitti, V. Latora, M. Marchiori, and A. Rapisarda, Error and attack tolerance of complex networks, *Physica A 340*, 2004, 388–394.

[65] M. Staroswiechi, G. Hoblos, and A. Aitouche, Sensor network design for fault tolerant estimation, *International Journal of Adaptive Control and Signal Processing*, January 2004, 55–72.

[66] S. Misra and G. Xue, Efficient anonymity schemes for clustered wireless sensor networks, *International Journal of Sensor Networks*, 1(1/2), 2006, 50–63.

[67] Y. Quyang, Z. Le, Y. Xu, N. Triandopoulos, S. Zhang, J. Ford, and F. Makedon, Providing anonymity in wireless sensor networks, in *Proc. IEEE International Conference on Pervasive Services*, July 2007, pp. 145–148.

[68] W. He, X. Liu, H. Nguyen, K. Nahrstedt, and T. Abdelzaher, PDA: Privacy-preserving data aggregation in wireless sensor networks, in *Proc. IEEE INFOCOM'07*, May 2007, Anchorage, Alaska.

[69] Y. Jian, S. Chen, Z. Zhang, and L. Zhang, Protecting receiver-location privacy in wireless sensor networks, in *Proc. 26th Annual IEEE Conference on Computer Communications (INFOCOM'07)*, Anchorage, Alaska, May 2007.

[70] Y. Zhang, W. Liu, W. Lou, and Y. Fang, Mask: Anonymous on-demand routing in mobile ad hoc networks, *IEEE Transactions on Wireless Communications*, 5(9), September 2006, 2376–2385.

[71] Y. Xi, L. Schwiebert, and W. Shi, Preserving source location privacy in monitoring-based wireless sensor networks, in *Proc. 20th International Parallel and Distributed Processing Symposium (IPDPS 2006)*, April 25–29, 2006, Rhodes Island, Greece.

[72] Z. Benenson, N. Gedicke, and O. Raivio, Realizing robust user authentication in sensor networks, in *Real-World Wireless Sensor Networks (REALWSN)*, June 2005, Stockholm.

[73] S. Banerjee and D. Mukhopadhyay, Symmetric key based authenticated querying in wireless sensor networks, in *Proc. 1st International Conference on Integrated Internet Ad Hoc and Sensor Networks*, May 2006, France.

[74] Q. Wang and T. Zhang, Sec-SNMP: Policy-based security management for sensor networks, in *Proc. The International Conference on Security and Cryptography (SECRYPT'08)*, July 2008, pp. 222–226.

[75] C. Jiang, B. Li, and H. Xu, An efficient scheme for user authentication in wireless sensor networks, in *Proc. 21st International Conference on Advanced Information Networking and Applications Workshops (AINAW'07)*, 1, 2007, pp. 438–442.

[76] K. Wong, Y. Zheng, J. Cao, and S. Wang, A dynamic user authentication scheme for wireless sensor networks, in *Proc. IEEE International Conference on Sensor Networks, Ubiquitous, and Trustworthy Computing (SUTC'06)*, 1, 2006, pp. 244–251.

[77] S. Almajali and T. Elrad, Remote dynamic policy deployment for sensor networks using application transparent approach, in *Workshop on Building Software for Sensor Networks (OOPSLA '06)*, October 2006.

[78] W. Lee, A. Datta, and R. Cardell-Oliver, WinMS: Wireless sensor network-management system, an adaptive policy-based management for wireless sensor networks, Technical Report UWA-CSSE-06-001, The University of Western Australia, June 2006.

[79] L. Ruiz, J. Nogueira, and A. Loureiro, MANNA: A management architecture for wireless sensor networks, *IEEE Communications Magazine*, 41(2), February 2003, 116–125.

15 Intrusion Detection in Wireless Sensor Networks

Thanassis Giannetsos, Ioannis Krontiris,
Tassos Dimitriou, and Felix C. Freiling

CONTENTS

Sensor networks are highly vulnerable to attacks due to the nature of the wireless media. There are several proposed protocols for authentication and encryption that prevent unauthorized nodes from accessing the network and the transferred information. However, a second line of defense is needed, as a broad range of attacks can be launched from compromised nodes that appear as legitimate members of the network. An intrusion detection system (IDS) can detect the misbehavior of such nodes and notify other nodes in the network to take necessary measurements. In this chapter, we discuss the general design principles of such systems for sensor networks, their requirements, and available approaches. Then, we present an architecture of a distributed IDS, in which, even though nodes do not have a global view of the network, they can still collaborate with each other and successfully detect an intrusion. Finally, we show how such a system can be implemented in TinyOS, which components and interfaces are needed, and what is the resulting overhead imposed.

15.1   INTRODUCTION

During recent years, wireless sensor networks have found several applications, ranging from military to civilian and commercial uses and it is expected that their adoption will spread more in the future. What makes sensor networks attractive is that they can operate unattended and without the help of any infrastructure or interaction with a human. They are able to operate under these constraints for many years, because of their low-energy requirements and new developed algorithms that allow them to adapt autonomously to any environmental changes.

However, it is exactly this unattended nature of sensor networks and the limited resources of their nodes that makes them susceptible to attacks. Their inadequate physical protection makes them receptive to being captured, compromised, and hijacked [1]. Thus any cryptographic material they contain can be used by adversaries to perform attacks from within the network and such attacks are much harder to detect and prevent. Moreover, the use of wireless links makes things even easier for adversaries, as they enable a wide class of attacks ranging from passive eavesdropping to active interfering. In addition to that the adversary can use powerful laptops with high-energy and long-range communication capabilities to attack the network, so, one can realize that designing a secure protocol for sensor networks is a nontrivial task.

Several intrusion prevention techniques have been introduced for sensor networks over the last few years. Key management protocols as well as encryption and authentication algorithms have been extensively studied aiming to protect information from being revealed to an unauthorized party and guarantee its integral delivery to the base station. Other specific services, like localization, aggregation, cluster formation, and time synchronization have also been secured under certain conditions [2–4]. Some security protocols have also been designed with the goal of protecting a sensor network against specific attacks, like selective forwarding [5], sinkhole [6], or wormhole attacks [7].

However, intrusion prevention techniques do not always guarantee the protection of the network. No matter how many defenses are inserted in a sensor network, an adversary can always find a weak point to exploit in order to break in. Besides, these techniques are designed to secure specific loopholes created by specific protocols. This does not exclude clever adversaries from finding new ways to achieve their goals, especially in systems like sensor networks with inherent vulnerabilities. That is why we refer to intrusion prevention as the first line of defense.

To truly secure sensor networks we also need a second line of defense: an intrusion detection system (IDS) that can detect third party break-in attempts, even if this particular attack has not been experienced before. If the intruder is detected soon enough, one can take any appropriate measures before any damage is done or any data is compromised. An effective IDS can also help us design better prevention mechanisms, by collecting information about intrusion techniques.

In this chapter, we discuss the process of designing an IDS for sensor networks. We present the parameters that one has to take under consideration, the different techniques and architectures that are appropriate for such networks, and the requirements that such a system should satisfy. Then we focus on an IDS architecture that relies on embedded preprogrammed policies and a coordinated, cooperative behavior resulting in a most effective way to gain maximum advantage against adversaries. The architecture of the system, the underline protocol, as well as its implementation in TinyOS are discussed.

15.2   DESIGNING AN IDS FOR SENSOR NETWORKS

In intrusion detection, we wish to provide an automated mechanism that identifies the source of an attack and generates an alarm to notify the network or the administrator, so that appropriate preventive actions can take place. As an attack, we consider any set of actions that target the computing or networking resources of our system. Attackers may be using an external system without authorization or have legitimate access to our system but are abusing their privileges (i.e., an insider attack). It is important to realize here that the IDS comes into the picture after an intrusion attempt has occurred. It does not try to prevent these attempts in the first place.

15.2.1 INTRUSION DETECTION TECHNIQUES

To detect an intruder, we need to use a model of intrusion detection. We need to know what an IDS system should look out for. In particular, an IDS system must be able to distinguish between normal and abnormal activities to discover malicious attempts in time. However, this can be difficult since many behavior patterns can be unpredictable and unclear. There are three main techniques that an intrusion detection system can use to classify actions [8]:

- *Misuse detection.* In misuse detection or signature-based detection systems [9,10], the observed behavior is compared with known attack patterns (signatures). So, action patterns that may pose a security threat must be defined and given to the system. The misuse detection system tries to recognize any "bad" behavior according to these patterns. Any action that is not clearly prohibited is allowed. The main disadvantage of such systems is that they cannot detect novel attacks. Someone must continuously update the attack signature database. Another difficulty is that signatures must be written in a way to encompass all possible variations of the pertinent attack, and yet avoid flagging nonintrusive activity as an intrusive one.
- *Anomaly detection.* Anomaly detection [11] overcomes the limitations of misuse detection by focusing on normal behaviors, rather than attack behaviors. This technique first describes what constitutes a "normal" behavior (usually established by automated training) and then flags as intrusion attempts any activities varying from this behavior by a statistically significant amount. In this way there is a considerable possibility to detect novel attacks as intrusions. There are two problems associated with this approach: First, a system can exhibit legitimate but previously unseen behavior. This would lead to a substantial false alarm rate, where anomalous activities that are not intrusive are flagged as intrusive. Second, and even worse, an intrusion that does not exhibit anomalous behavior may not be detected, resulting in false negatives.
- *Specification-based detection.* Specification-based detection [12,13] tries to combine the strengths of misuse and anomaly detection. It is based on deviations from normal behavior. However, in this case, the normal behavior is not defined by machine learning techniques and training. It is based on manually defined specifications that describe what is a correct operation and monitors any behavior with respect to these constraints. In this way, legitimate but previously unseen behaviors will not cause a high false alarm rate, as in the anomaly detection approach. Also, since it is based on deviations from legitimate behaviors, it can still detect previously unknown attacks. On the other side, the development of detailed specifications by humans can be time-consuming and bare the inherent risk that certain attacks may pass undetected.

Caution must be taken when applying the anomaly detection technique in sensor networks. It is not easy to define what is a "normal behavior" in such networks, as they usually adapt to variations in their environment or according to other parameters, such as the remaining battery level. So, these legitimate changes of behavior may easily be mistaken from the IDS as intrusion attempts. Moreover, sensor networks cannot bear the overhead of automatic training, due to their low-energy resources. Specification-based detection seems the most appropriate approach in this case, if one can design appropriate rules that cover as broad range of attacks as possible.

15.2.2 INTRUSION DETECTION ARCHITECTURES

Traditionally, intrusion detection systems for fixed networks were divided into two categories: host-based and network-based. The host-based architecture was the first architecture to be explored in intrusion detection. A host-based intrusion detection system (HIDS) is designed to monitor, detect, and respond to system activity and attacks on a given host (node). Any decision made is based on information collected at that host by reviewing audit logs for suspicious activity. This contradicts the distributed nature of sensor networks and makes it impossible to detect network attacks. A network-based architecture is clearly more appropriate here.

Network-based intrusion detection systems (NIDS) use raw network packets as the data source. A NIDS typically listens on the network, and captures and examines individual packets in real time. It can analyze the entire packet, not just the header. In wired networks, active scanning of packets from NIDS is usually done at specific traffic concentration points, such as switches, routers, or gateways. On the other hand, wireless sensor networks do not have such "bottlenecks." Any node can act as a router and traffic is usually distributed for load balancing purposes. So, it is impossible to monitor the traffic at certain points. To apply a network-based intrusion detection architecture, an IDS client must be installed in several nodes.

15.2.3 Decision-Making Techniques

IDSs can be further classified according to the decision-making techniques that they use to detect and initiate a response to an intrusion attempt. This decision can be made either collaboratively or independently by the nodes.

Since the nature of sensor networks is distributed and most of the services provided require cooperation of other nodes, it is only natural that intrusion detection should also be done in a cooperative manner. In this case, every node participates in intrusion detection and response by having an IDS client installed on them. Each node is responsible for detecting attempts of intrusion locally. If an anomaly is detected by a node with weak evidence, or if the evidence is inconclusive, then a cooperative mechanism is initiated with the neighboring nodes to take a global intrusion detection action. Such a mechanism is described in Ref. [14] for ad hoc networks, where nodes use a majority-based distributed intrusion detection procedure. More sophisticated cooperative decision-making schemes may use mobile agents [15,16] or fuzzy logic [17] to better support the decision process.

When designing a cooperative decision-making mechanism for intrusion detection in sensor networks, one should take under consideration the fact that a node can be compromised and hence, send falsified data to its neighbors trying to affect the decision. So, one must be skeptical as to which nodes should trust. The fact that it is difficult for an adversary to compromise the majority of the nodes in a specific neighborhood can play an important role here. Moreover, a cooperative mechanism has to consider the bandwidth and energy resources of the nodes. The nodes cannot exchange security data and intrusion alerts without considering the energy that has to be spent for sending, receiving, and processing these messages.

In an independent decision-making system, there are certain nodes that have the task to perform the decision-making functionality. They collect intrusion and anomalous activity evidences from other nodes and based on them they can make decision about network-level intrusions. The rest of the nodes do not participate in this decision. In such architectures, the decision-making nodes can attract the interest of an attacker, since their elimination would leave the network undefended. Furthermore, the information that they process is limited, since it originates from specific nodes. Another disadvantage of such approaches is that they restrict computation-intensive analysis of over-all network security state to a few key nodes. Their special mission of processing the information from other nodes and deciding on intrusion attempts results in an extra processing overhead, which may quickly lead to their energy exhaustion, unless different nodes are dynamically elected periodically.

15.3 WATCHDOG APPROACH

As in Section 15.2.2, to apply a NIDS in sensor networks, packet monitoring should take place in several nodes of the network. In this section, we look in more detail at a technique that can be used for packet monitoring, called the watchdog approach [18].

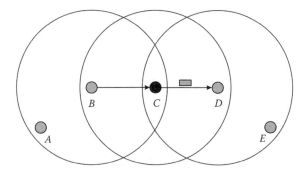

FIGURE 15.1 Node B is selectively forwarding packets to node C. Node A promiscuously listens to node B's transmissions.

The watchdog approach relies on the broadcast nature of the wireless communications and the fact that sensors are usually densely deployed. Each packet transmitted in the network is not only received by the sender and the receiver, but also from a set of neighboring nodes within the sender's radio range. Normally these nodes would discard the packet, since they are not the intended receivers, but for intrusion detection this can be used as a valuable audit source. Hence, a node can activate its IDS agent and monitor the packets sent by its neighbors, by overhearing them. However, this is not always adequate to draw safe conclusions on the behavior of the monitored node.

There are certain concerns that arise in this case which will be highlighted by way of an example. In the setting shown in Figure 15.1, suppose that a packet should follow the path $A \rightarrow B \rightarrow C \rightarrow D$. Now, suppose that C is compromised and exhibits a malicious behavior, selectively dropping packets. There are three cases, arising from the wireless nature of communications, where having a node B monitoring node C cannot result in a successful detection of node C:

1. Node C forwards its packet and node A sends a packet to B at the same time. Then a collision occurs at B. Node B cannot be certain which packets caused this collision, so it cannot conclude on C's behavior.
2. Node C forwards its packet to node D at the same time that node E makes a transmission. Then a collision occurs at D, which cannot be detected by B. Node B thinks that C has successfully forwarded its packet and therefore, C can skip retransmitting the packet, without being detected.
3. Node C forwards its packet to node D at the same time that D makes a transmission. Then a collision occurs at D. Again, node B thinks that C has successfully forwarded its packet, even though it never reached node D.

From the above cases we can conclude that only one watchdog is not always enough to detect an attack, so this approach should involve information from more nodes. Then these nodes could cooperate and exchange their partial views to draw their final conclusions. In Section 15.6, we describe an IDS based on this observation.

Furthermore, to detect certain attacks, it is not enough to monitor just one node, but rather a link, meaning the packets transmitted by the nodes at both of its ends. For example, to detect selective forwarding, a watchdog should be able to overhear packets arriving at a node and transmitted by that node. So, if we want to see whether a node B forwards packets sent by node A, we must activate a watchdog that resides within the intersection of A's and B's radio range. For example, in Figure 15.2, the nodes A, C, D, and E can be watchdogs for the communication between A and B.

Finally, one could argue that the watchdog approach increases the energy consumption of the nodes, since they have to overhear packets not destined for them. However, let us note that in most

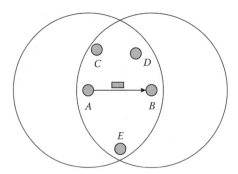

FIGURE 15.2   Nodes *A, C, D,* and *E* can be watchdogs of the link *A* → *B*.

radio stacks of today's sensor platforms each node receives packets sent by neighboring nodes anyway. They cannot know if a packet is addressed to them unless they receive it and check the destination field. So, the only overhead imposed to the nodes is any further processing of the packet.

15.4   EXISTING APPROACHES

Several proposed architectures of intrusion detection systems already exist for ad hoc networks. The first scheme to be proposed was introduced in Ref. [14], which is a distributed and cooperative IDS model, where every node in the network participates in the detection process. Another architecture, called local intrusion detection system (LIDS), was introduced in Ref. [19], which utilizes mobile agents on each of the nodes. These agents are used to collect and process data on remote hosts and transfer the results back to their home nodes, or migrate to another node for further investigation. Also based on mobile agents is the IDS proposed in Ref. [20]. The agents are categorized as monitoring, decision-making, and action agents. All nodes accommodate host-based monitoring agents but only a few nodes chosen by a distributed algorithm host agents with network monitoring and decision capabilities.

    These IDS architectures for ad hoc networks cannot be applied directly to sensor networks. The differences in the nature of the two kinds of networks impose different requirements, which forces us to redesign new solutions. A first attempt to apply anomaly detection in sensor networks is presented in Ref. [21]. According to the author's proposed algorithm, there are some monitor nodes in the network, which are responsible for monitoring their neighbors looking for intruders. These nodes listen to messages in their radio range and store certain message fields that might be useful to the rule application phase. The rules concern simple observations, such as

- Message sending rate must be within some limits
- Payload of a forwarded message should not be altered
- Retransmission of a message must occur before a defined timeout
- Same message can only be retransmitted a limited number of times

    Then they try to detect some attacks, like message delay, repetition, data alteration, blackhole, and selective forwarding. It is concluded from the paper that the buffer size to store the monitored messages is an important factor that greatly affects the false positives number. Given the restricted memory available in motes, it turns out that the detection effectiveness is kept to lower levels.

    A similar approach is followed in Ref. [22], where each node has a fixed-size buffer to store the packets received from neighbors and their corresponding arrival time and received power. If its power is not within certain limits, the packet is characterized anomalous. An intrusion alert is raised if the rate at which anomalous packets are detected over the overall rate at which packets are received is above a given threshold. In this way, the authors claim that it is possible for a node to effectively identify an intruder impersonating a legitimate neighbor.

Loo et al. [23] and Bhuse and Gupta [24] describe two more IDSs, emphasizing on routing attacks in sensor networks. Both papers assume that routing protocols for ad hoc networks can also be applied to wireless sensor networks (WSNs): Loo et al. [23] assume the ad hoc on-demand distance vector (AODV) protocol while Bhuse and Gupta [24] use the destination-sequenced distance-vector (DSDV) and dynamic source routing (DSR) protocols. Then, specific characteristics of these protocols are used like "number of route requests received" to detect intruders. However, to the best of our knowledge, these routing protocols are not attractive for sensor networks and they have not been applied to any implementation that we are aware of.

In Ref. [25], the authors propose an IDS architecture where all nodes are loaded with an IDS agent. This agent is divided into two parts: local agents and global agents. Local agents are active in every node and are responsible for monitoring and analyzing only local sources of information. Global agents are active at only a subset of nodes. They are in charge of analyzing packets flowing in their immediate neighborhood. In order for the whole communication in the network to be covered by global agents, the global agents must be activated at the right nodes. For example, if clusters are used, the global agents will be activated at the cluster-heads. In case of a flat architecture, the authors propose another solution (called spontaneous watchdogs) that tries to activate only one global agent for a packet circulating in the network.

A completely different approach is presented in Ref. [26], where the authors assume a signature-based intrusion detection. This is the only work that takes a position against promiscuous monitoring and argue that detection should be based only on the analysis of packets that pass through a node. The problem then is to determine at which nodes should the IDS modules be placed, such that all the packets are inspected at least once. The proposed solution is based on the concepts of dominating set and minimum cut set and on the requirement that the nodes running the IDS module should be tamper resistant.

15.5 REQUIREMENTS OF IDS FOR WSN

To elaborate on the requirements that an IDS system for sensor networks should satisfy, one has to look at the specific characteristics of these networks. Each sensor node has limited communication and computational resources and a short radio range. Furthermore, each node is a weak unit that can be easily compromised by an adversary [1], who can then load malicious software to launch an insider attack.

In this context, a distributed architecture, based on node cooperation is a desirable solution. In particular, we require that an IDS system for sensor networks must satisfy the following properties [27]:

1. *Localize auditing.* An IDS for sensor networks must work with localized and partial audit data. In such networks there are no centralized points (apart from the base station) that can collect audit data for the entire network, so this approach fits the sensor networks paradigm. Dealing with partial data means that the IDS should also address the problem of high false-alarm rate.
2. *Minimize resources.* An IDS for sensor networks should utilize a small amount of resources. The wireless network does not have stable connections and physical resources of network and devices, such as bandwidth and power, are limited. Disconnection can happen at any time. In addition, the communication between nodes for intrusion detection purposes should not take too much of the available bandwidth.
3. *Trust no node.* In a collaborative IDS, the nodes cannot assume that other participant nodes can be trusted. Unlike wired networks, sensor nodes can be easily compromised. These

nodes may behave normally with respect to the routing of the information in order to avoid being detected by the IDS. However, they can expose a malicious behavior to obstruct the successful detection of another intruder node. Therefore, in cooperative algorithms, the IDS must assume that no node can be fully trusted.

4. *Be truly distributed.* The process of data collection and analysis should be performed on a number of locations, to distribute the load of the intrusion detection. The distributed approach also applies to execution of the detection algorithm and alert correlation.

5. *Support addition of new nodes.* In practice it is likely that a sensor network will be populated with more nodes after its deployment. An IDS should be able to support this operation and distinguish it from an attack (e.g., wormhole attack) that has the same effect.

6. *Be secure.* An IDS should be able to withstand a hostile attack against itself. Compromising a monitoring node and controlling the behavior of the embedded IDS agent should not enable an adversary to revoke a legitimate node from the network, or keep another intruder node undetected.

15.6 DISTRIBUTED IDS FOR SENSOR NETWORKS

An IDS system should be both distributed and cooperative to suit the needs of wireless sensor networks. In this section, we present the architecture of such an IDS [27,28], which moves towards a general approach and focuses more on the collaboration of the nodes than on the detection of specific attacks. We emphasize both on the conceptual modules and their interconnections, as well as on the algorithmic part of the intrusion detection.

In this architecture, an IDS agent is installed in all sensor nodes. The agent runs independently from the application, monitoring communication activity within the radio range of the host node. Neighboring nodes collaborate with each other to make safer conclusions about any suspicious activity in their area. As a result, the nodes collectively form an IDS system to defend the sensor network.

The general functionality of the IDS agent can be described as follows:

- *Network monitoring*: Every agent performs packet monitoring in their immediate neighborhood collecting audit data.
- *Decision making*: Using this audit data, every agent decides on the intrusion threat level on a host-based basis. Then they publish their findings to their neighbors and make the final collective decision.
- *Action*: Every agent has a response mechanism that allow it to respond to an intrusion situation.

Based on these functions, we build the architecture of the IDS client based on the conceptual modules shown in Figure 15.3. Each module is responsible for a specific function, which we describe in the sections below. The IDS clients are identical in each node and they can broadcast messages for clients in neighboring nodes to listen. The communication among the clients allows us to use a distributed algorithm for the final decision on the intrusion threat.

Before presenting the architecture more analytically, let us note that for what follows, we do not assume a unit-disk graph model for the network. Instead, we consider a realistic representation for the communication model, where the range can be affected by various reasons and change from one transmission to the next. We also consider unreliable links and unpredictable delays for the wireless links. When a node transmits a packet, it does not know which nodes successfully received the message, since the message authenication code (MAC) layer of the receivers does not send any acknowledgments or requests for retransmissions. A node may miss to receive a message, either because a collision occurs or because its radio is not available at the time of the transmission.

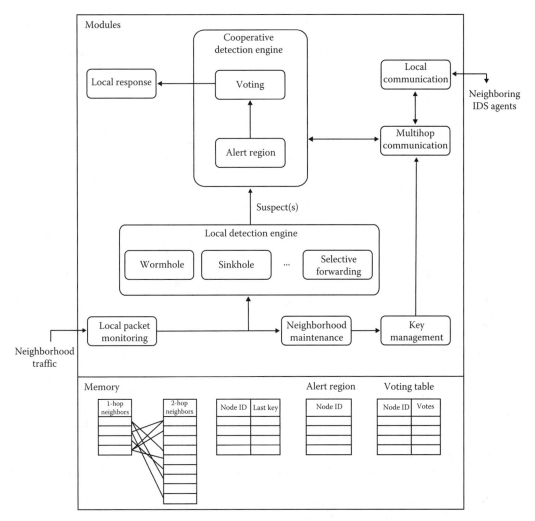

FIGURE 15.3 IDS architecture. The modules and their interconnections are shown in the upper part of the figure, while the necessary structures to be stored in the mote's memory are shown in the lower part.

15.6.1 LOCAL PACKET MONITORING MODULE

This module gathers audit data to be provided to the local detection module. Audit data in a sensor networks IDS can be the communication activities within its radio range. This data can be collected by listening promiscuously to neighboring nodes' transmissions. By promiscuously we mean that since another node is within range, the data collection module can overhear communications originating from that node.

15.6.2 NBPERIMETER MODULE

This module is responsible for maintaining consistent information about 1-hop and 2-hop neighbors of the nodes. Information about 2-hop neighbors is needed because, as we will see, the detection process involves the communication of the nodes which are neighbors of the (yet unknown) attacker but they might be 2-hops away from each other. During an initialization phase that takes place immediately after the deployment of the network, the NbPerimeter module broadcasts the node ID and the IDs of the node's immediate neighbors within a packet that has a TTL field equal to 1, meaning that each packet will be forwarded just once by the sender's 1-hop neighbors. The discovered neighborhood information is stored in a table, which we call the 2-hops neighborhood table. We make the reasonable assumption here that the time required for the completion of the initialization phase is smaller than the time needed by an adversary to compromise a sensor node during deployment.

15.6.3 KEY MANAGEMENT MODULE

After the deployment of the sensor network, the KeyManagement module of the node generates a one-way key chain of length n, using a preassigned unique secret key K_n. A one-way key chain [29] $(K_0, K_1, \ldots, K_{n-1}, K_n)$ is an ordered list of cryptographic keys generated by successively applying a one-way hash function F to the key seed K_n, such as $K_j = F(K_{j+1})$, for $j = n - 1 \ldots 0$. Therefore, any key K_j is a commitment to all subsequent keys K_i, $i > j$. In our implementation, SHA-1 hashing is used for the production of the key chain. As the last step in the initialization phase, the KeyManagement module in each node announces the resulted K_0 to all of its 1-hop and 2-hop neighbors.

The KeyManagement module also stores the corresponding information for the neighboring nodes (up to 2-hops), i.e., the node IDs and their keys. This information needs to remain consistent and up-to-date during the lifetime of the network. So, the KeyManagement module updates the corresponding key every time a node publishes a new one from its key chain. But, also, the topology can change, as nodes may be removed or added, and for that reason, the KeyManagement module is linked with the NbPerimeter module and is informed for the new or deleted nodes.

15.6.4 LOCAL DETECTION ENGINE

This module collects the audit data and analyzes it according to some given rules. A set of rules is provided for each attack, and whenever one or more rules are satisfied, a local alert is produced by the module. Whether a rule is satisfied or not does not just depend on information from the intercepted packets, but also on information from the 2-hop neighborhood table or information from past observed behavior. We elaborate more on this in Section 15.7.

When we say that the LocalDetectionEngine of a node s produces an alert, what we mean is that it simply outputs some set $D(s)$ of suspected nodes. The size of $D(s)$ depends on the quality of the alert module and the nature of the attack. If $|D(s)| = 1$, then the sensor has identified the source of the attack and the IDS protocol ends here. Most often however, $D(s)$ will contain a larger set of neighbors or may even be equal to the whole neighborhood of s, which we will denote as $N(s)$.

Generally speaking, $D(s)$ will be a subset of $N(s)$. In practice, given their myopic vision, sensor nodes might be able to realize that an attack is taking place, but they will not be able to identify the attacker directly. They can only suspect some of their neighbors. By communicating this information to the other nodes and collaborating with them is what can make them conclude on the attacker's identity. The rest of the modules, which are the core of the IDS and are discussed in the following sections, provide the functionality necessary for this collaboration.

Under an intrusion, several nodes at the vicinity of the attack will detect it through their local detection engine. Note however that not necessarily all neighbors of the attacker will notice something wrong. That depends merely on the nature of the attack and the constructed rules in the detection engine. If the LocalDetectionEngine of node s outputs a list of suspected nodes, we call s an alerted node.

15.6.5 VOTING MODULE

During the voting phase each alerted node sends its vote to all the other alerted nodes and respectively collects their votes. Let us denote the message that bears the vote from node s as $m_v(s)$. Each vote consists of the nodes suspected by the sender, so for node s,

$$m_v(s) = id||D(s).$$

Node s "signs" its vote calculating the MAC with the next key K_j from its one-way key chain, and broadcasts

$$m_v(s), MAC_{K_j}(m_v(s)).$$

Following that, it sets a timer T_v to expire after time τ_v. During that time it waits to receive the votes of the rest of the alerted nodes and buffers them, as it has to wait for the key publishing phase to authenticate them. The vote of each alerted node needs to reach all other alerted nodes, which means that we need a forwarding mechanism. Since the messages are signed with a key known only to the sender, the attacker cannot change the votes. However, we make no assumptions about the behavior of the attacking node. This means that the attacker may refuse to forward votes from its neighbors, so they must be forwarded through other paths, bypassing the attacker. Note that these paths can be more than two hops.

To ensure that the votes propagate to all alerted nodes, we follow a broadcast message-suppression protocol, similar to sensor protocols for information via negotiation (SPIN) [30]. When an alerted node receives a vote, it advertises it, by broadcasting an ADV message. Upon receiving an ADV, each neighboring node checks to see whether it already has received or requested the advertised vote. If not, it sets a random timer T_{req} to expire, uniformly chosen from a predetermined interval. When the timer expires, the node sends a REQ message requesting the specific vote, unless it has overheard a similar REQ from another node. In the latter case, it cancels its own request, as it is redundant.

Algorithm 1: The *Voting* algorithm

Data: Alerted node IDs
Result: Vector of collected votes
begin
    Create $m_v(s) = id||D(s)$;
    Calculate $MAC_{K_j}(m_v(s))$ using K_j;
    Broadcast $m_v(s), MAC_{K_j}(m_v(s))$;
    Set timer $T_v = \tau_v$;
    while $![T_v(expired)]$ **do**
        if *receive* $m_v(q)$ **then**
            Store $m_v(q)$ and corresponding MAC;
            Broadcast $m_{adv}(q)$; // ADV message
        end
        if $[$*receive* $m_{req}(q)$ $]$ **then**
            Broadcast $m_v(q), MAC_{K_j}(m_v(q))$;
        end
        if $[$*receive* $m_{adv}(q)$ $]\&\&[$ *don't have* $m_v(q)]$ **then**
            Start timer T_{req};
        end
        while $![T_{req}(expired)]$ **do**
            Register overheard $m_{req}(q)$;
        end
        if $[T_{req}(expired)]$ **then**
            if $!$*overheard* $m_{req}(q)$ **then**
                Broadcast $m_{req}(q)$; // REQ message
            end
        end
    end
end

Next each node broadcasts the next key of its hash chain, K_j, which was used to sign the vote. When a node receives the disclosed key, it can easily verify the correctness of the key by checking whether K_j generates the previous one through the application of F. If the key is correct, it replaces the old commitment K_{j-1} with the new one in its memory. The node can now use the key to verify the signature of the corresponding vote stored in its buffer from the previous phase. If this process is successful, it accepts the vote as authentic.

We allow sufficient time for the nodes to exchange their keys by setting a timer T_p. This timer is initialized just after a node publishes its own key and it is set to expire at time τ_p. During this time period, the nodes follow the same ADV-REQ scheme that we described for the exchange of votes. That is, when an alerted node acquires a key, it advertises it to its neighbors and they request it sending the corresponding REQ message. In this way, the keys can propagate to all the alerted nodes, even if the attacker does not participate in the process.

When the timer expires, the nodes move to the final step of processing the votes and exposing the attacker. In the case where a key has been missed, the corresponding vote is discarded. The code for this phase is given in Algorithm 2.

Algorithm 2: The *Publish Key* algorithm
***

Data: Buffer of received votes
Result: Attacker's ID
begin
    Broadcast key K_j;
    Set timer $T_p = \tau_p$;
    while $![T_p(expired)]$ **do**
        if *receive K_i* **then**
            if *Verify(K_i) && Authenticate(m_v^i)* **then**
                Store K_i;
                Broadcast $m_{adv}(K_i)$;
            end
            else
                Discard m_v;
            end
        end
        if $[receive\ m_{req}(K_i)\ ]$ **then**
            Forward the requested key K_i;
        end
        if $[receive\ m_{adv}(K_i)\ ]\&\&[\ don't\ have\ K_i]$ **then**
            Start timer T_{req};
        end
        while $![T_{req}(expired)]$ **do**
            Register overheard $m_{req}(K_i)$;
        end
        if $[T_{req}(expired)]$ **then**
            if $!overheard\ m_{req}(K_i)$ **then**
                Broadcast $m_{req}(K_i)$;
            end
        end
    end
end

Since nodes are not time synchronized, and some nodes may start publishing their keys while others are still in the voting phase, we need to consider "man in the middle" attacks. When a node sends its vote, an attacker may withhold it until that node publishes its key. Then it can change the vote, sign it again with the new key, and forward it to the next alerted node. Following this, the attacker also forwards the key, and the receiver will be able to verify the signature and accept the fake vote as authentic.

An explicit defense against the previous attack would be to require the nodes to be loosely synchronized as in μTESLA [31]. Here, however, we have decided to keep things simple and deal with this problem implicitly by relying on residual paths among the nodes (although we plan to investigate the synchronization approach and consider its possible benefits). As votes are forwarded by all nodes, even if an attacker refuses to forward a vote, it will arrive to the intended recipients via other paths. We also take some additional measures in our algorithm having a node accepting a vote only while it has not published its own key and it has not received the key from the node that sends the vote.

When each alerted node $s_1, s_2, \ldots, s_n$ has collected and authenticated the votes from all the other alerted nodes, it will have knowledge of all the corresponding suspect lists, $D(s_1), D(s_2), \ldots, D(s_n)$, itself included. Then it applies a local operator on these lists which will produce the final intrusion detection result, i.e., the attacker's ID. In particular it applies a count operator which counts the number of times δ_i each node i appears in the suspect lists, or else the number of votes it collects. All alerted nodes will reach the same result, since they all apply the same operator on the same sets.

15.6.6   LOCAL RESPONSE MODULE

Once the network is aware that an intrusion has taken place and have detected the compromised area, appropriate actions are taken by the `LocalResponse` module. The first action is to cut off the intruder as much as possible and isolate the compromised nodes. After that, proper operation of the network must be restored. This may include changes in the routing paths, updates of the cryptographic material (keys, etc.), or restoring part of the system using redundant information distributed in other parts of the network. Depending on the confidence and the type of the attack, we categorize the response to two types:

- *Direct response*: Excluding the suspect node from any paths and forcing regeneration of new cryptographic keys with the rest of the neighbors.
- *Indirect response*: Notifying the base station about the intruder or reducing the quality estimation for the link to that node, so that it will gradually loose its path reliability.

IDS in other types of networks always report an intrusion alert to a human, who takes the final action. Correctly, this approach is usually neglected in WSN IDS literature. Sensor networks should (and they actually are) able to demonstrate an autonomic behavior, taking advantage of their inherent redundancy and distributed nature. Autonomic behavior means that any response to an intrusion attempt is performed without human intervention and within finite time.

15.7   DETECTING ATTACKS

The architecture that we described in the previous section is a general architecture for detecting intruders in a sensor network. In this section, we take a closer look at an example of how this schema can be used to detect a specific attack, namely the Sinkhole attack. We use this example to show how the local detection module generates alerts based on the messages that it monitors and which rules one should built to analyze these messages. A more detailed analysis on the intrusion detection of Sinkhole attack can be found in Ref. [28].

The Sinkhole attack [32] is one of the main attacks against sensor networks routing protocols. A compromised node can draw all or as much as possible traffic from a particular area, by making itself look attractive to the surrounding nodes with respect to the routing metric. In particular, the compromised node will try to persuade its neighbors to change their current parents and choose the Sinkhole node as their new one, by trying to make these parents look like much less attractive than itself. For modern routing protocols, which use more sophisticated routing metrics, such as link quality [33], launching a Sinkhole attack is not trivial, but still possible.

The method followed by routing protocols to allow nodes advertise their link quality to their neighbors is the use of periodic packets, called route update packets. Then, to launch a Sinkhole attack, an adversary has to change the link quality estimates sent by the nodes, within the route update packets. To do that, the attacker listens to the route update messages from its neighbors, alters them and replays them, impersonating the original sender.

So, we can build a rule in the local detection engine to capture such kind of activity and trigger an alert. The intuition is that route update packets should originate only from their legitimate sender and that nodes should defend against impersonation attacks.

Rule: "For each overhead route update packet check the sender field, which must be equal to one of the neighboring nodes. If this is not the case, produce an alert."

Note that a node, which detects an anomaly according to the above rule, can infer the existence of an attacker, but it has not enough information to conclude on the attacker's identity (the sender field of the packet is altered). The only conclusion it can draw is that the attacker is one of the neighboring nodes, since the route update packets are only broadcasted locally. Then, the node needs to rely on the cooperative detection engine to reduce the candidates down to one node. So, the suspect list produced by the local detection engine will be the neighborhood of the node.

Then the cooperative detection engine will take over. Since the attacker broadcasted a packet impersonating another node, the alerted nodes are going to be neighbors of the attacker. Therefore, the attacker's ID will be in all suspect lists. By exchanging these lists during the voting phase, and applying the majority rule, nodes are looking to find which nodes are common within the sets, i.e., which node collects most of the votes. If only one node holds the majority, then that node is the attacker. Experiments and simulations presented in Ref. [28] show that more than 75 percent of the attacker's neighbors will manage to successfully detect it, meaning that they will end up with only one node ID as the result of the cooperative detection process.

15.8 IMPLEMENTATION OF IDS IN TINYOS

In this section, we present the implementation details of the IDS described in this chapter. The goal is to show that such a system for sensor networks is lightweight enough to be a viable and realistic solution from implementation and real deployment perspectives. We emphasize on the components that the IDS is made of, along with the proposed interface language and components compatibility. Component-based design is increasingly viewed as the cornerstone of software engineering. It has become a necessity for networked embedded systems, where hardware platforms come in great variety and evolve extremely rapidly.

15.8.1 MODULES AND INTERFACES

The current development of the IDS protocol builds on Moteiv Telos motes—a popular architecture in the sensor network research community. It features the 8 MHz TI MSP430 microcontroller, a 16 bit RISC processor that is well known for its low-energy consumption. Yet, even though the implementation is tested on the Telos motes, all the components are designed with adequate generality such that porting them to different sensor platforms should yield similar performance results.

Figure 15.4 shows the layered approach of System Network Architecture (SNA) and where the IDS exists with respect to the other layers. The bottom layers represent reusable mechanisms that are performed on the communication link interconnections for the best transmission scheduling. Routing, error detection and correction and flow control are performed in the Network and Transport

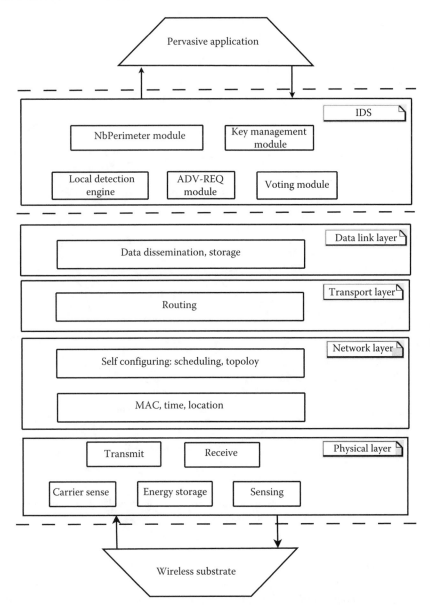

FIGURE 15.4 Embedded IDS in sensor network architecture.

layer. The IDS is placed on top of these layers and is self-contained, so that it can be easily modified without unduly affecting other layers. Finally, the Application layer represents programs developed by users that run on sensor motes. This application level technology does not introduce any limitations on the underlying infrastructure or the routing protocols.

The IDS consists of a set of interconnected modules (Figure 15.5) that is scheduled by a simple FIFO-based nonpreemptive scheduler. Modules communicate with each other through commands and events. Commands propagate downwards; they are issued by higher level modules to lower level ones. Events propagate upwards; they are signaled by lower level modules and handled by higher level ones.

The implementation of the IDS implicates parts of the TinyOS network stack and specifically the `GenericComm` module, which provides commands for the transmission and reception of Active

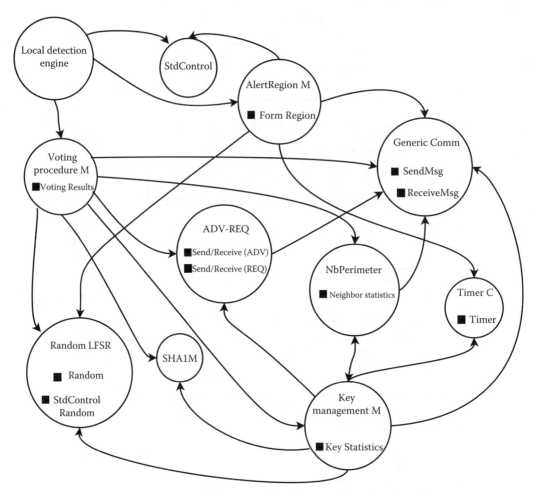

FIGURE 15.5 IDS modules and interfaces. Each line begins at the module that uses an interface and ends at the module that provides the corresponding interface.

Messages. GenericComm directs all messages to AMStandard, a defined Active Message layer that forwards messages to all needed modules. Messages are classified as route updates (sent by the routing protocol), neighborhood IDs, keys, or votes. All messages are sent using the current TinyOS packet format for Telos motes CC2420 radio, so changing of the default packet structure is not needed.

15.8.2 MEMORY AND COMPUTATIONAL REQUIREMENTS

The memory footprint of the IDS that we described in Section 15.6 is an important measure of its feasibility and usefulness on limited memory constrained sensor nodes. The total memory footprint is composed of the memory footprint of the compiled code, which is present in ROM, and the memory footprint of the data memory required to run the code. An IDS for constrained devices must be compact in terms of both code size and RAM usage, to leave room for applications running on top of the system. Table 15.1 lists the memory footprint of the modules, compiled for the MSP430 microcontroller.

The largest module in terms of RAM footprint in Table 15.1 is the Key Management module. This is because the Key Management module contains statically allocated tables for the neighbors

TABLE 15.1
Size of the Compiled Code (in Bytes)

Module	RAM Usage	Code Size
Neighborhood discovery	136	968
Exchange of keys	216	4060
Reliability (ADV-REQ)	104	32
Voting	159	4844
Total	615	9904

and their keys. In terms of ROM, the largest module is the voting module, since it has the most lines of code. In total, the IDS consumes 615 bytes of RAM and 9,904 bytes of code memory. This leaves enough space in the mote's memory for user applications. For example, the total RAM available in Telos motes is 10 KB.

15.8.3 EXPERIMENTS

To evaluate the implementation of the IDS, we can see its performance in a real environment. In particular we deployed several nodes in random topologies on the floor of an office building. We set a node to be the "attacker" and we gradually incremented the number of its neighbors to form larger alert regions. For each alert region size, we repeated the experiment for 20 different random topologies.

The experiments here are performed by having the motes running a typical monitoring application. In particular, we loaded the Delta application, where the motes report environmental measurements to the base station every 5 s. We also deployed the MultihopLQI protocol at the routing layer, which is an updated version of the MintRoute protocol [33] for the Chipcon CC2420 radio. We tuned it to send control packets every 5 s. In this way we can see how well the IDS functions, even under the presence of traffic on other layers.

Figure 15.6 depicts the communication cost of the protocol measured in packets sent by a node. In particular, we broke it down to the packets exchanged for the voting phase and the publish key phase (as a total of exchanging the votes, ADV, REQ, and keys). As it is expected, the number of packets exchanged in the two phases are the same, since the protocol does not change; only the content of the packets does. For small number of alerted nodes the cost is only about 12 packets, while for more dense regions the cost still remains low (19 packets). This is the total communication cost per attack and involves only the nodes in the alert region. It is also measured as a mean time averaged on different random topologies. The number of packets depends on the topology and the number of the alerted nodes, as these parameters determine the number of votes and keys circulated among them.

Figure 15.7 expresses the percentage of costs for computation and communication for the publish key phase. We can see that most of the overhead arises from the transmission of data rather than from any computational costs. This overhead for the communication is due to the inherent inability of TinyOS to receive the next packet before finishing the processing of the current one. In the implementation, upon receiving a key, the node has to verify if it is a valid one before accepting it. To save memory space, we do not buffer the key for later processing, but rather we authenticate it on the fly. Meanwhile, TinyOS cannot receive the next key. That is why we had to include a random delay so that nodes publish their keys in different time instances. This delay, although experimentally minimized, contributes significantly to the results of Figure 15.7.

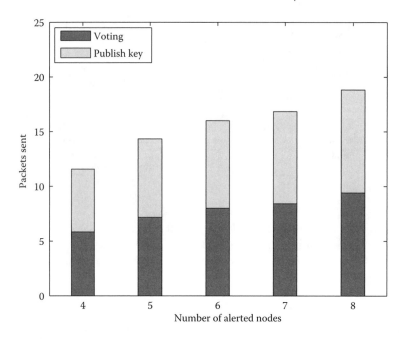

FIGURE 15.6   Measured communication cost for different number of alerted nodes.

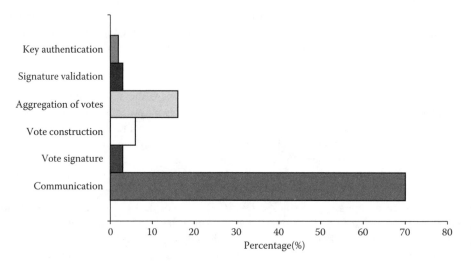

FIGURE 15.7   Costs of computation and communication in terms of time for the publish key phase.

15.9   CONCLUSIONS

In this chapter, we discussed the problem of intrusion detection in sensor networks in a formal and abstract manner, without focusing on underlying protocols or detection of specific attacks. We described an IDS that uses a large number of autonomous, but localized, cooperating agents to detect an attacker. The nodes use coordinated surveillance by incorporating inter-agent communication and distributed computing in decision making to collaboratively infer the identity of the attacker from a set of suspicious nodes.

The system is not based on any reputation system, as that would raise considerably the energy requirements. To the solution presented in this chapter achieves to minimize energy consumption by allowing the arbitrary behavior of the nodes. Instead it is based on the power of the majority to detect the misbehaving nodes. This approach is quite popular in sensor networks security protocols. If the honest nodes are more than the faulty nodes (which is usually the case), the intruder can be detected, even if some nodes are lying. We do not need to consider a procedure that would hold these faulty nodes responsible, when they are making false accusations. The only thing we need to ensure is that the protocol is protected from spoofing and that votes from honest nodes are delivered unchanged to their destinations.

The IDS we discussed is novel but most importantly realistic considering the current state-of-the-art in wireless sensor networks. The demonstrated implementation details show that it is lightweight enough to run on sensor nodes, requiring only limited computational and memory resources. This proves that studying the problem of IDS in sensor networks is a viable approach and with further research it can provide even more attractive and efficient solutions for securing such networks.

REFERENCES

[1] A. Becher, Z. Benenson, and M. Dornseif. Tampering with motes: Real-world physical attacks on wireless sensor networks. In *Proceedings of the 3rd International Conference on Security in Pervasive Computing (SPC)*, pp. 104–118, York, UK, April 2006.

[2] L. Lazos and R. Poovendran. Serloc: Robust localization for wireless sensor networks. *ACM Transactions on Sensor Networks*, 1(1):73–100, 2005.

[3] T. Dimitriou and I. Krontiris. Secure in-network processing in sensor networks. *Security in Sensor Networks*, pp. 275–290. CRC Press, Boca Raton, FL, 2006.

[4] S. Ganeriwal, S. Capkun, C.-C. Han, and M. Srivastava. Secure time synchronization service for sensor networks. In *Proceedings of the 4th ACM workshop on Wireless Security (WiSe '05)*, pp. 97–106, Cologne, Germany, 2005.

[5] B. Yu and B. Xiao. Detecting selective forwarding attacks in wireless sensor networks. In *Proceedings of the 20th International Parallel and Distributed Processing Symposium (SSN2006 Workshop)*, pp. 1–8, Rhodes, Greece, April 2006.

[6] E.C.H. Ngai, J. Liu, and M.R. Lyu. On the intruder detection for sinkhole attack in wireless sensor networks. In *Proceedings of the IEEE International Conference on Communications (ICC '06)*, Istanbul, Turkey, June 2006.

[7] Y.-C. Hu, A. Perrig, and D.B. Johnson. Packet leashes: A defense against wormhole attacks in wireless ad hoc networks. In *Proceedings of the Twenty-Second Annual Joint Conference of the IEEE Computer and Communications Societies (INFOCOM 2003)*, San Francisco, CA, April 2003.

[8] S. Axelsson. Intrusion detection systems: A survey and taxonomy. Technical Report 99-15, Department of Computer Engineering, Chalmers University of Technology, Gothenburg, Sweden, March 2000.

[9] K. Ilgun, R.A. Kemmerer, and P.A. Porras. State transition analysis: A rule-based intrusion detection approach. *Software Engineering*, 21(3):181–199, 1995.

[10] U. Lindqvist and P.A. Porras. Detecting computer and network misuse through the production-based expert system toolset (p-BEST). In *IEEE Symposium on Security and Privacy*, pp. 146–161, 1999.

[11] H.S. Javitz and A. Valdes. The NIDES statistical component: Description and justification. Annual report, Computer Science Laboratory, SRI International, Menlo Park, CA, March 1994.

[12] C. Ko, P. Brutch, J. Rowe, G. Tsafnat, and K.N. Levitt. System health and intrusion monitoring using a hierarchy of constraints. In *RAID '00: Proceedings of the 4th International Symposium on Recent Advances in Intrusion Detection*, pp. 190–204, Toulouse, France, 2001.

[13] C. Ko, M. Ruschitzka, and K. Levitt. Execution monitoring of security-critical programs in distributed systems: A specification-based approach. In *SP '97: Proceedings of the 1997 IEEE Symposium on Security and Privacy*, pp. 175–187, Oakland, CA, 1997.

[14] Y. Zhang, W. Lee, and Y.-A. Huang. Intrusion detection techniques for mobile wireless networks. *Wireless Networks Journal*, 9(5):545–556, 2003.

[15] P. Albers, O. Camp, J.-M. Percher, B. Jouga, L. Mé, and R. Puttini. Security in ad hoc networks: A general intrusion detection architecture enhancing trust based approaches. In *Proceedings of the First International Workshop on Wireless Information Systems (WIS-2002)*, pp. 1–12, Ciudad Real, Spain, April 2002.

[16] O. Kachirski and R. Guha. Effective intrusion detection using multiple sensors in wireless ad hoc networks. In *Proceedings of the 36th Hawaii International Conference on System Sciences (HICSS '03)*, p. 57, BigIsland, HI, January 2003.

[17] A. Siraj, S. Bridges, and R. Vaughn. Fuzzy cognitive maps for decision support in an intelligent intrusion detection system. In *IFSA World Congress and 20th North American Fuzzy Information Processing Society (NAFIPS) International Conference*, Volume 4, pp. 2165–2170, Vancouver, Canada, July 2001.

[18] S. Marti, T.J. Giuli, K. Lai, and M. Baker. Mitigating routing misbehavior in mobile ad hoc networks. In *Proceedings of the 6th Annual International Conference on Mobile Computing and Networking (MobiCom '00)*, pp. 255–265, Boston, MA, 2000.

[19] P. Albers, O. Camp, J.-M. Percher, B. Jouga, L. Mé, and R.S. Puttini. Security in ad hoc networks: A general intrusion detection architecture enhancing trust based approaches. In *Wireless Information Systems*, pp. 1–12, Ciudad Real, Spain, 2002.

[20] O. Kachirski and R. Guha. Intrusion detection using mobile agents in wireless ad hoc networks. In *KMN '02: Proceedings of the IEEE Workshop on Knowledge Media Networking*, p. 153, 2002.

[21] A.P. da Silva, M. Martins, B. Rocha, A. Loureiro, L. Ruiz, and H.C. Wong. Decentralized intrusion detection in wireless sensor networks. In *Proceedings of the 1st ACM International Workshop on Quality of Service & Security in Wireless and Mobile Networks (Q2SWinet '05)*, pp. 16–23. ACM Press, Montreal, Canada, October 2005.

[22] I. Onat and A. Miri. An intrusion detection system for wireless sensor networks. In *Proceedings of the IEEE International Conference on Wireless and Mobile Computing, Networking and Communications*, Volume 3, pp. 253–259, Montreal, Canada, August 2005.

[23] C.E. Loo, M.Y. Ng, C. Leckie, and M. Palaniswami. Intrusion detection for routing attacks in sensor networks. *International Journal of Distributed Sensor Networks*, 2(4), pp. 313–332, December 2006.

[24] V. Bhuse and A. Gupta. Anomaly intrusion detection in wireless sensor networks. *Journal of High Speed Networks*, 15(1):33–51, 2006.

[25] R. Roman, J. Zhou, and J. Lopez. Applying intrusion detection systems to wireless sensor networks. In *Proceedings of IEEE Consumer Communications and Networking Conference (CCNC '06)*, pp. 640–644, Las Vegas, NV, January 2006.

[26] F. Anjum, D. Subhadrabandhu, S. Sarkar, and R. Shetty. On optimal placement of intrusion detection modules in sensor networks. In *BROADNETS '04: Proceedings of the First International Conference on Broadband Networks (BROADNETS'04)*, pp. 690–699, San Jose, CA, 2004.

[27] I. Krontiris, T. Dimitriou, and T. Giannetsos. LIDeA: A distributed lightweight intrusion detection architecture for sensor networks. In *Proceedings of the 4th International Conference on Security and Privacy for Communication (SECURE COMM'08)*, Istanbul, Turkey, September 2008.

[28] I. Krontiris, Z. Benenson, T. Giannetsos, F. Freiling, and T. Dimitriou. In *6th European Conference on Wireless Sensor Networks (EWSN'09)*, Cork, Ireland, February 2009.

[29] L. Lamport. Password authentication with insecure communication. *Communications of the ACM*, 24(11):770–772, 1981.

[30] J. Kulik, W. Heinzelman, and H. Balakrishnan. Negotiation-based protocols for disseminating information in wireless sensor networks. *Wireless Networks*, 8(2/3):169–185, 2002.

[31] A. Perrig, R. Szewczyk, J.D. Tygar, V. Wen, and D.E. Culler. Spins: Security protocols for sensor networks. *Wireless Networks*, 8(5):521–534, 2002.

[32] C. Karlof and D. Wagner. Secure routing in wireless sensor networks: Attacks and countermeasures. *AdHoc Networks Journal*, 1(2–3):293–315, September 2003.

[33] A. Woo, T. Tong, and D. Culler. Taming the underlying challenges of reliable multihop routing in sensor networks. In *SenSys '03: Proceedings of the 1st International Conference on Embedded Networked Sensor Systems*, pp. 14–27, 2003.

16 Key Establishment in Wireless Sensor Networks

Ioannis Chatzigiannakis and Elisavet Konstantinou

CONTENTS

In this chapter, we will consider *key establishment protocols* for *wireless sensor networks*. Several protocols have been proposed in the literature for the establishment of a shared group key for wired networks. The choice of a protocol depends whether the key is established by one of the participants (and then transported to the other[s]) or agreed among the participants, and on the underlying cryptographic mechanisms (symmetric or asymmetric). Clearly, the design of key establishment protocols for sensor networks must deal with different problems and challenges that do not exist in wired networks. To name a few, wireless links are particularly vulnerable to eavesdropping, and that sensor devices can be captured (and the secrets they contain can be compromised); in many upcoming wireless sensor networks, nodes cannot rely on the presence of an online trusted server (whereas most standardized authentication and key establishment protocols do rely on such a server).

In particular, we consider five distributed group key establishment protocols. Each of these protocols applies a different algorithmic technique that makes it more suitable for (1) static sensor networks, (2) sensor networks where nodes enter sleep mode (i.e., dynamic with low rate of updates on the connectivity graph), and (3) fully dynamic networks where nodes may even be mobile. On the other hand, the common factor for all five protocols is that they can be applied in dynamic groups (where members can be excluded or added) and provide forward and backward secrecy. All these protocols are based on the Diffie–Hellman key exchange algorithm and constitute natural extensions of it in the multiparty case.

16.1   INTRODUCTION

Group key management mainly includes activities for the establishment and the maintenance of a group key. Secure group communication requires scalable and efficient group membership with appropriate access control measures to protect data and to cope with potential compromises. A secret key for data encryption must be distributed with a secure and efficient way to all members of the group. Another important requirement of group key management protocols is key freshness. A key is fresh if it can be guaranteed to be new. Moreover, the shared group key must be known only to the members of the group. Four important cryptographic properties must be encountered in group key agreement [31,35]. Assume that a group key is changed m times and the sequence of successive keys is $\mathcal{K} = \{K_0, \ldots, K_m\}$.

Computational group key secrecy: It guarantees that it is computational infeasible for any passive adversary to discover any group key $K_i \in \mathcal{K}$ for all i.

Decisional group key secrecy: It ensures that there is no information leakage other than public blinded key information.

Key independence: It guarantees that a passive adversary who knows a proper subset of group keys cannot discover any other of the remaining keys. Key independence can be decomposed into forward secrecy and backward secrecy. Forward secrecy guarantees that a passive adversary who knows a contiguous subset of old group keys cannot discover any subsequent group key. Backward secrecy guarantees that a passive adversary who knows a contiguous subset of group keys cannot discover preceding group key.

Group key establishment can be either centralized or distributed. In the first case, a member of the group is responsible for the generation and the distribution of the key. In distributed group key establishment all group members contribute to the generation of the key. Clearly, the second approach is suited for sensor networks because problems with centralized trust and the existence of single point of failure can be avoided. In this chapter, we consider distributed group key establishment protocols [7,14,29,45] which can be applied in dynamic groups (where members can be excluded or added) and provide forward and backward secrecy. Moreover, all these protocols are based on the Diffie–Hellman key exchange algorithm [21] and constitute natural extensions of it in the multiparty case.

Many cryptographic protocols have been developed to provide security for group communication [5,12,13,27]. Unfortunately, most of these protocols either require a particular structure for the network (that is neither desired nor available in ad hoc networks) or are resource intensive.

Most group key establishment protocols are based on generalizations of Diffie–Hellman key exchange protocol [21]. The first attempt for the construction of such protocols was made by Ingemarsson, Tang, and Wong [29] that arrange the participants in a standard form like a logical ring via a synchronous start-up phase. The protocol completes in $n - 1$ rounds, where n is the number of the participants.

Burmester and Desmedt presented in Ref. [14] a more efficient scheme which requires only two rounds. However, the protocol's disadvantage is that (i) every participant must perform $n + 1$ exponentiations and (ii) communication is based on concurrent broadcasts that lead to high number of collisions, a situation very common in wireless sensor networks that affects performance [17]. Moreover, the authors do not provide a proof of security (in the stronger sense of semantic security). Recently, Katz and Yung [30] proposed a more general framework that provides a formal proof of security for this protocol. In Hypercube protocol [7] the participants in the network are arranged in a logical hypercube. This topology decreases the number of transmitted data and exponentiation operations, but still the protocol is very demanding for use in sensor networks.

One of the most efficient protocols in the literature for group key management is the third protocol GDH.3 of Steiner, Tsudik, and Waidner presented in Ref. [45]. This protocol requires serial execution of computations that makes it inefficient for highly dynamic networks with large number of nodes. More precisely, this protocol may not be a good choice for a dynamically evolving ad hoc environment because the last node in the protocol's computation would have to know the whole structure of the network.

A performance analysis of all the above mentioned protocols is presented in Refs. [3,4] which clearly shows the superiority of GDH.3 protocol in the number of transmitted data and exponentiation operations required. In particular, the number of messages and exponentiations is linear to the number of the participants in the protocols, while for all other protocols are of order $n \log n$ or n^2.

A very efficient protocol is also presented in Ref. [31]. In this recent work, a logical key tree structure is used to improve the scalability of the key agreement protocol. Any device can calculate the group key if it knows all the keys in its co-path. This requirement makes the protocol quite expensive in storage memory that is critical for sensor networks. For these reasons, we believe that the simplicity and the limited memory requirements of GDH.3 protocol make it more suitable and applicable in sensor networks. Moreover, the recent papers of Bresson et al. [10,11] were the first to present a formal model of security for group authenticated key exchange and the first to give rigorous proofs of security for particular protocols.

Based on the above, we distinguish a category of protocols (e.g., [14,29,45]) which rely on communication primitives that provide global ordering of the devices, e.g., such as a (virtual) ring-based topology and enable many-to-many message exchanges. In fixed infrastructure-based networks, such communication primitives can be provided by the fixed part (i.e., base stations). However, in wireless sensor networks the fixed infrastructure is sparse (or even nonexisting), making it difficult (or even impossible) to implement such primitives via external coordination. Certainly one can assume that the participating devices are capable of transmitting at long ranges, allowing them to communicate directly with each other. Still, in the light of the dense deployment of sensor devices, a traditional single hop communication scheme consumes a lot of power compared to distributed short-range hop-by-hop propagation [28]. In addition, multi-hop communication can effectively overcome some of the signal propagation effects in long-distance wireless transmissions and may help to smoothly adjust propagation around obstacles. Finally, the low-energy transmission in hop-by-hop propagation may enhance security, protecting from undesired discovery of the data propagation operation.

In this direction, a number of protocols have been presented (e.g., [18,22,31]) that construct a connectivity-related distributed tree structure representing the topology of the network by a series of message passes. The network is viewed as a dynamically changing, directed graph, with devices as vertices and *edges* (*virtual links*) between vertices corresponding to devices that can currently communicate. Each device is required to store some small amount of information regarding the data structure (i.e., not the complete graph) and uses this information for group key establishment. Because these protocols do not require that all group members communicate directly with each other via long-range transmissions, the silent adversary will only be able to listen to a limited number of messages given its actual physical location.

Communication in wireless sensor networks usually occurs in ad hoc manner [9] and is subject to frequent, unpredictable changes: network connectivity changes by time as devices adjust their radio range [33] and duty cycle [32], sensor devices die and new sensor devices may be added to the network. To guarantee the correctness of the process of group key establishment, the protocols are required to exchange messages to update the data structure to reflect the changes to the topology of the network. If the rate of connectivity changes is low ("quasistatic" networks) or medium, then adaptive algorithmic techniques can apply, e.g., like the work of Ref. [37]. However, if the rate is high, the devices end up exchanging large amounts of information (that waste the wireless medium and the energy resources) and might even fail to react fast enough forcing the group key establishment to fail.

16.1.1 Modeling Assumptions

In the sequel we present five key establishment protocols [14,18,19,31,45] using elliptic curves terminology. We abstract the technological specifications of existing wireless sensor systems [1] as a system consisting of n devices connected through bidirectional channels allowing direct communication between pairs of neighbor processes linked by a channel. We assume that devices have distinct identities, and they know their own identities together with those of their neighbors. To simplify the presentation of the protocols, we here assume that channels are safe, that is, messages are delivered without loss or alteration after a finite delay, but they do not need to follow a first-in-first-out rule. Moreover, the protocols can be distributed (e.g., a structure-based algorithm, see [18,45]) or centralized (i.e., by a controlling center). We here consider general wireless sensor networks that are composed by a base station and simple devices where no kind of hierarchy exists. The base station represents the authorities of this remote surveillance system (i.e., where the wireless sensors report), has very large storage and data process capabilities and is usually a gateway to another network (i.e., the internet). Typically, the sensors are deployed around the area of the base station and form groups given the needs of the base station. Data flow in our network is group-wise within a group of sensor devices and the base station.

In Section 16.8 we implement (in nesC code running in TinyOS) safe communication demonstrating that the underlaying technology can fulfill these assumptions.

16.2 PRELIMINARIES OF ELLIPTIC CURVE THEORY

In this section we review some basic concepts regarding elliptic curves and their definition over finite fields. The interested reader may find additional information in e.g., [8,42]. We also assume familiarity with elementary number theory (see e.g., [16]).

The elliptic curves are usually defined over *binary fields* F_{2^m} ($m \geq 1$), or over *prime fields* F_p, $p > 3$. In the experimental results we used elliptic curves defined over prime fields. An *elliptic curve* $E(F_p)$ over a finite field F_p, where $p > 3$ and prime, is the set of points $(x, y) \in F_p$ (represented by affine coordinates) which satisfy the equation

$$y^2 = x^3 + ax + b \tag{16.1}$$

and a, $b \in F_p$ are such that $4a^3 + 27b^2 \neq 0$. The set of solutions (x, y) of Equation 16.1 together with a point $\mathcal{O}$, called the *point at infinity*, and a special addition operation define an Abelian group, called the *Elliptic Curve group*. The point $\mathcal{O}$ acts as the identity element (for details on how the addition is defined see [8,42]).

The *order m* of an elliptic curve is the number of the points in $E(F_p)$. The *order of a point P* is the smallest positive integer n for which $nP = \mathcal{O}$. Application of Langrange's theorem (see e.g., [16]) on $E(F_p)$, gives that the order of a point $P \in E(F_p)$ always divides the order of the elliptic curve group, so $mP = \mathcal{O}$ for any point $P \in E(F_p)$, which implies that the order of a point cannot exceed the order of the elliptic curve.

The security of elliptic curve cryptosystems is based on the difficulty of solving the discrete logarithm problem (DLP) on the EC group. The elliptic curve discrete logarithm problem (ECDLP) is about determining the least positive integer k which satisfies the equation $Q = kP$ for two given points Q and P on the elliptic curve group. A user A in an elliptic curve cryptosystem can choose a random integer $0 < k < p - 1$ and send Q to a user B with whom he wants to communicate secretly. A's public key is Q and his private key is k. Then an encryption algorithm can be applied (e.g., ElGamal encryption [24]) so that B can encrypt the message he wishes to send to A with the public key Q and A will decrypt it using his private key k.

The elliptic curve Diffie–Hellman algorithm [21] is based on the difficulty of solving the discrete logarithm problem in an elliptic curve group. The algorithm is as follows. Let A and B be two entities that wish to share a secret key. Both A and B agree a priori on an elliptic curve group, a generator

P of this group and generate a pair of private/public key $(k_A, Q_A = k_A P)$ and $(k_B, Q_B = k_B P)$, respectively. Then A sends Q_A to B and B sends Q_B to A. A computes the value $S = k_A Q_B$ and B the value $S = k_B Q_A$, where S is now their shared secret key. For an appropriately chosen elliptic curve group, an adversary who observes Q_A and Q_B cannot find the shared point S. The only weakness of this algorithm is that there must be an authentication process between A and B so that there is a guarantee that every entity is who he claims to be.

There is a family of protocols that are referred as "natural" extensions of the original, 2-party DH key exchange to n parties. Like in the 2-party case, all participants $M_1, \ldots, M_n$ agree on an elliptic curve group and a generator P. Each member M_i chooses randomly a value k_i and the final shared key among all the members will be equal to $Q_n = k_1 \cdots k_n P$. It can be proven that any adversary who observes any subproduct $k_1, \ldots, k_{i-1}, k_{i+1}, \ldots, k_n P$, cannot compute the shared key Q_n [45]. In particular, in Ref. [45] is proven that if a 2-party key is indistinguishable from a random value, the same is true for n-party keys.

16.3 BURMESTER–DESMEDT PROTOCOL

In 1994, Burmester and Desmedt proposed an efficient protocol for the establishment of a common secret key among the members of a group which requires a constant number of rounds [14]. Suppose that n users $M_1, \ldots, M_n$ wish to establish a common group key. The steps of the protocol are the following where the indices are taken modulo n so that member M_0 is M_n and member M_{n+1} is M_1.

16.3.1 BD Protocol

1. In the first stage, every group member M_i generates a random secret value k_i and broadcasts the point $Q_i = k_i P$ where P is the base point.
2. Each group member M_i computes and broadcasts the point $X_i = k_i(Q_{i+1} - Q_{i-1})$.
3. In the last stage, every group member M_i computes the group key as $K = nk_iQ_{i-1} + (n-1)X_i + (n-2)X_{i+1} + \cdots + X_{i+n-2}$.

If all group members follow the above steps, they will compute the same group key $K = (k_1 k_2 + k_2 k_3 + \cdots + k_n k_1)P$. For example, suppose that four group members M_1, M_2, M_3, and M_4 wish to establish a common secret key. They compute and broadcast the points $Q_1 = k_1 P$, $Q_2 = k_2 P$, $Q_3 = k_3 P$, and $Q_4 = k_4 P$. Then, M_1 broadcasts the point $X_1 = k_1(Q_2 - Q_4)$, M_2 broadcasts $X_2 = k_2(Q_3 - Q_1)$, M_3 broadcasts the point $X_3 = k_3(Q_4 - Q_2)$ and, finally M_4 the point $X_4 = k_4(Q_1 - Q_3)$. Every group member now knows all points Q_i, X_i and using its secret value k_i can compute the group key. For instance, M_1 computes $K = 4k_1Q_4 + 3X_1 + 2X_2 + X_3 = k_1 k_2 P + k_2 k_3 P + k_3 k_4 P + k_4 k_1 P$.

The security of the protocol is based on the decisional Diffie–Hellman problem. The protocol is unauthenticated and consequently is secure only against passive adversaries. The authors of Ref. [14] provided the full security proof of their protocol later in Ref. [15]. Recently, Katz and Yung [30] proposed a more general framework that provides a formal proof of security for this protocol. They also proposed a scalable compiler which transforms an unauthenticated group key agreement protocol into an authenticated group key agreement protocol preserving in the same time the forward secrecy of the original protocol. The modification that Katz and Yung proposed to this protocol adds one more round, two signature generations and $2n - 2$ signature verifications.

The main advantage of this protocol is that it completes in only two rounds. However, the protocol's disadvantage is that (i) every participant must perform $n + 1$ exponentiations and (ii) communication is based on concurrent broadcasts that lead to high number of collisions, a situation very common in wireless sensor networks that affect performance [17]. Another disadvantage of the protocol is that for any new dynamic event that may happen in the group (either join or leave) the protocol has to be executed again from the start.

Many variants of the Burmester–Desmedt protocol have been proposed. Two of the most recent variants are [20,23]. In Ref. [20] the authors proposed a bilinear variant of the Burmester–Desmedt protocol whose security relies on the Computational Diffie–Hellman problem in the random oracle model. Another variant was presented in Ref. [23] which improves on the efficiency and flexibility of the original Burmester–Desmedt protocol.

16.4 GROUP DIFFIE–HELLMAN PROTOCOLS

A class of three, generic n-party protocols was presented in Ref. [45], namely GDH.1, GDH.2, and GDH.3 protocols. Here we will present the elliptic curve analog of the protocols. Suppose that every member in the group has agreed on the use of the same elliptic curve parameters. The number of participants is n and we will denote by M_i the ith participant.

16.4.1 GDH.1 PROTOCOL

1. In the first stage every group member M_i generates a random secret value k_i. The M_1 participant selects a base point P and sends to M_2 the point $Q_1 = k_1 P$. Then M_2 sends to M_3 the points $(Q_1 = k_1 P, Q_2 = k_1 k_2 P)$, member M_3 sends to M_4 the points $(Q_1 = k_1 P, Q_2 = k_1 k_2 P, Q_3 = k_1 k_2 k_3 P)$ and so on until the protocol reaches member M_n. In this stage, every member of the group performs one scalar multiplication and the message sent by M_i to M_{i+1} contains i intermediate values.

2. Group member M_n computes the point $Q_n = k_1 k_2 \cdots k_n P$, which is the intended group key. Then, he sends to M_{n-1} the points $(k_n P, k_1 k_n P, k_1 k_2 k_n P, \ldots, k_1 k_2 \cdots k_{n-2} k_n P)$. In other words, he multiplies all the points he had received from M_{n-1} in the previous stage with k_n, removes the formed group key Q_n and adds the point $k_n P$.

3. In the following stage every group member M_i, $i \in [1, n-1]$ multiplies all points which were received from M_{i+1} with k_i, computes the group key Q_n, removes it from the set of points and sends the rest of them to group member M_{i-1}. In this stage every member M_i performs i scalar multiplications and the message he sends to M_{i-1} contains $i - 1$ points.

For example, assume that the size of the group is $n = 5$. Then, in the first stage, M_1 sends to M_2 the point $k_1 P$, M_2 to M_3 the set $(k_1 P, k_1 k_2 P)$, M_3 to M_4 the set $(k_1 P, k_1 k_2 P, k_1 k_2 k_3 P)$, and M_4 to M_5 the set $(k_1 P, k_1 k_2 P, k_1 k_2 k_3 P, k_1 k_2 k_3 k_4 P)$. Then, member M_5 calculates the final group key $Q_5 = k_1 k_2 k_3 k_4 k_5 P$, and sends to M_4 the message $k_5 P, k_1 k_5 P, k_1 k_2 k_5 P, k_1 k_2 k_3 k_5 P$. M_4 can compute now the group key by multiplying the last value of the message with its secret value k_4. Then, he sends to M_3 the set $k_4 k_5 P, k_1 k_4 k_5 P, k_1 k_2 k_4 k_5 P$ who calculates the group key and sends to M_2 the points $k_3 k_4 k_5 P, k_1 k_3 k_4 k_5 P$. Finally, M_2 acquires the key and forward to M_1 the point $k_2 k_3 k_4 k_5 P$.

In summary, GDH.1 protocol requires $2(n-1)$ rounds and $2(n-1)$ messages are sent. Computationally, the total number of scalar multiplications required in the protocol are $(n+3)n/2 - 1$, while every member M_i has to compute $i + 1$ scalar multiplications (except from M_n who computes n).

One of the main drawbacks of GDH.1 protocol is the large number of rounds. To reduce the number of rounds, the authors of Ref. [45] modified GDH.1 and proposed the GDH.2 protocol:

16.4.2 GDH.2 PROTOCOL

1. Similarly to GDH.1, in the first stage every group member M_i generates a random secret value k_i. In round i, $1 \leq i \leq n$, member M_i sends to M_{i+1} the points $\{(k_1 k_2 \cdots k_i / k_j) P \,|\, j \in [1, i]\}$ and $k_1 k_2 \cdots k_i P$. For example, M_3 sends to M_4 the points $(k_1 k_2 k_3 P, k_1 k_2 P, k_1 k_3 P, k_2 k_3 P)$. In this stage, every member M_i of the group performs i scalar multiplication and the message sent by M_i to M_{i+1} contains $i + 1$ intermediate values.

2. In the second stage, member M_n computes the point $Q_n = k_1 k_2 \cdots k_n P$, which is the intended group key and broadcasts the values $\{(k_1 k_2 \cdots k_n / k_i) P \,|\, i \in [1, n]\}$ to the rest of the users. In this way, every group member M_i can compute the group key by multiplying the point $(k_1 k_2 \cdots k_n / k_i) P$ with its secret value k_i.

GDH.2 protocol is executed in n rounds, its total computational cost is $(n+3)n/2-1$ multiplications and every member M_i computes $i + 1$ scalar multiplications (like in GDH.1 protocol).

16.4.3 GDH.3 PROTOCOL

1. In the first stage every group member M_i generates a random secret value k_i. The M_1 participant selects a point P and sends to M_2 the point $Q_1 = k_1P$. Then M_2 sends to M_3 the point $Q_2 = k_1k_2P$ and so on until the protocol reaches member M_{n-1}. Notice here that the protocol must pass only one time from every participant.
2. Group member M_{n-1} computes the point $Q_{n-1} = k_1k_2\cdots k_{n-1}P$ and sends it to all M_i, with $i \in [1,n]$.
3. In the third stage every group member $M_i, i \in [1, n-1]$ computes a point $G_i = k_i^{-1}Q_{n-1}$ and sends it to the last group member M_n.
4. M_n calculates the values k_nG_i and send them to the corresponding members M_i.

After these stages, every group member M_i can calculate the group key $Q_n = k_1k_2\cdots k_nP$ by multiplying the value k_nG_i with its secret number k_i. Despite its efficiency, the disadvantage of GDH.3 protocol is that it does not offer symmetric operation, because all the participants in the protocol do not perform the same number of operations. If the number of the participants is large, then the computational effort in member M_n can be devastating for its energy.

16.5 TREE-BASED GROUP DIFFIE–HELLMAN (TGDH) PROTOCOL

In Ref. [31], an efficient group key agreement protocol was presented that arranges the group members in a binary tree structure. For example, four members M_1, M_2, M_3, and M_4 will be arranged in a binary tree as shown in Figure 16.1. The tree should be kept balanced. Every node in the tree is labeled with a pair $<l, u>$, where l is the level in which the node belongs to and u is a number indicating the place of the node in the particular level. Obviously, $0 \le l \le h$ where h is the height of the tree and $0 \le u \le 2^l - 1$. Each node $<l, u>$ is associated with a secret key $k_{<l,u>}$ and a public key $K_{<l,u>} = k_{<l,u>}P$. Initially, all group members generate their secret keys and compute their public keys $K_{<h,u>}$. The public keys $K_{<l,u>}$ of the upper levels are computed using Diffie–Hellman key exchange between the left and right child of the particular node $<l, u>$, that is nodes $<l+1, 2u>$ and $<l+1, 2u+1>$. The secret key $k_{<l,u>}$ is computed by $k_{<l,u>} = map(k_{<l+1,2u>}K_{<l+1,2u+1>})$ or $k_{<l,u>} = map(k_{<l+1,2u+1>}K_{<l+1,2u>})$ where $map()$ is a function which maps an elliptic curve point to its x-coordinate. The final secret group key is $k_{<0,0>}$. Clearly, this key can be computed by every group member if he knows all public keys in the tree.

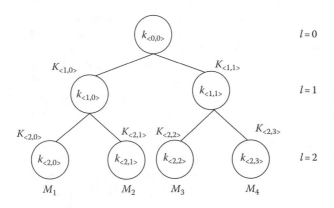

FIGURE 16.1 TGDH binary tree ($n = 4$).

The original description of TGDH [31] does not specify clearly the setup phase of the protocol, but it focuses on the group membership events. However, we can summarize the previously mentioned steps in the following setup procedure [38]:

16.5.1 TGDH PROTOCOL

1. In the first stage every group member M_i generates a random secret value $k_{<l_i,u_i>}$ and broadcasts the point $K_{<l_i,u_i>} = k_{<l_i,u_i>}P$ where P is the base point.
2. Each group member M_i computes the point $K_{<l,u>} = k_{<l,u>}P$, where $l = l_i - 1$ and $u = \lfloor \frac{u_i}{2} \rfloor$ and the rightmost member of the (sub)tree rooted at node $<l, u>$ broadcasts the point $K_{<l,u>}$.
3. All members of the group repeat step 2 with $l = l - 1$ and $u = \lfloor \frac{u}{2} \rfloor$ until they compute the group key $k_{<0,0>}$.

Consider, for example, a group with four members arranged in the tree structure of Figure 16.1. In the first step, the members of the group compute and broadcast the points $K_{<2,0>}$, $K_{<2,1>}$, $K_{<2,2>}$, and $K_{<2,3>}$. Then, members M_1 and M_2 compute the value $k_{<1,0>} = map(k_{<2,0>}K_{<2,1>}) = map(k_{<2,1>}K_{<2,0>})$ using the Diffie–Hellman key exchange protocol and member M_2 (the rightmost member of the subtree) broadcasts the point $K_{<1,0>} = k_{<1,0>}P$. Similarly, members M_3 and M_4 compute the value $k_{<1,1>}$ and M_4 broadcasts the point $K_{<1,1>}$ to the network. Finally, all members can compute the group key $k_{<0,0>} = map(k_{<1,0>}K_{<1,1>}) = map(k_{<1,1>}K_{<1,0>})$. Concluding, the main disadvantage (like in the case of Burmester–Desmedt protocol) of this protocol is that it requires many broadcasts that lead to high number of collisions, a situation very common in wireless sensor networks that affect performance [17].

The tree-based structure of the protocol eases the handling of additive (such as join or merge) and subtractive (such as leave or partition) events. In additive events, the new member or the rightmost member of the new tree broadcasts their own public keys to the network. Then, all nodes update the tree structure and the rightmost member M_s of the updated tree changes its secret value $k_{<l_s,u_s>}$, computes the new secret and public keys in its path up to the root and broadcasts them to rest of the group members. A similar procedure is followed in the case of subtractive events.

16.6 DISTRIBUTED SEQUENTIAL TRAVERSAL PROTOCOL

The protocols presented above rely on communication primitives that provide global ordering of the devices (stage 1), e.g., such as a (virtual) ring-based topology and enable many-to-many message exchanges (stages 2 and 3). In fixed infrastructure-based networks, such communication primitives can be provided by the fixed part (i.e., base stations). However, in wireless sensor networks the fixed infrastructure is sparse (or even nonexisting), making it difficult (or even impossible) to implement such primitives via external coordination. Certainly one can assume that the participating devices are capable of transmitting at long ranges, allowing them to communicate directly with each other. Still, in the light of the dense deployment of sensor devices close to each other, a traditional single-hop communication scheme consumes a lot of power when compared to distributed short-range hop-by-hop propagation [28]. In addition, multi-hop communication can effectively overcome some of the signal propagation effects in long-distance wireless transmissions and may help to smoothly adjust propagation around obstacles. Finally, the low-energy transmission in hop-by-hop propagation may enhance security, protecting from undesired discovery of the data propagation operation.

In this section we present a distributed protocol that does not require many-to-many message exchanges as in the case of the second and third stages of the GDH.3 protocol and does not rely on any global ordering of the devices. This protocol was presented in Ref. [18] and is based on the observation that in the first stage of the GDH protocols, group member M_n can compute the shared group key Q_n by acquiring the point Q_{n-1} from M_{n-1} and multiply it with its secret value k_n. Moreover, the points $Q_i = k_1 k_2 \cdots k_i P$ which are generated by each group member M_i can be used as their public keys while their private keys are the values k_i. Using this observation, the protocol totally avoids the third and fourth stages of GDH.3 protocol.

In particular, a distributed sequential traversal algorithm is used that visits each participant of the group to build the shared secret key (starting from M_1 and reaching M_n). This is similar to the first stage of GDH protocols, without necessarily requiring direct communication among all the group members and still guaranteeing that each participant is visited only one time. When the traversal is finished and all participants are visited, the protocol reaches member M_n that calculates the point Q_n, the shared secret key. Finally, in order for M_n to communicate the shared secret key to all the participants of the group, it repeats the distributed traversal but in reverse order. M_n now encrypts Q_n with M_{n-1}'s public key Q_{n-1} and sends it to M_{n-1}. M_{n-1} can decrypt the message with his private key k_{n-1}, acquire the secret value Q_n, encrypt it with the public key of M_{n-2} and send the result to M_{n-2}. The same process will be followed by M_{n-2} and so on, until the protocol reaches member M_1.

Initial group formation: Based on the initial assignment of the sensors in a given group M (defined by the base station), every group member M_i generates a random secret value k_i while the group leader M_1 selects a point P and calculates the point $Q_1 = k_1 P$ which is its public key. Then, based on a distributed sequential traversal algorithm that visits all participants at least once, and particularly a distributed *depth-first* traversal algorithm, it sends Q_1 to M_2 (i.e., a neighboring group member) via special SEARCH$<M_2, Q_1>$ message. When M_2 receives the message, it becomes active for the first time, it becomes visited and defines participant M_1 as its father (we say M_2 joins the traversal). Moreover, when M_2 is active, it calculates the point $Q_2 = k_1 k_2 P$ (which is the public key of M_2) and shifts the control to a nonvisited neighbor M_3, through a special SEARCH$<M_3, Q_2>$ message.

When the protocol reaches member M_u with all its neighbors being visited, M_u encrypts Q_u with M_{u-1}'s public key Q_{u-1} and sends it to M_{u-1} (its father) using a special PARENT$<M_{u-1},$ encrypt$(Q_u, Q_{u-1})>$ message. The function encrypt is defined as encrypt(data, key). M_{u-1} can decrypt the message with his private key k_{u-1}, acquire the secret value Q_u and either continue the distributed sequential traversal by shifting the control to the next nonvisited neighbor (if any), again through a special SEARCH$<M_v, Q_u>$ or if all its neighbors have been visited, send a PARENT$<M_{u-2},$ encrypt$(Q_u, Q_{u-2})>$ message to its parent. This process continues until the protocol reaches the last member of the group M_n that calculates the point Q_n, the shared secret key. In a similar way with M_u, M_n encrypts Q_n with M_{n-1}'s public key Q_{n-1} and sends it to M_{n-1} (its father) using a special PARENT$<M_{n-1},$ encrypt$(Q_n, Q_{n-1})>$ message.

In the case where M_u has more than one children (in the *depth-first* virtual tree), upon receiving the PARENT message from the last child (let this be M_v) that contains the shared key, it first sends the PARENT$<M_{u-1},$ encrypt$(Q_n, Q_{u-1})>$ message to its father and then informs its children by sending the special message UPDATE$< M_i,$ encrypt$(Q_n, Q_u)>$. The child M_i that receives an UPDATE$<M_i,$ encrypt$(Q_n, Q_u)>$ message, decrypts it to acquire the secret key and forward it to its child via a UPDATE$<M_{i+1},$ encrypt$(Q_n, Q_i)>$ message. This ensures that the shared key will traverse the *depth-first* virtual tree all the way up to the group leader M_1 but also reach all the nodes that belong to a branch of the tree.

Implementation of this traversal technique requires for the active process to know exactly which of its neighbors are visited. To do so, the distributed algorithm apart from the SEARCH and PARENT message uses a special VISITED message that allows each visited process to inform its neighbors (by broadcasting the message) that it has joined the traversal.

This protocol essentially builds a *depth-first* search spanning tree of a network, given a distinguished node as its root (i.e., M_1) and its correctness essentially follows from the correctness of the sequential depth first search (DFS) algorithm, because there is no concurrency in the execution of this algorithm [6].

Handling JoinGroup events: When a new member M_{n+1} wants to join a group, it must first be authenticated by the base station and get an ID and then contact the group leader (via the nearest group member M_u and through the virtual tree structure) a JOIN$<M_u, M_{n+1}>$ message.

The group leader replies by sending the old group key Q_n (again via the tree structure). Then M_{n+1} generates a random value k_{n+1}, computes the new group key $Q_{n+1} = k_{n+1}Q_n$ and sends it back to the group leader. Finally, the group leader sends an UPDATE message to all group members, again by using the virtual tree. The need to contact the group leader is necessary in cases of more than one nodes joining the group simultaneously, in which case, the group leader delays the UPDATE message until all new nodes have joined.

Handling LeaveGroup events: In the case that a member M_u leaves the group, the group leader generates a random value $\overline{k_n}$ and computes a new group key $\overline{Q_n} = \overline{k_n}Q_n$. Then, as in the case of the *JoinGroup* event, it informs all group members about the new shared by sending an UPDATE message using the virtual tree. However, because the removal of the old member will disrupt the tree structure, the children of M_u must now contact the parent of M_u and update the tree structure. If this is not possible, i.e., because the parent of M_u cannot directly communicate with the children of M_u, the handling of the event fails, and M_1 is signaled to restart the *depth-first* tree construction and generate a totally new shared key.

Handling MergeGroup events: When a group M' of nodes want to join group M, the procedure followed essentially expands the *depth-first* search tree to include the members of M'. The group leader of M' contacts the leader of M (via the nearest group member of M, M_u) by sending a MERGE$<M_u, M'_1>$. The group leader M_1 replies by sending the old group key Q_n (again via the tree structure) to M'_1 that is now denoted as M_{n+1}. When the message reaches M_{n+1}, the distributed sequential traversal algorithm continues as if the members of M' where unvisited members of M. In this sense, M_{n+1} computes a new random value k_{n+1}, calculates the point $Q_{n+1} = k_1 k_2 \cdots k_{n+1}P$ (which is now the new public key of M_{n+1}) and shifts the control to the next nonvisited neighbor M_{n+2} (an old member of M'), through a SEARCH$<M_{n+2}, Q_{n+2}>$ message. When all the old members of M' have been visited, the new shared key is propagated to the merged group through the use of the *PARENT* and *UPDATE*.

Handling PartitionGroup events: Instead of trying to compute a new group key for the two resulting groups, when a *PartitionGroup* event is signaled the protocol simply reconstructs the *depth-first* search spanning tree and generates a new shared key for each group.

Periodic group maintenance: We here note that to handle the above events, the virtual tree can degenerate into a spanning tree that no longer fulfills the *depth-first* search criteria. Therefore, to balance the tree and also to guarantee key freshness the group leader periodically restarts the *depth-first* search and generates a new shared key.

Discussion: This group key protocol satisfies the first two cryptographic properties mentioned in Section 16.1 and in particular the *JoinGroup* event accomplishes forward secrecy while backward secrecy is guaranteed by the *LeaveGroup* event. Regarding the computational group key secrecy this is satisfied because, if an adversary silently overhears radio communication and captures data, he cannot discover the group key as it is computationally infeasible to find any secret value k_i from the transmitted data (he has to solve an ECDL problem). Additionally, because the protocol does not require that all group members communicate directly with each other via long-range transmissions, the silent adversary will only be able to listen to a limited number of messages given its actual physical location.

All group events are handled in $O(n)$ time (as in the case of GDH.3, assuming that a bounded number of retransmissions are required due to collisions) and require $O(n)$ message exchanges (again similar to GDH.3, although it is expected that $2n$ less messages need to be transmitted in the network). However, in contrast to GDH.3, this protocol evenly distributes energy consumption among the participants as each device has similar roles in terms of required computations and communication exchanges (energy-wise, the two most demanding events). Balancing the energy dissipation among the sensors in the network avoids the early energy depletion of certain sensors (i.e., in GDH.3 participant M_{n-1}) and thus increases the lifetime of the system by preventing from early network disconnection [43]. In contrast to the GDH.3 protocol, this protocol does not assign different roles to the participating devices nor requires some of them to transmit more messages than others. The distributed sequential traversal ensures that the devices consume more or less equal amounts of energy as they perform the same number of events and communication exchanges leading to better energy balance.

16.7   RANDOM TRAVERSAL PROTOCOL

We now present a distributed protocol that totally avoids constructing and maintaining a distributed structure that reflects the topology of the network, does not require any strong communication primitive and relies only on simple short-range hop-by-hop message exchanges. This protocol [19] is particularly suitable for dense ad hoc networks where the topology is subject to frequent and unpredictable changes. The devices are coordinated in a distributed manner using minimum communication overhead through the use of a mobile agent (software, mobile code) that traverses the network. The protocol imposes minimum communication overhead as the mobile agent is of small size (it fits in a single packet). The mobile agent moves randomly through the devices of the network and no specific traversal strategy (e.g., strictly sequential, in both directions) or global information (e.g., such as the structure of the network) is required. This software agent (mobile code) is responsible for constructing a common key among the participants of the group (by taking into account the contribution of all members of the group) and for the delivery of the established key to all participants.

Initially, all the participants of the group execute a protocol that generates a unique group ID, an initial secret shared key and calculates the group size. This constitutes the setup phase of the protocol. This information is delivered to all participants. The shared key is used to encrypt the information exchanged between any two parties to authenticate each other. We assume that there is no structure that reflects the topology of the network and thus, there cannot be any guarantee that a member truly belongs to the network. No further global information is available to the participants. Note that the initial keys are used only for a short period of time (that we estimate) after which they are replaced by the keys established during the first round of execution of our protocol.

The protocol is executed in two stages. At the *first stage* all the sensor nodes contribute their random information to construct a shared secret key, which is shared to all nodes at the second stage. We suppose that every member in the group has agreed on the use of the same elliptic curve parameters, the number of participants is n and we will denote by M_i the ith participant visited by the mobile agent. The final shared group key will be equal to $Q_n = k_n k_{n-1} \cdots k_1 P$ where k_i are random values selected from every group member M_i and P is a fixed point in the elliptic curve. More precisely, the two stages of our protocol are as follows.

First stage: This stage starts by activating participant M_1. M_1 selects a point P, generates a random value k_1 and calculates the point $Q_1 = k_1 P$. Then, he constructs a mobile agent in which he puts the point Q_1, encrypts the agent with the shared key and transmits it to a random neighbor. Suppose that this neighbor is participant M_2. M_2 decrypts agent's data, acquires point Q_1, generates a random value k_2 and computes the point $Q_2 = k_2 Q_1$. Then, he updates agent's information with the point Q_2, encrypts the agent and sends it to a random neighbor. The same process is followed by every member M_i that is reached by the agent.

Continuing this way, the mobile agent will eventually reach the last unvisited device. This device will be the first to calculate the shared key $Q_n = k_n k_{n-1} \cdots k_1 P$. We here note that the encryption and decryption of agent's critical information is accomplished using as key the secret shared point Q_{old} which had been generated and shared at the previous round of our protocol.

Second stage: At this stage the produced secret key is injected back into the network with a new random walk, to inform all participants about the new key. Suppose that the last unvisited participant of the first stage was M_n. M_n constructs a mobile agent in which he puts the point $Q_{n-1} = k_n^{-1} Q_n$ and transmits the agent to a random neighbor M_i. Member M_i will multiply the point Q_{n-1} with his secret value k_i^{-1} and updating the agent's information will send it back to M_n. M_n will multiply the agent's value with k_n, send it again to M_i who will now be able to acquire the shared value Q_n by multiplying agent's context with k_i. Now M_i is ready to send the agent to another participant M_j by following the same process. If an eavesdropper wants to reveal the shared secret key, he will have to know one of the produced random values k_is. Note that the encryption and decryption of agent's critical information in the first stage can be avoided, provided that every participant of the group can authenticate its neighbors.

This scheme is strong enough for an environment where group membership is highly dynamic, key lifetime is short, and little data is available for cryptanalysis. The protocol produces a shared key using a contributory component from each group member. This component is incorporated along with the other members' components forming the shared group key. Because the group key must be changed periodically due to security requirements, each member must have the ability to quickly generate a random value and perform a decryption operation to decode the agent's message. We now show how to guarantee the fulfillment of the cryptographic properties presented in Section 16.1 using the main membership events. Moreover, these procedures can be applied in all cases of our protocol, i.e., either there is one agent or more. The main membership events include single member addition, single member deletion, group merge, and group partition. In all cases, the members that are added or deleted must first be authenticated by the base station using their ID.

A join event occurs when a single member wants to join the existing group. In this case, the new member generates a random value k_{new}, contacts the nearest group member and computes the new group key $Q_{\text{new}} = k_{\text{new}} Q_{\text{old}}$. Then the new group key is shared among all the members by the agent.

A leave event occurs when a member wishes to leave the group, or is forced to leave it. To handle this event, a random member of the group is chosen to generate a random value and calculate a new shared key by multiplying this value with the old key. This participant will then generate a new mobile agent to distribute the new key to all existing members.

A group merge event occurs when multiple potential members want to join an existing group. To handle this event, we have to develop a MERGE procedure in which a new agent is generated and initialized with the original group's shared key. Then the new agent must visit all new members and include them to the old group. After the establishment of the new shared key, the agent continues the walk to the participants of the expanded group.

A group partition event occurs when multiple members leave the group with or without forming their own subgroup. To handle this event, a random member is chosen in each partitioned subgroup (e.g., using a leader election algorithm, see [37]), it generates a new random value and computes the new group key for the partitioned subgroup by multiplying this value with the old group key. These new subgroup keys are delivered to all members of each subgroup by the corresponding agents.

Let us now consider some correctness issues of the protocol, i.e., some fundamental properties for any protocol that tries to establish a common shared key among the participants of a group. We assume here that the duty cycle of the sensor devices of the network are determined by application protocols (i.e., when to enter sleep mode and when to wake-up) and that the decision is independent of the motion of the mobile agent (i.e., we exclude the case where the devices are deliberately trying to avoid the mobile agent, or enter sleep mode when the mobile agent is located in the device). Moreover, we assume that the sensor devices of the network have sufficient power to support communication. Recall that we assume that channels are safe (messages are delivered without loss or alteration after a finite delay).

In such a case, the mobile agent will eventually meet all the participants of the group with probability 1. In fact, and by using the Borel–Cantelli lemmas for infinite sequences of trials [40], given an unbounded period of (global) time (not necessarily known to the sensor devices) the mobile agent will meet the devices *infinitely often* with probability 1 (because the events of meeting the devices are mutually independent and the sum of their probabilities diverges). This guarantees that the mobile agent will meet all the participants and collect their contribution to the shared key and, then, correctly distribute the established key within the group.

To estimate the time-efficiency of the protocol we model wireless sensor networks by abstracting away the physical layer details as a graph $G(V, E)$ in Euclidean space. The set V is the set of all devices. The set E contains an edge from device u to v if u can directly transmit to v (this can be determined by considering path loss and the signal-to-noise ratio). We refer to G as the *transmission graph*. Under this model, we assume that the mobile agent employed by our protocol does a *continuous time random walk* on $G(V, E)$, without loss of generality (if it is a discrete time random walk, all results will transfer easily). We define the random walk of the mobile agent on G that induces a continuous time Markov chain M_G as follows: The states of M_G are the vertices of G and they are finite. Let s_t denote the state of M_G at time t. Given that $s_t = u$, $u \in V$, the probability that $s_{t+dt} = v$, $v \in V$, is $p(u, v) \cdot dt$ where

$$p(u, v) = \begin{cases} \frac{1}{d(u)} & \text{if } (u, v) \in E \\ 0 & \text{otherwise} \end{cases}$$

and $d(u)$ is the degree of vertex u.

We know (Theorem 6.8, [39]) that the cover time of the random walk on G, i.e., the time required for the walk that started at vertex i to visit all vertices of G, is bounded from above by $2m(n - 1)$, where $m = |E|$. This can also be expressed as $\delta(G)n(n - 1)$, where $\delta(G)$ is the average degree of a vertex.

Now, by assuming that the sensor devices are random uniformly distributed on the area, the density of the network can be calculated according to Ref. [12] as $\mu(R) = n\pi R^2/A$, where R is the transmission range of the devices and A is the size of the area covered by the sensor network. Basically, $\mu(R)$ gives the number of sensor devices within the transmission radius of each device in region A. In our transmission graph model, this implies that the average degree $\delta(G) = \mu(R)$. Therefore the time required for the mobile agent to visit all devices (to collect their contributions) and then revisit all devices (to deliver the calculated key) is equal to two times the cover time of the random walk on G. Thus the time is bounded from above by $2 \cdot (n^2\pi R^2/A) \cdot (n - 1)$, implying an $O(n^3)$ time efficiency. Note that Theorem 6.8 gives the same $O(n^3)$ upper bound for the complete graph K_n, whereas it is known that in such graphs the cover time is $\Theta(n \log n)$. In this sense, this time-efficiency bound can be further improved.

If we assume a more controlled sensor deployment strategy [2], such as a lattice-shaped network (e.g., where $\delta(G) = 4$), then the resulting transmission graph will be a regular graph and the corresponding cover times drop to $O(n^2)$.

16.8 ALGORITHMIC ENGINEERING

There is generally a considerable gap between the theoretical results and the implemented protocols. Advancements have been made in the physical hardware level, embedded software in the sensor devices, systems for secure sensing applications and fundamental research in new communication and security paradigms. Although these research attempts have been conducted in parallel, in most cases they were also done in isolation, making it difficult to converge toward a unified global framework. Most currently deployed solutions lack the necessary sophistication, innovation, and efficiency, while state-of-the-art foundational approaches are often too abstract, missing a satisfactory accuracy of important technological details and specifications. To be effective and to produce applicable results, it is important to encourage interaction and bridge the gap between fundamental approaches and technological/practical solutions.

The development of a wireless sensor network application requires a significant amount of resources not only in hardware but also in software development. In most of the traditional systems and network architectures the software development process is more or less standardized or at least enhanced in several ways. There are specifications and standardized protocols, well-defined application programming interfaces, and software libraries that programmers can build upon. Such facilities are absent in new and emerging technologies, thus the means to develop applications are limited. Developers frequently need to reinvent the wheel porting applicable components from other paradigms or implementing new solutions from scratch. On the other hand, in emerging paradigms and new networking architectures, there are many open problems and design aspects that are still under active research, new algorithmic solutions as well as software and hardware components are developed. In such systems ideally, scientific research progresses through an iterative process; new concepts are modeled and analyzed with theoretic tools. The results of the theoretic research are transubstantiated into algorithms and protocols that are implemented usually in some mainstream programming language through an algorithmic engineering process. Algorithms and protocols are evaluated through simulation or deployed in experimental testbeds; through the experimental evaluation, flaws in the initial design, or unanticipated problems are traced, thus providing incentive to the theorists to improve and develop new analytical methods and models.

For this reason, it is expected that a systematic theory of *distributed algorithm engineering* will be developed. The term distributed algorithm engineering was first introduced in Ref. [44] and essentially it involves the considerable effort required to convert theoretically efficient and correct distributed algorithms to effective, robust and easily used software implementations on a simulated or real distributed environment, usually accompanied by thorough experimentation, fine-tuning and testing. Such a conversion process may lead to improved distributed algorithms through the experimental discovery of behaviors and properties that were not exploited in the initial theoretical version of the algorithm.

To evaluate the suitability of the protocols in wireless sensor networks we here consider the implementation of the protocols on the MICA2 mote architecture [1]. Currently, these devices represent the state of the art in wireless sensor networks technology based on commercial off-the-shelf hardware components and offer an 8-bit, 7.3 MHz ATmega 128L processor, 4 kB of primary memory (RAM) and 128 kB of program space (ROM) and 512 kB secondary memory (EEPROM) and a ChipCon CC1000 radio capable of transmitting at 38.4 kBps powered by 2 AA batteries.

16.8.1 Implementing Elliptic Curve Cryptography

To demonstrate the applicability of the protocols presented in this chapter, we used the elliptic curve version of Diffie–Hellman problem [21]. The reason is that elliptic curve cryptosystems use much smaller keys than conventional, discrete logarithm-based cryptosystems (a 160-bit key in an elliptic curve cryptosystem provides equivalent security with a 1024-bit key in a conventional cryptosystem). This fact makes elliptic curves the only reasonable choice for sensor networks, where the resources are very limited. Moreover, recent research has shown that public key cryptography based on elliptic curves is feasible to be used in sensor networks [25,26,36].

In software, based on the nesC programming language, the elliptic curve cryptography module EccM [36], implemented specifically for TinyOS, allows to represent and carry out basic operations with multiprecision integers of 160-bit size. Given this particular selection of hardware/software we can evaluate the running times for generating random secret values k_i, multiplying the secret values with a point P and encrypting/decrypting them based on a given set of public/private keys. The running times shown in Table 16.1 were measured by the MICA2 device using the SysTime, TinyOS component that provides a 32-bit system time based on the available hardware clock. The results indicate that elliptic curves implementation is feasible in sensor devices, as time to perform an encryption and decryption averages out to 74.481 and 38.365 s.

TABLE 16.1

Running Times of EccM-2.0 for Operations Using 163-Bit Multiprecision Integers

	Addition	Multiplication	Random	Encryption	Decryption
Running time (s)	2.250	36.114	0.22	74.481	38.365

16.8.2 PROTOCOL EVALUATION

Given the above running times for performing the necessary cryptography operations, we can investigate the performance of the key establishment protocols via simulation. The experimental evaluation is conducted with Power-TOSSIM [41] that simulates the wireless network at the bit level, using TinyOS component implementations almost identical to the MICA2 CC1000-based radio stack. Note that the amount of time spent executing instructions is not captured by TOSSIM. Regarding the generation of various different physical topologies of wireless sensor networks TOSSIM provides the LossyBuilder tool [34].

The evaluation presented here considers two different types of network topologies: (i) lattice-shaped networks of $n = \sqrt{n} \times \sqrt{n}$ devices, where the spacing of the devices is set to 45 ft and (ii) random uniform networks of n devices deployed in a square area of 50×50 ft. For both types of topologies the network size is set to $n = [16,25,36,49,64]$. The transmission range of the devices is set to 50 ft. Thus in the lattice-shaped networks, the maximum degree is 4 and for the random uniform networks, the resulting transmission graph is fully connected and the degree is $n - 1$.

All protocols rely on certain assumptions on the underlaying network. To provide the necessary high-level primitives for the protocol to be operational two additional modules are required: (i) the Reckon module that provides information on the identities of the neighboring nodes (that belong in the same group with the node) based on short HELLO messages and (ii) the SafeSend module that guarantees that messages are finally delivered to their destinations by periodically retransmitting messages until the destination confirms their safe reception. The GDH protocols require all devices to communicate directly with M_{n-1}. To be able to provide this primitive, a basic algorithm (the Flood module) is used that simply floods the messages in the network until they are received by their final destination. We here note that the Bcast module provided by TinyOS implements many-to-1 multi-hop routing and is thus unsuitable. Finally, because in GDH the devices need to know identities of all the group members (so that stage 1 can be carried out) we also implemented the ReckonGlobal module that provides information on the identities of all the nodes (that belong in the same group with the node) based on a simple flooding protocol of HELLO messages.

Given the above configuration, the performance of GDH.3, distributed sequential traversal (or DFS-based) and random traversal (or agent-based) protocols is evaluated. In our evaluation we included only these protocols as we believe that they are the most suitable for wireless sensor networks. Note that all implementations follow closely the protocol descriptions, are simple and have limited memory requirements.

As a starting point, let us consider the effect of network size on the communication efficiency of the protocols. The total number of transmissions performed by the devices when executing each protocol is shown in Figure 16.2. In this graph, the results for lattice-shaped networks are included. Similar results hold for the random uniform networks considered, see Figures 16.3 and 16.4. The results indicate that the random traversal, agent-based, protocol performs a large number of transmissions compared to the other protocols. This is explained by the fact that while the agent is moving randomly within the network, each device may be visited multiple times until all the network is covered. On the other hand, the other two protocols build a tree-based structure and thus improve

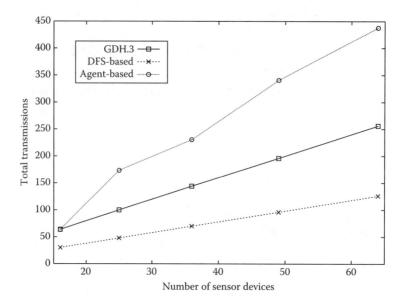

FIGURE 16.2   Total transmissions when executing each protocol, for lattice-shaped networks.

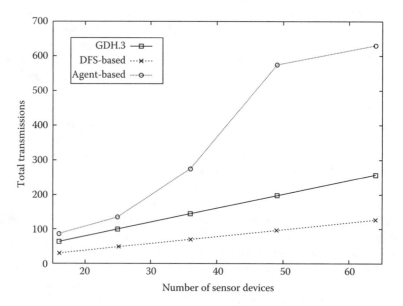

FIGURE 16.3   Total transmissions when executing each protocol, for single-hop networks.

communication as each device is visited only a couple of times. Therefore, as the network increases in size, although the performance of all protocols increases, the random traversal protocol performs significantly more message transmissions than the other.

To have a better view on the energy efficiency of the protocols, consider Figure 16.5 that depicts the power consumed by each device separately. Based on these results, the random traversal protocol consumes more energy than the other two, it manages to evenly distribute energy consumption among the participants. This is a result of the random motion of the agent that does not discriminate devices

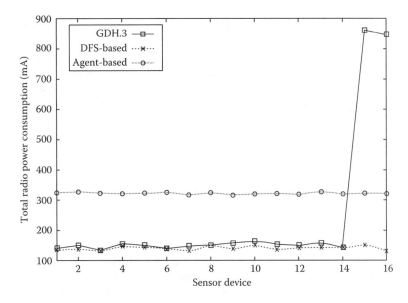

FIGURE 16.4 Power consumption of radio equipment (mA) for each device separately, when executing each protocol, for single-hop networks and $n = 16$.

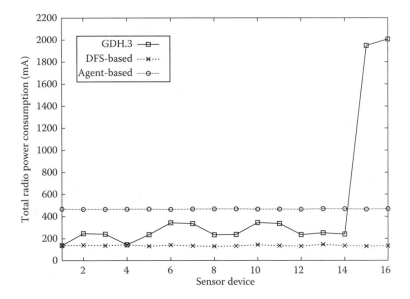

FIGURE 16.5 Power consumption of radio equipment (mA) for each device separately, when executing each protocol, for lattice-shaped networks and $n = 16$.

but more or less visits each device the same number of times (on average). This behavior is similar to the one achieved by the distributed sequential traversal protocol that similarly assigns the same roles in terms of required computations and communication exchanges (energy-wise, the two most demanding events). On the other hand, in the GDH protocol, the energy consumed by devices 15 and 16 is extremely higher than the other devices; these devices may "die" much earlier than the others, especially if the protocol is executed multiple times.

We conclude by evaluating the time efficiency of the protocols. We observe that the time required for a point multiplication (used by all three protocols) dominates the overall execution time. In this sense, because all three protocols require $O(n)$ point multiplication operations, their execution times are more or less the same. However, more importantly we observe that GDH.3 and the distributed sequential traversal protocol operate based on a data structure that reflects the topology of the network, to be correct and establish a common key, the network must remain fixed during the whole protocol execution. So, even for small networks, we must guarantee that during the execution of these protocols, no device enters sleep mode, or is destroyed, and no additional devices are deployed in the network, otherwise the procedure must be repeated leading to further delays and certainly further power consumption.

16.9 CONCLUSIONS AND OPEN ISSUES

This chapter presented five group key agreement protocols which can be used in wireless sensor networks. All protocols studied require that the participants are active throughout the establishment of the common key. It is realistic to consider that some of the devices might stop (due to a failure, or due to capture by an adversary). A future research direction is to come up with distributed protocols that tolerate stopping failures by establishing a common key even if a fraction of the devices fails.

Another characteristic property of all protocols is the sequential traversal of the group members. A future research direction is to parallelize this process and improve the time efficiency of the system. What is more, the approach of using an agent performing a random walk over the participants of the group seems to overcome fault tolerance issues in dynamic environments. On the other hand, the agent process of visiting every member of the group takes a considerable amount of time compared to other protocols. An interesting research direction will be the improvement of the protocol's time efficiency using a different methodology or more agents.

REFERENCES

[1] Crossbow Technology, Inc. MICA2 notes. http://www.xbow.com/Products/productsdetails.aspx?sid= 72, 2005.

[2] I. Akyildiz, Y. Sankarasubramaniam, W. Su, and E. Cayirci, Wireless sensor networks: A Survey, *Journal of Computer Networks*, 38:393–422, 2002.

[3] Y. Amir, Y. Kim, C. Nita-Rotaru, and G. Tsudik, On the performance of group key agreement protocols, *ACM Transactions on Information and System Security*, 7(3):457–488, 2004.

[4] E. Anton and O. Duarte, Performance analysis of group key establishment protocols in ad hoc networks, Technical report GTA-03-06, Universidade Federal do Rio de Janeiro, Brazil, 2006.

[5] G. Ateniese, M. Steiner, and G. Tsudik, Determining the optimal configuration for the zone routing protocol, *IEEE Journal on Selected Areas in Communications*, 18(4):1–13, 2000.

[6] H. Attiya and J. Welch, *Distributed Computing: Fundamentals, Simulations and Advanced Topics*, McGraw-Hill, Berkshire, U.K., 1998.

[7] K. Becker and U. Wille, Communication complexity of group key distribution, *5th ACM Conference on Computer and Communications Security (CCS 1998)*, pp. 1–6. ACM Press, San Francisco, CA, 1998.

[8] I. Blake, G. Seroussi, and N. Smart, Elliptic curves in cryptography, Technical report, London Mathematical Society Lecture Note Series 265, Cambridge University Press, 1999.

[9] A. Boukerche and S. Nikoletseas, Chapter 2: Protocols for data propagation in wireless sensor networks: A survey, in *Wireless Communications Systems and Networks*, Kluwer Academic Publishers, New York, 2004.

[10] E. Bresson, O. Chevassut, and D. Pointcheval, Dynamic group Diffie–Hellman key exchange under standard assumptions, *Eurocrypt 2001*, pp. 321–336. Springer-Verlag, Lecture Notes in Computer Science 2332, Innsbruck, Austria, 2002.

[11] E. Bresson, O. Chevassut, D. Pointcheval, and J. Quisquater, Provably authenticated group Diffie–Hellman key exchange, *ACM-CCS 2001*, pp. 255–264. ACM Press, New York, 2001.

[12] N. Bulusu, D. Estrin, L. Girod, and J. Heidemann, Scalable coordination for wireless sensor networks: Self-configuring localization systems, *International Symposium on Communication Theory and Applications (ISCTA 2001)*, Ambleside, U.K., 2001.

[13] M. Burmester and Y. Desmedt, Efficient and secure conference-key distribution, *Security Protocols Workshop*, pp. 119–129. Springer-Verlag, Lecture Notes in Computer Science 1189, Cambridge, U.K., 1996.

[14] M. Burmester and Y. Desmedt, "A secure and efficient conference key distribution system", Advances in Cryptology (EUROCRYPT 1994), pages 275–286. Springer-Verlag, Lecture Notes in Computer Science 950, Perugia, Italy, 1994.

[15] M. Burmester and Y. Desmedt, A secure and scalable group key exchange system, *Information Processing Letters*, 94(3):137–143, 2005.

[16] D. Burton, *Elementary Number Theory*, McGraw-Hill, 4th edn., Berkshire, U.K., 1998.

[17] I. Chatzigiannakis, A. Kinalis, and S. Nikoletseas, Wireless sensor networks protocols for efficient collision avoidance in multi-path data propagation, *ACM Workshop on Performance Evaluation of Wireless Ad Hoc Sensor, and Ubiquitous Networks (PE-WASUN 2004)*, pp. 8–16, Montreal, Canada, 2004.

[18] I. Chatzigiannakis, E. Konstantinou, V. Liagkou, and P. Spirakis, Design, analysis and performance evaluation of group key establishment in wireless sensor networks, *ICALP Workshop on Cryptography for Ad hoc Networks—WCAN, Electronic Notes in Theoretical Computer Science*, Vol. 2, Venice, Italy, 2006.

[19] I. Chatzigiannakis, E. Konstantinou, V. Liagkou, and P. Spirakis, Agent-based distributed group key establishment in wireless sensor networks, *Proceedings of the 3rd IEEE International Workshop on Trust, Security and Privacy for Ubiquitous Computing–TSPUC 2007*, Helsinki, Finland, 2007. ISBN 1-4244-0992-6.

[20] K. Y. Choi, J. Y. Hwang, and D. H. Lee, Efficient ID-based group key agreement with bilinear maps, *Public Key Cryptography—PKC' 04*, pp. 130–144. Lecture Notes in Computer Science 2947, Singapore, 2004.

[21] W. Diffie and M. Hellman, New directions in cryptography, *IEEE Transactions on Information Theory*, 22:644–654, 1976.

[22] T. Dimitriou, Securing communication trees in sensor networks, *2nd International Workshop on Algorithmic Aspects of Wireless Sensor Networks (ALGOSENSORS 2006)*. Springer-Verlag, Lecture Notes in Computer Science, Turku, Finland, 2006.

[23] R. Dutta and R. Barua, Constant round dynamic group key agreement, *8th International Conference in Information Security—ISC' 05*, pp. 74–88. Lecture Notes in Computer Science 3650, Singapore, 2005.

[24] T. ElGamal, A public key cryptosystem and a signature scheme based on discrete logarithms, *IEEE Transactions on Information Theory*, 31:469–472, 1985.

[25] G. Gaubatz, J. Kaps, and B. Sunar, Public key cryptography in sensor networks—revisited, *1st European Workshop on Security in Ad-Hoc and Sensor Networks (ESAS 2004)*, pp. 2–18. Springer-Verlag, Lecture Notes in Computer Science 3313, Heidelberg, Germany, 2004.

[26] N. Gura, A. Pate, A. Wander, H. Eberle, and S. Shantz, Comparing elliptic curve cryptography and RSA on 8-bit CPUs, *Cryptographic Hardware and Embedded Systems (CHES 2004)*, pp. 119–132. Springer-Verlag, Lecture Notes in Computer Science 3156, Cambridge, MA, 2004.

[27] L. Harn and T. Kiesler, Authenticated group key distribution scheme for a large distributed network, *IEEE Symposium on Security and Privacy*, pp. 300–309, 1989.

[28] W. R. Heinzelman, A. Chandrakasan, and H. Balakrishnan, Energy-efficient communication protocol for wireless microsensor networks, *33rd IEEE Hawaii International Conference on System Sciences (HICSS 2000)*, p. 8020, Maui, Hawaii, 2000.

[29] I. Ingemarsson, D. Tang, and C. Wong, A conference key distribution system, *IEEE Transactions on Information Theory*, 28:714–720, 1982.

[30] J. Katz and M. Yung, Scalable Protocols for authenticated group key exchange, *Advances in Cryptology—CRYPTO 2003*, pp. 110–125. Lecture Notes in Computer Science 2729, Santa Barbara, CA, 2003.

[31] Y. Kim, A. Perrig, and G. Tsudik, Tree-based group key agreement, *ACM Transactions on Information and System Security*, 7(1):60–96, 2004.

[32] P. Leone, L. Moraru, O. Powell, and J. Rolim, A localization algorithm for wireless ad-hoc sensor networks with traffic overhead minimization by emission inhibition, *2nd International Workshop on Algorithmic Aspects of Wireless Sensor Networks (ALGOSENSORS 2006)*. Springer-Verlag, Lecture Notes in Computer Science, Turku, Finland, 2006.

[33] P. Leone, S. Nikoletseas, and J. Rolim, An adaptive blind algorithm for energy balanced data propagation in wireless sensor networks, *1st IEEE/ACM International Conference on Distributed Computing in Sensor Systems (DCOSS 2005)*, pp. 35–48. Springer-Verlag, Lecture Notes in Computer Science 3650, Marina del Rey, CA, 2005.

[34] P. Levis, N. Lee, M. Welsh, and D. Culler, Tossim: Accurate and scalable simulation of entire tinyos applications, *1st ACM International Conference On Embedded Networked Sensor Systems (SENSYS 2003)*, pp. 126–137, Los Angeles, CA, 2003.

[35] L. Liao, Group key agreement for ad hoc networks, Master's thesis, Ruhr-University Bochum, Germany, 2005.

[36] D. Malan, M. Welsh, and M. Smith, A public-key infrastructure for key distribution in tinyos based on elliptic curve cryptography, *1st IEEE International Conference on Sensor and Ad Hoc Communications and Networks (SECON 2004)*, Santa Clara, CA, pp. 71–80, 2004.

[37] N. Malpani, J. Welch, and N. Vaidya, Leader election algorithms for mobile ad hoc networks, *3rd ACM Annual Symposium on Discrete Algorithms and Models for Mobility (DIALM 2000)*, pp. 96–103, Boston, MA, 2000.

[38] M. Manulis, Contributory group key agreement protocols, revisited for mobile ad-hoc groups, *Proceedings of the 2nd IEEE International Conference on Mobile Ad-hoc and Sensor Systems—MASS'05*, pp. 811–818, Washington, DC, 2005.

[39] R. Motwani and P. Raghavan, *Randomized Algorithms*, Cambridge University Press, U.K., 1995.

[40] S. M. Ross, *Stochastic Processes*, John Wiley & Sons, New York, 1996.

[41] V. Shnayder, M. Hempstead, B. Chen, G. Allen, and M. Welsh, Simulating the power consumption of large-scale sensor network applications, *2nd ACM International Conference on Embedded Networked Sensor Systems (SENSYS 2004)*, pp. 188–200, Baltimore, MD, 2004.

[42] J. Silverman, *The Arithmetic of Elliptic Curves*, Springer-Verlag, New York, 1986.

[43] M. Singh and V. Prasanna, Energy-optimal and energy-balanced sorting in a single-hop wireless sensor network, *1st IEEE International Conference on Pervasive Computing and Communications (PERCOM 2003)*, pp. 50–59, Dallas-Fort Worth, TX, 2003.

[44] P. Spirakis and C. Zaroliagis, Distributed algorithm engineering, *Experimental Algorithmics—From Algorithm Design to Robust and Efficient Software*, Springer-Verlag, Berlin, pp. 229–278, 2002.

[45] M. Steiner, G. Tsudik, and M. Waidner, Diffie–Hellman key distribution extended to group communication, *3rd ACM Conference on Computer and Communications Security (CCS 1996)*, pp. 31–37. ACM Press, New Delhi, India, 1996.

17 Malicious Node Detection in Wireless Sensor Networks

Yu Chen, Hao Chen, and Wei-Shinn Ku

CONTENTS

Recent advancements in microelectromechanical systems (MEMS), low-power and highly integrated electronic devices have led to the development and wide application of wireless sensor networks (WSNs) [6,12,13]. Due to the unattended nature of sensor networks, an attacker could launch various attacks and even compromise sensor devices without being detected [4]. Security in WSNs has been one of the hottest topics in research community [2]. A sensor network should be robust against attacks, and if an attack succeeds, its impact should be minimized. In other words, compromising a single sensor node should not crash the entire network. In this chapter, we introduce the malicious node detection problem under the framework of Byzantine General Problem and discuss the boundary conditions to be taken into account. Then, we present the state-of-the-art solutions to this problem and discuss their advantages and constraints in detail. Through this analysis readers can learn the trade-offs and key points in designing security solutions for WSN. This chapter concludes by identifying some open research issues and the research directions in the near future.

17.1   INTRODUCTION

In contrast to traditional wireless networks, special security and performance issues have to be carefully considered for sensor networks [14,15]. WSNs are often deployed in a hostile environment and work without human supervision; any individual node could be easily compromised or destroyed by the adversary. It is critical to detect and isolate the misbehaved nodes, such as compromised nodes or out-of-function nodes, to avoid being misled by the falsified information injected by the adversary or being disturbed by garbage data. In fact, the problem of coping with failure of one or more components in a computing system has been addressed abstractly as Byzantine General Problem [10]. For example, in cognitive radio network, Byzantine problem in spectrum sensing is also investigated [3]. However, the traditional solutions cannot be implemented in WSNs directly because of the constraints that arise from battery lifetime, smaller memory space, and limited computing capability [16,20,21]. Special security and performance issues have to be carefully considered for sensor networks [9,17].

In this chapter, the state-of-the-art for malicious node detection techniques in WSNs is presented. First the classic Byzantine General Problem is analyzed. Then, we introduce four novel malicious nodes detection schemes in detail: (1) trust-node based solution, (2) locationized anomaly detection (LAD), (3) signal strength-based solution, and (4) weighted-trust evaluation (WTE) based solution. Finally, we wrap up this chapter with a discussion of open research issues.

17.2   BYZANTINE GENERAL PROBLEM: WHO IS THE TRAITOR?

Considering a reliable distributed system, it is desired to be able to handle the failure of one or more of its components. The failure parts often behave in a way the system is not expected to do, for example, out of function part may send information that is conflicting to other normal working peers. The problem of handling this type of failure is expressed as the Byzantine General Problem [10], and the problem it tries to address is very close to the question of malicious nodes detection in WSNs.

Suppose that several divisions of the Byzantine army are preparing to attack an enemy city, each division is under the command of its own general. The generals can communicate with each other by messengers. After observing the enemy, the generals must reach a common plan for the next action. However, there are some generals who are traitors. They would like to prevent the loyal generals from achieving an agreement. Therefore, those loyal generals need an approach to guarantee that

- All loyal generals reach the same agreement no matter what the traitors do.
- Small number of traitors should not be able to mislead the loyal generals to adopt a wrong plan.

Although the Byzantine General Problem looks simple, the fact indicates it is difficult. Earlier research has pointed out that if each general can only send an oral message, in order to cope with m traitors at least $3m + 1$ generals are required [10]. Here, it is the traitor's ability to lie that makes Byzantine General Problem so difficult. However, if we allow each general to send unforgeable signed message, as long as the number of loyal generals is larger, a correct decision will be achieved. This implies that general information security techniques are helpful in solving the Byzantine General Problem, such as information encryption, digital signature, authentication, etc.

In WSNs, due to the constraints of computing power, storage memory space, and limited battery life, the general solutions cannot be adopted directly. At least a light-weighted version of these secure schemes is desired. However, some of the complexity is mandatory to prevent them from being broken easily. Hence, the simplification may imply that a weakened robustness can be achieved. Then, considering the oral message only scenario in Byzantine General Problem, we need more healthy nodes to fight against malicious nodes. Obviously, a tradeoff between the cost and the robustness need to be considered. In this chapter, we introduce four different solutions. Among them, the tradeoff is clearly presented.

17.3 TRUST NODE-BASED SOLUTION

Luo et al. [11] indicated that WSNs do not have a clear defense line due to its infrastructureless nature. Therefore, each individual node needs to be prepared for encounters with an adversary. A self-securing approach based on the collaboration among multiple trust nodes is proposed, which is able to provide authentication services to each node in the network.

17.3.1 SYSTEM MODELS

Considering the constraints existing both in the hardware and software configurations in WSNs, the trust-node based solution is proposed to be deployed in such a network:

- Communication channel: Bandwidth-constrained, error-prone, and insecure.
- Network topology: A dynamic network topology with n nodes, where the number n may keep changing as nodes join, leave, or die.
- Infrastructureless: Neither physical nor logical infrastructure support, no assured reliability in multihop packet forwarding.
- Node identifier: Each node has a unique nonzero ID, and has an approach to find its immediate (one hop) neighbors.
- Network degree: Each node has at least k one-hop neighbors.
- Mobility: Each node can move with a maximum speed as S_{max}.

Another important assumption is that each node is equipped with some local detection schemes to detect the misbehaviors among its immediate neighbors. Where the rationale is that although intrusion detection in WSN is more difficult than in wired networks, it is easier and practical to detect misbehaviors among one-hop peers.

This scheme is designed to address two types of attacks: the denial-of-service (DoS) attacks and node break-ins. For the convenience, we categorize the adversaries in two models as proposed in Ref. [11]:

- Model I: The adversary cannot gain access or control over k or more nodes during the whole lifetime of the network.
- Model II: The lifetime of the network is divided into intervals of length T, the adversary cannot gain access or control over k or more nodes during any time interval T.

17.3.2 LOCALIZED TRUST MODEL

In this localized trust model, an entity is considered trustful if any k trusted entities claim so within a certain time T_{cert}. Considering the unreliable communication channel and the mobility of the nodes, typically these k entities are chosen among the entities' immediate neighbors. This implies that as long as the node is trusted by its local community, it is trusted globally. Meanwhile, T_{cert} describes the time-varying characteristics of a trust relationship.

Therefore, how to choose the two parameters k and T_{cert} is critical. To some degree, k describes the strength of trustiness, it could be set in two different ways. The first is to set k as a global constant, it functions as a threshold used by each individual node across the network. The second option is to set k as a location-dependent variable. Obviously, the former approach is simple and easy to system administrator, however, it cannot reflect the variation of sensor topology, i.e., density. The second option gives flexibility, however, it would lead to an inconsistency in the trust criteria across the whole network. In Ref. [18] the first option is adopted: if the number of neighbors of a node is less than k, the authors allow it to move to a larger community or wait for a new node to join.

An extended polynomial secret-sharing technique is proposed, which is more scalable and robust to fight against the adversaries of model II. This security architecture is essentially built using RSA-based public key pairs. Besides the system key pair, each node v_i also holds an individual RSA key pair. To certify its personal keys, each node needs to hold a certificate $cert_i$, which indicates that the key pair node v_i holds during the time interval $[t, t + T]$. The certificate is valid only if it is signed by system secret key.

In this architecture, nodes without valid certificate will be denied from access to any network resources. While a node moves to a new position, it exchanges its certificate with new neighbors. Authenticated neighboring nodes help each other handle packets and monitor peers to detect intruders. In addition, the certificate of an individual node needs to be updated periodically with a new one issued by k immediate neighbors; each of the neighbors issue a partial certificate if they believe that the updating request is from a well-behaved node. This implies that each node can obtain a new certificate only if it is trusted by at least k neighbors. Otherwise, it will be isolated from the whole network. In turn, a validate certificate represents the trust from a coalition of k nodes.

17.3.3 DISCUSSIONS

Performance of this scheme has been verified through both network simulator ns-2 and UNIX platform. Four metrics are used to evaluate the performance:

- *Success ratio*: The number of successful certification services over the total number of requests.
- *Average delay*: The average latency for a node to request a certification service.
- *Overhead*: Communication overhead in bytes.
- *Convergence time*: The time required to complete a shared update.

The experimental results achieved a success ratio over 95 percent as the network size grows from 30 to 100 nodes, and the delay remains almost constant as size of the network grows. This implies a very nice scalability. The only issue that impacts the success ratio is the density of sensor node in the network. The low-node density makes a node difficult to get k neighbors to collect enough partial certificate updates. The overhead grows linearly as the node number increases.

Although the performance of the solution is satisfactory in simulation experiments, computation power of each individual node is critical to the performance. As an RSA based scheme integrated public key technique and polynomial secret sharing method, it is robust enough to most intrusion detection or finding malicious nodes. However, it is not suitable to WSNs. In the experiments, the processing power required is as much as that of Pentium 75, PentiumII 300, or PentiumIII 500. This is far beyond the embedded processor used in any normal sensor node currently in use.

17.4 LAD: LOCALIZATION ANOMALY DETECTION

It will be of great importance if sensors can discover whether their derived location is correct or not. Discovering the correct locations of sensors plays a critical role in this kind of WSN applications. Though beacon scheme is the most popular scheme being applied for localization in WSN, this section proposes one of beaconless scheme deployment knowledge-based scheme which can more efficiently detect localization anomalies. By leveraging the methodologies from the intrusion detection field, a scheme called localization anomaly detection (LAD) has been defined [5].

17.4.1 MODELING OF THE DEPLOYMENT KNOWLEDGE

In the context of model analysis, we assume that sensor nodes are static once they are deployed. Two points are essential to each sensor node: deployment point and resident point. The deployment point

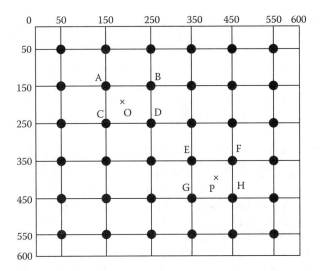

FIGURE 17.1 Illustration of grid deployment.

is defined as the location where this sensor is to be deployed, but this is not the location where this sensor finally resides. The location where the sensor finally resides is defined as resident point.

In practice, it is quite common that nodes are deployed in groups, i.e., a group of sensors are deployed at a single deployment point, and the probability distribution functions (PDFs) of the final resident points of all the sensors from the same group are the same. For example, N sensor nodes in a group-based deployment are going to be deployed. They are divided into n equal-size groups so that each group G_i, for $i = 1, 2, \ldots, n$ is deployed to the deployment point with index i. For simplicity, G_i can also be used to represent the deployment point, and (x_i, y_i) represents its coordinates. These N deployment points are arranged in a grid, Figure 17.1 illustrates such an example. During deployment, the resident point of node k in group G_i follows the Gaussian distribution:

$$f_k^i(x, y | k \in G_i) = f(x - x_i, y - y_i) \tag{17.1}$$

Assume that the probability that a node from group G_i can land at a location l distance from the deployment point of G_i follows a Gaussian distribution. That is

$$f_R(l | n_i \in G_i) = \frac{1}{2\pi^2} e^{-\frac{l^2}{2\sigma^2}} \tag{17.2}$$

where
  R is the wireless transmission range
  σ is the standard deviation of the Gaussian distribution

Similarly, an important concept of $g(z)$ is introduced.

$g(z | n_i \in G_i)$ is defined as the probability that the sensor node n_i from group i resides within the neighborhood of a sensor that is z distance from the deployment point of group G. Based on geometry knowledge, following formula can be derived:

$$g(z | n_i \in G_i) = 1\{z < R\} \left[1 - e^{-\frac{(R-z)^2}{2\sigma^2}} \right] + \int_{|z-R|}^{z+R} f_R(l | n_i \in G_i) \cdot 2l \cos^{-1}\left(\frac{l^2 + z^2 - R^2}{2lz} \right) dl \tag{17.3}$$

Where $1\{\cdot\}$ is the set indicator function: the value of $1\{\cdot\}$ is 1 when the evaluated condition is true, and 0 otherwise. For simplicity, $g(z)$ will be used to represent $g(z|n_i \in G_i)$ in the rest of literature. Due to complexity of the equation above, it is impossible to compute $g(z)$ in sensor networks. Instead, precalculated values of $g(z)$ will be stored in memory for looking up, and this operation takes constant time. More specifically, the range of z will be divided into ω equal-size sub-ranges, and store the $g(z)$ values for these $\omega + 1$ dividing points into a table for further simplification. In order to meet the level of accuracy, ω needs not to be too large.

17.4.2 LOCALIZATION ANOMALY DETECTION PROBLEM

Actually, the analysis of LAD problem can be decomposed into two consecutive phases: phase 1 is called the localization phase where sensors derive their locations, phase 2 is called the detection phase where sensors verify whether the derived locations are correct or not. Since the technology presented here is only focusing on the phase 2, a piecewise analysis will be efficient. An assumption is made to support the analysis in phase 2. The assumption is that all sensors have already derived a location in localization phase, and ready for further consistence inspection.

The key for consistence lies on the distance between actual point and derived point of a sensor node. In the following, $|L_1 - L_2|$ will be used to represent the distance between two locations L_1 and L_2. Three related concepts will be introduced for gradual approach of detection issue. The localization error is defined as the distance between L_e and L_a, where $L_e = (x_e, y_e)$ represents the location that a sensor node derived via certain localization scheme in phase 1, and $L_a = (x_a, y_a)$ represents the actual location of sensor node detected in phase 2.

Due to the cross talk or other disadvantage, localization errors are inevitable. Real sensor network applications should have capabilities to tolerate these errors. The maximum tolerable error (MTE) is the value of error that a sensor network can tolerate, and it is application dependent. An anomaly is defined as a phenomenon where the localization error is greater than the MTE, i.e., $|L_e - L_a| >$ MTE. The great such error is, the more successful the attacks are. An attack causing $|L_e - L_a| = 120$ leads to more severe damage than an attack that causes $|L_e - L_a| = 60$.

Finally, an anomaly called D-anomaly is introduced, it can be represented as: $|L_e - L_a| > D$, where D is called the Degree of Damage, and it is chosen by attackers based on their targeted errors. However, the actual location L_a is not observable, which leads the comparison between $|L_1 - L_2|$ and D is impossible. Up to this step, the rest things to do for anomaly detection have been pretty clear, which include developing an observable metric A and its corresponding threshold (called the detection threshold). Theoretically, a metric A should satisfy the property that metric A is larger than the detection threshold if and only if $|L_e - L_a| > D$. Unfortunately, such a metric, if exists at all, is difficult to find. Finally, such kind of metric is turned to the heuristic way trying to satisfy the above property as much as possible. The side effects of heuristic approach is the introduction of false positive or false negative. A good metric A should be able to minimize both of them.

17.4.3 DETECTING LOCALIZATION ANOMALIES USING DEPLOYMENT KNOWLEDGE

A concrete example can well describe the key idea of LAD scheme based on deployment knowledge. Assuming a WSN is well deployed in the area as shown in Figure 17.1. Sensor v is actually at the location O, but its derived location from phase 1 indicates that it locates at P due to certain adversaries. According to its accrual location, v is supposed to observe many neighbors from groups A, B, C, or D, other than observing neighbors from groups E, F, G, or H around point P. The farther apart O and P are, the more different their observations are. Because v's observation at actual location O is known and v's expected observation at P can be calculated using the deployment knowledge, the consistency can be easily obtained by comparing both observations.

Theoretically, each sensor broadcasts its group ID to its neighbors and counts the numbers of neighbors from G_i, for $(i = 1, \ldots, n)$ after sensors are deployed. Assume a sensor finds out that it has $O_1, \ldots, O_n$ neighbors from groups $G_1, \ldots, G_n$, which can be represented as $O = (O_1, \ldots, O_n)$.

Due to different locations, the observations of sensors are variable. This property can be used for consistency verification. On the other hand, a sensor can derive the expected observations and the likelihood of its actual observations, based on the estimated location $L_e = (x_e, y_e)$ and the deployment knowledge. Three metrics have been proposed to evaluate the efficiency of anomaly detection.

- Difference metric: Let $L_e = (x_e, y_e)$ be v's estimated location that has been derived using certain localization scheme in phase 1 and $O = (O_1, \ldots, O_n)$ be v's actual observation. V's expected observation can be defined as $\mu = (\mu_1, \ldots, \mu_n)$ based on location L_e. μ_i can be computed using the equation

$$\mu_i = m \cdot g_i(L_e) = g(\sqrt{(x_e - x_i)^2 + (y_e - y_i)^2}) \qquad (17.4)$$

  Using the difference between the expected observation μ and the actual observation O, the DM is defined as

$$DM = \sum_{i=1}^{n} |o_i - \mu_i| \qquad (17.5)$$

- Add-all metric: This metric utilizes the property of union for evaluation. The union of the observations o and μ is defined as $t = (t_1, \ldots, t_n)$, where t_i is defined as

$$t_i = \max(o_i, \mu_i) \quad \text{for } (i = 1, \ldots, n) \qquad (17.6)$$

  The rationale of Add-all metric is that the more different O and P are, the higher the value of $|t|$. It can be defined as

$$AM = \sum_{i=1}^{n} \max(o_i, \mu_i) \qquad (17.7)$$

- Probability metric : When a sensor node sees o_i sensors from group i, based on its estimated location, the sensor can calculate how likely it can have o_i neighbors from group i. If the probability is too small, it indicates a potential anomaly. Given the number m of nodes deployed in each group and the PDF function of the deployment, the probability that exactly o_i nodes from group G_i can be observed by v. The probability that the node v at the location L_e has exactly o_i neighbors from group i can be computed using the following equation:

$$PM = P_r(X_i = o_i | L_e) = \binom{m}{o_i} (g_i(L_e))^{o_i} (1 - g_i(L_e))^{(m-o_i)} \qquad (17.8)$$

  If any of the $P_r(X_i = o_i | L_e)$, for $i = 1, \ldots, n$ is less than a threshold, the detection will raise an alarm for the indication of abnormal.

The threshold value selection is very important. Due to the lack of actual network experiments, in Ref. [5] some simulation data is collected to train the system, then derives the detection threshold from the training results.

The simulation results have also shown that the LAD scheme is more effective in detecting attacks with higher degree of damage, and the difference metric performs the best among the three metrics.

17.5 SIGNAL STRENGTH-BASED SOLUTION

Pires Junior et al. [8] proposed to identify malicious nodes by detecting the inconsistency between the signal strength and the location of the sender. By comparing the received signal strength and calculated value, this mechanism is capable of addressing the Hello flooding and wormhole attacks. This section describes the principle and the protocol for detection of suspicious nodes.

17.5.1   System Model

This solution is proposed to be used in static homogeneous WSNs, in which all the nodes are of same hardware and software configuration, and the communication is symmetric. In addition, all nodes are uniquely identified and are aware of their own geographic position. For the purpose of authentication, the position information and individual identifier are included in each message. It is also assumed that each message is encrypted.

Regarding the strength of signals propagating in the network, it follows a well-defined model to describe the received signal power as a function of the distance between the receiver and the sender, plus some other parameters. For instance, the Two-Ray Ground propagation model [7] assumes that a signal from the sender does not arrive at the receiver through a unique straight line, but eventually also through a reflection in the ground:

$$P_\text{r} = \frac{P_\text{t} \times G_\text{t} \times G_\text{r} \times h_\text{t}^2 \times h_\text{r}^2}{d^4 \times L} \tag{17.9}$$

where
   P_r is the received signal power (Watt)
   P_t is the transmission power (Watt)
   G_t is the transmitter antenna gain
   G_r is the receiver antenna gain
   h_t is the transmitter antenna height (meter)
   h_r is the receiver antenna height (meter)
   d is the distance between the transmitter and the receiver (meter)
   L is the system loss (constant)

Only when the arriving signal power is higher than a preset threshold P_m the signal is detectable by the receiving node.

Therefore, when any node in the WSN received a message, which carries the sender's identifier and the position information, the receiver will compare the actual signal strength with the theoretical signal power calculated using the position information carried in the message. The mismatch implies the message is suspicious. When multiple nodes report suspicious against the same source, the sender is marked as malicious.

More concretely, a node can place a received message into one of the two categories: suspicious or unsuspicious, depending on whether the node thinks the transmission is malicious.

17.5.2   Suspicious Node Detection

Based on the model mentioned above, each individual node in the WSN is capable of detecting suspicious peers by comparing two values when it hears any communication. The first is the calculated signal strength, which is obtained according to the position and identifier of the sender, the antenna parameters, etc. The second is the actual signal strength received by the receiver.

If there is no anomaly behavior, these two values should be close to equal. However, if there is malicious node(s), and a Hello flooding or wormhole attack is undergoing, an obvious difference would be observed.

Taking advantage of this fact, a malicious node detection scheme is suggested and a protocol has been designed to specify the implementation of this detection. In this scheme, each node maintains a status table, in which the "reputation" of peers in the WSN is recorded. The reputation of each individual peer is stored in a table entry indexed using its unique ID, and it is described using the number of times this peer had been voted as "suspicious" and the number of times as "unsuspicious."

Everytime the suspiciousness of a received message has been checked, the receiver updates the local reputation table accordingly. If the difference between the actual signal strength and the calculated signal strength beyond certain preset threshold, the "suspicious" counter of the sender is increased by one; otherwise, the "unsuspicious" counter is increased. A suspicious node information dissemination protocol (SNIDP) has been proposed to specify the further actions each node need to take based on these two counters.

The operations specified by SNIDP are as follows:

- On receiving a suspicious message, the receiver (node A) broadcasts the ID of the sender (node S) to its neighbors, telling them that node S is suspicious.
- On receiving the broadcasting query from node A, if the neighbors of node A (say, node B) have also received the message from node S, Node B will respond according to the current status of its local reputation table. If the count of suspicious is larger than the count of unsuspicious, node B responds with suspicious; otherwise, responds with unsuspicious.
- Node A collects all the replies and updates its reputation table accordingly: For each voting of suspicious it receives, node A increases its local suspicious counter for node S by one; same operation is done for each unsuspicious voting.

There are some interesting issues worthy of further discussion regarding the design of this simple protocol. First of all, note that the node B's response is also a broadcast. This implies that all its neighbors will receive the responding message. Some of B's neighbors are neither node A's neighbor nor node S's neighbor. They will also update their table accordingly. One question is, does this incur a flooding in the network? In Ref. [8], authors specified that the SNIDP is executed only when a suspicious message is detected to avoid unnecessary overhead.

The second concern is why do we need the voting procedure? Why cannot node A make decision directly? The argument is to avoid being fooled by malicious node who disseminates false information against a legal node, which may potentially lead to denial-of-service attacks against a regular node. The multiple voting makes the scheme more robust against such type of lie.

17.5.3 Performance Evaluation

The performance of this scheme has been evaluated through simulation experiments [8]. Two different scenarios have been considered: unfocused and focused. In an unfocused case, all sensor nodes, including the malicious one, are turned on at the same time and broadcast one Hello message each. Malicious node adopts a different signal power but does not choose any victim purposely. In contrast, in a focused case, malicious nodes start sending message after the network initiate phase and a target victim is chosen. In the second case, the malicious nodes try to avoid being detected by mimicking the behavior of normal nodes and calculating the signal strength it should use accordingly.

The simulation reveals the impact of a set of parameters on the detection rate of this scheme: network density (number of node in a 179×179 m^2 field), transmission power multiplier, maximum ratio difference, message check probability, and SNIDP overhead. The experiment results show that in the unfocused case the detection rate is insensitive to network density; however, the focused attacks can fool some nodes when the density is low (less than 50 nodes). Another observation is lower message check probability leads to a much lower malicious node detection rate under unfocused attacks. Particularly, when the message check probability is less than 0.3, the detection rate drops below 50 percent. However, the change of message check probability has no influence to the detection rate under focused attack scenario. There is no clear explanation why this happens.

The overhead of the protocol is evaluated using the average number of message sent and received. This is highly dependent on the density and size of the network and there is no efficient approach to lower the overhead, particularly in case of a large size network.

17.6   WTE: WEIGHTED TRUST EVALUATION

In this section, we introduce a weighted-trust evaluation (WTE) based scheme [1] that detects the compromised/malicious nodes by monitoring the reported data. It is a light-weighted algorithm with little overhead. Considering the scalability and flexibility, a hierarchical network architecture is adopted. Research work reported in Ref. [4] is similar to WTE, however, it is based on aggregation of sampled data analysis. This requires much more computational power, which leads to higher overhead and more power consumption.

17.6.1   NETWORK ARCHITECTURE

Figure 17.2 presents the network architecture in which the WTE scheme is implemented. It is a three-layer hierarchical network architecture consists of three types of sensor nodes, which are similar to the architecture adopted by Zhao et al. [19]:

- Low-power common sensor node (SN) with limited functionality
- High-power forwarding nodes (FNs), which forward the data obtained by sensor nodes to upper layer
- Access point (AP) or base station (BS), which route data from the WSN into the wired network infrastructure

   In contrast to sensor nodes in flat ad hoc sensor networks, sensor nodes in the lowest layer of this hierarchical network do not offer multihop routing capability to its neighbors. A number of SNs are organized as a group and controlled by a higher layer node, the FN. Therefore, each sensor node only communicates with its FN and provides information such as sensor reading to its FN. FNs are located in the second layer on top of the sensor node layer and offer multihop routing capability to SNs or other FNs.

   Each FN has two wireless interfaces, one communicates with lower layer nodes (SNs) which belong to its management and the other connects to higher layer nodes: access points (APs). APs are located in the highest layer in the network, and have both wireless and wired interfaces. APs provide multihop routing for packets from SNs and FNs within radio range, in addition, APs route data to the wired network. APs also forward control information from wired network to FNs and SNs.

   This hierarchical network can also be considered as a distributed information aggregation system. SNs gather information and report to its FN. Based on the information collected from SNs, FNs

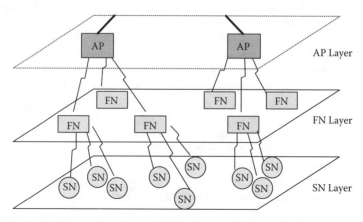

FIGURE 17.2   Hierarchical network architecture.

compute the aggregation result and commit the information to APs. However, since SNs may be compromised and report fake information, it is important for FNs to verify the correctness of the information. Similarly, it is also desired that APs possess the ability of verifying the committed information.

17.6.2 MALICIOUS NODE DETECTION BASED ON WTE

As mentioned earlier, sensor nodes in sensor networks are usually deployed in hostile environments such as battlefield, thus a sensor node may be compromised or out of function, and then provides wrong information that may mislead the whole network. This problem is called as the Byzantine problem. For example, a compromised sensor node (malicious node) can constantly report incorrect information to higher layers. The aggregator (FN or AP) in higher layer may make a wrong aggregation result due to the effect of the malicious node. It is therefore an important issue in sensor networks to detect malicious nodes in spite of such Byzantine problem.

As the first step toward the solution to the problem, we model it into a weight-based network as shown in Figure 17.3. The network is adaptive in the architecture between a group of SNs and its FN. As shown in the figure, a weight W is assigned to each sensor node. The FN collects all information provided by SNs and calculates aggregation result using weight assigned to each SN:

$$E = \sum_{n=0}^{N} W_n \times U_n \tag{17.10}$$

where
 E is the aggregation result
 W_n is the weight ranging from 0 to 1

A big concern is about the definition of sensor nodes output U_n. In practice, the output information U_n may be false or true information or kind of continuous numbers such as temperature reading. Thus the definition of output U_n is usually depending on the application where the sensor network is used.

The next important issue is to update the weight of each sensor node based on the correctness of information reported. Updating the weight has two purposes. First, if one sensor node is subject to malicious node and frequently sends its report inconsistent with the final decision, its weight is likely to decrease. Then if sensor nodes weight is lower than a specific threshold, we can identify it as a malicious node. Second, the weight decides how much a report may contribute to the final

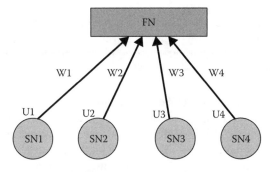

FIGURE 17.3 Weighted aggregation illustration.

decision. This is reasonable since if the report from a sensor node tends to be incorrect, it should be counted less in a final decision.

This logic is reflected in the following equation:

$$W_n = W_n - \theta \times r(n), \quad \text{if } (U_n \neq E) \tag{17.11}$$

where θ is a weight penalty. When a sensor node's output is not consistent with the final result, its weight is reduced by the weight penalty θ. Moreover $r(n)$ is the ratio of the number of the nodes sending different report in its neighbors to the number of all neighbor nodes.

An optimal θ value is essential in the WTE mechanism since it affects the detection time and the accuracy of the algorithm. In addition, due to various definitions of output information U_n mentioned above, the consistence determination which decides whether a nodes output is consistent with final result is also application dependent. For example, it is easy to determine the consistence for a false or true output. However, for a continues number of U_n like temperature reading, the probability distribution function could be used to determine the consistence.

Furthermore, a normalization operation as described in the following equation is used to guarantee the weight kept in the range from 0 to 1.

$$W_n = \frac{W_n}{\max(W_1, \dots, W_n)} \tag{17.12}$$

Based on updated weights, the forwarding node identifies a node as a malicious node if its weight is lower than a specific threshold. This detection algorithm can be widely used in different types of sensor networks. For example, the number of sensor nodes can vary in the algorithm, which makes it suitable for very large and small networks. However, the description of sensor node output and updating scaling factor which are depending on the applied application need to be determined carefully to achieve efficient and high accuracy detection.

17.6.3   WTE PERFORMANCE

Atakli et al. [1] studied the performance of WTE-based malicious node detection scheme through intensive simulation experiments. In the simulation, the detection algorithm is deployed at a forwarding node to monitor all sensor nodes under the control of the forwarding node, and the detection is performed at every cycle which is a basic time unit of the simulation.

Assume that a sensor node is compromised randomly by the attack at a specific probability of every cycle, referred to attack probability, and then this malicious node keeps reporting the opposite information after being compromised. For example, a malicious node always sends alarm although the aggregation result computed from other sensor nodes is no alarm. Meanwhile, a normal sensor node may also send alarm when real alarm occurs. This case also occurs randomly at a different probability referred to alarm probability.

Let's assume that sensor nodes are densely deployed to monitor certain target. In contrast to malicious nodes, if a normal node started sending alarm, its neighbor nodes would also start to send alarm after a short delay time. Furthermore, normal nodes sending with alarm stop sending alarm after a certain cycles. The node which is detected or misdetected as a malicious node is isolated from the whole processing.

Three metrics are defined to evaluate the performance of the detection algorithm. The response time, which is the average detection cycles of correctly detected malicious nodes shows how fast malicious nodes can be detected. The detection ratio, which is the ratio of correctly detected nodes to all malicious nodes. The third measure is misdetection ratio which is the ratio of misdetected nodes to all detected nodes including correctly detected and misdetected nodes. Short response time and high detection ratio are desired as well as low misdetection ratios.

The impact of parameters and several critical design issues such as scalability have been considered. The increasing weight penalty decreases the response time and improves the detection ration. However, the misdetection ratio also increases as weight penalty increasing, especially for 0.08 and above. Considering response time, detection and misdetection ratios comprehensively, the experiment results in Ref. [1] show that the weight penalties in range (0.05–0.1) are optimal values. Experiment also shows that WTE scheme is applicable to both small- and large-sized WSNs due to its good scalability.

17.7 SUMMARY

Although the security problem in WSNs has attracted the attention of many researchers in the network security community, there is still a lots of work to be done to make WSNs accepted widely by the public and be adopted in more application areas. Malicious node detection/isolation is one really desired capability, particularly for WSNs deployed in hostile environments, i.e., battlefield.

In this chapter, we introduced four different approaches to detect malicious nodes in WSNs. Each of them addresses the problem under certain circumstances, there is no silver bullet yet.

These schemes have been verified mostly through simulation, there exists a big question that how effective and correct these solutions could be in reality. Below are impending tasks in the near future:

- Physical testbed construction: Many researchers have pointed out that it is difficult to simulate the power consumption and the behavior of wireless communication channels in WSNs. Obvious differences have been observed between the simulation results and the behavior of a physically deployed WSN in real world. Some researcher has already proposed to construct large scale testbed using real sensor nodes [1].
- Methods discussed in this chapter are only with preliminary results of simulation experiment, more factors that affect the performance of them are to be investigated. Further systematic study against them are necessary before they are mature enough to be deployed in real world.
- As WSN nodes are much vulnerable due to their resource constraints, the security overhead sensor nodes can afford is limited. Therefore, light-weighted design is desired. Researchers are looking for novel secure solutions which are light-weighted but robust enough.

REFERENCES

[1] I. Atakli, H. Hu, Y. Chen, W.-S. Ku, and Z. Su, Malicious node detection in wireless sensor networks using weighted trust evaluation, *The ACM/SIGSIM Symposium on Simulation of System Security (SSSS'08)*, Ottawa, Canada, April 14, 2008.
[2] E. Ayday, F. Delgosha, and F. Fekri, Location-aware security services for wireless sensor networks using network coding, Infocom, May 2007.
[3] R. Chen, J.M. Park, and K. Bian, Robust distributed spectrum sensing in cognitive radio networks, Technical Report TR-ECE-06-07, Department of Electrical and Computer Engineering, Virginia Tech., July 2006.
[4] D.-I. Curiac, O. Banias, F. Dragan, C. Volosencu, and O. Dranga, Malicious node detection in wireless sensor networks using an autoregression technique, *The 3rd International Conference on Networking and Services (ICNS07)*, June 19–25, 2007, Athens, Greece.
[5] W. Du, L. Fang, and P. Ning, LAD: Localization anomaly detection for wireless sensor networks, *The 19th International Parallel and Distributed Priocessing Symposium (IPDPS05)*, April 3–8, 2005, Denver, Colorado.
[6] D. Estrin, R. Govindan, J. Heidemann, and S. Kumar, Next century challenges: Scalable coordination in sensor networks, MOBICOM, August 1999.
[7] Y. Hu, A. Perrig, and D. Johnson, Packet leashes: A defense against wormhole attacks in wireless ad hoc networks, *IEEE INFOCOM'03*, San Francisco, California, March 30–April 3, 2003.

[8] W.R. Pires, Jr., T.H. de Paula Figueriredo, H.C. Wong, and A.A.F. Loureiro, Malicious node detection in wireless sensor networks, *The 18th International Parallel and Distributed Processing Symposium (IPDPS04)*, April 26–30, 2004, Santa Fe, New Mexico.

[9] C. Karlof, N. Sastry, and D. Wagner, TinySec: A link layer security architecture for wireless sensor networks, ACM Sensys, November 2004.

[10] L. Lamport, R. Shostak, and M. Pease, The Byzantine generals problem, *ACM Transactions on Programming Languages and Systems*, 4(3), July 1982.

[11] H. Luo, P. Zerfos, J. Kong, S. Lu, and L. Zhang, Self-securing ad hoc wireless networks, *IEEE ISCC (IEEE Symposium on Computers and Communications) 2002*, Italy.

[12] B. Przydatek, D. Song, and A. Perrig, SIA: Secure information aggregation in sensor networks, *Proceedings of the 1st International Conference on Embedded Networked Sensor Systems*, November 05–07, 2003, Los Angeles, California.

[13] S. D. Servetto, From small sensor networks to sensor networks, EmNets 2006, May 2006.

[14] M. Tubaishat and S. Madria, Sensor networks: An overview, *IEEE Potentials*, 22(2), 20–23, April 2003.

[15] M.A.M. Vieira, D.C. da Silva Jr., C.N. Coelho Jr., and J.M. da Mata, Survey on wireless sensor network devices, Emerging Technologies and Factory Automation (ETFA03), September 2003.

[16] W. Ye, F. Silva, and J. Heidemann, Ultra-low duty cycle MAC with scheduled channel polling, in *Proceedings of the 4th ACM Conference on Embedded Networked Sensor Systems (SenSys)*, November 2006, Boulder, Colorado.

[17] Y. Yu, B. Krishnamachari, and V.K. Prasanna, Energy-latency tradeoffs for data gathering in wireless sensor networks, IEEE Infocom'04, Hong Kong, March 7–11, 2004.

[18] Y. Zhang and W. Lee, Intrusion detection in wireless ad hoc networks, *the sixth Annual International Conference on mobile Computing and networking* (ACM MOBICOM), Boston, Massachusetts, August 6–11, 2000.

[19] S. Zhao, K. Tepe, I. Seskar, and D. Raychaudhuri, Routing protocols for self-organizing hierarchical ad-hoc wireless networks, *Proceedings of the IEEE Sarnoff Symposium*, March 2003, Trenton, New Jersey.

[20] L. Zhou and Z.J. Haas, Securing ad hoc networks, *IEEE Network Special Issue on Network Security*, 13(6), 24–30, November 1999.

[21] S. Zhu, S. Setia, and S. Jajodia, LEAP: Efficient security mechanisms for large-scale distributed sensor networks, CCS'03, October 2003.

18 Jamming in Wireless Sensor Networks

Aristides Mpitziopoulos, Damianos Gavalas,
Charalampos Konstantopoulos,
and Grammati Pantziou

CONTENTS

The objective of this chapter is to provide a general overview of the critical issue of jamming in wireless sensor networks (WSNs) and cover all the relevant work, providing the interested researcher pointers for open research issues in this field. Jamming represents the most serious security threat in the field of WSNs, as can easily put out of order even WSNs that utilize strong high-layer security mechanisms, simply because it is often ignored in the initial WSN design. Law et al. [12] conclude that with typical WSN systems in use today no effective measures against link-layer jamming are possible. This chapter begins with a brief overview of the communication protocols typically used in WSN deployments. The next section highlights the characteristics of contemporary WSNs that make them susceptible to jamming attacks along with the various types of jamming which can be exercised against WSNs. Typical countermeasures against jamming are also detailed. Furthermore, the key ideas of existing security mechanisms against jamming attacks in WSNs are reviewed, focusing on their respective advantages/disadvantages. This chapter concludes by highlighting open research issues with respect to the defence against jamming attacks in sensor networks.

18.1   INTRODUCTION

WSNs [2] are used in many applications which often include the monitoring and recording of sensitive information. Hence, their critical importance raises major security concerns. Jamming is defined as the act of intentionally directing electromagnetic energy toward a communication system to disrupt or prevent signal transmission [1]. In the context of WSNs, jamming is the type of attack which interferes with the radio frequencies used by network nodes [35]. In the event that an attacker uses a rather powerful jamming source, disruptions of WSNs' proper function are likely to occur. As a result, the use of countermeasures against jamming in WSN environments is of immense importance, especially taking into account that WSNs suffer from many constraints including low computation capability, limited memory and energy resources, susceptibility to physical capture, and the use of insecure wireless communication channels.

Jamming attacks may be viewed as a special case of denial-of-service (DoS) attacks. Wood and Stankovic define DoS attack as "any event that diminishes or eliminates a network's capacity to perform its expected function" [41]. Typically, DoS prevents or inhibits the normal use or management of communications through flooding a network with "useless" information. In a jamming attack the radio frequency (RF) signal emitted by the jammer corresponds to the "useless" information received by all sensor nodes. This signal can be white noise or any signal that resembles network traffic.

18.2   COMMUNICATION IN WSNs

A WSN is usually composed of hundreds or even thousands of sensor nodes. These sensor nodes are often randomly deployed in the field and form an infrastructure-less network. Each node is capable of collecting data and routing it back to the processing element (PE) via ad hoc connections with neighbor sensor nodes. A sensor consists of five basic parts: a sensing unit, a central processing unit (CPU), storage unit, a transceiver unit, and a power unit [2]. It may also have additional application-dependent components attached such as location finding system (GPS), mobilizer, and power generator.

18.2.1   COMMUNICATION PROTOCOL STACK

The protocol stack used in sensor nodes contains physical, data link, network, transport, and application layers defined as follows [2]:

- Physical layer: Responsible for frequency selection, carrier frequency generation, signal deflection, data encryption,* and modulation. This is the layer that suffers the most damage from radio jamming attacks.
- Data link layer (DLL): Responsible for the multiplexing of data streams, data frame detection, medium access control (MAC), data encryption, and error control; as well as ensuring reliable point-to-point and point-to-multipoint connections. This layer and more specific MAC are heavily damaged by link-layer jamming. In link-layer jamming [10,12], sophisticated jammers can take advantage of the DLL to achieve energy-efficient jamming. Compared to radio jamming, link-layer jamming offers better energy efficiency.
- Network layer: Responsible for specifying the assignment of addresses and how packets are forwarded.
- Transport layer: Responsible for the reliable transport of packets and data encryption.
- Application layer: Responsible for specifying how the data are requested and provided for both individual sensor nodes and interactions with the end user.

18.2.2 COMMUNICATION PROTOCOLS AND STANDARDS USED IN WSNs

A considerable percentage of the nodes currently used in WSN environments comply with the ZigBee communications protocol. ZigBee protocol minimizes the time the radio functions so as to reduce power consumption. All ZigBee devices are required to comply with the IEEE 802.15.4-2003 or IEEE 802.15.4-2006 Low-Rate Wireless Personal Area Network (WPAN) standard. The standard only specifies the lower protocol layers—the physical layer (PHY) and the MAC portion of the DLL. The standard's specified operation is in the unlicensed 2.4 GHz, 902–928 MHz (North America), and 868 MHz (Europe) ISM (industrial, scientific, and medical) bands. In the 2.4 GHz band, there are 16 ZigBee channels (for both 2003 and 2006 versions of IEEE 802.15.4) with each channel occupying 3 MHz of wireless spectrum and 5 MHz channel spacing. The center frequency for each channel can be calculated as $FC = (2400 + 5 * k)$ MHz, where $k = 1, 2, \ldots, 16$. In 902–928 MHz, there are 10 channels (extended to 30 in 2006) with 2 MHz channel spacing and in 868 MHz 1 channel (extended to 2 in 2006).

The radios use direct-sequence spread spectrum (DSSS) coding in which the transmitted signal takes up more bandwidth than the information signal that is being modulated. In IEEE 802.15.4-2003 two physical layers are specified: BPSK in the 868 and 902–928 MHz and orthogonal O-QPSK that transmits two bits per-symbol in the 2.4 GHz band. The raw, over-the-air data rate is 250 kbit/s per channel in the 2.4 GHz band, 40 kbit/s per channel in 902–928 MHz, and 20 kbit/s per channel in the 868 MHz band. The 2006 revision improves the maximum data rates of the 868/915 MHz bands bringing them up to support 100 and 250 kbit/s as well. Moreover, it defines four physical layers depending on the modulation method used: BPSK and O-QPSK in 868/915 MHz band, O-QPSK in 2.4 GHz band, and a combination of binary keying and amplitude shift keying for 868/915 MHz band. Transmission range for both versions is between 10 and 75 m (33,246 ft), although it is heavily dependent on the particular environment where the nodes are deployed. The maximum output power of the radios is generally 0 dBm (1 mW).

18.3 VULNERABILITIES OF TODAY WSNs THAT MAKE THEM SUSCEPTIBLE TO JAMMING

The above discussion makes clear that a node that follows the IEEE 802.15.4 communications protocol (2003 or 2006 revision) may connect to the network via a limited number of frequencies (16 channels in 2.4 GHz band (2400–2483.5 MHZ), 10 channels (30 for 2006) in 902–928 MHz and 1 channel (3 for 2006) in 868.3 MHz). In addition, taking into account the maximum output power of

---

* Data encryption more commonly is done at the data link or transport layers.

the radio of a node (0 dBm), it becomes apparent that an attacker could easily jam a WSN (with the use of small power output) and disrupt sensor nodes communication. The main limitation of the above-mentioned protocols is that they have not been originally designed taking radio jamming into account. WSN nodes design also presents the same limitation. Other types of widely utilized motes such as Mica-2 [6] are even more susceptible to jamming since they use a limited number of frequencies (support of only lower 868/916 bands and not 2.4 GHz band) for communication. Thus, with typical WSNs in use today is very difficult to take effective measures against jamming, which raises a major security issue. A significant number of research works suggest addressing jamming attacks utilizing hardware used in contemporary WSNs [19,20,41,44–46], while other works suggest new design requirements of nodes that can effectively defend jamming attacks [18]. The main advantage of the former is that the implementation is much cheaper and straightforward and the compatibility with currently available hardware; yet, they cannot easily cope with heavy jamming attacks. On the other hand, suggesting new design requirements for nodes, jamming attacks can be addressed more efficiently. However, the implementation of these nodes requires a significant amount of research, while the associated cost is highly increased. In addition, this approach offers no compatibility with existing hardware.

18.4 DEFINITION OF JAMMING, JAMMING TECHNIQUES, AND TYPES OF JAMMERS

Jamming is defined as the emission of radio signals aiming at disturbing the transceivers' operation [1]. The key point in successful jamming attacks is signal-to-noise ratio (SNR), $SNR = P_{signal}/P_{noise}$ where P is the average power. Noise simply represents the undesirable accidental fluctuation of electromagnetic spectrum collected by the antenna. Jamming can be considered effective if $SNR < 1$. Existing jamming techniques are described below.

18.4.1 SPOT JAMMING

The most popular jamming method is the spot jamming wherein the attacker directs all its transmitting power on a single frequency that the target uses with the same modulation and enough power to override the original signal. Spot jamming is usually very powerful, but since it jams a single frequency each time it may be easily avoided by changing to another frequency.

18.4.2 SWEEP JAMMING

In sweep jamming, a jammer's full power shifts rapidly from one frequency to another. While this method of jamming has the advantage of being able to jam multiple frequencies in quick succession, it does not affect them all at the same time, and thus limits the effectiveness of this type of jamming. However, in a WSN environment, it is likely to cause considerable packet loss and retransmissions and, thereby, consume valuable energy resources.

18.4.3 BARRAGE JAMMING

In barrage jamming, a range of frequencies is jammed at the same time. Its main advantage is that it is able to jam multiple frequencies at once with enough power to decrease the SNR ratio of the enemy receivers. However, as the range of the jammed frequencies grows bigger the output power of the jamming is reduced proportionally.

18.4.4 DECEPTIVE JAMMING

In this type of jamming, the signal emitted by the adversary is not random but resembles network traffic. Deceptive jamming is used when the adversary does not want to reveal her existence.

By flooding the WSN with fake data she can deceive the network's defensive mechanisms (if any) and complete her task without leaving any traces. Deceptive jamming is a very dangerous type of attack as it cannot be easily detected and has the potential to flood the PE with useless or fake data that will mislead the WSN's operator and occupy the available bandwidth used by legitimate nodes.

18.4.5 TYPES OF JAMMERS

There are several types of jammers that may be used against WSNs. Xu et al. in Ref. [45] propose generic jammer models, namely (1) constant jammer, (2) deceptive jammer, (3) random jammer, and (4) reactive jammer.

Constant jammer emits continuous radio signals in the wireless medium (Figure 18.1). The signals that she emits are totally random. They do not follow any underlying MAC protocol and are just random bits. This type of jammer aims at keeping the channel busy and disrupting nodes' communication or causing interference to nodes that have already commenced data transfers and corrupt their packets.

The deceptive jammer uses deceptive jamming techniques (see Section 18.4.4) to attack the WSN (Figure 18.1).

The random jammer sleeps for a random time t_s and jams for a random time t_j (Figure 18.2). The type of jamming used can be of any kind depending on the situation. Also by changing t_s and t_j we can achieve different levels of effectiveness and power saving.

The reactive jammer (see Figure 18.2) listens for activity on the channel, and in case of activity, immediately sends out a random signal to collide with the existing signal on the channel. As a result the transmitted packets of data will be corrupted.

According to Xu et al. [45], the constant jammers, deceptive jammers, and reactive jammers are effective jammers in that they can cause the packet delivery ratio to fall to 0 or almost 0, if they are placed within a suitable distance from the victim nodes. However, these jammers are also energy-inefficient, meaning they would exhaust their energy sooner than their victims would supposed they are energy-constrained. Although random jammers save energy by sleeping, they are less effective. With respect to energy-efficiency, defeating energy-efficient jamming (DEEJAM) [43] presents a reactive design that is relatively energy efficient.

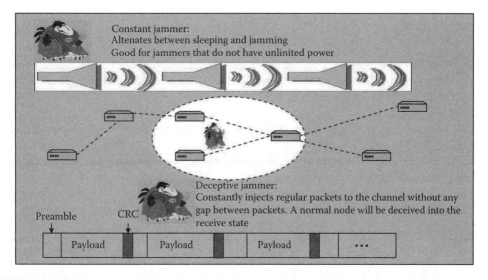

FIGURE 18.1 Constant and deceptive jammers.

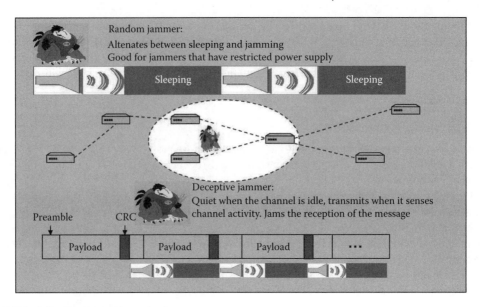

FIGURE 18.2   Random and reactive jammers.

18.5   COUNTERMEASURES AGAINST JAMMING

In this section we present countermeasures that deal with possible radio jamming scenarios.

18.5.1   REGULATED TRANSMITTED POWER

Using low transmitted power the discovery probability from an attacker decreases (an attacker must locate first the target before transmitting jamming signal). Higher transmitted power implies higher resistance against jamming because a stronger jamming signal is needed to overcome the original signal. A considerable percentage of sensor nodes currently used in contemporary WSNs (e.g., Sunspots [37]) possess the capability to change the output power of their transmitter.

18.5.2   FREQUENCY-HOPPING SPREAD SPECTRUM

Frequency-hopping spread spectrum (FHSS) is a spread-spectrum method of transmitting radio signals by rapidly switching a carrier among many frequency channels, using a shared algorithm known both to the transmitter and the receiver. FHSS brings forward many advantages in WSN environments:

- It minimizes unauthorized interception and jamming of radio transmission between the nodes.
- The SNR required for the carrier relative to the background decreases as a wider range of frequencies is used for transmission.
- It deals effectively with the multipath effect.*
- Multiple WSNs can coexist in the same area without causing interference problems.

---

* Multipath in wireless telecommunications is the propagation phenomenon that results in radio signals reaching the receiving antenna through two or more paths due to reflections of the original signal [5,13]. FHSS, when the hop rate is quite fast, eliminates the multipath effect because when the receiver receives the original signal it immediately changes frequency, thus the ghost of the original signal (harmonic signal) is not received at all.

One of the main drawbacks of frequency-hopping is that the overall bandwidth required is much wider than that required to transmit the same data using a single carrier frequency. However, transmission in each frequency lasts for a very limited period of time so the frequency is not occupied for long.

18.5.3 DIRECT-SEQUENCE SPREAD SPECTRUM

DSSS transmissions are performed by multiplying the data (RF carrier) being transmitted and a pseudonoise (PN) digital signal. This PN digital signal is a pseudorandom sequence of 1 and −1 values, at a frequency (chip rate[†]) much higher than that of the original signal. This process causes the RF signal to be replaced with a very wide bandwidth signal with the spectral equivalent of a noise signal; however, this noise can be filtered out at the receiving end to recover the original data, through multiplying the incoming RF signal with the same PN modulated carrier.

The first three of the above-mentioned FHSS advantages also apply to DSSS. Furthermore, the processing applied to the original signal by DSSS makes it difficult to the attacker to descramble the transmitted RF carrier and recover the original signal (the key factor is the chip rate which in 802.15.4 is 2 Mchip/s). Also since the transmitted signal of DSSS resembles white noise, radio direction finding of the transmitting source is a difficult task.

18.5.4 HYBRID FHSS/DSSS

Hybrid FHSS/DSSS communication between WSN nodes represents a promising antijamming measure. In general terms, direct-sequence systems achieve their processing gains through interference attenuation using a wider bandwidth for signal transmission, while frequency-hopping systems through interference avoidance. Consequently, using both these two modulations, resistance to jamming may be highly increased. Also Hybrid FHSS/DSSS compared to standard FHSS or DSSS modulation provides better low-probability-of-detection/low-probability-of-interception (LPD/LPI) properties. Fairly specialized interception equipment is required to mirror the frequency changes uninvited. It is stressed though that both the frequency sequence and the PN code of DSSS should be known to recover the original signal. Thus Hybrid FHSS/DSSS improves the ability to combat the near–far problem which arises in DSSS communications schemes. Another welcome feature is the ability to adapt to a variety of channel problems.

18.5.5 ULTRAWIDE BAND TECHNOLOGY

Ultrawide band (UWB) technology is a modulation technique based on transmitting very short pulses [40] on a large spectrum of a frequency band simultaneously. This renders the transmitted signal very hard to be intercepted/jammed and also resistant to multipath effects. In the context of WSNs, UWB can provide many advantages. The research work of Oppermann et al. [23] promises low power and low-cost wide-deployment of sensor networks. In addition, UWB-based sensor networks guarantee more accurate localization and prolonged battery lifetime. The IEEE standard for UWB, 802.15.3.a, is under development.

18.5.6 ANTENNA POLARIZATION

The polarization of an antenna is the orientation of the electric field of the radio wave with respect to the earths' surface and is determined by the physical structure of the antenna and its orientation. The antenna polarization of a nodes' radio unit plays a significant role in jamming environments. For line-of-sight communications (mainly used in WSNs) for which polarization can be relied upon, it can make a significant difference in signal quality to have the transmitter and receiver using the same polarization. Thus, an antenna with circular polarization will not receive signals that are circularly

---

[†] Chip represents a single bit of a pseudonoise sequence while chip rate is the rate at which chips are sent.

polarized as for the opposite direction. Furthermore, there will be 3 dB loss from a linear polarized antenna that receives signals circularly polarized; the same also stands vice-versa. Hence, if the nodes of a WSN are capable of changing the polarization of their antennas they will be able to effectively defend in jamming environments when they sense interference. One problem is that the nodes must inform first each other about the change of their antenna's polarization, otherwise communication among peers will be interrupted. A method to overcome this problem is to program the nodes when they sense interference or lack of network connectivity, to change to specific polarizations until they establish reliable link to the network. The change of nodes' polarization of a WSN hinders the jamming process because it makes necessary to use specialized jamming equipment with the capability to change its signal polarization rapidly during the jamming.

18.5.7 DIRECTIONAL TRANSMISSION

Today's sensor nodes typically use omnidirectional antennas. The use of directional antennas could dramatically improve jamming tolerance in WSNs. In general, directional antennas/transmission provide better protection against eavesdropping, detection, and jamming than omnidirectional transmission [21,29,36]. A directional antenna transmits or receives radio waves only from one particular direction unlike the omnidirectional antenna that transmits and receives radio waves from all directions in the same time. This feature allows increased transmission performance, more receiving sensitivity and reduced interference from unwanted sources (e.g., jammers) compared to omnidirectional antennas.

The main problems with directional transmission are (1) the requirement of a more sophisticated MAC protocol [3,9] and (2) multipath routing becomes more complex [14,31].

Two types of directional antennas are commonly used in wireless ad hoc networks (sensor networks are not included): sectored and beamforming. Sectored antennas have multiple fixed antenna elements, pointed in different directions that can often operate independently. On the other hand, beamforming antennas have multiple antenna elements that work in tandem to transmit or receive in different directions. Beamforming antennas can be electronically switched or steered. Switched beamforming antennas can select one from a set of predefined beams by shifting the phase of each antenna element's signal by a precalculated amount. Steered beamforming antennas are more dynamic in nature since the main antenna lobe can be directed in any desired direction. They have increased performance and cost compared to switched beamforming antennas.

In the context of wireless sensor networks, Noubir [22] proposed the use of sectored antennas as more resistant to jamming; however, the specific antennas are not yet available.

18.6   PROPOSED SECURITY SCHEMES AGAINST JAMMING IN WSNs

Securing WSNs against jamming attacks is an issue of immense importance. In the following subsections we review security schemes proposed in the WSN literature to address this issue. We propose taxonomy that categorizes these approaches in

- Proactive
- Reactive
- Mobile agent-based

The relevant advantages and disadvantages of each method are highlighted and evaluated.

18.6.1 PROACTIVE COUNTERMEASURES

The role of proactive countermeasures is rather to make a WSN immune to jamming attacks rather than to react in the event of such incident. Proactive countermeasures can be classified in software (e.g., algorithms for the detection of jamming or encryption of transmitted packets) and combined software–hardware countermeasures (e.g., encryption of packets and FHSS transmission in parallel).

18.6.1.1 Proactive Software Countermeasures

This class of countermeasures makes use of software approaches (e.g., specialized algorithms and MAC protocols) to help a node survive jamming attacks or detect them and then employ other defensive mechanisms.

18.6.1.1.1 The Feasibility of Launching and Detecting Jamming Attacks in WSNs

In Ref. [45] Xu et al. claim that understanding the nature of jamming attacks is critical to assuring the operation of wireless networks. They present four different jammer attack models that may be employed against a WSN and they explore various techniques for detecting jamming attacks in WSNs. To improve detection, they introduce the notion of consistency checking, where the packet delivery ratio is used to classify a radio link as having poor utility, and then a consistency check is performed to classify whether poor link quality is due to jamming. They propose two enhanced detection algorithms: one employing signal strength for consistency check, and one employing location information for consistency check.

The advantage of this work is that knowing and identifying the adversary (e.g., the jammer type), the WSN can deal with the problem more efficiently. However, the authors focus on the analysis and detection of jamming signals and they do not deal with effective countermeasures against jamming.

18.6.1.1.2 Energy-Efficient Link-Layer Jamming Attacks against WSN MAC Protocols

Law et al. in Ref. [12] explore energy-efficient attacks against the DLL. They examine three representative MAC protocols, S-MAC, L-MAC, and B-MAC and they derive several jamming attacks that allow an adversary to jam the above MAC protocols in energy-efficient manner: (1) they are effective on encrypted packets, (2) they are as effective as constant/deceptive/reactive jamming, and (3) they are more energy-efficient than random jamming or reactive jamming.

The authors also discuss potential countermeasures against the proposed jamming attacks for each of the above-mentioned MAC protocols. For S-MAC they propose the use of high duty cycle as a partial countermeasure to energy-efficient link-layer jamming and for L-MAC shorter data packets. Finally, they suggest the use of L-MAC as a better choice compared to other MAC protocols in terms of resistance against link-layer jamming.

The main conclusion drawn from this work is that with typical WSN systems in use today no effective measures against link-layer jamming are possible. Regarding WSNs that require increased security against link-layer jamming the authors propose (1) the encryption of link-layer packets to ensure a high entry barrier for jammers, (2) the use of spread spectrum hardware, (3) the use of time division multiple access (TDMA) protocol, and (4) the use of randomized intervals.

The strong points of this work lie in the in-depth exploration of jamming attacks against MAC protocols and the suggestion of several countermeasures. Mainly software countermeasures are proposed and tested. The only drawback is that although the use of spread spectrum hardware is suggested, such solution is neither tested nor simulated; also, the authors do not suggest a specific type of spread spectrum technique (e.g., FHSS, DSSS, or UWB). It is quite obvious that even the less sophisticated jamming attack could put out of order sensor nodes that use only one communication channel, whatever MAC protocol these nodes use.

18.6.1.1.3 Radio Interference Detection in Wireless Sensor Networks

Radio interference relations among the nodes of a WSN and the design of a radio interference detection protocol (RID) are discussed in Ref. [47]. The main purpose of RID and its variation RID-Basic (RID-B) is to detect run-time radio interferences relations among nodes. The interference detection results are used to design real collision-free TDMA protocols.

The basic idea of RID is that a transmitter broadcasts a high-power detection packet (HD packet), and subsequently a normal-power detection packet (ND packet). This is called an HD–ND detection sequence. The receiver uses the HD–ND detection sequence to estimate the transmitter's interference strength. An HD packet includes the transmitter's ID. The receiver estimates possible interference caused by the transmitter by sensing the power level of the transmitter's ND packet.

After the HD–ND detection, each node propagates the detected interference information to its neighbor nodes, and then uses this information to figure out all collision cases within the system.

Briefly RID comprises three phases: (1) HD–ND detection, (2) information sharing, and (3) interference calculation. RID-B is a simpler and more lightweight version of RID (it lacks phase 3).

The main advantage of RID and RID-B is that in simulated WSNs combined with the TDMA protocol the packet delivery ratio reaches 100 percent while TDMA alone has packet loss up to 60 percent, in high packet traffic.

The main problem with RID and RID-B is the extended usage of bandwidth and energy of sensor nodes in the constrained environment of WSNs (especially in the case of RID). Furthermore only interference from adjacent sensor nodes is efficiently addressed. Jamming from external sources is not investigated; hence RID remains highly vulnerable to jamming attacks.

18.6.1.1.4 DEEJAM: Defeating Energy-Efficient Jamming

Wood et al. in Ref. [43] propose DEEJAM, a new MAC-Layer protocol for defending against stealthy jammers using IEEE 802.15.4-based hardware. The general design approach of this protocol is to hide messages from a jammer, evade its search, and reduce the impact of messages that are corrupted anyway.

In this work the authors provide the definition, implementation, and evaluation of four jamming attack classes and the suitable countermeasures against these attacks which are all combined in DEEJAM protocol. They also assume that the attacker uses the same or similar hardware as WSN nodes in terms of capability, energy capacity, and complexity. For more sophisticated attackers they suggest other approaches [42,46] more suitable than DEEJAM. The jamming attacks and the relevant countermeasures proposed in this work are

- Interrupt jamming in which the jammer transmits only when its radio captures a multibyte preamble and a Start of Frame Delimeter (SFD) sequence (see Figure 18.3). The capture of the above sequence is an indication for the jammer that a packet of data follows, so she initiates jamming. The defense proposed for this type of jammer is "Frame Masking," where the sender and the receiver node agree on a secret pseudorandom sequence for the SFD in each packet. As a result, no interrupt will be signaled to begin jamming. However, against a constant jammer this method cannot defend effectively.
- Activity jamming illustrated in Figure 18.4 where the attacker is trying to detect radio activity by periodically sampling the radio signal strength indicator (RSSI) or the radio's clear-channel assessment (CCA) input (if available). When RSSI is above a programmable threshold the jammer initiates her attack. The proposed defense for this type of jamming is frequency hopping, since the attacker can only sample RSSI only for the channel on which its radio listens to.

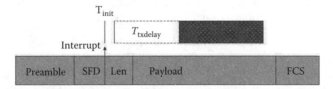

FIGURE 18.3 Interrupt jamming of packet, triggered by SFD reception. Fcs, frame check sequence; Len, length of payload; T_{init}, time needed for initialization of necessary state or radio chip registers; and $T_{txdelay}$, time imposed by the radio hardware for switching on transmitter circuits and oscillator stabilization.

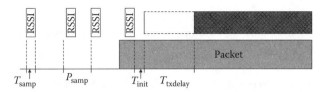

FIGURE 18.4 Activity jamming attack. RSSI in a single channel is periodically sampled every P_{samp} for T_{sam} time and jamming begins upon packet detection.

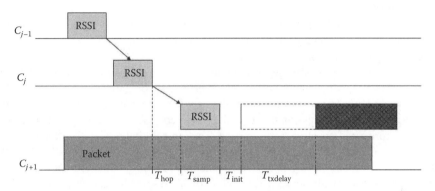

FIGURE 18.5 Channel scanning for activity, followed by jamming of the message found on C_{j+1} channel.

- Scan jamming illustrated in Figure 18.5 is the appropriate attack against frequency hopping. However, the attacker must have a significant higher hopping rate from the victim. In this type of attack, the attacker samples each channel as briefly as possible to determine if activity is present (the same method used in activity jamming). If she discovers radio activity, jamming is immediately initiated. The authors for this type of attack suggest packet fragmentation as the most suitable countermeasure. Scan jamming is most successful if the transmitted messages are long enough to be intercepted (the required transmission time is longer and this gives more time to the attacker for channel scanning). Using packet fragmentation a node breaks the transmitted packet into fragments which are transmitted separately on different channel and with different SFDs. If the fragments are short enough, an attacker's reactive jamming message does not start until after the sender has finished transmitting and hopped to another channel.

- Pulse jamming where the attacker is blindly jamming on short pulses on a single channel. This will corrupt the fragments of a packet that would use the current channel and taking into account that corrupting even one fragment is sufficient to cause the drop of the entire packet, pulse jamming can achieve excellent results in WSNs with limited communication channels available for frequency hopping. Redundant encoding is the proposed countermeasure, wherein the receiver is able to recover from one or more corrupted fragments. However, there is an increased cost in energy and bandwidth usage since transmission redundancy is occurred.

To conclude, DEEJAM utilizes all above-mentioned defensive mechanisms to protect a WSN against jamming from adversaries that use hardware with same capabilities as the deployed nodes. The main advantage is that it is compatible with existing nodes' hardware (no hardware modification is needed); the authors have also provided evidence of its effectiveness via simulations on MICAz nodes [6]. However, as the authors already noted against a powerful and more sophisticated jammer DEEJAM cannot effectively defend the WSN and the most probable scenario is that an adversary will use more advanced hardware compared to that of the nodes'. Another drawback is the overhead that DEEJAM requires to operate and the increased computational and energy cost in the already resource constrained nodes of a WSN.

18.6.1.2 Proactive Software–Hardware Combined Countermeasures

The category of proactive countermeasures focuses on the design of innovative antijamming hardware along with the proposal of algorithms that will utilize the new properties of the hardware.

18.6.1.2.1 Hermes II node

Mpitziopoulos and Gavalas [18] outline the design specifications of a prototype node, named Hermes II that effectively defends jamming attacks. Hermes II uses a hybrid FHSS/DSSS scheme as the main countermeasure against jamming.

The band that the authors propose for communication among Hermes nodes is the unlicensed 5 GHz band (5650–5925 MHz) which suffers less interference compared to the heavily used 2.4 GHz band. Also the increase in frequency has as a result a narrower and more directional transmitted signal, however with decreased transmission range. In the proposed 5 GHz band, there is 275 MHz of bandwidth available for spread-spectrum transmission. The same digital modulation that incorporated in ZigBee is used (O-QPSK for 5 GHz band). Hermes uses 55 frequency channels for FHSS with 5 MHz of bandwidth each available for DSSS. Each channel uses DSSS modulation with 270 KHz modulating (prespreading) bandwidth and 5 MHz total (two-sided) spread-spectrum signal bandwidth and so a 12.67 dB processing gain. The resulting raw, over-the-air data rate is approximately 252 Kbps per channel in the 5 GHz band. Hermes II nodes are able to perform up to 100.000 frequency hops/s, so is rather difficult for contemporary fast follow jammers to hop along and jam in the same time.

The sequence of channels used is determined by a channel sequence generation algorithm that uses a secret key as a seed. This secret key is derived from a secret word, known only to the nodes and the PE, using password-based key derivation function 2 (PBKDF2) [30]. Even if the secret word leaks, the adversary is not able to compromise the security of the entire WSN because she would not know the derivation function parameters of the secret key (salt,* iterations and key length). The encryption key can change upon PE request, at specific time intervals or in arbitrary fashion, depending to nodes computational power and energy availability. The authors propose—for security reasons—to "hard-code" the initial secret word onto nodes prior to the WSN deployment; however, the PE is able to modify it at any time. The generation of the channels' sequence is done with the use of the Mersenne Twister or MT19937 algorithm [16] combined with a hash function (SHA-1) for enhanced security [34]. Hermes II node also incorporates a DSSS chip with 5 MHz chip rate.

For the synchronization of communication nodes, the authors suggest the use of the flooding time synchronization protocol (FSTP) [15] in a predefined frequency channel. FTSP is designed especially for sensor networks and guarantees average precision of 0.5 μs/hop in a multihop scenario. It uses low communication bandwidth and exhibits proven robustness against node and link failures.

The main advantage of Hermes II is that it deals with the problem of jamming using a powerful hardware scheme, capable of defending a WSN even against the most sophisticated and with unlimited power resources jammers. In simulations that the authors have made Hermes II nodes guarantee a satisfactory packet even in heavily jammed environments where the attacker(s) are able to jam the entire area, the WSN is deployed, with barrage jamming, as opposed to typical sensor nodes communication schemes. However, all these advantages come with an increased cost.

As the authors state, the implementation of Hermes nodes is not a straightforward task due to the technologies incorporated, hence a significant amount of research is needed in various fields. First, a radio unit that complies with the Hermes II standards needs to be designed along with a

---

* Salt is a seed value used in the encryption of a plaintext password to expand the number of possible resulting cipher texts from a given plaintext. The use of a salt value is a defensive measure used to protect encrypted passwords against dictionary attacks.

new communication protocol that uses the 5 GHz band. Even if Hermes node is finally implemented the cost is expected to be considerably high and for sure not affordable for common WSNs. Only WSNs that will be used for applications that need extra high security (e.g., military) could afford the increased cost of Hermes nodes.

18.6.1.2.2   How to Secure a Wireless Sensor Network

Law and Havinga in Ref. [11] deal with many WSN security aspects including jamming on the physical and DLL.

In physical layer, they concentrate on spread-spectrum modulations as ideally countermeasures. They suggest the use of FHSS instead of DSSS because the latter requires more circuitry (higher cost) to implement is more energy-demanding and more sensitive to environmental effects [17,28]; on the contrary they claim that the hop rate in an FHSS system is typically much lower than the chip rate in a DSSS system, resulting in lower energy usage [17,39].

They do not recommend DSSS transceivers in WSNs but only the use of FHSS transceivers. For the latter they suggest the use of quaternary/binary frequency-shift keying (FSK) for data modulation [17,27] and a hop rate between 500 and 1000 hops/s [38]. Also they state that although a maximum hop rate of 1000 hops/s is not able to cope with most sophisticated jammers [33], however, for most of cases 500–1000 hops/s should be a practical compromise. For transceivers that do not support spread spectrum they recommend the channel surfing method by Xu et al. [46].

Finally for jamming on DLL, in absence of effective countermeasures they propose the use of TDMA protocols like L-MAC [7] as more resistant to jamming attacks.

The main advantage of this work is the recommendation of specific hardware that can cope with jamming clearly more efficiently than software countermeasures. However FHSS alone, as also noted by the authors, is not able to deal with contemporary fast-follower military jammers, which are able of jamming FHSS communications that perform even thousands of hops/s [33]. Furthermore, the disapproval of DSSS transceivers is not really justified since DSSS presents many advantages against jamming attacks [26]. For instance, since the transmitted signal of DSSS resembles white noise, crucial information for an adversary as the interception of the transmission channel and radio direction finding of the transmitting source become difficult. Also the use of UWB transceivers [23], which are used in many military systems as an effective countermeasure against jamming, is not considered as a possible countermeasure by the authors.

Regarding the countermeasures against physical layer it is stated that for transceivers not supporting spread spectrum, the channel surfing method by Xu et al. [46] is recommended. Channel surfing though, as detailed in a previous section, is in effect an adaptive form of FHSS, so it uses some type of spread spectrum hardware that allows switching to a different, orthogonal frequency $\pm\delta$ away when it discovers the current frequency is jammed.

18.6.2   Reactive Countermeasures

The main characteristic of reactive countermeasures is that they enable reaction only upon the incident of a jamming attack sensed by the WSN nodes. Reactive countermeasures can be further classified into software and combined software–hardware, similarly to proactive countermeasures.

18.6.2.1   Reactive Software Countermeasures

This category employs the use of exclusively software-based solutions to defend against jamming attacks. Similarly to proactive software countermeasures, this category is also compatible with IEEE 802.15.4-based hardware that is most often used in today's WSN.

18.6.2.1.1   JAM: A Jammed-Area Mapping Service for Sensor Networks

JAM algorithm enables the detection and mapping of jammed regions (Figure 18.6) to increase network efficiency [42]. Data is then simply routed around the jammed regions. The output of the jamming detection module is a JAMMED or UNJAMMED message broadcast to the node's

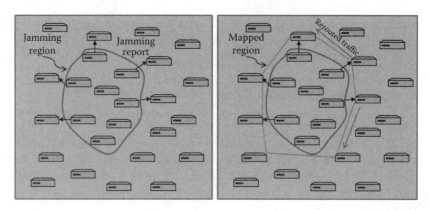

FIGURE 18.6   Detection and mapping of jammed region.

neighbors. However, as the authors stress, a jammed node would not be able to send any messages, since almost all MAC protocols require a carrier sense to indicate a clear channel in order to have clearance for transmission. To deal with this problem, MAC must provide a way to override carrier-sense so as to allow broadcasting a brief, high-priority, unacknowledged message. However, this could lead to the energy exhaustion of the jammed nodes in case of link-layer jamming. The mapping service that the authors propose can provide the following benefits:

- Feedback to routing and directory services
- An effective abstraction at a higher-level than local congestion, failed neighbors and broken routes
- Support for avoiding the region by network-controlled vehicles, military assets, and emergency personnel
- Reports to a base-station for further jamming localization
- Aid to power management strategies for nodes inside and around jammed regions

In cases that jamming attacks are restricted to small portions of the WSN, JAM has beneficial results for the robustness and functionality of the network. However, this method exhibits several shortcomings: first, it cannot practically defend in the scenario that the attacker jams the entire WSN or a significant percentage of nodes; second, in the case that the attacker targets some specific nodes (e.g., those that guard a security entrance) to obstruct their data transmission, again this technique fails to protect the attacked sensor nodes.

18.6.2.2   Reactive Software–Hardware Combined Countermeasures

This type of countermeasures requires the use of specialized hardware along with proposed software solutions.

18.6.2.2.1   Channel Surfing and Spatial Retreat

Xu et al. [46] proposed two evasion strategies against constant jammers: channel surfing and spatial retreat. Channel surfing is essentially an adaptive form of FHSS. Instead of continuously hopping from frequency to frequency, a node only switches to a different, orthogonal frequency $\pm\delta$ away when it discovers the current frequency is being jammed. The δ value can be determined by experiments, e.g., for Berkeley motes, δ is found to be multiples of 800 kHz [46]. The frequency hopping pattern suggested in Ref. [46] is $C(n + 1) = C(n) + 1 \bmod M$. The authors admit the predictability of this pattern and have suggested using of a preshared secret between the communicating parties.

Finally, a node can detect a jammed medium, if the packet delivery ratio is low while the received signal strength is high [45].

Spatial retreat, which can be applied only upon sensor nodes that have the capability of mobility, is an algorithm according to which two nodes move in Manhattan distances to escape from a jammed region. The key to the success of this strategy, as the authors state, is to decide where the participants should move and how should they coordinate their movements. The main shortcoming of the two above-mentioned strategies is that they are effective only against constant jammers and they have no results against more intelligent or follow-on jammers.

18.6.2.2.2 Wormhole-Based Antijamming Techniques in Sensor Networks

Wormholes, until recently, were considered as a threat for a WSN [21,24,32]. However, Cagalj et al. [4] proposed a reactive antijamming scheme for WSNs using wormholes. The basic idea is that jammed nodes use channel diversity to establish a communication with another node outside the jammed area. The authors propose three types of wormholes:

- Wired pair of sensors: In this solution, the authors propose the construction of a WSN with certain number of pairs of sensor nodes, each connected through a wire. The wired sensors are also equipped with wireless transceivers. To have a large probability that an arbitrary pair forms a wormhole from the exposure region to the area not affected by jamming, it might be required to create a large number of wired pairs. In large-scale WSNs this solution is very expensive and needs a large amount of time for the deployment of the sensor nodes. Also the scenario that an adversary simply tries to cut the wires before the attack should be taken into account.
- Frequency hopping pairs: This solution enables the creation of pairs by using frequency hopping techniques, like Bluetooth. All pairs are being deployed by wireless links and can afford longer links between pairs. A problem that arises with this solution is the synchronization among the nodes that use FHSS. This problem can be addressed with the use of a time synchronization protocol (e.g., FTSP [15]). Another one problem is that some nodes must be equipped with special and more sophisticated transceivers, which increases significantly the cost of the WSN. Also the deployment of the WSN becomes more complex and time-consuming. Finally, FHSS alone is not an effective countermeasure against fast-follower jammers [33].
- Uncoordinated channel-hopping: In this solution, the authors seek to create probabilistic wormholes by using sensor nodes capable of hopping between radio channels that ideally span a large frequency band. Unlike the previously mentioned solution, individual packets are transmitted on a single channel. As a result, the hops between the channels are much slower. That simplifies the adversary's work since she has more time in her disposal to recover the channel in use and jam it. Finally, the problems of frequency hopping pairs are also applied in this solution.

From the above, we conclude that wormholes may be an interesting idea to defend against jamming attacks but a different approach must be taken. In a large-scale attack or against sophisticated fast-follower jammers the proposed countermeasures cannot efficiently protect the WSN.

18.6.3 MOBILE AGENT-BASED SOLUTIONS

In this class of antijamming approaches enables mobile agents (MAs) to enhance the survivability of WSNs. The term MA [25] refers to an autonomous program with the ability to move from host to host and act on behalf of users towards the completion of an assigned task.

18.6.3.1 Jamming Attack Detection and Countermeasures in WSNs Using Ant System

Muraleedharan and Osadciw [20] propose the use of ant system algorithm as an effective countermeasure against jamming attacks in a WSN. In effect, ants may be viewed as a type of MAs.

An initial set of ants traverse through the nodes in a random manner and once they reach their destinations, they deposit pheromone on trails as a means of communication indirectly with the other ants. The amount of pheromone left by the previous ant agents increases the probability that the same route is taken during the current iteration. Parameters such as hops, energy, distance, packet loss, SNR, bit error rate (BER), and packet delivery affect the probability of selecting a specific path or solution. Also pheromone evaporation over time prevents suboptimal solutions from dominating in the beginning.

The main advantage of the ant solution is that the ant agents spread into the network and continuously try to find optimal and jamming-free routes for data transferring, taking into account crucial node and network parameters (e.g., nodes' remaining energy, packet loss, SNR). In a large WSN, they have a clear advantage over other antijamming solutions since they can adapt more easily to the jammed environment. An ant agent can remain in a node when this is under attack and move to an adjacent node upon detecting a "clear" communication channel (jamming pause).

Unfortunately, this system has not been tested in large-scale simulated WSNs (simulations have been conducted in topologies comprising 16 nodes), hence its scalability is questionable. Also the extra computational and energy cost required by ants is not evaluated. Furthermore, the authors omitted information on how quickly the "pheromone" trails are able to react to nimble attackers. Finally, in the case that a considerable proportion of WSN nodes are jammed then ants will probably fail to guarantee the uninterrupted network's operation.

18.6.3.2 JAID: An Algorithm for Data Fusion and Jamming Avoidance on WSNs

In Ref. [19] Mpitziopoulos et al. describe the critical role MAs can play in the field of security and robustness of a WSN in addition to data fusion. They propose the jam avoidance itinerary design (JAID) algorithm. The design objective of JAID algorithm is twofold: (1) to calculate near-optimal routes for MAs that incrementally fuse the data as they visit the nodes; (2) in the face of jamming attacks against the WSN, to modify the itineraries of the MAs so to avoid the jammed area(s) while not disrupting the efficient data dissemination from working sensors.

To meet the second objective, the PE uses the JAM algorithm [42] to map the jammed area(s) and identify the problematic nodes.* Furthermore, it executes queries in specific-time intervals so as to be informed as soon as they resume function. Assuming that not the entire WSN is affected, the MAs are scheduled not to visit the jammed nodes. Instead, they visit nodes in the perimeter of the jammed area(s) that are not affected to avoid the security risk and thus the collapse of the WSN. If the number of jammed nodes is below a specific threshold, JAID only modifies the prejamming scheduled itineraries ("connects" the cut off nodes to jam-free nodes) to increase the algorithm's promptness. Otherwise, JAID reconstructs the agent itineraries excluding the jammed area(s).

The authors evaluated the performance of JAID in simulated topologies, examining the scenario wherein multiple jammers initiate jamming attacks against a WSN consisted of 100 nodes randomly deployed in a field (Figure 18.7). They assume that jammers have limited jamming range, not covering the entire WSN because; the authors admit that in case the whole WSN is jammed, no algorithmic solution could effectively defend the network.

---

* Simulation results in Ref. [42] in a WSN composed of 121 sensor nodes proved that the mapping activity varies from 1.5 s for moderately connected networks to just over 5 s for the largest jammed region. This is fast enough to allow a reasonable real-time response to jamming in the current sensor network.

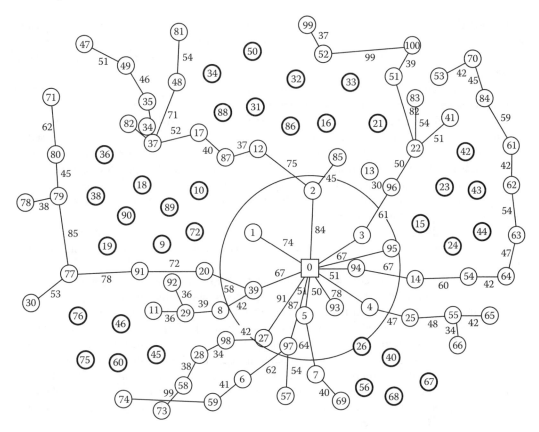

FIGURE 18.7 Delphi-based simulation of JAID algorithm in case of jamming. The PE is denoted with a square and the nodes with the thin and thick outlined circles. The big circle around PE is its effective transmission range. The thick outlined circles represent the jammed nodes.

The MAs used in JAID have the advantage of excluding jammed and problematic nodes from their itineraries and efficiently deliver the data from the working nodes to the PE. Furthermore JAID calculates near-optimal routes for MAs minimizing the energy cost needed for transmission.

The drawback of JAID is that it cannot defend the WSN in case the jammer(s) exercise efficient attack against all nodes simultaneously. However, if a WSN consisted of nodes that utilize specialized spread-spectrum radio units (e.g., Hermes II) and used JAID the best of the two worlds could be combined (effective defense even against heavy jamming attacks while maintaining the advantages of JAID, i.e., lower energy consumption, high responsiveness, and enhanced security).

18.7 COMPARISON OF ANTIJAMMING APPROACHES

Proactive countermeasures are performed in the background, even in jamming-free environments; typically, they cannot be initiated, stopped, or resumed on demand. Hence, they enable instant response against jamming at the expense of increased computational and energy cost upon the resource constrained sensor nodes. Thus, they defend more efficiently against stealth jamming attacks which may pass undetected for a significant period from a reactive countermeasure.

Proactive software countermeasures pose no requirement for specialized hardware; they only utilize the capabilities of IEEE 802.15.4-compliant hardware. Hence, they are compatible with existing nodes' hardware and can be applied on any contemporary WSN. On the contrary, proactive software–hardware combined countermeasures, since they imply new nodes design, are typically associated with prolonged implementation time and relatively high implementation cost. However, countermeasures of the particular category demonstrate improved results compared to alternative methods, namely superior resistance against jamming.

Reactive countermeasures need reduced computational and energy cost compared to proactive countermeasures but in the case of stealth or deceptive jamming there is a great possibility for delayed sensing of jamming. Reactive software countermeasures category, similarly to proactive software countermeasures, is also compatible with IEEE 802.15.4-based hardware that is most often used in todays WSN. The approaches belonging to the category of combined reactive software–hardware countermeasures, because of the need for specialized hardware along with proposed software solutions suffer considerable impact (1) on the implementation cost of the sensor nodes, and (2) in the deployment complexity of the WSN. On the other hand, reactive software–hardware combined countermeasures present higher effectiveness.

Finally, mobile agent-based solutions do not require the use of specialized hardware. However, in conjunction with spread-spectrum hardware their antijamming properties can be significantly improved.

Table 18.1 lists all the works reviewed in this chapter. We focus on the countermeasures that have been extensively analyzed and evaluated (general-purposed and undocumented countermeasures, e.g., the use of spread spectrum hardware, are not listed). Furthermore we assume an efficient number of constant jammers with unlimited power supply that perform spot jamming attacks upon large-scale WSNs. We assume that not all WSN nodes are jammed at a same time. In the "defense" column we evaluate the level of defense each countermeasure provides against the above-mentioned jamming scenario while in "compatibility with existing hardware" column we report if the proposed countermeasures are compatible with existing hardware or need a specialized hardware platform. Finally, in "expected implementation/deployment cost" column we evaluate the implementation and deployment cost of each countermeasure.

18.8 OPEN RESEARCH ISSUES

The constraints of contemporary sensor nodes resources (e.g., limited energy, computation and communication capabilities) and the fact that they are often deployed in insecure or even hostile terrains underline their susceptibility to jamming attacks. Therefore, the problem of jamming on the Physical and Link Data layers of WSNs has been a subject of intense research during the last few years. However, there are still many open research issues outlined below:

- UWB tranceivers: Despite the proposal of several spread-spectrum schemes for defense against jamming attacks, the usage of UWB radio units has not extensively examined, although UWB exhibits many advantages against jamming.
- MAs: The use of mobile agents for defending against jamming attacks is a partially unexplored and promising method. Currently, only two works address jamming in WSNs with the use of MAs [19,20]. The unique characteristics of mobile agents could be explored to intensify their benefit upon WSNs under jamming attacks (e.g., the fact that traveling agents can temporary remain on their current position and return to the PE with their collected data when they sense clear terrain).
- A new communication protocol that uses the 5 GHz free band, which suffers less interference compared to the heavily used 2.4 GHz band, needs to be proposed and designed.
- The antennas that sensor nodes currently use are omnidirectional. The design requirements and the implementation of alternative, interference and jamming-resistant antennas need to be devised.

TABLE 18.1
Characteristics and Features of Proposed Antijamming Schemes

	Proposed Countermeasures against Jamming	Type of Countermeasures	Defence Effectiveness	Compatibility with Existing Hardware	Expected Implementation Deployment Cost
Feasibility of launching and detecting jamming attacks in WSNs	• Detection of jamming using signal strength	Proactive software	Low	Yes	Low
	• Detection of jamming using location information	Proactive software	Low	Yes	Low
Energy-efficient link-layer jamming attacks against WSN MAC protocols	• S-MAC: High duty cycle	Proactive software	Low	Yes	Medium
	• L-MAC: Shorter data packets	Proactive software	Low	Yes	Medium
	• Encryption of link-layer packets	Proactive software	Low	Yes	Low
	• TDMA protocol	Proactive software	Low	Yes	Low
	• Transmission in randomized intervals	Proactive software	Low	Yes	Low
Radio interference detection protocol (RID)	a. High-normal packet detection b. Information sharing c. Interference calculation	Proactive software	Low	Yes	Medium
DEEJAM: Defeating energy-efficient jamming	a. Frame masking b. Frequency hopping c. Packet fragmentation d. Redundant encoding	Proactive software	Medium	Yes	High

(continued)

TABLE 18.1 (continued)
Characteristics and Features of Proposed Antijamming Schemes

Proposed Countermeasures against Jamming	Type of Countermeasures	Defence Effectiveness	Compatibility with Existing Hardware	Expected Implementation Deployment Cost
Hermes II node	Proactive software–hardware	High	No	High
How to secure a wireless sensor network by Law and Havinga	Proactive software–hardware	Medium	Partial	High
• FHSS				
• L-MAC	Proactive software	Low	Yes	Medium
JAM: A jammed-area mapping service for sensor networks	Reactive software	Medium	Yes	Medium
a. Detection of jamming				
b. Mapping of jammed area				
Channel surfing (adaptive FHSS)	Reactive software–hardware	Medium	Partial	High
• Channel surfing and spatial retreat				
• Spatial retreat	Reactive software–hardware	Medium	Partial	High
Wormhole-based Antijamming techniques in sensor networks	Reactive software–hardware	High	Partial	High
• Wired pair of sensors				
• Frequency hopping pairs	Reactive software–hardware	Medium	Partial	High
• Uncoordinated channel-hopping	Reactive software–hardware	Medium	Partial	High
Jamming attack detection and countermeasures in WNS using ant system	Mobile agent based	Medium	Yes	Medium
Ants (mobile agents)				
JAID: An algorithm for data fusion and Jamming avoidance on WSNs	Mobile agent based	Medium	Yes	Medium
Mobile agents				

18.9 CONCLUSION

This chapter reviewed the main aspects of wireless sensor network security against jamming attacks: vulnerabilities of today's WSNs, types of jammers and attacks, and effective countermeasures against jamming.

It also classifies the research works that deal with jamming in WSNs in three main categories: proactive, reactive, and mobile agent based, highlighting their relevant positive aspects and shortcomings. Furthermore it highlights open research issues in the field of jamming in WSNs. In the near future, the wider adoption and usage of WSN technologies in military and monitoring applications is expected to bring out the immense importance of this security issue.

REFERENCES

[1] D. L. Adamy and D. Adamy, *EW 102: A Second Course in Electronic Warfare*, Artech House, Inc., Norwood, MA, 2004.

[2] F. Akyildiz, W. Su, Y. Sankarasubramaniam, and E. Cayirci, A survey on sensor networks, *IEEE Communications Magazine*, pp. 102–114, August 2002.

[3] S. Bandyopadhyay, K. Hasuike, S. Horisawa, and S. Tawara, An adaptive MAC and directional routing protocol for ad hoc wireless network using directional ESPAR antenna, in *Proceedings of the ACM Symposium on Mobile Ad Hoc Networking & Computing 2001 (MOBIHOC 2001)*, Long Beach, CA, pp. 243–246, October 2001.

[4] M. Cagalj, S. Capkun, and J. P. Hubaux, Wormhole-based anti-jamming techniques in sensor networks, *IEEE Transactions on Mobile Computing*, 6(1), 100–114, May 2006.

[5] R. J. Cramer, M. Z. Win, and R. A. Scholtz, Impulse radio multipath characteristics and diversity reception, in *the Proceedings of the IEEE International Conference on Communications ICC'98*, pp. 1650–1654, 1998.

[6] Crossbow Technology Inc., http://www.xbow.com/.

[7] L.V. Hoesel and P. Havinga, A lightweight medium access protocol (LMAC) for wireless sensor networks: Reducing preamble transmissions and transceiver state switches, in *Proceedings of the First International Conference on Networked Sensing Systems*, Tokyo, 2004.

[8] Y. Hu, A. Perrig, and D. Johnson, Packet leashes: A defense against wormhole attacks in wireless networks, in *Proceedings of the IEEE INFOCOM 2003*, vol. 3, pp. 1976–1986, 2003.

[9] Y. B. Ko, V. Shankarkumar, and N. H. Vaidya, Medium access control protocols using directional antennas in ad hoc networks, in *Proceedings of the IEEE INFOCOM 2000*, vol. 1, pp. 13–21, March 2000.

[10] Y. Law, P. Hartel, J. den Hartog, and P. Havinga, Link-layer jamming attacks on S-MAC, in *Proceedings of the 2nd European Workshop on Wireless Sensor Networks (EWSN 2005)*, pp. 217–225, 2005 [Online]. Available at: http://ieeexplore.ieee.org/iel5/9875/31391/01462013.pdf.

[11] Y. Law and P. Havinga, How to secure a wireless sensor network, in *the Proceedings of the 2005 International Conference on Intelligent Sensors, Sensor Networks and Information Processing*, pp. 89–95, December 2005.

[12] Y. Law, L. V. Hoesel, J. Doumen, P. Hartel, and P. Havinga, Energy-efficient link-layer jamming attacks against wireless sensor network MAC protocols, in *Proceedings of the Third ACM Workshop on Security of Ad Hoc and Sensor Networks (SASN 2005)*, ACM Press, pp. 76–78, 2005.

[13] H. Lee, B. Han, Y. Shin, and S. Im, Multipath characteristics of impulse radio channels, in *Proceedings of the IEEE Vehicular Technology Conference*, pp. 2487–2491, 2000.

[14] Y. Li and H. Man, Analysis of multipath routing for ad hoc networks using directional antennas, in *the Proceedings of the IEEE 60th Vehicular Technology Conference*, vol. 4, pp. 2759–2763, September 2004.

[15] M. Maroti, B. Kusy, G. Simon, and A. Ledeczi, The flooding time synchronization protocol, Technical Report ISIS-04-501, Institute for Software Integrated Systems, Vanderbilt University, Nashville, TN, 2004 [Online]. Available at: http://www.eecs.harvard.edu/mdw/course/cs263/papers/ftsp-sensys04.pdf.

[16] M. Matsumoto and T. Nishimura, Mersenne Twister: A 623-dimensionally equidistributed uniform pseudo-random number generator, *ACM Transactions on Modeling and Computer Simulation*, 8(1), 3–30, 1998.

[17] J. Min, Analysis and design of a frequency-hopped spread-spectrum transceiver for wireless personal communications, PhD dissertation, University of California, Los Angeles, CA, 1995 [Online]. Available at: http://www.icsl.ucla.edu/aagroup/PDF files/tcvr-arch.pdf.

[18] A. Mpitziopoulos and D. Gavalas, Countermeasures against radio jamming attacks in wireless sensor networks, *International Journal of Computer Research*, 15(1), pp. 5–22, March 2007.

[19] A. Mpitziopoulos, D. Gavalas, C. Konstantopoulos, and G. Pantziou, JAID: An algorithm for data fusion and jamming avoidance on sensor networks, Technical Report, TR-2007-09-10, Department of Cultural Technology & Communication, University of the Aegean, Greece, September 2007 [Online]. Available at: http://www.aegean.gr/culturaltec/dgavalas/TR/TR-2007-09-10.pdf.

[20] R. Muraleedharan and L. Osadciw, Jamming attack detection and countermeasures in wireless sensor network using ant system, *2006 SPIE Symposium on Defense and Security*, Orlando, FL, April 2006.

[21] C. S. R. Murthy and B. S. Manoj, Transport layer and security protocols for ad hoc wireless networks, in *Ad Hoc Wireless Networks: Architectures and Protocols*. Prentice Hall PTR, NJ, May 2004.

[22] G. Noubir, On connectivity in ad hoc networks under jamming using directional antennas and mobility, in *Proceedings of the Wired/Wireless Internet Communications conference, Lecture Notes in Computer Science*, 2957, pp. 186–200, 2004.

[23] I. Oppermann, L. Stoica, A. Rabbachin, Z. Shelby, and J. Haapola, Uwb wireless sensor networks: Uwen- a practical example, *IEEE Communications Magazine*, 42(12), 27–32, December 2004.

[24] P. Papadimitratos and Z. Haas, Secure routing for mobile ad hoc networks, in *Proceedings of the SCS Communication Networks and Distributed Systems Modeling and Simulation*, 2002.

[25] V. Pham and A. Karmouch, Mobile software agents: An overview, *IEEE Communications Magazine*, 36(7), 26–37, 1998.

[26] R. L. Pickholtz, D. L. Schilling, and L. B. Milstein, Theory of spread spectrum communications—A tutorial, *IEEE Transactions on Communications*, 20(5), 855–884, 1982.

[27] S. Pollin, B. Bougard, R. Mangharam, L. V. der Perre, F. Catthoor, R. Rajkumar, and I. Moerman, Optimizing transmission and shutdown for energy-efficient packet scheduling in sensor networks in *Proceedings of the Second European Workshop on Wireless Sensor Networks (EWSN 2005)*, IEEE, pp. 290–301, February 2005.

[28] G. J. Pottie and L. P. Clare, Wireless integrated network sensors: toward low-cost and robust self-organizing security networks, in *Sensors, C3I, Information, and Training Technologies for Law Enforcement*, ser. SPIE Proceedings, vol. 3577, pp. 86–95, 1999 [Online]. Available at: http://wins.rsc.rockwell.com/publications/spie3577-12.pdf.

[29] R. Ramanathan, On the performance of ad hoc networks with beamforming antennas, *ACM International Symposium on Mobile Ad Hoc Networking and Computing (MobiHoc'01)*, Long Beach, CA, pp. 95–105, October 2001.

[30] RFC 2898, PKCS #5: Password-based cryptography specification version 2.0, 2000, http://rfc.net/rfc2898.html.

[31] S. Roy, S. Bandyopadhyay, T. Ueda, and K. Hasuike, Multipath routing in ad hoc wireless networks with omni directional and directional antenna: A comparative study, in *Proceedings of 4th International Workshop of Distributed Computing, Mobile and Wireless Computing, IWDC 2002*, Calcutta, India, pp. 184–191, December 2002.

[32] K. Sanzgiri, B. Dahill, B. Levine, C Shields, and E. Belding-Royer, A secure routing protocol for ad hoc networks, in *Proceedings of the 10th IEEE International Conference on Network Protocols*, pp. 78–87, November 2002.

[33] D. C. Schleher, Electronic warfare in the information age. Artech House, Inc., Norwood, MA, July 1999.

[34] B. Schneier. *Applied Cryptography : Protocols, Algorithms, and Source Code in C*. Wiley, 2nd edn., 1995.

[35] E. Shi and A. Perrig, Designing secure sensor networks, *Wireless Communications Magazine*, 11(6), 38–43, December 2004.

[36] A. Spyropoulos and C. S. Raghavendra. Energy efficient communications in ad hoc networks using directional antennas, in *Proceedings of IEEE Conference on Computer Communications (INFOCOM'02)*, vol. 1, pp. 220–228, New York, June 2002.

[37] SunSpotWorld. http://www.sunspotworld.com/.

[38] D. J. Torrieri, Fundamental limitations on repeater jamming of frequency-hopping communications, *IEEE Journal on Selected Areas in Communications*, 7(4), 569–575, May 1989.

[39] K. Tovmark, Chipcon Application Note AN014,Frequency hopping systems (Rev. 1.0), Chipcon AS, Mar. 2002 [Online]. Available: http://www.chipcon.com/files/AN 014 Frequency Hopping Systems 1 0.pdf.

[40] UWB-wikipedia. http://en.wikipedia.org/wiki/Ultra wideband.

[41] A. D. Wood and J. A. Stankovic, Denial of service in sensor networks, *Computer*, 35(10), 54–62, 2002.

[42] A. D. Wood, J. A. Stankovic, and S. H. Son, JAM: A jammed-area mapping service for sensor networks, in *Proceedings of the 24th IEEE Real-Time Systems Symposium (RTSS'2003)*, pp. 286–297, 2003.

[43] A. D. Wood, J. A. Stankovic, and G. Zhou, DEEJAM: Defeating energy-efficient jamming, in IEEE 802.15.4-based wireless networks, in *the Proceedings of the 4th Annual IEEE Communications Society Conference on Sensor, Mesh and Ad Hoc Communications and Networks (SECON)*, pp. 60–69, San Diego, CA, June 2007.

[44] W. Xu, K. Ma, W. Trappe, and Y. Zhang, Jamming sensor networks: Attack and defense strategies, *IEEE Network Magazine*, 20, 41–47, 2006.

[45] W. Xu, W. Trappe, Y. Zhang, and T. Wood, The feasibility of launching and detecting jamming attacks in wireless networks, in *Proceedings of the 6th ACM international symposium on Mobile Ad Hoc Networking and Computing*, pp. 46–57, 2005.

[46] W. Xu, T. Wood, W. Trappe, and Y. Zhang. Channel surfing and spatial retreats: Defenses against wireless denial of service, in *Proceedings of the 2004 ACM workshop on Wireless security*, pp. 80–89, New York, 2004.

[47] G. Zhou, T. He, J. A. Stankovic, and T. F. Abdelzaher, RID: Radio interference detection in wireless sensor networks, in *Proceedings of the IEEE INFOCOM' 2005*, vol. 2, pp. 891–901, 2005.

19 Concealed Data Aggregation for Wireless Sensor Networks

Aldar C.-F. Chan

CONTENTS

Wireless sensor networks are multihop networks composed of tiny devices with limited computation and energy capacities. For such devices, data transmission is a very energy-consuming operation. To achieve a reasonably long lifetime for a wireless sensor network, data aggregation is a widely adopted approach for answering queries from end users so as to minimize the number of bits sent by each device. End-to-end privacy and aggregate integrity are the two main security goals for data aggregation. In this chapter, the state of the art of secure data aggregation for wireless sensor networks is studied. The operation of the data aggregation paradigm is described. To precisely capture the criteria for securing data aggregation in wireless sensor networks, rigorous cryptographic models for end-to-end privacy and aggregate integrity are presented. The impossibility of achieving end-to-end aggregate integrity noninteractively without calling back aggregating nodes is discussed. The security of existing schemes is then discussed with reference to these security models.

19.1   INTRODUCTION

Wireless sensor networks are envisioned to be economic solutions to an increasing number of applications. These applications include environmental monitoring (such as temperature, humidity, and seismic activities), real-time traffic monitoring, military surveillance and homeland security, mining and industrial disaster detection. A sensor network may consist of hundreds or even thousands of low-cost sensors, each of which forms an information source sensing and collecting data from the environment for a given task. In a typical sensor network, there is one or more sinks which are much more powerful devices acting as a bridge between the sensors and the outside world (or the end user of the sensed data). The sink distributes queries from the end user to the sensors and collects data from the sensors and relays them back to the end user. In addition to sensing, the sensors also form a data delivery infrastructure, in which each sensor not only merely sends its data and receives query instructions directly but also forwards instructions and data to and from other sensors.

Regardless of the application, most sensor networks have two notable properties in common. First, the network's overall goal is typically to reach a collective conclusion regarding the outside environment which requires the sensors to detect a common phenomenon; consequently, there is relatively high redundancy in the sensors' raw data; thus, reporting raw data is often unnecessary but resource-consuming. Second, individual sensors have severely limited computation, communication and power (battery) resources; besides, a typical sensor has a disproportionately high cost of transmitting information as compared to performing local computation. It is thus very inefficient for every sensor node to report raw data back as each data packet could traverse many hops to reach the sink and each intermediate node may need to forward (in a query response) several packets, one for each downstream node. Recently, many data aggregation protocols have been proposed to fuse raw data en route to reduce the communication cost and energy consumption in data forwarding, thereby extending the lifetime of a sensor network. With aggregation, each intermediate node only needs to transmit a single packet of the data aggregate, compared to multiple packets of raw data when aggregation is not used.

Although possibly extending the lifetime of a typical sensor network, aggregation also undesirably opens new vulnerabilities to false data injection attacks and poses new challenges to privacy protection. Sensor nodes are usually deployed in open and unattended environment, thus making it easy for an adversary to physically capture a sensor node. An adversary can obtain confidential information, say, the secret keys, from a compromised sensor and reprogram it with malicious code.

The compromised node may then report an arbitrary, false fusion result, causing the final aggregate value to deviate far from its true measurement value. This attack can become more damaging when multiple compromised nodes collude in injecting false data. On the other hand, aggregation becomes problematic when end-to-end privacy between sensors and the sink is required as most conventional encryption schemes do not support ciphertext aggregation; hop-by-hop encryption is a straightforward solution. In hop-by-hop encryption, each sensor shares a common secret key with each of its neighbors. When it receives ciphertexts from downstream, it first decrypts the ciphertexts, then aggregates the data in plaintext, and finally re-encrypts the aggregation result with a different key and forwards it upstream. However, in the presence of compromised nodes, encrypting data on a hop-by-hop basis is insecure.

The problem of providing end-to-end privacy in data aggregation is usually called concealed data aggregation (CDA) due to Westhoff et al. [36]. In CDA, multiple source nodes send encrypted data to a sink along a concast tree with ciphertext aggregation performed en route; the aggregation is performed in the encrypted domain.

In general, end-to-end privacy and aggregate integrity are the two main security goals of secure data aggregation (SDA) in wireless sensor networks. Informally speaking, end-to-end privacy (regardless of information leakage due to the correlation among sensor measurements) ensures that nobody other than the sink could learn considerable information about the final aggregate even if he might control any subset of sensor nodes. Whereas, aggregate integrity assures that any manipulation of the final aggregate by an adversary beyond what is achievable through direct injection of data at compromised nodes under his control will be detected at the sink. Due to the subtlety of security mechanism design and the unpredictability of attack techniques, defining the security goals and requirements of a protocol or system is usually nontrivial. In this chapter, the reductionist's approach of "provably security" [2], a widely adopted paradigm in modern cryptography, is used to define the security notions of SDA.

In this chapter, security issues arising in the data aggregation paradigm of wireless sensor networks are studied. First, the motivation and typical operation of data aggregation are presented. Then cryptographic models to define the security goals of SDA are given. The impossibility of achieving end-to-end aggregate integrity noninteractively without calling back aggregating nodes by the sink is also discussed. The security of a number of schemes in the literature is then discussed.

19.2   DATA AGGREGATION

Sensor nodes come in various shapes and forms; however, they are generally assumed to be resource-limited with respect to computation power, storage, memory and, especially, battery life. A popular example is the Berkeley mote [24]. One common feature among sensor nodes is the disproportionately high cost of transmitting information as compared to performing local computation. For example, a Berkeley mote spends approximately the same amount of energy to compute 800 instructions as it does in sending a single bit of data [24]. It thus becomes essential to reduce the number of bits forwarded by an intermediate node to extend the entire network's lifetime.

One natural approach to reduce data transmission by each individual sensor involves aggregating sensor data as they propagate along the path from the sensors to the sink. In this approach, the sensors (in response to a query from the outside world) send their sensed values to the aggregating nodes. Each aggregating node then condenses the data prior to sending them on. In terms of bandwidth and energy consumption, aggregation is beneficial as long as the aggregation process is not too computationally intensive and is accompanied by a data length reduction, that is, the resulting aggregate has a smaller length compared to the total length of the input data incorporated into it. For instance, no energy reduction can be achieved in multiplicative aggregation due to the violation of the latter condition. In multiplicative aggregation, the aggregate is the product of all the input data and the length of the aggregate increases linearly with the number of inputs because multiplying n pieces of data, each of length l, results in a product of length nl.

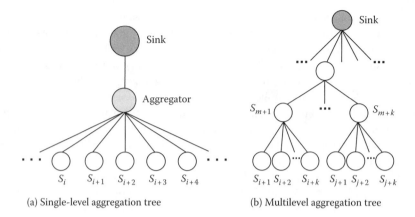

(a) Single-level aggregation tree   (b) Multilevel aggregation tree

FIGURE 19.1   Concast trees for data aggregation.

Although the aggregating nodes can either be special (more powerful) nodes or regular sensor nodes, it is more common to assume that all sensor nodes are potential aggregating nodes as is done in this chapter. In this setting, because each sensor node has very limited capabilities, aggregation must be simple and not involve any expensive or complex computations. Ideally, it should only require a few simple arithmetic operations such as additions and multiplications.

During a typical data aggregation process, when a sensor network receives an aggregate query from the outside world through the sink, sensor nodes organized themselves into a concast tree rooted at the sink—the so-called aggregation tree. Each nonleaf node in the aggregation tree acts as an aggregating node fusing data passed on from their child nodes in the aggregation tree and forwarding only the aggregation result to its parent node. Despite the existence of schemes considering only a single-level aggregation tree such as [31], in practice, only a multilevel tree [5,8,15,20,24,36,37] is considered useful. Figure 19.1 depicts a single-level and a multilevel aggregation tree.

From the viewpoint of information theory, data aggregation is a lossy data compression process in which all the individual sensor readings are lost in the aggregation operation. As a result, data aggregation is not applicable to some applications, for instance, perimeter control, because only individual sensor readings are of interest. However, many application scenarios that monitor an entire microenvironment (e.g., temperature or seismic activity) do not require information from individual sensors but, instead, put more emphasis on statistical aggregates/quantities, such as mean, median, and variance. To illustrate the application and the efficiency gain from data aggregation, additive aggregation is considered as follows.

19.2.1 EXAMPLE: ADDITIVE AGGREGATION

With additive aggregation, each sensor node sums all values, say, x_i (each l-bit long), received from its k child nodes (in the sink-rooted aggregation tree) and forwards only the sum to its parent node. As a result, it sends only one packet of length $(l+k)$-bit instead of k packets of length l-bit each. This drastically reduces the transmission load of a sensor node from lk-bits to $(l+k)$-bits, thus extending its lifetime. Suppose, eventually, the sink obtains the sum of all values sent by n sensor nodes. By dividing the sum by n (which is the total number of contributing sensor nodes), the sink can compute the average on the measured data.

This simple aggregation is very efficient because each aggregating node only performs k arithmetic additions. Note that an aggregating node is assumed to have its own measurement to contribute; thus k additions are needed. It is also robust because there is no requirement for all sensors to participate as long as the sink gets the total number of sensors that actually provide a measurement.

With additive aggregation, it is possible to compute useful statistical quantities such as variance, standard deviation, and any other moments on the measured data. For example, in case of variance, each aggregating node not only computes the sum, $S = \sum_{i=1}^{k} x_i$, of the individual values sent by its k children, but also the sum of their squares: $V = \sum_{i=1}^{k} x_i^2$. Eventually, the sink obtains two values: the sum of the actual samples S_n which it can use to compute the mean and the sum of the squares V_n which it can use to compute the variance using the following equation:

$$\text{Var} = E(x^2) - E(x)^2,$$

where $E(x^2) = \left(\sum_{i=1}^{n} x_i^2 \right)/n = V_n/n$, $E(x) = \left(\sum_{i=1}^{n} x_i \right)/n = S_n/n$, and n is the number of sensor nodes contributing to the two final aggregates S_n and V_n.

19.3   SECURITY MODELS FOR SECURE DATA AGGREGATION

End-to-end privacy and aggregate integrity are the two fundamental goals an SDA scheme needs to accomplish. Security models for both end-to-end privacy and aggregate integrity of SDA are given in this chapter; the privacy goal is expressed by means of an abstract primitive called CDA, and the aggregate integrity goal by means of another called aggregate message authentication code (AMAC). Through the consideration of these two primitives, a separation of the privacy goal and the aggregate integrity goal of SDA is possible.

19.3.1   OVERVIEW OF SDA SECURITY MODELS

To get a proper SDA design and analyze its security completely (i.e., no speculated vulnerability is missed out in the security evaluation), it is necessary to specify the goals of privacy and aggregate integrity precisely in terms of the assurances and functionalities a design is supposed to provide, the assumed capabilities of an expected attacker to be guarded against and the resources or information supposed to be available to the attacker.

Informally speaking, in a typical SDA scenario, an attacker is assumed to be global, that is, able to monitor any location in the sensor network or even the entire network. Furthermore, the attacker is supposed to be able to corrupt a subset of sensor nodes, to read the internal state of some sensors, and to control some sensors to withhold, inject, or modify data in the network.

The privacy goal is two-fold. First, the privacy of the data has to be guaranteed end-to-end, that is, only the sink could learn about the final aggregation result and only a negligible amount of information about the final aggregate should be leaked out to any eavesdropper or node along the path. Each node should only have knowledge about its data, but no information about the data of other nodes. Second, to reduce communication overhead, the data from different source nodes has to be efficiently combined by intermediate aggregating nodes along the path. Nevertheless, these aggregating nodes should not learn any information about the final aggregate in an ideal scheme.

The goal of aggregate integrity comprises correctness and soundness. The correctness goal ensures that if all sensor nodes act honestly, the sink should accept the resulting aggregate as a correct aggregate. The soundness goal assures that any deviation of the final aggregate from the desired value (beyond what can result from an attacker injecting data directly into the network through compromised nodes) will be detected at the sink even though the attacker is allowed to collude with a subset (but not all) of the sensor nodes, control the messages sensed by some sensor nodes and observe all the information flows in the network. That is, it is guaranteed that an attacker cannot manipulate or modify the portion of the final aggregate contributed by honest nodes without being detected by the sink.

Although the above discussion offers good insights and intuitions for understanding the desired security goals, a useful SDA scheme needs to be achieved; it is not sufficiently rigorous in cryptographic practice. "Provable Security" [2] is the most widely adopted approach for modeling and analyzing the security of a protocol in cryptography; the details of this approach will be given in Section 19.3.2. The advantage of bringing in this approach to model the security of SDA is that the resulting security model and analysis, if done properly, usually provide a clearer, more comprehensive, and precise specification of the essential security goals, and more importantly a security guarantee against not only known attacks but also unknown attacks (as long as they are within the specified model). Besides, this approach has established many well and widely studied security notions, each of which provides a reasonably wide coverage of known attacks in practice; in fact, many of these notions are widely adopted as the standard models for different classes of security protocols (e.g., indistinguishability against chosen ciphertext attacks [CCA] in encryption schemes). It is thus more likely to get a proper and correct security model for SDA by refining these standard notions to cover the salient features of data aggregation.

Although there has been a solid foundation in cryptography for both private-key [22,23,33] and public-key [10,17,18,28] encryption, as well as, for both message authentication codes [3,21] and digital signatures [17,19], to obtain security models tailored for SDA, a refinement to the standard security models is needed to cover the following salient features in the data aggregation scenario: First, an SDA scheme can be based on private-key or public-key cryptography. That is, the encryption function or the tag verification function of an SDA scheme could be public or private. Second, SDA is a many-to-one (multisender, single-receiver) cryptosystem although cryptosystems in the literature are either one-to-one [3,18,21,22] or one-to-many [13,19,34]. Because the verification of a digital signature can be done by anyone having the signer's public key, a digital signature scheme can be considered as an one-to-many authentication scheme. Third, SDA includes the aggregation functionality whose adversary model needs new definitions for both the privacy and aggregate integrity goals.

Natural extensions of the standard security notions for encryption, namely, semantic security and indistinguishability against CCA, and for NAC and digital signatures, namely, existential unforgeability against chosen message attacks, are given as the security models for SDA privacy (in Section 19.5) and SDA aggregate integrity (in Section 19.6), respectively. In these models, only the essence and functionality of data aggregation relevant to security analysis are extracted and this information is represented as two ideal, abstract primitives, namely, CDA and AMAC. That is, a separation of the privacy goal and the aggregate integrity goal is adopted. These primitives are artificial constructs to show what an ideal scheme for the security goal in question should look like (CDA for privacy and AMAC for aggregate integrity). Hence, only algorithmic input–output relations are specified in these primitives without including concrete algorithmic procedures or routines. As usual, the desired security goals of CDA and AMAC are described in terms of a simulation game.

19.3.2 BACKGROUND ON PROVABLE SECURITY AND REDUCTION ANALYSIS OF CRYPTOGRAPHIC PROTOCOLS

A cryptographic protocol is usually accompanied with a security goal which varies depending on the application scenarios. On the other hand, all the designs of cryptographic protocols are based on some computational problem assumed to be difficult, more precisely, unsolvable by any known polynomial time algorithm. To analyze whether a given design achieves its goal, the following needs to be specified:

1. What is considered as a break of security or a failure for a design to achieve its goal?
2. What are the capabilities of an adversary to be guarded against and what resources and information are available to him?
3. What is the underlying computational assumption?
4. What functionalities are supposed to be provided by the design?

Obviously, the first two items could be viewed as a specification of the problem of breaking the security of a given protocol, which is an equivalent alternative to precisely specify the security goal. As a result, they are usually combined together in the form of a simulation game in the context of cryptographic design and analysis. Instead of directly specifying the goal of a protocol, the goal of the adversary (the achievement of which is considered as a break of the security goal of the protocol) is usually given. Overall, the simulation game specification of a security notion of any protocol includes a list of possible types of information that could be obtained from the allowed adversary interaction/queries, a challenge (which can be a computational or decisional problem) for the adversary and a description of what is considered as a win for the adversary; usually, winning the challenge is equivalent to that the adversary breaks the security property or notion in question and the probability (for a computational challenge) or advantage (for a decisional challenge) of winning a challenge is quantified as how successful an adversary is in breaking the corresponding security property.

The required functionalities are usually abstracted as a list of algorithms called syntax. These algorithmic specifications usually only include input–output relations, algorithm nature (deterministic vs. probabilistic), and resources assumed (say, polynomial time). But no information about how the algorithms are implemented is given in the syntax.

19.3.2.1 Proof by Contradiction

In all protocols, it could easily be seen that a given design is secure only if the underlying computational problem is hard. For instance, nobody could recover the plaintext from a single given ElGamal ciphertext only if the discrete logarithm problem is hard (the necessary condition for the ElGamal scheme [12] to be secure). In other words, the hardness of the underlying computational problem is necessary for the security of a given protocol. It is thus natural to ask whether the hardness of such a computational problem is sufficient to guarantee the security of a given protocol (i.e., its goal achieved). Proving this sufficient condition directly is usually hard; instead, proving by contradiction is the widely adopted approach to show that a protocol achieves its security goal.

Using the standard contrapositive argument, the logic statement "statement p implies statement q" is equivalent to another statement "statement q implies statement p" (where p denotes the negation of statement p). In the context of security analysis, this means if the statement "breaking the security of a given design (subject to a specified adversary capability model) implies solving the underlying computational problem" can be shown, then it is equivalent to claim the statement: "if the underlying computational problem is hard, than the given design is secure in the given adversary capability model."

This kind of proof by contradiction is the standard argument in complexity theory or cryptographic analysis, called reduction. Suppose the possible adversary interaction with the system is neglected for the moment. A reduction proof usually starts by assuming there exists an imaginary, efficient adversary algorithm which can break the security of a given design in polynomial time and then shows this adversary algorithm can also solve the underlying computational problem (with slightly more effort) in polynomial time. That is, in a typical reduction argument, an efficient algorithm (using the assumed-to-exist adversary algorithm as a component) is constructed to solve the underlying computational problem. The adversary algorithm is imaginary because whether it exists or not is unknown but the constructed algorithm is based on the assumption that such an adversary algorithm exists. Consequently, a reduction proof is a conditional proof hinged on the assumed difficulty of the underlying computational problem.

19.3.2.2 Modeling Adversary Interaction

In the discussion so far, an adversary (say $\mathcal{A}$) is assumed to be given an input problem instance (i.e., breaking the security of a protocol) to solve but no other additional information, and the only assumption on the capability of $\mathcal{A}$ is that it is probabilistic polynomial time (PPT). However, this does not reflect the complete scenario in real systems. In the real scenario, an adversary could obtain additional information before solving the problem in question; that is, the adversary could have a higher capability or be given additional information it can leverage on due to the interaction with the actual system. For instance, before attempting to decrypt a given ciphertext, an adversary could possibly have obtained other plaintext–ciphertext pairs which give him additional information to leverage on. That is, depending on the allowed adversary interaction with the legitimate users or the system, the problem of breaking the security of a protocol could be an easier problem to an adversary in the real attack environment.

To reflect this additional information an adversary can leverage on, if the imaginary adversary $\mathcal{A}$ (assumed to be successful in breaking the security of a protocol) is to be used as a subroutine to solve the underlying computational problem, a simulation of the allowed interaction is needed to make $\mathcal{A}$ have the same view as it is in the real attack. As a result, to formulate the problem, it is necessary to specify what interaction or possible additional information an adversary is allowed in a real attack as an assumption of its capability in addition to it being PPT. That is why the security goal of a protocol is usually specified by a simulation game in which how an adversary is allowed to interact in the protocol (in terms of oracle access) and what is considered as a win by the adversary are described.

Besides, no concrete interaction that an adversary can make is assumed; instead, the focus is on what information he can obtain through such interaction. To do so, it is a usual practice to formulate the interaction as an imaginary oracle access. That is, the model only specifies what an adversary can get from an oracle and how this access is possible is beyond the scope of most security analysis/proofs in provable security. For example, an oracle access could be like the following: an adversary obtains a decryption of a ciphertext of his choice (the CCA) through accessing a decryption oracle; how he can do so does not matter to cryptographic modeling and analysis.

Hence, a typical security model in essence defines a class of attacks (both known and unknown) that a design can guard against. If a construction is shown to be secure in a given model, it would actually remain secure to all future attacks which can access no more information (in types and quantities) than what is specified by the oracle access in the model; the only cause for the security of such a protocol to be broken in the specified model is that an algorithm capable to solve the underlying computational problem in polynomial time has been discovered.

19.4   NOTATIONS

The notations for algorithms and probabilistic experiments that originate in [18,19] are adopted in this chapter. A detailed exposition can be found there and a brief introduction is given below.

Let $z \leftarrow A(x, y, \ldots)$ denote the experiment of running probabilistic algorithm A on inputs $x, y, \ldots$ generating output z. Then $\{A(x, y, \ldots)\}$ denotes the probability distribution induced by the output of A. The notations $x \leftarrow \mathcal{D}$ and $x \in_R \mathcal{D}$ are equivalent and mean randomly picking a sample x from the probability distribution $\mathcal{D}$; if no probability function is specified for $\mathcal{D}$, it is assumed that x is uniformly picked from the sample space. As usual, $\mathbb{N}$ denotes the set of nonnegative integers, and PPT is "probabilistic polynomial time." An empty set is always denoted by ϕ.

19.5   CDA SECURITY MODEL

The privacy goal of SDA is considered in this section using an abstract primitive called CDA. Recall that the cryptographic model of a security notion consists of two parts: a syntax (modeling the operation and functionalities of a protocol) and a simulation game (modeling the security goal of a protocol and the capability of an adversary and resources he can make use of). The CDA security model discussed here is taken from Ref. [6].

19.5.1 SYNTAX OF CDA

A typical CDA scheme includes a sink R and a set U of n source nodes (which are sensor nodes) where $U = \{s_i : 1 \leq i \leq n\}$. Denote the set of source identities by ID; in the simplest case, ID $= [1, n]$. In the following discussion, $hdr \subseteq$ ID is a header indicating the source nodes contributing to an encrypted aggregate. Given a security parameter λ, a CDA scheme consists of the following polynomial time algorithms:

Key Generation (KG): Let $\mathrm{KG}(1^\lambda, n) \rightarrow (dk, ek_1, ek_2, \ldots, ek_n)$ be a probabilistic algorithm. Then, ek_i (with $1 \leq i \leq n$) is the encryption key assigned to source node s_i and dk is the corresponding decryption key given to the sink R.

Encryption (E): $\mathrm{E}_{ek_i}(m_i) \rightarrow (hdr_i, c_i)$ is a probabilistic encryption algorithm taking a plaintext m_i and an encryption key ek_i as input to generate a ciphertext c_i and a header $hdr_i \subset$ ID. Here, hdr_i indicates the identity of the source node performing the encryption; if the identity is i, then $hdr_i = \{i\}$.

The encryption function is sometimes denoted by $\mathrm{E}_{ek_i}(m_i; r)$ to explicitly show by a string r the random coins used in the encryption process.

Decryption (D): Given an encrypted aggregate c and its header $hdr \subseteq$ ID (which indicates the source nodes included in the aggregation), $\mathrm{D}_{dk}(hdr, c) \rightarrow m/\perp$ is a deterministic algorithm which takes the decryption key dk, hdr, and c as inputs and returns the plaintext aggregate m or possibly $\perp$ if c is an invalid ciphertext.

Aggregation (Agg): With a specified aggregation function f, the aggregation algorithm $\mathrm{Agg}_f(hdr_i, hdr_j, c_i, c_j) \rightarrow (hdr_l, c_l)$ aggregates two encrypted aggregates c_i and c_j with headers hdr_i and hdr_j, respectively (where $hdr_i \cap hdr_j = \phi$) to create a combined aggregate c_l and a new header $hdr_l = hdr_i \cup hdr_j$. Suppose c_i and c_j are the ciphertexts for plaintext aggregates m_i and m_j, respectively. The output c_l is the ciphertext for the aggregate $f(m_i, m_j)$, namely, $\mathrm{D}_{dk}(hdr_l, c_l) \rightarrow f(m_i, m_j)$.

Note that the aggregation algorithm does not need the decryption key dk or any of the encryption keys ek_i as input; it is a public algorithm.

Depending on constructions, the aggregation function f could be any associative function, for instance, f could be the sum, multiplicative product, max, etc. Leveraging on the associativity property, this chapter abuses the notation: the composition of multiple copies of f is simply denoted by $f(m_1, m_2, \ldots, m_i)$ irrespective of the order of aggregation and it is called the f-aggregate on $m_1, m_2, \ldots, m_i$; to be precise, it should be written as $f(f(f(m_1, m_2), \ldots), m_i)$ with a certain aggregation order.

It is intentional to include the description of the header hdr in the above definition so as to make the CDA security model as general as possible (to cover schemes requiring headers in their operations). Nonetheless, generating headers or including headers as input to algorithms should not be treated as a requirement in the actual construction or implementation of CDA algorithms. For constructions which do not need headers (such as the generic construction given in Section 19.7.1), all hdr's can simply be treated as an empty set ϕ in the security model and the discussions still apply.

19.5.1.1 Typical CDA Operation

The operation of CDA runs as follows. In the initialization stage, the sink R runs KG to generate a set of encryption keys $\{ek_i : 1 \leq i \leq n\}$ and the corresponding decryption key dk and distributes each one of the encryption keys to the corresponding source, say, ek_i to s_i. Depending on constructions, the encryption keys ek_i could be private or public, but the decryption key dk has to be private in all cases.

At a certain instant, the sink selects a subset $S \subseteq U$ of the n sources to report their data. Each $s_i \in S$ uses its encryption key ek_i to encrypt its data represented by the plaintext m_i, giving a ciphertext c_i. The model does not pose restrictions on whether global or local random coins should be used for encryption. If each source generates its random coins individually, the random coins are said to be local; if the random coins are chosen by the sink and broadcast to all source nodes, they are global. Global random coins are usually public. When global random coins are used, no restriction is posed in the model about the reuse of randomness despite that, in practice, each global random coin is treated as nonce, that is, used once only.

Usually, the source nodes form a concast aggregation tree over which the encrypted data are sent. To save communication cost, aggregation is done en route to the sink whenever possible. When a node s_i in the aggregation tree receives x ciphertexts, say $(hrd_{i_1}, c_{i_1}), \ldots, (hdr_{i_x}, c_{i_x})$, from its child nodes (with identities $i_1, \ldots, i_x \in S$), it aggregates these ciphertexts along with its own ciphertext (hdr_i, c_i) by running Agg_f successively. Note that it is possible that some of these ciphertexts are already the encryption of aggregated data rather than the encryption of a single plaintext. The concast tree structure ensures that any pair of these headers have an empty intersection. Suppose $c_{i_1}, \ldots, c_{i_x}$ are the ciphertexts for the plaintext aggregates $m_{i_1}, \ldots, m_{i_x}$. The resulting ciphertext is (hdr_l, c_l) where $hdr_l = hdr_{i_1} \cup \cdots \cup hdr_{i_x} \cup hdr_i$ and c_l is the encryption of the aggregate $f(m_{i_1}, \ldots, m_{i_x}, m_i)$.

Eventually, a single encrypted aggregate c_{sink} arrives at the sink. The sink can then apply the decryption algorithm to c_{sink} to get back the plaintext aggregate $f(\ldots, m_i, \ldots)$ with $s_i \in S$. The CDA scheme is usually required to be correct in the sense that when the encryption and decryption are performed with matched keys and correct headers and all the aggregation is run properly, the decryption should give back an f-aggregate of all the data applied to the encryption.

19.5.2 Security Notion of CDA

Two types of oracle queries (adversary interaction with the system) are allowed in the CDA security model, namely, the encryption oracle $\mathcal{O}_E$ and the decryption oracle $\mathcal{O}_D$. Their details are as follows:

Encryption oracle $\mathcal{O}_E(i, m)$: For fixed encryption and decryption keys, on input an encryption query $\langle i, m \rangle$, the encryption oracle retrieves s_i's encryption key ek_i and runs the encryption algorithm on m and replies with the ciphertext $E_{ek_i}(m; r)$ and its header hdr. In case global random coins are used, the random coins r are part of the query input to $\mathcal{O}_E$.

Decryption oracle $\mathcal{O}_D(hdr, c)$: For fixed encryption and decryption keys, on input a decryption query $\langle hdr, c \rangle$ (where $hdr \subseteq \mathrm{ID}$), the decryption oracle retrieves the decryption key dk and runs the decryption algorithm D and sends the result $D_{dk}(hdr, c)$ as the reply.

The encryption oracle is needed in the security model because the encryption algorithm in some CDA could use private keys, for example [5,36]. In case the encryption algorithm does not use any secret information, an adversary can freely generate the ciphertext on any message of his choice without relying on the encryption oracle.

Shown below are three security notions for CDA privacy, namely, security against CCA, semantic security (equivalent to indistinguishability against chosen plaintext attacks), and one-wayness. All these notions are natural extensions of well-known notions in encryption schemes.

19.5.2.1 Security against Chosen Ciphertext Attacks

Security (more precisely, indistinguishability) against adaptive chosen ciphertext attacks (IND-CCA2) is defined by the following game played between a challenger and an adversary, assuming there is a set U of n source nodes. If no PPT adversary, even in collusion with at most t compromised node (with $t < n$), can win the game with nonnegligible advantage (as defined below), the CDA scheme is said to be t-secure. Note that the adversary is allowed to freely choose parameters n and t.

DEFINITION 1

A CDA scheme is t-secure (indistinguishable) against adaptive CCA if the advantage of winning the following game is negligible in the security parameter λ for all PPT adversaries.

Collusion choice: The adversary chooses to corrupt t source nodes. Denote the set of these t corrupted nodes and the set of their identities by S' and I', respectively.

Setup: The challenger runs KG to generate a decryption key dk and n encryption keys $\{ek_i : 1 \leq i \leq n\}$, and gives the subset of t encryption keys $\{ek_j : s_j \in S'\}$ to the adversary but keeps the decryption key dk and the other $(n-t)$ encryption keys $\{ek_j : s_j \in U \backslash S'\}$.

Query 1: The adversary can issue to the challenger two types of queries
- Encryption query $\langle i_j, m_j \rangle$. The challenger responds with $E_{e_{i_j}}(m_j)$.
- Decryption query $\langle hdr_j, c_j \rangle$. The challenger responds with $D_{dk}(hdr_j, c_j)$.

In case global random coins are used, the adversary is allowed to choose and submit his choices of random coins for both encryption and decryption queries. Depending on whether the encryption keys are kept secret, the encryption queries may or may not be needed.

Challenge: Once the adversary decides that the first query phase is over, it selects a subset S of d source nodes (whose identities are in the set I) such that $|S \backslash S'| > 0$, and outputs two different sets of plaintexts $M_0 = \{m_{0k} : k \in I\}$ and $M_1 = \{m_{1k} : k \in I\}$ to be challenged. The only constraint is that the two resulting plaintext aggregates x_0 and x_1 are not equal where $x_0 = f(\ldots, m_{0k}, \ldots)$ and $x_1 = f(\ldots, m_{1k}, \ldots)$.

The challenger flips a coin $b \in \{0, 1\}$ to select between x_0 and x_1. The challenger then encrypts each $m_{bk} \in M_b$ with ek_k and aggregates the resulting ciphertexts in the set $\{E_{ek_k}(m_{bk}) : k \in I\}$ to form the ciphertext C of the aggregate, that is, $D_{dk}(I, C) = x_b$, and gives (I, C) as a challenge to the adversary.

In case global random coins are used for encryption, the challenger chooses and passes them to the adversary. If a nonce is used, the global random coins should be chosen different from those used in the Query 1 phase and no query on them should be allowed in the Query 2 phase.

Query 2: The adversary is allowed to make more queries (both encryption and decryption) as previously done in Query 1 phase but no decryption query can be made on the challenged ciphertext C. Nevertheless, the adversary can still make a decryption query on a header corresponding to the set S except that the ciphertext has to be chosen different from the challenged ciphertext C.

Guess: Finally, the adversary outputs a guess $b' \in \{0, 1\}$ for b.

Result: The adversary wins the game if $b' = b$. The advantage of the adversary is defined as $Adv_A = \left| Pr[b' = b] - \frac{1}{2} \right|$.

Note that in CDA what the adversary is interested in is the information about the final aggregate. Consequently, in the above game, the adversary is asked to distinguish between the ciphertexts of two different aggregates x_0 and x_1 as the challenge, rather than to distinguish two different sets of plaintexts M_0 and M_1. By picking elements for M_0 and M_1, the adversary is essentially free to choose x_0 and x_1. Allowing the adversary to choose the two sets M_0, M_1 is to give him more flexibility in launching attacks. When an adversary cannot distinguish between the ciphertexts of two different aggregates (of his choice) with probability of success nonnegligibly greater than $1/2$, this means, in essence, he can learn no information about an aggregate from its ciphertext.

Although IND-CCA2 security is the most desired property for any protocol with privacy as its design goal, such as encryption, usually only a handful of constructions can achieve it for a given problem. In fact, it can be shown that IND-CCA2 security (as defined above) could not be achieved in the CDA problem. The reason is that the needed functionality of ciphertext aggregation in CDA would allow an adversary to create a new valid ciphertext for making a query in Query 2 phase and the result would allow him to win the challenge.

19.5.2.2 Semantic Security

Semantic security, which is equivalent to indistinguishability against chosen plaintext attacks (IND-CPA), is defined by the same game as in the definition of security against chosen ciphertext attacks except that no query to the decryption oracle $\mathcal{O}_D$ is allowed. Similar to the definition of IND-CCA2, a CDA scheme is said to be t-secure when it can still achieve IND-CPA security against a PPT adversary corrupting at most t compromised nodes.

For a CDA scheme to be useful, it should at least achieve semantic security. In the notion of semantic security, the main resource for an adversary is the encryption oracle $\mathcal{O}_E$. In some schemes like [5,36], the adversary may not know the encryption keys, meaning he might not have access to the encryption oracle in the real environment. Nevertheless, in sensor networks, he is able to obtain the encryption of any plaintext of his choice by manipulating the sensing environment and recording the sensed value using his own sensors. Hence, chosen plaintext attacks are still a real threat to CDA.

19.5.2.3 One-Wayness

One-wayness is the weakest possible security notion for encryption. A CDA scheme is t-secure with respect to one-wayness if no PPT attacker, corrupting at most t nodes, should be able, with non-negligible probability of success, to recover the plaintext aggregate matching a given ciphertext. To define one-wayness more formally, the same game for defining IND-CCA2 is used except that no query is allowed and the adversary can make no choice in the challenge phase but is given a ciphertext of a certain aggregate x (encrypted using at least one encryption key not held by the adversary) and asked to recover x.

19.6 AMAC SECURITY MODEL

The security model of AMAC considered here is a natural extension of that for MAC [3,21] commonly used in one-to-one communication. A typical MAC scheme is a two-tuple (MAC, VER) where MAC takes the secret key k (shared between the communicating parties) and message m as input to return a tag $t = \text{MAC}_k(m)$. $\text{VER}_k(m, t)$ taking the key k, the message m and the tag t as input returns either 1 (if t is a valid tag for m) or 0 (otherwise). For correctness, $\text{VER}_k(m, \text{MAC}_k(m)) = 1$. The security requirement for MAC schemes is in essence the same as that for digital signatures, namely, existential unforgeability against chosen message attacks [3,19]. In details, the secret key k is kept secret from an adversary. The adversary is allowed to query tags t_j for messages m_j of his choice, which can be done adaptively. Let $\mathcal{M}$ denote the set of all messages m_j queried by the adversary. The adversary breaks the scheme if he is able to find a message $m \notin \mathcal{M}$ and a valid tag t such that $\text{VER}_k(m, t) = 1$.

19.6.1 Syntax of AMAC

The setting for AMAC is the same as that of CDA, with one sink R and a set U of n source nodes. As before, let $U = \{s_i : 1 \leq i \leq n\}$ and the set of source identities $\text{ID} = [1, n]$. Same as in CDA, $hdr \subseteq \text{ID}$ is a header indicating the source nodes contributing to an aggregate. Note that in AMAC, the aggregation is done in the plaintext domain (compared to the encrypted domain aggregation in CDA) as the aggregate integrity goal is isolated from the privacy goal in the consideration of AMAC. It can be seen that CDA and AMAC can be integrated together to achieve both security goals. Without loss of generality, the aggregation function $f(\cdot)$ is assumed to be associative. For a security parameter λ, an AMAC consists of three polynomial time algorithms as follows.

Key generation (KG): Let $KG(1^\lambda, n) \to (k_1, k_2, \ldots, k_n)$ be a probabilistic algorithm. Then, k_i (with $1 \le i \le n$) is the secret key used to generate a verification tag. The key k_i is held by node i and the sink possesses all k_i's.

Tag generation (MAC): $MAC_{k_i}(m_i) \to tag_i$ takes a secret key k_i and a message m_i as input to generate a verification tag tag_i for m_i. The message sent out from node i is of the form (hdr_i, m_i, tag_i) where $hdr_i = \{i\}$.

Tag verification (Ver): Let m be an aggregate of messages $m_1, m_2, \ldots, m_i, \ldots$ and hdr be the set of all contributing identities. Then $Ver_{k_1, k_2, \ldots, k_i, \ldots}(m, tag_1, tag_2, \ldots, tag_i, \ldots) \to 0/1$ takes the aggregate m and the verification tag tag_i and secret key k_i for each $i \in hdr$ and outputs 1 if m is a correct aggregate (i.e., $m = f(m_1, m_2, \ldots, m_i, \ldots)$) and 0 otherwise.

The verification is also denoted as $Ver_{\{k_i: i \in hdr\}}(m, \{tag_i : i \in hdr\})$ in the following discussion.

Note that no aggregation algorithm is specified in AMAC; the aggregation is done in plaintext, just the same as in usual aggregation. When an aggregating node with identity z receives two measurement values and their tags from downstream, say, $(\{x\}, m_x, tag_x)$ and $(\{y\}, m_y, tag_y.)$, it would pass $(\{x, y, z\}, f(m_x, m_y, m_z), tag_x, tag_y, tag_z)$ as the aggregation result to its parent where m_z and tag_z are its own measurement and the corresponding tag, respectively. Aggregation of verification tags is not considered here. So all tags are needed in the verification. If a secure AMAC without tag aggregation cannot be achieved, then it also implies that a secure AMAC with tag aggregation cannot be achieved either.

The correctness requirement of AMAC is as follows:

$$Ver_{k_1, k_2, \ldots, k_i, \ldots}(f(m_1, m_2, \ldots, m_i, \ldots), MAC_{k_1}(m_1), MAC_{k_2}(m_2), \ldots, MAC_{k_i}(m_i), \ldots) = 1.$$

19.6.1.1 Typical AMAC Operation

The majority of AMAC operation is the same as that in CDA. In the initialization phase, the sink generates n secret keys k_is and gives k_i to node i. To respond to a query, each reporting sensor node $s_i \in S \subseteq U$ sends in a data-tag pair (m_i, tag_i) and its identity $hdr_i = \{i\}$ where $tag_i = MAC_{k_i}(m_i)$. That is, tag_i is supposed to be a verification tag for message m_i generated by node i. As these data-tag pairs and the corresponding headers go upstream along the aggregation tree, aggregation is performed on m_i leaving tag_i intact; a combined header would be forwarded as in CDA. Hence, eventually the sink would receive m and all the tags in $\{tag_i = MAC_{k_i}(m_i) : s_i \in S\}$ and the header corresponding to S where m is supposed to be an f-aggregate of all messages in $\{m_i : s_i \in S\}$. The sink runs the verification algorithm Ver to check the validity of m. If the security goal of AMAC is achieved, the only manipulation an adversary can do with the compromised nodes is to inject data directly through the compromised nodes. The adversary cannot modify the data portion contributed by all the honest nodes without being detected by the sink.

19.6.2 Security Notion of AMAC

MAC is usually regarded as the symmetric analog of digital signatures. The AMAC security model is actually a refinement of the notion of existential unforgeability against chosen message attacks [19], a well-known notion for digital signatures.

Two types of oracle queries are allowed in the security model, namely, the tag generation oracle $\mathcal{O}_T$ and the tag verification oracle $\mathcal{O}_V$. Their details are as follows:

Tag generation oracle $\mathcal{O}_T(i, m)$: For fixed secret keys, on input a tag generation query $\langle i, m \rangle$, the tag generation oracle retrieves key k_i and runs the tag generation algorithm on m and replies with the tag $MAC_{k_i}(m)$.

Tag verification oracle $\mathcal{O}_V(hdr, m, T = \{tag_i : i \in hdr\})$: For fixed secret keys, on input a tag verification query $\langle hdr, m, T \rangle$ (where $hdr \subseteq$ ID and $T = \{tag_i : i \in hdr\}$), the tag verification oracle retrieves the keys $\{k_i : i \in hdr\}$ and runs the tag verification algorithm Ver and replies with the result $\text{Ver}_{\{k_i : i \in hdr\}}(m, T)$.

19.6.2.1 Existential Unforgeability against Chosen Message Attacks

As in CDA, a simulation game is used to define the notion of existential unforgeability against chosen message attacks.

DEFINITION 2

An AMAC scheme is t-secure (unforgeable) against chosen message attacks if the advantage of winning the following game is negligible in the security parameter λ for all PPT adversaries.

Collusion choice: The adversary chooses to corrupt t source nodes. Denote the set of these t corrupted nodes and the set of their identities by S' and I', respectively.

Setup: The challenger runs KG to generate n secret keys $\{k_i : 1 \le i \le n\}$, and gives the subset of t keys $\{k_j : s_j \in S'\}$ to the adversary but keeps the other $(n - t)$ keys $\{k_j : s_j \in U \setminus S'\}$.

Query: The adversary can issue to the challenger two types of queries

- Tag generation query $\langle i_j, m_j \rangle$. The challenger responds by returning the tag $\text{MAC}_{k_{i_j}}(m_j)$.
- Tag verification query $\langle hdr_j, m_j, T_j \rangle$. The challenger responds by returning the verification result $\text{Ver}_{\{k_l : l \in hdr_j\}}(m_j, T_j)$.

The adversary can issue queries of both types until he decides to make a guess. Let $\mathcal{T}$ denote the set of all tag generation queries made by the adversary, that is, $\mathcal{T} = \{\langle i_j, m_j \rangle\}$. Then $\mathcal{M} = \{f(\ldots, m_{j_x}, m_{j_y}, \ldots) : \langle i_{j_x}, m_{j_x} \rangle, \langle i_{j_y}, m_{j_y} \rangle \in \mathcal{T}; i_{j_x} \ne i_{j_y}, \forall x, \forall y\}$, that is, $\mathcal{M}$ is the set of all possible aggregate values which can be obtained from aggregating messages in $\mathcal{T}$.

Guess: Finally, the adversary outputs an aggregate and the associated verification tags, that is, a tuple $\langle hdr, m, T = \{tag_i : i \in hdr\} \rangle$ where $hdr \setminus S' \ne \phi$ and $m \notin \mathcal{M}$.

Result: The adversary wins the game if $\text{Ver}_{\{k_l : l \in hdr\}}(m, T) = 1$. The advantage of the adversary is defined as the probability that the adversary wins the game.

Although the above security notion for AMAC is a natural one inspired by the notion of existential unforgeability against chosen message attacks which is widely adopted in digital signatures and MAC, no AMAC construction seems to be able to achieve this notion [7]. The explanation is that if such a secure AMAC scheme exists, it can be used as a component to convert any semantically secure CDA scheme into one which is IND-CCA2 secure as defined in Section 19.5. Semantically secure CDA exists under some computational assumptions, for example, the CMT scheme [5] (assuming a pseudorandom function exists) or the generic construction given in Ref. [6] (assuming semantically secure public-key homomorphic encryption exists), but an IND-CCA2 secure CDA scheme seems not to exist. Hence, there is a contradiction and a secure AMAC as defined should not exist if no IND-CCA2 secure CDA exists. It should be emphasized that the AMAC security model in Definition 2 is not a contrived one designed to show negative results. The security notion defined above is not achievable probably because supporting strong unforgeability and allowing the aggregation functionality in AMAC are in conflict; that is, the security model in Definition 2 is self-contradictory in some sense. It is not conclusive whether a secure AMAC scheme of some form can be achieved. For instance, relaxing the security model, say, posing more restriction on the adversary, may result in an achievable security notion for AMAC. It is thus fair to say providing end-to-end aggregate integrity in SDA remains an open problem for research.

19.7 CONSTRUCTIONS OF CDA

A number of CDA constructions are described in this section. Most CDA constructions in the literature are additive (i.e., the aggregation operation is addition), partly because it can compute the common statistical quantities like mean and partly because constructing CDA schemes for other aggregation operations seems to be a difficult task. In fact, not much has been done to construct CDA schemes with other aggregation operations. A CDA scheme for computing MAX/MIN is proposed in Ref. [1] but its security is too weak to draw sufficient interest for wide usage in practice and its design will not be discussed here. On the other hand, because IND-CCA2 secure CDA may not exist. Hence, only semantically secure CDA schemes are discussed.

19.7.1 GENERIC CDA CONSTRUCTION [6]

A semantically secure CDA (using local random coins) can be constructed from any semantically secure public-key homomorphic encryption. Note that an asymmetric-key homomorphic encryption is necessary for the CDA construction to achieve semantic security in the presence of compromised nodes.

19.7.1.1 Public-Key Homomorphic Encryption

A public-key homomorphic encryption scheme is a four-tuple (KG, E, D, A). The key generation algorithm KG receives the security parameter 1^λ as input and outputs a pair of public and private keys (pk, sk). E and D are the encryption and decryption algorithms. Given a plaintext x and random coins r, the ciphertext is $E_{pk}(x; r)$ and $D_{sk}(E_{pk}(x; r)) = x$. The homomorphic property allows one to operate on the ciphertexts using the poly-time algorithm A without first decrypting them; more specifically, for any x, y, r_x, r_y, A can generate from $E_{pk}(x; r_x)$ and $E_{pk}(y; r_y)$, a new ciphertext of the form $E_{pk}(x \otimes y; s)$ for some s. The operator $\otimes$ could be addition, multiplication, or others depending on specific schemes; for instance, it is multiplication for RSA [32] or ElGamal [12], and addition for Paillier [29].

19.7.1.2 Concealed Data Aggregation from Public-Key Homomorphic Encryption

Assume there are n source nodes in total. Suppose there exists a semantically secure public-key homomorphic encryption scheme $(KG^{HE}, E^{HE}, D^{HE}, A^{HE})$ with homomorphism on operator $\otimes$. A semantically secure CDA scheme, tolerating up to $n - 1$ compromised nodes, with aggregation function of the form: $f(m_i, m_j) = m_i \otimes m_j$, can be constructed as follows:

Key generation (KG): Run $KG^{HE}(1^\lambda)$ to generate (pk, sk). Set the CDA decryption key $dk = sk$ and each one of the CDA encryption keys to be pk, that is, $ek_i = pk, \forall i \in [1, n]$.

Encryption (E): Given a plaintext data m_i, toss the random coins r_i needed for E^{HE} and output $c_i = E_{pk}^{HE}(m_i; r_i)$. Set the header $hdr_i = \phi$. Output (hdr_i, c_i).

Decryption (D): Given an encrypted aggregate c and its header hdr, run D^{HE} using the private key sk to decrypt c and output $x = D_{sk}^{HE}(c)$ as the plaintext aggregate.

Aggregation (Agg): Given two CDA ciphertexts (hdr_i, c_i) and (hdr_j, c_j), the aggregation can be done using the homomorphic property of the encryption scheme. Generate $c_l = A^{HE}(c_i, c_j)$ and $hdr_l = hdr_i \cup hdr_j$. Output (hdr_l, c_l).

For a total of n source nodes, the above CDA construction is semantically secure against any collusion of at most $n-1$ compromised nodes, assuming that the underlying homomorphic encryption scheme is semantically secure.

One point to note about this generic CDA construction is that in practice there may be no efficiency gain (in terms of energy saving) through aggregating en route unless the network is pretty large (with thousands of nodes) because all known public-key homomorphic encryption schemes (such as Paillier [29] and RSA [32]) have large ciphertexts. For example, the ciphertext of a typical Paillier scheme is 2048 bits long for 2^{80} security. In small networks, the generic CDA construction cannot beat the trivial scheme in which each sensor node sends a distinct ciphertext along the concast tree toward the sink without any aggregation done en route.

19.7.2 WGA [36]

WGA uses Domingo-Ferrer's symmetric-key homomorphic encryption [11] as a building block. Each source node uses the same encryption key ek and the sink's decryption key $dk = ek$. When there is no compromised node, if the underlying symmetric-key homomorphic encryption is semantically secure, then WGA achieves semantic security.

However, as few as one node is compromised, the adversary knows the decryption key and can gain the knowledge of all future aggregates by just passive eavesdropping, that is, not even one-wayness can be achieved if there exist compromised nodes.

19.7.3 CMT [5]

CMT can be considered as a practical modification of the Vernam cipher or one-time pad [35] to allow plaintext addition to be done in the ciphertext domain. Basically, there are two modifications. First, the exclusive-OR operation is replaced by an addition operation. By choosing a proper modulus, multiplicative aggregation is also possible in CMT. CMT can achieve either additive or multiplicative aggregation but not both at the same time. Second, instead of uniformly picking a key at random from the key space, the key is generated by a certain deterministic algorithm (with an unknown seed) such as a pseudorandom function [16]. As a result, the information-theoretic security (which requires the key be at least as long as the plaintext) in the Vernam cipher is replaced with a security guarantee in the computational-complexity theoretic setting in CMT.

The operation of the CMT scheme is as follows: Let p be a large enough integer used as the modulus. Assume the key length is λ bits. Then p could be 2^{λ}. Besides, global random coins are used in CMT, that is, the sink chooses and broadcasts a public nonce to all nodes. In the following description, let $F = \{F_{\lambda}\}_{\lambda \in \mathbb{N}}$ be a pseudorandom function family where $F_{\lambda} = \{f_s : \{0,1\}^{\lambda} \to \{0,1\}^{\lambda}\}_{s \in \{0,1\}^{\lambda}}$ is a collection of functions indexed by a key $s \in \{0,1\}^{\lambda}$. Loosely speaking, given a function f_s from a pseudorandom function ensemble with unknown key s, any PPT distinguishing procedure allowed to get the values of $f_s(\cdot)$ at (polynomially many) arguments of its choice should not be able to tell (with nonnegligible advantage in λ) whether the answer of a new query (with the argument not queried before) is supplied by f_s or randomly picked from $\{0,1\}^{\lambda}$.

Key generation (KG): Randomly pick $K \in \{0,1\}^{\lambda}$ and set it as the decryption key dk. For each $i \in [1, n]$, $ek_i = f_K(i)$ is the encryption key for source node s_i with identity i.

Encryption (E): Given an encryption key ek_i, a plaintext data m_i and a broadcast nonce r from the sink, output $c_i = (m_i + f_{ek_i}(r)) \bmod p$. Set the header $hdr_i = \{i\}$. Output (hdr_i, c_i). Note: each r has to be used once only.

Decryption (D): Given the ciphertext (hdr, c) of an aggregate and a nonce r used in the encryption, generate $ek_i = f_K(i), \forall i \in hdr$. Output the plaintext aggregate $x = \left(c - \sum_{i \in hdr} f_{ek_i}(r)\right) \bmod p$.

Aggregation (Agg): Given two CDA ciphertexts (hdr_i, c_i) and (hdr_j, c_j), compute $c_l = (c_i + c_j) \bmod p$ and $hdr_l = hdr_i \cup hdr_j$ and output (hdr_l, c_l).

The CMT scheme is semantically secure against any collusion with at most $n-1$ compromised nodes if a pseudorandom function is used. There are various constructions of conjectured pseudorandom functions [3,21,26,27] which can be used for this purpose. Because the output length of these functions is usually much larger than the data size, a hash function $h(\cdot)$ is thus needed to match the lengths; Ref. [6] shows that the only requirement on h to preserve the security of CDA is that the output of h follows a uniform distribution if the input is uniformly distributed.

19.8 PROVIDING AGGREGATE INTEGRITY

As discussed in Section 19.6.2, providing end-to-end aggregate integrity may not be achievable in data aggregation. It should be emphasized that "end-to-end" actually means no interaction (i.e., no callback to the aggregating nodes after the aggregation result reaches the sink) is needed. Relaxing the security requirement or using stronger assumptions is a typical means to achieve aggregate integrity protection in wireless sensor networks. This section reviews some directions in this aspect and existing work includes Refs. [8,20,31,37].

19.8.1 Hu and Evans (HE) [20]

HE [20] shows a way to relax the security requirement by giving an in-network data aggregation scheme over a multitier tree which is secure as long as no two compromised nodes are parent-and-child of each other. In practice, it is necessary to guard against multiple compromised nodes. The HE scheme relies on a broadcast authentication scheme called μTESLA[30]. μTESLA is in essence a symmetric-key MAC scheme with delayed tag verification by sending the secret key at a delay after the message and tag are sent.

HE scheme assumes that only the leaf nodes have data to send and each node shares a distinct secret key with the sink from which a temporary key can be computed for each query session. In each query session, the temporary keys are used to generate verification tags. Each aggregation is done in two levels consisting of child nodes, parent nodes, and grandparent nodes. A parent node passes on all child nodes' data and their tags to the grandparent without any aggregation done; along with these, the parent node computes and sends a tag for the aggregate of the child nodes' data. That is, the aggregation tag is sent but not the aggregate itself. The grandparent node then computes the aggregates on its grandchildren's data, one aggregate for each branch/parent, and passes on these results along with the tags it obtains from the parent nodes. Like what the parent nodes do, the grandparent node computes tags for all aggregates it sends upstream. When the final aggregate reaches the sink, the sink discloses all the temporary keys for tag verification. With these keys, each grandparent node can verify whether the parent nodes have passed on correct data from the child nodes and identify malicious parent nodes. The major limitation of this scheme is it can only tolerate up to one malicious node per aggregation group; in other words, it fails if a child node and its parent node collude.

19.8.2 Przdatek, Song, and Perrig (PSP) [31]

Another means to lower the requirement is to allow only one level of aggregation [31]. There is only a single aggregating node (usually the sink) in PSP, just like the scenario depicted in Figure 19.1a. The aggregating node collects authenticated raw data from all sensor nodes, and then computes an aggregate on the raw data together with a commitment to the data based on a Merkle hash-tree [25]. These are sent to the end user directly or via a sink. This scheme assumes the bandwidth between the aggregating node/sink and the end user is the main bottleneck; hence, it attempts to minimize this bandwidth consumption while providing a means to detect a compromised aggregating node. Reference [31] proposes several algorithms in this setting to compute a fairly wide range of aggregates, including average, median, and MIN/MAX. However, the benefit brought about by aggregation could become insignificant in such a setting.

19.8.3 Yang, Wang, Zhu, and Cao (YWZC) [37] and Buttyán, Schaffer, and Vajda (BSV) [4]

The knowledge of the statistical distribution of the sensor measurements (partially or completely) would provide an important piece of information for a scheme to detect malicious manipulation of the aggregation result. The most straightforward technique is to perform hypothesis testing (on whether the aggregate has been tampered with) at the sink over the received aggregate if complete knowledge of the statistical distribution of sensor data is assumed. Refs. [4,37] assume some knowledge on the sensor data distribution to detect any deviation of the aggregation result that is too far away from a certain threshold. The deviation could be due to malicious manipulation, temporal variability of the quantity being measured, or measurement errors. But all such deviations are rejected in Refs. [4,37].

YWZC and BSV use different outlier elimination algorithms to detect or filter out deviated data and to identify suspected malicious nodes. YWZC uses the maximum normalized residual test, also known as the Grubb's test, although BSV uses random sample consensus. YWZC divides an aggregation tree into sub-trees, each of which reports its aggregate to the sink directly, and the outlier algorithm is run on the received aggregates to identify suspected sub-trees. Whereas, BSV runs tests on the overall aggregate.

19.8.4 CHAN, PERRIG, AND SONG (CPS) [8]

Similar to zero knowledge proof, interaction is a powerful means to attest the correctness of data aggregation. When callbacks to the aggregating node is possible, it is possible to construct efficient schemes for verifying the correctness of an aggregation process. The HE and YWZC schemes are examples of this; for instance, the YWZC scheme makes use of a commit-and-attest mechanism requesting nodes in a suspected sub-tree to prove they perform the aggregation properly.

CPS [8] also gives a commit-and-attest mechanism. This scheme only works for additive aggregation but, unlike [37], does not assume any knowledge of the statistical distribution of the data. The idea is to require a node to commit to its choice of intermediate aggregation result during the process and then request sensor nodes to verify that their contributions to the final aggregate are correctly incorporated by their parents after the aggregate reaches the sink.

19.9 CONCLUSIONS

In this chapter, the security of data aggregation in wireless sensor networks is studied. Although data aggregation is an essential paradigm to reduce energy consumption so as to maximize the lifetime of a sensor network, it also opens up new opportunities for an attacker and poses privacy/secrecy problems. This chapter gives a rigorous treatment to the security of data aggregation, discussing the main goals of an SDA scheme, namely, end-to-end privacy and aggregate integrity. Security models for end-to-end privacy and aggregate integrity are given. Impossibility results are discussed. A survey on existing constructions in the literature is presented.

REFERENCES

[1] M. Acharya, J. Girao, and D. Westhoff, Secure comparison of encrypted data in wireless sensor networks, in *Proc. WiOpt'05*, p. 47–53, April 2005.

[2] M. Bellare, Practice-oriented provable-security, in *Proc. First International Workshop on Information Security (ISW'97)*, LNCS, Springer-Verlag, Vol. 1396, pp. 221–231, 1998.

[3] M. Bellare, R. Canetti, and H. Krawczyk, Keying hash functions for message authentication, in *Advances in Cryptology—CRYPTO 1996*, LNCS, Springer-Verlag, Vol. 1109, pp. 1–15, 1996.

[4] L. Buttyán, P. Schaffer, and I, Vajda, RANBAR: RANSAC-based resilient aggregation in sensor networks, in *Proc. Fourth ACM Workshop on Security of Ad Hoc and Sensor Networks (SASN'06)*, pp. 83–90, October 2006.

[5] C. Castelluccia, E. Mykletun, and G. Tsudik, Efficient aggregation of encrypted data in wireless sensor networks, in *Proc. Second International Conference on Mobile and Ubiquitous Systems (MobiQuitous'05)*, pp. 109–117, July 2005.

[6] A.C.-F. Chan and C. Castelluccia, On the privacy of concealed data aggregation, in *Proc. 12th European Symposium on Research in Computer Security (ESORICS 2007)*, LNCS, Springer-Verlag, Vol. 4734, pp. 390–405, September 2007.

[7] A.C.-F. Chan and C. Castelluccia, On the (Im)possibility of aggregate message authentication codes, in *Proc. IEEE International Symposium on Information Theory (ISIT 2008)*, July 2008.

[8] H. Chan, A. Perrig, and D. Song, Secure hierarchical in-network aggregation in sensor networks, in *Proc. 13th ACM Conference on Computer and Communications Security (CCS 2006)*, pp. 278–287, October 2006.

[9] B. Chevallier-Mames, P. Paillier, and D. Pointcheval, Encoding-free ElGamal encryption without random oracles, in *Public Key Cryptography (PKC 2006)*, *LNCS*, Springer-Verlag, Vol. 3958, pp. 24–26, 2006.

[10] D. Dolev, C. Dwork, and M. Naor, Nonmalleable cryptography, *SIAM Journal on Computing*, 30(2):391–437, 2000.

[11] J. Domingo-Ferrer, A provably secure additive and multiplicative privacy homomorphism, in *Proc. 5th International Conference on Information Security (ISC 2002)*, *LNCS*, Springer-Verlag, Vol. 2433, pp. 471–483, September 2002.

[12] T. ElGamal, A public key cryptosystem and a signature scheme based on discrete logarithms, *IEEE Transactions on Information Theory*, IT-30(4):469–472, July 1985.

[13] A. Fiat and M. Naor, Broadcast encryption, in *Advances in Cryptology—CRYPTO 1993*, *LNCS*, Springer-Verlag, Vol. 773, pp. 480–491, 1994.

[14] R. Gennaro, H. Krawczyk, and T. Rabin, Secure hashed Diffie–Hellman over non-DDH groups, in *Advances in Cryptology—EUROCRYPT 2004*, *LNCS*, Springer-Verlag, Vol. 3027, pp. 361–381, 2004.

[15] J. Girao, D. Westhoff, and M. Schneider, CDA: Concealed data aggregation in wireless sensor networks, in *Proc. IEEE International Conference on Communication (ICC'05)*, pp. 3044–3049, May 2005.

[16] O. Goldreich. *Foundations of Cryptography: Part 1*. Cambridge University Press, Cambridge, UK, 2001.

[17] O. Goldreich. *Foundations of Cryptography: Part 2*. Cambridge University Press, Cambridge, UK, 2004.

[18] S. Goldwasser and S. Micali, Probabilistic encryption, *Journal of Computer and System Sciences*, 28(2):270–299, 1984.

[19] S. Goldwasser, S. Micali, and R. Rivest, A secure signature scheme secure against adaptive chosen-message attacks, *SIAM Journal on Computing*, 17(2):281–308, 1988.

[20] L. Hu and D. Evans, Secure aggregation for wireless networks, in *Proc. Application and the Internet Workshop 2003*, pp. 384–391, January 2003.

[21] T. Iwata and K. Kurosawa, OMAC: One-key CBC MAC, in *Fast Software Encryption (FSE 2003)*, *LNCS*, Springer-Verlag, Vol. 2887, pp. 129–153, 2003.

[22] J. Katz and M. Yung, Characterization of security notions for probabilistic private-key encryption, *Journal of Cryptology*, 19(1):67–95, 2006.

[23] M. Luby. *Pseudorandomness and Cryptographic Applications*. Princeton University Press, Princeton, NJ, 1996.

[24] S.R. Madden, M.J. Franklin, J.M. Hellerstein and W. Hong, TAG: A tiny aggregation service for ad-hoc sensor networks, in *Proc. Fifth Annual Symposium on Operating Systems Design and Implementation*, pp. 131–146, 2002.

[25] R. Merkle, A digital signature based on a conventional encryption function, in *Advances in Cryptology—CRYPTO 1987*, *LNCS*, Springer-Verlag, Vol. 293, pp. 369–378, 1987.

[26] M. Naor and O. Reingold, Number-theoretic constructions of efficient pseudo-random functions, in *Proc. IEEE Symposium on Foundations on Computer Science (FOCS'97)*, pp. 458–467, 1997.

[27] M. Naor, O. Reingold, and A. Rosen, Pseudorandom functions and factoring, *SIAM Journal on Computing*, 31(5):1383–1404, 2002.

[28] M. Naor and M. Yung, Public-key cryptosystems provably secure against chosen-ciphertext attacks, in *Proc. ACM Symposium on Theory of Computing (STOC 1990)*, pp. 427–437, 1990.

[29] P. Paillier, Public-key cryptosystems based on composite degree residuosity classes, in *Advances in Cryptology—EUROCRYPT 1999*, *LNCS*, Springer-Verlag, Vol. 1592, pp. 223–238, 1999.

[30] A. Perrig, R. Szewczyk, V. Wen, D. Culler and D. Tygar, SPINS: Security protocols for sensor networks, *Wireless Networks*, 8(5):521–534, September 2002.

[31] B. Przydatek, D. Song, and A. Perrig, SIA: Secure information aggregation in sensor networks, in *Proc. First International Conference on Embedded Networked Sensor Systems*, pp. 255–265, 2003.

[32] R. Rivest, A. Shamir, and L. Adleman, A method for obtaining digital signatures and publickey cryptosystems, *Communications of ACM*, 21(2):120–126, February 1978.

[33] C.E. Shannon, Communication theory of secrecy systems, *Bell Systems Technical Journal*, 28:656–715, 1949.

[34] V. Shoup and R. Gennaro, Securing threshold cryptosystems against chosen ciphertext attack, *Journal of Cryptology*, 15(2):75–96, 2002.

[35] G.S. Vernam, Cipher printing telegraph systems for secret wire and radio telegraphic communications, *Journal of the American Institute of Electrical Engineers*, 45:105–115, 1926. See also US patent #1,310,719.

[36] D. Westhoff, J. Girao, and M. Acharya, Concealed data aggregation for reverse multicast traffic in sensor networks: Encryption, key distribution, and routing adaption, *IEEE Transactions on Mobile Computing*, 5(10):1417–1431, 2006.

[37] Y. Yang, X. Wang, S. Zhu, and G. Cao, SDAP: A secure hop-by-hop data aggregation protocol for sensor networks, in *Proc. 7th ACM International Symposium on Mobile Ad Hoc Networking and Computing (MobiHoc 2006)*, pp. 356–367, 2006.

20 Message Authentication in Surveillance Networks

Raymond Sbrusch and T. Andrew Yang

CONTENTS

Wireless sensor networks (WSNs) simplify the collection and analysis of data from multiple locations. The self-organization capabilities of wireless sensor networks enable rapid deployment of target tracking and intrusion detection applications in hostile environments. However, sensor networks deployed in adversarial environments must be fortified against attacks. This chapter examines the threats against WSNs and surveys countermeasures that protect their communications with origin integrity and data integrity. This chapter also provides an analytical framework to solve the security problem in WSNs deployed for surveillance and target tracking.

20.1 INTRODUCTION

WSN simplify the collection and analysis of data from multiple locations. Target tracking and perimeter intrusion detection applications benefit from the ad hoc deployment and self-organization capabilities of WSNs. However, sensor networks deployed in hostile environments must be fortified against attacks by adversaries. This chapter examines the constraints that make WSN surveillance challenging and evaluates algorithms that provide origin integrity and data integrity for WSNs.

This chapter develops a framework for evaluating authentication mechanisms to protect target-tracking methods for routing and self-organization of a WSN. This includes authentication of sent messages to assure that they have not been altered (aka message integrity), and authentication of the sender to assure that the messages are not forged (aka origin integrity). The process begins with a survey of security threats and risk mitigation strategies common to all WSNs. This survey includes mainly attacks against origin and message integrity, as well as those against confidentiality and availability. Selection of an elegant authentication solution requires a survey of current unicast and broadcast message authentication protocols for WSNs. The protocols will be contrasted to facilitate selection of the most appropriate one for target tracking WSNs.

20.1.1 CHAPTER OUTLINE

Applications of WSNs span across diverse fields from military surveillance across enemy lines to monitoring of avian breeding habits in sensitive wildlife habitats. Sensor networks simplify the simultaneous collection and organization of data from multiple locations, which may be unreachable, inhospitable, or even hostile environments. The wired counterparts of these sensors have been utilized in industrial and military applications for decades, but the simple constraint of stringing cables to the monitored site limits the reach of this communication mechanism. Merging wireless communications with sensor network capabilities enables rapid deployment and reduces the cost of the infrastructure. However, adopting wireless communications introduces a new set of challenges. The first contribution of the chapter is an expanded risk analysis of WSNs used for target tracking, including examination of resource constraints and other challenges.

The chapter then follows the evolution of target-tracking methodologies from conventional protocols of direct communication (DC) to tree topologies organized for network longevity. Target-tracking techniques may focus on intruders penetrating a border in a two-dimensional space or may try to provide comprehensive coverage of a detection region. These techniques commonly impose a trade-off between accuracy of the tracking algorithm and efficient use of sensor resources, including energy, computation power, and communication bandwidths. The naive surveillance strategy delivers high accuracy by enabling the sensing mechanism on every node in the network and forcing each node to send alerts directly to the base station. This strategy reduces network longevity and increases contention for the radio channel. Advanced strategies organize nodes into clusters with child nodes detecting an intrusion and intermediary nodes relaying messages between children and the base station. This chapter illustrates how methods, including scalable tracking using networked sensors (STUN) [1], low energy adaptive clustering hierarchy (LEACH) [2], optimized communication and organization (OCO) [3], and guard duty alarming technique (GDAT) [4], balance network longevity and detection accuracy.

Secure WSN deployments remain elusive because of resource constraints; however, operation of sensor networks in hostile locations requires secure, authenticated communication. This chapter categorizes threats against WSNs and surveys authentication protocols that can be employed to mitigate threats against WSNs. The survey tracks the evolution of sensor network authentication from widely adopted standards, including security protocols for sensor networks (SPINS) [5] and TinySec [6], to more comprehensive strategies. The chapter concludes comparing power utilization characteristics of the security standards.

20.2 WIRELESS SENSOR NETWORKS

The term WSN describes an association among miniaturized embedded communication devices that monitor and analyze their surrounding environment. The network is composed of many tiny nodes sometimes referred to as motes [7]. A node is made up of the microcontroller, the sensor(s), the radio communication component, and a power source. Wireless sensor nodes range in size from a few millimeters to the size of a handheld computer. Regardless of size, sensor nodes share common constraints. This section identifies the unique challenges of WSNs and leads into a survey of organization, communication, and tracking algorithms.

20.2.1 CHARACTERISTICS OF WIRELESS SENSOR NETWORKS

WSNs are deployed for a diverse variety of applications, each characterized by a unique set of requirements. Romer and Mattern [8] organized a workshop with the European Science Foundation to classify application domains, define hardware and software requirements for each domain, and determine how to coordinate research into WSNs. Their results counter the general conviction that the most common scope for research into WSNs centers on military applications, which involve large-scale ad hoc networks with homogeneity among tiny, resource-constrained, immobile sensor nodes. The results uncover the inadequacy of this rigid definition. The participants in the workshop extract 14 characteristics of commercial and research networks. These characteristics consist of deployment, mobility, resources, cost, energy, heterogeneity, mobility, infrastructure, topology, coverage, connectivity, size, lifetime, and quality of service. These attributes influence design decisions, especially those related to security. They will be employed to characterize target-tracking WSNs in this chapter.

Although the classical sensor network consisted of homogeneous devices, contemporary sensor networks incorporate modular design and make use of heterogeneous nodes that fulfill unique requirements. For example, some nodes include a GPS sensor that other nodes can query to determine their location. Others may include interfaces to the Internet through satellite or cellular communications. Although radio frequency is the most common communication modality, information can also be transmitted via laser, sound, and diffuse light.

These communication capabilities support an assortment of network infrastructures. In a basic infrastructure-organized network, nodes can only communicate with a base station. The opposite is true in an ad hoc network where there is no base station or communication infrastructure. In this case, each node can communicate with any other node. The communication infrastructure influences network topology. In some cases, each node must be within the radio range of any other node because messages can only travel across a single hop. Networks organized into a graph-like topology allow routing of messages across multiple hops.

Some applications can achieve their goals with a network of sparsely deployed sensors. Others require a densely populated network with redundant nodes available. Network topology and coverage requirements determine the network size. Networks may range in size from thousands of nodes to only a few. Target-tracking networks such as OCO and GDAT have evolved to embrace dense populations as a means to improve efficiency while maintaining accuracy.

20.2.2 Challenges of Target Tracking

Target-tracking WSNs face a unique set of challenges when compared to sensor networks deployed for other applications such as building automation or habitat monitoring. Threats against the network increase with the hostility of the surveilled area and range from simple node destruction to sophisticated network-based exploits. The sensor network must be able to monitor intrusions and securely deliver alerts even if a large percentage of the nodes have been compromised.

20.2.2.1 Resource Constraints

Secure and reliable wireless deployments remain elusive because of four prohibitive constraints. WSNs are characterized by limited computational abilities, small amounts of permanent and temporary storage, and finite energy resources. The wireless communication medium introduces reliability and security risks that, if left unresolved, preclude the use of WSNs in adversarial locations. Conventional security mechanisms in use on the Internet are usually not applicable to WSNs because of the limited resources of the nodes. Sensor network packets commonly contain 32 bytes or less [9]. Conventional security mechanisms that add 16 bytes of network overhead are inappropriate for use in WSNs because they will quickly drain the sensor power source. On some platforms, each byte transmitted consumes about as much power as executing 4000–5000 instructions [6].

Besides the requirement of conserving energy, a WSN deployed for target tracking must provide a robust communication channel, real-time alerting, resistance to tampering and stealth. This paradox forces network designers to exploit ways to improve efficiency while maintaining a high level of accuracy by, for example, incorporating redundancy, clustering, etc. into the network.

20.2.2.2 Physical Threats

Deployment of a sensor network in a safe environment can be carefully planned and controlled. Safer environments can benefit from more deliberate node placement. For example, in the Twenty-nine Palms experiment [7], researchers at U.C. Berkeley deployed wireless sensor nodes to track the path of tanks along a predetermined path. A small number of sensors were carefully dropped from an unmanned aerial vehicle at points near the tank path. The sensors detected the presence of the tanks using a magnetometer and sent alerts over a 916.5-MHz radio channel to the base station. This application detected the arrival of the tank and calculated its speed and direction as it moved along the path.

In contrast, the challenges commence as soon as sensor nodes are deployed in a hostile environment, where they become subject to numerous threats. An adversary can physically destroy or displace nodes or launch network attacks against node communications. External forces such as explosions or earthquakes can unpredictably relocate nodes. Aerial dispersion results in a highly random coverage pattern with coverage gaps in some areas and excessive redundancy in others. Once sensors are deployed in an adversary's domain, there are few chances to modify the coverage patterns or refresh dying nodes.

20.3 TARGET-TRACKING METHODS

Detection and tracking of objects moving through a surveilled region remains in the forefront of WSN research. In WSNs deployed for tracking objects, the effectiveness of organization and communication methods depends on many factors, including the node deployment tactic, the layout of the surveyed landscape, and the path of the intruder. Careful selection of appropriate communication and organization protocols can assure longer life for the network. Target-tracking strategies can be evaluated on how they organize to form a network, sense intruders, communicate results to the base station, and reorganize to handle exhausted or damaged nodes. This section surveys target-tracking methods and shows the trend in target-tracking research toward improving efficiency and network longevity.

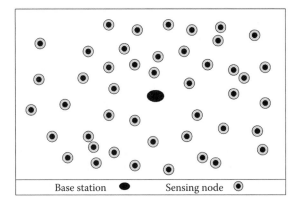

Base station ●   Sensing node ◉

FIGURE 20.1   DC network.

20.3.1   SINGLE-HOP COMMUNICATION

The simplest organization and communication strategy, known as DC [2], requires each node to transmit alerts directly to the base station. This single-hop strategy suffers from rapid depletion of resources. In DC, each sensor node monitors the environment with its sensing module and transmits alerts directly to the base station over a wireless channel. Although this delivers the most accurate detection of intrusions from any attack vector, DC faces a number of setbacks.

Figure 20.1 illustrates a DC network composed of sensing nodes and a base station. Each sensor node must have a radio transmitter powerful enough to reach the base station. This limits the effective range of the network and forces the base station to allocate enough radio channels for communication with all sensors. Each sensor node is required to maintain an activated sensor module throughout its lifetime. This module continuously draws power from the node battery. Because nodes may have overlapping sensor coverage areas, redundant alerts may be sent to the base station, again unnecessarily depleting the battery. While theoretically effective at tracking targets, DC proves inefficient and impractical for use in real-world applications.

20.3.2   HIERARCHY TREES

Organizing nodes into hierarchical network topologies or clusters can increase the coverage area and prune redundant alerts. In a tree-structured network organization, senor nodes occupy graph vertices, and graph edges signify DC links between nodes. STUN [1], for example, organizes its sensor nodes into a tree-structured network; it aims to track a large number of objects as they move through the surveilled region. STUN organizes sensing nodes into a linear graph. Leaf nodes at the bottom of the STUN tree function as sensing nodes. Nodes at intermediate levels relay messages from sensing nodes to the base station at the root of the tree. Intermediary nodes store information about the presence of detected objects. When leaf nodes send detection messages toward the base station, the intermediate nodes compare the alerts to the information they already recorded and drop the messages if they are redundant. Pruning redundant alerts lowers communication costs.

Although hierarchy trees are an improvement over DC, they may have their own deficiencies. STUN, for instance, does not account for the physical proximity of the sensor nodes when building the hierarchy tree; adjacency of nodes in STUN does not necessarily imply physical nearness of the nodes. Sensor networks can gain more efficiency and accuracy by accounting for the physical proximity of the sensor nodes after they are dispersed. Another potential problem with tree-structured methods is the need to form multihop communications between the sensing nodes and the base station (as the root of the tree). When the height of the tree increases, the cost of communications also increases, leading to poor scalability when the sensor network is deployed in a large area.

20.3.3 CLUSTERING METHODS

Clustering methods capitalize on the ability to detect node proximity when forming the hierarchy tree. Like a hierarchy tree, clustering algorithms prune redundant messages as they are sent to the base station. They also conserve energy by assuring low-cost radio communication between sensor nodes and intermediate nodes. The LEACH [2] illustrates how knowledge of actual node proximity can improve efficiency of hierarchy tree-based organizations. During the LEACH setup phase, nodes elect themselves to be local cluster heads and broadcast messages to their neighbors advertising their status.

The cluster heads act as intermediate nodes and relay messages between a neighborhood of leaf nodes and the base station. To qualify for the role of cluster head, a node must have enough power to relay messages to the base station. Sensor nodes analyze cluster-head advertisements to determine the cost of wireless communication with the cluster head. They choose which neighborhood they want to join by selecting the cluster head that requires the least radio transmission power. Figure 20.2 shows a LEACH network organized into nine neighborhoods. The role of cluster head consumes more energy than the role of sensor node, thus a LEACH network periodically repeats the setup process, electing new nodes to the cluster-head position.

LEACH distributes the cost of serving as cluster head among all nodes in the network, thus increasing the lifetime of the network. Like STUN, LEACH still lacks knowledge of the surveilled terrain. Because any node could theoretically elect itself as the cluster head, LEACH requires that all nodes initially have enough radio power to communicate directly with the base, even if they are on the perimeter of the surveilled region. LEACH also suffers from gaps in coverage because the cluster-head election process lacks knowledge of the surveilled region. The election process only evaluates the strength of the node's radio, not the coverage patterns of neighboring nodes. As with other proposals introduced until now, all nodes in the network activate their sensing component and their radio interface.

20.3.4 ENERGY-EFFICIENT CLUSTERING

Some WSN target-tracking approaches assume that an intruder will enter the detection region from the perimeter. These techniques conserve energy by allowing interior nodes to sleep until an intrusion occurs, thus increasing the longevity of the network. However, this strategy leaves the interior of the network vulnerable to attacks that bypass the perimeter, such as aerial or insider attacks. Rababaah and Shirkhodaie [4] propose a clustering and tracking technique that significantly increases efficiency comparable to basic clustering techniques while maintaining sensing throughout the detection region. This method, called GDAT, is inspired by military practice of rotating guard duty. While most soldiers

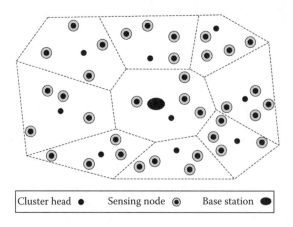

FIGURE 20.2   LEACH network organization.

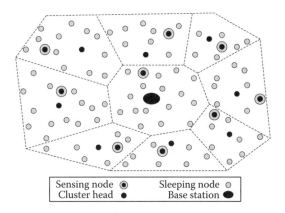

FIGURE 20.3   GDAT organization.

sleep, one soldier is ordered to remain on guard duty for an assigned period. In GDAT, one sensing node is on guard, actively sensing while its other cluster members sleep.

GDAT requires two types of nodes: head nodes and sensing nodes. Head nodes are provisioned with both strong radios to communicate with the base station and with global positioning receivers to append location information to intrusion alerts. Sensing nodes are equipped with intrusion detection devices and only require sufficient power to communicate with the head nodes.

Following network deployment, both head nodes and sensing nodes are awake. Head nodes broadcast advertisements requesting that sensing nodes adopt them as their cluster head. Sensing nodes accept the first offer they receive, record the ID of the head node, and send an acceptance message to the head node. Each head node can accept 32 sensing nodes into its cluster.

The tracking phase begins once clustering is complete. The cluster head assigns one sensing node to guard duty for a small interval and instructs the other sensing nodes to sleep. The sensing node can reside anywhere within the cluster. When the sensing node detects an intrusion, it alerts the head node, which appends GPS coordinates to the alert and forwards it to the base station. Following the intrusion, the sensing node will remain on duty until its shift is complete. At that point, the head node will assign another sensing node to guard duty. Figure 20.3 highlights the advantage of GDAT over the LEACH network depicted in Figure 20.2. In LEACH, all nodes drain energy with their activated sensors and radios. In GDAT, the majority of nodes sleep.

Important assumptions affect the success of GDAT. Most significantly, GDAT assumes that head nodes and sensors will be normally distributed across a detection region following a probability distribution function. Factors such as rugged terrain, battle, or natural forces can adversely influence the distribution of head nodes and sensing nodes. In comparison to image-processing techniques, which will be described in the following subsection, GDAT leaves the perimeter porous. Even though it delivers higher accuracy throughout the network, it may not detect an intruder crossing the perimeter as quickly as an image-processing technique. The authors indicate that they will include image-processing techniques in future GDAT research.

20.3.5   IMAGE-PROCESSING TECHNIQUES

Image-processing techniques have been shown to be a more efficient and accurate target-tracking method than conventional graph-based methods [3,10]. Image-processing techniques map physical node location onto a grid representing the coverage region and then assign nodes on an occupation based on their location in the grid. This improves efficiency in part by activating the sensing components of perimeter nodes and placing redundant nodes in an energy preserving sleep state until the perimeter is breached. The OCO method [3,10], for example, efficiently secures the perimeter of the detection region and reorganizes the network when a node is damaged or lost.

The OCO method segments a sensor network's lifetime into four phases. Upon mote deployment, the sensor network enters a position collection phase. The base station broadcasts a message to all nodes and requests that they report their ID and position. The base's neighbors acknowledge the request and broadcast a request for the ID and position of all their neighbors. The process repeats until all nodes are accounted for. The base station maps each node's unique ID and location onto a map representing the surveilled region.

The next phase, processing, selects the minimal set of sensor nodes required to cover the detection region. It achieves this with a three-step process for finding nonredundant nodes. OCO defines a redundant node as one whose sensing radius overlaps with the sensing radius of other nodes in the network. First, the base station initializes an empty list of nonredundant nodes and adds itself to the list. It then analyzes the map to identify sensor nodes with nonoverlapping coverage and adds them to the list. The base station then identifies areas of the graph that are not covered by a sensor node from the list. It assigns a nearby node responsibility for that region. All nodes not on the list after this step are considered redundant.

Identification of perimeter nodes (a.k.a. border nodes) and construction of the hierarchy tree also occur in the processing phase. The image-processor at the base station analyzes each pixel in the coverage map to determine if it is covered by a node from the nonredundant node list. It then assigns pixels in the map that are covered by a node of value 1. Regions of the graph that are not covered by a sensor node are assigned the value 0. When the image processor locates a pixel p with value 1, it checks the value of neighboring pixels. If any neighbor has a value of 0, then the node encompassing pixel p is considered a border node.

Once border nodes are defined, the base station must find the shortest path from itself to each of the active nodes, including border and intermediate nodes. Border nodes can route alerts to the base station by relaying them through multiple interior forwarding nodes, as shown in Figure 20.4. After the path is found, the base station broadcasts a message activating the sensor and the radio modules on the border nodes. The base station instructs redundant nodes to enter a sleeping state where their radio and sensor modules are temporarily inactive. Forwarding nodes deactivate their sensor module, yet keep their radio receiver on.

Once the network topology is defined, the network enters the target-tracking phase. Intruders are assumed to enter the surveilled area from outside the perimeter. When a node detects an intruder, it sends an alert to the base station via the forwarding nodes. It will continue to send periodic alerts to the base station although the intruder remains within its sensing radius. The sensing node will also

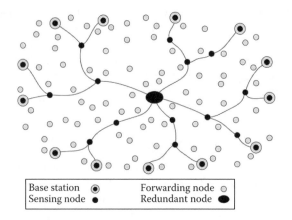

FIGURE 20.4   OCO network topology.

alert its neighbors of the intrusion. The neighbor notification strategy depends on the capabilities of the alerting sensor. Nodes capable of tracking multiple objects will only send two alerts to its neighbors; one when an intruder enters the coverage area and another when the intruder exits. A node that can only track one object at a time will periodically broadcast alerts to its neighbors while an intruder is present. When the neighbors receive an alert, they activate their sensor modules and try to determine if the intruding object has entered their sensing radius.

The maintenance phase of OCO manages reconfiguration of the network when a node is destroyed, moved, or depleted of power. To detect damaged nodes, status messages are periodically exchanged between parents and children. When a node fails to send the status message on time, its peer assumes that it has been damaged. The node that was expecting to receive the status message will notify the base that its peer has vanished. When nodes move, a repositioned node will broadcast an alert stating that it has been moved. Any nodes receiving this alert will forward the message to the base. When a node detects that its power level has fallen below a predefined threshold, it sends a notification toward the base. When the base station receives notification of damage, depleted power, or movement of a node, it will trigger a local reorganization algorithm that reorganizes the affected area to cover the gap left by the affected node(s). The reorganization algorithm may activate one or more of the redundant nodes.

OCO was evaluated in a network simulator against DC and LEACH. The simulation was run under various scenarios, respectively, with no intruders, one intruder, and multiple intruder objects. As exhibited in the simulation experiments, OCO has several strengths. When no intruders are present, an OCO network will outlast a LEACH network. An OCO network with multiple intruders will reach a constant rate of energy dissipation regardless of the number of intruders. The simulation shows that OCO is nearly as accurate as DC in detecting intruders penetrating the border. OCO surpasses the other two algorithms in terms of least cost per detected object.

20.4   SECURITY IN WSN FOR SURVEILLANCE

Security risks in WSNs include threats to the confidentiality, integrity, and availability of the system. Security mechanisms used on the Internet are not easily adaptable to sensor networks because of the limited resources of the sensors and the ad hoc nature of the networks. The adoption of efficient algorithms to mitigate security risks has not kept pace with the rate of miniaturization of the sensor devices. This section underscores the challenges of securing sensor network communications and illustrates general attacks against sensor networks.

20.4.1   SECURITY GOALS

Security assessments of any application focus on the five fundamental tenets of information security: confidentiality, origin integrity, data integrity, nonrepudiation, and availability. The definitions used in this section are derived from [11,12].

Confidentiality means the concealment of information from unauthorized entities. Mechanisms used to achieve confidentiality include access control mechanisms and cryptography. Cryptography scrambles, or encrypts, data to generate ciphertext unintelligible to any unauthorized viewer. The data can be made comprehensible to an authorized viewer who knows the secret key. Semantic security implies a stronger guarantee of confidentiality. Semantic security requires that repeated encryption of a message M would yield unique ciphertext each round. This limits the ability of an eavesdropper to interpret the plaintext even after observing multiple encryptions of the same message. Use of initialization vectors (IVs) seeded with a counter or a nonrepeating nonce provides semantic security.

Origin integrity, also known as authentication, refers to the trustworthiness of the source of data. It means that the receiver of a message can trust that the sender of the message is truthfully who it claims. An intruder should not be able to send a fabricated message and have it treated as a legitimate message from a trusted peer. Data integrity means that the user of the data can trust that the content of the information has not been changed in any way by an unauthorized intruder or improperly modified by an authorized user. Because similar mechanisms provide origin integrity and data integrity, they are commonly grouped under the moniker integrity. Integrity overshadows other security goals because of its influence on the reliability of the system and its output. In a robust WSN, the information contained in a message holds a lower priority than the integrity and authenticity of the message. For example, child nodes commonly send HELLO messages to parents to inform them that they are still active. Concealing this message is less important than assuring that it originated from a legitimate node and not an impostor who had perhaps compromised or destroyed that node.

Nonrepudiation means that the sender of a message should not be able to deny later that he ever sent that message. In the predigital world, one achieved nonrepudiation with a simple handwritten signature. In cryptography, it implies that authentication and data integrity can be certified with a high level of assurance and it cannot later be refuted. Nonrepudiation is a critical security service and must be guaranteed in applications that involve financial and business transactions, where accountability of actions is critical to ensure success of the applications. Digital signatures provide nonrepudiation.

Availability implies that an authorized user should be able to use the information or resource as required. In a WSN, the wireless communication link must remain available for the network to sustain operations.

Security literature commonly condenses the five security goals into the C.I.A. triad, signifying confidentiality, integrity, and availability.

20.4.2 CHALLENGES

The lack of efficient authenticated messaging exposes all layers of the sensor network protocol stack to potential compromise. Without link-layer authentication, an attacker may inject unauthorized packets into the network. This may be used to introduce collisions and force legitimate nodes into an infinite waiting state [13]. Network layer attacks against routing protocols give the attacker the ability to cause routing loops, delay messages, or selectively drop messages [14]. WSNs deployed for tracking targets provide valuable application layer notifications about the location of the target. Without authentication, the attacker can perpetrate attacks such as dropping intruder notifications, spoofing intruder notifications to create a diversion, or forcing the entire network into a continual state of reorganization.

In WSNs, the need for integrity surpasses all other security goals. Data integrity and authentication create a foundation for a highly available and trustworthy network. Although many authentication schemes have been conceived for WSNs, none of them is a panacea. Algorithms for unicast message authentication, for example, do not meet the requirements for authenticating broadcast messages. Similarly, algorithms that mimic the asymmetry of public key systems by dividing time into slots violate the real-time constraints of intrusion notification systems. Results in Ref. [15] show that the best solution tailors the authentication mechanism to the requirements of the application.

20.4.3 ATTACKS AGAINST SENSOR NETWORKS

Wood and Stankovic [13] provide a comprehensive survey of attacks against WSNs and describe strategies that have been used to reduce their impact. The discussions of attacks against sensor networks in the rest of this section are mainly derived from their work. The analysis classifies attacks following the OSI reference model for network protocol design.

Physical tampering poses a threat to sensors. If sensors are distributed in an unprotected area, an attacker could destroy the nodes or collect the sensors, analyze the electronics, and steal cryptographic keys. This complicates the process of bootstrapping newly deployed sensors with cryptographic keying material. To protect against this, sensors must be tamper-proof or they must erase all permanent and temporary storage when compromised. Secure key-rotation mechanisms can also mitigate the threat of stolen cryptographic keys.

Jamming attacks against wireless radio frequencies affect the availability of the network. Although it is most efficient to program sensors to communicate on one specific wireless frequency, an attacker could easily broadcast a more powerful signal on the same frequency and introduce interference into the communications channel. Spread spectrum technologies such as frequency-hopping spread spectrum alleviate the impact of jamming; however, complex channel hopping patterns reduce battery life. Nodes could also try to detect jamming and sleep until the jamming stops, resulting in a temporary, self-induced denial-of-service (DoS).

Link-layer protocols face similarly challenging threats. Attackers can introduce collisions that force communicating nodes to retransmit frames. Following a collision, a node must back off and wait for the channel to clear before attempting to resend. The attacker can continually introduce collisions until the victim runs out of power. Although error-detecting mechanisms suffice for common transmission errors, they do not reduce the influence of maliciously generated collisions. Collisions maliciously injected near the end of a legitimate frame rapidly exhaust the resources of the legitimate node. Authentication cannot alleviate these physical and link-layer attacks; other control measures are needed to counter these attacks.

Network layer attacks take advantage of the ad hoc organization of WSNs. Any node in the network can become a router, forwarding traffic from one node to another. Tree-structured WSNs, for example, typically designate some of the sensor nodes as forwarding nodes because of their location on the coverage map. By manipulating routing information, the attacker can shape the flow of traffic. The simplest attack compromises a routing node and forces it to drop messages, creating a network black hole. The attacker can also selectively delay messages routed by the compromised node. In a wormhole attack, the adversary tunnels messages destined for one part of the network through a path under enemy control. Wormhole attacks facilitates eavesdropping, message replay, or disconnection of a segment of the network.

One technique to create black holes circumvents the way routing protocols organize the network. Nodes typically adopt the router that broadcasts route advertisements with the strongest radio signal. This strategy reduces the energy required for a node to communicate with its default router. An attacker can manipulate this strategy to convince legitimate nodes that it requires the least communication overhead. One way to accomplish this is with HELLO floods [14]. To perpetrate a HELLO flood, the attacker repeatedly broadcasts HELLO messages at higher power than every other node. HELLO floods can be used by an attacker to persuade the network nodes that it is a legitimate neighbor and a reliable next hop. An attacker can also corrupt shared routing tables by spoofing, manipulating, or replaying routing messages.

Internet style attacks have their analogue in WSNs. Misdirection attacks, such as the Internet smurf attack, work in sensor networks. The attacker can send multiple messages to broadcast addresses with a source address forged to the intended victim's address. The broadcast responses will overwhelm the victim, flood its communication channel, and drain its power. Filtering the legitimate messages from the responses in a smurf attack requires a hierarchy not present in many WSN routing protocols. A similar attack, called a Sybil attack, targets systems that select peers based on their reputation. In a Sybil attack, the adversary sends a large number of fabricated messages that appear to be forwarded from other nodes. Legitimate nodes begin to trust the attacker because it seems to fairly route traffic. The legitimate nodes will eventually adopt the adversarial node as their router.

Transport-layer protocols provide end-to-end connectivity between nodes. Sequencing, such as that used in the Transmission Control Protocol (TCP), improves the reliability of the connection. Protocols that use sequencing may succumb to DoS attacks. The classic TCP SYN flood applies to sensor networks. An adversary can flood the victim with synchronization requests and limit the ability for other nodes to communicate with the victim. One solution limits the number of synchronization requests accepted, but this limits both adversaries and allies. Client puzzles, a more complex solution, require the client to make a commitment to the server before it is allowed to initiate a conversation. When the client initiates a connection, the server will respond with a puzzle that the client must solve. The client must complete the puzzle and send the answer to the server before the server will accept a full connection. Although this solution protects the server from SYN floods, it may harm allies that have fewer computational resources than the adversary does.

Origin authentication and message integrity can mitigate attacks at the network layer and above. Threats such as spoofing or fabrication of routing information justify the need for origin and data integrity of even the simplest HELLO messages.

20.5 SURVEY OF MESSAGE AUTHENTICATION PROTOCOLS

This section summarizes some of the most relevant proposals that integrate origin integrity and data integrity into WSN communications. Each proposal possesses unique qualities that influence its applicability to various sensor network applications. Many combine techniques for origin integrity and message integrity with other security goals, such as confidentiality or replay protection. However, these features may consume excessive processor, storage, or energy resources. This section concludes by contrasting the cost of each mechanism.

A number of requirements frame the analysis of authentication protocols for protecting WSNs deployed for target tracking. First, an authentication protocol should be resistant to node compromise by allowing secure key management. The protocol may provide an integrated key-rotation mechanism or allow for key rotation by an external module.

In addition, the protocol must have low computation overhead for both the sender and the recipient of a message. This analysis measures computation overhead in terms of the number of processor instructions, the amount of code memory, and the amount of data memory of the major authentication protocols. The protocol must also require low communication overhead. On some mote platforms, the radio transceiver consumes more energy than the processor.

Finally, messages supporting the authentication protocol must function in an unreliable network. In target-tracking networks, alarms are time-sensitive. Thus, the protocol should support the ability to immediately authenticate a message upon receipt.

20.5.1 CRYPTOGRAPHIC CONSTRUCTS

The authentication proposals presented later in this section differentiate themselves on factors including key management, packet format, and selection of block ciphers and modes of operation. This section presents the cryptographic foundation for the comparison of the proposals.

20.5.1.1 Conventional Authentication

The roots of message integrity begin with cryptographic checksums, also known as hashes. These checksum functions take a message and condense it into a smaller message digest [11]. The simplest example, the parity bit, counts the number of 1-bits in a message to produce a checksum of 1-bit in length. Strong cryptographic hash functions must possess three desirable properties. First, the hash must be easy to compute, not consuming significant computational resources. Second, it should be computationally infeasible to reverse the hash function. This means that given the result of the hash $h(M)$, one should not be able to determine M. Ideally, two distinct messages, when hashed, should yield two distinct checksums. However, according to the pigeonhole principle, there is a chance that two distinct messages, M and M', produce the same hash value, $h(M) = h(M')$. This condition, known as a collision, can be exploited to defeat hash functions [16]. The MD5 [17] and SHA-1 [18] hash functions are employed in several security applications and protocols. MD5 condenses a message into a hash of 16 bytes. SHA-1 condenses a message into a 20-byte hash. Both MD5 and SHA-1 have been proven susceptible to collisions [16,19].

Hash functions provide a level of message integrity between communicating peers. A sender prepares a message M and calculates the checksum $x = h(M)$. It then sends the checksum along with the message to the recipient. When the recipient receives message M, he can recalculate the

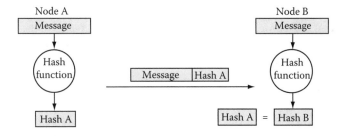

FIGURE 20.5   Hash function.

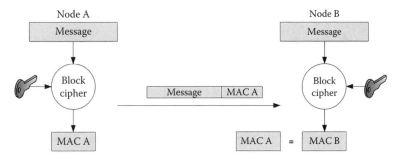

FIGURE 20.6   Message authentication code.

checksum on the received message *M*. If the checksum appended to the message matches the checksum calculated by the recipient, then the recipient can be assured of message integrity. Figure 20.5 illustrates this process.

Cryptographic checksums cannot provide assurance that messages arrive without modification or that they originate from an authentic sender. Because an attacker may know the hashing algorithm in use, an attacker could simply replace message *M* with message M', compute the hash $x' = h(M')$, and send the concatenation of the message M' and the hash x'. The recipient will calculate the hash of M', which will match the x' sent by the attacker. Thus, the recipient cannot validate that authenticity of the message. Message authentication codes (MACs), an instantiation of hashes that uses a unique key, provide both the data integrity of checksums and origin integrity provided by a secret key. Both the sender and the receiver should share the key. If an adversary learns the secret key, the hashing function is compromised.

As illustrated in Figure 20.6, a MAC is constructed by encrypting a message with a block cipher in cipher block chaining (CBC) or cipher feedback modes (CFB) [12]. Use of the CBC mode to generate a MAC is commonly known as CBC-MAC. Many WSN authentication mechanisms employ CBC-MAC. However, the CBC-MAC operation has been shown to be insecure for variable length messages [20]. The U.S. NIST has approved an implementation of CBC-MAC that securely authenticates variable length messages. This mode, known as CMAC [21,22], has not yet been reviewed in WSN research.

20.5.1.2   Unicast versus Broadcast Authentication

Unicast authentication provides the assurance of origin integrity when a message is sent from one sender to one receiver. A MAC, generated by the sender/creator of the message by using a secret key, can be used to ensure origin integrity. For unicast messages, static symmetric (shared) key cryptography meets the requirements because the two peers are trusted not to leak the key. The speed and efficiency of symmetric key cryptography suit the constraints of wireless motes.

Broadcast authentication ensures that multiple recipients of a message can verify its origin integrity. If using MACs to ensure broadcast authentication, all recipients of the message must share the symmetric key. The unique challenge for broadcast authentication involves the management of that shared key. If the key is broadcast to potential recipients, an adversary could eavesdrop on the key broadcast, capture the key, and generate a legitimate MAC for a forged message. Public key cryptography solves the problem of securely sharing a key for conventional Internet computing systems. However, public key cryptosystems consume far too much storage, computation, and bandwidth resources to be applied in WSNs. The research surveyed throughout this section introduces mechanisms that either replicate the asymmetry of public key cryptography or provide efficient key rotation.

20.5.1.3 Block Ciphers

Symmetric key cryptography consists of two categories of ciphers: block ciphers and stream ciphers. Stream ciphers operate on a single bit or byte at a time. Block ciphers operate on groups of bits called blocks [12]. Common block ciphers considered for WSNs accept block sizes of 32, 64, and 128 bits. Authentication mechanisms typically employ block ciphers because they can be used to generate MAC. Three block ciphers stand out when analyzing cryptographic ciphers for sensor nodes: RC5 [23], Skipjack [24], and AES [25].

The cost of these ciphers on wireless sensor nodes has been scrutinized in [26–29]. Roman et al. evaluated the purported benefits of implementing block ciphers in hardware. They reveal that software implementations suffice because the greatest cost arises from communication overhead, not processing or storage. In Roman's studies, the Skipjack cipher excels at providing efficient block cipher operations because of its speed and simple key schedule. Roman et al. show that Skipjack has an average encryption time of 48 ms/B. Law further improved its implementation to reach 25 ms/B. Despite its efficiency, the small 80 bit Skipjack key limits the algorithm's resistance to attack. The stronger AES algorithm, based on the Rijndael cipher [30], accepts key sizes up to 256 bits. Law and Choi separately show that AES has an average encryption time of 50 ms/B. This additional processing overhead can bottleneck high bandwidth communications. However, AES may be appropriate for target-tracking applications because they do not require high bandwidth. Grosschaedl et al. show that existing implementations of AES can be easily optimized to further reduce energy and storage consumption. Table 20.1 summarizes the block and key sizes of common block ciphers.

20.5.2 SPINS

In Ref. [5], Perrig et al. propose a suite of security protocols called SPINS, which provides message authentication, data integrity, confidentiality, and replay protection. The SPINS suite includes two protocols: SNEP and μTESLA. SNEP provides unicast authentication, confidentiality, and replay protection. μTESLA offers a solution for authenticated broadcast messaging in sensor networks.

TABLE 20.1
Common Block Ciphers

Cipher	Key Size (b)	Block Size (b)
AES	128/192/256	128
RC5	0–2040	32/64/128
RC6	128/192/256	128
Twofish	128/192/256	128
Skipjack	80	64
XTEA	128	64

These two protocols specifically address self-organizing WSNs that have a multihop routing topology. The proposal secures three communication paths: single node to the base station, base station to a single node, and base station broadcast to all nodes. SPINS do not offer a solution for node-originated broadcast messages.

20.5.2.1 SNEP

The SNEP component of SPINS packages data authentication, protection against replay, and weak data freshness into one cryptographic protocol. SNEP provides origin authentication and message integrity by appending an 8-byte MAC to the ciphertext. It generates the MAC by running the plaintext message and the counter through the RC5 block cipher in CBC-MAC mode. Because the MAC provides message integrity, the CRC field (two bytes) is dropped. Thus, total packet length increases only by 6 bytes.

SNEP requires communicating endpoints to maintain a shared counter as a mechanism to prevent replay attacks. Sensor network protocols commonly avoid counters because they increase communication overhead. SNEP initializes the counter when two nodes begin communicating and increments it after each communication block, thus eliminating the need to send it along with each message. Although this minimizes communication overhead, it requires the nodes to set aside additional memory for the counter.

The counter aids both in the encryption process and in the calculation of MAC. Use of the counter in the encryption process provides semantic security, which means that repeated encryption of a message M would yield unique ciphertext each round. Semantic security limits the ability of an eavesdropper to interpret the plaintext even after observing multiple encryptions of the same message. Use of the counter in calculation of the MAC provides protection against replay attacks. The counter also helps enforce weak message freshness. Message freshness describes the level of assurance that a message to a node A was created by node B in response to a request from node A. The authors employ a nonce in cases where strong message freshness is required. Node A generates the nonce and sends it in the request to node B. Node B implicitly returns the nonce to node A by encoding it into the MAC.

20.5.2.2 μTESLA

Although SNEP protects unicast messaging, secure broadcasting messages in a WSN remain a complex problem. In Internet communications, asymmetric or public-key cryptography typically serves as the foundation for broadcast authentication. When applied to wireless sensor networks, public key cryptography requires too many computation cycles and too much storage for sensor nodes. On the other hand, symmetric key cryptography requires strong protection for the shared key. Once the shared key is disclosed, an impostor could use it to spoof messages from the legitimate sender. As part of SPINS, μTESLA manages this problem by using a chain of symmetric keys that are periodically rotated. This mechanism works best with broadcast messages sent by a base station. For a node to send broadcast messages, an intermediary must intervene.

To accomplish base station originated broadcast authentication, the base station first randomly selects a key, K_n. It then successively applies a one-way hash function F to K_n, generating keys K_n through K_0. The base station rotates keys on a design specific interval.

To send an authenticated message, the base station computes a MAC on the packet with a key that is secret at that point in time. That is, when the base station broadcasts a message M_i, it appends to the message a MAC that is generated with key K_i, which is secret at that time. Sensor nodes that receive the message M_i may not be able to immediately authenticate the message until the next interval, when the base station broadcasts the new key, K_{i+1}. When the recipient receives $K(i + 1)$, it can calculate K_i by applying the one-way hash function to K_{i+1}, i.e., $K_i = F(K_{i+1})$, and then verify the MAC with key K_i. Note that this use of key rotation based on time intervals requires the base station and the nodes to be loosely time synchronized. If nodes loose synchronization with the base station, validation of the MAC will fail.

In SPINS, wireless sensor nodes cannot send authenticated broadcast messages without the assistance of the base station. The authors propose two strategies. First, the node can send the message to the base station and allow the base station to broadcast it. Second, the node can broadcast the message and let the base station manage distribution of keys.

The implementation of both SNEP and μTESLA calls upon the RC5 cipher. For message authentication and generation of random numbers, the authors employ CBC-MAC. Encryption and decryption occur with RC5 in counter (CTR) mode, a mode of operation that turns a block cipher into a stream cipher. Because the same function provides encryption and decryption, the algorithm preserves code memory. Other advantages of RC5 include its efficiency, its avoidance of multiplication, and its small lookup tables. Rijndael and other block ciphers require complex calculations and memory to store large lookup tables.

SPINS addresses the security requirements for a sensor network that does not have real-time constraints. This system provides confidentiality for unicast messages, and message authentication for both unicast and broadcast messages. The system minimizes utilization of energy, storage, and bandwidth resources. However, the SPINS model does not efficiently handle node-originated broadcast authentication, nor does it scale to all sensor network topologies. Most importantly, SPINS do not provide a solution for instantaneous authentication, which is required for alerts in WSNs deployed for target tracking, such as OCO.

20.5.3 REGULAR AND PREDICTABLE TIMES AND LOW ENTROPY AUTHENTICATION

μTESLA requires the recipient to wait for an interval of time before it can determine the key to authenticate a previously received message. This limits its usefulness in sensor networks with real-time constraints such as target-tracking networks. Intrusion notifications may occur irregularly, but need to be authenticated immediately. When keys are sent more frequently than they are used, the recipients expend computation cycles working through the key chain. Luk et al. [15] introduce regular and predictable times (RPT) and low entropy authentication (LEA) as solutions to these issues.

They begin by suggesting three modifications to μTESLA that make it more suitable for infrequent but urgent alarms. The simplest modification reduces the key disclosure interval. However, this approach wastes energy because the recipients must process the more frequent key distribution messages. Another solution proposes publication of multiple keys in a single message. A third solution replaces the one-way hash chain with a hash tree. A tree with N keys will grow to a height of $d = \log_2(N)$. Only d values must be sent with each message to verify that it is authentic. Hash chains can be combined with hash trees, providing a link from one chain to the next. The leaf node of one tree can be used to compute the hash chain appended to the subsequent branch. There are two advantages to this strategy. First, messages can be authenticated using the one-way hash of the last known key as with standard μTESLA. Second, if no messages were sent in a long period, the recipient can use the hash tree leaves as a shortcut through the hash chain.

The RPT protocol efficiently authenticates messages sent at RPT [15]. RPT breaks time into short intervals and assigns a key from a one-way key chain to each interval. The sender calculates δ, the sum of the maximum propagation delay and the maximum time synchronization error. To send a message, the sender first broadcasts only the MAC of the message. The sender then waits δ, and sends the message and the key used to generate the MAC. The receiver verifies that the key is still fresh, and then verifies that the MAC of the message matches the MAC originally broadcast by the sender. This allows for sparse key rotation and reduces communication overhead caused by key distribution. This approach benefits circumstances where the message contents are known well in advance, and some procedure requires that the message be sent regularly and on schedule. For example, consider time synchronization signals. A base station may send a synchronization and key rotation message every day at noon. With standard µTESLA, the message could not be authenticated until a key rotation the following day.

The second protocol introduced by Luk et al., LEA [15], provides security for short message that change infrequently. LEA evolved from one-time signatures like the Merkle–Winternitz construction. One-time signatures include private and public keying information connected by a one-way hash function. The function starts with a pseudorandom private key at the root of a directed acyclic graph. Two edges connect this private key to two vertices: the one-way hash of the key and a checksum of the key. These two vertices are again repeatedly hashed so that the length of the chain is sufficient to sign the message. A message of x bits requires a chain of $2x$ values long. At the end of the chain, the hash of the key and the hash of the checksum are concatenated to form the public key.

As with other asymmetric key algorithms, the sender of a message signs the message with their private key. The length of the signature depends upon the length of the message to be signed, thus one-time signatures such as Merkle–Winternitz suit short messages with low entropy. Although this protocol efficiently generates and verifies signatures, it does not scale well for signatures of long messages. Additionally, one-time signatures require a unique, authentic public key for each message. This challenge can be overcome, however, by distributing the public keys far in advance of the signed message. The sender could potentially send a set of n keys to sign the next n messages. Luk et al. suggest using their own RPT protocol for key predistribution. This approach suffers because it requires the recipient to store n keys until they are actually used.

20.5.4 TINYSEC

TinySec [6] aims to satisfy three security goals: origin integrity, message integrity, and message confidentiality. It achieves these goals while limiting the impact on computation, memory usage, and bandwidth. The TinySec approach to securing WSNs provides message integrity and confidentiality in a way that facilitates integration into sensor network applications. TinySec provides two modes of operation: TinySec with origin and message authentication only (TinySec-Auth) and TinySec with authentication and encryption (TinySec-AE). TinySec, the first proposal to make symmetric key encryption primitives available to sensor nodes [26], now comes bundled in the Tiny operating system (TinyOS) for wireless sensor nodes. This eases integration into application development and increases the likelihood of deployment of secure WSNs.

TinySec makes use of MACs to provide authentication and message integrity. The sender and the receiver share a secret key to cryptographically sign messages before their delivery and to validate messages upon receipt. The recipient validates the signature to detect tampering or damage incurred during transit. TinySec authenticates a packet with a 4-byte MAC, which authenticates the following TinyOS fields: destination address, active messages (AM) type, length, and the payload. This MAC replaces the CRC field at the end of a TinyOS packet because a MAC can detect both malicious changes and transmission errors. Use of a shared authentication key allows the TinyOS Group ID field to be dropped. Because TinySec drops these fields, TinySec-Auth only results in 1 byte of additional communication overhead.

Use of a 4-byte MAC illustrates the trade-off between risk and the cost of a security counter-measure. MACs usually range from 8 bytes, as in SPINS, to 16 bytes in length. The TinySec authors claim that a 4-byte MAC meets the requirements for WSNs. This length gives adversaries a 1 in 232 chance of successfully brute forcing a MAC for a particular message. To succeed in this forgery, the adversary must send messages to the target recipient. The limited wireless data transfer rate on WSNs only allows for 40 forgery attempts per second. It would take over 20 months to send all 232 possible MAC combinations. The recipient sensor would likely have run out of power before this attack could complete. As a compensating control, nodes should alert the base station when the rate of MAC failures exceeds a predefined threshold.

The TinySec MAC evolved from the CBC-MAC block cipher mode of operation. Bellare, Kilian, and Rogaway [20] have demonstrated that this construction fails to secure variably sized messages.

They suggest three alternatives for generating MACs of variably sized messages. The variant used by TinySec XORs the encryption of the message length with the first plaintext block. The authors considered the RC5 and Skipjack block ciphers the most appropriate ciphers for embedded micro-controllers. They implement TinySec with Skipjack because RC5 requires additional RAM to store a precomputed key schedule.

TinySec-AE offers both basic confidentiality and semantic security, preventing adversaries from gaining partial knowledge of the plaintext by observing repeated encryption of the same message. Semantic security requires use of IVs, which increase diversity in the plaintext. Target-tracking networks, like LEACH, GDAT, and OCO, commonly exhibit low message entropy. WSN that require confidentiality often include mechanisms to provide semantic security. TinySec uses the CBC mode of operation with the Skipjack cipher to provide confidentiality. Unlike SPINS, TinySec views keying mechanisms and cryptographic functions modularly. This enables use of a variety of keying mechanisms within TinySec, including a simple networkwide shared key.

Security mechanisms must be easy to integrate into an application; otherwise, they will not see widespread deployment. The security, performance, and usability of TinySec can be enabled in sensor network applications by setting a compile-time flag in a TinyOS makefile. Developers can select TinySec-Auth or TinySec-AE and tune the level of security around their application's require-ments. TinySec simplifies integration with a design focused on transparency and portability. TinySec achieves transparency by assuring that there is consistency between standard network APIs and net-work APIs that enable security. This facilitates upgrades of legacy applications to support security countermeasures. Implementation of TinySec as a link-layer module makes this transparency possi-ble. The TinyOS radio stack sends all radio events to the TinySec module. In support of portability, TinySec aims to run on a variety of platforms that support TinyOS.

20.5.5 MiniSec

The research team that introduced SPINS and μTESLA propose a WSN security architecture called MiniSec [31], which improves efficiency in delivering authentication and confidentiality to WSNs. MiniSec secures both unicast and broadcast communication. MiniSec's most significant contribution rests in its use of the same block cipher operation to provide confidentiality and authentication. MiniSec also introduces a novel semantic security strategy that only requires transmission of a fragment of the IV. MiniSec capitalizes on advancements in mote hardware, leveraging increasingly available mote memory to defend against replay attacks.

MiniSec proposes significant energy savings for WSNs that require both confidentiality and integrity. TinySec minimizes memory and energy use by omitting replay protection and using a single network shared key. In contrast, ZigBee takes the opposite approach and maintains a high level of security by sending an 8-byte IV [32]. The authors claim that MiniSec consumes one-third the memory of TinySec-AE, the authenticated encryption mode of TinySec.

MiniSec employs a block cipher mode known as offset code book (OCB), a block cipher mode of operation developed by Phillip Rogaway [33]. OCB provides authentication and encryption in one pass. In comparison, TinySec and ZigBee require two block cipher passes: one for confidential-ity and another for authentication. OCB takes in the message, the key, and a nonrepeating nonce, and concurrently generates the ciphertext and a tag used for authentication. The nonce provides semantic security, assuring that any two identical plaintext messages encrypted with the same key yield different ciphertext. The tag functions as a MAC. OCB does not cause ciphertext expansion; it produces ciphertext the same length as the plaintext. The MiniSec authors port Rogaway's OCB implementation into 4000 lines of nesC code. Their implementation consumes 874 bytes of RAM and 16 kB of code memory.

MiniSec is composed of two schemes, MiniSec-U for unicast messaging, and MiniSec-B for broadcast messaging. The two schemes differ in the way they handle counters for replay protection.

As in the SNEP protocol discussed previously, MiniSec-U requires each receiver to maintain a counter for each sender.

MiniSec-U reduces the cost of transmitting counters. On the Telos platform, radio transmissions consume the most energy. Sending a single byte consumes as much energy as executing about 4720 instructions. TinySec conserves radio energy by only sending the last few bits (the last bits [LB] value) of the 64-bit counter. A developer can select an LB value based on the potential for dropped packets. A low LB value can be used in environments with less potential for interference, thus reducing the communication overhead. The protocol requires memory to store two keys and two counters for each pair of communicating nodes, one each per direction.

Counters cannot be used in broadcast communication because of the complexity of keeping the counters synchronized among multiple nodes. OCB provides confidentiality and authentication for broadcast communication, but MiniSec requires a novel approach to defend against replay attacks. One proposal recommends the use of a sliding window. The tactic splits time into a series of finite epochs. Estimated network latency dictates the duration of each epoch. A unique ID assigned to each epoch serves as the nonce. When a node receives an encrypted message, it deciphers the message with both the current epoch ID and the previous epoch ID. If neither ID yields successful decipherment, the node flags the packet as a potential replay.

With this approach, a packet sent early in an epoch can be replayed throughout the epoch. MiniSec-B makes use of Bloom filters to defend against replay attacks. Bloom filters and loose time synchronization help defend against this type of replay attack. Bloom filters allow nodes to efficiently store a fingerprint of received messages into an array and quickly query the array to determine if the message has already been seen. Bloom filters guarantee that replayed messages will be detected. However, this strategy may flag some legitimate new messages as replayed messages. Bloom filters detect replayed messages within an epoch. Each receiving node maintains two Bloom filters: one for the current epoch and one for the previous epoch. Upon receipt of a packet, the recipient first validates the packet by performing OCB decryption. If this succeeds, the recipient performs a test to determine if the packet was replayed. The receiver first queries the Bloom filter for the packet. If the receiver finds the packet in the Bloom filter, it considers the packet a replay. If the receiver does not find the packet, it considers the packet fresh and adds it to the Bloom filter. This strategy will detect all replay attacks within the epoch.

MiniSec provides a high level of confidentiality and integrity with less overhead than its predecessors do. Implementation details such as packet length, key length, and authentication tag influence the level of security. MiniSec shares the following fields with TinyOS: length, frame control, sequence number, destination PAN, destination address, and active message AM type. Like TinySec-AE, MiniSec adds a 2-byte source address and replaces the 2-byte CRC with a 4-byte tag. Use of cryptographic keys replaces the functionality of the TinyOS group ID, thus this field is dropped. Overall, MiniSec adds three bytes to the standard TinyOS packet. Providing both authentication and encryption, MiniSec consumes one-third the energy of TinySec-AE.

Like TinySec, MiniSec employs Skipjack cipher with block size of 64 bits as the underlying block cipher. OCB requires the nonce to be the same length as the block size, so MiniSec uses a 64-bit counter. This monotonically increasing counter guarantees semantic security. MiniSec follows the Skipjack standard and requires 80-bit keys.

MiniSec achieves notable efficiency while maintaining a high level of security. It reduces the transmission overhead of TinySec-AE by two bytes and it reduces computational overhead by employing OCB. This strategy elevates MiniSec as one of the most efficient algorithms to provide confidentiality and integrity to WSNs.

20.5.6 AMSECURE

Although TinySec and MiniSec evaluated the cost of software-based cryptography, the AMSecure [34] proposal describes implementation of hardware-accelerated cryptography in TinyOS.

They insert a cryptographic acceleration hardware module wired between the AM module and the radio. When transmitting or receiving an AMSecure message, the AMSecure module performs the cryptographic operations, and then sends the payload to the other layers.

AMSecure offers the four cryptographic modes specified in IEEE 802.15.4 [35]: no cryptography, authentication-only with CBC-MAC, encryption-only with CTR mode, and authenticated encryption with CCM (CTR with CBC-MAC). The system appropriates between 1 and 9 bytes of the 29 byte TinyOS payload, depending on the cryptography utilized. Unlike its predecessors, AMSecure bases all cryptographic operations on the AES block cipher. Like TinySec, AMSecure preserves backward-compatibility with legacy TinyOS application by controlling activation of cryptography with a compile-time flag. This flag distinguishes standard TinyOS AM from AMSecure messages.

The proposal summarizes the processing and communication overhead of AMSecure. By using a hardware cryptographic module, AMSecure message processing time is kept to a low, predictable level. Because it offloads cryptographic operations, the overhead decreases as payload length increases. With a standard 29-byte payload, addition of authentication and encryption result in 20 percent message overhead. This drops to approximately 8 percent with a 100-byte payload. For a 90-byte payload, the receive processing time increases from about 500 ms for a standard TinyOS message to 1750 ms for an authenticated, encrypted AMSecure message. This represents a significant improvement over the 50 ms/B rate documented by Law and Choi [28,29].

20.5.7 INTERLEAVED AUTHENTICATION

As illustrated previously in this chapter, WSNs face significant exposure to node compromise. The compromise could be as simple as physical destruction of a node or sophisticated enough to allow an attacker to manipulate an active, compromised node. An adversary who overtakes a legitimate, active node may use this node to inject false information into the system. Secure key rotation alone cannot defend against this threat. Compromising an active node gives the attacker access to keying material, thus the attacker has the ability to calculate legitimate MACs. These reputedly authentic messages can misguide the base station, divert intrusion alerts, and deplete network resources.

When analyzing security of WSN protocols, Zhu et al. [36] assume that nodes will be compromised. Furthermore, compromised nodes may be exploited by the attacker to launch other attacks, like injecting false data into the packets. They present an interleaved hop-by-hop authentication method that can detect data injection attacks. The method adds an additional layer of security on top of SPINS, TinySec, and others by defining a threshold for the number of compromised nodes that a network can endure. They define a value t representing the maximum number of compromised nodes tolerated during an attack, and require $t + 1$ nodes to send an authenticated reports of an event. The scheme guarantees that if no more than t nodes are compromised, the network will detect and drop falsely injected data. A designer can adjust the value t based on the threat of node compromise. This strategy defends against collusion among compromised nodes.

The proposal by Zhu et al. requires a network topology organized into clusters, with a subset of nodes acting as cluster heads. The cluster must include at least $t + 1$ nodes, including the cluster head. The cluster may reside multiple hops away from the base station. Nodes within the network can send unicast messages up and down the tree, and broadcast messages to their neighbors. Nodes share a master key with the base station and have the ability to establish shared keys with most of their one-hop neighbors.

The scheme relies on an association of nodes that reside $t + 1$ hops apart on the path to the base station. They refer to a node's peer closer to the base station as the upper associated node and the peer closer to the leaf nodes as the lower associated node. Upper associated nodes validate the MAC appended to messages from their lower associated peers. Each message may carry as many as $t + 1$ MACs as it travels from leaf nodes to the base station. A validation failure on any of the MACs will cause the message to be dropped. As long as the number of compromised nodes remains below the value t, the system can detect false data injection.

Five unique phases comprise the hop-by-hop authentication technique, including initialization and deployment, association discovery, report endorsement, en-route filtering, and base-station verification.

During node initialization and deployment, the key server loads each node with a unique ID and a unique key that node shares with the base station. The node derives an authentication key from the encryption key it shares with the base station. The key server then can use one of many key establishment algorithms to initiate network key distribution. This enables nodes to establish shared keys with their neighbors.

The association discovery phase allows nodes to discover their upper and lower associated peers. The base station kicks off the process by broadcasting a HELLO message. Each node that receives the broadcast checks for the ID of the node $t + 1$ hops up the tree, replaces that ID with its own, and rebroadcasts the modified hello. This provides an upper bound of $t + 1$ node IDs attached to the HELLO message. A receiving node records the ID of the node $t + 1$ hops up the tree as its upper associated node. Note that nodes less than $t + 1$ hops from the base station do not have an upper associated node. When the hello message reaches a cluster, the cluster head assigns its leaves to upper associated nodes. The hello message can be authenticated with a broadcast authentication scheme such as μTESLA.

After the cluster head notifies its leaves of their peers, it sends an acknowledgment back to the base station. The lower associated nodes authenticate the acknowledgment with the pairwise key they share with their upper associated node. Along with the MAC, the acknowledgment includes the node IDs of the cluster head and the leaf nodes. As the acknowledgment is returned up the tree toward the base station, upper associated nodes learn the node ID of their lower associated node. They replace the ID of their lower associated node with their own ID and forward the acknowledgment back up the tree. In cases where upper nodes have branches to multiple clusters, they record cluster IDs and nodes in a table.

When nodes witness an event, they send a report to the base station. This hop-by-hop proposal requires $t + 1$ nodes to witness an event and endorse a report of the event. As the endorsement moves upstream, the IDs of the nodes that witnessed the events and authentication codes will be appended to it. Each node computes two MACs for the event. The individual MAC is computed using the node's key with the base station. The pairwise MAC is computed using the node's pairwise key with its upper associated node. If the cluster head can authenticate a report from all its leaf nodes, it compresses the individual MACs by XORing its individual MAC with the individual MACs from the leaf nodes. The pairwise MACs are not compressed.

Upper level nodes that receive this report must authenticate it using their pairwise key with their lower associated node. If authentication succeeds, the node will extract the MAC from its lower associated node and append its own. If authentication fails, the message will be dropped. This in-route filtering assures that, as long as no more than t nodes are compromised, falsely injected data will be detected and dropped.

Once the alert reaches the base station, the base station performs its own verification. It extracts the event data and node IDs from the report. It then computes its own compressed MAC on the event data using its keys shared with the nodes in the node list. If the MAC matches the compressed MAC in the report, the alert is considered valid.

Because WSN are susceptible to damage, this interleaved authentication proposal includes maintenance techniques. One strategy proposes piggybacking association discovery messages on base station beacons such as those sent in TinyOS. Nodes accept the first beacon they receive as their parent node. Because these beacons are sent every epoch, it is possible for nodes to change parents every epoch. Although this strategy is satisfactory for dynamic networks, it is costly for networks that do not change frequently. A less costly base station initiated strategy has the parent change only if the node determines that any of the nodes in the beacon from its parent have changed. Repair can also be initiated locally when nodes detect failure of a neighbor. The proposed technique requires nodes to use GPS or similar technology to determine the physical location of their neighbors. When a node detects that its parent has failed, it will send a REPAIR message to the first node counterclockwise from the edge between itself and its deceased parent. It will then exchange messages with this node to learn the IDs of the upstream nodes and rebuild any broken node associations.

The interleaved authentication proposal provides a higher level of security than simple pairwise authentication between neighboring nodes. The base station can trust that reports from leaf nodes are authentic based on the MAC computed with the pairwise key between itself and the leaf node. Intermediary nodes can authenticate reports based on pairwise keys with their lower associated nodes. As long as $t + 1$ nodes agree on an event, the system will detect a falsely injected report sent by t nodes or less. This high level of security requires a high level of node redundancy.

20.5.8 COMPARISON OF IMPLEMENTATIONS

This section aims to evaluate security countermeasures so that a network owner can balance the cost of security with the risks of a target-tracking system. Superficially, an authentication-only solution like TinySec-Auth appears to adequately secure the system without significantly affecting network longevity. Many target-tracking methods really do not require confidentiality. However, all of the solutions deserve a comparative analysis. Many of the proposals discussed above in this section evaluate overhead caused by communication costs, processing costs, and speed of cryptographic operations. This section will compare the costs of these proposals side-by-side. Changes in mote technology make this analysis complex because the different radios and processors exhibit different power consumption behavior. Table 20.2 summarizes the platforms on which the preceding proposals were evaluated.

Table 20.3 illustrates the communication overhead required by proposals that integrate security into a standard TinyOS packet. The last column shows the portion of total packet overhead dedicated to security. Some protocols immediately stand out for their high communication overhead. The additional 8-byte MAC in the SPINS method SNEP, for example, clearly exceeds the communication overhead required by its successors such as TinySec and MiniSec. This large MAC significantly

TABLE 20.2
Summary of Mote Platforms Used by the Various Methods

Proposal	Platform	Processor	Speed	Architecture	Radio Hardware
SNEP	Smart Dust	Unknown	4 MHz	8 bit	916 MHz
RPT, LEA	Moteiv Telos	TI MSP430	8 MHz	16 bit	Unknown
TinySec	Mica2, Mica2dot	ATMega128	8 MHz	8 bit	Chipcon CC1000
MiniSec	Moteiv Telos	TI MSP430	8 MHz	16 bit	Chipcon CC2420
AMSecure	MicaZ	ATMega128	8 MHz	8 bit	Chipcon CC2420

TABLE 20.3
Increase in Packet Length (bytes)

Proposal	Payload	Total Overhead	Total Size	Security Overhead
TinyOS	24	12	36	0
SNEP (SPINS)	24	18	42	6
TinySec-Auth	24	13	37	1
TinySec-AE	24	17	41	5
MiniSec	24	15	39	3

reduces the chance of message forgery, but other proposals agree that the excessive length provides security for low bandwidth WSNs at the cost of unnecessarily high overhead. The RPT and LEA proposals seem appealing: RPT can secure the regular and predictable health messages sent between parents and descendants; LEA can effectively protect the low entropy intrusion alerts sent from a border node to its parent. However, their proposal does not specify integration of RPT and LEA into one implementation. Such integration of various methods is needed to create a holistic solution of securing the various messages exchanged in a target-tracking WSN.

As a final drawback, the RPT and LEA proposals do not include analysis of the energy consumption characteristics of the protocols. A more current protocol, AMSecure, seems promising because it advocates the use of new standards like IEEE 802.15.4 and the NIST approved AES block cipher. With hardware-supported cryptography, AMSecure can quickly encrypt a packet in 1750 ms. However, to develop a WSN by adopting this hardware-based solution requires that sensor devices supporting AMSecure be available to the developers.

The preceding discussion leaves TinySec and MiniSec as the two most promising, low overhead contenders for securing a target-tracking network. Both employ well-documented, standards-based cryptographic block ciphers: CBC-MAC and OCB. Both proposals make available the nesC source code used in their evaluation. One can compare the costs of the two proposals simply by evaluating the increase in packet length. Although authentication-only TinySec-Auth increases packets by 1-byte, the authenticated-encryption TinySec-AE and MiniSec increase packet length by 5 bytes and 3 bytes, respectively. These increases take a direct toll on the energy resource because the radio must be turned on longer when transmitting or receiving longer packets. The cost of sending a packet involves more than just the transmission of bits of data and a header. Longer packets also increase communication latency and consume valuable processor cycles.

The TinySec and MiniSec authors both empirically evaluated the communication cost associated with their respective proposal. Three security options were implemented in TinySec, including TinyOS, TinySec-Auth, and TinySec-AE [6]. Four security options were implemented in MiniSec, including TinyOS, SNEP, TinySec-AE, and MiniSec [31]. The TinySec authors [6] evaluated energy consumed by an 8-bit processor when transmitting packets through a Chipcon CC1000 radio. They sampled current drawn by a transmitter when sending a packet with a 24-byte payload. The MiniSec authors [31] selected a more contemporary mote model, with a 16-bit processor and the Chipcon CC2420. Although their actual energy consumption measurements differ significantly from those in TinySec, the percent increase appears similar. TinySec-Auth clearly keeps the cost of authentication lower than its authenticated encryption counterparts do.

20.6   CONCLUSION

When used for detecting intrusions and tracking targets, WSNs must provide a high assurance of trust. Modification or fabrication of alerts introduces false positive conditions that reduce the trustworthiness of the network. Forgery of routing messages may enable an attacker to occupy key roles within the network, allowing the attacker to control the flow of information toward the base station. If a target-tracking WSN is deployed in an arena where it is subject to these attacks, the value of its service will be diminished. False alarms lead to operational inefficiencies in any environment. In defense applications, manipulation of alarms may lead to loss of valuable assets or even loss of life. A layer of authentication wrapped around WSNs mitigates these risks.

This chapter summarized generic attacks against WSNs. DC, LEACH, and GDAT all share an application layer alerting process vulnerable to message forgery. LEACH, GDAT, and OCO possess self-organization capabilities that expose vulnerable network routing services. Both the network organization and the intrusion notification functions need protection. A brief survey of related literature shows that there has been little convergence of security protocols with target-tracking applications. Perhaps this originates from the conflict between the increased costs of security measures with the network longevity goals of target-tracking networks. Each of the security mechanisms surveyed in this chapter requires a trade-off between functionality, efficiency, and ease of deployment with protection of the network's assets.

The survey conducted as part of this chapter tracked the evolution of two disparate fields in sensor network research: target-tracking applications and authentication protocols. Although both areas of research strive for efficiency, the addition of security to a target-tracking mechanism increases energy consumption. This increase contradicts the quest of target-tracking research to uncover maximum efficiency. Target-tracking methods such as GDAT and OCO improve network longevity by forcing the majority of nodes to sleep. These techniques demonstrate superiority in efficiency while maintaining a high level of accuracy in tracking intruders. Sensor network authentication protocols similarly strive for efficiency by reducing computation and communication costs. Classic proposals such as TinySec and SPINS integrate easily with any target-tracking method. However, the application designer must carefully evaluate the security requirements of an application. Blanket coverage of all communications with a security countermeasure unnecessarily reduces the ability of efficient target-tracking mechanisms to provide long-lasting networks.

Although the simplest authentication mechanisms, like TinySec and MiniSec, provide maximum efficiency, they leave vulnerabilities exposed. For example, Zhu et al. [36] demonstrate that simple peer-to-peer authentication protocols like TinySec-Auth suffer fatal flaws in the presence of node compromise. An attacker can overtake an active node, gain access to keying material, and falsely inject messages into the network. Providing protection for this level of attack will require even more energy consuming components, such as secure key rotation and interleaved authentication.

ACKNOWLEDGMENTS

The authors are partially supported by the UHCL Faculty Research Support Fund, the Institute of Space Systems Operations (ISSO), the Texas State Advanced Research Program (ARP), and the National Science Foundation (NSF).

REFERENCES

[1] Kung, H. T. and Vlah, D., Efficient location tracking using sensor networks, *IEEE Wireless Communications and Networking, WCNC*, 3, 1954–1961 vol. 3, 2003.

[2] Heinzelman, W. R., Chandrakasan, A., and Balakrishnan, H., Energy-efficient communication protocol for wireless microsensor networks, *Proceedings of the 33rd Annual Hawaii International Conference on System Sciences*, p. 10, 2000.

[3] Tran, S. P. and Yang, T. A., OCO: Optimized communication & organization for target tracking in wireless sensor networks, *IEEE International Conference on Sensor Networks, Ubiquitous, and Trustworthy Computing*, 1, 428–435, 2006.

[4] Rababaah, H. and Shirkhodaie, A., Guard duty alarming technique (GDAT): A novel scheduling approach for target-tracking in large-scale distributed sensor networks, *IEEE International Conference on System of Systems Engineering*, pp. 1-6, 2007.

[5] Perrig, A., Szewczyk, R., Tygar, J. D., Wen, V., and Culler, D. E. SPINS: Security protocols for sensor networks, *Wireless Networks*, 8, 521–534, 2002.

[6] Karlof, C., Sastry, N., and Wagner, D., TinySec: A link layer security architecture for wireless sensor networks, *Proceedings of the 2nd International Conference on Embedded Networked Sensor Systems*, pp. 162–175, 2004.

[7] Pister, K. S. J., et al., 29 Palms fixed/mobile experiment: Tracking vehicles with a UAV delivered sensor network, 2001. http://robotics.eecs.berkeley.edu/~pister/29Palms0103/, November 10, 2008.

[8] Romer, K. and Mattern, F., The design space of wireless sensor networks, *IEEE Wireless Communications*, 11, 54–61, 2004.

[9] Hill, J., Szewczyk, R., Woo, A., Hollar, S., Culler, D., and Pister, K., System architecture directions for networked sensors, *ACM SIGPLAN Notices*, 35, 93–104, 2000.

[10] Tran, S. P., Yang, T. A., Cao, D., and Nguyen, T. A., OCO: An efficient method for tracking objects in wireless sensor networks, Working paper, 2006. http://sce.uhcl.edu/yang/research/OCO%205-4-2006.pdf.

[11] Bishop, M., *Computer Security: Art and Science*. Boston, MA: Addison–Wesley, 2003.

[12] Schneier, B., *Applied Cryptography: Protocols, Algorithms, and Source Code in C*, 2nd ed. New York: Wiley, 1996.

[13] Wood, A. D. and Stankovic, J. A., Denial of service in sensor networks, *IEEE Computer*, 35, 54–62, 2002.

[14] Wood, A. D., Fang, L., Stankovic, J. A., and He, T., SIGF: A family of configurable, secure routing protocols for wireless sensor networks, *Proceedings of the Fourth ACM Workshop on Security of Ad Hoc and Sensor Networks*, pp. 35–48, 2006.

[15] Luk, M., Perrig, A., and Whillock, B., Seven cardinal properties of sensor network broadcast authentication, *Proceedings of the Fourth ACM Workshop on Security of Ad Hoc and Sensor Networks*, pp. 147–156, 2006.

[16] Wang, X. and Yu, H., How to break MD5 and other hash functions, *Advances in Cryptology, EURO-CRYPT 2005, Proceedings of the 24th Annual International Conference on the Theory and Applications of Cryptographic Techniques*, Cramer, R. (Ed.), Springer, pp. 19–35, 2005.

[17] Rivest, R., The MD5 Message-Digest Algorithm, RFC 1321, *IETF*, 1992.

[18] Eastlake, D. and Jones, P., US Secure Hash Algorithm 1 (SHA1), RFC 3174, *IETF*, 2001.

[19] Wang, X., Yin, Y. L., and Yu, H., Collision search attacks on SHA1, *Crypto 2004*, August 15–19, 2004.

[20] Bellare, M., Kilian, J., and Rogaway, P., The security of the cipher block chaining message authentication code, *Journal of Computer and System Sciences*, 61, 362–399, 2000.

[21] NIST, *Recommendation for Block Cipher Modes of Operation: The CMAC Mode for Authentication*, 2005.

[22] Michail, H. E., Kakarountas, A. P., Selimis, G., and Goutis, C. E., Throughput optimization of the Cipher Message Authentication Code, *15th International Conference on Digital Signal Processing*, 495–498, 2007.

[23] Rivest, R. and Baldwin, R., The RC5, RC5-CBC, RC5-CBC-Pad, and RC5-CTS Algorithms, RFC 2040, IETF, 1996.

[24] NIST, Skipjack and KEA algorithm specifications, 1998.

[25] FIPS 197, Advanced encryption standard (AES), 2001.

[26] Roman, R., Alcaraz, C., and Lopez, J., A survey of cryptographic primitives and implementations for hardware-constrained sensor network nodes, *Mobile Networks and Applications*, 12, 231–244, 2007.

[27] Grosschaedl, J., Tillich, S., Rechberger, C., Hofmann, M. A., and Medwed, M., Energy evaluation of software implementations of block ciphers under memory constraints, in *Design, Automation & Test in Europe Conference & Exhibition, 2007. DATE '07*, pp. 1–6, 2007.

[28] Law, Y. W., Doumen, J., and Hartel, P., Survey and benchmark of block ciphers for wireless sensor networks, *ACM Transactions on Sensor Networks*, 2, 65–93, 2006.

[29] Choi, K. J. and Song, J. -I., Investigation of feasible cryptographic algorithms for wireless sensor networks, *The 8th International Conference on Advanced Communication Technology, ICACT*, 2, 3, 2006.

[30] Daemen, J. and Rijmen, V., The block cipher Rijndael, *Proceedings of the International Conference on Smart Card Research and Applications*, 2000.

[31] Luk, M., Mezzour, G., Perrig, A., and Gilgor, V., MiniSec: A secure sensor network communication architecture, *Proceedings of the 6th International Conference on Information Processing in Sensor Networks*, pp. 479–488, April 25–27, 2007.

[32] Zigbee Alliance. Zigbee specification. Technical Report Document 053474r06, 2005.

[33] Rogaway, P., Bellare, M., Black, J., and Krovetz, T., OCB: A block-cipher mode of operation for efficient authenticated encryption, *Proceedings of the 8th ACM Conference on Computer and Communications Security*, 2001.

[34] Wood, A. D. and Stankovic, J. A., AMSecure: Secure link-layer communication in TinyOS for IEEE 802.15.4-based wireless sensor networks, *Proceedings of the 4th International Conference on Embedded Networked Sensor Systems*, 2006.

[35] IEEE 802.15.4-2003. *Wireless MAC and PHY Specifications for Low Rate Wireless Personal Area Networks (LR-WPANs)*, 2003.

[36] Zhu, S., Setia, S., Jajodia, S., and Peng Ning, A., An interleaved hop-by-hop authentication scheme for filtering of injected false data in sensor networks, *IEEE Symposium on Security and Privacy*, pp. 259–271, 2004.

21 A Broadcasting Authentication Protocol with DoS and Fault Tolerance for Wireless Ad Hoc Networks

Yixin Jiang, Minghui Shi, Xuemin (Sherman) Shen, Chuang Lin, and Xiaowen Chu

CONTENTS

In this chapter, a novel authentication protocol is proposed, which satisfies both security and reliability requirements for group communications in ad hoc networks. The security features include identity anonymity and intracability, one-way session key/identity refreshment, and data privacy.

The reliability features include efficient denial-of-service (DoS) tolerance for forged refreshment messages and fault tolerance for lost messages recovery. The theoretical and the simulative results demonstrate that the proposed protocol is robust and efficient under severe DoS attack and poor wireless channel quality.

21.1 INTRODUCTION

Wireless ad hoc networks have attracted great intension in both academia and industry due to their unique characteristics and wide application scenarios [1]. They consist of mobile nodes which communicate with each other through wireless medium without fixed infrastructure. The key advantages include easy and fast deployment, decreased dependence on infrastructure, etc. Therefore, wireless ad hoc networks are widely used in emergency operations, such as search and rescue, policing and fire fighting, and military operations, etc. In those applications, group communications, as a growing application area in mobile communications, are preferred in many cases to keep the privacy of information for each onsite units and reduce the wireless traffic load. As shown in Figure 21.1, the mobile nodes from two units in the ad hoc network form two communication groups. In such aforementioned applications, there is usually at least one officer leading each unit. We define the corresponding node as commander (*CMD*) node, which is in charge of issuing secret certificate to group nodes.

A secure group session guarantees that only legal nodes share a common key which can be used in the session. The concept of traditional conference key distribution was first proposed in Ref. [2], and further studied in Refs. [3–8], which is not quite suitable for ad hoc group communications scenario. The protocol in Ref. [3] provides a basic secure key distribution protocol for mobile networks. The schemes in Refs. [4,5] are for active nodes to dynamically join or leave an ongoing group session. Two cryptosystems used in the schemes are not friendly for the mobile devices. In Ref. [6], the protocol does not offer identity anonymity so that an intruder can easily obtain real identity of a node by message interception and trace its mobility and current location. The lightweight protocol [7] lacks key refreshment mechanism so that the communication may be compromised by using a stale key. The impaired ad hoc communication environment and other various attacks from the Internet, such as DoS attacks, need be considered carefully, which otherwise may lead to protocol failure if nodes cannot communicate with the *CMD* due to communication interruption.

It is important that the confidentiality and authenticity mechanism is available in ad hoc group communications to prevent various illegal intrusions [1,9,10]. The intrusions include traditionally

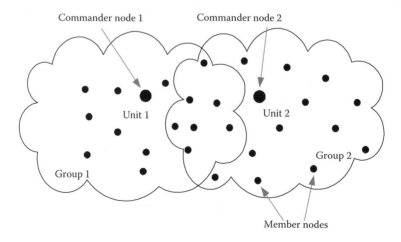

FIGURE 21.1   Ad hoc group communication architecture.

known attacks such as impersonation, conversation eavesdropping, mobile user's mobility tracing, and more severe DoS attack, which can diminish or black out a network's capacity. The main plausible ways for DoS attacks [11] include signal jamming in the physical layer and packet collision/exhaustion in the link layer. In this paper, we focus on the DoS attacks in the link layer. To address the aforesaid security issues, in this chapter, a DoS and fault-tolerant authentication protocol for ad hoc group communications is proposed, which features the following notable properties:

- Identity anonymity to protect nodes' identity and mobility information from tracking with dynamic pseudoidentity (*PID*)
- Forward secrecy so that the communication key (*CK*) and nodes' *PID* can be refreshed with implicit authentication in a one-way manner
- DoS tolerance for forged *CK* renewal message without relying on message retransmissions or acknowledgments (ACKs)
- Fault tolerance for the lost *CK*s
- Seamless *CK&PID* renewal without disrupting ongoing data transmissions
- Robustness to the clock skews among nodes

The proposed protocol also takes into account the resource constraints in the mobile devices by minimizing the computation overhead. Because of its implicit authentication capability of the *CK&PID* refreshment mechanism, the proposed protocol can work well under impaired wireless environment, without using message retransmission or ACKs. Therefore, the communication overhead is lightweight. Demonstrated by the performance analysis and simulation results, the proposed protocol can effectively tolerate high channel loss rate and DoS attacks, which are of particular importance in the emergency and military applications.

The rest of the chapter is organized as follows. In Section 21.2, the proposed authentication protocol for ad hoc group communications is introduced in detail. In Sections 21.3 and 21.4, the security and the performance analysis are presented, respectively, followed by conclusion given in Section 21.5.

21.2 PROPOSED AUTHENTICATION PROTOCOL

Figure 21.2 shows the messages used in the protocol between the nodes and the *CMD*. The *InitConfKey* message initiates or reinitiates refreshment parameters. It is sent to all nodes in the initial phase. The *CMD* uses the *RefreshKey* message to periodically broadcast the next *CK* in the key sequence to nodes. The nodes employ the *RequestKey* message to explicitly request the current *CK* in the key sequence. This message is generated by a node when it fails to receive *CK*s

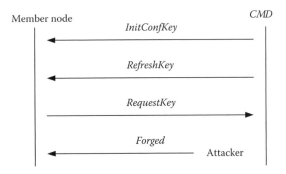

FIGURE 21.2 Message flows between *CMD* and group nodes.

TABLE 21.1
Notations

T_k	Member node
CMD	Commander node
t	Time stamp
ID_k	Identity of mobile node T_k
PID_k	Pseudoidentity of mobile node T_k
CK	Group Communication Key
CK_I	Integrity key derived from CK
CK_E	Data encryption key derived from CK
$\oplus$	Bitwise XOR operation
$\|$	Concatenation operation
$f(.)$	Key generating function
$H(.)$	One-way hash function
$E_k(.)$	Symmetric encryption with key k
$D_k(.)$	Symmetric decryption with key k
MIC	Message integrity code generated by integrity key CK_I
MK	Master key used to encrypt *InitConfKey* and *RefreshKey*

over key renewal intervals. We assume that the nodes may also receive forged *CMD* messages sent by attackers. For clarity, we list the related notations used in the rest of the chapter in Table 21.1.

21.2.1 ARCHITECTURE

The architecture of the proposed protocol primarily consists of four modules: DoS-tolerant module, fault-tolerant module, *CK&PID* switch module, and the stream encryption/decryption and integrity check module. As shown in Figure 21.3, the DoS-tolerant module provides two-phase DoS-tolerant

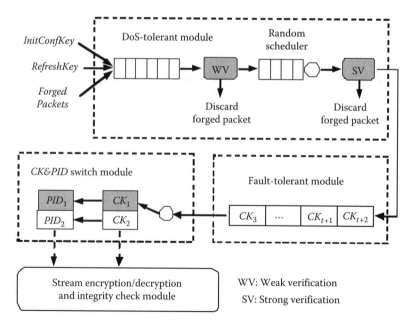

FIGURE 21.3   Architecture of proposed protocol.

authentication, weak verification (WV) and strong verification (SV), to filter out the forged packets efficiently. The *CMD* precomputes key sequence of *CK*s by utilizing a one-way hash function, which is similar to that of S/KEY [12]. Each *CK* is distributed to all nodes before it is used for encryption or decryption. The authenticity of the received *CK* is verified by using the other prestored *CK*s, and the missed *CK*s can be recovered from the new *CK*s. The new *PID* is computed based on the *CK* and its previous *PID*.

The DoS-tolerant module protects the *RefreshKey* message from DoS attack by using two-stage verification. The WV filters out a large number of forged messages by executing fast authentication with simple computation. And the SV executes strict authentication with a little more complex computation to drop a few forged messages which have passed the WV phase.

The fault-tolerant module provides a robust and reliable mechanism for tolerating the packet loss in the impaired wireless channel. On receiving an authentic *RefreshKey* message, each node can automatically recover the lost *CK&PID* without requesting the *CMD* to retransmit the lost message. The fault-tolerant feature relies on the distinctive property of the cryptographic one-way hash function, which is also used in TESLA [13–15] and LiSP [16]. The proposed protocol can improve (1) efficiency because each node only buffers the constant number of keys, whereas TESLA is required to buffer all the received messages until the node receives an authentic message; and (2) reliability because DoS-tolerance mechanism is offered while it is not considered in LiSP protocol.

The *CK&PID* switch module computes the new *PID* and seamlessly refreshes *CK&PID* without disrupting ongoing data transmission. To accomplish the functions, two key-slots, which can be operated concurrently, are set up in each node. When the *CK&PID* in one key-slot is being used for data encryption or decryption, the received new *CK* in the key sequence will be stored in the other key-slot. At the middle point of the refreshment interval, the node switches to the new *CK* key in the other key-slot.

Finally, the stream encryption/decryption and integrity check module guarantees data privacy, and also provides an enhanced security to resist key stream reuse attacks.

21.2.2　FORWARD SECRECY

Forward secrecy is used to assure the refreshment of *CK&PID* and offer a base for implementing DoS and fault-tolerant mechanisms. As shown in Figure 21.4, forward secrecy is ensured in three aspects: one-way *CK* refreshment, one-way *PID* refreshment, and one-way data privacy. To ensure the forward secrecy in *CK* refreshment, the proposed scheme offers an *MK* used by the *CMD* to encrypt

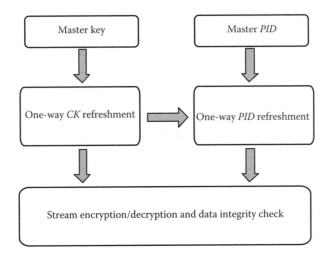

FIGURE 21.4　Forward secrecy: one-way *CK&PID* renewal and data privacy.

the *InitConfKey* or the *RefreshKey* message containing the temporal *CK*, which is used to encrypt or decrypt data. Similarly, to assure the forward secrecy in the *PID* renewal, it also defines a master *PID* and a temporal *PID*. The temporal *PID* is derived from master *PID* and its corresponding *CK*. The data privacy is also endowed with the forward secrecy, because we use the temporal *CK&PID* as the seeds to compute the block cipher and message integrity code. Hence, the key stream collisions can be efficiently avoided due to the forward secrecy.

Forward secrecy is based on a one-way hash chain, which is generated by a one-way function H. A one-way hash chain is a sequence of hash values, $x_n, x_{n-1}, \ldots, x_0$, such that $\forall j : 0 < j \leq n, x_{j-1} = H(x_j)$. Thus, there exists the following linear derivative relation:

$$x_1 = H(x_2) = \cdots = H^{j-1}(x_j) = \cdots = H^{n-2}(x_{n-1}) = H^{n-1}(x_n) \quad (1 < j \leq n).$$

In the proposed protocol, all temporal *CK*s are derived from a one-way hash function H and belong to one key sequence. In the initial phase, the *CMD* needs to precompute a one-way key sequence $\{CK_i | i = 1, 2, \ldots, n\}$, where n is reasonably large. The *CMD* selects CK_n as the last key in the key chain and repeatedly performs the hash function H to compute all the rest of keys as $CK_i = H(CK_{i+1})$, $0 \leq i \leq n - 1$. Each key CK_i will be distributed to all nodes by the *CMD* at ith time interval. With this one-way function, given CK_j in the key chain, anybody can compute the previous keys CK_i, $0 \leq i \leq j$, but cannot compute any of other keys CK_i, $j + 1 \leq i \leq n$. Similarly, all temporal *CK*s also form the following linear derivative relation:

$$CK_1 = H(CK_2) = \cdots = H^{j-1}(CK_j) = \cdots = H^{n-2}(CK_{n-1}) = H^{n-1}(CK_n) \quad (1 < j \leq n).$$

Given that the temporal $PID_{k,j}$ $(1 \leq j \leq n)$ is defined as the function of $PID_{k,j-1}$ and CK_j, for node T_k, the corresponding one-way *PID* chain can be derived as follows:

$$PID_{k,1} \Leftarrow \cdots \Leftarrow PID_{k,j} \Leftarrow \cdots \Leftarrow PID_{k,n-1} \Leftarrow PID_{k,n} \quad (1 < j \leq n).$$

Thus, the *PID* can be renewed with *CK* synchronously. Based on the mentioned linear derivative relations, the forward secrecy provides three significant security properties: (1) the identity anonymity mechanism is enhanced, because a dynamic *PID* can protect a node's location information from being tracked more efficiently than a long-term static *PID*; (2) the key stream reuse attacks are avoided, because the periodic or dynamic *CK&PID* is used to generate the stream cipher; and (3) forward secrecy leads to a solution to implement the important DoS and fault tolerance in the proposed protocol.

21.2.3 MUTUAL AUTHENTICATION PROTOCOL

When the head node (chairman node, also as the *CMD*) intends to start on a group session, it firstly initiates mutual authentication protocol (MAP). In this phase, the *CMD* sets up an *MK*, and then uses the *MK* to encrypt the *InitConfKey* message which includes the length t of key buffer for *CK*s, an initial CK, and the key refreshment period T_{refresh}. The message is securely broadcasted to each node. Then, at each interval T_{refresh}, the *CMD* uses *MK* to encrypt a *RefreshKey* message that contains the next *CK* in the precomputed key sequence. All the refreshment messages will be securely broadcasted or unicasted to each node.

Before mutual authentication, a node communicates with the *CMD*, and submits its identity to the *CMD*. The *CMD* generates a secret sufficiently long, e.g., 256 bits, random number N_i for each T_i, computes a pseudonym identity PID_i for using Equation 21.1, and records the mapping relation between PID_i and N_i.

$$PID_i = H\left(N_i || ID_{CMD}\right) \oplus ID_i \oplus ID_{CMD}, \tag{21.1}$$

where

"⊕" denotes XOR operation

ID_{CMD} is the identity of the *CMD*

H is a one-way hash function

Then, the *CMD* delivers PID_i to T_i in a secure way. With this secret-splitting mechanism, the real identity ID_i is concealed in PID_i. The *CMD* also shares a long-term secret key $s_i = f(ID_i)$ with T_i, where f is a key generating function.

In the following, we describe MAP according to the order of message exchanges, and discuss the security goals which can be achieved in each message (Figure 21.5).

1. The chairman T_i chooses a random r_1 and computes its long-term key s_1 by $s_1 = f(ID_1)$. Then, T_1 uses s_1 to encrypt $(t_1||s_1||r_1||ID_2||\ldots||ID_m)$ and sends $\{PID_1,\ E_{s_1}(t_1||\ s_1||\ r_1||\ ID_2||\ \ldots||\ ID_m)\}$ to the *CMD*.

2. On receiving the message from T_1, the *CMD* derives the real identity of node T_1 by computing

$$ID_1 = PID_1 \oplus h(N_1||ID_{CMD}) \oplus ID_{CMD}. \tag{21.2}$$

The *CMD* can retrieve corresponding shared key s_1 and decrypt $E_{s_1}(t_1||\ s_1||\ r_1||\ ID_2||\ \ldots||\ ID_m)$. Then, the *CMD* verifies the authenticity of s_1 and the time stamp t_1. If it is true, the *CMD* calls the other user ID_i, $i = 2,\ldots,m$. All the keys s_i $(i = 1,\ldots,m)$ are precomputed by the *CMD*.

3. Each node T_i, $i = 2, 3, \ldots, m$, does the same as T_i in step 1. The node T_i chooses a random r_i, computes the long-term key s_i as $s_i = f(ID_i)$, uses s_i to encrypt $\{t_i||s_i||r_i\}$, and sends the message $\{PID_i, E_{s_i}(t_i||s_i||r_i)\}$ to the *CMD*.

4. On receiving the message from T_i, the *CMD* extracts the real identity ID_i of node T_i by computing

$$ID_i = PID_i \oplus h(N_i||ID_{CMD}) \oplus ID_{CMD}. \tag{21.3}$$

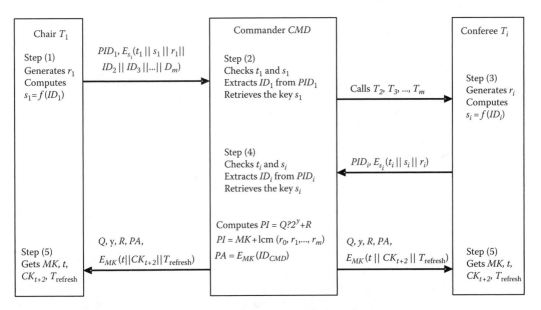

FIGURE 21.5 Mutual authentication protocol for the proposed teleconference.

Then, the *CMD* can retrieve corresponding shared key s_i, and further decrypt $E_{s_i}(t_i||s_i||r_i)$. Next, the *CMD* checks the authenticity of key s_i and t_i. If true, the *CMD* precomputes a key sequence by using a one-way hash function H, where n is chosen to be reasonably large and each CK_i satisfies $CK_i = H(CK_{i+1})$, or $CK_i = H^{n-i}(CK_n)$. The *CMD* selects a nonzero random r_0, and computes *PI* and *PA* by

$$PI = MK + \text{lcm}(r_0, r_1, \ldots, r_m), \quad (21.4)$$

$$PA = E_{MK}(ID_{CMD}), \quad (21.5)$$

where $\text{lcm}(r_0, r_1, \ldots, r_m)$ denotes the least common multiple function and $MK(= CK_0)$ is the master key of the group session. Then, at time t_{start}, the *CMD* broadcasts the following message to nodes T_i, $i = 1, 2, \ldots, m$

$$CMD \rightarrow T_i : \{PI, PA, InitConfKey\},$$

where *InitConfKey* denotes $E_{MK}(t||CK_{t+2}||T_{refresh})$, where MK and CK_{t+2} satisfy $MK = CK_0 = H^{t+2}(CK_{t+2})$.

5. According to the received message, T_i gets which is given by

$$MK = PI \bmod(r_i). \quad (21.6)$$

Then T_i verifies the validity of MK by checking if $PA = E_{MK}(ID_{CMD})$. If it holds, T_i gets $\{t, CK_{t+2}, T_{refresh}\}$ by decrypting *InitConfKey*. The detailed corresponding procedures are given by Algorithm 1.

Algorithm 1: Initial group communication session parameters

---

1. **function** *Init_Conf_Key* ( ) {
2.   **if** (*InitConfKey* message received) {
3.   Compute $MK = PI \bmod(r_i), E_{MK}(ID_{CMD})$;
4.   **if** $PA \neq E_{MK}(ID_{CMD})$ **return** ERROR;
5.   **if** (*InitConfKey* message received) {
6.   Allocate a key buffer of length t ($kb[1], \ldots, kb[t]$), and two key-slots ($ks[1], ks[2]$);
7.   **for** ($i = 1; i \leq t - 1; i++$) **do** $kb[i] = H^{t-i}(CK_{t+s})$;
8. $ks[2] = H^t(CK_{t+s})$, $ks[1] = H^{t+1}(CK_{t+s})$;
9. $CK_W = H^{t+2}(CK_{t+s})$;
10.   Set key $ks[1]$ for data encryption;
11.   Set *RefreshKeyTimer* to $T_{refresh}/2$;
12.   }
13. }

Figure 21.6 shows how the node copies *CK* sequence into its key buffer and key-slots, and switches the active key once receiving CK_{i+2}. Algorithm 2 gives the right-shift process of automatic key renewal at the midpoint of each interval $T_{refresh}$. Each node maintains two variables e and CK_W. Sentry CK_W tracks the most recently outdated *CK*, and traces the number of *CK* that the node failed to receive correctly.

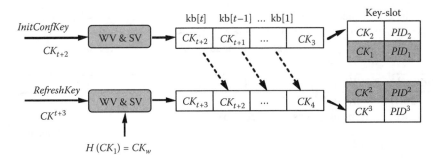

FIGURE 21.6   Initial setup and *CK&PID* refreshment mechanisms.

Algorithm 2: Refresh key timer

---

1. **function** *Refresh_Key_Timer* ( ) {
2.   **if** (*RefreshKeyTimer* triggered) {
3.   Right-shift the key buffer and key-slot;
4. $e + +$;
5. $CK_w = $ {the inactive key in key slots};
6.   }
7.   Set active *CK*s in key slots;
8.   Set *RefreshKeyTimer* to $T_{refresh}/2$;
9.   **if** ($e = t$), send *RequestKey* message to *CMD*;
10. }

21.2.4   ONE-WAY *CK&PID* RENEWAL

Forward secrecy requires that the *CK&PID* be refreshed in a one-way manner. One-way *CK* renewal guarantees that the *CMD* can update the *CK* at regular intervals. One-way *PID* renewal allows each node to renew its *PID* frequently and reduces the risk that it uses a compromised *PID* to communicate with the *CMD*.

If the *RefreshKey* message is broadcast every interval $T_{refresh}$, an intruder may predict when to launch an attack. Thus, it is much easier for the intruder to disrupt such messages by initiating the DoS attacks. To refrain from such attack, the *CMD* should send the *RefreshKey* packets randomly or in a pseudorandom ways that cannot be predicted by an attacker.

Assume that the MAP phase completes at time t_{init}. To resist DoS attack, the *CMD* broadcasts the *RefreshKey* message with CK_{i+t+2}, $i = 0, \ldots, n - t - 2$ for the *i*th *CK&PID* renewal to all nodes at the time randomly chosen from the interval $[t_{init} + i \cdot T_{refresh} - \delta, t_{init} + i \cdot T_{refresh} + \delta]$, $\delta < T_{refresh}/4$, i.e.,

$$CMD \rightarrow T_j (j = 1, \ldots, m) : \{CK_i, E_{MK}(CK_{i+t+2}||CK_{i+1})\},$$

where
    CK_{i+1} is the active encryption key at the time when the *RefreshKey* message is broadcast
    CK_i is the outdated *CK* in the *CK* sequence

To provide the DoS tolerance, CK_i is used for WV and $E_{MK}(CK_{i+t+2}||CK_{i+1})$ is used for SV.

On receiving the *RefreshKey* message, each participant deals with this message according to Algorithm 3. Figure 21.6 illustrates how to initialize and refresh the *CK&PID*. Due to one-way property of the CK sequence, the $RefreshKey$ message does not need message authentication code, because the receiver can verify if the received CK belongs to the same key sequences as those stored in the key buffer. Such implicit authentication notably decreases the message size.

Algorithm 3: $CK\&PID$ refreshment for node

---

1. **function** $Refresh\_CK\&PID$ ( ) {
2.   **while** ($Refreshkey$ message received) {
3.   **if** $CK_i \neq CW_w$ {   // weak verification;
4.   Discard this message;
5.   **skip**;
6.   }
7.   Decrypt $Refreshkey$ to get CK_{i+t+2} ;
8. $CK_w = \{$**the inactive key in key slots**$\}$;
9.   Right-shift $kb[1] = CK_{i+3}$ to the inactive key slot;
10.   Computing $PID_{k,i+3} = H(PID_{k,i+2}||CK_{i+3})$;
11.   **for** $(i = 2; i \leq t; i++)$ do $kb[i] \to kb[i-1]$;
12.   **if** $(e \neq 0)$ {   //there are lost CKs;
13.   Recover the lost CKs by Algorithm 4;
14. $CK_{i+t+2} \to kb[t]$;
15.   }
16.   }
17. }

In the following, we discuss the PID renewal mechanism. Though in the MAP phase, basic identity anonymity is provided by using PID_i for T_i instead of its real identity ID_i, there are still some security issues to be concerned. For instance, even T_i never reveals ID_i to parties other than the CMD, it does reveal its long-term PID_i during the session. Hence, illegal parties can track a node's location by PID_i, although they cannot obtain ID_i.

In the proposed protocol, the PID renewal is in progress with the CK renewal synchronously. For the jth refreshment, a node T_k can compute its new pseudonym identity $PID_{k,j}$ as

$$PID_{k,j} = H(PID_{k,j-1}||CK_j), \quad j = 1, 2, \ldots, n, \tag{21.7}$$

where $PID_{k,0}$ of T_k is equal to the initial PID in the MAP phase, i.e., $PID_{k,0} = PID_k$, $k = 1, 2, \ldots, m$. The PID will vary with CK_j. Hence, the PID of each node is updated with forward secrecy due to the one-way CK renewal.

The computation complexity of the refreshment algorithm is lightweight, because it is only requisite to broadcast the $RefreshKey$ messages and perform low-cost hash operations.

21.2.5 DoS-Tolerant Authentication

The $CK\&PID$ refreshment relies on the authenticity of the $RefreshKey$ messages, which makes the $RefreshKey$ messages attractive targets for the DoS attack. An attacker may send a large amount of forged messages to exhaust the nodes' buffer before they can verify the messages, and force them to drop some authentic messages.

An efficient way for an attacker to disrupt the $RefreshKey$ message is to jam the communication channel when the $RefreshKey$ messages are transmitted. If the attacker can predict the schedule of such messages, it would be much easier for the attacker to disrupt such message transmissions. Thus the CMD is required to send the $RefreshKey$ packets randomly or in a pseudorandom ways so that prediction is not feasible.

For each node, a packet filter is designed to fast verify the *RefreshKey* message, $\{CK_i, E_{MK}(CK_{i+t+2}||CK_{i+1})\}$. As shown in Figure 21.3, in the WV phase, the nodes perform a fast check to identify the forged messages, and try to discard most unintended forged messages. The "unintended" refers to those random packets used for jam purpose only, and the "intended" refers to those fake packets used for both fraud and jam. Upon receiving the *RefreshKey* message, each node first checks the authenticity of clear text CK_i in $\{CK_i, E_{MK}(CK_{i+t+2}||CK_{i+1})\}$. The messages that fail this test are discarded. The computation overhead for the WV is very low. Those intended forged messages that slip through the WV are removed in the SV phase. Compared with the WV, the SV performs strict check with hash computation.

To further improve the possibility that a node receives authentic *RefreshKey* packets, the node adopts a random selection policy to store and authenticate the incoming packets that pass the above WV.

Without loss of generality, assume that the length of the buffer at each node is m. During each time interval T_{refresh}, a node can save the first m copies of *RefreshKey* packets that pass the WV. Then, if a new copy is to be kept, the node randomly selects one of the m buffers and replaces the corresponding copy. For the kth copy ($k > m$), the node keeps it with the probability m/k. It is easy to verify that if a node receives n copies of *RefreshKey* packets, all copies have the same probability m/n to be kept in one of the buffers. The key issue is to make sure that all *RefreshKey* copies have the equal probability to be selected. Otherwise, an attacker who knows the refreshment rule may exploit the unequal probabilities and make a forged *RefreshKey* be chosen with high possibility. Therefore, each node verifies $E_{MK}(CK_{i+t+2}||CK_{i+1})$ for at most m times and $(m-1)/2$ on average. With random selection strategy, the probability that a node receives an authentic *RefreshKey* copy can be estimated as

$$P[\textit{RefreshKey} \text{ packet is authentic}] = 1 - p^m, \tag{21.8}$$

where

$$p = \frac{\#\text{forged copies}}{\#\text{total copies}}. \tag{21.9}$$

Therefore, the longer the buffers are, the more effective the random selection algorithm is. Due to the exponential form of Equation 21.8, a little longer buffer can remarkably improve the reliability of broadcasting the *RefreshKey* messages. To maximize the successful DoS attack, an attacker has to send as many forged copies as possible. Hence, such algorithm makes the DoS attack so difficult that the attacker would rather use signal jamming than directly attacking the nodes.

21.2.6 FAULT-TOLERANT KEY RECOVERY

The *RefreshKey* message can also be used to recover the lost CKs for fault-tolerant and key-recovery mechanism shown in Algorithm 4. Suppose that there are $r(\leq t)$ CKs reserved in the key buffer due to previous lost messages, i.e., there are $e = t - r$ empty slots in the key buffer. Let $\{CK'_r, \ldots, CK'_1\}$ denote these r CKs in the key buffer $\{kb[r], \ldots, kb[1]\}$, respectively. They also belong to the same key sequence, and satisfy $H(CK'_r) = CK'_{r-1}, \ldots, H(CK'_2) = CK'_1$.

Upon receiving a *RefreshKey* message with CK_k, each node checks if $H(CK_{i+1}) = CK_W$, where CK_W tracks the most recently outdated CK. If it is true, the node uses CK_{i+t+2} to recover the lost CKs in the same key sequence. Figure 21.7 illustrates the recovery of the lost $CK(s)$. Assume that a node receives a *RefreshKey* message with CK_{t+2}. Because $H(CK_{t+2}) = CK_{t+1}$ and $e = 0$, there is no message loss. However, the next two renewal messages are discarded because $H(CK^*_1) \neq CK_W$ and $H(CK^*_2) \neq CK_W$. Thus, there are $t - 2$ CKs in the key buffer. The node receives an authentic *RefreshKey* message with CK_{i+5}. Because $H(CK^*_3) = CK_W$, the node can recover the previous two lost CKs as $CK_{t+3} = H^2(CK_{t+5})$ and $CK_{t+2} = H^3(CK_{t+5})$.

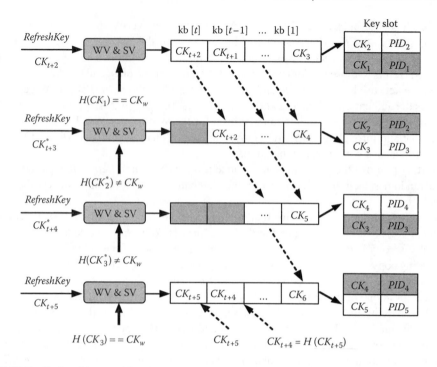

FIGURE 21.7   Fault-tolerant and key-recovery mechanisms.

Algorithm 4: Strong verification & CK recovery

   1. **function** $Recover\_CK$ ( ) {
   2.   **if** $H(CK_{i+1}) \neq CW_W$ {   // strong verification;
   3.   Discard $RefreshKey$ message $\{CK_i, E_{MK}(CK_{i+t+2}||CK_{i+1})\}$;
   4.   **return** FALSE;
   5.   }
   6.   **if** $(e \geq 1)$ **for** $(i = 1; i \leq e; i + +)$ **do** {
   7. $H^i(CK_k) \rightarrow kb[t - i + 1]$;
   8. $e = 0$;
   9.   };
 10.   **return** TRUE;
 11. }

21.2.7 Dynamic Participation

The dynamic participation mechanism, as a basic security requirement, is that any active participant can join or leave an onging group session while assuring the freshness of the key.

Node joining: When a participant T_{m+1} joins an onging session, T_{m+1} is required to obtain the permission from the *CMD* to join the session. The corresponding actions are described as follows:

1. T_1 sends the *CMD* the message $J = E_{s_1}(ID_{m+1}||t'||\text{JOIN})$ with its current PID_1, where t' is a time stamp and ID_{m+1} is the identity of the participant T_{m+1}.
2. The *CMD* decrypts J with s_1 to obtain t' and ID_{m+1}, and then it checks the validity of the time stamp t' and ID_{m+1}. If it is true, the *CMD* calls T_{m+1}.

3. T_{m+1} chooses a random r_{m+1}, and computes the secret key $s_{m+1} = f(ID_{m+1})$. It uses s_{m+1} to encrypt $\{t_{m+1}||s_{m+1}||r_{m+1}\}$ and sends the message $\{PID_{m+1}, E_{s_{m+1}}(t_{m+1}||s_{m+1}||r_{m+1})\}$ to the CMD.

4. On receiving the message from T_{m+1}, the CMD extracts ID_{m+1} as $ID_{m+1} = PID_{m+1} \oplus H(N||ID_{CMD}) \oplus ID_{CMD}$, computes s_{m+1}, and decrypts $E_{s_{m+1}}(t_{m+1}||s_{m+1}||r_{m+1})$ with s_{m+1}. Later, the CMD checks the authenticity of s_{m+1} and t_{m+1}. If it is true, the CMD calculates PI' by

$$PI' = MK + r_{m+1} \cdot s_{m+1},$$

where MK is the main key of the session. Finally, at time $t_{\text{init}} + i \cdot T_{\text{refresh}}$ (assume that the MAP phase ends at time t_{init}), the CMD sends following message to T_{m+1}

$$CMD \rightarrow T_{m+1} : \{PI', PA, InitConfKey'\},$$

where $InitConfKey' = E_{MK}(t||CK_{i+t+2}||T_{\text{refresh}})$.

5. T_{m+1} processes the $RefreshKey$ message according to Algorithm 1. That is, t_{m+1} computes MK as $MK = PI'\text{mod}(r_{m+1})$ and verifies its validity by checking if PA is equal to $E_{MK}(ID_{CMD})$. If it holds, T_{m+1} decrypts $InitConfKey'$ with MK. T_{m+1} joins the session.

Node leaving: When a node leaves an onging session, the CMD must update all of the previous CKs to assure the freshness of CK. Assume that node T_q has exited the session. The procedure of updating CK can be depicted as follows:

1. T_1 sends the message $Q = E_{s_1}(ID_q||t''||QUIT)$ to the CMD, where ID_q is the identity of T_q.

2. The CMD obtains t'' and ID_q by decrypting Q with s_1, and then it checks the validity of t''. If it is true, the CMD selects a new MK' and further calculates PI' and PA' as

$$PA' = E_{MK'}(ID_{CMD}),$$
$$PI' = CK_0' + lcm(r_0', r_1', \ldots, r_{q-1}', r_{q+1}', \ldots, r_m'),$$

where $r_i' = r_i + t'$, and t' denotes the current time. Then, the CMD broadcasts following message to the remaining nodes T_i, $i \neq q$

$$CMD \rightarrow T_i(i \neq q) : \{PI', InitConfKey'\},$$

where $InitConfKey' = E_{MK'}(t||CK_{t+2}||T_{\text{refresh}})$, and MK' and CK_{t+2} satisfy $MK' = H^{t+2}(CK_{t+2})$.

3. The rest of nodes $T_i(i \neq q)$ attain the MK' as $MK' = PI'\text{mod}(r_i + t')$ and verify the authenticity of MK' by checking $PA' = E_{MK'}(ID_{CMD})$. If true, they get $\{t, CK_{t+2}, T_{\text{refresh}}\}$ by decrypting $InitConfKey'$ and then execute the Algorithm 1 to reinitiate the system. The CMD updates the CKs and makes all previous CKs obsolete.

21.2.8 Reinitialization

The CMD needs to reinitialize the group session, if all n CKs in the CK sequence have been used up, or existing nodes have been compromised, or a node has explicitly requested CK because it has missed more than t CKs. In the former two scenarios, all nodes are forced to be reinitialized, while in the last case only the requested node needs to be reinitialized.

Specifically, in the first case, the CMD recomputes a new CK sequence $\{CK_i'|i = 1, 2, \ldots n\}$ and then broadcasts a new $InitConfKey$ message with CK_{t+2}' to all the nodes. In the second case, the steps of reinitialization are similar to those when a node leaves an onging session. For the third case, the CMD

only sends the requesting node an *InitConfKey* message with the current configuration parameter $\{t||CK_{i+t+2}||T_{\text{refresh}}\}$. Subsequently, this node can periodically renew the *CK&PID* by receiving the *RefreshKey* message.

21.2.9 ROBUSTNESS FOR CLOCK SKEWS

The proposed protocol is robust to clock skew among the nodes and the *CMD*. Let $m_i(T)$ denote the mapping from clock time to real time at node T_i. Then the clock skew between nodes A and B is denoted as $\lambda = |m_A(T) - m_B(T)|$, where T is clock time. To seamlessly renew the *CK&PID*, should satisfy $\lambda < T_{\text{refresh}}/2$, because each node will switch the active key to the new one at the midpoint of the renewal interval.

Assume that MAP ends at time t_{init}. Then at the ith renewal period $[t_{\text{init}} + (i - 1/2) \cdot T_{\text{refresh}}, t_{\text{init}} + (i+1/2) \cdot T_{\text{refresh}}]$, node A uses CK_{i+2} for data encryption, while node B still uses CK_{i+1} due to the clock skew between A and B. However, they can still successfully communicate with each other during this period, because both A and B hold the same decryption key pair $\{CK_{i+1}, CK_{i+2}\}$. Therefore, the proposed protocol can ensure the worst-case clock skew of $T_{\text{refresh}}/2$. For any two nodes A and B, the timing margin against clock skews should satisfy,

$$\max\{|m_A(T) - m_B(T)|, \forall A, B\} < T_{\text{refresh}}/2. \tag{21.10}$$

The proposed protocol can also tolerate the clock skew between the *CMD* and the nodes. For the ith *CK&PID* renewal, to resist DoS attacks, the *CMD* broadcast *RefreshKey* message at a time randomly chosen from the interval $[t_{\text{init}} + i \cdot T_{\text{refresh}} - \delta, t_{\text{init}} + i \cdot T_{\text{refresh}} + \delta]$, $\delta < T_{\text{refresh}}/4$. δ should satisfy for refreshing the *CK&PID* seamlessly. Under this constraint, the timing margin against clock skews between the *CMD* and any node B is denoted as

$$\max\{|m_{NC}(T) - m_B(T)|, \forall B\} < T_{\text{refresh}}/2 - 2\delta. \tag{21.11}$$

21.2.10 MESSAGE ENCRYPTION/DECRYPTION AND INTEGRITY

We also propose a privacy mechanism for the session between any two nodes (sender or receiver), which offers data confidentiality via encryption and data integrity via an integrity checker, message integrity code (MIC).

Figure 21.8 shows the frame format of a message, in which the *PID* identifies the pseudoidentity of the sender, *KeyID* indexes which *CK* in the two key-slots is active, and *IV* denotes the initialization vector. They are all sent unencrypted, only the data is encrypted and denoted by shaded part of the frame. Once the sender generates such a frame, it sends the frame to the receiver

$$sender \rightarrow receiver : \{Header||KeyID||PID||IV||Ciphertext||MIC\}.$$

IV, *PID*, and *KeyID* offer seeds for computing the block cipher and the *MIC*, and thus make encryption and decryption self-synchronous between the sender and the receiver. Because of the key renewal mechanism, the length of the *IV* field can be small, e.g., 32 bits. Typically, the *IV* is varied with each frame. The *KeyID* field is used to identify the *CK*, which is used to derive the integrity key *CK_I* and encryption key *CK_E*, respectively.

Header	KeyID	PID	IV	Cipher text	MIC

FIGURE 21.8 Message frame format.

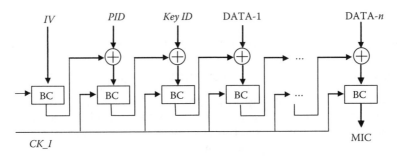

FIGURE 21.9 Message integrity check mechanisms (BC: block cipher).

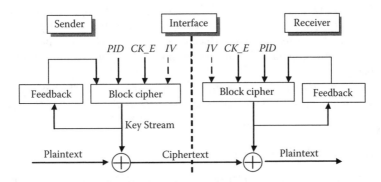

FIGURE 21.10 Message encryption and decryption mechanisms.

As shown in Figure 21.9, the *MIC* field is used to provide integrity mechanisms, which is computed from *KeyID, PID, IV,* and the ciphertext data by cipher block chaining (CBC) mode. The integrity key is *CK_I*. The *MIC* is created by using an *IV* that is fed into a cipher block and its output is XOR'd with selected elements from the frame header which is then fed into the next cipher block. The process continues over the remainder of the frame header until a 128-bit *MIC* is obtained.

To assure data confidentiality, data encryption involves bitwise modulo 2 addition of the output of a block stream cipher with the transmitted data. Figure 21.10 shows how the data is encrypted/decrypted between the sender and the receiver. The block cipher is seeded by the encryption key *CK_E, PID,* and *IV.* The output key stream is fed back to the block cipher. This process is repeated until the entire frame has been encrypted.

The block cipher takes the concatenation of *PID, IV, CK_E,* and previous key stream *KeyStream$_{i-1}$* as input. As a result, it outputs a new stream block *KeyStream$_i$*

$$KeyStream_i = BlockCipter(KeyStream_{i-1}, CK\_E, PID, IV).$$

Thus, if the plaintext is equally divided into n blocks, $\{PlainText_i | i = 1, 2, \ldots, n\}$, the ith ciphertext is generated as $CipherText_i = KeyStream_i \oplus PlainText_i$.

The proposed protocol ensures that the key steam will never be reused with the following measures: (1) a sender blends its own *PID* into the key steam to ensure that all the nodes sharing the *CK* use different key streams; (2) a sender increases its own *IV* by 1 for each message to avoid any repetition of key stream; and (3) updating *CK* periodically also guarantees that none of nodes reuse *IV*. Therefore, the proposed protocol addresses data integrity with a *MIC* value and confidentiality with symmetric CBC encryption.

21.3   SECURITY ANALYSIS

We show that the proposed protocol satisfies security requirements for ad hoc group communications.

21.3.1   IDENTITY ANONYMITY AND INTRACEABILITY

The security requirement for concealing nodes' location information is achieved by introducing a dynamic identity mechanism. This feature makes an intruder difficult to trace a particular user's location by intercepting the conversation. This protocol provides identity anonymity in all phases by replacing nodes' real identity with a pseudonym identity.

1. In the MAP phase, the real identity ID_i of T_i is replaced with PID_i. Because only the CMD knows the secret N_i and $H(N_i||ID_{CMD}) \oplus ID_{CMD}$, nobody except the CMD can obtain ID_i from PID_i by computing $ID_i = PID_i \oplus h(N_i||ID_{CMD}) \oplus ID_{CMD}$. Given that a tracker does not know $H(N_i||ID_{CMD}) \oplus ID_{CMD}$, it cannot get ID_i from the transmitted messages and then trace the location of a mobile node. Because each T_i's PID_i is computed using unique N_i, the legitimate T_i cannot compute another node T_i's by intercepting PID_k and further impersonate T_k.

2. In one-way $CK\&PID$ renewal phase, the identity anonymity is enhanced by one-way CK refreshment mechanism. A node can renew its PID as $PID_{k,j} = H(PID_{k,j-1}||CK_j)$, $j = 1, 2, \ldots, n$.

21.3.2   RESISTANCE TO RELAY ATTACK

A replay attack is a method that an intruder stores "stale" intercepted messages and retransmits them at a later time. An efficient measure against a replaying attack is to introduce time stamp t and lifetime L into the transmitted messages and set an expected valid time interval Δt for transmission delay.

All transmitted messages in each step of the proposed protocol scheme contain time stamps. According to the time stamp t and the interval Δt, the receiver can verify the validity of these messages by checking if $t - t_1 < \Delta t$ is true, where t_1 is the time stamp of a message while t is the current time when it is received. If this inequality holds, the message is valid. Otherwise, the CMD regards the message as a replaying message. This mechanism resists replaying attacks to a large extent.

21.3.3   PRIVACY OF GROUP CONVERSATION

The conversation information will be encrypted with CK. Hence, an intruder cannot know the conversation content without CK. To obtain CK in $InitConfKey$ or $RefreshKey$ message, an intruder must get the secret random r_i and then use it to calculate the MK as in Equation 21.6. However, $r_i(i = 1, 2, \ldots, m)$ is generated secretly by T_i. Nobody except T_i itself and the CMD know r_i. Hence, even $\{PID_1, E_{s_1}(t_1||s_1||r_1||ID_2||\ldots||ID_m)\}$ and $\{PI, PA\}$ can be intercepted, the intruder cannot obtain $r_i(i = 1, 2, \ldots, m)$ and compute $MK = PI \bmod r_i$, because it is impossible for him to get the key s_i unless it knows ID_i of T_i. Hence, the intruder is prohibited from taking CK and eavesdropping any communication content.

21.3.4   FORWARD SECRECY

The proposed protocol meets the security requirement for forward secrecy, because its key distribution mechanism can assure the one-way $CK\&PID$ refreshment by periodically or dynamically reconfiguring the protocol, when (1) a node joins or leaves an ongoing group communication session; (2) the lifetime of the keys is overdue; (3) all n CKs in the CK sequence have been used up; (4) existing nodes have been compromised; and (5) a node has definitely requested the CK, because more than t CKs is lost.

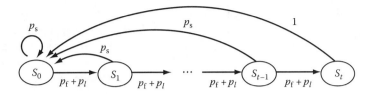

FIGURE 21.11   State transition diagram of each node.

21.4   PERFORMANCE ANALYSIS

In this session, the computation and communication overhead of the proposed protocol is analyzed. We quantify the cost of the communication and computation overhead when nodes renew the *CK&PID*, and the performance improvement because to the robust and reliable mechanism.

21.4.1   STEADY MARKOV STATE DISTRIBUTION

Figure 21.11 shows the distribution state of a node with a Markov chain. We assume that occurrence of the *CK* loss or authentication failure is random and mutually independent, and each node can finish the *RequestConfKey* operation within the interval T_{refresh}, if the key buffer of a node is full. Let state S_i denote that there are i *CK* authentication failures, and thus there are j empty slots in the key buffer. The state transition is triggered by three events: packet loss, *CK* authentication failure, and *CK* authentication success. Let

$$p_f = \Pr\{CK \text{ authentication fails}|CK \text{ is received}\},$$

$$p_s = \Pr\{CK \text{ authentication succeeds}|CK \text{ is received}\}.$$

Without loss of generality, we also assume that all transmitted messages (both legal and bogus packets) via the wireless channel have the same loss probability p_l, which is defined as $p_l = \Pr\{\text{message is lost}\}$. The assumption is reasonable because the wireless channel cannot distinguish the difference of the packets. According to Equations 21.8 and 21.9, we have $p_s = p_l \cdot (1 - p^m)$ and $p_f = p_l \cdot p^m$, where $p_f + p_s + p_l = 1$. p_l represents the channel condition. A high p_l means that a wireless channel is with high loss or error rate. p_f is imposed by the bogus packets, which leads to successful DoS attacks.

Let $P(k)$ denote the steady-state probability of state S_i that there are exactly k empty slots. According to the global balance equation, we have

$$\begin{cases} P(i) \cdot (p_s + p_f + p_l) = P(i-1) \cdot (p_f + p_l), & i = 1, \dots, t \\ P(0) \cdot (p_f + p_l) = P(t) + \sum_{i=1}^{t-1} P(i) \cdot p_s. \end{cases} \tag{21.12}$$

Considering $\sum_{k=0}^{t} P(k) = 1$, each $P(k)$ is derived as

$$\begin{cases} P(0) = (1 - \theta)/(1 - \theta^{t+1}) \\ P(i) = P(0) \cdot \theta^k, & k = 1, 2, \dots, t. \end{cases} \tag{21.13}$$

where $\theta = p_f + p_l$.

21.4.2   COMMUNICATION OVERHEAD

To evaluate the communication overhead between the *CMD* and a node, we normalize the expected communication overhead C_{comm} by the cost of transmitting *RefreshKey* messages. Let C_{init} and C_{refresh}

denote communication costs for sending the *InitConfKey* and the *RefreshKey* message, respectively. Let $\alpha = C_{\text{init}}/C_{\text{refresh}}$ be the ratio of communication cost of *InitConfKey* to that of *RefreshKey*. Clearly, $\alpha > 1$ because the *InitConfKey* message needs more bandwidth or resources than the *RefreshKey* message.

It is necessary for the *CMD* to transmit the *InitConfKey* message when the following events occur:

1. When a participant joins an onging group communication session, the *CMD* sends this new node the current configuration via an *InitConfkey* message.
2. When a participant leaves a group communication session, the *CMD* will revoke this node by broadcasting the *InitConfkey* message to all the other nodes.
3. When all n *CK*s have been used, the *CMD* recomputes a new key sequence $\{CK_i'|i = 1, 2, \ldots, n\}$ and broadcasts the *InitConfkey* message to all nodes.
4. A node has definitely requested the *CK*, because it missed more than t *RefreshKey* messages. For this event, the *CMD* sends an *InitConfkey* message containing the configuration parameters to the node.

Note that in cases 2 and 3, all nodes are required to be reinitialized, while in cases 1 and 4 the requesting node needs to be reinitialized by sending the *InitConfKey* message. Except for these cases, the *CMD* broadcasts the *RefreshKey* message periodically. So the expected communication cost of a node is

$$E[C_{\text{Comm}}] = C_{\text{init}} \cdot \left[\frac{1}{n} + P(t) + p_{\text{e}} + p_{\text{j}} \right] + C_{\text{refresh}} \cdot \sum_{k=0}^{t-1} \{P(k)\},$$

where p_{e} and p_{j} denote the probability of a node joining or leaving a group session, respectively. According to Equation 21.13, the communication cost can be normalized with C_{refresh} as

$$C_{\text{comm}} = \alpha \cdot \left[\frac{1}{n} + \theta^t \cdot P(0) + p_{\text{e}} + p_{\text{j}} \right] + \sum_{k=0}^{t-1} \{\theta^k \cdot P(0)\}. \tag{21.14}$$

If C_{comm} is close to 1, it indicates that most *RefreshKey* messages should work well. If C_{comm} is close to α, *RefreshKey* messages works less efficiently. To analyze the dynamics of the *CK* refreshment, we assume that the frequency of joining or leaving a group session is low so that Equation 21.14 can be approximated as

$$C_{\text{comm}} = \alpha \cdot \left[\frac{1}{n} + \theta^t \cdot P(0) \right] + \sum_{k=0}^{t-1} \theta^k \cdot P(0).$$

Figure 21.12 shows the function relation between C_{comm} and the key buffer length t, where $n = 500$, $p_1 = 0.05 \sim 0.45$, and $\alpha = 10$. The choice of α implies that the cost of transmitting and dealing with *InitConfKey* message is higher than that of transmitting *RefreshKey* message, because the packet size of *InitConfKey* is larger than that of *RefreshKey*. It can be seen that the key buffer length in each node determines the communication cost. A small t will lead to a high communication overhead while a large t can remarkably reduce the communication overhead. Figure 21.12 also shows that the proposed protocol is efficient in terms of communication overhead even under serious DoS attacks ($p_{\text{f}} \geq 0.25$) and packet loss ($p_1 \geq 0.25$), because the normalized communication cost C_{comm} approaches 1 if $t \geq 10$.

21.4.3 COMPUTATION OVERHEAD

The main computation overhead in nodes is the modular arithmetic per *InitConfKey* message and the hash computation per *RefreshKey* message. Let N_{m} and N_{h} denote the number of modular arithmetic

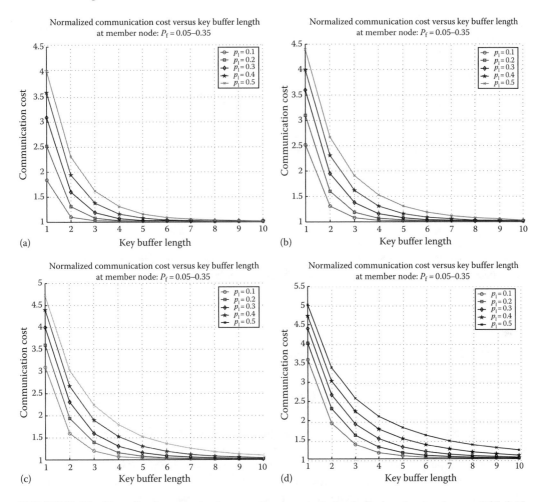

FIGURE 21.12   Normalized communication costs C_{comm} versus the key buffer length t at node: $p_l = 0.05\text{--}0.45$, $n = 500$, $\beta = 5$ at (a) $p_i = 0.05$, (b) $p_i = 0.15$, (c) $p_i = 0.25$, and (d) $p_i = 0.35$.

per *InitConfKey* message and hash computations per *RefreshKey* message, respectively. If there are exactly $k < t$ empty slots, N_h can be computed as

$$N_h = \begin{cases} 0, & \text{if } CK \text{ message is lost} \\ 1, & \text{if } CK \text{ strong authentication fails} \\ k+1, & \text{if } CK \text{ authentication succeeds.} \end{cases} \tag{21.15}$$

If all t slots in the key buffer are empty due to the CK authentication failure or message loss, the node can explicitly initiate a *RequestConfKey* message to obtain the new CK. Thus it needs to do $t + 1$ extra hash computations according to the received CK, and we can have

$$N_h = \begin{cases} t+1, & \text{if } CK \text{ message is lost} \\ t+2, & \text{if } CK \text{ strong authentication fails} \\ t+1, & \text{if } CK \text{ authentication succeeds.} \end{cases} \tag{21.16}$$

Therefore if there are k empty slots, the corresponding conditional expected value of N_h, can be derived as

$$E[N_h|k \text{ slots}] = \begin{cases} p_f + (k+1) \cdot p_s, & k < t \\ (t+2) \cdot p_f + (t+1) \cdot (p_s + p_l), & k = t. \end{cases} \quad (21.17)$$

and the expected value N_h of is calculated as

$$E[N_h] = \sum_{k=0}^{t} E[N_h|k \text{ slots}] \cdot P(k) = (t+1+p_f) \cdot P(0) \cdot \theta^t$$
$$+ \sum_{k=0}^{t-1} \{(k+1) \cdot (1-p_l) - k \cdot p_f\} \cdot P(0) \cdot \theta^k. \quad (21.18)$$

Assume that C_{hash} and $C_{modular}$ denote the cost of calculating a single modular arithmetic and hash operation, respectively. Let $\beta = C_{modular}/C_{hash}$. Similar to the evaluation of the communication costs, the expected computation costs of a node can be computed as follows:

$$E[C_{Comp}] = C_{hash} \cdot E[N_h] + C_{modular} \cdot \left[\frac{1}{n} + P(t) + p_e + p_j\right].$$

According to Equation 21.13, the computation cost of nodes can be normalized with $C_{modular}$ as

$$C_{comp} = E[N_h] + \beta \cdot \left(\frac{1}{n} + P(0) \cdot \theta^t + p_e + p_j\right). \quad (21.19)$$

To reduce the analysis complexity, assume that the frequency of joining or exiting a group session is low. So, p_e and p_j in Equation 21.19 can be ignored. Figures 21.13 and 21.14 show the normalized computation cost C_{comp} as the function of p_l and p_f ($n = 500$ and $\beta = 5$), respectively. The computation cost of each node is low, because each node only computes less than three hash functions per *CK&PID* renewal even in the worst case, e.g., where $p_l = 0.45$ and $p_f = 0.45$.

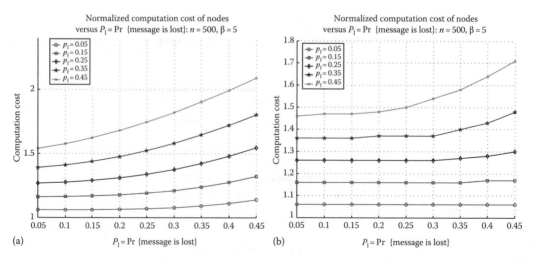

FIGURE 21.13 Normalized computation costs C_{comp} of nodes versus p_l: $n = 500$, $\beta = 5$ at (a) $t = 5$ and (b) $t = 10$.

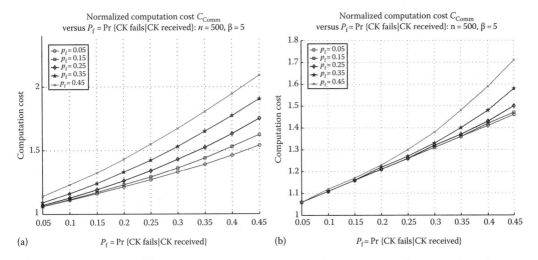

FIGURE 21.14 Normalized computation costs C_{comp} of nodes versus p_f: $n = 500$, $\beta = 5$ at (a) $t = 5$ and (b) $t = 10$.

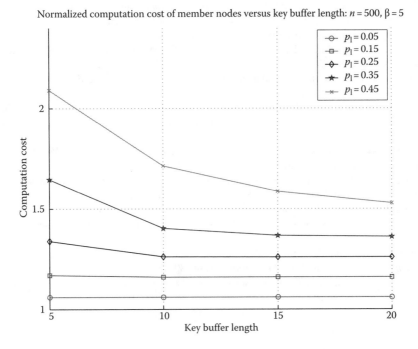

FIGURE 21.15 Normalized computation costs C_{comp} of nodes versus key buffer length t: $n = 500$, $\beta = 5$.

Figure 21.15 depicts C_{comp} as the function of the key buffer length t, where both p_f and p_l vary from 0.05 to 0.45. It can be seen that the desirable number of key buffer is $t \geq 15$ to keep communication and computation cost low, i.e., the normalized communication or computation cost is within the range of $1 \approx 1.5$. Therefore, the proposed protocol is efficient in terms of communication and computation overhead, even under heavy DoS attacks and high packet loss rate.

21.5   CONCLUSION

In this chapter, a novel secure and reliable authentication protocol has been developed for group communications in ad hoc networks. The protocol has several attractive features, such as identity anonymity and intraceability, one-way $CK\&PID$ refreshment, forward secrecy, and data privacy, etc. In addition, reliability enhancement features, such as DoS and fault tolerance and clock skews resistance, can efficiently improve the protocol robustness and enhance the system survivability.

REFERENCES

[1] D. Djenouri, L. Khelladi, and A. N. Badache, A survey of security issues in mobile ad hoc and sensor networks, *IEEE Communications Surveys & Tutorials*, 7, 2–28, 2005.

[2] I. Ingemarson, D. T. Tang, and C. K. Wong, A conference key distribution system, *IEEE Transactions on Information Theory*, IT-28, 714–720, 1982.

[3] M.-S. Hwang and W.-P. Yang, Conference key distribution schemes for secure digital mobile communications, *IEEE Journal on Selected Areas in Communications*, 13, 416–420, 1995.

[4] M.-S. Hwang, Dynamic participation in a secure conference scheme for mobile communications, *IEEE Transactions on Vehicular Technology*, 48, 1469–1474, 1999.

[5] S.-L. Ng, Comments on dynamic participation in a secure conference scheme for mobile communications, *IEEE Transactions on Vehicular Technology*, 50, 334–335, 2001.

[6] K.-F. Hwang and C.-C. Chang, A self-encryption mechanism for authentication of roaming and teleconference services, *IEEE Transactions on Wireless Communications*, 2, 400–407, 2003.

[7] X. Yi, C. K. Siew, C. H. Tan, and Y. Ye, A secure conference scheme for mobile communication, *IEEE Transactions on Wireless Communications*, 2, 1168–1177, 2003.

[8] Y. Jiang, C. Lin, M. Shi, and X. Shen, A self-encryption authentication protocol for teleconference services, *Proc. of IEEE Globecom 2005*, pp. 1706–1710, St. Louis, MO, 2005.

[9] D. Brown, Techniques for privacy and authentication in personal communication system, *IEEE Personal Communication Magazine*, 2, 6–10, 1995.

[10] Y. Jiang, C. Lin, M. Shi, and X. Shen, Hash-binary-tree based group key distribution with time-limited node revocation, in Y. Xiao and Y. Pan (Eds.), *Security in Distributed and Networking Systems*, World Scientific, 2007.

[11] A. D. Wood and J. A. Stankovic, Denial of service in sensor networks, *Computer*, 35, 54–62, 2002.

[12] N. Haller, The s/key one-time password system, RFC1760, IETF, 1995.

[13] A. Perrig, R. Szewczyk, V. Wen, D. Culler, and J. D. Tygar, SPINS: Security protocols for sensor networks, *Proc. of IEEE/ACM MobiCom '01*, pp. 189–199, Rome, Italy, 2001.

[14] A. Perrig, J. D. Tygar, D. Song, and R. Canetti, Efficient authentication and signing of multicast streams over lossy channels, *Proc. of the 2000 IEEE Symp. on Security and Privacy*, pp. 56–65, Berkeley, CA, 2000.

[15] D. Liu and P. Ning, Multilevel μTESLA: Broadcast authentication for distributed sensor networks, *ACM Transactions on Embedded Computing Systems*, 3, pp. 800–836, 2004.

[16] T. Park and K. G. Shin, LiSP: A lightweight security protocol for wireless sensor networks, *ACM Transactions on Embedded Computing Systems*, 3, pp. 634–660, 2004.

Part III

Security in Integerated RFID and WSN

22 Threats and Vulnerabilities of RFID and Beyond

Jaap-Henk Hoepman and Thijs Veugen

CONTENTS

The physical characteristics of radio-frequency identification (RFID) tags and the typical architecture of RFID-based systems introduce specific threats to the security and privacy of such systems. The privacy threats of RFID have received most attention so far, especially in the media. But the security threats are equally important in real applications.

This chapter discusses both security and privacy threats to RFID systems, in a balanced way. We first outline the general architecture of an RFID system, the stakeholders managing the different components, and identify the specific properties of RFID systems relevant for privacy and security. We then discuss threats aimed at the tag itself, at the wireless interface between tag and reader, and at the backoffice. We also discuss how these threats can lead to security or privacy breaches. We clarify the types of threats and their risks by discussing several cases of RFID systems (animal tracking, public transport, identity cards, and smart shops).

We also look to the future and discuss how a related technology, called near field communication (NFC), suffers from a few more threats of its own. Note that this chapter only deals with threats and risks, not possible countermeasures to mitigate these risks. These are discussed elsewhere in this book.

22.1   INTRODUCTION

Inspite of being a one-woman show, consumers against supermarket privacy invasion and numbering (CASPIAN)* ran by Katherine Albrecht [2] has made a tremendous impact on the public perception of radio-frequency identification (RFID) technology. Consistently called "spychips" by Albright, RFID tags have acquired a predominantly negative public image in a relatively short period of time. Campaigns against Benetton in 2003 and the METRO "Future Store" in Germany in 2004 received worldwide media attention, forcing these companies to change their plans.

The debates were often heated, and strained with emotion. As a result, although they were based on real and serious issues with the use of RFID, the threats to privacy were sometimes exaggerated. On the other hand, the security problems with the use of RFID were hardly mentioned at all. For a public debate this is not very surprising, because security affects only the business applying RFID, whereas privacy aspects affect the general public as a whole.

The purpose of this chapter is to give a balanced treatment of both privacy and security issues surrounding the use of RFID technology. This chapter is based on our own research† in this area [30], as well as numerous other general studies by, for instance, the Federal Office for Information Security (BSI) [5], the European Technology Assessment Group (ETAG) [8], and the National Institute for Standards and Technology (NIST) [25].

There is also a growing body of academic literature on this topic. Sarma et al. [28,29] were the first to discuss the security and privacy implications of RFID. Others surveyed the issues as well, all taking a slightly different angle, see for example [11,26,32]. For an extensive bibliography on RFID security and privacy, we refer to the Web site‡ maintained by Gildas Avoine.

22.1.1   SECURITY AND PRIVACY RELEVANT PROPERTIES OF RFID SYSTEMS

Certain aspects of RFID systems have a profound impact on the security and privacy of the systems that use them. Although the basics of RFID systems are discussed in depth elsewhere (see for instance the *RFID Handbook* [9]), we briefly summarize the security and privacy relevant properties of RFID systems here.

RFID tags (also called transponders) and readers communicate with each other through a wireless communication link (as specified by ISO 18000 [19]). This link can operate on different radio

* http://www.nocards.org
† http://www.pearl-project.org/
‡ http://lasecwww.epfl.ch/~gavoine/rfid/

frequencies (RFs), which influences the distance over which the tags can be read by a reader. This has an impact on the privacy of the RFID system. Typical settings are low frequency (LF) (125–134 kHz) with a reading range of 1 m, high frequency (HF) (13.56 MHz) and ultrahigh frequency (UHF) (868/916 MHz) with a reading range of up to 4 m, and microwave (2.45 or 5.8 GHz) with a reading range of up to 15 m, with recorded cases of 1 km [5] (in case of an active tag). The wireless channel itself also poses a point of attack.

Reading distance is also determined by the shape and size of the antenna, and the type of the tag. There are basically two types of tags: active tags that have their own source of power (e.g., a battery), and passive tags that receive their power from the reader when they are within range (either through capacitive coupling or through inductive coupling). Active tags have a longer reading range. Capacitive coupling implies close proximity of the tag and the reader, typically a few centimeters. ISO 14443 [17] standardizes these so-called Proximity tags. Vicinity tags (with reading distances between 10 cm up to 2 m) are standardized through ISO 15693 [18].

When many tags are in reading range of a single reader, an anticollision procedure must be executed to single out a particular tag to communicate with. This is usually based on the unique number of the tag. This number is used before any type of authentication or access control between reader and tag has been performed, and thus may pose a security or privacy risk.

Transponders are packaged in many different shapes and sizes. Smart labels are tags attached as overlays on paper, cardboard, plastic, or other flat surfaces. Tags can be embedded in glass cylinders (for small dimensions) or plastic sheath (to protect against moisture) for example. Larger packages can be used to achieve longer active ranges. Card-shaped transponders are tags embedded in a credit card shape piece of plastic, usually conforming to the dimensions mandated by ISO 7816 [21]. The packaging has a big influence on the visibility of the tag (i.e., whether it is obvious that the object is tagged) and on the strength of the "binding" between the tag and the object being tagged (i.e., whether the tag can easily be removed form the object). It is therefore an important parameter both for the privacy as well as the security of the RFID application in which it is used.

Finally, the tags (i.e., the chips) themselves also come in a wide variety of classes, with different capabilities in terms of processing power and storage space. The traditional low-end RFID tags (like the EPC* tags [15]) simply encode a product and item number that is broadcasted every time the tag is within reading range. Higher-end systems are really just like contactless smart cards: they have a few kB of ROM and RAM, and can execute complex protocols involving cryptographic operations. Typical examples are payment cards (like EMV†) or the new electronic passports [13]. Our focus in this chapter is on the more low-end tags, although the treatment is general enough to cover the whole range of RFID and RFID-like systems.

In the second half of this chapter we also discuss near-field communication (NFC), which is a more recent technology that subsumes RFID and adds certain extra capabilities that raise their own security and privacy issues.

22.1.2 Entities and Components in RFID Systems

When discussing (computer) security it is important to distinguish the devices under attack from the entities (one could say stakeholders) that use, control, or own them. Although, in general, the devices are the targets of an attack, the damage is usually not incurred by the devices themselves, but rather by the owners or users of these devices instead. They are the ones paying the cost: financial damages, bad press or loss of reputation (in case of a security breach), or the loss of control over one's personal information (in case of a privacy breach).

The typical components and entities of an RFID-based system are shown in Figure 22.1, and are briefly discussed below.

---

* http://www.epcglobalinc.org
† http://www.emvco.com/

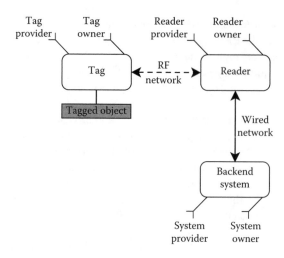

FIGURE 22.1   Entities and components in an RFID system.

22.1.2.1   Components

In general, we distinguish the following five different components in an RFID application:

- Tags
- Wireless RF network(s) between tags and readers
- Readers
- (Usually wired and IP-based) network(s) between readers and backend systems
- Backend systems

Note that the backend systems themselves could be quite complex. In a supermarket with several branches, the backend systems comprise the collection systems in the individual stores connected with the systems at the company headquarters. Typical Electronic Product Code (EPC)-based backend systems* also comprise an RFID middleware solution (e.g., Savant), an Object Naming Server (ONS), and a Physical Markup Language (PML) repository. We do not go into any more detail here, as the threats against the backend system and its components are not RFID specific.

22.1.2.2   Entities

Many different entities are involved in the development and deployment of an RFID system. So-called providers deliver the different hardware components and software applications. Owners deploy these applications or own particular pieces of hard- and software. Each of these entities has different, sometimes conflicting, requirements regarding the functionality, security, and privacy of the RFID system. They also control parts of the RFID system, and thereby influence whether certain requirements can actually be met. In malicious cases, certain entities may actively subvert the security and privacy of other entities.

- Providers of tags, readers, or backend systems
- Owners of tags, readers, or backend systems
- Owner of the application

---

* http://www.epcglobalinc.org

Note that we distinguish between the hardware platform (tags, readers, and backend systems) and the (software) application running on top of the hardware. Typically, several applications use the same pieces of hardware. We do not consider the providers or owners of networks separately. In the case of the wireless network between reader and tag they are effectively provided by and owned by the provider/owner of the reader. The network between reader and backend systems is generic.

22.1.3 OUTLINE

This chapter is organized as follows. In Section 22.2 we classify and describe the threats against an RFID system, and discuss how these threats affect the security and privacy of the system. We clarify these threats and their risks by discussing several cases of RFID systems (animal tracking, public transport, identity cards, and smart shops) (Figures 22.4 through 22.7) in Section 22.3. The successor to RFID, NFC, introduces some new threats that are discussed in Section 22.4. This chapter ends with some conclusions and an inventory of open issues in Section 22.5.

22.2 THREATS AND THEIR SECURITY AND PRIVACY CONSEQUENCES

Let us first be a little more precise about the terminology we use in this chapter (see Figure 22.2). A vulnerability is a certain weakness in the system that allows an attacker to gain illegal access to a resource. A control aims to limit the severity of a vulnerability, making attacks exploiting this vulnerability more difficult or impossible. A threat is the possibility to exploit a vulnerability in a certain way. An attack is the act of really trying to use a vulnerability in a certain way. A successful attack leads to an incident (or breach), which incurs a certain damage.

There are different ways to classify threats against a particular system. One way is to classify threats according to their purpose (doing physical harm, to incur financial damage, denial-of-service, undermine the trustworthiness of the target, etc.). Another way, and the way we follow here, is to classify threats according to their target (i.e., the tag, the reader, and the back office). This classification is depicted in Figure 22.3, and discussed in detail below.

After describing this classification, we discuss how these basic threats can be used to subvert the security and the privacy of the system and its owners.

22.2.1 THREAT CLASSIFICATION

As said earlier, we classify the threats according to the target at which they are aimed. We first discuss the tag, then the network between reader and tag, then the reader itself, then the network between reader and back office, and finally (briefly) the back office itself.

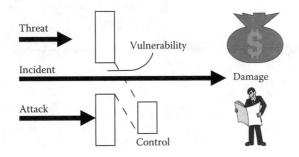

FIGURE 22.2 Threats, attacks, countermeasures, etc.

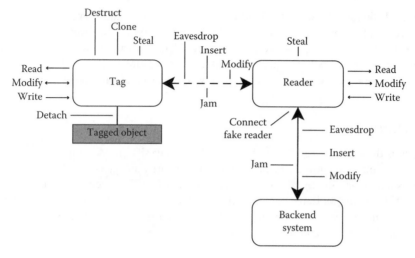

FIGURE 22.3   Threats against tag, reader, and backend systems.

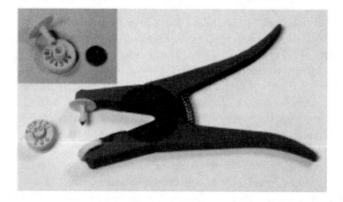

FIGURE 22.4   Animal tracking.

FIGURE 22.5   Public transport.

FIGURE 22.6   Identity cards.

FIGURE 22.7   Smart shopping.

22.2.1.1   On the Tag

The following threats to the tag exist:

Unauthorized read: Data on the tag is read by an unauthorized reader. For example, basic RFID tags have no form of access control and broadcast their ID whenever they come within range of a reader. This includes readers that were not meant (i.e., are unauthorized) to access this data.

Unauthorized write (aka falsification of contents): Data on the tag is overwritten by an unauthorized reader (note that even if the reader writes data, we call it a reader). An unauthorized write changes the data on the tag, leading to wrong or false data on the tag. This threat is particularly important if the tag stores other data about the object being tagged, besides the tag ID.

Unauthorized writes may also invalidate or disable tags, for instance if a nonexistent ID is written to the tag.

Unauthorized modification: When an unauthorized read is combined with an unauthorized write, an unauthorized modification occurs. The difference between a simple unauthorized write is that in case of a modification the new data can depend on old data on the tag, whereas this is not the case with an unauthorized write.

Cloning (falsification of identity—tag): A tag can be cloned by reading all its data and putting it on a new tag, making the new tag indistinguishable from the old tag. If the tag has no security features, this is as simple as copying the ID from the old tag to the new tag. If the old tag does have security features, the new tag must be able to simulate them to trick the reader into accepting the new tag as valid. Whether this is possible depends on the nature of the security features.

Destruction/deactivation: Tags can be physically destructed (e.g., through the use of an RFID zapper*). They can also be deactivated by sending an appropriate command to the tag (cf. the EPC kill command). The attack either results in the tag being completely invisible to a reader, or that the data on the tag (e.g., the tag ID) can no longer be read. This constitutes a denial-of-service attack.

Detaching the tag: Tags can also be removed from the object they are attached to, thus breaking the link between the tag and the object they belong to. In general this is hard for the tag to detect (unless it contains certain sensors). This is a high impact threat, because the whole point of using RFID tags was to be able to track the physical objects they were connected to.

Theft (of tag): The tag can be stolen from its owner (either with or without the object to which it was attached, or even before the tag got attached to an object in the first place). Tags may contain keys or other sensitive data that may be recoverable by hardware-based attacks, and that may prove to be valuable to attack other tags or readers. The ramifications of this threat are roughly the same as the deactivation threat, except that in this case the threat is imposed by an outsider instead of the owner.

22.2.1.2   On the Air Interface

The following threats on the air interface between reader and tag can be identified:

Eavesdropping: Listening into the communication between reader and tag is very easy, because the communication link is wireless. It should be noted that data sent from the reader to the tag is easier to intercept than the communication from the tag to the reader. This is because the reader-to-tag signal is much stronger than the tag-to-reader signal. Which signal can be overheard also depends on the location of the eavesdropper with respect to the reader and the tag, and on the presence of materials that block or deteriorate the RF signal.

Injection, and falsification of identity—reader: This threat comes in many different guises. A fake reader can set up a communication link with a genuine tag. With simple RFID systems there is no form of reader authentication, and therefore fake readers cannot be distinguished from genuine readers. This threat becomes less severe if some security mechanisms are in place, e.g., some form of authentication. Note, however, that in that case, readers may get stolen to retrieve the credentials or use the credentials on the stolen reader to access tags surreptitiously.

Similarly, a fake tag can connect to a genuine reader. This may feed inaccurate data through the reader into the back office. In fact, the theoretical possibility of embedding a virus on an RFID tag to compromise a reader (or the back office) has recently been discussed [27].

Unauthorized messages can also be sent on an existing communication session both to the reader or to the tag quite easily, again because the communication link is wireless. This type of attack is a bit more involved than eavesdropping though, because the correct signals

---

* https://events.ccc.de/congress/2005/static/r/f/i/RFID-Zapper(EN)_77f3.html

have to be generated at exactly the right time. Moreover, an attack may be detected by some of the parties if they see signals on the link that appear to be sent by them that they did not in fact send themselves.

Modification/relay: When eavesdropping and injection threats are combined, this yields a modification threat. In this case, messages in transit can be relayed or modified online. In fact, complex man-in-the-middle and relay attacks are possible, where signals from one tag are intercepted by a rogue reader, transmitted over the Internet over long distances to a rogue tag emulator that resends the messages to a unsuspecting reader [24].

Jamming/blocking: Data transfer via the air interface can be disrupted either by passive means (such as shielding by a piece of metal) or by active means (jamming the signal sent by the transmitter). This threat is actually used in certain cases to improve the privacy of the tag holders. So-called blocker tags [23] use this technique to simulate the presence of any number of tags, thus hiding the identity of the actual tags present in the reader area. The RFID Guardian [27] uses a different technique to cloak a particular tag. The Guardian jams responses from the tag back to the reader by sending a strong radio signal in the time slot where it knows the tag will respond. Again note that the options for jamming depend on the location of the jamming station with respect to the tag and the reader.

22.2.1.3   On the Reader

The reader itself suffers much less from RFID-specific threats. In fact, we only distinguish the following two threats.

Unauthorized access: read, write, or modify: Unauthorized access to the reader (either reading it, writing it, or modifying it) might be achieved through any of its interfaces (including the wireless link to the tag, the connection to the back office, and perhaps other connections to point-of-sale equipment, or its own user interface).

Theft (of reader): A reader may get stolen (especially in unattended settings). If the reader contains credentials (e.g., keys) to access certain tags, these tags may get compromised. If the reader contains credentials to access the back office systems, stealing the reader gains control over these back office systems.

22.2.1.4   On the Network between Reader and Backend Systems

Threats on the network between reader and backend systems are not specific to RFID systems, because they mostly employ standard network technology. The connections are mostly wired, and hence less easy to get access to. On the other hand, the networks may be large and span several continents in the case of an international company. In this case, the data carried over these networks cross many unknown and untrusted nodes. In any case, the threat analysis is the same as for any other large-scale wired Internet Protocol (IP)-based network.

22.2.1.5   On the Backend System

Threats on the backend system are mostly not specific to RFID systems. Certain RFID-specific components (such as the ONS and the PML repository) may be present though, and could be the focal point of an attack on an RFID system, either to obtain data about certain classes of tags, or to disrupt service by modifying data about tags.

It is worth pointing out that threats to the reader can extend to the backend systems as well: fake RFID tags may inject erroneous data into the system, corrupting databases. In theory, if the reader or the backend systems themselves do not perform rigorous sanity checks on their inputs, excessively long RFID data could cause crashes in the backend systems because of buffer overflows and the like. Similarly, viruses, worms, or Trojans could be injected into the backend system this way.

22.2.2   SECURITY BREACHES

The threats discussed above can result in security breaches in various ways. We describe how these threats can have an impact on the integrity and authenticity of the system, the confidentiality of the data and the availability of the overall system. Security affects mostly the owners of the overall RFID application, or the owners of the readers and or backend systems.

In Section 22.2.3 that follows, we will discuss how the basic threats have an effect on the privacy. This mostly affects the owners of the tags.

22.2.2.1   Integrity and Authenticity

Several attacks on the integrity and authenticity of the system are possible.

Cloned tags cause confusion about the true whereabouts of the associated objects, and hence constitute a threat to the integrity of the system. If tags are used for access control or ticketing purposes, cloned tags enable illegal access for the carrier of the cloned tag. This is especially a concern when tags are used for identification purposes (e.g., biometric passports [12]).

If tags contain additional data about the object they tag, modifying this data causes a discrepancy between the object tag and the information about that object stored on the tag. This could be a concern in for instance health care applications where tags contain essential health-related data of the person carrying the tag. If the tag identifier itself is modified, the object is essentially renamed and becomes untraceable.

Tags may get detached from the object they tag. This way, objects become untraceable. The removed tag can be used at other locations, for instance to gain illegal access, or to confuse the system about the location of the real object. In the latter case, the integrity of the data in the backend system is compromised.

False data may be injected in the networks, both through the wireless link between tag and reader as well as the network connecting the reader with the backend systems.

Stolen readers may be used to gain access and modify data on the backend system, or to modify data on tags that come into range. This was a real concern for the design of the European passport standard that wanted to make sure that only authorized readers operated at border inspection stations would be able to read private biometric data of the passports (see Ref. [12]).

22.2.2.2   Confidentiality

Confidentiality of the system is threatened either by attacking the network between the reader and the backend system or by attacking the network between the reader and the tag. The latter case is what concerns us most.

Reading out tags may convey a lot of information, and is not only a privacy issue. For instance, in a supply chain scenario, the possibility to read out tags will give a lot of information about logistic operations, or the contents of warehouses and the availability of items in stock. This may be valuable information for a competitor.

Reading out tags may also reveal the value of goods stored at a certain location. This information is useful to thieves, and applies not only to logistics. Reading out the contents of a purse (when bank notes are tagged [3,22,33]) is attractive to identify potential pickpocketing targets. Scanning the inventory of a house is useful for burglars (although in this case the maximal reading distance of tags limits the number of items "visible" from outside). And determining the nationality of passport holders may be valuable to certain groups of terrorists (for instance when using a so-called RFID-bomb that detonates only if people wearing passports of a certain nationality come in range*).

---

* See the overhyped YouTube video http://www.youtube.com/watch?v=-XXaqraF7pI for a graphic demonstration of this.

Tags can be read by placing rogue readers in areas of interest. They could be hidden to prevent detection. Tag data can also be read by eavesdropping on the communication between a genuine reader and the tag. In this case, shielding does not help as the owner of the tag has agreed to the tag being read out by the genuine reader.

22.2.2.3 Denial-of-Service

Denial-of-service attacks are generally aimed at the tag or the wireless link between tag and reader, because they are the easiest targets. We do note, however, that certain virus and buffer overflow attacks can be a serious threat to the availability of the reader and backend systems. Badly programmed readers, for instance, can crash when confronted with rogue tags that transmit more data than the reader expects.

Tags can be broken in a variety of ways. They can be destroyed by mechanical force or due to exposure to certain chemical components. Tags can be destroyed (over a certain distance) by electromagnetic fields or powerful inductive sparks. They can get deactivated by accepting a fake kill command, or when being overwritten with false data. Finally, tags can be removed from the objects they tag.

The tag to reader interface can be jammed with a radio signal with the same frequency of the reader. Separate jamming signals may be needed to jam the tag to reader communication channel as well. Certain materials (e.g., metal and water) may block or deteriorate the RF signal. Metal cages, or metal laminated wallets or bags shield the tag from any outside RF field.

22.2.3 PRIVACY BREACHES

So far we have discussed the security threats. It is now time to focus our attention on the privacy threats associated with RFID systems. In fact, this is, by many, perceived to be their main threat [2]. Incidents with the Metro Future Store* the Benetton campaign,[†] and many others have been brought into the spotlight by groups like CASPIAN. Similar incidents continue to get press coverage to this day. Fear of negative press has made the RFID industry more careful. It has made suppliers of RFID equipment concerned about the negative impact on sales, if a particularly privacy unfriendly RFID system should put the whole RFID industry in a bad daylight.

In this section we discuss how RFID systems pose a threat to the privacy of their users. As said before, this threat is mainly aimed at the owners of the tags.

Privacy is a surprisingly hard concept to define [10]. In fact, one can consider several types of privacy, like physical/bodily privacy, location privacy, privacy of the home, informational privacy, and others. Warren and Brandeis [31] defined privacy to be "the right to be let alone." Computer systems mostly touch on *informational privacy*:

> the right to control one's personal information, and the ability to determine if and how that information is obtained and used (even after the data was collected).

In other words: privacy is "informational self determination."

Privacy is, in general, also protected by law. In the Netherlands, for instance, the law defines the notion of "personal information" as information that can be traced back to a natural person (using only moderate effort). At a European level, we have the European Privacy Directive [6]. which states

> ... Member States shall protect the fundamental rights and freedoms of natural persons, and in particular their right to privacy with respect to the processing of personal data ...

---

* http://www.wired.com/techbiz/media/news/2004/02/62472
† http://www.boycottbenetton.com/

The problem with privacy and RFID is that, in the narrow legalistic sense, RFID tags hardly constitute personal data. EPC tags for instance only contain a product code and an item number, but no information that is directly linked to an individual. This means that, at least in the Netherlands, the law does not assume a privacy risk associated with RFID usage, and offers no protection to consumers. At the European level similar caveats apply, given the fact that the Directive defines:

> "personal data" shall mean any information relating to an identified or identifiable natural person ("data subject"); an identifiable person is one who can be identified, directly or indirectly, in particular by reference to an identification number or to one or more factors specific to his physical, physiological, mental, economic, cultural or social identity.

Yet consumers do perceive a real privacy threat, for several reasons.

Tags may reveal information about the objects carried by consumers. For instance, if clothes are routinely tagged and not disabled at the point of sale, these tags reveal what you are wearing. They may even reveal what size you have. The privacy threat arises because the tag can be linked to the person wearing it, by visually linking the data provided by the reader with the person being scanned.

Item-level tags (i.e., EPC tags that contain a unique item number as well as the product code) uniquely identify one object. If this object happens to be a personal item that is almost always worn (for example a wristwatch, or a mobile phone), your presence can be detected and followed anywhere. All it takes is to link the tag code once to a particular person. This constitutes a breach of location privacy. Mobile phones can (theoretically) be tracked in a similar fashion. However, global system for mobile communication (GSM) communication is encrypted, so to catch the International Mobile Equipment Identity (IMEI) or the International Mobile Subscriber Identity (IMSI) (two different unique numbers associated with the mobile handset or the SIM card) you need to set up you own base station. This is more involved than acquiring an RFID reader. On the other hand, GSM signals can be tracked from much further away.

But even if tags do not contain a unique item level code and only transmit a product code, consumers can be tracked. The reason is that the brand and type of a small collection of personal belongings (for instance your watch, your favorite clothes, a lighter, the brand of cigarettes you smoke, your perfume, etc.) uniquely determine you. In other words, such a profile is just as good as a unique item number to track a particular person.

This tracking of persons using a unique profile is certainly a real privacy threat. Unfortunately, however, it is one that is not very well understood, and therefore often dismissed by others. Moreover, the law does not provide any protection against this type of tracking and tracing of people.

22.3 CASE STUDIES

In this section we will focus on a number of selected cases. The cases that are considered are animal tracking, public transport, identity card, and smart shop. We describe the most important security risks specifically for these cases, and sometimes briefly mention solutions to the threats listed. Naturally, other chapters in this handbook go into further detail here.

22.3.1 ANIMAL TRACKING

Sometimes RFID tags are used to track animals like goats, sheep, or pets. This can be done externally (using an ear mark) or internally (below the skin or in the stomach). Such animals are tracked because it makes them easier to find in case a pet is lost, or when a goat or sheep has an infectious disease [4].

Most of the tags that are used are read-only, so the data on the tag cannot be modified unauthorized. Only rewritable tags would allow fraudulent behavior of the animal owner. This would disable the tracking of animals in case of infectious disease. But because all animals are also registered in a central database, such an action by the animal owner will not be effective.

A real risk in this case would be the physical destruction or detachment of the tag by the animal. This risk can be reduced by regularly checking the presence of tags on the animals. As mentioned before, the deliberate physical destruction by the owner will not be effective.

22.3.2 Public Transport

In several countries (e.g., the Netherlands) an electronic ticketing system is being developed to control the fair use of tickets for public transport [14]. These tickets contain RFID tags and some rewritable area for extra services and ease of use. Some tickets are anonymous.

Unauthorized modification of the contents of an electronic ticket is very critical, like changing the departure or arrival location, the billing information, or the time of use. When an electronic purse is used on the ticket, this purse could be misused.

The detachment of tag from a ticket is not a serious threat, because this usually means destruction of the ticket itself, or at least making it invalid. By putting an eavesdropping device in a gate or RFID reader, eavesdropping of communication could be used with a replay attack to spend tickets more than once, or to simply steal tickets. Although blocking and jamming will probably not lead to free tickets, it could be used to frustrate the public transport system.

22.3.3 Identity Card

Some new identity cards contain RFID chips to enable automatic reading of passports. Because unauthorized reading is a serious threat, an initial step is incorporated that uses an optical device that actually looks into the passport and reads some cryptographic key which is subsequently used to encrypt the session between identity card and RFID reader [12,13].

A deactivation function is usually not implemented in the identity card.

The tag is incorporated in the card in such a way that detachment of the tag will lead to destruction of the card. Destruction of a tag will lead to fraud investigation and will therefore not be very successful.

Eavesdropping is very relevant, so some kind of cryptographic measure is taken to encrypt the communication. It should not be possible to detect, e.g., the nationality of a person by reading his passport from a distance [1,12]. Or, even worse, to obtain his valid and approved fingerprints and facial image templates.

Blocking and jamming of the communication could be used to frustrate the system, but might be considered as sabotage and could be prohibited by law. On the other hand, tags may break, so the border inspection procedure should cater for the case that an unsuspecting citizen presents a passport with a nonfunctional RFID tag.

22.3.4 Smart Shop

An RFID tag is often seen as a successor to the bar code. The RFID tag on each product could provide new services to the customer. By linking to a Web site, information could be found about warranty, product ingredients, etc. [7,16].

Cheap articles will usually have cheap and read-only RFID tags. Only more expensive articles might have reusable and rewritable tags which makes them vulnerable to unauthorized modification of data. This could lead to wrongly priced items and extended warranties for example.

A deactivation function can be used to enable the customer to shut down the tag after the product leaves the store. On the other hand, this will prevent the extra services to be purchased. This issue is still an ongoing debate.

Tags could be destructed or detached by customers after purchasing. It will protect unauthorized reading of the contents and thereby their privacy. Unauthorized reading in a shop by using fake readers could be detected by suitable scanners.

Blocking and jamming are only used for frustrating the RFID system. Selectively blocking the reading of tags attached to some (expensive) goods in a shopping cart may be a successful strategy for shoplifting.

22.3.5 Conclusions

The analysis of the four cases shows that in each case the security risks and counter measures are very specific. To implement particular counter measures one should incorporate the entire system and architecture of which the RFID responder and reader are part. The results are summarized in Table 22.1.

The goal of the attacker varies over the four different cases which leads to the following conclusions:

Animal tracking: The main threat is the physical destruction of the responder by the animal. It will no longer be able to track animals that are infected or lost, which is essential to the system.
Public transport: The main target is to avoid illegal tickets, i.e., tickets for which is paid less than one is supposed to pay. This could be accomplished by a replay attack after eavesdropping on a communication, or by unauthorized modification of the contents or ID of a tag.
Identity card: Unauthorized modification is very harmful to the goal of electronic identity cards: proper authentication of persons. As with non-RFID-based identity cards, adequate measures are taken such that this risk is sufficiently reduced. A specific RFID-based threat is the unauthorized reading of the contents of a card. To avoid this, an initial optical step is introduced before electronically reading out the card.
Smart shop: Although the RFID tags offer interesting new services, the most important threat is the loss of privacy of the consumer that bought such a product. By deactivating the reading functionality, or completely removing the tag, this risk has disappeared. But it will interfere with the introduction of new services.

The risk of unauthorized modification of tags is relevant to most cases, but only when the tags are rewritable. Some security risks are only annoying or frustrating and will not lead to loss of object or money. The risk of detachment of a tag is often underrated and is highly relevant in some cases. Especially in the smart shop case the risk of privacy plays a role, a risk that forms a great part of the public debate nowadays.

TABLE 22.1
Overview of RFID Risks in Several Cases

Security Risk	Animal Tracking	Public Transport	Identity Card	Smart Shop
Falsification of ID and contents	Tags are read-only	Relevant	High impact	Relevant for rewritable variants
Deactivation and destruction	N/A	Procedural measures	Fraud	Consumer right
Detaching the tag	Possible by animal	Leads to physical destruction	Leads to physical destruction	Consumer right
Eavesdropping	N/A	Replay attack	Relevant	Privacy
Blocking and jamming	N/A	Sabotage	Sabotage	Sabotage
Falsifying	N/A	Replay attack	Relevant	Privacy reader ID

22.4 BEYOND RFID: NFC

NFC* is attracting more and more attention in the last few years. As it is a technology very similar to RFID, we wish to discuss the security and privacy issues of this technology here as well. Especially because NFC 'enjoys' certain threats that are not present in RFID.

NFC is quickly becoming the preferred technology for new mobile close-contact applications, such as smart posters, contactless payments, ticketing applications, and the like. Especially now that several mobile phones (like the Nokia 6131, and the Samsung X700) have become available that contain an NFC device, usually connected to a secure element.

With smart posters, for instance, the NFC device in the mobile phone reads an RFID tag embedded in a poster. Using the URL obtained from the tag, the Internet-enabled phone quickly accesses additional information from the Web site set up by the poster publisher. In a contactless payment scenario, the user pays for goods at the counter by placing his mobile phone close to the NFC-enabled payment terminal, thus using his mobile phone as a virtual wallet. Here the secure element is used to secure the transaction.

22.4.1 SHORT PRIMER INTO NFC

NFC is a wireless connectivity technology that enables short-range communication between electronic devices. NFC is similar to RFID, but the focus of NFC is broader than identification. The NFC interface and protocol are defined by the ISO18092 standard [20], also known as NFCIP-1 and ECMA340. NFC operates in the 13.56 MHz RF band, which it shares with other RFID technologies such as ISO 15963. The NFC protocol is compatible with the lower layers of the contactless smart card protocols ISO 14443, Mifare[†] and Felica.[‡] This means that NFC devices can read these types of cards, and may also be able to emulate such cards. The "near field" in NFC refers to the short operating distance of 0–10 cm.

Each NFC device implements the architecture as shown in Figure 22.8. All devices have an RF layer to be able to communicate over NFC, and can operate in different modes:

- *Peer-to-peer mode*: The NFC device communicates directly (peer-to-peer) with another NFC device.
- *Reader/writer mode*: The NFC device reads/writes and existing proximity card.
- *Card emulation mode (optional)*: The NFC devices emulate a proximity card, that can be read and written by existing proximity card readers.

We are now ready to discuss the threats that occur due to the use of NFC.

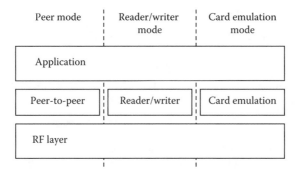

FIGURE 22.8 NFC device architecture.

---

* http://www.nfc-forum.org

† http://www.mifare.net/

‡ http://www.sony.net/Products/felica/index.html

22.4.2   RFID Security Threats Applicable to NFC

Because NFC is strongly related to RFID, the two technologies share a number of risks. Here we summarize these risks, for more detailed information see [4–6].

- Similar to RFID tags, NFC devices like mobile phones and NFC tags may be vulnerable to unauthorized reading of information stored on the NFC device or tag, using the NFC radio interface. This threat is especially important for NFC devices and tags that store data for payment applications, like credit card information.
- Another content-related threat is the falsification of content due to unauthorized writing to the data storage part of a device. Next, a falsified unique ID and other forged security information can trick an NFC device into accepting an emulated or cloned NFC device as the original. Just like other RFID tags, also NFC devices can be deactivated by physically destroying the NFC chip or any other critical part of the device (e.g., mechanically or using a microwave), and NFC tags can be detached from a tagged item and subsequently associated with a different item.
- With respect to communication, messages exchanged between NFC devices are vulnerable to eavesdropping as well, although the communication distance between two NFC devices is smaller than that of most other RFID technologies, reducing the risk of eavesdropping. Eavesdropping can lead to privacy infringement and replay attacks: repeating one side of the recorded communication to simulate that NFC device; and man-in-the-middle (relay) attacks: placing a rogue device in between the two communication NFC devices, such that that all communication goes unnoticed through this third device. Smartly modifying this communication could for example in payment systems lead to charging the wrong electronic wallet. Furthermore, communication may be jammed by using powerful transmitters or by shielding.
- Finally, an often overlooked aspect when discussing the security threats associated with RFID (that applies to NFC applications as well) is the interaction with backend systems. Attacks to the backend system can affect the entire NFC application, and are therefore a serious threat. Examples include rogue tags that contain viruses [7] or other RFID malware that spread in the backend system, tags that contain data that make the systems crash, e.g., due to their unexpected length, causing buffer overflows, or simply tags (or devices programmed to appear as tags) that (over)load the system with incorrect information.

22.4.3   All NFC Devices Are Readers/Writers

All NFC devices support the active communication mode. Therefore, all NFC devices are potential NFC readers. For example, when NFC has become a standard feature of mobile phones, many people will actually carry an NFC reader with them all the time. As a consequence, you could be surrounded by a crowd of potential readers, that are capable of communicating with the NFC devices and tags you carry with you, without your knowledge. This new feature of NFC creates a number of threats.

- People's privacy will be threatened, as certain information on NFC devices and tags can be read by everybody. This includes the unique ID of an NFC chip, which can always be read, but also nonsecured personal data stored on the device or tag.
- We expect to see mobile phones infected with malware that uses the NFC reader capability to scan and attack other NFC devices or tags in its vicinity. Possible attacks include unauthorized reading (confidential information like credit card data, and user names and

passwords) or unauthorized writing. The NFC interface also opens up a new channel to distribute viruses and other malware directed at the device containing the NFC itself.

- When carrying NFC readers has become the norm, the incentive to try to rewrite NFC tag data on, e.g., products in shops to get a cheaper price will increase.

22.4.4 ALL NFC DEVICES CAN EMULATE A TAG

Similarly, if many NFC devices implement card emulation mode (and whether they will is unclear at the moment), many people will be carrying a device that can be programmed to look like any conceivable NFC tag all the time. Again there may be the incentive to write software that allows one to emulate tags of cheap products, or to emulate access tickets for the cinema or access cards to office buildings and the like. The impact of this threat appears to be small due to the limited reading distance.

22.4.5 NFC NOT ENFORCED

Many organizations try to assure their customers that NFC is secure, as the reading distance of NFC is 10 cm at maximum. In fact this is only true for NFC readers that have "normal" antennas, such as the ones current mobile NFC phones are equipped with. However, an NFC reader with a more powerful antenna is not restricted to this distance.

Moreover, reading distances of 25 cm [8] and 40–50 cm [9] have been reported for NFC systems [10]. This distance is mostly limited by the fact that the NFC tag needs to be powered by the electromagnetic field emanating from the reader. For battery-powered NFC devices this is not an issue, so reading distances may actually be even larger. The consequence is that it is still possible to communicate from a greater distance with NFC devices and tags than organizations claim.

This induces two separate risks: first, the reading distance cannot be enforced by the tag alone and longer range eavesdropping is possible. But there is a separate risk of unwarranted perceived security. Organizations and individuals may feel that it is not necessary to secure data on NFC devices or tags that they carry with them, because they assume their data is secure from everything that is further away than 10 cm. Because this is not true, this perceived security threatens the actual security of everything that is accessible through an NFC interface.

22.4.6 NFC AS GATEWAY TO ATTACHED DEVICE

NFC as communication mechanism on more complex devices such as mobile phones or PDAs opens up a new entrance to these devices. The same happened with the introduction of Bluetooth on mobile devices, but an important difference is that NFC does not have Bluetooth's undiscoverable mode. On top of that, the GSM Association (GSMA) has standardized the NFC-SIM link (called SWP, for Single-Wire Protocol) that enables access through NFC to confidential data on the SIM even if the device is turned off or contains no batteries.

Therefore NFC creates (1) an always open-input channel to an NFC-enabled device, making the device extra vulnerable for unauthorized reading and hacking attacks; and (2) an always open-output channel, out of which applications may leak information to any passing writable NFC device or tag.

22.4.7 INTERFERENCE AMONG MULTIPLE NFC APPLICATIONS ON SINGLE DEVICE

Another impact of using one NFC interface on devices that contain multiple NFC applications is the risk of improper application separation. In such a system, all applications use the same interface to communicate with the outside world. This introduces the risk that one application is capable of interfering with or eavesdropping on the communication of another application that runs on the same device. Separation of applications (e.g., when running in a sandbox) is undermined in such a case.

Remember that many current NFC applications are high-value applications, such as credit cards and public transport ticketing. The threat of malware that is capable of gaining access to such high-value applications is not one that should be underestimated.

22.4.8 NO SECURITY STANDARD

As explained in Section 22.4.1, NFC is not equivalent to Mifare. NFC specifications do not provide a security standard or functions for authentication and access control, as for example for Bluetooth or WiFi, leaving it up to the application developer to properly secure data and communication on NFC devices or tags. Due to time constraints or lack of experience, the security of NFC applications may be badly implemented or even nonexistent, leaving the end-user vulnerable to all kinds of threats.

22.4.9 UNINTENDED PAIRING BETWEEN NFC DEVICES

Because NFC devices are intended to operate autonomously, they may connect to each other without explicit user interaction. This is cumbersome when multiple devices are in range: in that case, the device one connects to is not obvious. For example, in crowds (like the subway, on festivals) NFC mobile phones may connect to each other or to NFC (credit) cards in someone's wallet. This problem is similar to unintended Bluetooth pairing, but with a smaller likelihood due to the limited reading distance.

22.4.10 PRIVACY

Last, but not least, privacy issues are raised by the use of NFC as well. In fact, the previous paragraphs have already touched upon some of these. To summarize those for completeness, privacy is threatened by

- The possibility of unauthorized reading of tags,
- The fact that all NFC devices are potential readers,
- The fact that NFC is not enforced, and that
- NFC can act as a gateway to the attached device.

Furthermore, applications using NFC may impose further privacy infringements. They may collect and show a lot of data about their users without those users being aware of this. This is due to the fact that interaction between devices does not necessarily require explicit user interaction, and may not enforce a secure form of access control. For instance, in public transport, all details of a traveler's journeys can be stored. In payment applications, the fact that the user also carries a loyalty card may automatically be detected, even if the user would prefer not to show this loyalty card this time. The facts that users are not aware of which personal information is collected from their NFC and stored, and that users may be powerless to prevent this from happening, threaten their privacy.

22.5 CONCLUSIONS AND OPEN ISSUES

RFID suffers from a few specific threats, mainly due to the wireless link between the reader and the tag, and the importance of the physical connection between the tag and the object being tagged. This raises a couple of concerns with respect to the security and privacy of RFID applications that we have discussed in this chapter. NFC technology introduces a few more specific security and privacy problems, mainly because of the more active nature of NFC devices and the fact that they get embedded in highly mobile and personal devices like mobile phones.

Other chapters in this handbook discuss how to mitigate these threats and how to develop secure and privacy friendly RFID applications.

REFERENCES

[1] American Civil Liberties Union (ACLU). Naked data: HOW the U.S. ignored international concerns and pushed for radio chips in passports without security. An ACLU White paper, November 24, 2004.

[2] K. Albrecht and L. McIntyre. *Spychips: How Major Corporations and Government Plan to Track Your Every Move with RFID*. Nelson Current, Nashville, TN, 2005.

[3] G. Avoine. Privacy issues in RFID banknotes protection schemes. In *6th USENIX Smart Card Research and Advanced Application Conference (CARDIS)*, pp. 33–48, Toulouse, France, 2004.

[4] G. Becker. Animal identification and meat traceability. CRS Report for Congress, 2006.

[5] BSI. Security aspects and prospective applications of rfid systems, 2006.

[6] European Commission. Directive 95/46/ec of the European Parliament and of the council of 24 October 1995 on the protection of individuals with regard to the processing of personal data and on the free movement of such data, November 1995.

[7] UK RFID Council. A UK code of practice for the use of radio frequency identification (rfid) in retail outlets. April 12, 2006. Release 1.0.

[8] European Technology Assessment Group (ETAG). Rfid and identity management in everyday life. Technical Report IPOL/A/STOA/2006-22, European Parliament, Scientific Technology Options Assessment (STOA), 2007.

[9] K. Finkenzeller. *RFID-Handbook*, 2nd edn. Wiley & Sons, Chichester, 2003.

[10] C. Freid. Privacy. *Yale Law Journal*, 77(3):475–493, 1968.

[11] S.L. Garfinkel, A. Juels, and R. Pappu. Rfid privacy: An overview of problems and proposed solutions. *IEEE Security & Privacy*, 3(3):34–43, 2005.

[12] J.-H. Hoepman, E. Hubbers, B. Jacobs, M. Oostdijk, and R. Wichers Schreur. Crossing borders: Security and privacy issues of the European e-passport. In H. Yoshiura, K. Sakurai, K. Rannenberg, Y. Murayama, and S. Kawamura (Eds.), *1st Int. Workshop on Security*, Lecture Notes in Computer Science 4266, pp. 152–167, Kyoto, Japan, October 23–24, 2006. Springer.

[13] ICAO. PKI for machine readable travel documents offering ICC read-only access, version-1.1. Technical report, International Civil Aviation Organization (ICAO), 2004.

[14] IDTechEx. London—Transport for London tfl. oyster card UK. January 26, 2006.

[15] EPCglobal Inc. Epc radio-frequency identity protocols. class-1 generation-2 uhf rfid. protocol for communications at 860 MHz–960 MHz. http://www.epcglobalinc.org, 2004. Version 1.1.0.

[16] USA Strategies Inc. Rfid adoption in the retail industry, 2005.

[17] ISO 14443. ISO/IEC 14443 Identification cards—Contactless integrated circuit(s) cards—Proximity cards. Technical report, International Organisation for Standardisation (ISO) JTC 1/SC 17.

[18] ISO 15693. ISO/IEC 15693 Identification cards—Contactless integrated circuit(s) cards—Vicinity integrated circuit(s) card. Technical report, International Organisation for Standardisation (ISO) JTC 1/SC 17.

[19] ISO 18000. ISO/IEC 18000 Information technology—Radio frequency identification for item management. Technical report, International Organisation for Standardisation (ISO) JTC 1/SC 31.

[20] ISO 18092. ISO/IEC 18092 Information technology—Telecommunications and information exchange between systems—Near field communication. Technical report, International Organisation for Standardisation (ISO) JTC 1/SC 6.

[21] ISO 7816. ISO/IEC 7816 Identification cards—Integrated circuit(s) cards with contacts. Technical report, International Organisation for Standardisation (ISO) JTC 1/SC 17.

[22] A. Juels and R. Pappu. Squealing euros: Privacy protection in RFID-enabled banknotes. In R.N. Wright (Ed.), *7th Int. Conf. Financial Cryptography*, Lecture Notes in Computer Science 2742, pp. 103–121, Guadeloupe, French West Indies, January 27–30, 2003. Springer.

[23] A. Juels, R. Rivest, and M. Szydlo. The blocker tag: Selective blocking of RFID tags for consumer privacy. In V. Atluri and P. Liu (Eds.), *10th Int. Conf. on Computer and Communications Security*, pp. 103–111, Washington, DC, October 27–30, 2003. ACM.

[24] Z. Kfir and A. Wool. Picking virtual pockets using relay attacks on contactless smartcard. In *1st Int. Conf. on Security and Privacy for Emerging Areas in Communications Networks (SecureComm 2005)*, pp. 47–58, Athens, Greece, September 5–9, 2005.

[25] National Institute of Standards and Technology. Guidelines for securing radio frequency identification (rfid) systems, 2007. NIST Special Publication 800–98.

[26] M. Ohkubo, K. Suzuki, and S. Kinoshita. RFID privacy issues and technical challenges. *Communications of the ACM*, 48(9):66–71, 2005.

[27] M.R. Rieback, B. Crispo, and A.S. Tanenbaum. Is your cat infected with a computer virus? In *4th IEEE Int. Conf. on Pervasive Computing and Communications (PerCom)*, pp. 169–179, Pisa, Italy, March 13–17, 2006.

[28] S.E. Sarma, S.A. Weis, and D.W. Engels. RFID systems and security and privacy implications. In B.S. Kaliski Jr., K. Ko, and C. Paar (Eds.), *4th Workshop on Cryptographic Hardware and Embedded Systems*, Lecture Notes in Computer Science 2523, pp. 454–469, Redwood Shores, CA, August 13–15, 2002. Springer.

[29] S.E. Sarma, S.A. Weis, and D.W. Engels. Rfid systems, security & privacy implications (white paper). Technical Report MIT-AUTOID-WH-014, Auto-ID Center, MIT, Cambridge, MA, 2002.

[30] M. van Lieshout, L. Grossi, G. Spinelli, S. Helmus, L. Kool, L. Pennings, R. Stap, T. Veugen, B. van der Waaij, and C. Borean. Rfid technologies: Emerging issues, challenges and policy options. Technical Report EUR 22770 EN, DG Joint Research Centre, Institute for Prospective Technological Studies (IPTS), 2007. ISBN: 978-92-79-05695-6.

[31] S.D. Warren and L.D. Brandeis. The right to privacy. The implicit made explicit. *Harvard Law Review*, IV(5):193–220, 1890.

[32] S.A. Weis, S.E. Sarma, R.L. Rivest, and D.W. Engels. Security and privacy aspects of low-cost radio frequency identification systems. In D. Hutter, G. Müller, W. Stephan, and M. Ullmann (Eds.), *1st Int. Conf. on Security in Pervasive Computing*, Lecture Notes in Computer Science 2802, pp. 201–212, Boppard, Germany, March 12–14, 2003. Springer.

[33] J. Yoshida. Euro bank notes to embed rfid chips by 2005. EE Times, December 19, 2001.

23 Finite Field Arithmetic for RFID and Sensor Networks

José L. Imaña

CONTENTS

Embedded security is in constant evolvement and new applications are constantly emerging. Important examples are radio-frequency identification (RFID) tags and sensor nodes as they require efficient implementations of cryptographic protocols due to the limitations on number of gates, bandwith, power, etc.

Cryptographic protocols are based on finite field (prime or binary extension field) arithmetic. $GF(2^m)$ binary field arithmetic has many implementation advantages with respect to prime field arithmetic, and is used in the main cryptographic schemes [4] based on the advanced encryption standard (AES) and elliptic curve cryptography (ECC) [9]. For example, ECC based on $GF(2^m)$ is one of the cryptographic algorithms widely used today for its high security per bit length, less computational power, and memory requirement. Complexity reduction of $GF(2^m)$ arithmetic primitives leads to improved implementations of the corresponding cryptographic schemes. Therefore, efficient algorithms for arithmetic operations over $GF(2^m)$ are of fundamental interest for those applications where compact and low-power/energy architectures and implementations of cryptographic protocols are required, such as RFID and wireless sensor networks (WSNs). In this chapter, some of these efficient arithmetic algorithms over finite fields $GF(2^m)$ are reviewed.

23.1 INTRODUCTION

Embedded security and its applications are in constant evolvement. Sensor nodes and radio-frequency identification (RFID) tags are important examples. They require efficient implementations of cryptographic protocols due to the limitations on number of gates, bandwidth, power, etc.

WSNs typically consist of a wide number of tiny nodes communicating with a base station which collects the data from the nodes and communicates with the outside world. In this scenario, security in WSN is of special importance because a wide number of nodes are exposed in potentially hostile environments and if only one of these nodes is captured by the attackers, then the effect on the complete network can be disastrous. Therefore, various cryptographic services are required for WSN [6,28]. Furthermore, there is an evident necessity for low-cost and low-power architectures that can be used on sensor nodes [6,16,17,32,36,48,55].

RFID systems are used for the automatic retrieval of data about different objects (animals, goods, persons). The object is equipped with a small circuit (an RFID tag), from which stored information can be retrieved by a reader device. Each RFID system consists of a tag (attached to the object to identify) and a reader. The reader device can retrieve data from the RFID tag, and it may also be able to write data into the tag's memory. In RFID systems, data and energy are transmitted via radio frequency. Owing to these properties, RFID is a rapidly upcoming technology that is used in more and more applications where compact tags are mostly required. In 2006, already more than one billion RFID devices were shipped. These devices range from low-cost passive devices to complex actively powered devices.

Passive RFID devices essentially consist of a microchip that is attached to an antenna that is used to draw energy from an electromagnetic field of an RFID reader. Passive RFID devices are completely powered by the energy that is provided by this field [23]. Furthermore, this field is used for the communication between the reader and the device. The field's intensity is limited by national and international regulations, so the power consumption of the RFID tag also must fulfill limitations. For this reason, power-aware designing of the tag circuits is very important and necessary. Moreover, less power consumption leads to longer reader ranges where the tags can work with the available energy [13,24].

RFID systems can be used in a wide number of applications: animal tracking, toll systems, access control, secure money, and so on. Because of this variety of applications, security and privacy issues are fundamental concerns with this technology. However, enhanced security always comes with extra costs. Nowadays there exist passive RFID devices that implement proprietary cryptographic algorithms. These algorithms have been developed with the goal to provide security at very low implementation costs (power, time, area). However, often these proprietary algorithms have not the same cryptographic strength as standardized algorithms. To avoid these problems, there have been several efforts to develop low-cost implementations of elliptic curve cryptography (ECC) and advanced encryption standard (AES) for RFID devices [5,7,31,34,44,50,52,58,60].

ECC is one of the cryptographic algorithms widely used today in such applications due to its high security per bit length, less computational power, and memory requirements [9]. When implementing an elliptic curve system, a crucial consideration is how to implement the underlying finite field arithmetic. Finite fields of characteristic 2 (also known as *binary fields*) are very attractive to implementers due to their "carry-free" arithmetic, and the availability of different equivalent representations of the field, which can be adapted and optimized for the computational environment. That is the reason why ECC based on $GF(2^m)$ is widely used today. Furthermore, binary field $GF(2^m)$ arithmetic is also used in the AES.

The reduction of the complexity for $GF(2^m)$ arithmetic primitives leads to efficient implementations of the corresponding cryptographic schemes. Therefore, efficient algorithms for $GF(2^m)$ arithmetic operations are of fundamental interest for those applications where compact and low-power/energy architectures and implementations of cryptographic protocols are required, such as RFID and WSN. In this chapter, some of the main existing algorithms for arithmetic operations over finite fields $GF(2^m)$ and their efficient architectures will be reviewed.

23.2   MATHEMATICAL BACKGROUND

A *finite field* or *Galois field* denoted by $GF(q = p^m)$ is a field with characteristic p and q elements. Such a finite field exists for every prime p and positive integer m, and contains a subfield having p elements. This subfield is called the *ground field* of the original field. For every nonzero element $\alpha \in GF(q)$, the identity $\alpha^{q-1} = 1$ holds [35]. In cryptographic applications, the two most studied cases are $q = p$ (with p a prime) and $q = 2^m$. The former case, $GF(p)$, is denoted as *prime field*, whereas the latter one, $GF(2^m)$, is known as finite field of characteristic 2 or *binary extension field*, also denoted as $\mathcal{F}_{2^m}$.

A polynomial P of degree at least 1 in $GF(q)$ is said to be *irreducible* if it cannot be written as the product of two polynomials, each of positive degree. Let $P(x) = x^m + p_{m-1}x^{m-1} + \cdots + p_1x + p_0$ be an irreducible polynomial in the indeterminate x over $GF(2)$ of degree m, and let α be a root of $P(x)$, i.e., $P(\alpha) = 0$. Then, $P(x)$ can be used to construct a binary finite field $\mathcal{F} = GF(2^m)$ with $q = 2^m$ elements, where α itself is one of those elements. Furthermore, the set $\{1, \alpha, \alpha^2, \ldots, \alpha^{m-1}\}$ forms a basis for $GF(2^m)$ that is called the *polynomial* (or *canonical* or *standard*) *basis* of the field. Any element $A \in GF(2^m)$ can be expressed in this basis as $A = \sum_{i=0}^{m-1} a_i\alpha^i = a_0 + a_1\alpha + \cdots + a_{m-1}\alpha^{m-1}$, with $a_i \in GF(2), i = 0, 1, \ldots, m - 1$.

The order of an element $\beta \in \mathcal{F}$ is defined as the smallest positive integer s such that $\beta^s = 1$. Any finite field contains always at least one element, called a *primitive element*, with order $q - 1$ [35]. A polynomial $P(x)$ is said to be a *primitive polynomial* if any of its roots is a primitive element in $\mathcal{F}$. If $P(x)$ is primitive, then all the q elements of $\mathcal{F}$ can be expressed as the union of the zero element and the set of the first $q - 1$ powers of α. For $\mathcal{F} = GF(2^m)$ with $q = 2^m$ elements, the field $GF(2^m) = \{0, \alpha^1, \alpha^2, \ldots, \alpha^{2^m-2}, \alpha^{2^m-1} = 1\}$. Because $P(\alpha) = 0$, we have that $\alpha^m = p_{m-1}\alpha^{m-1} + \cdots + p_1\alpha + p_0$. Therefore, any element in $GF(2^m)$ can be expressed as an $(m - 1)$-degree polynomial in α, i.e., $GF(2^m) = \{a_{m-1}\alpha^{m-1} + \cdots + a_1\alpha + a_0 | a_i \in GF(2) \text{ with } 0 \leq i \leq m - 1\}$.

For $\alpha \in GF(2^m)$, $\{\alpha^{2^0}, \alpha^{2^1}, \ldots, \alpha^{2^{m-1}}\}$ is called a *normal basis* of $GF(2^m)$ over $GF(2)$ if $\alpha^{2^0}, \alpha^{2^1}, \ldots, \alpha^{2^{m-1}}$ are linearly independent [39,41,43,45–47]. If $\alpha \in GF(2^m)$, then the elements $(\alpha, \alpha^2, \alpha^{2^2}, \ldots, \alpha^{2^{m-1}})$ are called the *conjugates* of α with respect to $GF(2)$. There exists at least one normal basis for any m. Using a normal basis, any $\beta \in GF(2^m)$ is represented as a vector $(b_0, b_1, \ldots, b_{m-1})$ where $\beta = b_0\alpha^{2^0} + b_1\alpha^{2^1} + \cdots + b_{m-1}\alpha^{2^{m-1}}$ and $b_i \in \{0, 1\}$ for $0 \leq i \leq m - 1$. For any β and $\gamma \in GF(2^m)$, $(\beta + \gamma)^2 = \beta^2 + \gamma^2$ holds because $2\beta\gamma = 0$ [61]. From Fermat's theorem, i.e., $\beta^{2^m-1} = 1$, $\beta^{2^m} = \beta$ holds. Therefore, when $\beta = b_0\alpha^{2^0} + b_1\alpha^{2^1} + \cdots + b_{m-1}\alpha^{2^{m-1}}$, $\beta^2 = b_0\alpha^{2^1} + b_1\alpha^{2^2} + \cdots + b_{m-1}\alpha^{2^m} = b_{m-1}\alpha^{2^0} + b_0\alpha^{2^1} + \cdots + b_{m-2}\alpha^{2^{m-1}}$. Hence, in normal basis, if $\beta = (b_0, b_1, \ldots, b_{m-1})$, $\beta^2 = (b_{m-1}, b_0, \ldots, b_{m-2})$. In other words, squaring is carried out by a simple cyclic right shift. Furthermore, powering by 2^i can be carried out by an $(i \bmod m)$-bit cyclic right shift.

The definition of the *trace* function is fundamental to the concept of *duality* [35]. The *trace* $\mathrm{Tr}_{\mathcal{F}/\mathcal{K}}(\alpha)$ of an element α over $\mathcal{K}$ is defined by $\mathrm{Tr}_{\mathcal{F}/\mathcal{K}}(\alpha) = \alpha + \alpha^q + \cdots + \alpha^{q^{m-1}}$, where $\alpha \in \mathcal{F} = \mathcal{F}_{q^m}$ and $\mathcal{K} = \mathcal{F}_q$. It must be noted that the trace of α over $\mathcal{K}$ is the sum of the conjugates of α with respect to $\mathcal{K}$ and it is an element from $\mathcal{K}$. Let $\mathcal{K}$ be a finite field and let $\mathcal{F}$ be a finite extension of $\mathcal{K}$. Then, two bases $\{\alpha_1, \ldots, \alpha_m\}$ and $\{\beta_1, \ldots, \beta_m\}$ of $\mathcal{F}$ over $\mathcal{K}$ are said to be *dual bases* [3,10,12,40] if $\mathrm{Tr}_{\mathcal{F}/\mathcal{K}}(\alpha_i\beta_j) = \delta_{ij}$, for $1 \leq i, j \leq m$, where δ_{ij} is the Kronecker delta function, which is equal to 1 if $i = j$ and 0 otherwise. It can be proven that for any basis $\{\alpha_1, \ldots, \alpha_m\}$ of $\mathcal{F}$ over $\mathcal{K}$ there exists a dual basis $\{\beta_1, \ldots, \beta_m\}$.

Some special classes of irreducible polynomials are more convenient for the implementation of efficient binary finite field arithmetic. Amongst them, trinomials, pentanomials, and equally spaced-polynomials (ESPs) are the most important. *Trinomials* are polynomials with three nonzero coefficients of the form $P(x) = x^m + x^n + 1$. *Pentanomials* are polynomials with five nonzero coefficients, i.e., $P(x) = x^m + x^{k_2} + x^{k_1} + x^{k_0} + 1$. Finally, irreducible ESPs have the same space separation between two consecutive nonzero coefficients, i.e., $P(x) = x^m + x^{(k-1)d} + \cdots + x^{2d} + x^d + 1$, where $m = kd$. If $d = 1$, then ESPs turn into *all-one-polynomials* (AOPs), i.e., $P(x) = x^m + x^{m-1} + \cdots + x^2 + x + 1$. If $d = \frac{m}{2}$, then ESPs turn into *equally spaced-trinomials* (ESTs), i.e., $P(x) = x^m + x^{\frac{m}{2}} + 1$.

Arithmetic over a finite field of characteristic 2 (or a binary finite field) is essentially *modulo* arithmetic. The elements of the finite field $GF(2^m)$ are the polynomials $\{0, 1, \alpha, \alpha+1, \alpha^2, \alpha^2+1, \ldots, \alpha^{m-1}+ \alpha^{m-2} + \cdots + \alpha + 1\}$. The *addition* of such elements is carried out under modulo 2 arithmetic, i.e., is simply the addition of two polynomials where the coefficients are added in $GF(2)$. Therefore, the addition of two polynomials becomes the bitwise XOR of their binary representations. *Subtraction* is exactly the same as addition in modulo 2 arithmetic, so $1 - \alpha$ equals $1 + \alpha$. *Multiplication* is defined as the polynomial product of two operands followed by a reduction modulo the generating polynomial $P(x)$, and is considered the most important $GF(2^m)$ arithmetic operation because other important operations (such as inversion or exponentiation) are based on multiplication. *Inversion* of an element $\chi \in \mathcal{F}$ is the process to find an element $\chi^{-1} \in \mathcal{F}$ such that $\chi \cdot \chi^{-1} = 1 \bmod P(x)$.

Generally speaking, the arithmetic modules over $GF(2^m)$ can be classified into *bit-parallel* and *bit-serial* architectures. Bit-parallel systems process one whole word of the input sample in 1 clock cycle, and are ideal for high-speed applications when pipelined at bit-level. Bit-serial systems process 1 bit of the input sample in 1 clock cycle, and are area-efficient and suitable for low-speed applications. In applications which require moderate sample rates, these systems may be ineffective: bit-parallel systems may be faster than necessary and occupy a large amount of area, while that bit-serial systems may be too slow. For these reasons, *digit-serial* systems are very attractive for digital designers. These systems process multiple bits of the input word (the *digit-size*) in 1 clock cycle, where the digit-size can vary from 1 bit to the word-length to achieve trade-off between speed, area, and power consumption.

Efficient $GF(2^m)$ arithmetic algorithms and architectures have been proposed in the literature for the different bases of representation. Among them, the polynomial basis is more promising in the sense that it gives designers more freedom on irreducible polynomial selection and hardware optimization. For these reasons, we focus our review on polynomial basis arithmetic.

23.2.1 EXAMPLE

Let $\mathcal{F} = GF(2^4)$ be a binary finite field with defining primitive trinomial $P(x) = x^4 + x^3 + 1$. Let α be a root of $P(x)$, $P(\alpha) = 0$, hence $P(\alpha) = \alpha^4 + \alpha^3 + 1 = 0$. For binary field arithmetic, subtraction is equivalent to addition, therefore $\alpha^4 = \alpha^3 + 1$. Using this expression, the elements of $GF(2^4)$ represented in the polynomial basis $\{\alpha^3, \alpha^2, \alpha, 1\}$ are given in Table 23.1. For example, the element $\alpha^9 = \alpha \cdot \alpha^8 = \alpha \cdot (\alpha^3 + \alpha^2 + \alpha) = \alpha^4 + \alpha^3 + \alpha^2 = (\alpha^3 + 1) + \alpha^3 + \alpha^2 = 2 \cdot \alpha^3 + \alpha^2 + 1 = \alpha^2 + 1$, where $2 \cdot \alpha^3 = 0$ because this is a field with characteristic 2.

The addition of two elements is simply the bitwise XOR of their coordinates. For example, $\alpha^6 + \alpha^{10} = (\alpha^3 + \alpha^2 + \alpha + 1) + (\alpha^3 + \alpha) = 2 \cdot \alpha^3 + \alpha^2 + 2 \cdot \alpha + 1 = \alpha^2 + 1$. In binary notation, it can be written as $(1111) \oplus (1010) = (0101)$. The multiplication of two elements A and B is the field element $C = A \cdot B \bmod P(x)$. For example, $\alpha^8 \cdot \alpha^9 = (\alpha^3 + \alpha^2 + \alpha) \cdot (\alpha^2 + 1) = (\alpha^5 + \alpha^4 + \alpha^3) + (\alpha^3 + \alpha^2 + \alpha) = \alpha^5 + \alpha^4 + \alpha^2 + \alpha = (\alpha^3 + \alpha + 1) + (\alpha^3 + 1) + \alpha^2 + \alpha = \alpha^2$. In binary notation the product $\alpha^8 \cdot \alpha^9$ can therefore be written as $(1110) \cdot (0101) = (0100)$. It is clear that the multiplication depends on the generating polynomial $P(x)$ selected for the field. Therefore, the complexity of the arithmetic modules implementing this and other complex operations (such as inversion, exponentiation, etc.) will be different for different generating polynomials.

23.3 BINARY FINITE FIELD ARITHMETIC

Addition and subtraction are equivalent operations in $GF(2^m)$. Addition in binary finite fields is defined as polynomial addition and can be implemented as the XOR addition of the two m-bit operands. We give in the following several algorithms for multiplication and inversion with polynomial basis, although we center our review on multiplication because it can be considered as the most important arithmetic operation. Hardware architectures for implementation and their corresponding complexities are also outlined.

TABLE 23.1

Elements for the Field $GF(2^4)$ Defined by the Primitive Trinomial $P(x) = x^4 + x^3 + 1$

Element in $GF(2^4)$	Polynomial	Coordinates
0	0	(0000)
α	α	(0010)
α^2	α^2	(0100)
α^3	α^3	(1000)
α^4	$\alpha^3 + 1$	(1001)
α^5	$\alpha^3 + \alpha + 1$	(1011)
α^6	$\alpha^3 + \alpha^2 + \alpha + 1$	(1111)
α^7	$\alpha^2 + \alpha + 1$	(0111)
α^8	$\alpha^3 + \alpha^2 + \alpha$	(1110)
α^9	$\alpha^2 + 1$	(0101)
α^{10}	$\alpha^3 + \alpha$	(1010)
α^{11}	$\alpha^3 + \alpha^2 + 1$	(1101)
α^{12}	$\alpha + 1$	(0011)
α^{13}	$\alpha^2 + \alpha$	(0110)
α^{14}	$\alpha^3 + \alpha^2$	(1100)
α^{15}	1	(0001)

23.3.1 MULTIPLICATION

Let $A(\alpha) = \sum_{i=0}^{m-1} a_i \alpha^i$, $B(\alpha) = \sum_{i=0}^{m-1} b_i \alpha^i$, and $C(\alpha) = \sum_{i=0}^{m-1} c_i \alpha^i \in GF(2^m)$ and $P(\alpha)$ be the irreducible polynomial generating $GF(2^m)$. Multiplication over $GF(2^m)$ in polynomial basis $\{1, \alpha, \alpha^2, \ldots, \alpha^{m-1}\}$ is defined as the polynomial multiplication modulo the irreducible polynomial $P(\alpha)$, namely,

$$C(\alpha) = A(\alpha)B(\alpha) \bmod P(\alpha) \tag{23.1}$$

Several multipliers have been proposed in the literature with different complexities [15,19,20, 22,25,27,42,49,51,54,56,66]. Among the different approaches for computing the product $C(\alpha)$, *two-step multipliers*, *inter-leaved multiplication*, *matrix–vector multipliers*, *digit-serial multipliers*, and *Montgomery multipliers* can be considered.

23.3.1.1 Two-Step Multipliers

In the two-step multipliers, the field product $C(\alpha)$ given in Equation 23.1 can be obtained by first computing the polynomial product $D(\alpha)$ of degree at most $(2m - 2)$ given by

$$D(\alpha) = A(\alpha)B(\alpha) = \left(\sum_{i=0}^{m-1} a_i \alpha^i\right)\left(\sum_{i=0}^{m-1} b_i \alpha^i\right) = \sum_{k=0}^{2m-2} d_k \alpha^k \tag{23.2}$$

where $d_k = \sum_{i+j=k} a_i b_j$, $0 \leq i,j \leq m-1$, $0 \leq k \leq 2m-2$, followed by a reduction operation performed to obtain the $(m-1)$-degree polynomial $C(\alpha) = D(\alpha) \bmod P(\alpha)$.

Several techniques have been proposed in the literature for the polynomial multiplication [8,29,30]. Among them, one of the most important is the Karatsuba–Ofman algorithm that will be outlined in the following.

Karatsuba–Ofman Multiplier. The Karatsuba–Ofman algorithm [30,53] is a recursive method for efficient polynomial multiplication or efficient multiplication in positional number systems.

It is known [49] that two arbitrary polynomials in one variable of degree less or equal $m - 1$ with coefficients from a field $GF(2^m)$ can be multiplied with not more than m^2 multiplications in $GF(2^m)$ and $(m - 1)^2$ additions in $GF(2^m)$. The Karatsuba–Ofman algorithm provides a recursive algorithm which reduces the above multiplicative and additive (for large enough m) complexities. A Karatsuba–Ofman algorithm restricted to polynomials where $m = 2^t$ with t an integer, was presented in Ref. [49] and is outlined in the following.

Let $A(\alpha)$ and $B(\alpha)$ be two elements in $GF(2^m)$. We are interested in finding the product $D(\alpha) = A(\alpha)B(\alpha)$ with $\deg(D(\alpha)) \leq 2m - 2$. Both elements can be represented in the polynomial basis as

$$
\begin{aligned}
A &= \alpha^{\frac{m}{2}}(\alpha^{\frac{m}{2}-1}a_{m-1} + \cdots + a_{\frac{m}{2}}) + (\alpha^{\frac{m}{2}-1}a_{\frac{m}{2}-1} + \cdots + a_0) = \alpha^{\frac{m}{2}}A_h + A_l \\
B &= \alpha^{\frac{m}{2}}(\alpha^{\frac{m}{2}-1}b_{m-1} + \cdots + b_{\frac{m}{2}}) + (\alpha^{\frac{m}{2}-1}b_{\frac{m}{2}-1} + \cdots + b_0) = \alpha^{\frac{m}{2}}B_h + B_l
\end{aligned}
\tag{23.3}
$$

Using Equation 23.3, the polynomial product is given as

$$
D = \alpha^m A_h B_h + \alpha^{\frac{m}{2}}(A_h B_l + A_l B_h) + A_l B_l
\tag{23.4}
$$

Let us define the following auxiliary polynomials $M^{(1)}(\alpha)$:

$$
\begin{aligned}
M_0^{(1)}(\alpha) &= A_l(\alpha)B_l(\alpha) \\
M_1^{(1)}(\alpha) &= [A_l(\alpha) + A_h(\alpha)][B_l(\alpha) + B_h(\alpha)] \\
M_2^{(1)}(\alpha) &= A_h(\alpha)B_h(\alpha)
\end{aligned}
\tag{23.5}
$$

Then the product given in Equation 23.4 can be achieved by

$$
D = \alpha^m M_2^{(1)}(\alpha) + \alpha^{\frac{m}{2}}[M_1^{(1)}(\alpha) + M_0^{(1)}(\alpha) + M_2^{(1)}(\alpha)] + M_0^{(1)}(\alpha)
\tag{23.6}
$$

The algorithm becomes recursive if it is applied again to the polynomials given in Equation 23.5. The next iteration step splits the polynomials $A_l, B_l, A_h, B_h, (A_l + A_h)$, and $(B_l + B_h)$ again in half. With these newly halved polynomials, new auxiliary polynomials $M^{(2)}(\alpha)$ can be defined in a similar way to Equation 23.5. The algorithm eventually terminates after t steps. In the final step, the polynomials $M^{(t)}(\alpha)$ are degenerated into single coefficients. Because every step halves the number of coefficients, the algorithm terminates after $t = \log_2 m$ steps.

In Figure 23.1, a block diagram for a bit-parallel realization of the Karatsuba–Ofman algorithm over $GF(2^4)$ is shown [49]. In this case, the input polynomials $A(\alpha)$ and $B(\alpha)$ have degree 3 and the product $D(\alpha)$ has degree 6.

Reduction Modulo the Field Polynomial. The second step of the multiplication given in Equation 23.1 is the reduction mod $P(\alpha)$. The pure polynomial multiplication of two polynomials $A(\alpha)B(\alpha)$, both of degree $m - 1$, results in a product polynomial $D(\alpha)$ with $\deg(D(\alpha)) \leq 2m - 2$. To perform a multiplication in $GF(2^m)$, $D(\alpha)$ must be reduced modulo the irreducible polynomial $P(\alpha)$. The modulo operation will result in a polynomial $C(\alpha)$ with $deg(D(\alpha)) \leq m - 1$ which represents the desired field element:

$$
C(\alpha) = D(\alpha) \bmod P(\alpha) = (d_{2m-2}\alpha^{2m-2} + \cdots + d_0) \bmod P(\alpha) = c_{m-1}\alpha^{m-1} + \cdots + c_0 \tag{23.7}
$$

The reduction modulo $P(\alpha)$ can be considered as a linear mapping of the $2m - 1$ coefficients of $D(\alpha)$ into the m coefficients of $C(\alpha)$. This mapping can be represented in a matrix notation as follows [49]:

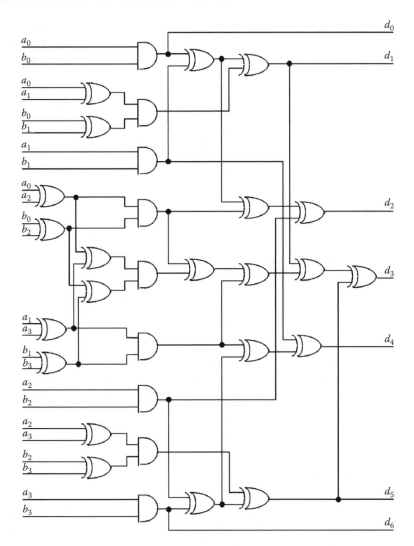

FIGURE 23.1 Bit-parallel implementation of Karatsuba–Ofman algorithm for polynomials of degree 3 over fields with characteristic 2.

$$
\begin{pmatrix} c_0 \\ c_1 \\ \vdots \\ c_{m-1} \end{pmatrix} =
\begin{pmatrix}
1 & 0 & \cdots & 0 & r_{0,0} & \cdots & r_{0,m-2} \\
0 & 1 & \cdots & 0 & r_{1,0} & \cdots & r_{1,m-2} \\
\vdots & \vdots & \ddots & \vdots & \vdots & \ddots & \vdots \\
0 & 0 & \cdots & 1 & r_{m-1,0} & \cdots & r_{m-1,m-2}
\end{pmatrix}
\begin{pmatrix} d_0 \\ \vdots \\ d_{m-1} \\ d_m \\ \vdots \\ d_{2m-2} \end{pmatrix}
\tag{23.8}
$$

The matrix on the right hand side of Equation 23.8 consists of a $(m \times n)$ identity matrix and a $(m \times m - 1)$ matrix $\mathbf{R}$ named *reduction matrix*. The $\mathbf{R}$ matrix is a function of only the selected irreducible polynomial $P(x) = x^m + p_{m-1}x^{m-1} + \cdots + p_0$ with a root α. Therefore, to every $P(x)$ a reduction matrix $\mathbf{R}$ is uniquely assigned. The $r_{j,i} \in GF(2)$ coefficients of $\mathbf{R}$ can be recursively computed in function of $P(x)$ as follows [49]:

$$
r_{j,i} = \begin{cases} p_j; & j = 0, \dots, m-1; \quad i = 0 \\ r_{j-1,i-1} + r_{m-1,i-1}r_{j,0}; & j = 0, \dots, m-1; \quad i = 1, \dots, m-2 \end{cases}
\tag{23.9}
$$

where $r_{j-1,i-1} = 0$ if $j = 0$. $\mathbf{R}$ is function of the selected irreducible polynomial. Therefore, choosing an appropriate reduction polynomial $P(x)$, like a *trinomial* or *pentanomial*, the complexity of this operation can be reduced, i.e., $P(x) = x^m + x^n + 1$ or $P(x) = x^m + x^{k_2} + x^{k_1} + x^{k_0} + 1$. Such polynomials are widely recommended in all the major standards [1,14,21].

Example

Reduction for $P(x) = x^4 + x^3 + 1$. For the generating irreducible trinomial $P(x) = x^4 + x^3 + 1$ for $GF(2^4)$ used in Table 23.1, the corresponding reduction matrix is

$$\mathbf{R} = \begin{pmatrix} 1 & 1 & 1 \\ 0 & 1 & 1 \\ 0 & 0 & 1 \\ 1 & 1 & 1 \end{pmatrix} \tag{23.10}$$

and the reduction operation given in Equation 23.8 is for this instance:

$$\begin{pmatrix} c_0 \\ c_1 \\ c_2 \\ c_3 \end{pmatrix} = \begin{pmatrix} d_0 + d_4 + d_5 + d_6 \\ d_1 + d_5 + d_6 \\ d_2 + d_6 \\ d_3 + d_4 + d_5 + d_6 \end{pmatrix} \tag{23.11}$$

which can be computed with 9 $GF(2)$ additions, i.e., with 9 XOR gates. For example, the polynomial multiplication of the two elements α^8 and $\alpha^9 \in GF(2^4)$ represented in polynomial basis $\{\alpha^3, \alpha^2, \alpha, 1\}$ is $\alpha^8 \cdot \alpha^9 = (\alpha^3 + \alpha^2 + \alpha) \cdot (\alpha^2 + 1) = \alpha^5 + \alpha^4 + \alpha^2 + \alpha = D(\alpha) = (d_6, d_5, d_4, d_3, d_2, d_1, d_0)$. Using Equation 23.11, $(c_3, c_2, c_1, c_0) = (0, 1, 0, 0) = C(\alpha) = \alpha^2$, as computed in the example given in Section 23.2.

23.3.1.2 Interleaved Multiplication

The simplest algorithm for $GF(2^m)$ multiplication is the shift-and-add method [29] with the reduction step interleaved [15,54]. For hardware, the shift-and-add method can be efficiently implemented and is suitable when area is constrained as in RFID and sensor networks. The multiplication of two elements $A(\alpha), B(\alpha) \in GF(2^m)$ can be expressed as

$$C(\alpha) = A(\alpha)B(\alpha) \bmod P(\alpha) = A(\alpha) \left(\sum_{i=0}^{m-1} b_i \alpha^i \right) \bmod P(\alpha)$$

$$= \left(\sum_{i=0}^{m-1} b_i A(\alpha) \alpha^i \right) \bmod P(\alpha) \tag{23.12}$$

Therefore, the product $C(\alpha)$ can be computed as

$$C(\alpha) = (b_0 A(\alpha) + b_1 A(\alpha)\alpha + b_2 A(\alpha)\alpha^2 + \cdots + b_{m-1} A(\alpha)\alpha^{m-1}) \bmod P(\alpha) \tag{23.13}$$

Equation 23.13 can be computed by processing $B(\alpha)$ from its least-significant bit to its most-significant bit. Then the shift-and-add method receives the name of *least-significant bit-serial* (LSB) multiplier. Thus, multiplication according to this scheme is given in algorithm 1 [15,54].

The product $D = \alpha \cdot A(\alpha)$ can be expressed as

$$D = \alpha(a_0 + a_1\alpha + \cdots + a_{m-1}\alpha^{m-1}) = a_0\alpha + a_1\alpha^2 + \cdots + a_{m-1}\alpha^m \tag{23.14}$$

Algorithm 1: LSB-first serial/parallel multiplier

Input: $A(\alpha), B(\alpha), P(\alpha)$
Output: $C(\alpha) = A(\alpha)B(\alpha) \bmod P(\alpha)$
1: $C = 0$;
2: **for** $i = 0$ **to** $m - 1$ **do**
3:     $C = b_i A + C$;
4:     $A = A\alpha \bmod P(\alpha)$;
5: **end for**
6: Return(C)

Using the fact that α is a primitive root of the irreducible polynomial $P(x)$, we have that $\alpha^m = p_0 + p_1\alpha + \cdots + p_{m-1}\alpha^{m-1}$, where p_is are the coefficients of the irreducible polynomial. Substituting this expression into Equation 23.14 we obtain

$$D = d_0 + d_1\alpha + \cdots + d_{m-1}\alpha^{m-1} \qquad (23.15)$$

where $d_0 = a_{m-1}p_0$ and $d_i = a_{i-1} + a_{m-1}p_i$, for $i = 1, 2, \ldots, m - 1$.

Using the above equations, the multiplier implementation of Algorithm 1 is given in Figure 23.2. The architecture shown in Figure 23.2 consists of two LFSRs plus extra circuitry [54]. Initially, the register C is set to zero, whereas the register in the upper part of Figure 23.2 is loaded with the m coefficients of the element A. Thereafter, when the clock is applied to the registers, the value of $A\alpha$ is computed. The coefficients of element B $(b_0, b_1, \ldots, b_{m-1})$ are then serially introduced in that order, generating the values $b_i A\alpha^i$, for $i = 0, 1, \ldots, m - 1$, which are accumulated in register C until all the coefficients of the product $c_0, c_1, \ldots, c_{m-1}$ are collected.

Multiplication can also be computed by processing $B(\alpha)$ from its most-significant bit to the least-significant bit. In such a case, the shift-and-add method receives the name of *most-significant*

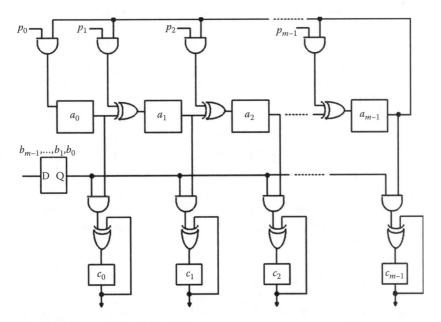

FIGURE 23.2 LSB-first serial/parallel multiplier.

bit-serial (MSB) multiplier, and the multiplication according to this scheme is given in the following algorithm [15].

Algorithm 2: MSB-first serial/parallel multiplier

Input: $A(\alpha), B(\alpha), P(\alpha)$
Output: $C(\alpha) = A(\alpha)B(\alpha) \bmod P(\alpha)$
1: $C = 0$;
2: **for** $i = m - 1$ **downto** 0 **do**
3:      $C = C\alpha \bmod P(\alpha) + b_i A$;
4: **end for**
5: Return(C)

It must be noted that in Step 3 of Algorithm 2, the computation of $C\alpha \bmod P(\alpha)$ and $b_i A$ can be performed in parallel because they are independent of each other. However, the value of C in each iteration depends on the value of C at the previous iteration and on the value of $b_i A$. Due to this dependency, the MSB-first multiplier has a longer critical path than the LSB-first multiplier. As conclusion, both LSB-first and MSB-first multipliers are suitable to RFID and sensor networks due to their constrained area.

23.3.1.3 Matrix--Vector Multipliers

The multiplication given in Equation 23.1 can also be described in terms of matrix–vector operations. There are mainly two different approaches based on matrix–vector operations for the computation of a field product. The first one is the previously described *two-step multiplier*, in which the polynomial multiplication is performed by any method and then the resulting product is reduced by using a reduction matrix. In the second approach, the polynomial multiplication and modular reduction parts are performed in a single step by using the so-called *Mastrovito* matrix. This second method was first proposed by Mastrovito [37,38] and is also known as *Mastrovito* multiplication.

First, we introduce a matrix notation [49] for the multiplication $C(\alpha) = A(\alpha)B(\alpha) \bmod P(\alpha)$ in the field $GF(2^m)$. All elements are binary polynomials of degree less than m:

$$c_{m-1}\alpha^{m-1} + \cdots + c_0 = (a_{m-1}\alpha^{m-1} + \cdots + a_0)(b_{m-1}\alpha^{m-1} + \cdots + b_0) \bmod P(\alpha) \quad (23.16)$$

The elements $B(\alpha)$ and $C(\alpha)$ can be also represented as column vectors with the polynomial coefficients. The matrix $\mathbf{Z} = h(A(\alpha), P(\alpha))$ can be introduced in such a way that the multiplication can also be described as

$$C = \begin{pmatrix} c_0 \\ c_1 \\ \vdots \\ c_{m-1} \end{pmatrix} = \mathbf{Z}B = \begin{pmatrix} h_{0,0} & \cdots & h_{0,m-1} \\ \vdots & \ddots & \vdots \\ h_{m-1,0} & \cdots & h_{m-1,m-1} \end{pmatrix} \begin{pmatrix} b_0 \\ b_1 \\ \vdots \\ b_{m-1} \end{pmatrix} \quad (23.17)$$

The $(m \times m)$ matrix $\mathbf{Z}$ is named *product matrix* or *Mastrovito matrix*. Its coefficients $h_{i,j} \in GF(2)$ depend recursively on the coefficients a_i and on the coefficients $f_{i,j}$ of the $\mathbf{F}$ matrix (introduced in Equation 23.19) as follows:

$$h_{i,j} = \begin{cases} a_i; & j = 0; \ i = 0, \ldots, m - 1 \\ u(i - j)a_{i-j} + \sum_{t=0}^{j-1} f_{j-1-t,i}a_{m-1-t}; & j, i = 0, 1, \ldots, m - 1; \ j \neq 0 \end{cases} \quad (23.18)$$

where the step function $u(\mu)$ is 1 for $\mu \geq 0$ or 0 for $\mu < 0$. The matrix–vector product given in Equation 23.17 describes the entire field multiplication. The $\mathbf{F}$ matrix required for the computation of $\mathbf{Z}$ is function of the irreducible polynomial $P(\alpha)$ of degree m. Its binary entries $f_{i,j}$ are defined such that

$$
\begin{pmatrix} \alpha^m \\ \alpha^{m+1} \\ \vdots \\ \alpha^{2m-2} \end{pmatrix} \equiv \begin{pmatrix} f_{0,0} & \cdots & f_{0,m-1} \\ f_{1,0} & \cdots & f_{1,m-1} \\ \vdots & \ddots & \vdots \\ f_{m-2,0} & \cdots & f_{m-2,m-1} \end{pmatrix} \begin{pmatrix} 1 \\ \alpha \\ \vdots \\ \alpha^{m-1} \end{pmatrix} \bmod P(\alpha) \qquad (23.19)
$$

The binary entries $f_{i,j}$ of $\mathbf{F}$ in Equation 23.19 can be computed recursively in function of the coefficients of the irreducible polynomial $P(x) = p_m + p_{m-1}x^{m-1} + \cdots + p_1x + p_0$ with a root α as follows:

$$
f_{i,j} = \begin{cases} f_{i-1,m-1}; & i = 1, \ldots, m-2; \quad j = 0 \\ f_{i-1,j-1} + f_{i-1,m-1}f_{0,j}; & i = 1, \ldots, m-2; \quad j = 1, \ldots, m-1 \end{cases} \qquad (23.20)
$$

where $f_{0,j} = p_j$.

It must be noted that the $\mathbf{F}$ matrix given by Equations 23.19 and 23.20 is equivalent to the *reduction* matrix $\mathbf{R}$ given in Equations 23.8 and 23.9 and used in the *two-step* multiplication. In fact, $\mathbf{R} = \mathbf{F}^T$.

The matrix–vector operation given in Equation 23.17 requires m^2 modulo 2 multiplications. Therefore, it can be proven [49] that the space complexity of a bit-parallel Mastrovito multiplier is m^2 AND gates and more than $(m^2 - 1)$ XOR gates (the equality, i.e., $(m^2 - 1)$ XOR gates corresponds to the irreducible trinomial $P(x) = x^m + x + 1$). The delay of the bit-parallel multiplier can also be upper bounded by $T \leq T_{\text{AND}} + 2T_{\text{XOR}} \lceil log_2 m \rceil$, where T_{AND} and T_{XOR} represent the delays of two-input AND and XOR gates, respectively.

Example

Multiplication for $P(x) = x^4 + x^3 + 1$. Let $P(x) = x^4 + x^3 + 1$ be the generating irreducible polynomial for $GF(2^4)$. The polynomials $x^4, x^5,$ and x^6 are given as

$$
\begin{aligned} x^4 &\equiv 1 + x^3 && \bmod P(x) = 1 + x^3 \\ x^5 &\equiv x + x^4 && \bmod P(x) = 1 + x + x^3 \\ x^6 &\equiv x^2 + x^5 && \bmod P(x) = 1 + x + x^2 + x^3 \end{aligned} \qquad (23.21)
$$

Equation 23.21 can be rewritten in matrix form to obtain the $\mathbf{F}$ matrix, also obtained using Equations 23.19 and 23.20, as follows:

$$
\begin{pmatrix} x^4 \\ x^5 \\ x^6 \end{pmatrix} \equiv \begin{pmatrix} 1 & 0 & 0 & 1 \\ 1 & 1 & 0 & 1 \\ 1 & 1 & 1 & 1 \end{pmatrix} \begin{pmatrix} 1 \\ x \\ x^2 \\ x^3 \end{pmatrix} \bmod x^4 + x^3 + 1 \qquad (23.22)
$$

The *product matrix* $\mathbf{Z}$ can be finally achieved using Equations 23.17 and 23.18:

$$
C = \mathbf{ZB} = \begin{pmatrix} a_0 & a_3 & a_2 + a_3 & a_1 + a_2 + a_3 \\ a_1 & a_0 & a_3 & a_2 + a_3 \\ a_2 & a_1 & a_0 & a_3 \\ a_3 & a_2 + a_3 & a_1 + a_2 + a_3 & a_0 + a_1 + a_2 + a_3 \end{pmatrix} \begin{pmatrix} b_0 \\ b_1 \\ b_2 \\ b_3 \end{pmatrix} \qquad (23.23)
$$

Several research papers use the Mastrovito scheme outlined above for different irreducible polynomials [19,20,22,25,51,56,66]. In most of them, the *decomposition* of the Mastrovito matrix $\mathbf{Z}$ in a sum of matrices is normally used. As an example, an efficient multiplication scheme was introduced in Ref. [51] for the computation of the product $C = \mathbf{Z}B$. In this approach, the *product* or *Mastrovito* matrix $\mathbf{Z}$ is decomposed as

$$\mathbf{Z} = \mathbf{L} + \mathbf{F}^{\mathrm{T}} \cdot \mathbf{U} = \mathbf{L} + \mathbf{R} \cdot \mathbf{U} \tag{23.24}$$

where $\mathbf{L}$ and $\mathbf{U}$ are the following $(m \times m)$ and $(m - 1 \times m)$ Toeplitz matrices, respectively:

$$\mathbf{L} = \begin{pmatrix} a_0 & 0 & 0 & 0 & \cdots & 0 \\ a_1 & a_0 & 0 & 0 & \cdots & 0 \\ a_2 & a_1 & a_0 & 0 & \cdots & 0 \\ \vdots & \vdots & \ddots & \ddots & \ddots & \vdots \\ a_{m-2} & a_{m-3} & \cdots & a_1 & a_0 & 0 \\ a_{m-1} & a_{m-2} & \cdots & a_2 & a_1 & a_0 \end{pmatrix} \tag{23.25}$$

$$\mathbf{U} = \begin{pmatrix} 0 & a_{m-1} & a_{m-2} & \cdots & a_2 & a_1 \\ 0 & 0 & a_{m-1} & \cdots & a_3 & a_2 \\ \vdots & \vdots & \ddots & \ddots & \vdots & \vdots \\ 0 & 0 & \cdots & 0 & a_{m-1} & a_{m-2} \\ 0 & 0 & \cdots & 0 & 0 & a_{m-1} \end{pmatrix} \tag{23.26}$$

and where $a_i's$ are the binary coordinates of A. The two vectors $D = \mathbf{L} \cdot B$ and $E = \mathbf{U} \cdot B$, which are functions of A and B, can be defined in such a way that the product C in $GF(2^m)$ can also be defined as [51]

$$C = D + \mathbf{F}^{\mathrm{T}} \cdot E = D + \mathbf{R} \cdot E \tag{23.27}$$

Using the above definitions, a bit-parallel architecture for polynomial basis multiplication over $GF(2^m)$ is shown in Figure 23.3. This architecture [51] is divided into two parts: IP-network and F-network. The IP-network has m blocks (denoted as $I_0, I_1, \ldots, I_{m-1}$) and generates the vectors D and E previously defined. For $0 \le i \le m - 2$, block I_i consists of two inner product cells, IP$(i + 1)$ and IP$(m - i - 1)$. The last block I_{m-1}, however, consists of only one such cell, IP(m). The cyclic shift block in the IP-network takes the vectors $(a_{m-1}, \ldots, a_2, a_1, a_0)$ into the vectors $(a_{m-2}, \ldots, a_1, a_0, a_{m-1})$.

In Figure 23.3, the F-network generates the product C using D and E. It consists of m binary trees of XOR gates $BTX_{0 \cdots m-1}$, where the number of XOR gates in BTX_i, $0 \le i \le m - 1$, is equal to the number of 1s in the ith column of the $\mathbf{F}$ matrix. The interconnection bus IB in Figure 23.3 has $m - 1$ lines equal to the number of $e_j s$ (coordinates of E vector).

The complexity of this multiplier depends on the irreducible polynomial $P(x)$ by means of the $\mathbf{F}$ matrix. In Ref. [51], the number of AND and XOR gates (#*AND* and #*XOR*, respectively) is proven to be #$AND = m^2$ and #$XOR \le (m - 1)^2 + H(\mathbf{F})$, where $H(\mathbf{F})$ is the Hamming weight of $\mathbf{F}$. The time delay is upper bounded by $T_C \le T_{\mathrm{AND}} + (\lceil \log_2 m \rceil + \lceil \log_2(\theta + 1) \rceil)T_{\mathrm{XOR}}$, where $\theta = \max\{H(\mathbf{f}_j), 0 \le j \le m - 1\}$ and $H(\mathbf{f}_j)$ is the Hamming weight of the column $\mathbf{f}_j$ of $\mathbf{F}$. Therefore, different complexities can be found for irreducible trinomials, pentanomials, ESPs, or AOPs [51].

23.3.1.4 Digit--Serial Multipliers

Bit-parallel $GF(2^m)$ multipliers have a hardware complexity that is proportional to m^2. Its area and power consumption greatly increase as the field order m increases, hence limiting the use of these multipliers on a single chip. Digit-serial multipliers [57] are best suited for systems requiring a trade-off between speed, area, and power consumption as in RFID and sensor networks. This is achieved by processing several coefficients of $B(\alpha)$ at the same time. The number of coefficients

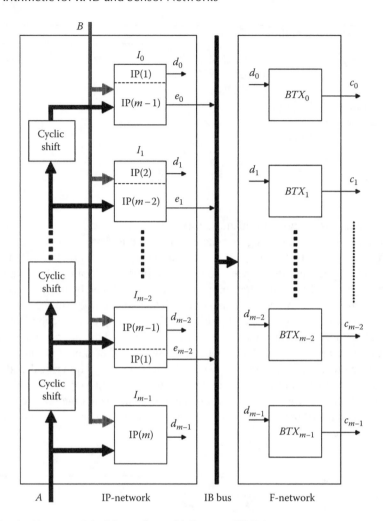

FIGURE 23.3 Architecture of the Mastrovito multiplier over $GF(2^m)$.

that are processed in parallel is defined to be the digit-size D. Therefore, the total number of digits in a polynomial of degree $m - 1$ will be $d = \lceil m/D \rceil$. Then the operand B can be rewritten as $B = \sum_{i=0}^{d-1} B_i \alpha^{Di}$, where the coefficients B_i are given as [57]

$$B_i = \sum_{j=0}^{D-1} b_{Di+j} \alpha^j, \ 0 \le i \le d - 1 \tag{23.28}$$

and where it is assumed that B is completed with zero coefficients in such a way that $b_i = 0$ for $m - 1 < i < d \cdot D$. Then

$$C(\alpha) = A(\alpha)B(\alpha) \bmod P(\alpha) = A(\alpha) \sum_{i=0}^{d-1} B_i \alpha^{Di} \bmod P(\alpha) \tag{23.29}$$

The *least-significant digit-serial* (LSD) multiplier is the generalization of the LSB-first multiplier in which the digits of $B(\alpha)$ are processed starting from its least-significant to its most-significant. The LSD-serial multiplication is given in the following algorithm [15,57]:

Algorithm 3: LSD-serial multiplier

Input: $A(\alpha) = \sum_{i=0}^{m-1} a_i \alpha^i, B(\alpha) = \sum_{i=0}^{\lceil \frac{m}{D} \rceil - 1} B_i \alpha^{Di}$
Output: $C(\alpha) = A(\alpha)B(\alpha) \bmod P(\alpha) = \sum_{i=0}^{m-1} c_i \alpha^i$
1: $C = 0$;
2: **for** $i = 0$ **to** $\lceil m/D \rceil - 1$ **do**
3:      $C = B_i A + C$;
4:      $A = A\alpha^D \bmod P(\alpha)$;
5: **end for**
6: Return(C)

---

where $a_i \in GF(2)$ and B_i is given in Equation 23.28. *Most-significant digit-serial* (MSD) multipliers can be deduced in a similar way as a generalization of the MSB-first multipliers.

The architecture of the LSD multiplier is given in Figure 23.4, where $A^{(i-1)}$ and $C^{(i-1)}$ represent A and C in step i, respectively [57]. There are two loops in this multiplier. The right loop performs the computation given in Step 4 in Algorithm 3, i.e., $A^{(i-1)}$ multiplied by α^D followed by mod $P(\alpha)$ to reduce the degree to be less than m. The partial product generator $\times$ (in Figure 23.4) computes $A^{(i-1)}B_{i-1}$ in iteration i, as in Step 3 in Algorithm 3. The left loop is an accumulator consisting of XOR gates and memory elements, where the partial products and the intermediate result $C^{(i-1)}$ are accumulated using a binary tree of XOR gates. Finally, the lower mod $P(\alpha)$ unit in Figure 23.4 reduces the degree of the result from $m + D - 2$ to less than m. The critical path computation time in this LSD-first multiplier is lower bounded by the dominant loop computation time, which is the time required for one mod $P(\alpha)$ reduction operation. The cycle time for one multiplication is $D(T_{XOR} + T_{AND})$, and the latency is given by $\lceil m/D \rceil + 1 - \delta(D, 1)$, where $\delta(D, 1)$ is 1 for $D = 1$ and 0 otherwise [57].

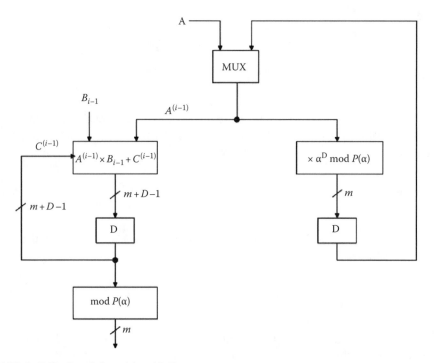

FIGURE 23.4   LSD-first digit-serial multiplier.

In LSD-serial multipliers, there are products in the form $A\alpha^D \bmod P(\alpha)$ which have to be reduced. As in LSB-first multipliers, modular reduction can be derived for general irreducible polynomials $P(\alpha)$. However, there exist certain types of polynomials that minimize the complexity of the reduction operation. The following two theorems were given in Ref. [57]:

1. Let $P(\alpha) = \alpha^m + p_k \alpha^k + \sum_{j=0}^{k-1} p_j \alpha^j$, with $k < m$, be an irreducible polynomial. For $t \le m - 1 - k$, α^{m+t} can be reduced to degree less than m in one step with the following equation:

$$\alpha^{m+t} \bmod P(\alpha) = p_k \alpha^{k+t} + \sum_{j=0}^{k-1} p_j \alpha^{j+t} \qquad (23.30)$$

2. For digit multipliers with digit-element size D, when $D \le m - k$, the intermediate results in Algorithm 3 can be reduced to degree less than m in one step.

From the above theorems the following conclusion can be achieved [15]: for a given irreducible polynomial $P(\alpha) = \alpha^m + p_k \alpha^k + \sum_{j=0}^{k-1} p_j \alpha^j$, the digit-element size D has to be chosen based on the value of k, i.e., the second highest degree in the irreducible polynomial. Furthermore, in Ref. [57] it was concluded that the complexity and energy consumption of the mod $P(\alpha)$ operation can be reduced by selecting those irreducible polynomials with smaller value of k and less Hamming weight. Therefore, the use of digit-serial multipliers with a suitable selection of the irreducible polynomials is of fundamental interest for RFID and WSN.

23.3.1.5 Montgomery Multipliers

Montgomery multiplication was first proposed for integer modular multiplication that can avoid trial division [42]. Later, it was extended to finite field multiplication over $GF(2^m)$ [27], where it was shown that the operation can be simplified if certain type of element $R(x)$ is selected. In the following, we give a brief review of the Montgomery multiplication in $GF(2^m)$ proposed in Ref. [27].

Let $P(x)$ be an irreducible polynomial over $GF(2)$ that defines the field $GF(2^m)$. Rather than computing Equation 23.2, the Montgomery multiplication calculates

$$C(x) = A(x)B(x)R^{-1}(x) \bmod P(x) \qquad (23.31)$$

where $R(x)$ is a fixed element and $gcd(R(x), P(x)) = 1$. One can find two polynomials $R^{-1}(x)$ and $P'(x)$ such that

$$R(x)R^{-1}(x) + P(x)P'(x) = 1 \qquad (23.32)$$

where $R^{-1}(x)$ is the inverse of $R(x)$ modulo $P(x)$. These two polynomials can be calculated with the extended Euclidean algorithm. In Ref. [27], $R(x)$ was selected to be $R(x) = x^m$ to have a modular reduction with high performance in the Montgomery multiplication algorithm, which is given in algorithm 4 [27,54]:

It can be shown [27,54] that Algorithm 4 has an associated area complexity of $2m^2$ two-input AND gates and $2m^2 - 3m - 1$ two-input XOR gates, whereas the time complexity is $3T_{AND} + (2\lceil \log_2 m \rceil + \lceil \log_2(m-1) \rceil)T_{XOR}$. Several architectures and implementations for Montgomery multiplication over $GF(2^m)$ have been reported in the literature for different irreducible polynomials $P(x)$. For example, bit-parallel Montgomery multipliers with reduced complexities were proposed in Ref. [62] for different trinomials.

Algorithm 4: Montgomery multiplication in $GF(2^m)$

Input: $A(x), B(x), R(x), P'(x)$
Output: $C(x) = A(x)B(x)R^{-1}(x) \bmod P(x)$
1: $T(x) = A(x)B(x)$;
2: $U(x) = T(x)P'(x) \bmod R(x)$;
3: $C(x) = [T(x) + U(x)P(x)]/R(x)$;
4: Return(C)

23.3.2 INVERSION

Besides multiplication, *inversion* and *division* are the other important operations over finite fields $GF(2^m)$. Division over $GF(2^m)$ means dividing a polynomial $A(x)$ by $B(x)$ modulo the irreducible polynomial $P(x)$. This operation is usually performed in two steps:

1. Find the inverse element $B^{-1}(x)$.
2. $A(x)/B(x) = A(x) \cdot B^{-1}(x)$.

Therefore, division is implemented as multiplication by the (*multiplicative*) *inverse*. The simplest technique for finding the inverse of any element $\alpha \in GF(2^m)$ uses table lookup. For smaller fields like $m < 8$ (used for Reed-Solomon codes), table lookup shows better performance in terms of chip area and computation time than algorithm-based methods. For larger fields, however, lookup-based implementations are highly inefficient.

Most popular methods for finite field inversion are based on *Fermat's little theorem* (mainly used for inversion in normal basis representations [11,59,61]) and on *Euclidean algorithm* and its derivatives [2,18,33,63–65]. Based on Fermat's little theorem and using an ingenious rearrangement of the required field operations, the *Itoh and Tsujii* multiplicative inverse algorithm was proposed in Ref. [26] that reduces the number of multiplications needed for calculating the inverse. Originally, the Itoh and Tsujii inversion algorithm was proposed to be used with normal basis representation. However, several variations have been reported showing that it can also be used with other field element representations [54].

Euclid's algorithm for polynomial calculates the greatest common divisor (GCD) polynomial of two polynomials $A(x)$ and $B(x)$. The algorithm can be extended for calculating the two

---

Algorithm 5: Extended Euclid's algorithm

Input: $A(x), B(x)$
Output: $R_{j-1}(x), U_{j-1}(x), W_{j-1}(x)$ *as the results.*
$(R_{j-1}(x) = GCD(A(x), B(x)) = U_{j-1}(x) \cdot A(x) + W_{j-1}(x) \cdot B(x))$
1: $R_{-1}(x) = B(x), U_{-1}(x) = 0, W_{-1}(x) = 1$;
1: $R_0(x) = A(x), U_0(x) = 1, W_0(x) = 0$;
1: $j = 0$;
2: **repeat**
3:     $j = j + 1$;
4:     $Q_j(x) = R_{j-2}(x) \div R_{j-1}(x)$;
5:     $R_j(x) = R_{j-2}(x) - Q_j(x) \cdot R_{j-1}(x)$;
6:     $U_j(x) = U_{j-2}(x) - Q_j(x) \cdot U_{j-1}(x)$;
7:     $W_j(x) = W_{j-2}(x) - Q_j(x) \cdot W_{j-1}(x)$;
6: **until** $R_j(x) = 0$;

Algorithm 6: Inversion algorithm

Input: $A(x), P(x)$
Output: $U(x)$ *as the result* $(U(x) = V(x) = A^{-1}(x))$
1: $S(x) = P(x), V(x) = 0$;
1: $R(x) = A(x), U(x) = 1$;
1: $\delta = 0$;
2: **for** $i = 1$ **to** $2m$ **do**
3:    **if** $r_m = 0$ **then**
4:       $R(x) = x \cdot R(x)$;
5:       $U(x) = x \cdot U(x) \bmod P(x)$;
6:       $\delta = \delta + 1$;
7:    **else**
8:       **if** $s_m = 1$ **then**
9:          $S(x) = S(x) - R(x)$;
10:          $V(x) = (V(x) - U(x)) \bmod P(x)$;
11:       **end if**
12:       $S(x) = x \cdot S(x)$;
13:       **if** $\delta = 0$ **then**
14:          $\{R(x) = S(x), S(x) = R(x)\}$;
15:          $\{U(x) = x \cdot V(x) \bmod P(x), V(x) = U(x)\}$;
16:          $\delta = 1$;
17:       **else**
18:          $U(x) = U(x)/x \bmod P(x)$;
19:          $\delta = \delta - 1$;
20:       **end if**
21:    **end if**
22: **end for**

---

polynomials $U(x)$ and $W(x)$ such that $GCD(A(x), B(x)) = U(x) \cdot A(x) + W(x) \cdot B(x)$. The extended Euclid's algorithm is given in Algorithm 5, where $\div$ denotes the operation that calculates a quotient polynomial [33].

Let $P(x)$ be the irreducible polynomial with degree m that defines the field, and $A(x)$ be the polynomial representation of an element in the field. Because the GCD polynomial of $A(x)$ and $P(x)$ is 1, the multiplicative inverse $A^{-1}(x)$ of $A(x)$ can be obtained as $U_{j-1}(x) \bmod P(x)$ by replacing $B(x)$ with $P(x)$ in Algorithm 5 as follows [33]: $GCD(A(x), B(x)) = U_{j-1}(x) \cdot A(x) + W_{j-1}(x) \cdot P(x)$, therefore $1 \equiv U_{j-1}(x) \cdot A(x) \pmod{P(x)}$ and finally $A^{-1}(x) \equiv U_{j-1}(x) \pmod{P(x)}$.

Using the above facts, an algorithm for inversion in $GF(2^m)$ based on the extended Euclid's algorithm and developed for VLSI implementation was proposed in Ref. [2]. The algorithm is given in Algorithm 6, where $\{X, Y\}$ means that the operations X and Y are performed in parallel; r_m and s_m are the most-significant coefficients of $R(x)$ and $S(x)$, respectively; δ is the difference of $deg^*(R(x))$ and $deg^*(S(x))$, where $deg^*(\cdot)$ denotes the upper bound of the degree of the proper one. Efficient hardware implementations of Algorithm 6 can be found in Refs. [2,33].

23.4 CONCLUSIONS

Security in RFID and sensor networks is of crucial importance. RFID tags and sensor nodes require efficient implementations of cryptographic schemes due to the limitations on number of gates, power, bandwidth, and so on. Cryptographic protocols are based on finite field (prime or binary extension field) arithmetic. $GF(2^m)$ binary field arithmetic is used in the main cryptographic schemes based on AES and ECC. Therefore, efficient algorithms for $GF(2^m)$ arithmetic operations with compact and low-power/energy architectures and implementations are of fundamental interest for RFID and WSN. Most important efficient algorithms for $GF(2^m)$ arithmetic suitable to hardware/software implementations have been outlined in this chapter.

REFERENCES

[1] American National Standards Institute, New York. *ANSI X9.62: Public Key Cryptography for the Financial Services Industry: The Elliptic Curve Digital Signature Algorithm (ECDSA)*, 1999.

[2] H. Brunner, A. Curiger, and M. Hofstetter, On computing multiplicative inverses in $GF(2^m)$, *IEEE Transactions on Computers*, 42(8), 1010–1015, August 1993.

[3] E.R. Berlekamp, Bit-serial Reed-Solomon encoders, *IEEE Transactions on Information Theory*, 28, 869–874, November 1982.

[4] T. Beth and D. Gollmann. Algorithm engineering for public key algorithm. *IEEE Journal on Selected Areas in Communications*, 7(4), 458–465, May 1989.

[5] L. Batina, J. Guajardo, T. Kerins, N. Mentens, P. Tuyls, and I. Verbauwhede. Public-key cryptography for RFID-tags. In *Fourth IEEE International Workshop on Pervasive Computing and Communication Security—PerSec 2007*, p. 6, 2007.

[6] L. Batina, N. Mentens, K. Sakiyama, B. Prenel, and I. Verbauwhede, Low-cost elliptic curve cryptography for wireless sensor networks, *ESAS 2006, LNCS 4357*, pp. 6–17, 2006.

[7] L. Batina, N. Mentens, K. Sakiyama, B. Preneel, and I. Verbauwhede. Public-key cryptography on the top of a needle. In *Proceedings of IEEE International Symposium on Circuits and Systems (ISCAS 2007), Special Session: Novel Cryptographic Architectures for Low-Cost RFID*, p. 4, 2007.

[8] A. Booth, A signed binary multiplication technique, *Quarterly Journal of Mechanics and Applied Mathematics*, 4(2), 236–240, 1951.

[9] I.F. Blake, G. Seroussi, and N.P. Smart, *Elliptic Curves in Cryptography*, Cambridge University Press, Cambridge, MA, 1999.

[10] S.T.J. Fenn, M. Benaissa, and D. Taylor, GF(2^m) multiplication and division over the dual basis, *IEEE Transactions on Computers*, 45(3), 319–327, March 1996.

[11] G.-L. Feng, A VLSI architecture for fast inversion in $GF(2^m)$, *IEEE Transactions on Computers*, 38(10), 1383–1386, October 1989.

[12] S.T.J. Fenn, Optimised algorithms and circuit architectures for performing finite field arithmetic in Reed-Solomon codecs, PhD thesis, University of Huddersfield, Huddersfield, England, 1993.

[13] K. Finkenzeller, RFID-Handbook, Carl Hanser Verlag München, Germany, 2nd edition, April 2003.

[14] National Institute for Standards and Technology, *FIPS 186-2: Digital Signature Standard (DSS). 186-2*, Gaithersburg, MD, February 2000. http://csrc.nist.gov/encryption.

[15] J. Guajardo, T. Güneysu, S.S. Kumar, C. Paar, and J. Pelzl, Efficient hardware implementation of finite fields with applications to cryptography. In J.L. Imana, Ed., *Acta Applicandae Mathematicae*, 93(1–3), pp. 75–118, Springer Verlag, the Netherlands, September 2006.

[16] G. Gaubatz, J.-P. Kaps, E. Öztürk, and B. Sunar. State of the art in ultra-low power public key cryptography for wireless sensor networks. In *IEEE Conference on Pervasive Computing and Communications Workshops (PerCom 2005 Workshops)*, pp. 146–150. IEEE Computer Society, March 8–12, Kauai, Hawaii, 2005.

[17] G. Gaubatz, J.-P. Kaps, and B. Sunar. Public key cryptography in sensor networks—revisited. In *1st European Workshop on Security in Ad-Hoc and Sensor Networks (ESAS 2004)*, Heidelberg, Germany, August 2004.

[18] J.-H. Guo and C.-L. Wang, Hardware-efficient systolic architecture for inversion and division in $GF(2^m)$, *IEE Proceedings—Computer and Digital Techniques*, 145(4), 272–278, July 1998.

[19] A. Halbutogullari and Ç.K. Koç, Mastrovito multiplier for general irreducible polynomials, *Applied Algebra, Algebraic Algorithms, and Error-Correcting Codes, LNCS 1719*, pp. 498–507, Springer-Verlag, Honolulu, HI, 1999.

[20] A. Halbutogullari and Ç.K. Koç, Mastrovito multiplier for general irreducible polynomials, *IEEE Transactions Computers*, 49(5), 503–518, May 2000.

[21] IEEE Computer Society Press, *IEEE P1363-2000: IEEE Standard Specifications for Public-Key Cryptography*, Silver Spring, MD, 2000.

[22] J.L. Imaña, R. Hermida, and F. Tirado, Low complexity bit-parallel multipliers based on a class of irreducible pentanomials, *IEEE Transactions on VLSI Systems*, 14(12), 1388–1393, December 2006.

[23] International Organization for Standardization (ISO). ISO/IEC 14443: Identification Cards—Contactless Integrated Circuit(s) Cards—Proximity Cards, 2000.

[24] International Organization for Standardization (ISO). ISO/IEC 18000-3: Information Technology AIDC Techniques—RFID for Item Management, March 2003.

[25] J.L. Imaña, J.M. Sánchez, and F. Tirado, Bit-parallel finite field multipliers for irreducible trinomials, *IEEE Transactions on Computers*, 55(5), 520–533, May 2006.

[26] T. Itoh and S. Tsujii, A fast algorithm for computing multiplicative inverses in $GF(2^m)$ using normal basis, *Information and Computation*, 78, 171–177, 1998.

[27] Ç.K. Koç and T. Acar, Montgomery multiplication in $GF(2^k)$, *Designs, Codes and Cryptography*, 14, 57–69, 1998.

[28] M. Keller and W. Marnane, Low power elliptic curve cryptography, *PATMOS 2007, LNCS 4644*, pp. 310–319, Göteborg, Sweden, 2007.

[29] D.E. Knuth, *The Art of Computer Programming: Seminumerical Algorithms*, Vol. 2, 2nd edition, Addison–Wesley, Reading, MA, 1981.

[30] A. Karatsuba and Y. Ofman, Multiplication of multidigit numbers on automata, *Soviet Physics-Doklady*, 7(7), 595–596, 1963.

[31] S. Kumar and C. Paar. Are standards compliant elliptic curve cryptosystems feasible on RFID? In *Proceedings of Workshop on RFID Security*, Graz, Austria, July 2006.

[32] J.-P. Kaps and B. Sunar. Energy comparison of AES and SHA-1 for ubiquitous computing. In E. Sha, Ed., *IFIP International Conference on Embedded and Ubiquitous Computing—EUC 2006, LNCS 4097*, pp. 372–381, Springer, Seoul, Korea, 2006.

[33] K. Kobayashi, N. Takagi, and K. Takagi, An algorithm for inversion in $GF(2^m)$ suitable for implementation using polynomial multiply instruction on GF(2), *Proceedings of 18th IEEE Symposium on Computer Arithmetic, ARITH-18*, pp. 105-112, Montpellier, France, June 2007.

[34] P. Leung, C. Choy, C. Chan, and K. Pun, An optimal normal basis elliptic curve cryptoprocessor for inductive RFID application, *ISCAS 2006*, pp. 309–312, Island of Kos, Greece, 2006.

[35] R. Lidl and H. Niederreiter, *Finite Fields*, Addison-Wesley, Reading, MA, 1983.

[36] Y.K. Lee and I. Verbauwhede, A compact architecture for montgomery elliptic curve scalar multiplication Processor. In *Workshop on Information Security Applications—WISA, LNCS 4867*, Jeju Island, Korea, 2007.

[37] E.D. Mastrovito, VLSI design for multiplication over finite fields $GF(2^m)$, *LNCS 357*, pp. 297–309, Springer-Verlag, Berlin, March 1989.

[38] E.D. Mastrovito, VLSI architectures for computation in Galois fields, PhD thesis, Department of Electrical Engineering Linköping, Linköping University, Sweden, 1991.

[39] A.J. Menezes (Ed.), *Applications of Finite Fields*, Kluwer Academic, Boston-London-Dordrecht, 1993.

[40] M. Morii, M. Kasahara, and D.L. Whiting, Efficient bit-serial multiplication and the discrete-time Wiener-Hopft equation over finite fields, *IEEE Transactions on Information Theory*, 35, 1177–1183, November 1989.

[41] J.L. Massey and J.K. Omura, Apparatus for finite field computation, *U.S. Patent Application*, pp. 21–40, 1984.

[42] P.L. Montgomery, Modular Multiplication without Trial Division, *Mathematics of Computation*, 44, 519–521, 1985.

[43] R. Mullin, I. Onyszchuk, S.A. Vanstone, and R. Wilson, Optimal normal bases in $GF(p^n)$, *Discrete Applied Mathematics*, 22, 149–161, 1989.

[44] M. McLoone and M.J.B. Robshaw, Public key cryptography and RFID tags. In M. Abe, Ed., *Topics in Cryptology—CT-RSA 2007, LNCS*, Springer, 2007.

[45] F.J. MacWilliams and N.J.A. Sloane, *The Theory of Error-Correcting Codes*, North-Holland, New York, 1977.

[46] A.M. Odlyzko, Discrete logarithms in finite fields and their cryptographyc significance, *LNCS 209*, 224–316, Springer, 1984.

[47] J.K. Omura and J.L. Massey, Computational method and apparatus for finite field arithmetic, U.S. Patent Number 4,587,627, May 1986.

[48] E. Özturk, B. Sunar, and E. Savas. Low-power elliptic curve cryptography using scaled modular arithmetic. In M. Joye and J.J. Quisquater, Eds., *Proceedings of 6th International Workshop on Cryptographic Hardware in Embedded Systems (CHES), LNCS 3156*, pp. 92–106. Springer-Verlag, Cambridge, MA, 2004.

[49] C. Paar, Efficient VLSI architectures for bit parallel computation in Galois fields, PhD thesis, Universität GH Essen, Germany, 1994.

[50] S. Peter and P. Langendörfer, An efficient polynomial multiplier in $GF(2^m)$ and its application to ECC designs, *DATE 2007*, pp. 1253–1258, 2007.

[51] A. Reyhani-Masoleh and A. Hasan, Low complexity bit parallel architectures for polynomial basis Multiplication over $GF(2^m)$, *IEEE Transactions on Computers*, 53(8), 945–959, August 2004.

[52] C. Ruland and T. Lohman, Digital signatures based on elliptic curves in RFIDs, *International Journal of Computer Science and Network Security*, 7(1), 275–281, January 2007.

[53] F. Rodríguez-Henríquez, G. Morales-Luna, N. Saqib, and N. Cruz-Cortés, Parallel Itoh-Tsujii multiplicative inversion algorithm for a special class of trinomials, Cryptology ePrint Archive, Report 2006/035, 2006. http://eprint.iacer.org/.

[54] F. Rodríguez-Henríquez, N. Saqib, A. Díaz-Pérez, and Ç.K. Koç, *Cryptographic Algorithms on Reconfigurable Hardware*, Springer, 2006.

[55] K. Sakiyama, L. Batina, N. Mentens, B. Preneel, and I. Verbauwhede, Small-footprint ALU for public-key processors for pervasive security. In *Proceedings of Workshop on RFID Security 2006*, Graz, Austria, p. 12, 2006.

[56] B. Sunar and Ç.K. Koç, Mastrovito multiplier for all trinomials, *IEEE Transactions on Computers*, 48(5), 522–527, May 1999.

[57] L. Song and K.K. Parhi, Low-energy digit-serial/parallel finite field multipliers, *Journal of VLSI Signal Processing*, 19(2), 149–166, 1998.

[58] B. Toiruul and K. Lee, An advanced mutual-authentication algorithm using AES for RFID systems, *International Journal of Computer Science and Network Security*, 6(9B), 156–162, September 2006.

[59] N. Takagi, J.-I. Yoshiki, and K. Takagi, A fast algorithm for multiplicative inversion in $GF(2^m)$ using normal basis, *IEEE Transactions on Computers*, 50(5), 394–398, May 2001.

[60] J. Wolkerstorfer, M. Feldhofer, and S. Dominikus, Strong authentication for RFID systems using the AES algorithm. In M. Joye and J.J. Quisquater, Eds., *Proceedings of 6th International Workshop on Cryptographic Hardware in Embedded Systems (CHES), LNCS 3156*, pp. 357–370. Springer-Verlag, Cambridge, MA, 2004.

[61] C.C. Wang, T.K. Truong, H.M. Shao, L.J. Deutsch, J.K. Omura, and I.S. Reed, VLSI architectures for computing multiplications and inverses in $GF(2^m)$, *IEEE Transactions on Computers*, C-34(8), 709–717, August 1985.

[62] H. Wu, Montgomery multiplier and squarer in $GF(2^m)$, *CHES 2000, LNCS 1965*, pp. 264–276, Worcester, MA, 2000.

[63] Z. Yan and D.V. Sarwate, New systolic architectures for inversion and division in $GF(2^m)$, *IEEE Transactions on Computers*, 52(11), 1514–1519, November 2003.

[64] Z. Yan, D.V. Sarwate, and Z. Liu, High-speed systolic architectures for finite field inversion, *INTEGRATION, the VLSI Journal*, 38, 383–398, 2005.

[65] Z. Yan, D.V. Sarwate, and Z. Liu, Hardware-efficient systolic architectures for inversion in $GF(2^m)$, *IEE Proceedings of Information Security*, 152(1), 31–45, October 2005.

[66] T. Zhang and K.K. Parhi, Systematic design of original and modified mastrovito multipliers for general irreducible polynomials, *IEEE Transactions on Computers*, 50(7), 734–749, July 2001.

24 Designing Secure Wireless Embedded Systems

Ilker Onat and Ali Miri

CONTENTS

Unlike a general-purpose computer, an embedded system is defined as a specialized device performing fewer but well-defined tasks. This chapter analyzes secure design methods for very low-power wireless embedded devices used in radio-frequency identification (RFID) and wireless sensor systems. Although design and operation principles are very simple, widespread use of RFID technology comes with complex security and privacy issues. Sensitive health-monitoring tasks and biometric data use are growing in low-power wireless applications, hence the demand for dependable secure operation is increasing. Active RFID tags and wireless sensor network (WSN) nodes are almost identical in technology yet few studies target the combined general class. The main focus of this work is the common security issues of RFID and wireless sensor systems. Both the individual sensor node and the RFID tag are treated under the main title of "Low-power wireless embedded device."

Operating principles and related protocol standards are summarized along with their different security implications on the system design. An overview of secure hardware and firmware design principles, major attacks, and countermeasures and related cryptographic algorithms are presented.

24.1   INTRODUCTION

The use of embedded systems is a ubiquitous part of daily life, from simple devices such as alarm clocks to more complex gadgets like cell phones. Because of application needs, embedded devices must generally be more reliable than the general-purpose computation systems. Security and reliability are therefore two essential pillars of embedded system design principles. Although generalized under the general title, embedded systems vary greatly in complexity, resources, connectivity, and security requirements. They can be designed to perform a few task occasionally during their whole lifetime with small batteries or they can be real-time systems with intense computational tasks controlling very important tools and processes that require high levels of security. In this work, we will concentrate on the security issues of very low-power and short-range embedded wireless systems, namely, passive and active RFID devices and wireless sensor network (WSN) nodes. The differences between wireless sensor nodes and RFID tags are getting less and less visible. In addition, there are now many implementations that use both technologies complementary to each other. Given the growing concerns of security and dependability, embedded system designers must pay greater attention to security in system design. Generally for the wireless embedded systems, specifically for the RFID systems, security is a major obstacle to the proliferation of the application areas. Wireless communications are inherently more open to abuse and security flaws. Networked or not, security challenges for these low-power wireless devices are more serious because of their resource constraints. These constraints limit the deployment of proven strong security schemes.

Whether they are battery powered as in wireless sensor nodes or active RFID tags, or powered by the reader as in passive RFID tags, energy limitation is an important constraint. Generally, cryptographic operations are computationally intense and consume high amounts of energy and require processors capable of running the firmware implementing these algorithms. The size of volatile and nonvolatile memory needed will increase with increasing cryptographic code size and number of operations. Secure hardware and software design with these constraints is a nontrivial design issue. Growing complexity in software and hardware increases vulnerable points in the system. Growing operating system and firmware size undermines security by increasing programming errors. Software testing cannot fully preview different scenarios of operation even in the moderate complexity systems.

Security requirements of embedded systems with varying degree of complexity are analyzed in Refs. [1–3]. All of these studies target the general embedded systems class. In any system, security is a function of available system resources. Providing security in the light of very limited system resources is the main design challenge for low-power wireless embedded devices which is the topic of this study. Section 24.2 overviews the RFID systems and their security issues. Section 24.3 is a summary of security solutions for WSNs. Security vulnerabilities of general low-power wireless embedded systems class in the light of various attacks and design methods against these attacks are analyzed in Section 24.4. One of the important challenges for embedded system designers is the selection of suitable components for their design. Low-power embedded system component selection principles are explained in Section 24.5. State of the art in the lightweight cryptographic algorithms for low-power embedded wireless systems is overviewed in Section 24.6. A short summary and concluding remarks for the chapter are presented in Section 24.7.

24.2   RFID SYSTEMS

An RFID system consists of two components: a transponder, or tag and a reader, or interrogator. The transponder carries the actual data and is attached to the object to be identified. Operating frequency of an RFID system is the frequency at which the reader transmits. RFID systems are categorized according to fundamental operating principles, tag complexity, operating frequency, range, and powering methods.

24.2.1 OPERATION PRINCIPLES OF RFID SYSTEMS

There are three main physical operation types for RFID tags: inductive coupling, electromagnetic backscatter, and close coupling.

Inductively coupled transponders receive energy from the reader-generated strong electromagnetic field which passes through the transponder's coil area. This electromagnetic field induces a current proportional to its strength (decreasing with distance), the coil area, and the number of windings, providing energy to the transponder. Majority of inductively coupled systems use either 30–300 kHz low-frequency or 3–30 MHz high-frequency ranges. They constitute about 90 percent of today's RFID systems [4]. Their range is less than 1 m. Inductively coupled systems use load modulation to transfer data from the transponder to the reader. A resonant transponder is a transponder with a self-resonant frequency same as the transmission frequency of the reader [4]. In load modulation, a resistor on a resonant transponder switched on and off according to data which effects the voltage across the reader and transmits data.

Electromagnetic backscatter transponders reflect back the electromagnetic waves created by the reader. The radiation power of the waves decreases, with the square of the distance from the source; therefore, a much weaker signal is returned to the reader by the passive backscatter transponder. Increasing frequency increases reflectivity; hence, this transponders use ultrahigh-frequency (UHF) range at 900 MHz or 2.4 GHz. Short wavelengths at these frequencies enable the construction of smaller antennas than the inductively coupled system coils. The UHF backscatter systems are also called long-range systems because they can transmit at up to 5 m apart from the reader.

Close coupling systems are powered through the magnetic field generated on the transponder coil when it is placed between the two windings of the reader carrying high-frequency alternating current. The transponder can be coupled up to 1 cm away from the reader. The frequency used is usually less than 30 MHz. Close coupling is used in contactless smart cards common in secure identification systems.

Because of the very limited useful energy that can be converted and used at the passive transponders, they can send information only to very limited distances. Active transponders, on the other hand, are very similar to WSN nodes with their activation methods being the only functional difference. A WSN node generally uses duty cycling (put to sleep and waken up periodically to check for the incoming message) whereas an active RFID tag is activated the same way as passive tags. Because there is no duty cycling, active tags have longer lifetimes compared to wireless sensor nodes. Active backscatter transponders have significantly higher ranges but their use is limited because of the maintenance and cost issues associated with battery use. The other transponder type is the semipassive transponders which use battery to retain memory contents or to do data processing; their radio functions supplied power through the reader as in passive readers.

24.2.2 SECURITY OF RFID SYSTEMS

Security and privacy breaches such as unauthorized reading and writing of the RFID tags can cause a wide range of problems from small monetary losses for the retailers to major disruptions in the supply chain. In many cases, the data contained in an RFID tag should be protected against unauthorized reading. This privacy problem is especially important for the RFID systems storing biometric data for authentication such as RFID enabled passports. With the growing number of such sensitive fields, security issues surrounding the RFID technology have become a major concern. RFID systems' security also depend on the bigger picture. Other components of the whole system, database, database management system, middleware have to be designed securely. The information derived from the tags has to be safely processed and transferred to processing centers. However, in this study we concentrate only on the embedded device-level security issues.

Different RFID technologies come with different problems and dictate different solutions. Device capabilities and available energy determine the use of security algorithms. In general, when security is concerned, the generalizations should clearly specify the target subarea. As a general rule, short-range systems can provide more power to transponders; hence, complicated security algorithms can be implemented. Low-power usage at the transponder increases the range of passive RFID systems. Higher transmit power and increased antenna gain are other factors affecting the range of an RFID system. If an RFID system has a higher range, it can be powered from far, thus can be read from far. The problem is also partly due to the fact that in long-range backscatter systems the power at the tag is very limited and it is a challenge to use full-scale cryptographic primitives on a general-purpose microcontroller powered by RF energy. Authentication and encryption operations, even in their simplest versions, require the capabilities of a microcontroller; hence, when such algorithms are implemented at a tag, enough power has to be transmitted to power the microcontroller running these algorithms.

It is not possible to talk about an operating system or a security software in RFID systems. Both devices merely contain communications and encryption units only. However, even bare-bone cryptography can be a challenge for RFID devices. Mainly because of the technology-oriented power limitation and application-oriented cost constraints, RFID tags usually contain a few thousand logic gates as computation elements. Especially in the case of UHF backscatter tags, until recently, it was not possible to equip them even with very low-power microcontrollers that will perform cryptographic operations. Many lightweight and minimalist protocols are proposed for this resource-constrained environment [5–8]. Recent studies such as Ref. [9] demonstrated the use of conventional strong cryptographic algorithms on a UHF RFID tag. Ultimately, the use of microcontrollers in the UHF tags is an efficient rectifier design problem that is able to collect and supply enough power to tag's microelectronic circuit and the ultralow-power microcontroller. All RFID devices contain nonvolatile memory in various forms. The unique identifiers of RFID tags are stored using nonvolatile memory. In addition, all the embedded code including network stack and security algorithms also have to be stored in the nonvolatile memory. Therefore, elaborate security algorithms increase the size of nonvolatile memory. Security solutions also require volatile memory tied to processor usage.

In Ref. [10], authors explain and demonstrate the possibility of application layer malware for RFID systems that was thought to threaten only more capable higher level systems. RFID viruses and worms, and other attacks such as sniffing, tracking, spoofing, replay, and denial-of-service create important vulnerabilities for RFID systems.

24.3 SECURITY OF WIRELESS SENSOR NETWORKS

IEEE 802.15.4 [11] is the standard for low-rate, low-power wireless personal area networks (WPAN) that define the physical and the medium access control layers. It supports 900 MHz and 2.4 GHz bands with data rates up to 250 kbps. ZigBee specification which is developed by the ZigBee industry alliance builds network and application layers on top of the 802.15.4 specified lower layers. 802.15.4 uses advanced encryption standard (AES) [12] in different modes to provide confidentiality and integrity. Following is a summary of the modes of this security suite:

- AES-CTR (AES-counter) mode: encryption
- AES-MIC (AES-message integrity code) mode: integrity (authentication)
- AES-CCM (AES-CTR with CBC-MAC) mode: encryption and integrity

Encryption and decryption are done by XORing data with a key stream produced from a secret key and a counter which is incremented for each data block. Integrity is achieved through cipher block chaining message authentication code (CBC-MAC) generated message integrity codes (MICs), thereby also providing authenticity. Each plaintext block is XORed with the previous ciphertext block before encryption and the last ciphertext is used as the MIC.

AES has a very high theoretical security and is proven to be a good overall design for various architectures. In Ref. [13], software implementation of various block ciphers are tested for their performance on an ultralow-power microcontroller. This study reveals that the overhead induced by AES may be high compared to other block ciphers in some microcontroller architectures. However, as the hardware implementation of AES gaining popularity and it is a standard requirement for 802.15.4 compatible transceivers, AES will remain as the dominant security mechanism for WSN nodes for years to come. In an AES enabled WSN, attacks will have to find vulnerabilities in the embedded implementation of the algorithm. In this study, we will concentrate on the attacks against a WSN node in the form of attacks against AES implemented in a wireless embedded system.

24.4 VULNERABILITIES OF WIRELESS EMBEDDED SYSTEMS AND DEFENSES

Small, low-power wireless devices are now an ubiquitous part of our daily lives. Almost all of these devices require some degree of security and privacy. Connectivity, especially in the wireless form, exposes embedded systems to intrusions and attacks. The advances in microelectronics and integrated circuit (IC) technology increase the capacities of these device. These increased capabilities also facilitated the implementation of powerful embedded security algorithms. However, these systems still carry vulnerabilities that can be exploited. Recent years have brought the possibility of exploiting these vulnerabilities through side-channels. Independent from the theoretical strength of the security algorithm, various types of side-channel information, not traditionally considered during algorithm designs, are now becoming weak links in the secure system design process. In this section, we will first give a summary of side-channel attacks in general, along with the characteristics and examples of such attacks against low-power wireless systems. Secure hardware and software design principles will be detailed next.

24.4.1 Side-Channel Attacks

Today's most common IC technology is the complementary metal-oxide-semiconductor (CMOS) technology. Without special precautions that will be discussed here, CMOS devices leak information about the operations they are performing. The attacks performed by exploiting this side-channel information can target a wide range of embedded processors performing cryptographic operations. A survey of side-channel attacks against smart cards is given in Ref. [14]. As discussed previously, IEEE 802.15.4 standard uses AES as the link layer encryption algorithm. Various studies [15–17] pointed out the vulnerabilities of AES against side-channel attacks. Sensor radios equipped with AES coprocessors are therefore vulnerable to side-channel attacks. Recently, side-channel vulnerabilities of RFID systems are pointed out in Ref. [18]. Side-channel attacks against low-power, low-cost devices are in general much more effective because the constraints prohibit the implementation of strong countermeasures. Still, there are various techniques to prevent such attacks that can be implemented in these devices, that are explained in the following sections.

24.4.1.1 Timing Attacks

Timing attacks use the precise time measurements from the specific implementations of cryptographic algorithms [19]. In many cases of the algorithm implementations, particular inputs use different amount of processing time that can be measured by the attacker and used to guess the cryptographic parameters used in the algorithm. Defenses against timing attacks include obfuscation of the variable execution time by equalizing or randomizing each algorithm run. These techniques are also called blinding techniques and they target the elimination of correlations between the key and the encryption time [19–21]. There are three main blinding techniques: randomization-based, equalization-based, and quantization-based. In randomization-based blinding, the defense includes adding a random number to be exponentiated with the encryption exponent before the ciphertext decryption. In equalization-based blinding, dummy operations are added to make the same operations equal time independent from the ciphertexts or other related parameters. In quantization-based blinding, all computations take a multiple of some predefined time quantum. In this technique, unless all decryptions last as long as the maximum decryption time, secret key information may still leak. For the present time, the most effective method for protecting against timing attacks is to use randomization-based blinding [21].

24.4.1.2 Power Analysis Attacks

These attacks collect and analyze the power consumption of the processors during cryptographic operations. Simple power analysis (SPA) attacks [22] measure the power consumption of the processor directly. Cryptographic operations usually require intense computation activity that can be observed by measuring the chip's power consumption. Crypto rounds and exponentiations can be differentiated; further analysis may also pinpoint instructions, multiplication, and squaring operations. Conditional branching can also be easily recognized by SPA. To protect against SPA, power consumption of different branches should be equalized. Other protection methods are software based and require code modification to make power consumption as uniform as possible at different branches. Crypto coprocessors are more resistant to SPA because of their rather uniform power consumption during different branches.

Differential power analysis (DPA) attacks [23] use the statistical differences between power consumption of the same hardware and algorithm section while processing different parameters. In DPA attacks, signal processing and error correction techniques are applied off-line over the collected data to obtain information about secret keys. Because DPA analysis is much more elaborate than SPA it is more difficult to prevent.

Various protection methods against DPA are classified in Ref. [24]. Simple balancing solutions are those that eliminate differences among the different events. These solutions are vulnerable against larger data sets and higher resolutions [24]. Other algorithmic solutions include randomly masking of the data before the encryption. However, such specific measures do not lay down a general methodology and they have to be adapted to each individual process.

Observable power dissipation in CMOS transistor circuits occurs mainly due to the switching activity produced by a change of input data. Therefore, CMOS circuits should switch independent of the processed data, or dissipate constant power. Hardware solutions such as Refs. [25–27] aim at not creating side-channel information by balancing power consumption at the gate level. Automatic routing and placement can introduce errors through various optimizations that create imbalances, for example, in wire lengths, producing power leakage. At the CMOS level, equalizing the capacitances across different gates by avoiding transistor size optimization is also useful in preventing the leakage [28]. In Ref. [29], an automated IC design flow aiming a constant power dissipation is introduced. Reference [30] explains a formal method for quantification and verification of hardware against side-channel vulnerabilities.

Recent successful power attacks revealed the vulnerabilities of various RFID system types. Reference [18] targets passive UHF backscatter tags and details a method of measuring the power consumed by a tag during computations. Different from a standard power analysis attack, this attack requires no physical contact with the tag. Different variations of the attack are described and the possibility of carrying out the attack even with a completely passive attacker is explained, which makes this type of attack very practical and hard to detect. The implications of this vulnerability is the requirement to build tags with cryptographic properties resistant to such power analysis attacks, which increases their cost. The attack described in Ref. [31] targets 13.56 MHz RFID devices equipped with AES coprocessors.

24.4.1.3 Electromagnetic Analysis Attacks

All electronic devices radiate electromagnetic waves directly dependent on their current usage. This radiation generally contains two main spectral elements: narrowband signals produced by the main clock, and broadband signals due to various sources of circuit elements and board layout, or noise. Narrowband clock-dependent emissions are more powerful than the wideband noise, and therefore they can easily be detected. Electromagnetic analysis (EMA) attacks are based on the detection and analysis of these narrowband clock-dependent emissions of processors. They are first introduced in Ref. [32]. In general, precise spectral signature is architecture dependent; however, there are common emission patterns among processors. When a coil is brought on top of the device under attack, the local signals on the chip can be detected which are later correlated with the power consumption.

A simple but effective defense against EMA attacks is the use of strong ferromagnetic metals in and around the processors that will absorb the emissions. Decreasing power consumption of the processors is another prevention mechanism. Low-power consuming designs radiate relatively less electromagnetic energy; therefore, they are naturally more resilient against EMA attacks. The use of asynchronous processors and blinding by current consumption equalization in hardware and software are other common methods. If each instruction has a specific electromagnetic signature, dummy instructions with reverse electromagnetic signature can be executed in parallel [32].

24.4.1.4 Fault Injection Attacks

A fault attack uses physical perturbations of system externals to cause transient or permanent faults. These faults can then be exploited to control or gain knowledge about the system under attack. Faults can be injected by radiation, temperature, or supply voltage variations; however, even random faults can also be used to obtain secrets. The method is first applied to factor RSA modulus in Ref. [33] then extended to other cryptosystems in Ref. [34]. Fault attacks may disrupt the availability of the subsystems, or may modify memory contents to gain control of the embedded system as shown in Ref. [35].

24.4.2 PHYSICAL SECURITY AND TAMPER RESISTANCE

Physical security measures protect a system against physical incursions. Low-power wireless embedded systems operate unattended at environmentally unsecure places; therefore, their physical security is as important as the computational security. Effective physical security decreases the probability information extraction even in the case of successful and noninvasive penetration. Many side-channel attacks described above can be classified as probe attacks. These attacks obtain information by observing or changing the inner workings of the attacked device to gather secrets. Probe attacks are classified as active when the attacker injects signals or information into a working system [1]. By changing the environmental conditions such as voltage or operating frequency, the contents of volatile and nonvolatile memory devices can be revealed [36]. If the probing points are not externally available, various machining methods can be used to cut and remove the protective layering. Manual, mechanical, water, laser, or chemical machining are major uncoating methods. Imprinting attacks try to obtain the information contained in the device by imprinting the CMOS transistor states. Radiation, temperature, and high-voltage imprinting are major methods. Imaging attacks use many imaging technologies such as x-ray, tomography, ultrasound, or electron microscope to analyze the static contents or dynamic state at the printed circuit board or IC level.

Defenses against physical attacks aim to prevent the reach of the attacker in the system. Tamper resistance measures try to prevent the attacker from reaching the secrets or inner working details of a system. Hard barriers are built from durable strong material hard to penetrate without destroying the device under attack. Chip coatings is a chemical method used to prevent probing attacks. Some coatings are designed such that their removal permanently destroys the device. Insulation methods use layers of insulators against imprinting or imaging attacks. Tamper resistance by itself is not a sufficient property. Tamper evident systems have the design feature of revealing whether or not a device is operated upon physically, even if the attacker performed a sophisticated operation without leaving much evidence. Most tamper evidence measures use sensitive materials that reveal tampering attempts. *Tamper response* measures are provisioned for alerting other protection subsystems against intrusions.

Unlike secure smartcards, most RFID tags are not tamper-resistant, though physical security is important for many RFID applications. The size and weight of a tamper resistance barrier is very important for light RFID tags. Various commercial solutions are proposed and currently on the market. Reference [37] is an example of a tamper-proof RFID tag that destroys itself when a tampering attempt is detected. It thereby prevents the use of the tags on counterfeit or substitute products. Other configurations include the continuation of the functioning but alerting the reader about the tampering attempt. Tamper resistance measures should be used as an additional security layer on top of other protection measures. Their use should be tailored according to application security needs, cost, and size of the embedded system.

24.4.3 SECURE FIRMWARE DESIGN AND MEMORY PROTECTION

Hardware and firmware are two major building blocks of an embedded system. Incorporation of secure design principles in firmware development process is necessary for a secure embedded system. Main phases of a firmware development life cycle are the requirements analysis, design, implementation, and testing. Principles of designing secure software are detailed in Saltzer's early work [38]. Following is the subset of these mechanisms adapted to our target environment.

Small, simple, and effective design: Security should be implemented and verified as simple as possible and it should be small in size compared to whole firmware size. Growing complexity generally increases the vulnerable points and probability of implementation errors. In low-power, low-cost wireless systems, this criterion is especially crucial.

Open design: The security of the algorithms should come from their tested strength instead of obscurity of code. Obfuscation-based protection should be avoided. Reviews by the research community is essential while selecting these algorithms.

A multilevel design validation strategy for secure firmware design is proposed in Ref. [39]. Most embedded systems are programmed with C/C++ language and therefore are vulnerable against buffer overflow attacks. An example buffer overflow attack for RFID systems is given in Ref. [10]. Various hardware and software protection and detection methods can be used against these attacks [40,41]. Adding instructions to check array bounds, pointer checks at runtime, pointer encryption, guarding the stack with hardware measures, and input bounding are some of them. The overheads introduced by such protection methods are major implementation problems in resource-constrained environments.

Firmware should also be protected against reverse engineering. If the executable code has been extracted from the embedded device, various powerful automated reverse engineering tools can be used to disassemble the corresponding source code. Reverse engineering generally involves two main steps [42]: disassembling machine code into assembly code and decompilation of the assembly code into higher level language. The protection methods in general aim to obfuscate one of these two main steps. An example of disassembly phase obfuscation with the insertion of dummy binary code is given in Ref. [42]. This method makes parsing more difficult by confusing disassemblers. Reference [43] details a method to hide decompilation phase by indirect memory references via obfuscated pointers. Reverse engineering gives attackers additional details about the inner workings of the embedded system to be exploited. In general, it is very difficult to protect against reverse engineering in architectures with fixed length instructions whereas much easier in variable length instruction architectures. New code retrieval attacks are developed [44] along with the new protection methods. The code protection should be done without impacting runtime performance or incrementing the binary size. For the RFID devices, a reverse engineering attack is detailed in Ref. [45].

In addition to secure firmware design, both volatile and nonvolatile memory devices should be protected against code retrieval attacks. The best solution to protect the memory against tampering attacks is utilizing tamper sensors and quickly and completely erasing the memory contents. Other security solutions can involve the use of secure memory management functions such as strong access control and antitearing algorithms as used in secure chip design of Refs. [46,47].

For higher level systems that use real-time operating systems, secure embedded design standards such as multiple independent levels of security (MILS) [48] divide the memory space into secure and nonsecure parts. Many real-time operating systems follow this standard in securing their code space. For WSN node design, the sensitive parts of the memory contents such as node ID, network key, or bootloader authentication key should be secured in the protected area against tampering for cases where a node is captured.

24.5 COMPONENT SELECTION FOR SECURE WIRELESS EMBEDDED SYSTEMS

The security of an embedded system as a whole is strongly dependent upon its subcomponents; hence, it is important to select proper building blocks. Main building blocks of an embedded system are various generic or Intellectual Property (IP)-based IC subsystems, mostly designed by other companies. Therefore, the hardware and software security vulnerabilities of these components have direct effects on the whole embedded system security. Here we give a general characteristics of the components that are vital for the security of low-power wireless embedded systems.

24.5.1 SECURE EMBEDDED MICROCONTROLLERS

Other than very simple RFID systems that operate with a few thousand logic gates, most wireless embedded systems contain a general-purpose microcontroller as the main computational element.

Many protection schemes such as voltage sensors, clock speed sensors, light sensors, metal layers, bus encryption, or password protection are used in high-end processors [49]. For secure computation in general-purpose processors operating at resource-constrained environments, dedicated hardware is the preferred solution because it can achieve better performance with lower energy consumption. Hence, at the low-end market, cryptoprocessors are generally implemented as crypto coprocessors in microcontrollers. Crypto coprocessors and pseudorandom number generators decrease the power consumption and increase the speed of cryptographic operations. An embedded crypto coprocessor is a tamper resistant processor that performs requested cryptographic operations using protected secrets. They are used in various applications requiring different levels of security. They can also support key generation on board. Unlike smart cards which are specifically designed to protect against most known attacks, the majority of low-cost, low-power microcontrollers offer only weak protection [31,45]. However, low-cost microcontroller designs are increasingly adopting secure design features from high-end secure microcontrollers.

Low-end crypto coprocessor are increasingly becoming a part of our everyday lives through low-power wireless devices and smart cards. AES crypto coprocessors are now implemented in all IEEE 802.15.4 compatible low-power sensor transceivers [50–52], various RFID devices, and smart cards [53–55]. These coprocessors are also used for digital signature and identification and can implement complex authentication and encryption schemes such as RSA and elliptic curve cryptography (ECC) [56]. A processor for low-power application should have low energy consumption and should fit in small chip area. It should be low cost and at the same time be robust against attacks. A processor also needs flexibility to changing program size in terms of volatile and nonvolatile memory capacity.

24.5.2 MICROCONTROLLER SELECTION FOR WSN NODES AND ACTIVE RFID TAGS

The microcontrollers used in active wireless devices should also have additional enhanced features. Because long device lifetime is the most important design goal, consuming low-power and providing various power-down modes is the most important feature of a processor that is used in such applications. There should also be enough memory space for network stack and security algorithms. The processor should also be low-cost, fast enough to execute time-critical tasks with a wide operating voltage range.

24.5.2.1 Duty Cycling Operation

To save energy, the low-power transceiver and processor of a battery powered sensor node have to be put into sleep (power-down) mode. The nodes wake up periodically and check the air for preamble. The periodic wakeup of the radio is the task of the processor. The periodic duty cycling may be enforced synchronously in the subnet, in which case precise timing of the wakeups is necessary. To provide such timing, a sleeping processor will need to keep its low-frequency clock (oscillator) running continuously. Therefore, low-jitter, low-frequency internal clock feature of a processor is essential. When a preamble is detected during the wakeup, a node wakes up with all peripherals and prepares to process the incoming packet stream. One of the main design decision in sensor network deployments is to build an efficient duty cycling mechanism that will provide maximum system lifetime while maintaining application requirements. In this context, the proper use of low-power modes and the programming API is also critical.

24.5.2.2 External Interrupts

If there is no duty cycling, external interrupts are used to wake up the node from the sleep state. The powered-down processor in this case is waken up with a general-purpose pin interrupt. Low voltage level interrupts capability and fast wakeup from the power-down modes is an important feature for microcontrollers that will be used in WSN nodes. Fast wakeups also decrease current consumption [57].

24.5.3 Radio Selection

Passive RFID circuits are simple transceivers that operate at short ranges. Depending on the applications, cost, frequency band, size, and form should be decided. Some applications may require higher reliability and faster response. The design should be modified according to application requirements and the budget. Unlike the rather simple transceiver circuitry of the passive tags, active tags and sensor nodes use full-featured low-power radios. These transceivers have to satisfy the following criteria:

Power consumption: A transceiver operates in four major states: sleeping, listening, receiving, and transmitting. In low-power WSN or active RFID transceivers, listening, receiving, and transmitting modes consume similar amount of power, on the order of 10–15 mA. In the IEEE 802.15.4 standard, transceivers spend most of their operational life in a sleep state; each device periodically wakes up and listens to the RF channel to determine whether a message is pending. This is done through the detection of a preamble. Because listening mode is as power consuming as transmitting and receiving, and it occurs periodically according to duty cycle, it is the major energy drain for duty cycled embedded designs. This is also referred to as idle listening.

Fast wakeup: Whether duty cycled or not, transceivers should leave sleep state as fast as possible to listening state so as not to miss further from the preamble.

Output power: High output power improves connectivity, decreases error rate but also drains the batteries faster, and may expose the network to outside. High output power also increases interference to other close frequency devices. According to application needs, a suitable output power should be selected. Programmable output power feature of a transceiver can be used to adapt the radio to different network and connectivity setups either statically before the deployment or dynamically during the operations.

Protocol stack and standards: A protocol stack consists of layers of abstraction. Each layer provides well-defined interfaces to neighboring layers and uses interfaces provided by them. Basically, a network protocol stack consists of four main layers: physical, medium access control (MAC), network layer, and application. If an IEEE 802.15.4 compliant radio is selected, physical and MAC layers embedded in the radio firmware can be readily used instead of developing, for example, an in-house collision avoidance scheme. Configuration options can be used for creating different, standard topologies. In general, the abstraction IEEE 802.15.4 provides in addressing, framing, channel access, acknowledgements and security is valuable for flexible designs. Standards also enforce specific data rates, frequency, and modulation schemes that increase interoperability of devices from different vendors.

24.6 LIGHTWEIGHT AUTHENTICATION AND ENCRYPTION ALGORITHMS

Many security algorithms designed for resourceful general-purpose computing devices cannot be used in small, battery powered devices. In this section, we give design principles and several examples of lightweight and efficient cryptographic algorithms. Most of these algorithms are the adaptations of strong cryptographic algorithms to resource-constrained wireless embedded systems. There are also brand new algorithms that are designed from scratch targeting only these systems. Design decisions about the algorithms are based on the trade-offs between cost, performance, resource consumption, and security.

Most of the lightweight applications trade theoretical security and speed with resource use. With decreasing resources, theoretical security of the algorithms decreases. Although determined attackers having time and computational resources can finally break some of these algorithms, they are in many cases sufficient deterrents and efficient protection mechanisms in the context of the low-power and low-cost wireless device being protected.

When implemented in hardware, in addition to high theoretical security, a lightweight algorithm should consume little power, should be implemented with small number of gates, and should be fast. For software implementations, small firmware size, small RAM requirement, and small number of clock cycles are the design goals. Symmetric and asymmetric algorithms have different uses. In general, public-key-based (asymmetric) algorithms consume 100 to 1000 times more clock cycles [58].

24.6.1 SYMMETRIC CIPHERS

Many symmetric ciphers are adapted. Reference [54] is a novel AES hardware implementation which encrypts a 128-bit block of data within 1000 clock cycles and with very low power consumption. The design targets 13.56 MHz RFID tags, but can be modified for other RFID standards. Reference [55] introduces another AES hardware core for 8-bit architecture which uses 128-bit keys. Compared to previous 8-bit implementations, this design provides a significantly higher throughput with corresponding area. Eight bit designs are important because 8-bit microcontrollers hold a major share of the embedded microcontroller market and they are dominant in the smart card industry. Reference [59] is a similar efficient core design for IEEE 802.15.4 WSN nodes. HIGHT [60] is a new block cipher with low-resource usage. It is specifically designed for WSN nodes and RFID tags. HIGHT requires approximately the same number of gates as AES but it is faster. Reference [61] is an efficient low-area hardware implementation of DES for RFID systems named DES lightweight extension (DESL). DESL uses a single S-box repeated eight times which decreases the area requirements significantly. The implementation is also efficient in number of clock cycles.

Very small-area stream ciphers can be alternatives to block ciphers in resource constrained environments. There are various efficient stream chipper designs. Reference [62] is a hardware implementation optimized block cipher-like stream cipher which can be implemented in small area. Reference [63] is a small-area hardware design with small gate count and power use that is suitable for RFID tags.

24.6.2 Asymmetric Ciphers

It is quite a challenge to implement public-key algorithms in resource-constrained environments. Because of their efficiency, ECC-based schemes are becoming popular for various low-power architectures. ECC uses smaller key lengths than other public-key algorithms to achieve the same given security level. It is based on the discrete logarithm problem over the points on an elliptic curve and it can be used to provide digital signatures and encryption. Because they are less resource constrained, WSNs can be more easily equipped with public-key schemes. TinyECC [64] is an ECC software library that provides configurable public-key system for WSNs. Reference [65] is a hardware/software codesign solution for RSA and ECC over GF(p) on an 8-bit microcontroller. Reference [66] presents another ECC implementation for 8-bit microcontrollers. Reference [56] discusses the use and feasibility of public-key algorithms for RFID systems and concludes that ECC-based asymmetric schemes can be efficiently run on RFID systems.

24.7 CONCLUSION

Designing secure embedded systems is a challenge in resource-constrained environments. Still it is possible to design dependable secure embedded systems if the various aspects of the design are in line with the secure design principles for embedded systems. In this study, we analyzed the adaptation of the general design rules to resource-constrained environments. We concentrated our analysis on the low-power, low-rate, short-range embedded systems. We also analyzed various specific novel techniques and algorithms that are developed against powerful attacks.

We studied the secure system design from both hardware and software perspective. Secure design today requires more than ever a holistic approach combining all abstraction layers around the same goal: system and communication security. Because of this, both software and hardware designs are increasingly becoming interrelated and dependent on each other. Ultimately, secure system design is a trade-off between capital resources to be invested for security and the actual value of the systems being protected. With the widespread use of low-power wireless embedded systems in critical applications, security vulnerabilities are no longer trivial secondary concerns but major design challenges for both hardware and firmware designers.

REFERENCES

[1] S.H. Weingart, Physical security devices for computer subsystems: A survey of attacks and defences, in *Proceedings of the Second International Workshop on Cryptographic Hardware and Embedded Systems, CHES'00*, pp. 302–317, 2000.

[2] S. Ravi, A. Raghunathan, P. Kocher, and S. Hattangady, Security in embedded systems: Design challenges, *Transactions on Embedded Computing Systems*, 3(3), pp. 461–491, 2004.

[3] P. Koopman, Embedded system security, *Computer*, 37(7), pp. 95–97, 2004.

[4] K. Finkenzeller, *RFID Handbook: Fundamentals and Applications in Contactless Smart Cards and Identification*. John Wiley & Sons, Inc., New York, 2003.

[5] P. Peris-Lopez, J.C. Hernandez-Castro, J.M. Estevez-Tapiador, and A. Ribagorda, EMAP: An efficient mutual authentication protocol for low-cost RFID tags, in *Proceedings of the OTM Federated Conferences and Workshop: IS Workshop, IS '06*, vol. 4277, pp. 352–361, November 2006.

[6] P. Peris-Lopez, J.C. Hernandez-Castro, J. Estevez-Tapiador, and A. Ribagorda, M2AP: A minimalist mutual-authentication protocol for low-cost RFID tags, in *Proceedings of the International Conference on Ubiquitous Intelligence and Computing, UIC 06*, vol. 4159, pp. 912–923, September 2006.

[7] C. Benoit, S. Canard, M. Girault, and H. Sibert, Low-cost cryptography for privacy in RFID systems, in *Proceedings of the International Conference on Smart Card Research and Advanced Applications, CARDIS '06*, 2006.

[8] Y. Cui, K. Kobara, K. Matsuura, and H. Imai, Lightweight asymmetric privacy-preserving authentication protocols secure against active attack, in *Proceedings of the International Workshop on Pervasive Computing and Communication Security, PerSec '07*, pp. 223–228, March 2007.

[9] H. Chae, D. Yeager, J. Smith, and K. Fu, Maximalist cryptography and computation on the WISP UHF RFID tag, in *Proceedings of the Conference on RFID Security*, July 2007.

[10] M.R. Rieback, B. Crispo, P.N.D. Simpson, and A.S. Tanenbaum, RFID malware: Design principles and examples, *Pervasive and Mobile Computing (PMC) Journal*, 2, 405–426, 2006.

[11] Wireless Medium Access Control (MAC) and Physical Layer (PHY) specifications for Low Rate Wireless Personal Area Networks (LR-WPANs), http://standards.ieee.org/ getieee802/download/802.15.4-2003.pdf.

[12] J. Daemen and V. Rijmen, *The Design of Rijndael: AES—The Advanced Encryption Standard*, 2002.

[13] Y.W. Law, J. Doumen, and P. Hartel, Survey and benchmark of block ciphers for wireless sensor networks, *ACM Transactions on Sensor Networks*, 2(1), 65–93, 2006.

[14] E. Hess, N. Janssen, B. Meyer, and T. Schuetze, Information leakage attacks against smart card implementations of cryptographic algorithms and countermeasures—A survey, in *the Proceedings of the Eurosmart Security Conference*, pp. 55–64, June 2000.

[15] S.B. Örs, F.Gürkaynak, E. Oswald, and B. Preneel, Power-analysis attack on an ASIC AES implementation, in *ITCC '04: Proceedings of the International Conference on Information Technology: Coding and Computing*, vol. 2, p. 546, 2004.

[16] K. Schramm, G. Leander, P. Felke, and C. Paar, A collision-attack on AES combining side channel- and differential-attack, in *Proceedings of the CHES '04*, pp. 163–175, 2004.

[17] S. Mangard and K. Schramm, Pinpointing the side-channel leakage of masked AES hardware implementations, in *Proceedings of the CHES '06*, pp. 76–90, 2006.

[18] Y. Oren and A. Shamir, Remote password extraction from RFID tags, *IEEE Transactions Computer*, 56(9), 1292–1296, 2007.

[19] P.C. Kocher, Timing attacks on implementations of Diffie-Hellman, RSA, DSS, and other systems, in *CRYPTO '96: Proceedings of the 16th Annual International Cryptology Conference on Advances in Cryptology*, pp. 104–113, 1996.

[20] W. Schindler, A timing attack against RSA with the Chinese Remainder Theorem, in *Proceedings of the Second International Workshop on Cryptographic Hardware and Embedded Systems, CHES '00*, pp. 109–124, 2000.

[21] D. Brumley and D. Boneh, Remote timing attacks are practical, in *Proceedings of the 12th Conference on USENIX Security Symposium, SSYM '03*, 2003.

[22] P. Kocher, J. Jaffe, and B. Jun, Introduction to differential power analysis and related attacks, tech. rep., Cryptography Research Inc., 1998.

[23] P.C. Kocher, J. Jaffe, and B. Jun, Differential power analysis, in *CRYPTO '99: Proceedings of the 19th Annual International Cryptology Conference on Advances in Cryptology*, pp. 388–397, 1999.

[24] S. Chari, C.S. Jutla, J.R. Rao, and P. Rohatgi, Towards sound approaches to counteract power-analysis attacks, in *CRYPTO '99: Proceedings of the 19th Annual International Cryptology Conference on Advances in Cryptology*, pp. 398–412, 1999.

[25] K. Tiri and I. Verbauwhede, Securing encryption algorithms against DPA at the logic level: Next generation smart card technology, in *Proceedings of the Cryptographic Hardware and Embedded Systems, CHES '03*, vol. 2779, pp. 125–136, 2003.

[26] S. Moore, R. Anderson, R. Mullins, and G. Taylor, Balanced selfchecking asynchronous logic for smart card applications, *Journal of Microprocessors and Microsystems*, 27(9), 421–430, 2003.

[27] K. Tiri and I. Verbauwhede, A logic level design methodology for a secure DPA Resistant ASIC or FPGA implementation, in *DATE '04: Proceedings of the Conference on Design, Automation and Test in Europe*, IEEE Computer Society, 2004.

[28] J. Fournier, S. Moore, H. Li, R. Mullins, and G. Taylor, Security evaluation of asynchronous circuits, in *Proceedings of the Cryptographic Hardware and Embedded Systems, CHES '03*, vol. 2779, pp. 137–151, 2003.

[29] K. Tiri and I. Verbauwhede, A digital design flow for secure integrated circuits, *IEEE Transactions on Computer-Aided Design of Integrated Circuits and Systems*, 25(7), 1197–1208, 2006.

[30] A.E. Cohen and K.K. Parhi, Side channel resistance quantification and verification, in *Proceedings of the IEEE International Conference on Electro/Information Technology*, pp. 130–134, 2007.

[31] M. Hutter, S. Mangard, and M. Feldhofer, Power and EM attacks on passive 13.56 MHz RFID devices, in *Proceedings of the Workshop on Cryptographic Hardware and Embedded Systems, CHES '07*, vol. 4727, pp. 320–333, 2007.

[32] J.-J. Quisquater and D. Samyde, ElectroMagnetic Analysis (EMA): Measures and counter-measures for smart cards, in *E-SMART '01: Proceedings of the International Conference on Research in Smart Cards*, pp. 200–210, 2001.

[33] D. Boneh, R. DeMillo, and R. Lipton, On the importance of checking cryptographic protocols for faults, in *Proceedings of Eurocrypt97*, pp. 37–51, 1997.

[34] E. Biham and A. Shamir, Differential fault analysis of secret key cryptosystems, in *CRYPTO '97: Proceedings of the 17th Annual International Cryptology Conference on Advances in Cryptology*, pp. 513–525, 1997.

[35] S. Govindavajhala and A. Appel, Using memory errors to attack a virtual machine, in *the Proceedings of IEEE Symposium on Security and Privacy*, pp. 154–165, 2003.

[36] R. Anderson and M. Kuhn, Tamper resistance—A cautionary note, in *the Proceedings of the Second Usenix Workshop on Electronic Commerce*, pp. 1–11, Nov. 1996.

[37] MIKOH, *Smart & Secure RFID tags*. http://www.mikoh.com/smartsecure.asp.

[38] J. Saltzer and M. Schroeder, The protection of information in computer systems. http://web.mit.edu/Saltzer/www/publications/protection.

[39] P. Schaumont, D. Hwang, S. Yang, and I. Verbauwhede, Multilevel design validation in a secure embedded system, *IEEE Transactions on Computers*, 55(11), 2006.

[40] Z. Shao, C. Xue, Q. Zhuge, M. Qiu, B. Xiao, and E.H.M. Sha, Security protection and checking for embedded system integration against buffer overflow attacks via hardware/software, *IEEE Transactions on Computers*, 55(4), 443–453, 2006.

[41] M.G. Grasser, Embedded security solution for digital safe-guard ecosystems, in *Proceedings of the Digital EcoSystems and Technologies Conference, DEST '07*, pp. 529–534, 2007.

[42] C. Linn and S. Debray, Obfuscation of executable code to improve resistance to static disassembly, in *CCS '03: Proceedings of the 10th ACM conference on Computer and Communications Security*, pp. 290–299, 2003.

[43] T. Ogiso, Y. Sakabe, M. Soshi, and A. Miyaji, Software obfuscation on a theoretical basis and its implementation, *IEICE Transactions on Fundamentals of Electronics, Communications and Computer Sciences*, E86-A(1), 176–186, 2005.

[44] C. Kruegel, W. Robertson, F. Valeur, and G. Vigna, Static disassembly of obfuscated binaries, in *the Proceedings of the 13th Conference on USENIX Security Symposium, SSYM '04*, 2004.

[45] S.C. Bono, M. Green, A. Stubblefield, A. Juels, A.D. Rubin, and M. Szydlo, Security analysis of a cryptographically-enabled RFID device, in *Proceedings of the 14th Conference on USENIX Security Symposium, SSYM '05*, 2005.

[46] ATMEL, *AT98SC008CT Secure Chip*. http://www.atmel.com/products/SecureASSP/.

[47] Dallas Semiconductor, *DS2432, 1k-Bit Protected 1-Wire EEPROM with SHA-1 Engine*. http://datasheets.maxim-ic.com/en/ds/DS2432.pdf.

[48] G. Uchenick and W. Vanfleet, Multiple independent levels of safety and security: High assurance architecture for MSLS/MLS, *Military Communications Conference, MILCOM*, vol. 1, pp. 610–614, 2005.

[49] R. Anderson, M. Bond, J. Clulow, and S. Skorobogatov, Cryptographic processors—A survey, *Proceedings of the IEEE*, 94(2), 357–369, Feb. 2006.

[50] Texas Instruments, *CC2420: Single-Chip 2.4 GHz IEEE 802.15.4 Compliant and ZigBee Ready RF Transceiver*. http://focus.ti.com/docs/prod/folders/print/cc2420.html.

[51] Ember, *EM260: ZigBee/802.15.4 Network Processor*. http://www.ember.com/pdf/120-1003-000M_EM260_Datasheet.pdf.

[52] Nordic Semiconductor, *nRF24LU1:Single Chip 2.4 GHz Transceiver*. http://www.nordicsemi.com/files/Product/data_sheet/nRF24LU1datasheet.pdf.

[53] NXP Semiconductors, *Secure Smart Card Controller Platform*. http://www.nxp.com/acrobat_download/other/identification/095710.pdf.

[54] M. Feldhofer, S. Dominikus, and J. Wolkerstorfer, Strong authentication for RFID systems using the AES algorithm, in *Proceedings of the Workshop on Cryptographic Hardware and Embedded Systems, CHES '04*, vol. 3156, pp. 357–370, 2004.

[55] P. Hamalainen, T. Alho, M. Hannikainen, and T.D. Hamalainen, Design and implementation of low-area and low-power AES encryption hardware core, in *DSD '06: Proceedings of the 9th EUROMICRO Conference on Digital System Design*, pp. 577–583, 2006.

[56] L. Batina, J. Guajardo, T. Kerins, N. Mentens, P. Tuyls, and I. Verbauwhede, Public-key cryptography for RFID-tags, in *International Workshop on Pervasive Computing and Communication Security, PerSec '07*, pp. 217–222, 2007.

[57] Texas Instruments, *Choosing An Ultralow-Power MCU*. http://focus.ti.com/lit/an/slaa207/slaa207.pdf.

[58] T. Eisenbarth, S. Kumar, C. Paar, A. Poschmann, and L. Uhsadel, A survey of lightweight-cryptography implementations, *IEEE Design and Test of Computers*, 24(6), 2007.

[59] P. Hamalainen, M. Hannikainen, and T. Hamalainen, Efficient hardware implementation of security processing for IEEE 802.15.4 wireless networks, *48th Midwest Symposium on Circuits and Systems*, pp. 484–487, 2005.

[60] D. Hong, J. Sung, S. Hong, J. Lim, S. Lee, B.-S. Koo, C. Lee, D. Changand, J. Lee, K. Jeong, H. Kim, J. Kim, and S. Chee, HIGHT: A new block cipher suitable for low-resource device, in *Proceedings of the Second International Workshop on Cryptographic Hardware and Embedded Systems, CHES '06*, pp. 46–59, 2006.

[61] A. Poschmann, G. Leander, K. Schramm, and C. Paar, New light-weight crypto algorithms for RFID, in *the Proceedings of IEEE International Symposium on Circuits and Systems, ISCAS '07*, pp. 1843–1846, 2007.

[62] C.D. Canniere and B. Preneel, Trivium: A stream cipher construction inspired by block cipher design principles, http://www.ecrypt.eu.org/stream/papersdir/2006/021.pdf.

[63] M. Hell, T. Johansson, A. Maximov, and W. Meier, A stream cipher proposal: Grain-128, *IEEE International Symposium on Information Theory*, pp. 1614–1618, 2006.

[64] A. Liu, P. Kampanakis, and P. Ning, *TinyECC: A Configurable Library for Elliptic Curve Cryptography in Wireless Sensor Networks*. http://cdl.csc.ncsu.edu/software/TinyECC/.

[65] K. Sakiyama, L. Batina, B. Preneel, and I. Verbauwhede, HW/SW co-design for public-key cryptosystems on the 8051 microcontroller, *Computer and Electrical Engineering*, 33(5–6), 324–332, 2007.

[66] S. Janssens, J. Thomas, W. Borremans, P. Gijsels, I. Verhauwhede, F. Vercauteren, B. Preneel, and J. Vandewalle, Hardware/software co-design of an elliptic curve public-key cryptosystem, in *Proceedings of the IEEE Workshop on Signal Processing Systems*, pp. 209–216, 2001.

Index

A

AACS, *see* Advanced access content system

Abstractions of integer arithmetics (AIA), 143–144

Access control, wireless sensor networks
  device authorization, 313–314
  user authentication, 314

Access policy enforcement service, 248–249

Acknowledgment spoofing, 296

Active attacks, 124

Active jamming, 41

Active tags
  definition of, 201
  functionality of, 101
  power source of, 122

Activity jamming, 384–385

Add-all metric, 367

Additive aggregation, 402–403; *see also* Data aggregation

Ad hoc group communication architecture
  DoS and fault-tolerant authentication protocol, 447
  mobile nodes, 446
  security requirement for, 460

Advanced access content system, 274

Advanced encryption standard (AES), 124
  8-bit operations, 125
  32-bit operations, 125
  128-bits operations, 126
  encryption and integrity, 512–513
  modes of, 512
  theoretical security, 513

Adversarial model, 86

AES-based authentication protocol, 115

Aggregate integrity
  goal of, 403–404
  manipulation of final aggregate, 401
  providing
    CPS, 416
    PSP and HE scheme for, 415
    YWZC and BSV, 415–416

Aggregation tree, 402; *see also* Data aggregation

AIA, *see* Abstractions of integer arithmetics

Algorithmic engineering
  elliptic curve cryptography, in TinyOS, 354
  software development process, 354

Algorithmic tamper-proof (ATP) model, 83

AMAC security model
  aggregation in, 410
  operation of, 411
  polynomial time algorithms of, 410–411
  security notion of, 412–413
  sink and source nodes, 410

Amplitude modulation, 59

Amplitude shift keying
  for amplitude modulation, 59
  carrier and input signal, 58
  constellation for, 59

AMSECURE, cryptographic modes of, 438

Angle-of-arrival (AoA), 180

Animal tracking
  physical destruction of tags, 481–482
  tag role in, 480

Anomaly detection, 323

Antenna arrays, 65–66

Antenna polarization
  in jamming environments, 382
  of nodes' radio unit, 381

Antenna selection, for receiver
  schematic representation of, 65
  signal quality, 66

Anticollision algorithms, effect of multi-tags on, 20–21

Anticollision protocols, 154

Antijamming schemes
  mobile agent-based solutions, *see* Mobile agent-based solutions, antijamming
  proactive countermeasures, *see* Proactive countermeasures
  reactive countermeasures, *see* Reactive countermeasures, antijamming

Antiskimming card, functions of, 37

Application layer
  protocol stack and, 376–377
  threats to, 296–297

Application software, functionality of, 53

Architectural-based solutions
  delegation
    based on tree of secrets, *see* Tree-based delegation
    pseudonyms, 227
    with two secret keys, 231–233
  distributed architecture, 227
    arrival protocol, roaming protocol, and departure protocol, 229
    randomized hash-locks, 228
    readers coverage, 228

Arrival protocol, 229

ASK, *see* Amplitude shift keying

Asymmetric encryption, 196

Attacker-perspective, threat model
  attacker types
    competitors, 260–261
    insiders and saboteurs, 261
  threats during
    trace data announcement, 262–263
    trace data creation and storage, 262
    trace data deletion, 264–265
    trace data retrieval, 263–264
    trace data search, 263

Attack system model,

Authentication
  active, 39
  based on
    asymmetric public-key system, 301
    key-chains, 302–303

Sensor nodes
 deployment point and resident point
  of, 365
 g(z) concept, 365–366
 of hierarchical network architecture, 370
 static, 364
 subversion of, 295
Session hijacking attacks, 250–251
Shift-and-add method, 496
Side-channel attacks
 on MAC protocols, 93
 physical privacy model and, 83
 prevention by multi-tags, 21
 as security vulnerabilities, 164
Signal strength-based solution
 performance evaluation of, 369
 in static homogeneous WSNs, 368
 suspicious node detection, 368–369
Single-hop communication, 423
Single-hop networks, 355, 357
Sinkhole attack, 296, 333–334
Skimming attacks, 37–38
Smart antennas
 beamforming problems, 67
 transmission patterns, 64
Smart RFID Keeper
 architecture of, 238–239
 attack model supporting, 241
 network model supporting
  offline AS, 241
  online AS, 240
 security analysis under
  DoS attacks, 251–252
  eavesdroppers, 251
  hijacking attack, 250–251
 security services of, 239
Smart shop, 481–482
Soft Defined Radios, 63
Software agent, 351
Spatial retreat, 389
Specification-based detection, 323
SPINS, 432
 SNEP component of, 433
 µTESLA component of, 433–434
Spoofing, 100
Spot jamming, 378
SRK, *see* Smart RFID Keeper
Static key management schemes
 location-based pairwise key predistribution
  scheme, 299
 polynomial pool-based pairwise key predistribution,
  298–299
 random key predistribution scheme, 298
Stream cipher, 124
Structure Query Language (SQL) injection, 45
STUN tree, 423; *see also* Hierarchy trees
Super high frequency/microwaves tags, 35
Suspicious node information dissemination protocol
  (SNIDP), 369
Sweep jamming, 378
Sybil attack, 295–296, 429

Symmetric ciphers
 AES, *see* Advanced encryption standard (AES)
 block cipher and stream cipher, 124
 CLEFIA, Clefia, SEA, and TEA, 127
 DES and its variants, 126
 grain cipher, 127–128, 130
 HIGHT, 127, 519
 present cipher, 127
 Profile-2, 130
 trivium, 128–129
Symmetric encryption, 196
Synchronization
 back-end database, 222–223, 225–226
 with mutual authentication, 223–225
 RFID systems, 67
 with universal time stamps, 226–227
System access points (SAP) and system exit points
  (SEP), 228
System availability, 31
System delay
 frame delay timing and FWT, 68
 FWI, 69
 synchronization and, 67
System model, based on life-cycle phases
 phase 4, and phase 5, 260
 phase 1, phase 2, and phase 3, 259

T

Tag ID/data, splitting of, 21
Tag variability experiments, 16
Tag-verifier authentication, messages exchanged for, 224
Tamper-resistantmicroprocessors, 39
Target tracking, wireless sensor networks
 challenges in, 422
 methods for, 422
  GDAT, 424–425
  image-processing techniques, *see* Image-processing
   techniques
  LEACH network, 424
  single-hop communication and hierarchy trees, 423
Tav-128 bits, 132, 145
TCP SYN flood, 429
"Terrorist"-resistant distance-bounding protocol,
  see Reid et al.'s protocol
µTESLA broadcast authentication, 302–303, 415
TGDH protocol
 public keys and secret group key, 347
 tree-based structure of, 348
Theft prevention, 23
ThingMagic circular antennas, 16
ThingMagic reader, 16
Threat model
 attacker-centric perspective against, *see*
  Attacker-perspective, threat model
 as basis for risk management, 264
  risk evaluation, 267
  risk identification, 264, 266–267
  risk response, 267–268
  threat list and countermeasures, 265